T0236154

Lecture Notes in Artificial Intelligence   10352

Subseries of Lecture Notes in Computer Science

More information about this series at http://www.springer.com/series/1244

Marzena Kryszkiewicz · Annalisa Appice
Dominik Ślęzak · Henryk Rybinski
Andrzej Skowron · Zbigniew W. Raś (Eds.)

Foundations of Intelligent Systems

23rd International Symposium, ISMIS 2017
Warsaw, Poland, June 26–29, 2017
Proceedings

 Springer

Editors

Marzena Kryszkiewicz
Institute of Computer Science
Warsaw University of Technology
Warsaw
Poland

Annalisa Appice
Department of Computer Science
University of Bari Aldo Moro
Bari
Italy

Dominik Ślęzak
Institute of Informatics
University of Warsaw
Warsaw
Poland

Henryk Rybinski
Institute of Computer Science
Warsaw University of Technology
Warsaw
Poland

Andrzej Skowron
Institute of Mathematics
Warsaw University
Warsaw
Poland

and

Systems Research Institute
Polish Academy of Sciences
Warsaw University
Warsaw
Poland

Zbigniew W. Raś
Department of Computer Science
University of North Carolina at Charlotte
Charlotte, NC
USA

ISSN 0302-9743   ISSN 1611-3349   (electronic)
Lecture Notes in Artificial Intelligence
ISBN 978-3-319-60437-4   ISBN 978-3-319-60438-1   (eBook)
DOI 10.1007/978-3-319-60438-1

Library of Congress Control Number: 2017943070

LNCS Sublibrary: SL7 – Artificial Intelligence

Printed on acid-free paper

This Springer imprint is published by Springer Nature
The registered company is Springer International Publishing AG
The registered company address is: Gewerbestrasse 11, 6330 Cham, Switzerland

Preface

This volume contains the papers selected for presentation at the 23rd International Symposium on Methodologies for Intelligent Systems (ISMIS 2017), which was held at Warsaw University of Technology, Warsaw, Poland, June 26–29, 2017. The symposium was organized by the Institute of Computer Science at Warsaw University of Technology, in cooperation with University of Warsaw, Poland, and University of Bari Aldo Moro, Italy.

ISMIS is a conference series that started in 1986. Held twice every three years, ISMIS provides an international forum for exchanging scientific, research, and technological achievements in building intelligent systems. In particular, major areas selected for ISMIS 2017 include both theoretical and practical aspects of machine learning, data mining methods, deep learning, bioinformatics and health informatics, intelligent information systems, knowledge-based systems, mining temporal, spatial and spatio-temporal data, text and Web mining. In addition, four special sessions were organized: namely, Special Session on Big Data Analytics and Stream Data Mining, Special Session on Granular and Soft Clustering for Data Science, Special Session on Knowledge Discovery with Formal Concept Analysis and Related Formalisms, and Special Session devoted to ISMIS 2017 Data Mining Competition on Trading Based on Recommendations, which was launched as a part of the conference. The ISMIS conference was accompanied by the Industrial Session on Data Science in Industry: Algorithms, Tools, and Applications.

ISMIS 2017 received 118 submissions that were carefully reviewed by three or more Program Committee members or external reviewers. Papers submitted to special sessions were subject to the same reviewing procedure as those submitted to regular sessions. After a rigorous reviewing process, 56 regular papers and 15 short papers were accepted for presentation at the conference and publication in the ISMIS 2017 proceedings volume. This volume also contains one invited paper and three abstracts by the plenary keynote speakers.

It is truly a pleasure to thank all people who helped this volume come into being and made ISMIS 2017 a successful and exciting event. In particular, we would like to express our appreciation for the work of the ISMIS 2017 Program Committee members and external reviewers who helped assure the high standard of accepted papers. We would like to thank all authors of ISMIS 2017, without whose high-quality contributions it would not have been possible to organize the conference. We are grateful to the organizers of special sessions of ISMIS 2017: Martin Atzmueller and Jerzy Stefanowski (Special Session on Big Data Analytics and Stream Data Mining), Pawan Lingras, Georg Peters, and Richard Weber (Special Session on Granular and Soft Clustering for Data Science), Davide Ciucci, Sergei Kuznetsov, and Amedeo Napoli (Special Session on Knowledge Discovery with Formal Concept Analysis and Related Formalisms), Andrzej Janusz and Kamil Żbikowski (Special Session devoted to ISMIS 2017 Data Mining Competition on Trading Based on Recommendations). We would also like to express our appreciation to the organizers of the accompanying events: Andrzej Janusz and Kamil Żbikowski, who successfully launched the competition; Robert Bembenik, Grzegorz Protaziuk, and Łukasz Skonieczny for organizing the Industrial Session and for their involvement in all organizational matters related to ISMIS 2017 as well as the creation and maintenance of the conference website. We are grateful to Bożenna Skalska, Joanna Konczak, and Anna Filipowska-Chlebek for their administrative work.

We also wish to express our thanks to Perfecto Herrera Boyer, Paola Mello, Marek Rusinkiewicz, and Osmar R. Zaïane for accepting our invitation to give plenary talks at ISMIS 2017. We would like to thank mBank S.A. for sponsoring the ISMIS 2017 Data Mining Competition, and Tipranks for providing data for this competition. Also, we are thankful to Alfred Hofmann of Springer for supporting the ISMIS 2017 Best Paper Award.

Moreover, our thanks are due to Alfred Hofmann for his continuous support and to Anna Kramer for her work on the proceedings.

We believe that the proceedings of ISMIS 2017 will become a valuable source of reference for your ongoing and future research activities.

June 2017   Marzena Kryszkiewicz
                                                            Annalisa Appice
                                                            Dominik Ślęzak
                                                            Henryk Rybinski
                                                            Andrzej Skowron
                                                            Zbigniew W. Raś

Organization

ISMIS 2017 was organized by the Institute of Computer Science, Warsaw University of Technology, Warsaw, Poland, in cooperation with University of Warsaw, Poland, and University of Bari Aldo Moro, Italy.

ISMIS 2017 Symposium Organizers

Symposium Chairs

Henryk Rybinski	Warsaw University of Technology, Poland
Andrzej Skowron	University of Warsaw, Poland and Polish Academy of Sciences, Poland

Program Chairs

Annalisa Appice	University of Bari Aldo Moro, Italy
Marzena Kryszkiewicz	Warsaw University of Technology, Poland
Dominik Ślęzak	University of Warsaw, Poland

Steering Committee Chair

Zbigniew W. Raś	University of North Carolina, USA and Warsaw University of Technology, Poland

Organizing Committee

Robert Bembenik	Warsaw University of Technology, Poland
Grzegorz Protaziuk	Warsaw University of Technology, Poland
Łukasz Skonieczny	Warsaw University of Technology, Poland

Steering Committee

Troels Andreasen	Roskilde University, Denmark
Jaime Carbonell	CMU, USA
Li Chen	Hong Kong Baptist University, Hong Kong, SAR China
Henning Christiansen	Roskilde University, Denmark
Juan Carlos Cubero	University of Granada, Spain
Floriana Esposito	University of Bari Aldo Moro, Italy
Alexander Felfernig	Graz University of Technology, Austria
Mohand-Saïd Hacid	Université Claude Bernard Lyon 1, France
Marzena Kryszkiewicz	Warsaw University of Technology, Poland
Jiming Liu	Hong Kong Baptist University, Hong Kong, SAR China

Olivier Pivert	IRISA-ENSSAT, University of Rennes 1, France
Henryk Rybinski	Warsaw University of Technology, Poland
Lorenza Saitta	Università del Piemonte Orientale, Italy
Andrzej Skowron	University of Warsaw, Poland and Polish Academy of Sciences, Poland
Dominik Ślęzak	University of Warsaw, Poland
Maria Zemankova	NSF, USA

Program Committee

ISMIS Regular Papers

Piotr Andruszkiewicz	Warsaw University of Technology, Poland
Annalisa Appice	University of Bari Aldo Moro, Italy
Jarek Arabas	Warsaw University of Technology, Poland
Martin Atzmueller	Tilburg University, The Netherlands
Robert Bembenik	Warsaw University of Technology, Poland
Salima Benbernou	Université Paris Descartes, France
Petr Berka	University of Economics, Prague, Czech Republic
Marenglen Biba	University of New York in Tirana, Albania
Maria Bielikova	Slovak University of Technology in Bratislava, Slovakia
Konstantinos Blekas	University of Ioannina, Greece
Jerzy Błaszczyński	Poznań University of Technology, Poland
Gloria Bordogna	National Research Council of Italy - CNR, Italy
Jose Borges	University of Porto, Portugal
Ivan Bratko	University of Ljubljana, Slovenia
Henrik Bulskov	Roskilde University, Denmark
Michelangelo Ceci	University of Bari Aldo Moro, Italy
Edward Chang	HTC Research, Taiwan
Jianhua Chen	Louisiana State University, USA
Silvia Chiusano	Polytechnic University of Turin, Italy
Davide Ciucci	University of Milano-Bicocca, Italy
Emmanuel Coquery	Université Lyon 1 - LIRIS, France
Germán Creamer	Stevens Institute of Technology, USA
Bruno Cremilleux	Université de Caen, France
Alfredo Cuzzocrea	University of Trieste, Italy
Claudia D'Amato	University of Bari Aldo Moro, Italy
Jeroen De Knijf	AVAST, The Netherlands
Marcilio De Souto	LIFO/University of Orleans, USA
Luigi Di Caro	University of Turin, Italy
Nicola Di Mauro	University of Bari Aldo Moro, Italy
Stephan Doerfel	University of Kassel, Germany
Brett Drury	National University of Ireland, Galway, Ireland
Wouter Duivesteijn	TU Eindhoven, The Netherlands
Inês Dutra	Universidade do Porto, Portugal

Christoph F. Eick	University of Houston, USA
Tapio Elomaa	Tampere University of Technology, Finland
Floriana Esposito	University of Bari Aldo Moro, Italy
Nicola Fanizzi	University of Bari Aldo Moro, Italy
Stefano Ferilli	University of Bari Aldo Moro, Italy
Sebastien Ferre	Université de Rennes 1, France
Carlos Ferreira	LIAAD INESC Porto LA, Portugal
Naoki Fukuta	Shizuoka University, Japan
Fabio Fumarola	Math & IT Research and Development, Italy
Mohamed Gaber	Birmingham City University, UK
Tomasz Gambin	Warsaw University of Technology, Poland
Paolo Garza	Polytechnic University of Turin, Italy
Piotr Gawrysiak	Warsaw University of Technology, Poland
Laura Giordano	University of Eastern Piedmont, Italy
Michael Granitzer	University of Passau, Germany
Jacek Grekow	Bialystok Technical University, Poland
Jerzy Grzymala-Busse	University of Kansas, USA
Hakim Hacid	Zayed University, United Arab Emirates
Mohand-Saïd Hacid	Université Claude Bernard Lyon 1, France
Allel Hadjali	LIAS/ENSMA, France
Mirsad Hadzikadic	UNC Charlotte, USA
Maria Halkidi	University of Piraeus, Greece
Shoji Hirano	Shimane University, Japan
Lothar Hotz	University of Hamburg, Germany
Dino Ienco	IRSTEA, France
Andrzej Janusz	University of Warsaw, Poland
Matthias Jarke	RWTH Aachen University, Germany
Szymon Jaroszewicz	Polish Academy of Sciences, Poland
Adam Jatowt	Kyoto University, Japan
Mehdi Kaytoue	LIRIS - INSA de Lyon, France
Matthias Klusch	DFKI, Germany
Mieczysław Kłopotek	Polish Academy of Sciences, Poland
Dragi Kocev	Jozef Stefan Institute, Slovenia
Jacek Koronacki	Polish Academy of Sciences, Poland
Bozena Kostek	Gdansk University of Technology, Poland
Lars Kotthoff	University of British Columbia, Canada
Marzena Kryszkiewicz	Warsaw University of Technology, Poland
Sergei O. Kuznetsov	National Research University Higher School of Economics, Russia
Mark Last	Ben-Gurion University of the Negev, Israel
Anne Laurent	LIRMM - UM, France
Dominique Laurent	Université Cergy-Pontoise, France
Marie-Jeanne Lesot	LIP6 - UPMC, France
Antoni Ligęza	AGH University of Science and Technology, Poland
Pawan Lingras	Saint Mary's University, Halifax, Canada
Corrado Loglisci	University of Bari, Italy

Christophe Rey	LIMOS - Université Clermont Auvergne, France
Rita P. Ribeiro	University of Porto, Portugal
Jadwiga Rogowska	University of Utah, USA
Przemysław Rokita	Warsaw University of Technology, Poland
André L.D. Rossi	Universidade Estadual Paulista Júlio de Mesquita Filho, Brazil
Salvatore Ruggieri	University of Pisa, Italy
Henryk Rybinski	Warsaw University of Technology, Poland
Dominik Ryzko	Warsaw University of Technology, Poland
Lorenza Saitta	University of Turin, Italy
Hiroshi Sakai	Kyushu Institute of Technology, Japan
Yacine Sam	Université François Rabelais Tours, France
Daniel Sanchez	University of Granada, Spain
Jose Salvador Sanchez	Universitat Jaume I, Spain
Yucel Saygin	Sabanci University, Turkey
Christoph Schommer	University of Luxembourg, Luxembourg
Matthias Schubert	Ludwig Maximilians University of Munich, Germany
Marian Scuturici	LIRIS-INSA de Lyon, France
Hamida Seba	University of Lyon 1, France
Nazha Selmaoui-Folcher	University of New Caledonia, Canada
Giovanni Semeraro	University of Bari Aldo Moro, Italy
Junming Shao	UESTC, China
Łukasz Skonieczny	Warsaw University of Technology, Poland
Andrzej Skowron	University of Warsaw, Poland and Polish Academy of Sciences, Poland
Dominik Ślęzak	University of Warsaw, Poland
Gregory Smits	IRISA-IUT de Lannion, France
Arnaud Soulet	Université François Rabelais Tours, France
Urszula Stanczyk	Silesian University of Technology, Poland
Jerzy Stefanowski	Poznan University of Technology, Poland
Luis Enrique Sucar	INAOE, Mexico
Marcin Sydow	Polish-Japanese Academy of Information Technology, Poland
Xiaohui Tao	University of Southern Queensland, Australia
Hadi Tork	University of Oslo, Norway
Vicenc Torra	University of Skövde, Sweden
Hakim Touati	Optym, USA
Brigitte Trousse	Inria, France
Shusaku Tsumoto	Shimane University, Japan
Theodoros Tzouramanis	University of the Aegean, Greece
Giorgio Valentini	University of Milan, Italy
Julien Velcin	Université de Lyon 2, France
Herna Viktor	University of Ottawa, Canada
Christel Vrain	LIFO - University of Orléans, France
Yang Wang	University of New South Wales, UK
Richard Weber	University of Chile Santiago, Chile

Alicja Wieczorkowska	Polish-Japanese Academy of Information Technology, Poland
Yiyu Yao	University of Regina, Canada
Jure Zabkar	University of Ljubljana, Slovenia
Slawomir Zadrozny	Polish Academy of Sciences, Poland
Wlodek Zadrozny	UNCC, USA
Bernard Zenko	Jozef Stefan Institute, Slovenia
Ying Zhao	Tsinghua University, Chia
Xiaofeng Zhu	Guangxi Normal University, China
Cezary Zielinski	Warsaw University of Technology, Poland
Kamil Żbikowski	Warsaw University of Technology & mBank S.A., Poland

Special Session on Big Data Analytics and Stream Data Mining

Martin Atzmueller (Chair)	Tilburg University, The Netherlands
Jerzy Stefanowski (Chair)	Poznan University of Technology, Poland
Petr Berka	University of Economics, Prague, Czech Republic
Dariusz Brzezinski	Poznan University of Technology, Poland
Bruno Cremilleux	Université de Caen, France
Nicola Di Mauro	University of Bari Aldo Moro, Italy
Wouter Duivesteijn	TU Eindhoven, The Netherlands
Mohamed Gaber	Birmingham City University, UK
Matthias Jarke	RWTH Aachen University, Germany
Szymon Jaroszewicz	Polish Academy of Sciences, Poland
Georg Krempl	Otto von Guericke University of Magdeburg, Germany
Mark Last	Ben-Gurion University of the Negev, Israel
Elio Masciari	ICAR-CNR, Italy
Ernestina Menasalvas	Universidad Politechnica de Madrid, Spain
Jose Salvador Sanchez	Universitat Jaume I, Spain
Herna Viktor	University of Ottawa, Canada
Michał Woźniak	Wroclaw University of Technology, Poland

Special Session on Granular and Soft Clustering for Data Science

Pawan Lingras (Chair)	Saint Mary's University, Halifax, Canada
Georg Peters (Chair)	Munich University of Applied Sciences, Germany, and Australian Catholic University, Australia
Richard Weber (Chair)	University of Chile Santiago, Chile

Special Session on Knowledge Discovery with Formal Concept Analysis and Related Formalisms

Davide Ciucci (Chair)	University of Milano-Bicocca, Italy
Sergei O. Kuznetsov (Chair)	National Research University Higher School of Economics, Russia
Amedeo Napoli (Chair)	LORIA/CNRS-Inria Nancy Grand Est-Université de Lorraine, France

Mehwish Alam	Université de Paris-Nord, France
Gabriela Arevalo	Universidad Nacional de Quilmes, Argentina
Jaume Baixeries	UPC Barcelona, Spain
Victor Codocedo	Inria Chile, Chile
Miguel Couceiro	LORIA Nancy/Université de Lorraine, France
Bruno Cremilleux	GREYC Caen/Université de Basse Normandie, France
Florent Domenach	University of Nicosia, Cyprus
Marianne Huchard	LIRMM/Université de Montpellier, France
Dmitry I. Ignatov	National Research University Higher School of Economics, Russia
Mehdi Kaytoue	INSA-LIRIS Lyon, France
Jan Konecny	Palacky University, Czech Republic
Nizar Messai	Université François Rabelais Tours, France
Jean-Marc Petit	INSA-LIRIS Lyon, France
Marc Plantevit	INSA-LIRIS Lyon, France
Pascal Poncelet	LIRMM/Université de Montpellier, France
Henri Prade	IRIT CNRS/Université de Toulouse, France
Artem Revenko	Technische Universität Dresden, Germany
Christian Sacarea	Babes-Bolyai University, Romania
Henry Soldano	Université de Paris-Nord, France
Arnaud Soulet	Université François Rabelais Tours, France
Diana Troanca	Babes-Bolyai University, Romania
Renato Vimiero	UFPE Recife, Brazil
Christel Vrain	LIFO/Université d'Orléans, France

Special Session on ISMIS 2017 Data Mining Competition on Trading Based on Recommendations

Andrzej Janusz (Chair)	University of Warsaw, Poland
Kamil Żbikowski (Chair)	Warsaw University of Technology and mBank S.A., Poland
Annalisa Appice	University of Bari Aldo Moro, Italy
Robert Bembenik	Warsaw University of Technology, Poland
Piotr Gawrysiak	Warsaw University of Technology, Poland
Marzena Kryszkiewicz	Warsaw University of Technology, Poland
Henryk Rybinski	Warsaw University of Technology, Poland
Łukasz Skonieczny	Warsaw University of Technology, Poland
Dominik Ślęzak	University of Warsaw, Poland

Additional Reviewers

Bioglio, Livio
Bradshaw, Stephen
Branco, Paula
Calefato, Fabio
Carbonnel, Jessie
Castelltort, Arnaud
Cerqueira, Vitor
Cheniki, Nasreddine
Chou, Jason
Ciecierski, Konrad
Fawagreh, Khaled

Garcia, Kemilly
Gašpar, Peter
Guarascio, Massimo
Haidar, Diana
Iaquinta, Leo
Impedovo, Angelo
Kaššák, Ondrej
Khiari, Jihed
Lippi, Marco
Mettouris, Christos
Pinto, Fábio

Polignano, Marco
Ritacco, Ettore
Rocha, Gil
Saidani, Fayçal Rédha
Saltos, Ramiro
Santos Costa, Vítor
Wu, Jake
Yao, Jingtao
Yao, Yiyu
Yeratziotis, Alexandros
Yu, Hong

Invited Talks

Elements of Musical Intelligence for the Next Generation of Digital Musical Tools

Perfecto Herrera-Boyer

Music Technology Group, Universitat Pompeu Fabra, Roc Boronat 138, 08018,
Barcelona, Spain
Departament de Sonologia, Escola Superior de Música de Catalunya, Padilla 155,
08013, Barcelona, Spain
perfecto.herrera@upf.edu

Extended Abstract

Musical intelligence can be observed in many diverse situations: while tapping to the beat, when recommending interesting tracks to a friend that we know well, when we play the right notes even the score has a hole in their place… Intelligence has frequently been presented linked to observable behavior. The infamous Turing test requires of overt (verbal) behavior. Deep blue played chess, again an observable behavior, as it was the case of IBM's Watson, the winner of the TV question-answering contest "Jeopardy". Maybe less known are Aaron (a painting system), Libratus (a poker player), or the Google AI system that writes sad poems. When considering intelligent musical systems, those showing observable behavior abound. But maybe there is more than that to deal with intelligence. Maybe the behavioral requirement is a residue of the "behavioral" perspective that ruled in the academia during some decades of the XXth century. In an influential, though speculative, work titled "On Intelligence" [2], we can read: "Ignoring what goes on in your head and focusing instead on behavior has been a large impediment to understanding intelligence and building intelligent machines". The idea of intelligence as a compact, monolithic construct, as long as the possibility to measure or detect it with a single type of test has also been challenged, among many other perspectives, by Gardner's theory of multiple intelligences [1].

In accordance with such critical views, musical intelligence starts up from basic building blocks like the capacity to entrain a beat, or to categorize a reduced list of pitches as related, or to detect recurring patterns (in time, in pitch, etc.). This is what some authors identify as "musicality" [3]. Although we do not then need to build full-fledged systems to observe musical intelligence, it can nevertheless be differently observed under different circumstances and in different scenarios. In the end, it will be in the mind of the observer, and no "bot versus human" test will be convincing enough or appropriate to conclude about that.

Which are the typical scenarios where we observe or induce intelligent music behavior? Among the earliest and more frequent "intelligent music systems" we find mostly music composition systems, which are developed with the goal to generate music scores with a minimal input from the users. It is usually the case that these systems are aimed at the "imitation" of a well-known composer (e.g., Bach), or at the generation of pieces in a given style. Other systems have been shown to compose in a highly idiomatic way, somehow inventing her own style (e.g., "Emily Howell"). A different flavor of musical intelligence can be found in music performance systems, capable of performing music in real time, accompanying one or more human players. In this category we also find systems that address "expressive playing" (i.e., playing as humans do, not robotically following just what is notated in a score, but making reasonable errors, adding grace notes, etc.). Music mixing systems are able to create musical programs by means of the concatenation of pre-existing musical pieces or its fragments, like disc-jockeys usually do and, also connected with the idea of "directing" music, there are intelligent systems that assist audio engineers in music mixing by proposing balances, equalization parameters, fades, etc. Finally, most of the intelligent music analysis systems do not target an overt behavior. They, instead, take the score or an audio file as input and transform the data into human-interpretable descriptions (see, for example, [4]). Content-based recommender systems (like Pandora radio) could be considered a special case of intelligent analysis systems capable of, again, generating observable behavior or even interaction with a user.

In this talk, after my overview of some remarkable intelligent music systems, I will argue about some necessary (but not sufficient) conditions for music systems to be considered intelligent (though, at the same time, I will question the utility of such notion for scientific or engineering purposes), such as:

- Multiple-faceted representation: the system manages different musical dimensions: rhythm, pitch, timbre, harmony, structure, expressivity...
- Multiple time-scale representation: the system manages music data using different temporal scopes (i.e., there are representations from onsets to sections, from notes to tonality).
- Predictive/Anticipating representation: the system manages representations that are not just current or past data but also representations of what is likely to happen next.
- Symbolic representation: the system manages abstract representations that can be easily mapped to human-understandable words or concepts.
- Pattern completion: the system is capable to work with incomplete information (like when a beat is missed because of noise, or when there is a printing defect in the music sheet).
- Interaction with the environment, where this environment consists of humans or other intelligent music systems.
- Experience-based learning: some of the internal representations can be modified by means of different learning processes (e.g. reinforcement, association, imitation...).

In connection with this perspective, the final minutes of my talk will be devoted to present some of the outcomes of the EU-funded project GiantSteps, which has been endeavoring on the development of intelligent musical agents and music interfaces that, in music creation contexts, functionally work as humble assistants to suggest or guide human users when they lack inspiration, or technical or musical knowledge.

References

1. Gardner, H.: Frames of Mind: The Theory of Multiple Intelligences. Basic Books, New York, (1993)
2. Hawkins, J.: On Intelligence. Times Books (2004)
3. Honing, H., ten Cate, C., Peretz, I., Trehub, S.E.: Without it no music: cognition, biology and evolution of musicality. Philos. Trans. R. Soc. B Biol. Sci. **370**(1664) (2015). doi:10.1098/rstb.2014.0088
4. Schedl, M., Yang, Y-H., and Herrera-Boyer, P.: Introduction to intelligent music systems and applications. ACM Trans. Intell. Syst. Technol. **8**(2), Article 17, 8 (2016). doi:10.1145/2991468

Security of Cyber-physical Systems

Marek Rusinkiewicz

New Jersey Institute of Technology, Newark, USA
marek.rusinkiewicz@njit.edu

Abstract. Cyberspace, the ubiquitous collection of interconnected IP networks and hosts, has become the nervous system of the country. Healthy functioning of Cyberspace is essential for the proper operation of numerous critical infrastructures, including telecommunication, energy distribution, and transportation. It is also necessary to support the ever expanding business infrastructure for commerce and banking. The increasing reliance on Cyberspace has been paralleled by a corresponding increase in the variety, frequency and impact of attacks from a range of assailants. Both commercial companies and government agencies face continuous and increasingly more sophisticated cyber-attacks ranging from data exfiltration and spear phishing to sophisticated worms and logic bombs. The targets include not only computer information systems, but also the network communication infrastructure and power grids.

In this talk, I will discuss protecting cyber-physical systems from attacks and argue that cyber security can significantly benefit from multidisciplinary research including Web Intelligence, Data Analytics and Network Science.

Classification by Association

Osmar R. Zaïane

University of Alberta, Edmonton, Canada
zaiane@ualberta.ca

Abstract. In machine learning the centre of attention these days is on Deep Learning. The spotlight has turned towards this re-branding of Neural Networks because of the tremendous success it recently demonstrated for image classification, speech recognition, text categorization and other applications where labeled data is plentiful. However, this is not necessarily true when data is not sufficient enough to optimize the hyperparameters. Alternatives are still needed in most applications and we advocate for rule-based classifiers such as associative classifiers. Associative classifiers piggyback on association rule mining to discover classification rules. The learned model is a set of rules that are easy to understand and can be edited. Established associative classification algorithms have shown to be very effective in handling categorical data such as text data. There were many improvements this last decade for rule representation, rule pruning, rule selection strategies, etc. Yet, most proposed approaches still suffer from the limitations of the support-confidence framework inherited from association rule mining. We will talk about the challenges in mining these classification rules, the progress reached using this paradigm and the last incarnation of associative classifiers that finds statistical dependency in the data to avoid the detection of meaningless rules and the omission of interesting ones while doing away with the restraining parameters support and confidence.

Contents

Invited Paper

Abductive Reasoning on Compliance Monitoring: Balancing Flexibility
and Regulation   3
   Federico Chesani, Paola Mello, and Marco Montali

Bioinformatics and Health Informatics

Multi-levels 3D Chromatin Interactions Prediction
Using Epigenomic Profiles   19
   Ziad Al Bkhetan and Dariusz Plewczynski

A Supervised Model for Predicting the Risk of Mortality
and Hospital Readmissions for Newly Admitted Patients   29
   Mamoun Almardini and Zbigniew W. Raś

An Expert System Approach to Eating Disorder Diagnosis   37
   Stefano Ferilli, Anna Maria Ferilli, Floriana Esposito,
   Domenico Redavid, and Sergio Angelastro

Multimodal System for Diagnosis and Polysensory Stimulation
of Subjects with Communication Disorders   47
   Adam Kurowski, Piotr Odya, Piotr Szczuko, Michał Lech,
   Paweł Spaleniak, Bożena Kostek, and Andrzej Czyżewski

Acute Kidney Injury Detection: An Alarm System to Improve
Early Treatment   57
   Ana Rita Nogueira, Carlos Abreu Ferreira, and João Gama

New Method of Calculating ^{SR}CM Chirality Measure   64
   Przemyslaw Szurmak and Jan Mulawka

Data Mining Methods

Selection of Initial Modes for Rough Possibilistic K-Modes Methods   77
   Asma Ammar and Zied Elouedi

Accelerating Greedy K-Medoids Clustering Algorithm with L_1 Distance
by Pivot Generation   87
   Takayasu Fushimi, Kazumi Saito, Tetsuo Ikeda, and Kazuhiro Kazama

On the Existence of Kernel Function for Kernel-Trick of k-Means   97
  Mieczysław A. Kłopotek

Using Network Analysis to Improve Nearest Neighbor Classification
of Non-network Data.   105
  Maciej Piernik, Dariusz Brzezinski, Tadeusz Morzy, and Mikolaj Morzy

An Accurate and Efficient Method to Detect Critical Links to Maintain
Information Flow in Network   116
  Kazumi Saito, Kouzou Ohara, Masahiro Kimura, and Hiroshi Motoda

Deep Learning

Continuous Embedding Spaces for Bank Transaction Data   129
  Ali Batuhan Dayioglugil and Yusuf Sinan Akgul

Shallow Reading with Deep Learning: Predicting Popularity
of Online Content Using only Its Title.   136
  Wojciech Stokowiec, Tomasz Trzciński, Krzysztof Wołk,
  Krzysztof Marasek, and Przemysław Rokita

Recurrent Neural Networks for Online Video Popularity Prediction   146
  Tomasz Trzciński, Paweł Andruszkiewicz, Tomasz Bocheński,
  and Przemysław Rokita

Intelligent Information Systems

User-Based Context Modeling for Music Recommender Systems   157
  Imen Ben Sassi, Sadok Ben Yahia, and Sehl Mellouli

Uncalibrated Visual Servo for the Remotely Operated Vehicle   168
  Chi-Cheng Cheng and Tsan-Chu Lu

Comparative Analysis of Musical Performances by Using
Emotion Tracking   175
  Jacek Grekow

Automated Web Services Composition with Iterated Services   185
  Alfredo Milani and Rajdeep Niyogi

On the Gradual Acceptability of Arguments in Bipolar Weighted
Argumentation Frameworks with Degrees of Trust   195
  Andrea Pazienza, Stefano Ferilli, and Floriana Esposito

Automatic Defect Detection by Classifying Aggregated Vehicular Behavior . . .   205
  Felix Richter, Oliver Hartkopp, and Dirk C. Mattfeld

Automatic Speech Recognition Adaptation to the IoT Domain
Dialogue System.   215
    Maciej Zembrzuski, Heesik Jeon, Joanna Marhula, Katarzyna Beksa,
    Szymon Sikorski, Tomasz Latkowski, and Paweł Bujnowski

Knowledge-Based Systems

Rule-Based Reasoning with Belief Structures   229
    Łukasz Białek, Barbara Dunin-Kęplicz, and Andrzej Szałas

Combining Machine Learning and Knowledge-Based Systems
for Summarizing Interviews   240
    Angel Luis Garrido, Oscar Cardiel, Andrea Aleyxendri,
    and Ruben Quilez

Validity of Automated Inferences in Mapping of Anatomical Ontologies   251
    Milko Krachunov, Peter Petrov, Maria Nisheva, and Dimitar Vassilev

An Experiment in Causal Structure Discovery. A Constraint
Programming Approach   261
    Antoni Ligęza

Machine Learning

Actively Balanced Bagging for Imbalanced Data.   271
    Jerzy Błaszczyński and Jerzy Stefanowski

A Comparison of Four Classification Systems Using Rule Sets Induced
from Incomplete Data Sets by Local Probabilistic Approximations   282
    Patrick G. Clark, Cheng Gao, and Jerzy W. Grzymala-Busse

Robust Learning in Expert Networks: A Comparative Analysis.   292
    Ashiqur R. KhudaBukhsh, Jaime G. Carbonell, and Peter J. Jansen

Efficient All Relevant Feature Selection with Random Ferns   302
    Miron Bartosz Kursa

Evaluating Difficulty of Multi-class Imbalanced Data   312
    Mateusz Lango, Krystyna Napierala, and Jerzy Stefanowski

Extending Logistic Regression Models with Factorization Machines   323
    Mark Pijnenburg and Wojtek Kowalczyk

Filtering Decision Rules with Continuous Attributes Governed
by Discretisation.   333
    Urszula Stańczyk

Mining Temporal, Spatial and Spatio-Temporal Data

OptiLocator: Discovering Optimum Location for a Business Using Spatial
Co-location Mining and Spatio-Temporal Data   347
   Robert Bembenik, Jacek Szwaj, and Grzegorz Protaziuk

Activity Recognition Model Based on GPS Data, Points of Interest
and User Profile   358
   Igor da Penha Natal, Rogerio de Avellar Campos Cordeiro,
   and Ana Cristina Bicharra Garcia

Extended Process Models for Activity Prediction   368
   Stefano Ferilli, Floriana Esposito, Domenico Redavid,
   and Sergio Angelastro

Automatic Defect Detection by One-Class Classification
on Raw Vehicle Sensor Data   378
   Julia Hofmockel, Felix Richter, and Eric Sax

Visualizing Switching Regimes Based on Multinomial Distribution
in Buzz Marketing Sites.   385
   Yuki Yamagishi and Kazumi Saito

"Serial" versus "Parallel": A Comparison of Spatio-Temporal
Clustering Approaches.   396
   Yongli Zhang, Sujing Wang, Amar Mani Aryal, and Christoph F. Eick

Time-Frequency Representations for Speed Change Classification:
A Pilot Study   404
   Alicja Wieczorkowska, Elżbieta Kubera, Danijel Koržinek,
   Tomasz Słowik, and Andrzej Kuranc

Text and Web Mining

Opinion Mining on Non-English Short Text   417
   Esra Akbas

Pathway Computation in Models Derived from Bio-Science Text Sources . . .   424
   Troels Andreasen, Henrik Bulskov, Per Anker Jensen,
   and Jørgen Fischer Nilsson

Semantic Enriched Short Text Clustering   435
   Marek Kozlowski and Henryk Rybinski

Exploiting Web Sites Structural and Content Features for Web
Pages Clustering   446
   Pasqua Fabiana Lanotte, Fabio Fumarola, Donato Malerba,
   and Michelangelo Ceci

Concept-Enhanced Multi-view Co-clustering of Document Data   457
    Valentina Rho and Ruggero G. Pensa

Big Data Analytics and Stream Data Mining

Scalable Framework for the Analysis of Population Structure
Using the Next Generation Sequencing Data   471
    *Anastasiia Hryhorzhevska, Marek Wiewiórka, Michał Okoniewski,
    and Tomasz Gambin*

Modification to K-Medoids and CLARA for Effective
Document Clustering.   481
    Phuong T. Nguyen, Kai Eckert, Azzurra Ragone, and Tommaso Di Noia

Supporting the Page-Hinkley Test with Empirical Mode Decomposition
for Change Detection.   492
    Raquel Sebastião and José Maria Fernandes

Co-training Semi-supervised Learning for Single-Target Regression
in Data Streams Using AMRules.   499
    Ricardo Sousa and João Gama

Time-Series Data Analytics Using Spark and Machine Learning   509
    Patcharee Thongtra and Alla Sapronova

Granular and Soft Clustering for Data Science

Scalable Machine Learning with Granulated Data Summaries:
A Case of Feature Selection   519
    Agnieszka Chądzyńska-Krasowska, Paweł Betliński, and Dominik Ślęzak

Clustering Ensemble for Prioritized Sampling Based on Average
and Rough Patterns   530
    *Matt Triff, Ilya Pavlovski, Zhixing Liu, Lori-Anne Morgan,
    and Pawan Lingras*

C&E Re-clustering: Reconstruction of Clustering Results
by Three-Way Strategy   540
    Pingxin Wang, Xibei Yang, and Yiyu Yao

Multi-criteria Based Three-Way Classifications with Game-Theoretic
Rough Sets.   550
    Yan Zhang and JingTao Yao

Theoretical Aspects of Formal Concept Analysis

A Formal Context for Acyclic Join Dependencies   563
    Jaume Baixeries

On Containment of Triclusters Collections Generated by Quantified
Box Operators   573
   Dmitrii Egurnov, Dmitry I. Ignatov, and Engelbert Mephu Nguifo

The Inescapable Relativity of Explicitly Represented Knowledge:
An FCA Perspective   580
   David Flater

Blocks of the Direct Product of Tolerance Relations   587
   Christian Jäkel and Stefan E. Schmidt

Viewing Morphisms Between Pattern Structures via Their Concept Lattices
and via Their Representations   597
   Lars Lumpe and Stefan E. Schmidt

Formal Concept Analysis for Knowledge Discovery

On-Demand Generation of AOC-Posets: Reducing the Complexity
of Conceptual Navigation   611
   Alexandre Bazin, Jessie Carbonnel, and Giacomo Kahn

From Meaningful Orderings in the Web of Data to Multi-level
Pattern Structures   622
   Quentin Brabant, Miguel Couceiro, Amedeo Napoli,
   and Justine Reynaud

On Locality Sensitive Hashing for Sampling Extent Generators   632
   Victor Codocedo and My Thao Tang

An Application of AOC-Posets: Indexing Large Corpuses
for Text Generation Under Constraints   642
   Alain Gutierrez, Michel Chein, Marianne Huchard,
   and Pierre Pompidor

On Neural Network Architecture Based on Concept Lattices   653
   Sergei O. Kuznetsov, Nurtas Makhazhanov, and Maxim Ushakov

Query-Based Versus Tree-Based Classification: Application
to Banking Data   664
   Alexey Masyutin and Yury Kashnitsky

Using Formal Concept Analysis for Checking the Structure of an Ontology
in LOD: The Example of DBpedia   674
   Pierre Monnin, Mario Lezoche, Amedeo Napoli, and Adrien Coulet

A Proposal for Classifying the Content of the Web of Data Based
on FCA and Pattern Structures   684
   Justine Reynaud, Mehwish Alam, Yannick Toussaint,
   and Amedeo Napoli

**ISMIS 2017 Data Mining Competition on Trading Based
on Recommendations**

ISMIS 2017 Data Mining Competition: Trading Based
on Recommendations.   697
   Mathurin Aché, Andrzej Janusz, Kamil Żbikowski, Dominik Ślęzak,
   Marzena Kryszkiewicz, Henryk Rybinski, and Piotr Gawrysiak

Predicting Stock Trends Based on Expert Recommendations
Using GRU/LSTM Neural Networks   708
   Przemyslaw Buczkowski

Using Recommendations for Trade Returns Prediction
with Machine Learning   718
   Ling Cen, Dymitr Ruta, and Andrzej Ruta

Heterogeneous Ensemble of Specialised Models - A Case Study
in Stock Market Recommendations   728
   Michał Kozielski, Katarzyna Dusza, Józef Flakus, Krzysztof Kozłowski,
   Sebastian Musiał, and Bartłomiej Szwej

Algorithmic Daily Trading Based on Experts' Recommendations   735
   Andrzej Ruta, Dymitr Ruta, and Ling Cen

Author Index   745

Invited Paper

Abductive Reasoning on Compliance Monitoring

Balancing Flexibility and Regulation

Federico Chesani[1]([✉]), Paola Mello[1]([✉]), and Marco Montali[2]

[1] University of Bologna, Bologna, Italy
{federico.chesani,paola.mello}@unibo.it
[2] Free University of Bozen–Bolzano, Bolzano, Italy
montali@inf.unibz.it

Abstract. Many emerging applications in Business Process Management, Clinical Guidelines, Service-Oriented and Multi-Agent Systems, are characterized by distribution, complex interaction and coordination dynamics. Such domains, apparently unrelated, all ask for a suitable tradeoff between flexibility and regulation. In this light, compliance checking emerged as an effective way to understand whether an observed course of interaction agrees with what is expected by a model of the system. In this paper, we single out a non exhaustive list of desiderata and challenges for compliance checking applied at runtime. We then argue that methods, tools and techniques of Computational Logic, and Abductive Reasoning in particular, can be fruitfully exploited to tackle all such challenges in a formally grounded, computationally effective way.

Keywords: Compliance monitoring · Abductive logic programming · Business Process Management · Multi-agent Systems

1 Introduction

Many real world scenarios in the context of the new digital revolution, such as Business Process Management (BPM), e-Health, Web-Services, Multi Agent Systems, Self Adaptive Systems, and Internet of Things and People, require interaction among different entities such as sensors, smart objects, human users, (soft-)bots and APIs. Traces of these interactions can be conceived as streams of data, such as transactional records, logs of process activities, messages produced by multiple, possibly heterogeneous and autonomous sources, and recorded as ordered sets of events. Such data represent a footprint of reality, and their processing provides the basis to understand the relationship between the expected and experienced courses of interaction.

In this light, among the many types of data analysis, of particular interest is *compliance checking*, which compares models and observed events in order to detect possible discrepancies or deviations. Compliance may be assessed at

---

This is an invited paper, related to the keynote presentation given by Prof. Mello.

© Springer International Publishing AG 2017
M. Kryszkiewicz et al. (Eds.): ISMIS 2017, LNAI 10352, pp. 3–16, 2017.
DOI: 10.1007/978-3-319-60438-1_1

different stages of the lifecycle of a system. In general, though, compliance of complex interacting system cannot be guaranteed by design, nor completely assessed through static verification techniques. On the one hand, it is in fact rarely the case that all interacting components are controlled by a single, centralized orchestrator (think about a service company interacting with external suppliers). On the other hand, compliance typically requires to analyze data that are only available while the system runs (think about a regulation imposing that two tasks have to be executed by the same person). This is why we focus on *runtime compliance checking*, that is, on the problem of assessing compliance of partial, evolving courses of execution.

A plethora of desirable features have been identified in the literature as key towards effective compliance checking. First and foremost, complex scenarios cannot be described by complete and detailed models of interaction. Instead, they call for models that reflect adaptivity to change, and that are able to deal with incomplete information, i.e., models that enjoy *flexibility*. At the same time, interaction has to be disciplined by expressing that the involved entities are expected to behave in agreement with regulations, norms, business rules, protocols and time constraints. This calls for models that incorporate the notion of *regulation*. Beside seeking a suitable tradeoff between flexibility and regulation, other fundamental features have been identified, including support to open and closed models, clear semantics, data and time constraints, and incomplete information.

In this paper, we single out a non exhaustive list of desiderata and challenges for compliance checking applied at runtime. We then show that knowledge representation and reasoning in Artificial Intelligence, and in particular Computational Logic methods based on abduction [32,34], constitute a formally grounded, computationally effective framework to model such complex systems, and provide compliance monitoring facilities. We ground the discussion on the the SCIFF Abductive Framework [6], on the one hand showing how SCIFF integrity constraints capture flexible, expressive interactions models, and on the other hand discussing how the SCIFF abductive proof procedure provides an effective computational mechanism to monitor compliance at runtime.

2   Desired Features for Compliance Frameworks

In compliance checking (see the *Process Mining Manifesto* [2]) a model describing the dynamics of a system is compared with the events related to its concrete executions, to check if the event traces agree with the model's prescriptions.

In an organizational setting, the model is typically constituted by a business process, and its execution traces contain events marking the execution of tasks. Similar abstractions can be found in a plethora of other domains. In multiagent systems, models correspond to interaction protocols, while events are the messages/utterances exchanged by the agents. In healthcare, medical guidelines and clinical pathways play the role of models, while events are represented by the actions performed by the involved healthcare practitioners. In service-oriented computing, dynamic models of interest are orchestrations/choreographies, while events correspond to service invocations. Finally, in an IoT setting events emerge from sensor and device data, while models aim at guaranteeing global properties that the overall systems should exhibit.

In all these complex scenarios, compliance violations and deviations from what is expected by the model usually require to promptly act, so as to bring the system back within the boundaries defined by the model, plan compensating activities, and/or determine sanctions. This, in turn, calls for runtime compliance checking (or compliance monitoring), in which compliance is assessed over finite, evolving event traces, as defined in [42]. As extensively argued in the literature (see, e.g., [2,42,48]), compliance monitoring poses a number of challenges. We recap some of the most important ones in the remainder of this section.

Trade-off between Flexibility and Compliance to Regulation. Complex scenarios usually cannot be represented by complete and detailed models of interaction, and demands a high level of flexibility to promptly adapt to change. At the same time, interactions must be disciplined so as to ensure that involved entities comply with external regulations, norms, business rules, data and time constraints. In [2] four dimensions are identified, relatively to mining in general, and compliance in particular: (a) fitness, (b) simplicity, (c) precision, and (d) generalization. Balancing between such quality dimensions is challenging and highly influences the compliance monitoring quality too. Discrepancies may be interpreted as errors in the observed, or vice-versa as an inadequacy of the model in describing reality (or a combination thereof).

Another approach to tackle flexibility comes from the defeasible interpretation of constraints and norms [28,43]. Requirements can vary from rigid constraints (prescriptions) to soft rules (recommendations). This leads to a non-crisp interpretation of the compliance check, since violation of soft-constraints does not imply non-compliance of the whole, and some constraints may override/defeat other constraints. This is subject of discussion, as pointed out in [20], that advocates the importance of a crisp compliance check: rather than considering violable constraints, new (better) requirements should be elicited.

Real applications often call for exceptions and peculiar situations that could not be directly incorporated in the model without undermining its understandability, nor be ignored. Borrowing a solution from the Object Oriented Programming, we could say that violations should be treated as exceptions, i.e., first class objects/events. Adaptation mechanisms to violations should be supported in the model, and the compliance checking task should take them into account as well.

Open vs Closed Interaction Models. In compliance checking, the model plays a fundamental role in describing the coordination of activities, in terms of required and/or forbidden interactions. Traditionally, procedural models have been adopted for explicitly enumerating strict compliance requirements (e.g., BPMN [31] and Yawl [3] for Business Process, AUML [9] for agents, GLARE [54] for Careflows). Procedural specifications are interpreted as "closed" models, i.e., what is not explicitly cited in the model is forbidden. This is contrasted by declarative, constraint-based models, which adopt an "open" approach, which supports all courses of execution that satisfy a given set of constraints, compactly and implicitly describing a variety of behaviours. This approach has been proven especially suitable in all those complex scenarios where high flexibility and adaptation are needed, i.e., where procedural approaches would become "spaghetti-like". Notable examples of declarative, open models are the ConDec language for business process management [49], the CLIMB approach to interaction [44], and social commitments in open multiagent systems [51,52,55]). Similar positions can be recognized in the area of programming languages, when Kowalski advocated the need for declarative languages (and for logic programming in particular) in its seminal work "Algorithm = Logic + Control" [35].

Comprehensive Formal Semantics. To a large extent, compliance checking techniques come with ad-hoc implementations that are not backed up by a formal semantics. This poses a twofold problem. First, how compliance is verified is not precisely described, raising concerns about the interpretation of the compliance results. Second, it is not possible to assess the generality of the approach, i.e., if it could be used to deal with compliance in different stages of the system lifecycle, or to solve other problems related to compliance.

E.g., compliance is tightly related to model consistency ("Does the model allows any interaction at all?"), so as to understand whether the model used to assess compliance is faulty. At runtime, compliance could be instrumental to prediction/recommendation ("Given a not-yet completed interaction, what about its possible future courses? What to do next so as to remain compliant?").

Beyond Control Flow. Control-flow aspects (i.e., dynamic constraints among tasks such as sequencing, alternatives, parallel executions [4]) have been extensively studied in the past, also in the context of compliance [50]. However, it is often advocated that more "semantic" constraints shall be considered as well. In [29,43], semantic constraints/rules focus not only on the control flow dimension, but also tackle the data associated to the execution of tasks. The same happens in [41], where structural, data-aware and temporal compliance is tackled. Notice that time-related aspects (such as deadlines, duration constraints, delays, etc.) can be considered a special case data-aware constraints, with the obvious difference that the temporal domain comes with its own specific axioms/properties (e.g., that time flows only forward). Interestingly, approaches coming from the area of computational logic proved particularly useful in the combination of such different ingredients, not only to handle compliance monitoring [45,46], but also to deal with other forms of reasoning and verification [27,53].

Mixing Procedural and Declarative Aspects. Works like [29,43] argue that properties related to data and time do not always fit well with control flow modelling. Consider the following general rule applied in medicine: "Do not administer a certain drug if the patient is allergic to any of its components". This rule applies to each step in a process execution, and would not be well captured in the control flow description of the process. Generally speaking, this is the problem of suitably mixing procedural and declarative knowledge, and it emerges also in the legal setting when it comes to so-called *regulatory compliance*

[10]. Logic-based frameworks, such as [12, 28, 29], show that these two seemingly contrasting approaches can be captured in a coherent and unified way.

Missing Information. Information about the model and the execution can be incomplete; some events can be missing (or not observable) or described in a partial and/or noisy, and/or uncertain way. Hence, the degree of discrepancy between the interaction model and the execution becomes hard to be evaluated, and compliance checking can be a difficult hard task.

Also temporal information might be affected by incompleteness, with consequences like, for example, being difficult to predict the duration of the tasks, as well as to check inter-tasks temporal constraints. In [21], for example, an ad-hoc algorithm is proposed to ensure dynamic controllability of a workflow (i.e., there exists at least one way to ensure the compliance of a workflow, independently of some task and their unknown duration).

Extraction of Meaningful Events. Complex systems can provide a huge number of low-level, meaningless information. Two issues emerge here. The first concerns the conceptual granularity of such events: quite often, meaningful events (w.r.t. compliance) are not traceable, but need to be derived by properly aggregating the low-level, recorded data. This problem is connected to the more general problem of *activity/event recognition* (cf., e.g., [8]). The second issue concerns the manipulation of large-scale streams of low-level events, which falls under the umbrella of *complex event processing* [40].

Scalability of Compliance Monitoring. Issues arise when compliance needs to be assessed on large systems generating "big event data". In [1], van der Aalst argues that scalability issues can be tamed through horizontal and/or vertical partitioning, so as to distribute the compliance monitoring task over a network of nodes (e.g. multi-core systems, grid infrastructures, or cloud environments).

3   Abductive Logic Programming in Action

We now introduce the paradigm of abductive logic programming, and show how it can be used to reason on compliance, obtaining a general, coherent framework in which all aspects mentioned in Sect. 2 can be suitably tackled.

3.1   Abductive Logic Programming

Abduction is a non-monotonic reasoning process, where facts are observed, and hypotheses are made to explain the observed facts [33]. Given a Logic Program $\mathcal{P}$ and an observation $\mathcal{G}$ (the goal), deductive reasoning tries to prove that $\mathcal{G}$ is a logic consequence of $\mathcal{P}$. However, this might not be possible due to some missing knowledge in $\mathcal{P}$. Abduction tackles this issue by looking for a set Δ of hypotheses such that $(\mathcal{P} \cup \Delta)$ logically entails the goal. An abductive framework provides the reasoning tool for formulating such hypotheses.

Typically, not all configurations of hypotheses make sense. To declaratively capture how hypotheses relate to each other, *integrity constraints* are employed.

A typical integrity constraint (IC) is a *denial*, expressing that two explanations are mutually exclusive. Given a set $\mathcal{IC}$ of integrity constraints, an abductive explanation Δ must be such that $(\mathcal{P} \cup \Delta)$ logical entails all ICs present in $\mathcal{IC}$.

Abductive Logic Programs (ALP) have been formalized in [32]. There, an ALP is defined as a triple $\langle \mathcal{KB}, \mathcal{A}, \mathcal{IC} \rangle$, where: *(i)* $\mathcal{KB}$ is a logic program, *(ii)* $\mathcal{A}$ is a set of abducible predicates, and *(iii)* $\mathcal{IC}$ a set of ICs. Given a goal $\mathcal{G}$, abductive reasoning looks for a set of literals $\Delta \subseteq \mathcal{A}$ such that they entail $\mathcal{G} \cup \mathcal{IC}$. The integration of constraint solving (ACLP) [22,32] enhances the practical utility of ALP by enriching the representation of the problem domain, and by empowering the computation of abductive explanations.

3.2 The SCIFF Framework

SCIFF [6] is a ACLP framework that extends Fung and Kowalski's IFF proof-procedure for ALP [26] towards compliance checking. Conceptually, this is done by introducing two types of special predicates, *happened events* and *expectations*.

A *happened event*, denoted by $\mathbf{H}(Ev, T)$, captures the idea that event Ev occurs at timestamp T. Ev is a logical term, and consequently may refer to a complex description of the event and its data. For example, fact

$$\mathbf{H}(administer\_drug(p42, paracetamol, mg, 500), 45).$$

indicates that at time 45, 500 mg of paracetamol have been given to the patient $p42$. A finite set of facts denoting happened events constitutes a *trace*.

Expectations capture the expected behaviour, and come with a positive or a negative flavour. A *positive expectation* $\mathbf{E}(Ev, T)$ states that event Ev is expected to happen at time T. Variables employed therein are quantified existentially. A *negative expectation* $\mathbf{EN}(Ev, T)$ states that event Ev is expected not to happen at time T. Variables employed therein are quantified universally. For example

$$\mathbf{E}(administer\_drug(p34, paracetamol, mg, 1000), T_1) \wedge T_1 < 78$$
$$\wedge \, \mathbf{EN}(administer\_drug(p67, paracetamol, mg, Q), T_2).$$

models that two expectations are in place. First, patient $p34$ is expected to receive 1 g of paracetamol within the deadline of 78. Second, patient $p67$ is expected to not receive any quantity of paracetamol at any time.

In SCIFF, models are specified through three components: (i) a knowledge base $\mathcal{KB}$, i.e. a set of (backward) rules *head $\leftarrow$ body*; (ii) a set $\mathcal{IC}$ of ICs, each of which is a (forward) rule of the form *body $\rightarrow$ head*; and (iii) a goal $\mathcal{G}$ (as in logic programming). The integrity constraints can be considered as reactive rules, i.e., when the body becomes true (because of the occurrence of the involved events), then the rule "fires" and the expectations in the head are generated. E.g., IC

$$\mathbf{H}(temperature(P, Temp), T_1) \wedge Temp > 38 \rightarrow$$
$$\mathbf{E}(administer\_drug(P, paracetamol, mg, 1000), T_2) \wedge T_2 > T_1 \wedge T_2 < T_1 + 10.$$

states that if a patient P is detected to have the temperature at a time T_1, then it is expected that P is given 1 g paracetamol, within ten time units after T_1.

Intuitively, SCIFF defines compliance as a sort of "hypothesis" confirmation relating happened events and expectations: a trace is compliant with a SCIFF specification if all expectations generated by combining the trace with the ICs are confirmed, i.e., for every positive expectation (**E**) there is a *corresponding* happened event (**H**) in the trace, and if for every negative expectation (**EN**) there is no *corresponding* happened event (**H**) in the trace. The *correspondence* between events and expectations is grounded on the notion of *unification* in Logic Programming [38]. Finally, SCIFF is inherently "open": everything that is not explicitly forbidden is indeed allowed. The interested reader might refer to [6] for a comprehensive description of the SCIFF Framework.

3.3   On the Suitability of SCIFF to Monitor Compliance

We now go through the different desired features for compliance frameworks discussed in Sect. 2, and argue how they can be tackled using SCIFF.

Formal Semantics, Flexibility, Open/Closed Models. Being SCIFF grounded on ALP, it natively comes with a formal declarative semantics, and with a corresponding (sound and complete) proof procedure to generate and confirm expectations. Furthermore, the notion of expectation confirmation briefly recalled in Sect. 3.2 allows us to lift the semantics of ALP to handle (runtime and offline) compliance [44]. Other reasoning tasks, such as model consistency and prediction/recommendation, can be supported as well (see our preliminary works in [14,15]). The definition of compliance rules in terms of logic-based specifications also allowed us to learn rules from traces, by realizing a form of discriminative mining through inductive logic programming techniques [16].

Notice that a number of concrete, end-user oriented languages for interaction/process modeling can be translated into SCIFF, and consequently directly inherit such compliance-related functionalities. E.g., we translated into SCIFF the constraint-based service interaction language DecSerFlow [47], and the AUML language for multiagent interaction protocols [5].

Notably, DecSerFlow and AUML are radically different in the way they approach interaction models, since they respectively adopt an open and closed approach to interaction. In Sect. 3.2, we argued that SCIFF adopts an open approach, so one might wonder how is it possible to capture closed models in SCIFF. We show this versatility by considering the sequencing between two events a and b, that is, the compliance rule indicating that whenever a occurs, then b should occur afterwards. In an open setting, any other event may happen before, during and after the sequence. The SCIFF formalization of this open interpretation is constituted by the IC:

$$\mathbf{H}(a, T_a) \rightarrow \mathbf{E}(b, T_b) \wedge T_b > T_a.$$

What if, instead, one would prefer a closed interpretation of the sequence, that a and b should occur next to each other? This could be declaratively captured by suitably exploiting temporal constraints, negative expectations, and logic programming variables in the following two ICs:

$$\mathbf{H}(a, T_a) \rightarrow \mathbf{E}(b, T_b) \wedge T_b > T_a \wedge \mathbf{EN}(X, T_x) \wedge T_a < T_x < T_b.$$
$$\mathbf{H}(b, T_b) \rightarrow \mathbf{E}(a, T_a) \wedge T_b > T_a \wedge \mathbf{EN}(X, T_x) \wedge T_a < T_x < T_b.$$

The first IC models again the sequence, but it also forbids the occurrence of any event between a and b. The second IC instead completes the notion of "chaining" between a and b, indicating that if b occurs, a should have occurred previously.

This modelling technique can be tuned so as to capture hybrid models combining openness and closeness. Finally, a generalized notion of closeness may also be realized by revisiting the declarative semantics of compliance provided by SCIFF, in particular expressing that a trace is compliant if *and only if* for every happened events there is a corresponding expectation, *and vice-versa*.

Beyond Control Flow. Since SCIFF is based on (constraint) logic programming, it inherits the expressive power needed to capture background, domain knowledge in the form of a knowledge base. In addition, events may be represented using complex, structured terms involving data, and data-aware constraints are seamlessly supported as constraints over variables used inside such terms. As shown above, special variables are used to explicitly denote timestamps, consequently allowing one to specify both qualitative (e.g., response, precedence, sequencing) and quantitative (e.g., delays, deadlines) temporal conditions by means of constraints involving such special variables.

Mixing Procedural and Declarative Aspects. SCIFF naturally lends itself to model declarative compliance rules, but procedural flow constructs can be captured as well [15,17]. Unfortunately, a direct combination of declarative and procedural specifications is not always satisfactory, since the two may impose conflicting requirements. For example, in the healthcare domain procedural guidelines are used to indicate a default treatment, and subtly interact with the background medical knowledge used to avoid harmful situations. E.g., a guideline for treating pneumonia may prescribe the administration of penicillin, which would conflict with the general clinical practice if the patient is allergic to penicillin. The conflict is resolved, in this case, by administering a different drug with equivalent effects. The mindful reader might object that this interaction can be tackled by explicitly incorporating alternative choices within the guideline. However, this approach would hinder the readability of the guideline, and may become unmanageable in general, since it requires to foresee, at modelling time, all possible subtle interactions between these two sources of knowledge.

A suitable way to manage this types of conflict while retaining understandability and flexibility is an open challenge. In [11] we separated the representation of the guideline and the basic medical knowledge, and assumed a user-defined priority of the medical knowledge with respect to the guideline prescriptions. This solution is peculiar to the medical field, and cannot be directly generalized. Relevant directions of research in this respect are approaches that assign different weights to rules, like defeasible logic, default logic, and variants of preference logic [7,30].

Missing Information. In a number of applications the collection of event data is far from being precise and complete: some events may not be monitorable, or may be monitored by error-prone systems, thus producing incomplete, corrupted and/or noisy traces. Such common situations require to suitably revise the notion of compliance, so as to distinguish noncompliant traces from traces that, due to logging issues, appear to be noncompliant. Thanks to its abductive semantics, SCIFF naturally lends itself to deal with this setting, as shown in [14,15]. There, a novel approach to compliance is presented, where beside yes/no answer, also the concept of weak compliance is explored. The idea is that a trace with missing information might be weak compliant with respect to a SCIFF specification, if it is possible to make a set of consistent hypotheses about the missing information. This approach simultaneously accounts for missing events, events with missing data (e.g., a logged event without the indication of "who" generated it), and events with missing time. Intuitively, missing events are handled by allowing SCIFF to hypothesise the occurrence of expected events, and to assess weak compliance by checking whether the overall set of formulated hypotheses is consistent. Traces with incomplete information are instead dealt with by allowing the happened events contained therein to include variables, and letting SCIFF formulating hypotheses on the values that such variables may take.

Of course, the problem of incomplete information in compliance may be tackled with other techniques, such as planning [25], and would also benefit from techniques that are able to tame uncertainty, such as fuzzy set theory, probabilistic reasoning, evolutionary computing, and machine learning.

Extraction of Meaningful Events. A well-established approach to complex event/activity recognition and processing in a declarative setting is that of the Event Calculus (EC) [36]. EC is a powerful logic-based framework to reason on the effects that events have on the "state" of the world. On the one hand, this mechanism can be used to infer high-level events from low level samples [8]. On the other hand, it can be exploited to reason on the effects that events have on the normative state of business interactions, which in turn can be used to relate compliance at the level of events with compliance at the level of contracts and commitments [19,55].

Notably, in [18] it has been shown that the EC can be formalized in SCIFF, with a twofold advantage. From the semantical point of view, event-based and state-based specifications are reconciled in a single logical framework. From the operational point of view, the SCIFF proof procedure provides a very effective form of runtime reactive reasoning for the EC, which is quite unique in the literature.

Scalability of Compliance Monitoring. The expressiveness of SCIFF, and the sophistication of its proof procedure, make it sensible to performance and scalability issues. An open challenge concerns the exploitation of modern parallelization and distribution architectures so as to improve the efficiency of compliance monitoring in SCIFF.

Interesting insights towards this goal come from the BPM field, where [1] has proposed an approach based on horizontal and vertical partitioning, to split traces and models and decompose the evaluation of compliance accordingly. Specifically, vertical partition assumes the existence of a special data slot, called *case identifier*, that is attached to all happened events and that is used to correlate and group them into different traces depending on the case. As an example, consider a loan management process: while the information system used to handle loans would log all events about all requested loans, such events may be separated into different traces on the basis of the loan identifier. This impacts on compliance, as each trace can be combined with the regulatory model of interest independently form the other traces, paving the way towards a direct parallelization/distribution of compliance monitoring.

In horizontal partitioning the regulatory model and the monitored traces are decomposed into sub-models and sub-traces, on the basis of the relevance of events to the different sub-models. This technique is effective if the overall framework guarantees some *compositionality* property, that is, if compliance of a trace with a regulatory model can be simply derived by checking compliance of its sub-traces with the corresponding sub-models. In general, the determination of sub-models and sub-traces so as to preserve compositionality is a difficult task, especially in the case of procedural models, while logic-based approaches naturally lend themselves to be decomposed.

Vertical and horizontal partitioning techniques have been reconstructed within SCIFF in [39], paving the way towards compliance monitoring via Map-Reduce. A similar approach has been investigated in [13], for the context of multiagent systems.

4   Conclusions

In this overview paper, we have introduced the important problem of compliance checking, focusing in particular on runtime compliance monitoring in the context of several application domains. We have argued that abductive reasoning, and in particular the SCIFF abductive logic programming framework, provides a comprehensive and rich approach to model sophisticated, flexible regulatory specifications, and to operationally monitor compliance. We have also outlined a number of challenges, how they can be tackled in SCIFF, and how they open interesting lines of research for the future.

It is worth recalling that checking compliance in SCIFF amounts to formulate hypotheses and abductive explanations that provide very valuable insights to human stakeholders. In this respect, this approach is fully in line with the recent debate on AI as "artificial" vs "augmented" intelligence.

Two additional open challenges are worth to be mentioned when it comes to aiding humans in compliance assessment. A first crucial challenge concerns how to aggregate and combine multiple alternative hypotheses formulated by SCIFF when monitoring compliance, so as to obtain a sort of integrated, compact explanation for the essential causes for noncompliance. Specifically, we are interested in understanding whether it is possible, in the context of compliance, to reconstruct a notion similar to that of *event structure* [23], which has been advocated as a unifying representation formalism for processing models and traces.

A second important challenge is how to exploit the formal foundations provided by SCIFF so as to go beyond compliance checking. As witnessed by recent works [24,37], there is a great interest towards more advanced techniques to estimate the "distance" between a regulatory model and a monitored trace, possibly also indicating how to realign/repair the trace and/or the model so as to restore compliance.

References

1. van der Aalst, W.M.P.: Distributed process discovery and conformance checking. In: de Lara, J., Zisman, A. (eds.) FASE 2012. LNCS, vol. 7212, pp. 1–25. Springer, Heidelberg (2012). doi:10.1007/978-3-642-28872-2_1
2. van der Aalst, W.M.P., et al.: Process mining manifesto. In: Daniel, F., Barkaoui, K., Dustdar, S. (eds.) BPM 2011. LNBIP, vol. 99, pp. 169–194. Springer, Heidelberg (2012). doi:10.1007/978-3-642-28108-2_19
3. van der Aalst, W.M.P., ter Hofstede, A.H.M.: YAWL: yet another workflow language. Inf. Syst. **30**(4), 245–275 (2005)
4. van der Aalst, W.M.P., ter Hofstede, A.H.M., Kiepuszewski, B., Barros, A.P.: Workflow patterns. Distrib. Parallel Databases **14**(1), 5–51 (2003)
5. Alberti, M., Chesani, F., Daolio, D., Gavanelli, M., Lamma, E., Mello, P., Torroni, P.: Specification and verification of agent interaction protocols in a logic-based system. Scalable Comput.: Pract. Exp. **8**(1), 1–13 (2007)
6. Alberti, M., Chesani, F., Gavanelli, M., Lamma, E., Mello, P., Torroni, P.: Verifiable agent interaction in abductive logic programming: the SCIFF framework. ACM Trans. Comput. Log. **9**(4), 29:1–29:43 (2008)
7. Antoniou, G., Billington, D., Governatori, G., Maher, M.J.: Representation results for defeasible logic. ACM Trans. Comput. Log. **2**(2), 255–287 (2001)
8. Artikis, A., Sergot, M.J., Paliouras, G.: An event calculus for event recognition. IEEE Trans. Knowl. Data Eng. **27**(4), 895–908 (2015)
9. Bauer, B., Cossentino, M., Cranefield, S., Huget, M.P., Kearney, K., Levy, R., Nodine, M., Odell, J., Cervenka, R., Turci, P., Zhu, H.: The FIPA Agent UML. http://www.auml.org/
10. Boella, G., Janssen, M., Hulstijn, J., Humphreys, L., van der Torre, L.W.N.: Managing legal interpretation in regulatory compliance. In: ICAIL 2013, pp. 23–32. ACM (2013)
11. Bottrighi, A., Chesani, F., Mello, P., Montali, M., Montani, S., Terenziani, P.: Conformance checking of executed clinical guidelines in presence of basic medical knowledge. In: Daniel, F., Barkaoui, K., Dustdar, S. (eds.) BPM 2011. LNBIP, vol. 100, pp. 200–211. Springer, Heidelberg (2012). doi:10.1007/978-3-642-28115-0_20
12. Bragaglia, S., Chesani, F., Mello, P., Montali, M.: Conformance verification of clinical guidelines in presence of computerized and human-enhanced processes. In: Hommersom, A., Lucas, P.J.F. (eds.) Foundations of Biomedical Knowledge Representation. LNCS, vol. 9521, pp. 81–106. Springer, Cham (2015). doi:10.1007/978-3-319-28007-3_6
13. Briola, D., Mascardi, V., Ancona, D.: Distributed runtime verification of JADE and Jason multiagent systems with prolog. In: CILC 2014, CEUR, vol. 1195, pp. 319–323 (2014)
14. Chesani, F., De Masellis, R., Di Francescomarino, C., Ghidini, C., Mello, P., Montali, M., Tessaris, S.: Abducing compliance of incomplete event logs. In: Adorni, G., Cagnoni, S., Gori, M., Maratea, M. (eds.) AI*IA 2016. LNCS, vol. 10037, pp. 208–222. Springer, Cham (2016). doi:10.1007/978-3-319-49130-1_16

15. Chesani, F., De Masellis, R., Francescomarino, C.D., Ghidini, C., Mello, P., Montali, M., Tessaris, S.: Abducing workflow traces: a general framework to manage incompleteness in business processes. In: ECAI 2016. Frontiers in Artificial Intelligence and Applications, vol. 285, pp. 1734–1735. IOS Press (2016)
16. Chesani, F., Lamma, E., Mello, P., Montali, M., Riguzzi, F., Storari, S.: Exploiting inductive logic programming techniques for declarative process mining. In: Jensen, K., Aalst, W.M.P. (eds.) Transactions on Petri Nets and Other Models of Concurrency II. LNCS, vol. 5460, pp. 278–295. Springer, Heidelberg (2009). doi:10.1007/978-3-642-00899-3_16
17. Chesani, F., Mello, P., Montali, M., Storari, S.: Testing careflow process execution conformance by translating a graphical language to computational logic. In: Bellazzi, R., Abu-Hanna, A., Hunter, J. (eds.) AIME 2007. LNCS, vol. 4594, pp. 479–488. Springer, Heidelberg (2007). doi:10.1007/978-3-540-73599-1_64
18. Chesani, F., Mello, P., Montali, M., Torroni, P.: A logic-based, reactive calculus of events. Fundam. Inform. **105**(1–2), 135–161 (2010)
19. Chesani, F., Mello, P., Montali, M., Torroni, P.: Representing and monitoring social commitments using the event calculus. Auton. Agents Multi-Agent Syst. **27**(1), 85–130 (2013)
20. Chopra, A.K.: Requirements-driven adaptation: compliance, context, uncertainty, and systems. In: RE@RunTime 2011, pp. 32–36. IEEE (2011)
21. Combi, C., Posenato, R.: Towards temporal controllabilities for workflow schemata. In: TIME 2010, pp. 129–136. IEEE (2010)
22. Denecker, M., Kakas, A.: Abduction in logic programming. In: Kakas, A.C., Sadri, F. (eds.) Computational Logic: Logic Programming and Beyond. LNCS, vol. 2407, pp. 402–436. Springer, Heidelberg (2002). doi:10.1007/3-540-45628-7_16
23. Dumas, M., García-Bañuelos, L.: Process mining reloaded: event structures as a unified representation of process models and event logs. In: Devillers, R., Valmari, A. (eds.) PETRI NETS 2015. LNCS, vol. 9115, pp. 33–48. Springer, Cham (2015). doi:10.1007/978-3-319-19488-2_2
24. Fahland, D., van der Aalst, W.M.P.: Model repair - aligning process models to reality. Inf. Syst. **47**, 220–243 (2015)
25. Francescomarino, C., Ghidini, C., Tessaris, S., Sandoval, I.V.: Completing workflow traces using action languages. In: Zdravkovic, J., Kirikova, M., Johannesson, P. (eds.) CAiSE 2015. LNCS, vol. 9097, pp. 314–330. Springer, Cham (2015). doi:10.1007/978-3-319-19069-3_20
26. Fung, T.H., Kowalski, R.A.: The IFF proof procedure for abductive logic programming. J. Log. Program. **33**(2), 151–165 (1997)
27. Giordano, L., Martelli, A., Spiotta, M., Dupré, D.T.: Business process verification with constraint temporal answer set programming. TPLP **13**(4–5), 641–655 (2013)
28. Governatori, G., Hashmi, M., Lam, H.-P., Villata, S., Palmirani, M.: Semantic business process regulatory compliance checking using LegalRuleML. In: Blomqvist, E., Ciancarini, P., Poggi, F., Vitali, F. (eds.) EKAW 2016. LNCS, vol. 10024, pp. 746–761. Springer, Cham (2016). doi:10.1007/978-3-319-49004-5_48
29. Governatori, G., Rotolo, A.: Norm compliance in business process modeling. In: Dean, M., Hall, J., Rotolo, A., Tabet, S. (eds.) RuleML 2010. LNCS, vol. 6403, pp. 194–209. Springer, Heidelberg (2010). doi:10.1007/978-3-642-16289-3_17
30. Greco, S., Trubitsyna, I., Zumpano, E.: On the semantics of logic programs with preferences. J. Artif. Intell. Res. (JAIR) **30**, 501–523 (2007)
31. Initiative, B.P.M.: Business process modeling notation. http://www.bpmn.org
32. Kakas, A.C., Kowalski, R.A., Toni, F.: Abductive logic programming. J. Log. Comput. **2**(6), 719–770 (1992). http://dx.doi.org/10.1093/logcom/2.6.719

33. Kakas, A.C., Mancarella, P.: Abduction and abductive logic programming. In: Logic Programming, Proceedings of ICPL, pp. 18–19 (1994)
34. Kakas, A.C., Michael, A., Mourlas, C.: ACLP: abductive constraint logic programming. J. Log. Program. **44**(1–3), 129–177 (2000)
35. Kowalski, R.A.: Algorithm = logic + control. Commun. ACM **22**(7), 424–436 (1979)
36. Kowalski, R.A., Sergot, M.J.: A logic-based calculus of events. New Gener. Comput. **4**(1), 67–95 (1986)
37. de Leoni, M., Maggi, F.M., van der Aalst, W.M.P.: An alignment-based framework to check the conformance of declarative process models and to preprocess event-log data. Inf. Syst. **47**, 258–277 (2015)
38. Lloyd, J.W.: Foundations of Logic Programming, 2nd edn. Springer, Berlin (1987)
39. Loreti, D., Chesani, F., Ciampolini, A., Mello, P.: Distributed compliance monitoring of business processes over mapreduce architectures. In: ICPE 2017, to appear
40. Luckham, D.C.: The Power of Events: An Introduction to Complex Event Processing in Distributed Enterprise Systems. Addison-Wesley, Boston (2001)
41. Ly, L.T., Knuplesch, D., Rinderle-Ma, S., Göser, K., Pfeifer, H., Reichert, M., Dadam, P.: SeaFlows toolset – compliance verification made easy for process-aware information systems. In: Soffer, P., Proper, E. (eds.) CAiSE Forum 2010. LNBIP, vol. 72, pp. 76–91. Springer, Heidelberg (2011). doi:10.1007/978-3-642-17722-4_6
42. Ly, L.T., Maggi, F.M., Montali, M., Rinderle-Ma, S., van der Aalst, W.M.P.: Compliance monitoring in business processes: functionalities, application, and tool-support. Inf. Syst. **54**, 209–234 (2015)
43. Ly, L.T., Rinderle-Ma, S., Göser, K., Dadam, P.: On enabling integrated process compliance with semantic constraints in process management systems - requirements, challenges, solutions. Inf. Syst. Front. **14**(2), 195–219 (2012)
44. Montali, M.: Specification and Verification of Declarative Open Interaction Models - A Logic-Based Approach. LNBIP, vol. 56. Springer, Heidelberg (2010). doi:10.1007/978-3-642-14538-4
45. Montali, M., Chesani, F., Mello, P., Maggi, F.M.: Towards data-aware constraints in declare. In: Proceedings of SAC 2013. ACM (2013)
46. Montali, M., Maggi, F.M., Chesani, F., Mello, P., van der Aalst, W.M.P.: Monitoring business constraints with the event calculus. ACM TIST **5**(1), 17:1–17:30 (2013)
47. Montali, M., Pesic, M., van der Aalst, W.M.P., Chesani, F., Mello, P., Storari, S.: Declarative specification and verification of service choreographies. TWEB **4**(1), 3:1–3:62 (2010)
48. Munoz-Gama, J.: Conformance Checking and Diagnosis in Process Mining - Comparing Observed and Modeled Processes. LNBIP, vol. 270. Springer, Heidelberg (2016). doi:10.1007/978-3-319-49451-7
49. Pesic, M., van der Aalst, W.M.P.: A declarative approach for flexible business processes management. In: Eder, J., Dustdar, S. (eds.) BPM 2006. LNCS, vol. 4103, pp. 169–180. Springer, Heidelberg (2006). doi:10.1007/11837862_18
50. Rozinat, A., van der Aalst, W.M.P.: Conformance checking of processes based on monitoring real behavior. Inf. Syst. **33**(1), 64–95 (2008)
51. Singh, M.P.: Agent communication languages: rethinking the principles. In: Huget, M.-P. (ed.) Communication in Multiagent Systems. LNCS, vol. 2650, pp. 37–50. Springer, Heidelberg (2003). doi:10.1007/978-3-540-44972-0_2
52. Singh, M.P., Chopra, A.K., Desai, N.: Commitment-based service-oriented architecture. IEEE Comput. **42**(11), 72–79 (2009)

53. Smith, F., Proietti, M.: Rule-based behavioral reasoning on semantic business processes. In: ICAART 2013. SciTePress (2013)
54. Terenziani, P., Raviola, P., Bruschi, O., Torchio, M., Marzuoli, M., Molino, G.: Representing knowledge levels in clinical guidelines. In: Horn, W., Shahar, Y., Lindberg, G., Andreassen, S., Wyatt, J. (eds.) AIMDM 1999. LNCS, vol. 1620, pp. 254–258. Springer, Heidelberg (1999). doi:10.1007/3-540-48720-4_28
55. Yolum, P., Singh, M.P.: Commitment machines. In: Meyer, J.-J.C., Tambe, M. (eds.) ATAL 2001. LNCS, vol. 2333, pp. 235–247. Springer, Heidelberg (2002). doi:10.1007/3-540-45448-9_17

Bioinformatics and Health Informatics

Multi-levels 3D Chromatin Interactions Prediction Using Epigenomic Profiles

Ziad Al Bkhetan[1,2,5,6] and Dariusz Plewczynski[1,3,4,7(✉)]

[1] Centre of New Technologies, Warsaw University, Warsaw, Poland
{z.albkhetan,d.plewczynski}@cent.uw.edu.pl
[2] Biology Department, Warsaw University, Warsaw, Poland
[3] Faculty of Pharmacy, Medical University of Warsaw, Warsaw, Poland
[4] Centre for Innovative Research, Medical University of Bialystok,
Białystok, Poland
[5] Department of Computing and Information Systems,
The University of Melbourne, Parkville, VIC, Australia
[6] Centre for Neural Engineering, The University of Melbourne,
Parkville, VIC, Australia
[7] Faculty of Mathematics and Information Science,
Warsaw University of Technology, Warsaw, Poland

Abstract. Identification of the higher-order genome organization has become a critical issue for better understanding of how one dimensional genomic information is being translated into biological functions. In this study, we present a supervised approach based on Random Forest classifier to predict genome-wide three-dimensional chromatin interactions in human cell lines using 1D epigenomics profiles. At the first level of our in silico procedure we build a large collection of machine learning predictors, each one targets single topologically associating domain (TAD). The results are collected and genome-wide prediction is performed at the second level of multi-scale statistical learning model. Initial tests show promising results confirming the previously reported studies. Results were compared with Hi-C and ChIA-PET experimental data to evaluate the quality of the predictors. The system achieved 0.9 for the area under ROC curve, and 0.86–0.89 for accuracy, sensitivity and specificity.

Keywords: 3D genome organization · 3D chromatin interactions · Epigenomics · Hi-C · ChIA-PET · CTCF motifs · Statistical learning · Physical interactions · Chromatin looping · Topologically associated domains · Chromatin contact domains · Random Forest

1 Introduction

The DNA of the mammalian cell is well organized together with histone proteins forming nucleosomes. It is packed into 10 nm chromatin fiber, that allows approximately two meters of human DNA [1] to be fitted in 6–10 μm diameter nucleus [2]. The biological function of a genome is influenced by major factors: one dimensional genomic data such as the DNA sequence of coding (genes) and non-coding (regulatory motifs) regions, further modifications of chromatin described as epigenomics; and

© Springer International Publishing AG 2017
M. Kryszkiewicz et al. (Eds.): ISMIS 2017, LNAI 10352, pp. 19–28, 2017.
DOI: 10.1007/978-3-319-60438-1_2

finally the three-dimensional structure of chromatin at the scale of loops, domains, chromosomal territories and compartments. In this contribution, we focus on the association between 1D epigenomic information and 3D interactions. Epigenetics sequence annotation could be categorized into DNA methylation and the post-translational modifications of the histone proteins (i.e., histone modifications) [4]. Epigenetics play a major role in gene activity regulation either by up-regulation, or down-regulation of the associated genes [5], while the genome 3D organization can actively organize the physical interaction of DNA regulatory elements (e.g. enhancers, promoters, etc.). The three-dimensional structure is defined by Chromatin Interactions (CI), similarly like in the case of proteins their contact maps determine their spatial conformation [6]. Detailed analysis of chromatin interactions that define the three-dimensional structure of the genome is still an open question that requires further investigations.

Hi-C experiments can detect chromatin interactions mediated by any protein but it requires high sequencing depth to obtain accurate results [3]. Chromatin Interaction Analysis by Paired-End Tag sequencing (ChIA-PET) experiments are used to detect subset of interactions mediated by specific proteins, such as CTCF and RNAPII [7, 8]. Those and some other methodologies for chromosomes conformation capture (3C) experiments are modified and specifically tailored to detect either local or global spatial interactions at unprecedented resolution. Yet, they are always affected by noise introducing false positive interactions, or by unavoidable systemic biases. Furthermore, many genome-wide high-resolution experiments (reaching 1 kb genomic size of interacting chromatin segments) were recently introduced to find these interactions such as in situ Hi-C [9], and new long-read ChIA-PET method using CTCF immunoprecipitation [8] that our study focused on. Each of them has its advantages and disadvantages, together with other newly introduced chromosome conformation capture (3C-based) approaches. The long-read ChIA-PET experimental data were accompanied by computational simulations [10] bridging the structural and functional interpretation of chromatin contacts and allowing biologists to visualize their data using specifically tailored bioinformatics web server [11]. Following the development of experimental and computational methods, there is a need for machine learning algorithms that can either predict the interactions at the higher spatial scales using accurate ChIP-seq experiments, or allow denoising of 3C-type of experimental data by identifying false interactions. One of such methods, EpiTensor [3]: unsupervised learning method introduced recently to predict the chromatin interactions, and detect high-resolution interactions at the genomic scale of 200 bp within the topologically associating domains (TADs) with the high accuracy. Yet, this method is yet not very effective in prediction of the physical interactions mediated by specific protein factors (like CTCF), and promoter-exon and exon-exon interactions. The Epitensor method together with earlier studies by various authors proved the possibility of predicting 3D chromatin interactions by analyzing 1D epigenomic data collected for the same cell line. But, does the epigenetics pattern differ across the whole collection of TADs in the same cell type? Does it differ between the chromosomes in the same cell type? Does it follow the same pattern at different levels (domains, chromosomes, compartments, whole-genomes, cell types) among different individuals? Translating these questions into more precise context: do the high resolution Hi-C and ChIA-PET interactions have the same epigenetics profiles in the interacting segments across domains, chromosomes, compartments, cell types, and finally individuals at the scale of whole populations? These are open questions of great importance that we are trying to approach by this preliminary study. Previous studies were able to predict genome-wide chromatin interactions based on epigenomic data [3, 12]. In our study, we applied supervised learning approach to predict chromatin interactions using 1D epigenomic profiles and the spatial distance between the segments involved in these interactions. The study aims to build multi-level machine learning predictors starting from specifically tailored statistical models trained separately on each level of topologically associating domains, chromosomes and compartments, then the whole genome level and each cell type meta-predictors. In addition, we aim at evaluating the in silico predictors trained on one specific cell type and tested on different one to see if the epigenomics profiles follow the same pattern in different cell types. In this preliminary study, we report interactions at the TADs level because intra-domain interactions are dominating in typical Hi-C or ChIA-PET experiments [3, 8, 9, 12]. It allows also to reduce the computational time needed to train the statistical models at the whole genome scale.

The data used in this study was obtained from high-throughput experiments conducted on the following cell lines: Human lymphoblast GM12878, Human lung fibroblasts IMR90, and Human embryonic stem cells H1.

2   Pipeline Description

2.1   Features Encoding

Two different features encoding schemes were applied to represent the genomic segments:

1. **Density Coverage Encoding:** The density coverage of all histone modifications was mapped to the genome using 1 kb resolution, the percentage of each histone modification was assigned to the related segment in the genome.
2. **Peaks Information Encoding:** MACS2 peaks calling method [13] was invoked for each histone modification assay, then the peak height and the distance between the peak summit and the center of the segment were assigned to each 1 kb genomic segment.

2.2   Data Preparation

Features encoding was applied for all genomic segments with the same resolution (1 kb or 5 kb). Both encoding methods were used in different experiments. All possible pairs formed from the genomic segments in the same TAD were mapped to interactions confirmed by either Hi-C, or ChIA-PET according to the applied experiment. These interactions were used to train the predictor then evaluate its performance. In the first phase we built one predictor for each single TAD, so all pairs belong to the same TAD were split into training and testing datasets. During data partitioning phase, all pairs related to any segment exist in one dataset either training or testing, but not in both.

2.3   Building Random Forest Classifier

We chose Random Forests Classifier to accomplish the prediction task, as it improves
the generalization accuracies by using group of decision trees working together. We
aim in the next steps at trying different classifiers to assess their performance in such
problems. The Random Forest classifier was trained on the training dataset for each
TAD, then evaluated on the testing dataset for the same TAD. Moreover, we tested
predictors on the same TADs but different cell lines to assess if the epigenomics follow
the same pattern in different cell types. Figure 1 illustrates the whole pipeline used in
the study.

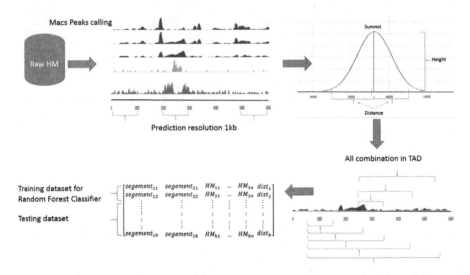

Fig. 1. The whole workflow when using peaks encoding. 1- MACS2 is invoked for all histone
modifications involved in the test. 2- Peaks are mapped to the genome at 1 kb resolution. 3-
Peaks height and the distance between the peak summit and the center of the segment are used to
represent each segment. 4- For each TAD, all possible combinations of segments are formed. 5-
These pairs were split into training and testing datasets (80–20%).

3   Results

The initial results obtained using above two features encoding methods for interacting
segments representation showed the better efficiency is when using peaks information.
That is why we suggest using this encoding scheme for the final genome-wide predictor. Nevertheless, we report here the features ranking for both encoding schemes, to
identify which features are present in both cases, and which ones are unique. First,
density coverage feature encoding was applied on IMR90 and H1 hESC cell lines,
using sixteen (16) histone modification assays in addition to DNA accessibility assay.
The distance between each pair anchors was also included in the features. All epigenomic data was mapped to the human genome version hg19 using 1 kb resolution.
Hi-C interactions were used for training and testing datasets as true interaction

representing spatial proximity. IMR90 in situ Hi-C interactions were obtained from reference [9], while Hi-C interactions for H1 hESC were obtained from reference [14]. We present results of this encoding scheme for selected domains (TADs) from chromosome 1. The domain-level predictors trained on H1 hESC cell line were tested on the same domains in IMR90 cell line, and vice versa. Figure 2 illustrates the most important features according to the Random Forest classifier. The distance between the pair's segments was the most important feature, which was expected as all Hi-C heatmaps show that the interactions are denser close to the diagonal. The rankings of features importance at the same TADs was almost identical in these two cell lines. This suggests that statistical learning trained on one cell type could predict the interactions in another cell type.

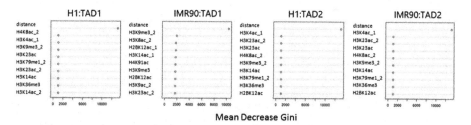

Fig. 2. Features importance in IMR90 and H1 hECS, using density coverage encoding scheme. TAD1 - chr1:43691938-46036881, TAD2 - chr1:3826747-5976883. Y-axis represent the histone modifications, the numeric suffix for each feature could be interpreted as *1*: the maximum value of left and right segments, *2*: refers to the minimum, ***none***: the multiplication of the left and right anchors values. X-axis refers to the mean decrease gini which represents the importance of the variable in data partitioning.

We calculated the confusion matrix, accuracy, precision, sensitivity and specificity for all predictors. Moreover, the ROC curve and the area under it were used to describe the performance of the predictors. Table 1 shows the area under the ROC curve obtained for five (5) different samples TADs located in Chromosome 1. Their locations are as follows: TAD1 - chr1:43691938-46036881, TAD2 - chr1:3826747-5976883, TAD3 - chr1:8975291-11021795, TAD4 - chr1:49037987-50814828, TAD5 - chr1:14421136-16068594. These results confirm that the epigenetics profiles have the same pattern among two tested different cell lines. The predictor trained on H1 cell line could detect IMR90 in situ Hi-C interactions with approximately the same accuracy as the predictor trained on IMR90. The same is also true for the predictor trained on IMR90. Figure 3 illustrates the ROC curve for the selected examples. The average accuracy for IMR90 cell line was 0.607 while for H1 hECS was 0.778. Figure 4 illustrates the distribution of the interactions according to the distance between the anchors. Secondly, we used peaks information to encode the epigenetic features. MACS2 peaks calling method was applied using the same parameters as in the reference [15]. We applied this scheme on IMR90, and GM12878 cell lines using in situ Hi-C data at 5 kb resolution from reference [9].

Table 1. The area under ROC curve for density coverage encoding predictors, columns headers represent training and testing cell types, the left side is the cell line where the predictor was trained, while the right side is the cell line where the predictor was tested.

TAD	H1 - H1	IMR90 -IMR90	H1 - IMR90	IMR90 - H1
TAD1	0.8582383	0.7997983	0.7990944	0.8580672
TAD2	0.820146	0.7555849	0.7551435	0.8205423
TAD3	0.8537698	0.7675929	0.7592675	0.8539412
TAD4	0.8936046	0.6441319	0.6456873	0.8941119
TAD5	0.8661579	0.7845425	0.7899033	0.8663949

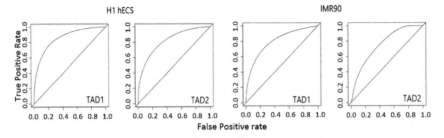

Fig. 3. ROC Curves for two different predictors applied on IMR90 and H1 hECS cell lines. TAD1 - chr1:43691938-46036881, TAD2 - chr1:3826747-5976883.

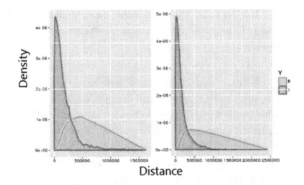

Fig. 4. Interaction density profiles as the function of the genomic distance between the pairs of interacting segments selected from GM12878 cell line. Y = 1 encodes the interaction. X-axis represents the genomic distance while Y-axis represents the density.

Figure 5 illustrates the important features according to the predictor for both cell lines GM12878, and IMR90. The average of the area under ROC curve obtained by these predictors was 0.9. Figure 6 illustrates the ROC curves for some samples from all chromosomes. Finally, the third in silico experiment used the predictor trained on GM12878 cell line with peaks encoding scheme and *in situ* Hi-C interactions to predict the ChIA-PET interactions for the same cell line. *In situ* Hi-C interactions represent spatial proximity between chromatin segments, whereas long-read ChIA-PET inter-actions are believed to be true physical interactions mediated by CTCF proteins. In general, number of *in situ* Hi-C loops reported in [9] is lower than ChIA-PET inter-actions recognized as statistically significant with calling score larger or equal to four paired-end reads for each interaction [8]. This discrepancy is caused by the better quality of loop calling in the case of CTCF ChIA-PET experiments, whereas unspecific *in situ* Hi-C uses non-trivial and complicated method to call loops from interaction heatmaps. Surprisingly, *in situ* Hi-C based machine learning predictor could recover in average about 66% of ChIA-PET strong interactions, resolving partially the above-described discrepancy. In the selected 150 TADs our method predicted approximately 53,416 physical interactions from the whole set of 80,112 CTCF interactions as reported in [8].

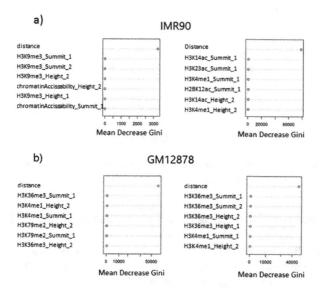

Fig. 5. Features importance in GM12878, IMR90 cell lines.

4   Data Resources

All Histone modification experiments were obtained from two on-line public resources as follows: H2BK12ac, H3K14ac, H3K18ac, H3K23ac, H3K27ac, H3K27me3, H3K36me3, H3K4ac, H3K4me1, H3K4me2, H3K4me3, H3K79me1, H3K9ac, H3K9me3, H4K8ac, H3K91ac, and DNase-seq for IMR90 and hESC H1 were downloaded from Roadmap Epigenomics project (NCBI). H3K04me1, H3K4me2, H3k4me3, H3K9ac, H3K9me3, H3K27ac, H3K27me3, H3K36me3, H3K79me2, H4K20me1 for GM12878 were downloaded from ENCODE Project.

In situ Hi-C Interactions Data for IMR90, and GM12878 were obtained from NCBI using accession number GSE63525 [9], then we normalized them using KR normal-ization vector obtained from the same study. Raw data for Hi-C interactions in hESC

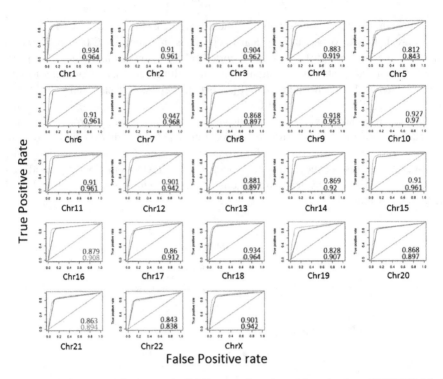

True Positive Rate

False Positive rate

Fig. 6. ROC Curves and the area under ROC curves for selected domains from GM12878 (blue) and IMR90 (green) cell lines chromosomes. (Color figure online)

H1 were obtained from NCBI using the accession number GSE43070 [14], then we created the heatmap using 5 kb resolution, and normalized it using the Vanilla Coverage Normalization method. At the end, liftover tool was used to transform the coordinates from hg18 to hg19. p-values for Hi-C interactions were calculated based on the normalized value of the interaction strength, and we considered the most important interactions as real interactions when p-value ≤ 0.05 while the rest considered as noise. Topologically associating domains coordinates were obtained from NCBI using accession number GSE63525 [9]. We used chromatin contact domains (CCDs) as domains genomic coordinates as found in GM12878 cell line using long-read ChIA-PET [8]. The genomic localization of three-dimensional domains are typically similar for all cell types, because TADs are highly conserved among different cell lines and organisms for mammals [16]. High-resolution ChIA-PET interactions were obtained from NCBI using the accession number GSE1806 [8].

5 Summary

The initial results obtained in this study confirmed the validity of using epigenomics profiles to predict chromatin interactions, the machine learning predictors performed very well for interactions prediction. Different evaluation methods were used to assess the performance of the predictors. The average accuracy, specificity, sensitivity and the area under ROC curve achieved by the predictors were above 0.85. TADs level predictors could identify the interaction for the same cell line and other cell lines. The tests showed that Peaks information coding could describe the interactions better than density coverage information, so in the successor studies we will focus on this method to build better predictors. More tests should be done to confirm many different cases. For TADs level predictors we will train the predictors on ChIA-PET interactions and check how many interactions we can recover using these predictors. Training a predictor on one cell line then test it on different individuals for the same cell type will be covered in the next tests. Multi-level predictors are also planned by assessing two approaches:

1. *Universal machine learning predictor* for each chromosome, trained on subset of the interactions from all TADs in the chromosome, then tested on the whole chromosome, and other chromosomes. Build one predictor for all cell lines trained on subset data from all chromosomes, and test it on the same and other cell lines.
2. Combining all predictors trained on the TADs in each chromosome to predict the interactions for each chromosome, then combine the chromosomes predictors to obtain cell line level predictor.

Acknowledgements. This work was supported by grants from the Polish National Science Centre (2014/15/B/ST6/05082), the European Cooperation in Science and Technology action (COST BM1405) and 1U54DK107967-01 "Nucleome Positioning System for Spatiotemporal Genome Organization and Regulation" grant within 4DNucleome NIH program.

References

1. McGraw-Hill Encyclopedia of Science and Technology. McGraw-Hill Education, New York (1997)
2. Bruce, A., Alexander, J., Julian, L., Martin, R., Keith, R., Peter, W.: Molecular Biology of the Cell. Garland Science, New York (2002)
3. Yun, Z., Zhao, C., Kai, Z., Mengchi, W., David, M., John, W.W., Bo, D., Nan, L., Lina, Z., Wei, W.: Constructing 3D interaction maps from 1D epigenomes. Nature Commun. **7** (2016). Article no. 10812. doi:10.1038/ncomms10812
4. Zubek, J., Stitzel, M.L., Ucar, D., Plewczynski, D.M.: Computational inference of H3K4me3 and H3K27ac domain length. PeerJ **4**(1), e1750 (2016)
5. Tyson DeAngelis, J., Farrington, W.J., Tollefsbol, T.O.: An overview of epigenetic assays. Mol. Biotechnol. **38**(2), 179–183 (2008)
6. Pietal, M.J., Bujnicki, J.M., Kozlowski, L.P.: GDFuzz3D: a method for protein 3D structure reconstruction from contact maps, based on a non-Euclidean distance function. Bioinformatics **31**(21), 3499–3505 (2014)
7. Fullwood, M.J., Liu, M.H., Pan, Y.F., Liu, J., Han, X., Mohamed, Y.B., Orlov, Y.L., Velkov, S., Ho, A., Mei, P.H., Chew, E.G.Y., Huang, Y.H., Welboren, W.-J., Han, Y., Ooi, H.-S., Pramila, N., Wansa, S.: An oestrogen-receptor-α-bound human chromatin interactome. Nature **462**(7269), 58–64 (2009)

8. Tang, Z., Luo, O.J., Li, X., Zheng, M., Zhu, J.J., Szalaj, P., Trzaskoma, P., Magalska, A., Wlodarczyk, J., Ruszczycki, B., Michalski, P., Piecuch, E.: CTCF-mediated human 3D genome architecture reveals chromatin topology for transcription. Cell **163**(7), 1611–1627 (2015)
9. Rao, S.S.P., Huntley, M.H., Durand, N.C., Stamenova, E.K., Bochkov, I.D., Robinson, J.T.: A 3D map of the human genome at kilobase resolution reveals principles of chromatin looping. Cell **159**(7), 1665–1680 (2014)
10. Przemysław, S., Zhonghui, T., Paul, M., Michal, J.P., Oscar, J.L., Michał, S., Xingwang, L., Kamen, R., Yijun, R., Dariusz, P.: An integrated 3-dimensional genome modeling engine for data-driven simulation of spatial genome organization. Genome Res. **26**(12), 1697–1709 (2016)
11. Szalaj, P., Michalski, P.J., Wróblewski, P., Tang, Z., Kadlof, M., Mazzocco, G., Ruan, Y., Plewczynski, D.: 3D-GNOME: an integrated web service for structural modeling of the 3D genome. Nucleic Acids Res. **44**(1), 288–293 (2016)
12. He, C., Li, G., Nadhir, D.M., Chen, Y., Wang, X., Zhang, M.Q.: Advances in computational ChIA-PET data analysis. Quant. Biol. **4**(3), 217–225 (2016)
13. Zhang, Y., Liu, T., Meyer, C.A., Eeckhoute, J., Johnson, D.S., Bernstein, B.E., Nussbaum, C., Myers, R.M., Brown, M., Li, W., Liu, X.S.: Model-based analysis of ChIP-seq (MACS). Genome Biol. **9**, R137 (2008)
14. Jin, F., Li, Y., Dixon, J.R., Selvaraj, S., Ye, Z., Lee, A.Y., Yen, C.-A., Schmitt, A.D., Espinoza, C.A., Ren, B.: A high-resolution map of the three-dimensional chromatin interactome in human cells. Nature **503**, 290–294 (2012)
15. Feng, J., Liu, T., Zhang, Y.: Using MACS to identify peaks from ChiP-seq data. Curr. Protoc. Bioinform. **14**, 1–14 (2011)
16. Dixon, J.R., Selvaraj, S., Yue, F., Kim, A., Li, Y., Shen, Y., Hu, M., Liu, J.S., Ren, B.: Topological domains in mammalian genomes identified by analysis of chromatin interactions. Nature **485**(7398), 376–380 (2012)

A Supervised Model for Predicting the Risk of Mortality and Hospital Readmissions for Newly Admitted Patients

Mamoun Almardini[1][(✉)] and Zbigniew W. Raś[1,2,3]

[1] College of Computing and Informatics, University of North Carolina, Charlotte, NC 28223, USA
{malmardi,ras}@uncc.edu
[2] Institute of Computer Science, Warsaw University of Technology, 00-665 Warsaw, Poland
[3] Polish-Japanese Academy of Information Technology, 02-008 Warsaw, Poland

Abstract. Mortality and hospital readmission are serious problems in healthcare, due to their negative impact on patients in specific and hospitals in general. Expenditure on healthcare has been increasing over the last few decades. This increase can be attributed to hospital readmissions; which is defined as a re-hospitalization of a patient after being discharged from a hospital within a short period of time. In this paper, we propose a system that is capable of using a supervised learning model to predict the status of patients when they get discharged from the hospital. For example, given a medical dataset, we would like to build a model that is capable of predicting with a high accuracy the likelihood that a newly admitted patient could be at risk of death or hospital readmission. For the purpose of prediction, we select five machine learning algorithms, namely Naïve Bayes, Decision Tree, Logistic Regression, Neural Networks, and Support Vector Machine. The results show the ability to predict the risk of mortality and hospital readmission with an accuracy of 82.78% and 63.59% respectively. The objective of the paper extends the scope of prediction to further include certain methods, such as feature engineering and boosting, to enhance the prediction accuracy.

Keywords: Hospital readmission prediction · Mortality prediction · Supervised model · Feature engineering · Boosting

1 Introduction

Recently, expenditure on healthcare has risen rapidly in the United States. According to [4], healthcare spending has been rising at twice the rate of growth of our income for the past 40 years. The projection of the growth rate in healthcare spending is 5.8% during the period 2014–2024, which means that the spending will rise to 5.4 trillion by 2024. This increase in healthcare spending can be attributed to several factors as listed by Price Waterhouse Coopers (PWC)

© Springer International Publishing AG 2017
M. Kryszkiewicz et al. (Eds.): ISMIS 2017, LNAI 10352, pp. 29–36, 2017.
DOI: 10.1007/978-3-319-60438-1_3

research institute: over-testing, processing claims, ignoring doctors orders, ineffective use of technology, hospital readmissions, medical errors, unnecessary ER visits, and hospital acquired infections [2]. Analyzing the reasons behind the increase in mortality and readmissions rates and reducing them can save lives and reduce the expenditure on healthcare. There are some published papers in the literature that apply machine learning algorithms in healthcare to predict the risk of mortality [5], readmission [6], and primary procedure [1]. In this paper we focus our interest to investigate a twofold problem that predicts the risk of mortality and hospital readmission for the newly admitted patients. We provide a comprehensive model that can include more medically diverse group of patients. That said, we built a supervised learning model to predict the status of patients when they get discharged. Several machine learning algorithms were used on the provided medical dataset to build an accurate classifier. In addition to that, feature filtering techniques and boosting were applied to enhance the prediction accuracy and utilize the processing performance.

2   HCUP Dataset Description

In this project, we mined the Florida State Inpatient Databases (SID) that are part of the Healthcare Cost and Utilization Project (HCUP). The SID dataset is primarily a state-level discharge data that is collected from non-federal community hospitals, which constitute the majority of hospitals in the USA. The SID include patients' demographic data, such as age, gender, and race. In addition to the demographic information, SID include patients' medical data, such as diagnoses, procedures, status of the patient, and the length of stay. The main table in the SID is the Core table, which is considered as the nucleus of the HCUP databases. The Core table contains over 280 features (attributes), however, many of those features are repeated with different values according to the patient's status. There are two types of coding schemes used in the Core table for labeling and formatting, which are the International Classification of Diseases, Ninth Revision, Clinical Modification (ICD-9-CM) and the Clinical Classifications Software (CCS). The ICD-9-CM coding is detailed and uses more codes to label the procedures and diagnoses. On the other hand, CCS is more generalized and a collapsed form of the ICD-9-CM. For example, there are 15,072 diagnosis categories and 3,948 procedure categories in the ICD-9-CM. On the other hand, CCS collapses these categories into a smaller number of more generalized categories, so there are 285 diagnosis categories and 231 procedures categories. For this work, we used the CCS coding scheme; as it provides more meaningful and descriptive presentation of the clinical categories. Table 1 shows the features that are used in this work.

3   Data Preprocessing

Data pre-processing is the most important step in machine learning. The better understanding and screening of the data helps in avoiding any misleading

Table 1. Description of the used core table features.

Features	Concepts
VisitLink	Patient identifier
DaysToEvent	Temporal visit ordering
LOS	Length of stay
DXCCSn	n^{th} Diagnosis, flexible feature
PRCCSn	n^{th} Procedure, meta-action
FEMALE	Indication of sex
RACE	Indication of race
AGE	Age in years at admission
DIED	Died during hospitalization

results in the steps to follow. There are several methods used to prepare the data depending on the nature of the data itself. In this work, we used five methods to prepare the data for the prediction step. These methods are: *cleaning, formatting, sampling, under − sampling*, and *aggregation*. In the following we provide an explanation of the steps underwent in each method:

- Cleaning: in this step we dealt with the missing instances in the dataset. That being said, we deleted some of them and replaced the remaining with zeros were applicable. The 0 in this case means that the patient does not exhibit a certain diagnosis or did not undergo a certain procedure.
- Formatting: most of the columns in the dataset are formatted in a way to include different codes for the diagnoses and procedures, which makes the header of the column insignificant in the prediction process (think about it as a bag-of-word). Therefore, we transposed the data in a way to create columns for all the possible diagnostic and procedure codes. In addition to that, we transformed the data into a binary representation to simplify the classification process and make it faster. At the end of this step, we ended up with 520 features.
- Sampling: the Core table in the SID includes a large number of features, which makes it very hard to process. Therefore, we have selected a subset of features that are relevant to the problem. By doing so, we increased the usage of the processing and memory capabilities.
- Under-Sampling: one of the problems we faced in the classification process is the skewed distribution of the classes. The number of the positive classes in the mortality and readmission datasets is way less than the negative classes. The classification of an imbalance data, as a result, will give us a non-realistic accuracy. In order to deal with the imbalanced dataset, we reduced the number of the majority classes (negative classes) to be close to the minority classes (positive classes).
- Aggregation: our dataset does not include a feature that gives us an indication whether a patient got readmitted or not. Therefore, we aggregated three

features ($VisitLink$, $DaysToEvent$, and LOS) into one feature that is used later as a class feature in the readmission prediction model. Following is the equation used to calculate the number of days between the discharge date and the next admission date:

$$RPeriod = DTE_2 - DTE_1 - LOS_1 \tag{1}$$

where $RPeriod$ refers to the number of days between the discharge date and the next admission date, DTE is a short for $DaysToEvent$ which refers to the number of days between any two consecutive admission dates, and LOS refers to the length of stay at the hospital. The subscript in the variable names indicates the visit number; 1 being the first visit and 2 being the second visit.

In order to consider that a patient had a readmission, the result of Eq. 1 should be less than or equal to 30 days for any two consecutive visits, as shown in Eq. 2.

$$Readmission = \begin{cases} Yes, & \text{RPeriod} \leq 30 \\ No, & \text{RPeriod} > 30 \end{cases} \tag{2}$$

4   Predicting the Risk of Mortality and Readmission

Several supervised learning algorithms are used to predict the mortality of patients while they are in hospice care and the likelihood of patients' readmission. The specific algorithms used are: Naïve Bayes, Decision Tree, Logistic Regression, Neural Network, and Support Vector Machine. These algorithms are selected due to their prevalent use in the field of machine learning and high level of support in the form of libraries and frameworks in various programming environments.

Overall, we were able to achieve a reasonably high accuracy in predicting the likelihood of mortality and hospital readmission. Table 2 shows the accuracies achieved by each algorithm on both the training and testing datasets for the mortality and readmission problems respectively. As can be noticed, the predictability of mortality is higher than the predictability of readmission. Moreover, the algorithms gave relatively close accuracies, however, some algorithms gave slightly better accuracies than others, such as neural networks and support vector machine, in both the mortality and readmission predictions. In addition to that, decision tree outperforms the other algorithms over the training dataset, which is foreseeable as decision trees tend to overfit the training data.

5   Feature Engineering

The high dimensionality of the dataset sometimes has a negative impact on the accuracy of the prediction model. Our dataset has 520 features, as described in Sect. 3, however, not all of them have the same impact on the prediction. Therefore, instead of using the whole set of features, we can select a subset that has significant impact on the prediction. We used three techniques to assign

Table 2. Predicting mortality and readmission accuracies

Algorithm	Mortality		Readmission	
	Training	Testing	Training	Testing
Naïve Bayes	87.43%	80.47%	62.76%	61.98%
Decision Tree	96.29%	80.34%	91.38%	56.92%
Logistic Regression	84.65%	65.21%	64.57%	63.59%
Neural Network	84.64%	82.78%	64.38%	63.1%
SVM	95.21%	82.68%	76.75%	62.71%

weights for the most significant features with respect to each prediction problem, which are: Entropy-Based, Chi-Squared, and Correlation filtering. Each of these techniques is explained briefly in Table 3.

Table 3. Descriptions of the feature engineering techniques

Filtering method	Description
Entropy-based filter	Entropy-based filtering is used to determine the overall relevance of a feature in predicting the decision class. This technique assigns high weights to the features that return the highest information gain; which is an indication of the homogeneity of the dataset
Chi-squared filter	Chi-squared filtering uses the chi-square statistical test to measure the independence of two events. Each feature is tested against the decision class feature. If the feature is found to be independent, then it is discarded. Otherwise, the feature is considered significant to the decision class feature
Correlation filter	Correlation filtering works similarly to the chi-squared filter, but in an opposite way. A feature is selected if it has a high correlation with the decision class feature. That being said, each filtering technique assigns weights differently

We applied the aforementioned feature filtering techniques on both the mortality and readmission data. We could reduce the number of features by [50–70]%. We selected the reduction percentage that gave us the best accuracy. The results did not show a significant improvement in the testing accuracy. However, there was a noticeable improvement in the processing performance. Another advantage is the ability to list the features that are significant in predicting the risk of mortality and readmissions as shown in Table 4.

Table 4. The most significant features in the mortality and readmission predictions. The prefix in the CCS code columns indicates whether the code refers to a procedure (PR) or a diagnosis (DX)

Readmission		Mortality	
Code	Description	Code	Description
DX196	Normal pregnancy and/or delivery	DX131	Respiratory failure; insufficiency; arrest (adult)
DX158	Chronic kidney disease	PR216	Respiratory intubation and mechanical ventilation
DX59	Deficiency and other anemia	DX249	Shock
DX99	Hypertension with complications and secondary hypertension	DX244	Other injuries and conditions due to external causes
PR137	Other procedures to assist delivery	DX2	Septicemia (except in labor)
PR58	Hemodialysis	DX107	Cardiac arrest and ventricular fibrillation
DX55	Fluid and electrolyte disorders	DX157	Acute and unspecified renal failure
PR134	134:Cesarean section	DX55	Fluid and electrolyte disorders
DX181	Other complications of pregnancy	DX122	Pneumonia (except that caused by tuberculosis or sexually transmitted disease)
DX195	Other complications of birth; puerperium affecting management of mother	DX106	Cardiac dysrhythmias

6   AdaBoost

The ultimate goal that we plan to achieve when dealing with classification algorithms is to build a prediction model, given a training dataset, that is able to generalize on a testing dataset and provide a high accuracy. One way to achieve that is by using boosting, which is simply defined as the process of building a highly accurate classifier model by combining a set of weak classifiers. Adaptive Boost (AdaBoost) algorithm, proposed by Freund and Schapire [3], is the very first implementation of boosting that maximizes the accuracy of a classifier by focusing on the points where the classifier does not perform well. The algorithm works by testing the model repeatedly using different portions of the training dataset. The selection of the data points, to be used in building the model, is based on a weight $(\omega_i(x, y))$ given to the points where the model performed poorly. This weight gives these data points advantage over the other data points in the selection process in the next iteration. The process keeps repeating n times in which the weight adapts in each iteration until a stronger classifier is built.

In order to study the effect of the AdaBoost on our dataset, we applied it on the decision tree classifier and noticed the improvement on the accuracy over the training and testing datasets. Table 5 shows a comparison before and after using AdaBoost on the readmission and mortality datasets. As we can notice, AdaBoost could build a stronger classifier that is able to better predict the class label. The accuracy has increased by 3% and 6% for the readmission and mortality predictions respectively. On the other hand, the accuracy of the training dataset was decreased, which means that the generated model does not overfit the dataset and can generalize on new observations.

Table 5. The effect of AdaBoost on the decision tree classifier

	Readmission		Mortality	
	Decision tree	AdaBoost	Decision tree	AdaBoost
Training accuracy	91.38%	60.57%	96.29%	86.31%
Testing accuracy	56.92%	59.39%	80.34%	86.29%

7 Conclusion

In this paper, we worked on a two-fold prediction problem where we applied a supervised model to predict the patients' mortality and the risk of hospital readmission. We built our prediction model using five algorithms: Naïve Bayes, Decision Tree, Logistic Regression, Neural Network, and Support Vector Machine. The highest accuracy we achieved in the readmission prediction was 82.78% using neural networks. On the other hand, the highest accuracy achieved in the mortality prediction was 63.59% using logistic regression. In addition to that, we applied three feature reduction techniques on the dataset and studied their impact on the accuracy and processing performance in prediction. Finally, we built a stronger classifier using AdaBoost on the decision tree classifier. In the future, we plan to apply deep learning on the two datasets and study how that would help in enhancing the prediction accuracy. Also, we will use some known and popular techniques to extract knowledge from imbalanced datasets.

References

1. Almardini, M., Hajja, A., Raś, Z.W., Clover, L., Olaleye, D.: Predicting the primary medical procedure through clustering of patients' diagnoses. In: Fifth International Workshop on New Frontiers in Mining Complex Patterns, nfMCP 2016. Springer, Heidelberg (2016)
2. Coopers, P.: The Price of Excess. Identifying Waste in Healthcare Spending (2006)
3. Freund, Y., Schapire, R.E.: A desicion-theoretic generalization of on-line learning and an application to boosting. In: Vitányi, P. (ed.) EuroCOLT 1995. LNCS, vol. 904, pp. 23–37. Springer, Heidelberg (1995). doi:10.1007/3-540-59119-2_166

4. Gorman, L.: Priceless: curing the healthcare crisis. Bus. Econ. **48**(1), 81–83 (2013)
5. Motwani, M., Dey, D., Berman, D.S., Germano, G., Achenbach, S., Al-Mallah, M.H., Andreini, D., Budoff, M.J., Cademartiri, F., Callister, T.Q., et al.: Machine learning for prediction of all-cause mortality in patients with suspected coronary artery disease: a 5-year multicentre prospective registry analysis. Eur. Heart J. **38**(7), 500–507 (2016)
6. Zolfaghar, K., Agarwal, J., Sistla, D., Chin, S.C., Basu Roy, S., Verbiest, N.: Risk-o-meter: an intelligent clinical risk calculator. In: ACM 19th SIGKDD International Conference on Knowledge Discovery and Data Mining, pp. 1518–1521 (2013)

An Expert System Approach to Eating Disorder Diagnosis

Stefano Ferilli[1,2]([envelope]), Anna Maria Ferilli[3], Floriana Esposito[1,2],
Domenico Redavid[4], and Sergio Angelastro[1]

[1] Dipartimento di Informatica, Università di Bari, Bari, Italy
{stefano.ferilli,floriana.esposito,sergio.angelastro}@uniba.it
[2] Centro Interdipartimentale di Logica e Applicazioni, Università di Bari, Bari, Italy
[3] Azienda Ulss 6 Euganea Regione del Veneto, Padova, Italy
annamaria.ferilli@aulss6.veneto.it
[4] Artificial Brain S.r.l., Bari, Italy
redavid@abrain.it

Abstract. Medical diagnosis in general is a hard task, requiring significant skill and expertise. Psychological diagnosis, in particular, is peculiar
for several reasons: since the illness is mental rather than physical, no
instrumental measurements can be done, more subjectivity is involved
in the diagnostic process, and there is more chance of comorbidity. Eating disorders, specifically, are a quite relevant kind of psychological illness. This paper proposes an Expert Systems-based solution to the above
issues. The Expert System was built upon a proprietary general inference
engine, that provides several features and reasoning strategies, allowing
to properly tune their exploitation through appropriate parameter settings. Qualitative analysis of the prototype revealed interesting insights,
and suggests further extensions and improvements.

Keywords: Eating disorders · Expert systems

1 Introduction and Motivation

Medical diagnosis in general is a hard task, due to the need to take into account
and correlate much, often partial or conflicting, information, based on a thorough
knowledge and understanding of the state-of-the-art in a given medical domain.
Often, specific diseases are consequences of, or strictly related to, more general
ones, and it is important to know about both in order to have an appropriate
account of the patient's status. The co-occurrence of many disorders is known
as *comorbidity*.

Compared to other medical domains, psychological diagnosis is peculiar for
several reasons. In this case, the therapist needs to know what actions to take
when facing a malaise, and this requires an understanding of the many factors
that, at different levels, affect the adaptive process of an individual. In turn,
this requires collecting, and reasoning on, a significant amount of information.

© Springer International Publishing AG 2017
M. Kryszkiewicz et al. (Eds.): ISMIS 2017, LNAI 10352, pp. 37–46, 2017.
DOI: 10.1007/978-3-319-60438-1_4

In addition to the usual complexity of medical diagnosis, it involves mental aspects that are, by themselves, hard to capture and partly subjective, both on the side of the patient and on the side of the therapist. E.g., since there are no physical measurements that can be done to identify a problem, all information must be gained by observing the patient's behavior and by collecting his thoughts using interviews and/or questionnaires to be filled, whose interpretation is not always agreed upon by all experts. Also, the questionnaires are typically made up of a fixed list of questions, to be answered by the patient in a specified order independently of the answers he already gave to previous questions. Moreover, the patient may be somehow scared by the therapist, and unwilling to explicitly express him all of his thoughts. This can be a relevant problem, because lacking, or —even worse— purposely distorted information may lead to a completely wrong and inappropriate diagnosis.

Eating disorders (EDs for short), in particular, are quite relevant in this landscape for several reasons. First of all, they are directly related to physical issues, and thus they may have a direct impact on the physical health of people. Moreover, the theoretical apparatus in this field [1] has recently become more complex, by identifying many more specific kinds of diseases than in the past, and thus introducing new challenges about how to properly discriminate very similar but different diseases. Last but not least, since eating disorders are often side effects of more basic mental diseases, the diagnostic search cannot be limited to a restricted domain. There is a pressing need to identify EDs as early as possible, because this allows to tackle the problems more easily than after the disorders have reached the full-blown stage, avoiding in this way the consequences of a full-blown disorder, and preventing its becoming chronic.

To tackle the above issues, this paper proposes a solution based on Expert Systems technology. After some cross-fertilization meetings held with a team of psychologists, Expert Systems were selected as a viable approach to this problem for the following reasons:

- by embedding formal and heuristic knowledge drawn from several textbooks and experts, it ensures a more objective viewpoint, both in acquiring the patient's information and in determining a diagnosis; this provides the therapist with a solid basis to devise further interviews, aimed at investigating relevant themes detected, and to properly evaluate all elements needed to put the malaise in context and to start a suitable treatment;
- the knowledge base may include several psychological questionnaires, and the sequence of questions that are posed to the patient can be suitably selected by the system based on the previously given answers, instead of being stuck fixed;
- it may be able to detect possible comorbidity by correlating symptoms to personality structures;
- it can quickly process a significant amount of information, establishing relationships which allow the psychologists to save time in getting to a diagnostic hypothesis; this, in turn, allows the therapists to start as soon as possible a treatment that is tailored to the specific case;

– it can be easily exploited by the user: after a quick description made by the therapist, the subject may be left alone with the system and freely answer to the questions (of course, under specific request of the patient, the therapist may be present and help him during the whole procedure, obviously being careful not to affect his answers).

The Expert System, called EDES (after Eating Disorder Expert System), was built upon a proprietary general inference engine, called GESSIE, that provides several features and reasoning strategies, allowing to properly tune their exploitation through appropriate parameter settings.

This paper is organized as follows. The next section presents GESSIE, while Sect. 3 describes the knowledge base structure and relevant features. Then, a qualitative evaluation of EDES is proposed in Sect. 4. Finally, Sect. 5 draws some conclusions and outlines future work issues.

2   GESSIE

GESSIE (an acronym for General Expert System Shell Inference Engine) is a general-purpose inference engine developed to support the implementation of a wide variety of expert systems. It was inspired by CLIPS [8], but designed to overcome some shortcomings of that environment and to be embedded with the Prolog language rather than C. It provides the following features:

Complex Preconditions involving any combination of AND, OR and NOT logical operators;
Forward Chaining reasoning, allowing to go from known facts to all their possible consequences[1];
Backward Chaining reasoning, allowing to prove a given hypothesis;
Explanation (upon request) of the reasoning followed to reach a conclusion, or of the reason why a question is posed during the interaction process;
Knowledge Base Modification through assertions and retractions of facts[2];
Conflict Resolution to express priority among incompatible conclusions;
Uncertainty Handling inspired by MYCIN's strategy [4], but applied to a more classical $[0, 1]$ confidence interval[3];
(De-)Fuzzification to handle vagueness with different, user-definable fuzzy functions;

---

[1] During forward reasoning, all current (complete or partial) instantiations of a knowledge item are maintained in a tree, where a child node represents an instantiation that extends the instantiation in the parent. This allows to keep the process of adding or extending instantiations within logarithmic complexity.

[2] This also allows to save time: by asserting relevant partial conclusions, driven by the meta-rules, the corresponding information is made immediately available to future deductions; by retracting useless knowledge items, less burden is placed on subsequent reasoning.

[3] While quite simple, this strategy is more intuitive and less computationally heavy than other approaches, e.g. the several probabilistic extensions of Prolog.

Knowledge Modularization to load/unload and combine, also dynamically, blocks of knowledge concerning different problem (sub-)domains[4];

Global Facts to define information for which only a single instance is allowed during the reasoning process;

Structure Definition to allow the representation of complex objects;

Predicate Indexing to quickly retrieve all information concerning a given property or relationship[5];

Prolog Embedding to directly exploit all the power and flexibility of the underlying Prolog engine.

Especially relevant, compared to CLIPS, is the native possibility to provide explanations, to use a backward chaining reasoning, and to handle uncertainty.

The core of the knowledge base formalism consists of two predicates:

fact(I, F, C) allows to express facts, where:

- I is a unique identifier for each fact;
- F is the fact, expressed as a Prolog atom;
- $C \in [0, 1]$ is the confidence with which the fact holds.

rule(I, H, B, P, C) allows to express rules, where:

- I is a unique identifier for each rule;
- H is the head of the rule, consisting of a single Prolog atom or of a Prolog list of atoms;
- B is the body of the rule, expressed as a combination of the following operators:
  - $and([A_1, \ldots, A_n])$ denotes the logical conjunction of $A_1, \ldots, A_n$;
  - $or([A_1, \ldots, A_n])$ denotes the logical (inclusive) disjunction of $A_1, \ldots, A_n$;
  - $not(A)$ denotes the negation of A;
  where the A's are Prolog atoms or nested *and*, *or* and *not* operators.
- $P > 0$ is the priority level of the rule, used for conflict resolution;
- $C \in [0, 1]$ is the confidence with which the fact holds.

Reserved predicates allow the knowledge engineer to exploit the various features provided by the inference engine. E.g., predicate unique/1 allows one to define global facts, by specifying that only one fact (i.e., instance) can be present at any time in the knowledge base for the predicate taken as argument. So, for instance, setting unique(patient_score(joe,Y)), if a fact patient_score(joe,36) is present in the knowledge base, and the assertion of fact patient_score(joe,39) is attempted, the former will be retracted and replaced by the latter.

A Web-service was developed to provide GESSIE reasoning facilities to non-technical users. Users of the service may use Java APIs that allow them to deploy projects (i.e., knowledge bases), run the system on a project, send facts and commands during an execution, get the outcomes, and ask for explanations.

---

[4] This allows to keep in the working memory only the parts of knowledge that are relevant to the current reasoning task, this way reducing the possible deductions and combinations and improving efficiency.

[5] This allows to keep constant the time to get all knowledge items associated to a given predicate.

3   Eating Disorder Expert System

To date, Expert Systems technology has not received much attention in the field of psychological diagnosis. In addition to the usual skepticism that many experts in any field have towards this technology, in psychology the involvement of subjective and non-physically measurable facts seems to suggest that the application of computer technology is unworthy. However, there are some examples against this assumption. SCIROPPO [3] is a support decision system for psychotherapeutic treatment. While chatting with the user about the patient's healthcare history, it processes the data and provides their possible consequences, the possible objectives for the therapist to set and a range of treatments that may be appropriate for the specific case. ALFA [7] is a software for the pharmacological therapy of panic disorder. It is based on KES, a commercial program that provides the inferential engine and lets the user enter the knowledge base, using a formalism that is very close to natural language. SEXPERT [2] is an expert system for the assessment and treatment of sexual disorders. It was conceived as an interactive system for computer-supported therapy and is based on techniques borrowed from both the Artificial Intelligence and the Intelligent Tutoring fields. ESQUIZOR [11] is an expert system designed to devise and test the effectiveness of neuroleptic treatments applied to schizophrenic subjects of paranoiac kind. It involves the use of an example-based decision tree in order to reach a conclusion. TRAUM [10] is a system developed to help people who is in charge of diagnosing and treating neuro-psychological disorders due to brain damage, stroke, or developmental delay. In addition to providing assistance, it can provide valid and accurate diagnosis, comparable to those of clinicians.

In this landscape, EDES was motivated by both the lack of an existing solution to the specific field of eating disorders, and the need to have a more flexible inference engine that provides the knowledge engineer with a wider range of inference tools.

Based on the DSM-V [1], EDES can handle the following main diseases:

1. *Anorexia Nervosa* (AN), consisting in a distorted perception of one's own body, seen as always overweight;
2. *Bulimia Nervosa* (BN), involving a morbid fear of gaining weight, characterized by gorging balanced by subsequent vomit, compulsive physical exercise, and consumption of laxatives and diuretics;
3. *Binge-Eating Disorder* (BED), characterized by gorging without balancing actions;
4. *Abstention/Restriction of food assumption*, characterized by indifference to food, selection of food based on sensory aspects (extreme sensitivity) or on brand, concern for negative consequences of eating, such as vomit or suffocation;
5. *Rumination Disorder*, consisting in bringing again ingested food to mouth through regurgitation, in order to chew it again, without the presence of pain or nausea;
6. *Geofagia*, i.e. the habit, in children, of eating non edible things;

plus some restricted forms or subtypes of the above: *Atypical Anorexia Nervosa* (despite a significant weight loss, the weight still stays within or above the normal range), *Anorexia Nervosa with restrictions, Anorexia Nervosa with gorging/elimination, Low-frequency and/or low-duration Bulimia Nervosa, Low-frequency and/or low-duration Binge-eating, Purgative Disorder* (frequent purgative behavior to modify body's weight and shape, when Binge-Eating is not present). We also included two disorders that, while not strictly classified as EDs by the DSM-V, are nevertheless associated to eating, and thus may be relevant to our study: *Reverse Anorexia* (characterized by a continuous and obsessive worrying for one's own muscular mass, even to health detriment) and *Orthorexia*, a form of abnormal attention to eating rules, to the choice of food and to its features. Many studies in literature note a frequent co-occurrence between EDs and Personality Disorders (PDs). They show a significant presence of PDs within EDs, or, *vice versa*, a strong presence of EDs in patiens having PDs. E.g., patients with EDs are more likely to show a simultaneous PD, compared to subjects without EDs [5]. Also, EDs are associated to specific clusters of PDs and to peculiar personality traits such as perfectionism, self-directivity and impulsiveness. Specifically, in EDs there is a prevalence of *avoidant, obsessive-compulsive* and *dependent* PD. In addition to including knowledge about all of the above, the system also embeds two well-known psychological questionnaires related to eating disorders: the Eating Attitudes Test, made up of 26 items (EAT-26) [6] and the Binge Eating Scale (BES) [9].

Overall, the knowledge base of this first prototype includes 93 domain rules, involving 45 concepts. Additional meta-rules are present to guide the reasoning strategy. Rules about the relationships among the disorders were extracted by the knowledge engineers directly from the literature and validated by the experts, while rules about the heuristics and reasoning to apply during a diagnosis were extracted by interviews and discussions with the experts. Different subsets of psychologists contributed to the development of the knowledge base for different disorders. For each disorder, first each expert was interviewed separately, then the knowledge extracted from him/her was submitted to the other members of the subgroup associated to that disorder for validation purposes. Since some knowledge items were changed during this validation process, a final wrap-up discussion with the entire subgroup in a plenary meeting was conducted.

The graphical interface of EDES proposes four stages. Stage 1, mandatory, asks the user for personal details (Name, Surname, Birth date, Sex), Height and Weight (needed to compute the user's Body Mass Index), and 6 questions aimed at a preliminary, very general identification of possible relevant problems. Stages 2 and 3 allow the user to (completely or partially) fill the EAT-26 and the BES questionnaires, respectively. Stage 4 is devoted to the interactive execution of the expert system: it provides a textual account of the interaction, along with four buttons (one to start the system, two for answering yes or no to the system's questions, and one to ask why a question or a diagnosis is made by the system); an account of the partial deductions of the system is also reported.

As regards the reasoning strategy, when the user starts the interactive phase, the information he/she has filled in the questionnaires is made available to the diagnostic reasoning, and is exploited by a preliminary forward chaining deduction to obtain all of its consequences. Then, backward chaining deduction is used in the interactive phase to obtain the final diagnosis, guided by the meta-rules to determine the priority of hypotheses to be checked. During this phase, in case of need, the system may ask specific questions taken from the part of questionnaires that was not filled by the user. The questions are devised so that a binary (yes/no) answer is expected by the user. So, uncertainty handling is currently limited to rules and deduced facts only. This keeps low the cognitive burden on the user, and makes the interaction faster.

4 Evaluation

The prototype of EDES was evaluated by means of controlled experiments run by the domain experts (psychologists) who contributed to its development. Specifically, they ran several times the system, each time pretending to be a patient with a given disorder, and answering accordingly to the questions posed by the system during its execution. While possibly introducing a bias, this approach has some justifications. First, it allowed the experts to cover all the disorders that the system should recognize, and to control the distribution of cases among the disorders. Second, in this way they knew the expected outcome, which allowed them to obtain a quantitative evaluation, based on the comparison of the system's diagnosis with the hypothesized disorder. Finally, they could check the reasoning followed by the system to reach a conclusion in some peculiar cases, aimed at a qualitative evaluation. Note, however, that the experts who simulated a given disorder never belonged to those that contributed to build the portion of knowledge base related to that disorder.

As regards the quantitative evaluation, 105 experiments were run, whose distribution and results for the different disorders are as follows:

1. Anorexia Nervosa (3 very minor, 3 minor, 3 moderate, 4 serious, 5 very serious): 1 serious diagnosed as very serious;
2. Atypical Anorexia: 5 not diagnosed, 1 diagnosed as Orthorexia;
3. Anorexia with gorging: 2 diagnosed as minor Anorexia, 2 diagnosed as very serious Anorexia;
4. Reverse Anorexia: 1 classified as 'not Anorexia Nervosa', 1 diagnosed as 'Atypical Anorexia';
5. Bulimia Nervosa (6 minor, 5 moderate, 7 serious, 1 very serious): 1 serious diagnosed as minor, the very serious one diagnosed as 'serious';
6. Binge-Eating Disorder (5 minor, 3 moderate, 8 serious, 3 very serious): 1 minor not diagnosed, 1 serious diagnosed as minor, 1 very serious diagnosed as serious;
7. Abstention/Restriction: 1 not diagnosed;
8. Purgative disorder: 2 not diagnosed, 1 classified as 'minor Anorexia';
9. Orthorexia;

10. Rumination disorder: 1 not diagnosed;
11. Geofagia: 1 not diagnosed.

Table 1 reports dataset and performance statistics, both for each disorder and overall. Row 'Cases' reports the number of cases on which EDES was run. Row 'No diagnosis' reports the number and percentage of cases on which it did not identify a disorder. Row 'Strict' reports the number and percentage of correct diagnoses only. Row 'Loose (extent)' reports the number of cases that can be considered acceptable, due to only the seriousness of the disorder being incorrect, and the cumulative percentage of accuracy with 'Strict'. Finally, row 'Loose (subtype)' reports the number of cases that can be considered acceptable, due to the disorder being mismatched with a very close one, and the cumulative percentage of accuracy with Strict and Loose (extent). It makes sense to consider 'Loose' cases as correct because they can point out the presence of a disorder with sufficient closeness to the correct diagnosis as to provide useful indications to the therapist. Performance falls below 80% for only 4 disorders out of 11 in the 'Strict' and 'Loose (extent)' settings, and for only 2 disorders in the 'Loose (subtype)' setting. The worst performance is on disorders #2, #3 and #8. However, #3 has acceptable 'loose' performance, and, according to the DSM-V, #8 is actually a particular form of the other disorders, and thus it is more complex to be classified. Column ('All - #2') reports the overall accuracy excluding disorder #2. All in all, we may conclude that EDES is usually able to diagnose the correct disorder and, when different extents are considered, also sufficiently accurate in identifying the correct extent. Most errors were caused by missing preconditions, due to the experts overlooking them or the knowledge engineers not being able to catch their relevance in the experts' accounts.

Table 1. Diagnostic accuracy

	#1	#2	#3	#4	#5	#6	#7	#8	#9	#10	#11	All	All - #2
Cases	20	10	5	5	19	19	5	5	5	5	7	105	95
No diagnosis %	0	5	0	0	0	1	1	2	0	1	1	11	6
	0.00	0.50	0.00	0.00	0.00	0.05	0.20	0.40	0.00	0.20	0.14	0.10	0.06
Strict %	19	4	1	3	17	16	4	2	5	4	6	81	77
	0.95	0.40	0.20	0.60	0.89	0.84	0.80	0.40	1.00	0.80	0.86	0.77	0.81
Loose (extent) cumulative %	1	0	0	0	2	2	0	0	0	0	0	5	5
	1.00	0.40	0.20	0.60	1.00	0.95	0.80	0.40	1.00	0.80	0.86	0.82	0.86
Loose (subtype) cumulative %	0	0	4	2	0	0	0	1	0	0	0	7	7
	1.00	0.40	1.00	1.00	1.00	0.95	0.80	0.60	1.00	0.80	0.86	0.89	1.00

Concerning the qualitative evaluation, let us report a couple of relevant cases.

1. One of the psychologists played the role of a bulimic patient. She kept answering the system's questions, committed to play that role, but the system's diagnosis was 'anorexia', which is actually the opposite disorder than bulimia.

So, she decided to ask for an explanation of the reasoning followed by the system, and discovered with much surprise that the answers she actually gave were more related to anorexia than they were to bulimia. While testing the system she got distracted and drifted to a different disorder than the originally planned one. This shows that the system may help in avoiding that the therapists drift toward wrong diagnoses due to distraction or stress.

2. An intern was asked to use the system. She is in fact overweight, but it is just due to an hormone disorder, and thus her personality traits are absolutely different than those of bulimic subjects. She likes eating, as many normal persons, more for the pleasure of being together than for the excessive pleasure of ingurgitating food. So, hers was a kind of disease more clinical than psychological, but knowledge associated to this kind of problems had not be entered in the knowledge base yet. The intern was asked to ask the questions without simulating any particular disorder, but just being herself. The final outcome of EDES was that she 'is not anorexic', which quite surprised the experts. While apparently strange, this diagnosis was in fact correct considering the knowledge available to the system. Indeed, EDES realized there had to be some problem, but it did not fit any of the known diseases. Rather than saying she was healthy, it tried to convey the presence of something by the outcome that she was not anorexic. This shows two things: first, that the system can handle to the best of its knowledge even cases that go beyond its realm of applicability; second, that it may point out in this way the need to extend its knowledge base with additional disorders.

As to runtime, the system always operated in real-time on a standard laptop, not showing noticeable delays between a user's answer and the next question. This suggests that its performance can scale, at least up to the foreseeable size of an extended knowledge base. It is also important for avoiding that the patient worry about having given a 'problematic' answer to the previous question.

5 Conclusions and Future Work

Psychological diagnosis adds several complexities to other kinds of medical diagnosis, which is in itself a hard task. It is more subjective, and is more related to malaise than to physical issues. So, there is a need to gain indirect information about the problem, and to be objective in doing this. Also, many factors (genetic, socio-cultural, psychological, biological ones) may be involved in the conditioning of a subject's adaptive process. In particular, in some aspects the diagnosis of eating disorders may be more relevant and challenging than other fields. This paper proposed and evaluated an Expert Systems-based solution to the above issues built upon a proprietary general inference engine. This is conceived as a tool that can be used directly by the therapist, or by the user alone, and then by the therapist to have an account of the situation before issuing a final diagnosis. While still at the prototype stage, the expert system has provided interesting outcomes that helped the psychologists to deal with several cases. This suggests that the solution is viable and worth further effort.

Future work on the expert system will involve refining and extending the knowledge base in several directions: including additional psychological questionnaires and knowledge about eating disorders and associated general psychological disorders; applying changes inspired by the outcomes of the system on real cases; adding new modules aimed at identifying a proper therapy for the diagnosed diseases. Some of these improvements might suggest, as a side effect, new features to be added to the inference engine, resulting in a cross-fertilization between EDES and GESSIE. E.g., a more theoretically founded probability handling feature, based on Bayesian Networks, is currently under development.

Acknowledgments. This work was partially funded by the Italian PON 2007–2013 project PON02_00563_3489339 'Puglia@Service'.

References

1. Diagnostic and Statistical Manual of Mental Disorders, Fifth Edition (DSM-V). American Psychiatric Association (2013)
2. Binik, Y., Servan-Schreiber, D., Freiwald, S., Hall, K.: Intelligent computer-based assessment and psychotherapy: an expert system for sexual dysfunction. J. Nerv. Ment. Dis. **176**(7), 387–400 (1988)
3. Brighetti, G., Contento, S., Cotti, M.: Esame di fattibilità di una banca dati (sistema esperto) sulle informazioni documentarie concernenti le attività psicoterapeutiche in Emilia Romagna. Rapporto conclusivo (1986, in Italian)
4. Buchanan, B.G., Shortliffe, E.H. (eds.): Rule-Based Expert Systems: The MYCIN Experiments of the Stanford Heuristic Programming Project. Addison Wesley, Boston (1984)
5. Dolan, B., Evans, C., Norton, K.: Disordered eating behavior and altitudes in female and male patients with personality disorders. Int. J. Eat. Disord. **8**, 17–27 (1994)
6. Garner, D., Garfinkel, P.: The eating attitudes test: an index of the symptoms of anorexia nervosa. Psychol. Med. **9**, 273–279 (1979)
7. Ghirlanda, L., et al.: ALFA: A.I. in medicina. un sistema esperto per il trattamento farmacologico del disturbo da attacchi di panico. Inform. Oggi **43**, 12–18 (1988). (in Italian)
8. Giarratano, J.C., Riley, G.D.: Expert Systems: Principles and Programming, 4th edn. Course Technology, Boston (2004)
9. Gormally, J., Black, S., Daston, S., Rardin, D.: The assessment of binge eating severity among obese persons. Addict. Behav. **7**, 47–55 (1982)
10. Grim-Haines, J., Parenté, R., Hoots, F., McNeely, J., Quattrone, W., Wriston, J.: Experimental validation of a neuropsychological expert system. Cogn. Technol. **11**, 4–9 (2006)
11. Madera-Carrillo, H., Loyo, J., Sanchez, L., Perez, M.: An expert system of prediction for the classification of schizophrenic patients: Esquizor. Rev. Mex. Psicol. **20**, 19–27 (2003)

Multimodal System for Diagnosis and Polysensory Stimulation of Subjects with Communication Disorders

Adam Kurowski[1]([✉]), Piotr Odya[1], Piotr Szczuko[1], Michał Lech[1], Paweł Spaleniak[1], Bożena Kostek[2], and Andrzej Czyżewski[1]

[1] Faculty of Electronics, Telecommunications and Informatics, Multimedia Systems Departament, Gdańsk University of Technology, Gabriela Narutowicza 11/12, Gdańsk, Poland
{adakurow, piotrod, szczuko, mlech, papol}@multimed.org,
ac@pg.gda.pl
[2] Faculty of Electronics, Telecommunications and Informatics, Audio Acoustics Laboratory, Gdańsk University of Technology, Gabriela Narutowicza 11/12, Gdańsk, Poland
bokostek@audioacoustics.com

Abstract. An experimental multimodal system, designed for polysensory diagnosis and stimulation of persons with impaired communication skills or even non-communicative subjects is presented. The user interface includes an eye tracking device and the EEG monitoring of the subject. Furthermore, the system consists of a device for objective hearing testing and an autostereoscopic projection system designed to stimulate subjects through their immersion in a virtual environment. Data analysis methods are described, and experiments associated with classification of mental states during listening exercises as well as audio-visual stimuli are presented and discussed. Feature extraction was based on discrete wavelet transformation and clustering employing the k-means algorithm was designed. All algorithms were implemented in the Python programming language with the use of Open Source libraries. Tests of the proposed system were performed in a Special School and Educational Center in Kościerzyna, Poland. Results and comparison with data gathered from the control group of healthy people are presented and discussed.

Keywords: Communication disorders · Special education · Polysensory stimulation · EEG · Electroencephalography · Multimodal interfaces · Signal segmentation · Data clustering

1 Introduction

Recent advances in technology made it possible to find novel ways to communicate with people who for some reason lack ability to effectively communicate with others. Moreover, in some situations there is a strong need for finding such unconventional ways of communication and diagnostics. For instance, recent studies show that each year in Europe approx. 300 people for every 100,000 people suffer from traumatic

© Springer International Publishing AG 2017
M. Kryszkiewicz et al. (Eds.): ISMIS 2017, LNAI 10352, pp. 47–56, 2017.
DOI: 10.1007/978-3-319-60438-1_5

brain injury (TBI), usually due to traffic accidents [1]. From diagnostic and rehabilitation points of view, the key problem is the lack of communication with such patients. In some cases it may lead to erroneous medical conclusions. The ratio of incorrect diagnosis amounts to as much as 40% [2]. This is not the only example where alternative ways of interacting with certain groups of people is needed. Another group associated with such kind of needs are people with impaired communication skills or behavioral issues, for instance due to developmental disorders such as autism or Asperger's syndrome. The polysensory stimulation and diagnosis system developed at the Multimedia Systems Department of Gdańsk University of Technology, is an innovative solution that was used in the context of diagnosis and rehabilitation of subjects considered to be a person in a vegetative state [3, 4]. However, in this work we would like to present the way of using it for diagnosis and therapy of persons with impaired communication skills. The stand integrates different technologies: eye-gaze tracking, analysis of bioelectrical activity of the subjects' brain and eye tracking. The paper presents the concept of the developed multimodal stand for organizing experiments in human-machine communication. Also techniques of post-processing of electroencephalographic (EEG) signals with the use of the Python programming language are described.

2 Multimodal Experimental Stand

The developed system assumes few phases of the subject's diagnosis and stimulation. The first one is an objective testing of hearing ability, then producing audio-visual stimuli, finally monitoring the brain neuronal activity, and tracking the subject's gaze. There is also possibility for a therapist to conduct exercises with a subject in order to train desired skills.

2.1 Hearing Assessment

The first step in evaluation of the subject is verifying one's hearing ability. Detailed data associated with each possible particular case of hearing impairment make it easier to prepare more efficient exercises and speed up a therapy. Therefore, in certain cases it is reasonable to conduct a hearing test to obtain additional data about subject's hearing even if therapists have gathered some related information, earlier.

It is assumed that the hearing test must first be conducted for persons with communication disorders, while the interaction with the subject is limited. Therefore the test must be carried out using an objective method. In the present study, auditory evoked potentials (ABR - Auditory Brainstem Response) were chosen for their numerous advantages: the method is non-invasive, painless and does not involve any complex preparation [5, 6]. ABR requires only that the subject is relaxed and not distracted by external stimuli. The main disadvantage of ABR is the duration of the examination. It depends on the number of test signals and their types. But it gives a possibility to obtain also information on the threshold of hearing. This knowledge may be useful during the preparation of auditory stimuli signals, for applying necessary

corrections, e.g. adjusting appropriately signal levels to the person's hearing characteristics. The equipment for the ABR method is similar to electroencephalograpy
(EEG), but due to technical differences (e.g. number and arrangement of electrodes and
signal acquisition parameters), it is not possible to use a typical EEG hardware, hence a
dedicated setup is applied. Both the ABR and EEG signal acquisition systems are
depicted in Fig. 1.

Fig. 1. Echodia Elios device allowing ABR measurement at the research stand, the electrode
mounted on the right mastoid and in-the-ear headset (the left picture), left-side electrodes and
device units visible on the table (the picture in the middle) and Emotiv INSIGHT used for
gathering EEG signals (the right picture).

2.2 Electroencephalography Monitoring

Electrical activity of neurons in the brain can be estimated in an indicative manner by
measuring changes in the local potentials on the person's scalp. This measurement
called the electroencephalogram typically uses from 5 to 14 electrodes, arranged
symmetrically on the head. In the current research, the following EEG devices were
used:

- ENOBIO8 helmet, made by Neuroelectronics, equipped with 8 electrodes,
- EPOC and INSIGHT models, made by Emotiv with 14 and 5 electrodes,
  respectively.

  Monitoring the potentials on the skin using non-professional EEG helmet might be
inaccurate because of the relatively large size of the area covered by a single electrode,
low reproducibility due to non-permanent electrode placement and differences between
individuals. However, for diagnostic assessment and multimedia applications such
simple devices can be employed. Above solutions offer monitoring of brain wave bands
from 0 to 31 Hz and determining the degree of concentration of the subject, as well as
the electrical activity in different brain areas. Raw EEG signals may be gathered from
the acquisition system and then be processed with the use of open source solutions like
Python programming language, which is equipped with a number of libraries dedicated
to scientific computations. A solution presented in this work is based on SciPy,
NumPy, Scikit Learn and PyWavelets libraries used for the purposes of numerical
calculations, machine learning and obtaining discrete wavelet transform (DWT) coefficients [7–10]. More details regarding the processing of the mentioned signals are
provided in the later part of this work. To monitor user attention and reaction to visual

stimuli presented on a screen, the user's gaze fixation point is captured, using an EyeX Controller by Tobii. It allows for recording the fixation point coordinates with more than 60 samples per second. EyeX can be mounted on a typical LCD monitor, a screen of a laptop, or on a tablet. The device allows for working comfortably in a wide range of distances from the screen 45–100 cm [11]. This allows for tracking parts of the screen which drew the most attention of the subject during performed exercises and correlate this fact with data collected by other means.

2.3   Post-processing Software

The stand enables to gather few types of data, associated with eye-gaze tracking, types of exercises performed by subjects and EEG signals collected during this process. Post-processing scripts are written in the Python programming language. Handling the eye tracker signal is simple and is based on visualizing trajectory of eye-gaze in the given ranges of time. The processing of the EEG signal is more complicated and divided into few steps. First, signals obtained from 5 electrodes of the headset are transformed with the use of independent component analysis (ICA), which allows for blind separation of multiple sources of electrodes signals. It is assumed that they are mixtures of signals originating from many sources placed on the scalp [12]. Signals were collected from the following standard electrode placements on the subject's scalp, according to the 10–20 system: AF3, AF4, T7, T8 and Pz. Next, signals obtained from the ICA module were split into segments of a desired length on which a feature extraction procedure was performed. The length of a frame is problem-dependent and in case of our research varied from 64 up to 512 samples. The following features were calculated for each of five channels of the EEG data, i.e.: means and variances of each level of discrete wavelet transform (DWT) coefficients. Multiple types of wavelets were used for this purpose on the basis of the literature review [13–15]. Wavelets types used were coiflets 1 and 2, Daubechies 1, 2 and 4 and symlet 9. Matrix of parameters obtained in this process was transferred to the k-means clustering algorithm which calculated fifteen data clusters used for classification of mental state associated with each segment of signal. A schematic representation of this algorithm is shown in Fig. 2.

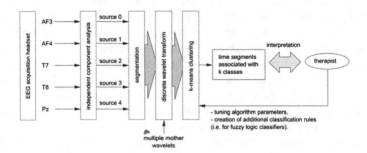

Fig. 2. The diagram of proposed architecture of EEG signal processing module. The obtained results may provide useful information for therapist regarding changes of mental states of tested subjects. This knowledge may be then utilized to improve the system performance.

The output data is visualized in the form of a plot in which ICA-processed EEG signal segment has its own shade of gray associated with a class assigned to it by the clustering algorithm. Such a type of graphical visualization makes it possible to trace changes of mental state of subjects during the exercises. It may be a valuable source of information for therapists who will be able to better understand what influences subjects' emotional state during the exercise performance.

2.4   Integration of the Stand Components and Initial Tests

All mentioned devices were integrated in a prototype of the experimental stand for a polysensory diagnosis and stimulation of subjects with communication disorders. It was equipped with an autostereoscopic screen, eye tracker device, EEG measurement headset and speakers. Eye-gaze tracking system and EEG device allow for the observation and recording of subject's reactions. The stimulation and the exercise procedure is managed by the therapist. Based on the development tools, provided by the manufacturers of EEG devices and eye-gaze tracking system, a set of software applications has been developed for simultaneous stimulation and recording of responses. This set consists of an application for the therapist that allows for managing various types and stages of the training, including a database of anonymized subjects' data with results, as well as the so-called player application, used as a polysensory interface for the subject. The player application provides an interaction between the user's action (gaze fixation point, EEG reaction) and the elements visible on the screen, the reproduced sound. The performance of all system components was initially validated with the participation of the healthy person without communications disorders. There were three sessions of system validation involving watching audiovisual stimuli such as music videos and watching flashing light displayed on the computer screen and states of relaxation with closed eyes. At this stage, the Emotiv INSIGHT 5-electrode headset was used. Examples of EEG signals captured during one of such sessions are depicted in Fig. 3. This stage of testing allowed for validating the signal pre-processing, feature extraction and segmentation procedures used for determining various mental states of a subject. The data enable also to assess quality of the signal recorded by all electrodes of the EEG device, as well as determine the difference in the brain activity during tasks completing and during the relaxation phase. It was assumed that the research study is conducted with the participation of subjects without problems in the auditory pathway (based on information from therapists or an ABR test described above), capable of responding to commands and able to execute verbal commands and use eye tracker. The data acquired during one of such sessions is presented in Fig. 3.

Some kinds of typical artifacts may be observed in this signal. First, the electrooculogram impulses may be identified in signals originating from AF3 and AF4 electrodes during the initial 110 s of the recording while the subject had the eyes opened and was blinking. Some spike-shaped signals are also present in the signals from all electrodes due to short instances when the electrode lost contact with the skin. This was because the Emotiv INSIGHT headset is equipped with dry electrodes and the loss of the skin contact is an issue with such headsets. Contrarily, this type of EEG acquisition hardware is easier and quicker to set up, and thanks to that, the prototyping,

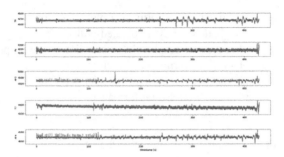

Fig. 3. Sample signals gathered with the use of the Emotiv INSIGHT headset.

correcting and tuning process of signal processing algorithms are also faster. The signal was processed by the algorithm described in the previous subsection. Results of those calculations are shown in Fig. 4.

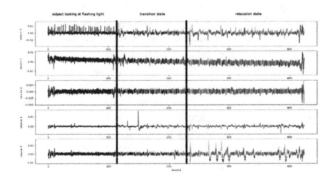

Fig. 4. Example of results of the proposed segmentation algorithm for two activities occurring one after another.

As can be seen in Fig. 4, some artifacts are separated into a single channel after processing. This is visible in the electrooculographic signals present in two electrodes: AF3 and AF4. After processing, the artifacts are present only in the signal coming from the source with the index of 0. Spike shaped signals were also extracted into two separate channels. Results may be even better, if the headset with more electrodes is to be used. Such arrangement would allow for obtaining more sources from the ICA processing stage. Coefficients calculated with the use of DWT made it possible to find two mental states of the subject with the transition state taking place between them. The first one is associated with the initial 110 s of the captured material marked with the darkest shade of gray. When the subject closed his eyes the electroocculogram signal disappeared and the clustering algorithm started to mark more and more frequent classes of the signal segments associated with brighter shades of gray. Finally, in the state of relaxation the initial class of segments is nearly non-existent which allows for distinguishing between two mental states of the subject during the conducted experiment.

3   Tests in Special School and Education Center in Kościerzyna

The hardware setup was tested also in real-world conditions in Special School and Education Center in Kościerzyna in Poland. There were differences between the prototype stand prepared earlier and the set installed in the school. A regular computer screen was used instead of an autostereoscopic monitor and Emotiv EPOC EEG headset was used instead of INSIGHT, however only signals from five electrodes were gathered. A set of language communication exercises and presentation of a video stimulus to three subjects were performed. Three children attending the early years of primary school, suffering from moderate autistic disorders, participated in these sessions. Exercises consisted of a test for understanding text heard and validity of the phoneme hearing. As a result, it was possible to quickly assess whether listening capabilities and ability to focus on given tasks of each subject are sufficient for proceeding to further stages of the investigation. In the beginning, each of them performed some exercises associated with phoneme hearing. The eye tracker was used instead of a traditional computer mouse controller at this stage of the study. Such an arrangement was already used in the earlier research studies [16, 17]. Next, the audio-visual stimulus was shown to them. The stimulus chosen for the presentation is a short excerpt of the motion picture adaptation of a children's story. The animation contains several plot twists which were considered by the therapist as capable to induce emotional responses in subjects. A sample of results of signal clustering and classification are depicted in Fig. 5. The experimental setup and a subject during the hearing exercise are shown in Fig. 6.

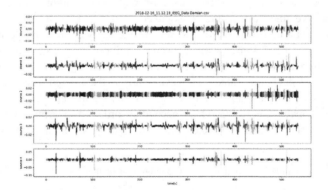

Fig. 5. Segmented EEG signals captured during the session in the special school in Kościerzyna, each mental state is associated with a different shade of gray.

The main problem during the acquisition of data in this case was the size of an EEG headset which is designed for heads of adults. Therefore subjects had to minimize their movements during the test which was not comfortable for them. Despite this recommendation, some artifacts originating from movements of headset are present in the signal and with only five electrodes it is a very hard task for the ICA decomposition to

Fig. 6. Experiment setup used for the in-field testing and acquisition of the EEG and eye tracker signals.

separate all noise from the useful signals. On the other hand, despite these unfavorable circumstances the clustering stage was capable of selecting some instances of different mental state of a subject, whose case is depicted in Fig. 6. It should be noted that such data are more difficult to interpret than those gathered from simple experiments conducted during the initial test stage. This is mostly due to the complexity of the stimulus. The selection of the subject's stimulation method is a difficult and significant decision while designing such a kind of experiment.

The analysis of the mean mental state duration was performed for three subjects from the facility in Kościerzyna and for the control group of three persons. It should be noted that persons from the control group were between 20 and 30 years' old. Each group was associated with six records – three associated with solving simple tasks with the use of the eye tracker and three associated with watching the same animation prepared for the school in Kościerzyna. Results of the analysis are depicted in Table 1.

A frame length of 512 samples was chosen. A difference between groups is visible in the case of maximal length of mental state in both the lengths of states and to less extent in standard deviation values. However, due to the limited size of a data set, differences presented in the Table 1 are not statistically significant if a t-variable test is concerned. Obviously, more research is needed in order to determine the statistical model of mental states. Also, more clear differences may be found, if the stimulation

Table 1. Statistic parameters associated with the lengths of mental states found by the clustering algorithm.

	Mean subject mental state duration [s]		Maximal mental state duration [s]	
	Subjects from Kościerzyna	Control group	Subjects from Kościerzyna	Control group
Mean value	4.45	4.52	7.73	9.45
Std. dev.	0.40	0.32	3.38	4.58

time is longer. In the literature, recordings up to several hours were used for the purpose of sleep state clustering [18]. Moreover, in the future, tests will be performed with a control group of the same age as subjects.

4 Conclusions and Future Work

Further research are necessary for developing robust and precise methods of identification classification of mental states and diagnosis of possible disorders of investigated subjects. In this work we presented a proposal of a hardware and software framework for performing such analyses. It was shown that analyses of data gathered from EEG and eye tracker may lead to interesting conclusions associated with mental state of investigated subjects identified and traced by a data clustering algorithm. Data obtained from this algorithm, which were fed with features extracted from EEG, may lead to identification of time intervals when, for instance, something interesting is happening for the subject or something draw his or her attention. Some differences in statistical properties of clustering algorithm outcomes for two groups of subjects involved into the study may be found, however, more research based on larger sets of data are necessary. In the future, we would like to implement more sophisticated algorithms of clustering, for example based on self-organizing maps, increase the number of analyzed EEG signals (all 14 signals from the EPOC headset), implement the variant of ICA allowing for identifying more sources than the available number of electrode signals and use statistical analysis to optimize the feature vector calculated for each segment of signal. All these additional efforts may provide a tool for therapists for more precise investigation of subjects' needs and therefore for enhancement of results of the performed therapy. Also tests on larger groups of subjects with stimulation of different levels of complexity are planned.

Acknowledgement. The project was funded by the National Science Centre on the basis of the decision number DEC-2013/11/B/ST8/04328.

References

1. Peeters, W., van den Brande, R., Polinder, S., Brazinova, A., Steyerberg, E.W., Lingsma, H.F., Maas, A.I.R.: Epidemiology of traumatic brain injury in Europe. Acta Neurochir. **157**(1), 1683–1696 (2015)
2. Schnakers, C., Vanhaudenhuyse, A., Giacino, J., Ventura, M., Boly, M., Majerus, S., Moonen, G., Laureys, S.: Diagnostic accuracy of the vegetative and minimally conscious state: clinical consensus versus standardized neurobehavioral assessment. BMC Neurol. **9**, 35 (2009)
3. Kunka, B., Sanner, T., Czyżewski, A., Kwiatkowska, A.: Consciousness study of subjects with unresponsive wakefulness syndrome employing multimodal interfaces. In: Proceedings of the International Conference on Brain Informatics and Health, Warsaw (2014)
4. Kunka, B., Czyżewski, A., Kwiatkowska, A.: Awarness evaluation of patients in vegetative state employing eye-gaze tracking system. Int. J. Artif. Intell. Tools **21**(2), 1240007 (2012)

5. Rodrigues, R.A., Busssiere, M., Froeschl, M., Nathan, H.J.: Auditory-evoked potentials during coma: do they improve our prediction of awakening in comatose patients? J. Critic. Care **29**, 93–100 (2014)
6. Skoe, E., Kraus, N.: Auditory brainstem response to complex sounds: a tutorial. Ear Hear. **31** (3), 302–324 (2010)
7. Jones, E., Oliphant, E., Peterson, P., et al.: SciPy: Open Source Scientific Tools for Python (2001) http://www.scipy.org. Accessed 04 Feb 2017
8. http://www.numpy.org/. Accessed 04 Feb 2017
9. Pedregosa, F., et al.: Scikit-learn: machine learning in Python. JMLR **12**, 2825–2830 (2011)
10. https://pywavelets.readthedocs.io. Accessed 04 Feb 2017
11. Tobii – EyeX Controller technical specification. www.tobii.com/xperience/products/ #Specification. Accessed 14 Sept 2016
12. Cichocki, A.: Blind signal processing methods for analyzing multichannel brain signals. Int. J. Bioelectromagnetism **6**(1), 22–27 (2014)
13. Daud, S., Sudirman, R.: Butterworth bandpass and stationary wavelet transform filter comparison for electroencephalography signal. In: 6th International Conference on Intelligent Systems, Modelling and Simulation, Kuala Lumpur (2015)
14. Al-Quazz, N., Ali, S., Ahmad, S., Islam, M., Escudero, J.: Selection of mother wavelet functions for multi-channel EEG signal analysis during a working memory task. Sensors **15**, 29015–29035 (2015). doi:10.3390/s151129015
15. Lin, S., Huang, Z.: Feature extraction of P300 s in EEG signal with discrete wavelet transform and fisher criterion. In: 2015 8th International Conference on Biomedical Engineering and Informatics, Shenyang (2015). doi:10.1109/BMEI.2015.7401500
16. Czyżewski, A., Łopatka, K., Kunka, B., Rybacki, R., Kostek, B.: Speech synthesis controlled by eye gazing. In: 129th Convention of the Audio Engineering Society, preprint 8165, San Francisco, USA (2010)
17. Kostek, B.: Music information retrieval in music repositories. In: Skowron, A., Suraj, Z. (eds.) Rough Sets and Intelligent Systems - Professor Zdzisław Pawlak in Memoriam. Intelligent Systems Reference Library, vol. 42, pp. 463–489. Springer, Heidelberg (2013)
18. Güneş, S., Polat, K., Yosunkaya, S., Efficient sleep stage recognition system based on EEG signal using k-means clustering based feature weighting. Expert Syst. Appl. **37**, 7922–7928 (2010). DOI:http://doi.org/10.1016/j.eswa.2010.04.043

Acute Kidney Injury Detection: An Alarm System to Improve Early Treatment

Ana Rita Nogueira$^{(\boxtimes)}$, Carlos Abreu Ferreira, and João Gama

LIAAD - INESC TEC, Rua Dr. Roberto Frias, 4200-65 Porto, Portugal
ardn@inesctec.pt
https://www.inesctec.pt/

Abstract. This work aims to help in the correct and early diagnosis of the acute kidney injury, through the application of data mining techniques. The main goal is to be implemented in Intensive Care Units (ICUs) as an alarm system, to assist health professionals in the diagnosis of this disease. These techniques will predict the future state of the patients, based on his current medical state and the type of ICU.

Through the comparison of three different approaches (*Markov Chain Model, Markov Chain Model ICU Specialists* and *Random Forest*), we came to the conclusion that the best method is the *Markov Chain Model ICU Specialists*.

Keywords: AKI · Markov Chain Model · RIFLE and Intensive Care Unit

1 Introduction

The acute kidney disease or AKI, is a disease that affects the kidneys, characterized by the rapid deterioration of these organs, usually associated with a pre-existing critical illness [1].

The purpose of this paper is to address the problem of correctly predicting the illness path in various patients by creating an alarm system to work as an early warning to health professionals in intensive cares units, so the necessary measures can be taken in order prevent further damage to the kidneys.

One possible approach is to apply data mining techniques, in which patients records are analyzed and used to predict disease progression for each patient. The patient's evolution is related to the comorbidities and consequently to the type of Intensive Care Unit where they are treated in [2]. For instance, a patient with a surgical heart problem will be treated at Cardiac Surgery Recovery ICU. It is expected that the disease's evolution will be different according to the patient's current state and the hospitalization unit. Taking this, we can apply these data mining techniques to each ICU to maximize the accuracy of the prediction. This paper is organized in four sections: Sect. 2 describes the related work in the AKI disease. Section 3 describes the AKI disease, the criteria used to classify it, the dataset used and the distribution of the data. Section 4 describes the approaches implemented and Sect. 5 the results obtained on the tests.

© Springer International Publishing AG 2017
M. Kryszkiewicz et al. (Eds.): ISMIS 2017, LNAI 10352, pp. 57–63, 2017.
DOI: 10.1007/978-3-319-60438-1_6

2   Related Work

In the recent years, several researchers had the opportunity to work in this subject. Back in 2013, Cruz et al. [3] applied Bayesian Networks to the AKI problem. In this case they used Weka and GeNIe as development tools to implement this algorithm and as data they used the MIMIC II database, of *PhysioNet*.

In 2016, Kate et al. [4] studied different methods (Logistic Regression, Support Vector Machines, Decision Trees and Naive Bayes) in a population of hospitalized older adults (over 60 year old). The objective of this study was to find a way to detect if a patient is going to develop AKI and also detect if a patient already has it.

3   Acute Kidney Disease

Acute kidney disease or AKI is a disease that affects the kidneys, and is described as a rapid decrease of the kidney function, specifically at the elimination of the toxins produced by the organism [1]. Depending on the progression and severity of the disease, a patient can expect to get a full recovery, develop a chronic disease, with the need or not of kidney transplant, or even death [2]. The best way to prevent any permanent damage to the kidneys is to prevent the progression to more severe stages.

For classifying patients considering the severity of their state and act according to that, several criteria can be applied. Currently, the most commonly used is the RIFLE criteria [2], since it has been shown to have the best results in several types of patients [5]. This criteria is divided in levels and outcomes (Table 1). The levels represent the stages the patients with this disease can be at (Risk, Injury and Failure). The outcomes are the representation of the patient's future, if the disease persists: kidney loss in four weeks or end stage kidney disease (ESKD) in three months. The diagnosis with the RIFLE criteria can be made through two different methods: by combining serum creatinine and the glomerular filtration rate (GFR), through measuring the variance between the patient's normal values and the current values, or by measuring the urine output through a tabulated period of time.

Table 1. RIFLE criteria

	Serum creatinine criteria	Urine output criteria
Risk	Increase in creatinine $\geq 1.5X$ or decrease in GFR $\geq 25\%$	<0.5 mL/kg/h for ≥ 6 h
Injury	Increase in creatinine $\geq 2.0X$ or decrease in GFR $\geq 50\%$	<0.5 mL/kg/h for ≥ 12 h
Failure	creatinine $\geq 354\,\mu$mol/L with an acute rise of $\geq 44\,\mu$mol/L	<0.3 mL/kg/h ≥ 24 h or anuria ≥ 12 h

Table 2. Patient's data example

ID	Time	Age	Gender	ICU type	Creatinine	GFR	Initial creatinine	AKI Stage
132592	t1	35	Female	Medical ICU	2	28,35355546	0,86089906	2
	t2				1,7	34,20251844		1

3.1 Data Description

To solve this problem, data from the challenge *"Predicting Mortality of ICU Patients: the PhysioNet/Computing in Cardiology Challenge 2012"* from PhysioNet was used. This dataset is contains demographic data and by blood, urine, heart rate and body temperature indicators.

Using this information as the base, a new dataset was created. This new dataset has only data from patients with at least one creatinine measurement. Besides the information from the original dataset other attributes specific to the AKI problem were computed from it (glomerular filtration rate and initial creatinine). One important measure to correctly diagnose the AKI disease and to predict the disease stage is the Glomerular Filtration Rate or GFR (1). Combined with the creatinine, is the basis of the diagnosis, according to the RIFLE criteria.

$$GFR = 175 \times Scr^{-1.154} \times Age^{-0.203}[\times 0.742 \ (if\ female)] \tag{1}$$

In this equation, if the patient is male, the 0.742 is replaced by 1. Another important calculated value is the initial creatinine measure. This value is required because, according to the RIFLE criteria, the AKI stage is obtained through the difference between the normal creatinine (or GFR) and the measured creatinine (or GFR). If the real normal creatinine value is not provided it can be calculated. In this case is attributed to the patient a normal GFR (75 mg/dL) [2] and by using a derivation of the GFR Eq. (2), the normal creatinine value is extracted:

$$Scr = \frac{75}{175 \times Age^{-0.203}[\times 0.742 \ (if\ female)]}^{\left(\frac{1}{-1.154}\right)} \tag{2}$$

Data Analysis

As stated previously, the dataset used is a modified version of the *PhysioNet* dataset, more focused on the AKI disease. The new dataset is composed by data from 6558 different patients, with an average of 1.19 different stages per patient. Each patient as a set of measurements similar to Table 2. Since the patients are in ICUs and therefore are frequently monitored, the time interval between each measurement is regular.

By analyzing Table 3, which represents the distribution of the RIFLE criteria stages in the dataset, we can verify that 66% of the patients do not show any symptoms of the disease, 14% are in risk, 9% can develop an injury, 11% of the patients can have kidney failure. The distribution of the progression of the disease is mostly stationary, that is, 65% of the patients in a stage (*Normal, Risk, Injury* or *Failure*) tends to maintain it during his stay in the ICU, as shown in Table 4.

Table 3. Distribution of AKI stages

	Percentage
Normal	66%
Risk	14%
Injury	9%
Failure	11%

Table 4. Distribution of the disease progression

	Percentage
Regressed	17%
Progressed	18%
Kept state	65%

4   Methodology

To develop the alarm system, there are already several techniques for predicting events. The general idea of this approach is, knowing the patient's current stage, as well as the patient's clinical records, try to predict the AKI progression. Within the extensive list of algorithms that can be used for solving this problem, the following were explored: *Random Forest* [6] and *Markov Chain Model* [7].

In addition to these two algorithms, an adaptation of the *Markov Chain Model* algorithm was developed. This adaptation divides the dataset into four more restrained *Markov Chain Models*, considering the type of ICU where the patient is hospitalized. These approaches will be explained in more detail in the next subsections.

4.1   Markov Chain Model

The Markov Chain Model algorithm is a stochastic process that uses a sequence of random variables based on the Markov property [8]. This property states that, to obtain the X_{n+1} state, we only need the previous state X_n, as it's shown in the Eq. (3) [7].

$$Pr(X_{n+1} = x_{n+1}|X_1 = x_1, X_2 = x_2, \ldots, X_n = x_n) = Pr(X_{n+1} = x_{n+1}|X_n = x_n) \quad (3)$$

The system changes from one state to the next one, by computing the transition probability [8]. This probability of transitioning from state i to j is given by Eq. (4).

$$p_{ij} = Pr(X_1 = s_j|X_0 = s_i), \text{ being } S = \{s_1, s_2, \ldots, s_n\} \text{ the set of possible states} \quad (4)$$

This probability can be represented in a matrix, where each pair (i, j) is the element in position i, j [7].

Applying this to the AKI problem, we have four different states: *Normal, Risk, Injury* and *Failure*. These are the RIFLE criteria stages plus the normal stage, where the patient, despite being in an ICU, does not show any symptoms of having AKI (Table 5). The parameters of the model were obtained using a maximum likelihood estimate. The transition probability matrix shows that in some cases, specifically in the transition from *Risk*, it's more likely that the disease regress, than to progress or maintain its state (Table 5).

Table 5. AKI Markov's transition probabilities

	Normal	Risk	Injury	Failure
Normal	83.31%	14.25%	2.10%	0.35%
Risk	53.71%	30.41%	14.96%	0.93%
Injury	10.34%	23.46%	49.01%	17.19%
Failure	1.31%	3.54%	18.24%	76.90%

Table 6. AKI Markov's Specialists *"Coronary Care Unit"* transition probabilities

	Normal	Risk	Injury	Failure
Normal	86.30%	11.30%	1.68%	0.72%
Risk	36.77%	45.16%	16.77%	1.29%
Injury	4%	21.60%	59.20%	15.20%
Failure	0%	4.12%	15.46%	80.41%

4.2 Markov Chain Model Specialists

The *Markov Chain Model Specialists* are an adaptation of the *Markov Chain Model* algorithm into the medical environment. These specialists are normal *Markov Chain Model*, trained to predict the next AKI stage for a specific type of ICU. In the used dataset, there are four different ICU types: Coronary Care Unit, Cardiac Surgery Recovery Unit, Medical ICU and Surgical Unit. Each ICU has different patients and specific diseases that influence the AKI progression. Having one *Markov Chain Model* specialist specific to each ICU type may improve the overall performance.

We can verify that the statement above is correct by analyzing Table 6: this table represents the transition probabilities for patients admitted in the *"Coronary Care Unit"* unit. In this table the *Risk* stage has a higher probability of maintaining the same stage than to transit to other stage, which differs from what is shown in Table 5, where the whole population of patients is represented.

5   Results and Discussion

To evaluate the *Markov Chain Model Specialists* and make a comparative study, we designed the following configuration of experiments: we compared the performance of *Markov Chain Model*, *Specialist* and *Random Forest* algorithms using 10-fold cross validation, to divide the data of all 6558 patients. The 10-fold were organized so that all the records belonging to each patient appeared in only one fold. The *Random Forest* algorithm used is the standard implementation and consists of five trees. This number was obtained by testing different numbers of trees (2 to 100). Then the results were compared. The set of trees with the highest number of similar results was the chosen one.

Since the problem in question is a multiclass problem, the results from the test were arranged in a confusion matrix [9]. After creating the confusion matrix, we analyze the results by computing the performance measures: precision, recall (5) and f-measure (6) [10]:

$$Precision = \frac{TP}{TP + FP} \qquad Recall = \frac{TP}{TP + FN} \qquad (5)$$

$$F - Measure = 2 \times \frac{Precision \times Recall}{Precision + Recall} \qquad (6)$$

Table 7. Results comparison

	Precision (%)	Recall (%)	F-measure (%)
Random Forest	65,86	58,87	62,17
Markov Chain Model	58,28	61,6	59,89
Markov Chain Model Specialists	71,08	63,65	67,16

If we analyze Table 7, that represents the performance measures obtained in the tests, we can see the approach with the better overall performance is the *Markov Chain Model Specialists*. This is explained by the Markov's transition probabilities in Table 5. The probability of a patient transiting from *Risk* to *Normal* stage is higher than the other probabilities. However, the probability of maintaining the stage *Risk* is lower than changing to *Normal*, but still high, compared to transitioning to other stages. This is reflected in the results: whenever the current stage is *Risk* the *Markov Chain Model* predicts that the next stage will be *Normal*, since it's the one with higher probability. This is the reason why the specialists approach has better results: by splitting the patients by type of ICU, the computed transition probabilities differ in each of the four ICUs and consequently, it restrains patients with more similar diagnostics than in the basic *Markov Chain Model*.

Besides the *Markov Chain Model*, the ICU Specialists had also better results than the *Random Forest* algorithm, which is an algorithm known to have excellent results in the majority of the problems [11]. This was also achieved by only exploring one attribute of the dataset (*Disease Stage*), while in *Random Forest* all the attributes present in the dataset were explored to create the model. Despite obtaining a worse result than the ICU specialists, the *Random Forest* algorithm still achieves a better result than the traditional approach of the *Markov Chains Model*. This proves that the next state depends, not only from the previous sate, but also from other factors present in the dataset.

6   Conclusions

The acute kidney disease or AKI is a disease of rapid progression, which manifests in ICUs as consequence of a previous disease. Within the implemented approaches, the *Markov Chain Model Specialists* approach showed to have the best results, beating the normal *Markov Chain Model* and *Random Forest*, models known to obtain excellent results in the majority of the problems. This fact occurs, as stated before, due to the grouping of the patients by ICU type, that restrains the outcome of the chain. This results in a more accurate prediction, based on common inputs. In the future, we are working in new and different approaches to solve the AKI early detection problem and exploring other attributes to create other more specific specialists. We are also discussing with a doctor in a Portuguese hospital the implementation of an alarm system that uses the *Markov Chain Model Specialist* and cost analysis of such implementation.

Acknowledgments. This work is supported by the *NanoSTIMA Project: Macro-to-Nano Human Sensing: Towards Integrated Multimodal Health Monitoring and Analytics/NORTE-01-0145-FEDER-000016* which is financed by the *North Portugal Regional Operational Programme (NORTE 2020)*, under the *PORTUGAL 2020 Partnership Agreement*, and through the *European Regional Development Fund (ERDF)*.

References

1. Ostermann, M., Joannidis, M.: Acute kidney injury 2016: diagnosis and diagnostic workup. Crit. Care **20**, 299 (2016)
2. Hoste, E.A.J., Clermont, G., Kersten, A., Venkataraman, R., Angus, D.C., De Bacquer, D., Kellum, J.A.: RIFLE criteria for acute kidney injury are associated with hospital mortality in critically ill patients: cohort analysis. Crit. Care **10**, R73 (2006)
3. Cruz, H., Grasnick, B., Dinger, H., Bier, F., Meinel, C.: Early detection of acute kidney injury with Bayesian networks. (2013)
4. Kate, R.J., Perez, R.M., Mazumdar, D., Pasupathy, K.S., Nilakantan, V.: Prediction and detection models for acute kidney injury in hospitalized older adults. BMC Med. Inform. Decis. Making **16**, 39 (2016)
5. Bagshaw, S.M., George, C., Bellomo, R.: A comparison of the RIFLE and AKIN criteria for acute kidney injury in critically ill patients. Nephrol. Dial. Transplant. **23**, 1569–1574 (2008)
6. Breiman, L.: Random forests. Mach. Learn. **45**, 5–32 (2001)
7. Spedicato, G.A., Signorelli, M.: The R package "markovchain": Easily Handling Discrete Markov Chains in R. CRAN (2014)
8. Ye, N.: A markov chain model of temporal behavior for anomaly detection. Proc. 2000 IEEE Syst. Man, Cybern. Inf. Assur. Secur. Work. 171–174 (2000)
9. Ferri, C., Hernández-Orallo, J., Salido, M.A.: Volume under the ROC surface for multi-class problems. In: Lavrač, N., Gamberger, D., Blockeel, H., Todorovski, L. (eds.) ECML 2003. LNCS (LNAI), vol. 2837, pp. 108–120. Springer, Heidelberg (2003). doi:10.1007/978-3-540-39857-8_12
10. Sokolova, M., Japkowicz, N., Szpakowicz, S.: LNAI 4304 - Beyond Accuracy, F-Score and ROC: A Family of Discriminant Measures for Performance Evaluation (2006)
11. Wainberg, M., Alipanahi, B., Frey, B.J.: Are random forests truly the best classifiers? J. Mach. Learn. Res. **17**, 1–5 (2016)

New Method of Calculating ^{SR}CM Chirality Measure

Przemyslaw Szurmak[✉] and Jan Mulawka

Artificial Intelligence Division, Warsaw University of Technology, Warsaw, Poland
przemyslaw@szurmak.pl, jmulawka@elka.pw.edu.pl
http://szurmak.pl

Abstract. Bioinformatics plays an important role in natural sciences. One of its branches – Computer-Aided Drug Design (CADD) – gives practical insights for designing and discovering of novel – better and safer – drugs. The CADD encompasses many different techniques like docking, virtual screening and quantitative structure-activity relationships (QSAR). The latter deals with building equations relating drug activities and their structures represented by variables called molecular descriptors. An important and promising type of such descriptors are Sinister-Rectus Chirality Measures (SRCM). However, the only so far available software for ^{SR}CM calculation is very slow, and this impedes wider application of ^{SR}CM by QSAR community. Therefore, an attempt to develop a novel algorithm for calculation of ^{SR}CM (using Genetic Algorithm) and to implement it in an efficient and modern computer program was made. The result of these efforts is Chirmes. Performed tests have shown that Chirmes gives correct results of ^{SR}CM calculations and performs way faster than the so far available software does. The paper describes first chemical and computational background behind the tackled problem. Then details of the implementation are presented, along with the test results and future prospects.

Keywords: Computer-aided drug design · QSAR · SRCM · Genetic algorithms · Cuda

1 Introduction

Although the XX century witnessed unprecedented development of medicine and chemical pharmacology, there are still many diseases for which safe and efficient therapies are lacking. The efforts for finding them are the domain of an interdisciplinary research field called medicinal chemistry. In recent years, computer-aided drug design (CADD) has become an integral part of this research. CADD encompasses a number of methodologies that allow medicinal chemists to model the behaviour of drugs and their molecular targets. Thus it facilitates rational directing of expensive laboratory work. One of CADD techniques is Quantitative Structure-Activity Relationship (QSAR) analysis. The aim of QSAR study is to find a mathematical (quantitative) relationship between chemical structure and the medicinal activity. To this aim, equations are built that are of general structure:

© Springer International Publishing AG 2017
M. Kryszkiewicz et al. (Eds.): ISMIS 2017, LNAI 10352, pp. 64–73, 2017.
DOI: 10.1007/978-3-319-60438-1_7

$$activity = f(structure) \qquad (1)$$

Here, the structure is expressed as molecular descriptors, that is variables that describe a certain aspect of molecular structure., e.g. number of atoms of a given kind, number of flexible bonds, energy of molecular orbitals etc. [7] The choice of descriptors and the way they are related to activity may be knowledge-based or supported by special methods [2]. The obtained equations - if they are of good statistical quality - may be used to explain the observed experimental findings and to predict the behaviour of novel - yet unsynthesized molecules. In such a case, QSAR allows to save money and time and to increase the rate of drug discovery. Many drug molecules bear a special property - chirality. This means that they are not superposable on their mirror image. The chirality of drug molecules is an important and well-known problem in medicinal chemistry since some chiral molecules exhibit different activity and/or toxicity depending on which mirror image (left-handed or right-handed enantiomer) they are. There appeared an idea to use chirality - described quantitatively by Sinister-Rectus Chirality Measures - for QSAR modelling. The descriptor has been successfully applied several times [5], but a lack of a fast tool to compute the descriptor seriously hampered the use of this variable in QSAR modelling. The aim of our research was to fill this gap by designing and implementing a novel method for calculating Sinister-Rectus Chirality Measures.

This paper is organized as follows. Section 2 describes theory behind chirality measures. Section 3 discusses implementation details of the developed solution. Sections 4 presents the results of tests made using Chirmes, with comparison to Chimea. Finally, Sect. 5 presents our summary.

2 Theory

2.1 Chirality Measures

Chirality is a property of a three-dimensional shape of a molecule. The degree to which a drug is efficient is strongly connected to spatial fit of a drug molecule and its molecular target. Thus chirality can be used to describe structure of molecule. Chirality measures may be of use here. They are variables that describe how much chiral a molecule is, or in other words (according to the IUPAC definition [1]): how much non-superposable on its mirror image it is.

Out of many known chirality measures, this paper is focused on Sinister-Rectus (^{SR}CM) chirality measures [3,4]. They are calculated as follows: a mirror image of an analysed molecule is generated and then superimposed onto original molecule structure so that the superposition is optimal. The goodness of fit is rated based on normalized cartesian sum of distances between corresponding atoms in original and mirror molecules weighted depending on chosen property space. Mathematical equation describing ^{SR}CM reads:

$$^{SR}CM(A) = \frac{1}{a}min(\sum_{i=1}^{n} w_i d_i) \tag{2}$$

where a is normalization component (molecular mass), d_i – distance, w_i – weight (most often, property assigned to selected atom like electrical charge or atom mass), n – number of atoms in molecule.

A most important issue during ^{SR}CM equation solving is to find optimal superposition that would minimize the ^{SR}CM value. For achiral molecules their mirror images should be superposed ideally on original structures ($^{SR}CM = 0$), in case of chiral molecules the situation is more complicated though. Such molecule has infinite number of possible bad superpositions. If we also take into consideration that a typical molecule can contain up to several hundreds of atoms it can be clearly seen that a problem domain in such a case is enormous thus choosing a proper algorithm and implementing it in an efficient way plays a key role for solving the problem.

^{SR}CM chirality measures have been already used in a real-world scientific research, including studies on chiral heterofullerenes [4], modelling activity of steroids binding to sex-hormone binding globuline [3] and in analysis of Vibrational Circular Dichroism spectra [6]. Other chirality measures were also used to model activity of pain relief drugs, acetylcholinesterase inhibitors, behaviour of amino acids in plate chromatography, or for explaining of catalytic activity. Such numerous and versatile examples of applications show potential and necessity of developing software to calculate such descriptors in an efficient way which will help spreading usage of chirality measures in bioinformatics (especially in computer aided drug design) and in general – computational chemistry.

3   Implementation

3.1   General Application Structure

The main goal of the presented research was to develop an effective algorithm to calculate ^{SR}CM chirality measure and to implement such an algorithm as a working desktop application. The algorithmic problem is related to optimization problem during calculation of ^{SR}CM. After generating mirror image of molecules, the best superposition of mirror and base molecule needs to be found so that the value of (2) defining ^{SR}CM measure is the lowest.

During the problem discussion, the authors decided to apply genetic algorithms (GA). Main reason behind usage of GA was ease of implementation and well known as very universal and flexible method for problem solving, also in computer drug design. The detailed description is given in sections: Sect. 3.2 and 3.3.

One of the main targets for created software was performance. Because of that, the whole application was developed using C++ in its most recent specification - C++14.

Also, experiments were made to find out if implementation on GPU will gain performance increase. To allow simple integration of GPU working code with rest of application whole program was divided to separate modules which can work independently. Another advantage of such approach is ease of allowing application to work on different operating systems such as Apple Mac OS, Microsoft Windows and Linux.

Developed software was named Chirmes from words **Chir**ality **Me**asures. Figure 1 shows modular structure of application divided by work phases.

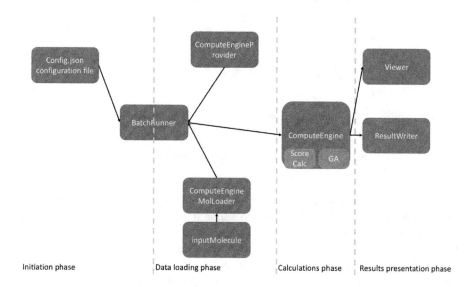

Fig. 1. Chirmes typical workflow chart with developed modules

First step of the work flow is loading configuration file which is then passed to *BatchRunner* module which. Separate class *ComputeEngineMolLoader*, loads input molecules and allocates (through helper *ComputeEngineProvider* object) chosen version of main calculation unit called *ComputeEngine* (implementation of *Abstract Factory* pattern). During loading of each molecule (from standard chemistry description file formats, more than 100 are supported) mirror images are created. Instance of *ComputeEngine* using implemented genetic algorithm (shown in Fig. 2) tries to find optimal superposition of molecule and its mirror image and then calculates ^{SR}CM value (presented in Fig. 3). Description of both processes can be found in Sects. 3.2 and 3.3.

3.2 Genetic Algorithm

Solving optimization problem (superposition of molecule and its mirror image) during calculation of ^{SR}CM is the most important part of Chirmes application. After loading molecule and its mirror image, a genetic algorithm is used, through series of rotation and translation applied to molecules to find combinations giving lowest value of chirality measure.

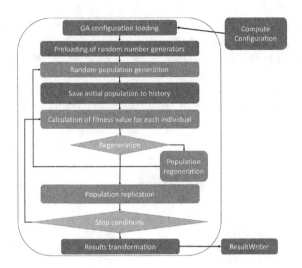

Fig. 2. Implemented genetic algorithm scheme

The application implements genetic algorithm in a typical form, overview of which is presented in Fig. 2. In order to tailor it better for the given problem, small improvements were made compared to the original idea of genetic algorithm. Gene coding is not binary but using floating point numbers[1] because they are mapping much better spatial coordinates for rotation, translation and, at the same time, providing better precision for such problem. Also, usage of binary coding would impose usage of much more complicated crossover and mutation operators to ensure correct solution domain (not every numeric value has meaning for spatial rotation or coordinate).

In order to achieve best performance, each gene (x, y, z for rotation and for translation) is coded using 32 bit floating point number instead of double precision. Main benefit of such approach is possibility to use vectorization support (SSE/AVX) provided by *Eigen* library which allows to make two times more calculations using single precision numbers than with double precision numbers at the same time. To prepare transformation matrix that converts genotype into phenotype following equation is used (3):

$$matrix = (translation * translationFromOrigin \\ * rotation * translationToOrigin) \tag{3}$$

Another modification of original genetic algorithm was implemented in process of random population generation. Right from the start values generated using *Mersenne Twister 19937* algorithm[2] are limited to only those having physical meaning – for example, translation components should not make

---

[1] Usage of real numbers was inspired by [8].
[2] Period of algorithm is $2^{19937} - 1$ which is more than enough for generating random population. Moreover, it is well known and optimized.

absolute distance between corresponding atoms greater than distance between geometrical center of molecule and its mirror image.

After generation of random population *Compute Engine* calculates value of chirality measure for each individual Fig. 3 (described in Sect. 3.3).

Next step is conditional population regeneration which is also another modification comparing to original genetic algorithm. This operator was introduced to respond more effectively in case of poor improvement of best individual score comparing to previous algorithm iterations. When such situation is discovered operator takes best individual from current population and, depending on chosen algorithm settings and current population situation, puts it into new population, created randomly from scratch without memory about previous iterations.

The operations mentioned above are important, however the essence of GA are three following steps: 1. Selection 2. Crossover 3. Mutation. They are responsible for exchange of genetic information between individuals which is why algorithm is able to find proper solution.

Out of many known selection methods, tournament selection was chosen as the one that is efficient enough and allows easily to control selective pressure through size of tournament. High value of this parameter decreases diversity of population[3].

In Chirmes two methods of crossover were implemented, one with fixed exchange point (between genes in genotype) and another one with random exchange point. After that, with random probability of occurence, mutation operator is used (independently for each child from crossover). During mutation, the application determines first whether mutate rotation or translation and then, which component should be changed – x, y or z. After randomly selecting mutation type new value is assigned, either by adding some random value to rotation or by generating new translation value in domain prepared in random population generation phase.

These operations are repeated for each individual from population to get completely new that replaces existing one for next algorithm iteration.

3.3   Chirality Measure Calculations

Fitness function (chrirality measure) in Chirmes is defined by Eq. (2). In the first step, new temporal individual is created by multiplying transformation matrix (inherited from genotype) with atoms positions matrix. Having that the application enters the most computationally intensive part – calculation of chirality measure. It is presented in Fig. 3. Initially, distance matrix between corresponding atoms from mirror image and original molecule is calculated. Currently, the application uses only distances in geometrical space without taking into consideration chemical or physical properties of molecule – it is planned to be implemented in future versions. In next step iterative search of smallest distance sum

---

[3] Strong selective pressure make algorithm to convergence to early, from the other side, weak pressure searching less effective. It is important to find correct bias between too strong and too weak pressure [8].

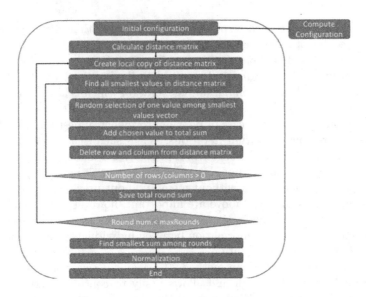

Fig. 3. Algorithm for fitness function calculating – chirality measure

between atoms is being performed. Each iteration consists of following parts: 1. finding smallest values in local distance matrix copy, 2. random selection from one of them. Because several equally small values can be found whole process needs to be repeated multiple times, 3. chosen distance is added to general sum of distances. Row and column where this distance was located in matrix are deleted from local copy, 4. whole process is continued until local matrix will be empty. In last step the algorithm finds smallest value among all calculated in all rounds and its normalization through division by number of molecules in analysed molecule.

4   Results

4.1   Achiral Molecules Test

The basic test for the novel method of calculating chirality measures is to check if ^{SR}CM values calculated for achiral molecules are zero. These molecules are identical with their mirror image thus in most optimal superposition atoms of input compound and their mirrored version are on the exactly same positions so final value of optimization method will be null.

In order to verify if Chirmes fulfill this requirement, a test with four achiral compounds with different size was performed, also in comparison to CHIMEA application. Results are shown in Table 1.

It can be seen that Chirmes finds values very close to ideal 0.0000. Deviation from this value is relatively small, moreover it is known that usually chiral compounds has ^{SR}CM measure in a range of 0.100–0.200. Error is even more

Table 1. ^{SR}CM values calculated for achiral molecules by Chirmes and Chimea

n	Molecule	Atoms no	^{SR}CM value	Calculation time [ms]	^{SR}CM value, CHIMEA	Calculation time, CHIMEA[s]
1	Fullerene	60	0.0067	132884	0.0007	3480
2	Tetraphenyl	44	0.0018	49360	0.0018	600
3	Butane	14	0.0043	4479	0.0005	120
4	Etane	8	0.0020	1819	0.0017	60

negligible when compounds are presented visually using built in visualisation module - in all four cases atoms were perfectly superimposed on their mirror images. What is even more important, achieving similar level of accuracy with CHIMEA takes more than an hour while using Chirmes it took about three minutes.

4.2 Chiral Molecules Test

Further, a comparison between values of ^{SR}CM calculated by CHIMEA and Chirmes for eleven chiral compounds was made. Results are presented by Table 2.

Table 2. Results gathered for ^{SR}CM values from CHIMEA and Chirmes

n	Molecule	CHIMEA	Chirmes	Deviation [%]
1	5-androstenediol	0.6777	0.7051	4
2	3β-androstanediol	0.6962	0.7756	11
3	5.6-didehydroizoandrosterone	0.7281	0.7944	9
4	5α-dihydrotestosterone	0.7512	0.7828	4
5	3α-androstanediol	0.7689	0.8139	6
6	androsterone	0.8154	0.8978	10
7	epitestosterone	0.8188	0.8372	2
8	testosterone	0.8435	0.8727	3
9	4-androstenedione	0.8464	0.9758	15
10	5β-dihydrotestosterone	0.848	1.0453	23
11	4-androstenediol	0.8568	0.8686	1

Percentage deviation of values calculated by Chirmes is in range of 1 to 23 percent. However results from CHIMEA and Chirmes are correlated and correlation rate is $R = 0.86$. Therefore, despite quite high error for some compounds made by Chirmes, results are correct for QSAR results application. In QSAR most important problem is to find good mapping of interrelationships in set of many compounds, not only about absolute values.

As in test from previous section here also Chirmes was significantly quicker than CHIMEA. To calculate results for all eleven compounds it took only 16 min for CHIMEA comparing to less then 2 min for Chirmes which is almost ten times better.

4.3   Chirmes Usage in Drug Research

To find out about practical advantages of developed application, test verifications were made in Mossakowski Medical Research Centre Polish Academy of Sciences which are shortly described below.

Values of chirality measure for 11 steroids shown in Sect. 4.2 were used to build QSAR model describing affinity of those molecules to androgen receptor. It needs to be explained that androgen receptor is a protein which connects with testosterone and causes production and maintenance of male sexual characteristics. Moreover, it is responsible for building bones, muscles as well as muscles strength. Androgen receptor is important target for drugs assisting in muscle recovery in case of serious diseases or surgery.

Hypothesis, which was verified thanks to calculated QSAR model, says that presence, character of molecule ending elements and shape of whole molecule is most important for successful bonding of steroids with androgen receptor.

After QSAR modelling following equation describing relationship between molecule affinity $(log(RBA))$ and character of ending elements (presented as partial molecular charges $q3$ i $q17$) and general shape of molecule described by ^{SR}CM chirality measure was developed

$$
\begin{aligned}
log(RBA) = 4.1(\pm 3.1) + 9.2(\pm 2.5) * q3 - 7.0(\pm 2.3) * q17 \\
-3.3(\pm 1.2) * {}^{SR}CM, r = 0.83, n = 11
\end{aligned} \tag{4}
$$

where $log(RBA)$ – relative binding affinity; $q3$ i $q17$ – electrical charges of c3 and c17 carbon atoms. It can be clearly seen that correlation ratio between test and experimental data for this equation (which is only a preliminary model) is at acceptable level ($R = 0.83$) thus it is a good base for more detailed analysis.

Again, it should be emphasized that all these ^{SR}CM values were calculated more then 10 times faster then using previous CHIMEA application which will be especially important for calculating ^{SR}CM for larger molecules. Time gain achieved by Chirmes is very promising for real-world use of the developed application in CADD scientific research. Preliminary scalability tests performed on larger molecules show a similar (or even higher) improvement, as compared to CHIMEA. Their results are beyond the scope of this paper and they will be presented in the following works.

5 Summary

The main goal of the presented research was achieved. A new method for efficient calculation of ^{SR}CM chirality measures by usage of genetic algorithms was developed and implemented as a computer application.

As it is presented in Sect. 4 usage of Chirmes gives significant performance gain (comparing to existing CHIMEA software) with necessary level of correctness. Thanks to possibility of customisation of all parameters in genetic algorithm it is possible to achieve even better results after tuning several GA parameters.

Another advantage of Chirmes is handling of more then 100 well used formats of chemistry related files and multiplatform availability.

During works on Chirmes authors also analysed possibility of porting Chirmes into highly parallel environments such as CUDA. Even though time frame for this paper was too short to prepare fully functional CUDA implementation quick proof of concept application showed potential to gain even higher performance increase then with usage of regular CPU.

The application developed in the presented research is an important step forward in bringing chirality measures to the mainstream of Computer-Aided Drug Design. Chirmes advantages shows that is a good answer for real needs of chemistry science. In biochemistry there is still a lot of space for use of computer aided computations with addition of artificial intelligence and new hardware solutions. Developed software opens new chances for further and intensive computers use in chemistry.

References

1. Blackwell Scientific Publication: Chirality. IUPAC. Compendium of Chemical Terminology (1997)
2. Gonzalez, M.P., Teran, C., Saiz-Urra, L., Teijeira, M.: Variable selection methods in QSAR: an overview. Curr. Top. Med. Chem. **8**(18), 1606–1627 (2008)
3. Jamroz, M.H., Rode, J.E., Ostrowski, S., Lipinski, P.F.J., Dobrowolski, J.C.: Chirality measures of α-amino acids. J. Chem. Inf. Model. **6**(52), 1462–1479 (2012)
4. Jamroz, M., et al.: On stability, chirality measures, and theoretical VCD spectra of the chiral C58X2 fullerenes (X = N, B). J. Phys. Chem. A **1**(116), 631–643 (2012)
5. Nguyen, L.A., He, H., Pham-Huy, C.: Chiral drugs: an overview. Int. J. Biomed. Sci. **2**(2), 85–100 (2006)
6. Lipinski, P., Dobrowolski, J.: Local chirality measures in QSPR: IR and VCD spectroscopy. RSC Adv. **87**(3), 47047–47055 (2014)
7. Todeschini, R., Consonni, V., Mannhold, R., Kubinyi, H., Timmerman, H.: Handbook of Molecular Descriptors. Methods and Principles in Medicinal Chemistry, Wiley (2008)
8. Michalewicz, Z.: Genetic Algorithms + Data Structures = Evolution Programs. Springer-Verlag, Berlin (1996)

Summary

Data Mining Methods

Selection of Initial Modes for Rough Possibilistic K-Modes Methods

Asma Ammar$^{(\boxtimes)}$ and Zied Elouedi

LARODEC, Institut Supérieur de Gestion de Tunis, Université de Tunis,
41 Avenue de la Liberté, 2000 Le Bardo, Tunisia
asmaammarbr@gmail.com, zied.elouedi@gmx.fr

Abstract. This paper proposes two new improvements of the k-modes algorithm under possibilistic and rough frameworks. These new versions of the k-modes deal with uncertainty in the values of attributes and in the belonging of objects to several clusters and handle rough clusters using possibility and rough set theories. In fact, both of the k-modes under possibilistic framework (KM-PF) and the k-modes using possibility and rough set theories (KM-PR) provide successful results when clustering categorical values of attributes under uncertain frameworks. However, they use a random selection of the initial modes which can have a bad impact on the final results. As a good selection of the initial modes can make better the clustering results, our aim through this paper is to propose new methods that improve the KM-PF and KM-PR methods through the selection of initial modes. Besides, we will study and analyze the impact of the good selection of the initial modes on both KM-PF and KM-PR. The test of the new methods that select the initial modes from the most dissimilar objects shows the improvement made in terms of accuracy, iteration number, and execution time.

Keywords: K-modes · Initial modes · Possibility theory · Rough set theory · Clustering

1 Introduction

The k-modes clustering algorithm [8] aims to cluster objects with categorical values of attributes to k clusters under certain framework. This method provides interesting results when dealing with this kind of data. However, this standard version of the k-modes method (SKM) has different limitations essentially its inability to handle uncertainty and the random selection of the k initial modes. In [1,2], authors proposed two new soft versions of the k-modes method that handle uncertainty. The k-modes under possibilistic framework KM-PF [2] uses the possibility theory to deal with uncertainty in the values of attributes and in the belonging of objects to several clusters. Besides, the k-modes using possibility and rough set theories KM-PR handles the uncertainty in the same way of the KM-PF using possibility theory and takes profits of the rough set theory by detecting rough clusters. However, both of the KM-PF and KM-PR selects

© Springer International Publishing AG 2017
M. Kryszkiewicz et al. (Eds.): ISMIS 2017, LNAI 10352, pp. 77–86, 2017.
DOI: 10.1007/978-3-319-60438-1_8

randomly the initial modes which can affect the final partition. This latter could be unstable. As a result, our aim through this paper is to develop two new methods that improve the KM-PF and KM-PR. We propose the KMPF-SIM and KMPR-SIM to mean respectively the KM-PF and the KM-PR with selection of initial modes.

The rest of the paper is structured as follows: Sect. 2 provides an overview of the possibility and rough set theories. Section 3 presents the background concerning the k-modes method. Section 4 details the main parameters of the proposed methods. Section 5 shows and analyzes experiments.

2   Possibility and Rough Set Theories

2.1   Possibility Theory

Possibility theory is a well-known uncertainty theory devoted to handle incomplete and uncertain states of knowledge [14]. It has been developed by several researchers including Dubois and Prade [3]. This theory has been widely applied in many fields such as clustering [2,12,13].

One fundamental concept in possibility theory consists of the possibility distribution function denoted by π.

Assume Ω as the universe of discourse containing several states of the world such as $\Omega = \{W_1, W_2, ..., W_n\}$. A possibility distribution is defined as the mapping from the universe of discourse Ω to the interval $L = [0, 1]$ which represents the possibilistic scale.

Several similarity and dissimilarity measures were proposed in possibility theory in order to compare pieces of uncertain information. We can mention the information affinity [4] defined by: $InfoAff(\pi_1, \pi_2) = 1 - 0.5[D(\pi_1, \pi_2) + Inc(\pi_1, \pi_2)]$. Where $D(\pi_1, \pi_2) = \frac{1}{n}\sum_{i=1}^{n}|\pi_1(W_i) - \pi_2(W_i)|$ is the Manhattan distance, $Inc(\pi_1, \pi_2) = 1 - max(\pi_1(W_i) Conj \pi_2(W_i))$ is the inconsistency measure and the conjunctive "Conj" is chosen as the minimum operator.

2.2   Rough Set Theory

Rough set theory was proposed in the early 1980s by Pawlak [10,11]. In rough set theory, data are represented through an *information table*. A decision table contains two types of attributes mainly *condition* and *decision* attributes. An information system is defined as a system $IS = (U, A)$ where U and A are finite and nonempty sets. U is the universe containing a set of objects and A is the set of attributes. Each $a \in A$ has a domain of attribute V_a defined by an information function $f_a : U \rightarrow V_a$. The approximation of sets is a fundamental concept in rough set theory. Given an $IS = (U, A)$, $B \subseteq A$ and $Y \subseteq U$. The set Y is described through the values of attribute from B by defining two sets named the B-upper approximation $\overline{B}(Y)$ $(\overline{B}(Y) = \bigcup\{B(y) : B(y) \cap Y \neq \phi\})$ and the B-lower approximation $\underline{B}(Y)$ $(\underline{B}(Y) = \bigcup\{B(y) : B(y) \subseteq Y\})$. Here $y \in U$, and $B(y)$ is a subset of U such that $\forall x \in B(y)$ values of attributes in B are the same. The B-boundary region is defined by $BR(Y) = \overline{B}(Y) - \underline{B}(Y)$.

3   K-Modes Method

The k-modes method [8,9] has been developed to handle large categorical datasets. It uses as parameters the simple matching dissimilarity measure and a frequency based function. The simple matching is defined by $d(X_1, Y_1) = \sum_{t=1}^{m} \delta(x_{1t}, y_{1t})$. Note that $\delta(x_{1t}, y_{1t}) = 0$ if $x_{1t} = y_{1t}$ and $\delta(x_{1t}, y_{1t}) = 1$. The frequency based function consists of selecting the most frequent value as the new value of the mode.

In order to avoid the non-uniqueness of the cluster mode and to deal with uncertain datasets, several improved versions of the k-modes method under possibilistic and rough frameworks are proposed. They are detailed in the following subsection.

4   Selection of Initial Modes for the KM-PF and KM-PR

The KM-PF [2] and KM-PR [1] are two uncertain versions of the k-modes methods. The KM-PF uses the possibility theory in order to handle uncertain values of attributes through possibility distributions. It also defines possibilitic membership degrees to represent the degree of belonging of each object to several clusters.

The KM-PR takes also profits of the possibility theory by handling the uncertainty as the KM-PF does. In addition, it introduces the rough set theory and uses the approximation of sets in order to compute the boundary region and detect rough clusters.

However, both of KM-PF and KM-PR methods select randomly the initial modes. They do not have a method that can provide the best final partitions. The obtained results could be unstable or not optimized.

The following subsections presents the improved versions of these uncertain methods i.e., KM-PF and KM-PR. These latter apply a new method for the selection of the initial modes by detecting the most dissimilar objects.

4.1   K-Modes Under Possibilistic Framework with Selection of Initial Modes

The k-modes under possibilistic framework based on the selection of initial modes denoted by KMPF-SIM is an improvement of the KM-PF method [2]. The new method combines the possibility theory with the k-modes method in order to handle two levels of uncertainty. It is considered as a soft version of the k-modes since it allows the belonging of each object to k available clusters. Besides, KMPF-SIM selects the initial modes from the most dissimilar objects. Hence, it avoids the main limitation of the k-modes method.

The KMPF-SIM uses the following parameters:

– An uncertain training set: containing attributes' values that may be certain and/or uncertain. Each value of attribute takes a possibility degree from $[0, 1]$.

– The selection of the initial modes: is based on the most dissimilar objects. The KMPF-SIM computes the similarity between each pair of objects of the training set using the information affinity measure $InfoAff$. Here, we handle two cases. If $k=2$ i.e., our aim is to cluster the training instances into 2-partitions, we choose as initial modes the most dissimilar pair of object. In other words, the objects that have the lowest $InfoAff$ value. Otherwise, if $k > 2$, we select the two initial modes as defined in the first case (i.e. $k = 2$). Then, we compute the $InfoAff$ between these two initial modes and the remaining objects (i.e. we do not consider objects that are already selected as the 2 initial modes). For each object, we use the mean operator in order to compute the average of the similarity with the two modes. Finally, we choose the remaining modes (i.e. $(k - 2)$ modes) from the objects that have the lowest average. The following formula presents how to choose the modes:
$$Min[\frac{\sum_{j=1}^{2} InfoAff(mode_j, O_1)}{2}, \frac{\sum_{j=1}^{2} InfoAff(mode_j, O_2)}{2}, ..., \frac{\sum_{j=1}^{2} InfoAff(mode_j, O_i)}{2},$$
$$\frac{\sum_{j=1}^{2} InfoAff(mode_j, O_{n-2})}{2}].$$

– The possibilistic similarity measure applied is based on the InfoAff [4]: It consists of a modified version of the information affinity and denoted by $IAf(X_1, X_2) = \frac{\sum_{j=1}^{m} InfoAff(\pi_{1j}, \pi_{2j})}{m}$. Note that m presents the number of attributes and π_{1j} and π_{2j} are the possibility distributions respectively of the objects X_1 and X_2 relative to the attribute j.

– The possibilistic membership degree ω_{ij}: is obtained by computing the possibilistic similarity measure. It represents the degree of belonging of an object X_i to the cluster j. This membership degree $\omega_{ij} \in [0, 1]$ is equal to 0 when there is no similarity between the object i and the cluster j. ω_{ij} is equal to 1 if the object i and the mode of cluster j have the same values.

– The update of the cluster mode: is based on the degree of belonging of each object to the k clusters i.e., ω_{ij}. It also depends on the initial possibility degrees of the training instances [2].

4.2   K-modes Using Possibility and Rough Set Theories with Selection of Initial Modes

The k-modes using both possibility and rough set theories and based on the selection of initial modes from the most dissimilar objects is a modified version of the KM-PR [1]. Our proposal denoted by KMPR-SIM differs from the k-modes and from the KM-PR by avoiding the random choice of the k initial modes in the clustering task. Besides, it keeps the advantages of the KM-PR by handling three levels of uncertainty. First, KMPR-SIM deals with uncertain values of attributes using possibility degrees. Then, it computes for each object the possibilistic membership degree. It uses possibility to handles these two uncertainty. Finally, it applies the rough set theory to detect boundary region and hence, rough clusters. The KMPR-SIM parameters are detailed as follows:

– Uncertain database: The KMPR-SIM uses an uncertain training set where there is both certain and/or uncertain values. A possibility distribution is defined for each attribute value relative to each object.

- The selection of the initial modes: is the same method used for the KMPF-SIM. It consists of selecting the k initial modes from the most dissimilar objects using the information affinity measure.
- The possibilistic similarity measure: In the KMPR-SIM, each value of attributes for both modes and instances may be uncertain. As a result, the KMPR-SIM uses the same possibilistic similarity measure defined for the KMPF-SIM.
- The possibilistic membership degree ω_{ij}: It characterizes the degree of belonging of each object i of the training set to the cluster j. It is deduced from the possibilistic similarity measure.
- The update of the modes: the KMPR-SIM takes into consideration the possibilistic distributions relative to the modes and the instances when updating the modes. It uses the same method defined for the KM-PR [1].
- The detection of peripheral objects: After computing the final possibilistic memberships of instances, the detection of boundary region is needed using rough set theory. In order to specify the peripheral objects, we introduce a new parameter that consist of the ratio (R) [5,6]. This latter determines if an object belongs to the upper or the lower region. The R is defined by $R_{ij} = \frac{\max \omega_i}{\omega_{ij}}$. After computing R for all objects, we set a threshold $T \geq 1$ [5] [6]. Then, the R_{ij} is compared to T. If $R_{ij} \leq T$, the object i is considered as a member of the upper bound of the cluster j. If an object i belongs to the upper bound of only one cluster j, i necessarily belongs to its lower bound.

5   Experiments

5.1   The Framework

In order to test and compare our new methods to the standard k-modes (SKM) and to KM-PF and KM-PR methods, we use UCI Machine Learning Repository [7]. The databases are described in Table 1.

Table 1. Description of the used data sets

Databases	#Instances	#Attributes	#Classes
Shuttle landing control (SL)	15	6	2
Balloons (Bal)	20	4	2
Post-operative patient (PO)	90	8	3
Congressional voting (CV)	435	16	2
Balance sscale (BS)	625	4	3
Tic-tac-toe endgame (TE)	958	9	2
Solar-flare (SF)	1389	10	3
Car evaluation (CE)	1728	6	4

5.2   Artificial Creation of Uncertain Data Sets

There are different data sets that contain uncertain values of attributes relative to objects. These uncertain values are due to some errors or lack of knowledge. The proposed algorithms of the KMPF-SIM and KMPR-SIM replace uncertain values by possibility degrees given by expert as follows:

1. For certain attribute values: the true value i.e., known with certainty is replaced by the possibility degree 1. Remaining values take the value of 0.
2. For uncertain attribute values: The true value also takes the degree 1. The remaining values take possibilistic degrees from $]0, 1]$.

5.3   Evaluation Criteria

In order to evaluate our proposals, we use as evaluation criteria *the accuracy* [8] $AC = \frac{\sum_{j=1}^{k} a_j}{n}$, with a_j is the correctly classified objects and n the total number of objects. As we deal with soft methods where an object can belong to several clusters with respect to its possibilistic degree, an object is considered as correctly classified if it belongs to the correct cluster with a possibilistic degree more than a threshold $Th1 \in [0, 1]$. In our case, we set $Th1 = 0.5$. In addition, we use the *the iteration number (IN)* that indicates the number of iterations needed in the main program and *the execution time (ET)* that represents the time taken in the whole program to cluster objects.

5.4   Experimental Results

In this section, we present and analyze the results of our proposals using the evaluation criteria. We have two main cases corresponding to the values of attributes that are artificially created. We will compare our new methods i.e., KMPF-SIM and KMPR-SIM to the SKM, KM-PF, and KM-PR.

1. Certain case: where values of attributes are defined from $\{0, 1\}$. Table 2 shows the results of KMPF-SIM and KMPR-SIM compared to the SKM, KM-PF, and KM-PR using the AC, IN, and ET.

    From Table 2, we can remark that the KMPF-SIM and KMPR-SIM improve the clustering results. This is obvious for the first criterion i.e., *accuracy*. Besides, we notice that the KMPR-SIM method provides the best results. For example, for the Car Evaluation database (CE) the AC is equal to *0.93* for the KMPR-SIM and it is equal to *0.92* for the KMPF-SIM.

    Moving to *the iteration number* and *the execution time*, we notice that the KMPR-SIM also provides good clustering results by avoiding the random choice of the initial modes. The KMPR-SIM and KMPF-SIM generate the same number of iteration as the SKM, KM-PF, and KM-PR. In addition, the new methods are very close to the other methods in terms of *ET*. We also remark that the KMPR-SIM and KMPF-SIM needed in some cases more time to detect the most dissimilar objects for the initial modes.

Table 2. The KMPF-SIM and KMPR-SIM compared to the SKM, KM-PF, and KM-PR in the certain case

	SL	Bal	*PO*	*CV*	*BS*	*TE*	*SF*	*CE*
SKM								
AC	0.61	0.52	0.684	0.825	0.785	0.513	0.87	0.795
IN	8	9	11	12	13	12	14	11
ET/s	12.431	14.551	17.238	29.662	37.819	128.989	2661.634	3248.613
KM-PF								
AC	0.71	0.74	0.749	0.91	0.834	0.625	0.932	0.897
IN	2	3	6	3	2	3	6	4
ET/s	2.3	0.9	1.4	6.7	8.51	40.63	55.39	89.63
KM-PR								
AC	0.73	0.75	0.77	0.92	0.88	0.67	0.94	0.91
IN	2	3	6	3	2	3	6	4
ET/s	2.92	1.76	1.91	7.19	8.81	41.93	56.03	89.91
KMPF-SIM								
AC	**0.72**	**0.75**	**0.77**	**0.92**	**0.88**	**0.65**	**0.95**	**0.92**
IN	2	3	6	3	2	3	6	4
ET/s	2.3	0.8	1.5	6.6	8.52	40.62	55.41	89.81
KMPR-SIM								
AC	**0.74**	**0.75**	**0.78**	**0.92**	**0.89**	**0.7**	**0.95**	**0.93**
IN	2	3	6	3	2	3	6	4
ET/s	2.92	1.74	1.93	7.17	8.82	41.92	56.1	89.97

2. Uncertain case: where the values of attributes of the used databases have possibilistic values from $]0, 1]$.

Tables 3 and 4 provide the final results of our proposals using the three evaluation criteria. Note that, in this uncertain case, we cannot apply the SKM method that is why we present in the following tables (i.e. Tables 3 and 4) only the KMPF-SIM and KMPR-SIM compared to the KM-PF and KM-PR.

In Table 3, we introduce two new parameters, A_u and P_d. $0\% \leq A_u \leq 100\%$ corresponds to the percentage of attributes with uncertain values from the database. $P_d \in [0, 1]$ presents the possibilistic degree of the uncertain value.

It is obvious from Table 3 that the KMPF-SIM and KMPR-SIM methods improve the results in terms of accuracy in the uncertain case. Besides, we remark that the KMPR-SIM produces the best results for all databases. The accuracy reaches its maximum for the Solar-Flare (SF) database by 0.97 for the KMPR-SIM. However, it is equal to 0.95, 0.922, and 0.912 respectively for KMPF-SIM, KM-PR, and KM-PF.

Table 3. The AC of the KMPF-SIM and KMPR-SIM vs. the KM-PF and KM-PR in the uncertain case

	SL	Bal	PO	CV	BS	TE	SF	CE
$A_u < 50\%$ and $0 < P_d < 0.5$								
KM-PF	0.659	0.64	0.73	0.88	0.79	0.589	0.875	0.87
KM-PR	0.66	0.643	0.739	0.89	0.8	0.6	0.89	0.88
KMPF-SIM	0.66	0.65	0.74	0.89	0.81	0.62	0.9	0.88
KMPR-SIM	0.68	0.66	0.74	0.91	0.83	0.65	0.92	0.92
$A_u < 50\%$ and $0.5 \leq P_d \leq 1$								
KM-PF	0.647	0.651	0.71	0.85	0.8	0.58	0.89	0.83
KM-PR	0.65	0.652	0.72	0.86	0.81	0.59	0.892	0.84
KMPF-SIM	0.65	0.654	0.75	0.87	0.82	0.61	0.92	0.86
KMPR-SIM	0.67	0.66	0.77	0.89	0.85	0.63	0.93	0.91
$A_u \geq 50\%$ and $0 < P_d < 0.5$								
KM-PF	0.735	0.811	0.784	0.921	0.87	0.63	0.912	0.9
KM-PR	0.74	0.82	0.787	0.929	0.878	0.64	0.922	0.91
KMPF-SIM	0.75	0.82	0.79	0.93	0.89	0.66	0.95	0.92
KMPR-SIM	0.79	0.85	0.8	0.95	0.89	0.68	0.97	0.95
$A_u \geq 50\%$ and $0.5 \leq P_d \leq 1$								
KM-PF	0.71	0.73	0.72	0.85	0.8	0.62	0.87	0.88
KM-PR	0.72	0.738	0.73	0.861	0.817	0.63	0.883	0.892
KMPF-SIM	0.73	0.75	0.75	0.88	0.83	0.66	0.89	0.92
KMPR-SIM	0.75	0.76	0.76	0.89	0.85	0.69	0.93	0.93

Table 4. The number of iterations and the execution time in second of the KMPF-SIM and KMPR-SIM vs. the KM-PF and KM-PR

	SL	Bal	PO	CV	BS	TE	SF	CE
KM-PF								
IN	3	3	8	4	2	6	8	4
ET/s	2.02	0.672	0.95	6.97	9.653	41.31	56.781	90.3
KM-PR								
IN	3	3	8	4	2	6	8	4
ET/s	2.12	0.69	0.99	7.01	9.67	41.42	56.79	90.5
KMPF-SIM								
IN	3	3	8	4	2	6	8	4
ET/s	2.1	0.74	1.12	7.14	9.81	42.61	68.71	99.14
KMPR-SIM								
IN	3	3	8	4	2	6	8	4
ET/s	2.06	0.81	1.52	7.15	10.8	40.31	67.2	96.3

Moving to the *iteration number*, we remark that all methods have the same IN. However, the KMPF-SIM and KMPR-SIM are very close to the SKM, KM-PF, and KM-PR in terms of execution time. In fact, our proposals need more time to get the initial modes.

Generally, the KMPF-SIM and KMPR-SIM improve the clustering results of the SKM, KM-PF, and KM-PR by getting stable partitions and avoiding the random selection of initial modes.

6   Conclusion

In this paper, we have focused on the problem of the random selection of the initial modes for two uncertain clustering methods. In fact, we have proposed two new versions of the k-modes method that are based on possibility and rough set theories. These new methods denoted by KMPF-SIM (k-modes under possibilistic framework with selection of initial modes) and KMPR-SIM (k-modes using possibility and rough set theories with selection of initial modes) handled the uncertainty in the values of attributes and in the belonging of objects to several clusters. In addition, they avoided the random choice and selected the initial modes from the most dissimilar objects. For the evaluation of our new methods, we used UCI machine learning repository databases and three main evaluation criteria namely the accuracy, the iteration number, and the execution time. Experiments proved that the KMPF-SIM and KMPR-SIM methods improved the clustering results since they avoided the random selection of initial modes.

References

1. Ammar, A., Elouedi, Z., Lingras, P.: The k-modes method using possibility and rough set theories. In: Proceedings of the IFSA World Congress and NAFIPS Annual Meeting, IFSA/NAFIPS. IEEE (2013)
2. Ammar, A., Elouedi, Z., Lingras, P.: The K-modes method under possibilistic framework. In: Zaïane, O.R., Zilles, S. (eds.) AI 2013. LNCS (LNAI), vol. 7884, pp. 211–217. Springer, Heidelberg (2013). doi:10.1007/978-3-642-38457-8_18
3. Dubois, D., Prade, H.: Possibility Theory: An Approach to Computerized Processing of Uncertainty. Plenium Press, New York (1988)
4. Jenhani, I., Ben Amor, N., Elouedi, Z., Benferhat, S., Mellouli, K.: Information affinity: a new similarity measure for possibilistic uncertain information. In: Mellouli, K. (ed.) ECSQARU 2007. LNCS (LNAI), vol. 4724, pp. 840–852. Springer, Heidelberg (2007). doi:10.1007/978-3-540-75256-1_73
5. Joshi, M., Lingras, P., Rao, C.R.: Correlating fuzzy and rough clustering. Fundamenta Informaticae **115**, 233–246 (2012)
6. Lingras, P., Nimse, S., Darkunde, N., Muley, A.: Soft clustering from crisp clustering using granulation for mobile call mining. In: Proceedings of the GrC 2011: International Conference on Granular Computing, pp. 410–416 (2011)
7. Murphy, M.P., Aha, D.W.: UCI repository databases (1996) http://www.ics.uci.edu/mlearn

8. Huang, Z.: Extensions to the k-means algorithm for clustering large data sets with categorical values. Data Mining Knowl. Discov. **2**, 283–304 (1998)
9. Huang, Z., Ng, M.K.: A note on k-modes clustering. J. Classif. **20**, 257–261 (2003)
10. Pawlak, Z.: Rough sets. Int. J. Inf. Comput. Sci. **11**, 341–356 (1982)
11. Pawlak, Z.: Rough Sets: Theoretical Aspects of Reasoning About Data. Kluwer Academic Publishers, Berlin (1992)
12. Tanaka, H., Guo, P.: Possibilistic Data Analysis for Operations Research. Physica-Verlag, Heidelberg (1999)
13. Viattchenin, D.A.: A Heuristic Approach to Possibilistic Clustering: Algorithms and Applications. Springer Publishing Company, Incorporated, Heidelberg (2013)
14. Zadeh, L.A.: Fuzzy sets as a basis for a theory of possibility. Fuzzy Sets Syst. **1**, 3–28 (1978)

Accelerating Greedy K-Medoids Clustering Algorithm with L_1 Distance by Pivot Generation

Takayasu Fushimi[1]($^{\boxtimes}$), Kazumi Saito[2], Tetsuo Ikeda[2], and Kazuhiro Kazama[3]

[1] Tokyo University of Technology, Tokyo, Japan
takayasu.fushimi@gmail.com
[2] University of Shizuoka, Shizuoka, Japan
{k-saito,t-ikeda}@u-shizuoka-ken.ac.jp
[3] Wakayama University, Wakayama, Japan
kazama@ingrid.org

Abstract. With the explosive increase of multimedia objects represented as high-dimensional vectors, clustering techniques for these objects have received much attention in recent years. However, clustering methods usually require a large amount of computational cost when calculating the distances between these objects. In this paper, for accelerating the greedy K-medoids clustering algorithm with L_1 distance, we propose a new method consisting of the fast first medoid selection, lazy evaluation, and pivot pruning techniques, where the efficiency of the pivot construction is enhanced by our new pivot generation method called PGM2. In our experiments using real image datasets where each object is represented as a high-dimensional vector and L_1 distance is recommended as their dissimilarity, we show that our proposed method achieved a reasonably high acceleration performance.

1 Introduction

With the explosive increase of multimedia data objects on the Internet, techniques for clustering these objects into groups according to their distances have received much attention in recent years [1]. As one of the most prominent clustering techniques, we focus on the greedy K-medoids clustering algorithm, which is theoretically guaranteed to produce unique high quality solutions [2] unlike the standard K-means algorithm. However, it usually requires a large amount of computational cost for large-scale datasets. Assume that objects are represented as high-dimensional feature vectors, and let N and S be the number of objects and the dimension of the feature vectors, respectively. Then, after calculating the distances of these vectors with computational cost of $O(N^2 S)$, we can run the K-medoids algorithm with computational cost of $O(KN^2)$, in case that we have enough memory space for storing all of the $N(N-1)/2$ distances. However, in case that the number N of these vectors is too large and memory space is not enough, we need to re-calculate most of the $N(N-1)/2$ distances for each greedy step, thus it amounts to computational cost of $O(KN^2 S)$. Here note that in our experiments shown later, sample values for these variables are $K = 100$,

© Springer International Publishing AG 2017
M. Kryszkiewicz et al. (Eds.): ISMIS 2017, LNAI 10352, pp. 87–96, 2017.
DOI: 10.1007/978-3-319-60438-1_9

$N = 1,000,000$, and $S = 340$, and it takes quite long computation time by the original K-medoids algorithm.

In order to reduce the number of distance calculations without changing the final results obtained by the original greedy K-medoids algorithm, we employ an approach combining the lazy evaluation technique [2] and the pivot pruning technique [3]. As for the latter technique, we focus on a pivot generation algorithm called the PGM (Pivot Generation based on Manhattan distance) method [4]. Here note that adopting L_1 (Manhattan) distance can be a more natural choice than L_2 in many situations, for instance, dissimilarity between histograms of color images [5] and so on. The main advantage of the PGM method is that each element of a pivot vector can be independently and optimally updated at each iteration with low computational cost. Moreover, it is guaranteed that the PGM method converges within a finite number of iterations. However, after calculating distances between every pair of object and pivot, the PGM method requires computational cost of $O(N^2)$ for calculating its objective function. In order to reduce the corresponding computational cost to $O(N \log N)$, we propose a novel pivot generation method called PGM2 (Pivot Generation on Manhattan distance using 2 pivots), which calculate the objective function efficiently in case that the number of pivots is 2.

In this paper, we focus on clustering a large number of objects under the case that memory space is not enough, and propose a new method for accelerating the greedy K-medoids algorithm with L_1 distance by combining the fast first medoid selection, lazy evaluation and pivot pruning techniques. In our experiments using the CoPhIR image search dataset [6] where L_1 distance is standardly adopted, we evaluate the performance of our proposed method in terms of computational efficiency.

2 Related Work

One of the simplest K-medoids algorithms is PAM [7], which iteratively optimizes the K medoid objects by swapping selected medoid and non-medoid ones. However, because of its high computational complexity, PAM is not practical for large datasets. Park and Jun proposed a fast algorithm of K-medoids, which calculates and stores the distances between every object pair [8]. Thus, a large amount of memory space is required. Although several sampling algorithms have been proposed for clustering of a large scale dataset, they compute their clustering results based on an approximated objective function consisting of some sampling objects selected from the original dataset [9]. These algorithms might produce significantly poor results depending on the sampling number. Paterlini et al. [10] proposed a fast algorithm of K-medoids by using pivots, which efficiently selects initial medoids and accelerate the convergence of iterative steps. In this paper, we focus on the greedy K-medoids algorithm. It is guaranteed that we can obtain a unique greedy solution with reasonably high quality, because of submodularity of the objective function.

3 Preliminaries

Let $\mathcal{X} = \{\mathbf{x}_1, \cdots, \mathbf{x}_N\}$ be a given set of object vectors, and we denote the L_1 distance between two objects $\mathbf{x}_n$ and $\mathbf{x}_m$ as $d(\mathbf{x}_n, \mathbf{x}_m) = \sum_{s=1}^{S} |x_{n,s} - x_{m,s}|$, where $x_{n,s}$ and S represent the s-th value of the object vector $\mathbf{x}_n$ and the dimension of vectors, respectively. The K-medoids problem is formulated as a problem of selecting the set of K objects, denoted as $\mathcal{R} \subset \mathcal{X}$, which minimizes the following objective function:

$$F(\mathcal{R}; \mathcal{X}) = \sum_{n=1}^{N} \min_{\mathbf{r}_k \in \mathcal{R}} d(\mathbf{x}_n, \mathbf{r}_k). \tag{1}$$

The selected objects belonging to $\mathcal{R}$ are referred to as medoids. Hereafter, $F(\mathcal{R}; \mathcal{X})$ is abbreviated as $F(\mathcal{R})$ for simplicity.

In order to minimize the objective function $F(\mathcal{R})$, we successively select each medoid by evaluating every candidate object $\mathbf{x}_n$ according to a greedy algorithm that uses the following marginal gain $g(\mathbf{x}_n; \mathcal{R})$ fixing the set $\mathcal{R}$ of the already selected medoids:

$$g(\mathbf{x}_n; \mathcal{R}) = F(\mathcal{R} \cup \{\mathbf{x}_n\}) - F(\mathcal{R}) = \sum_{\mathbf{x}_m \in \mathcal{X} \setminus \mathcal{R}} \min\{d(\mathbf{x}_m, \mathbf{x}_n) - \mu(\mathbf{x}_m; \mathcal{R}), 0\}, \tag{2}$$

where $\mu(\mathbf{x}_m; \mathcal{R}) = \min_{\mathbf{r} \in \mathcal{R}} d(\mathbf{x}_m, \mathbf{r})$ if $\mathcal{R} \neq \emptyset$; otherwise $\mu(\mathbf{x}_m; \emptyset) = 0$. Then, we can summarize the greedy algorithm as follows: after initializing $k \leftarrow 1$ and $\mathcal{R}_0 \leftarrow \emptyset$, we repeatedly select and add each medoid by $\hat{\mathbf{r}}_k = \arg\min_{\mathbf{x}_n \in \mathcal{X} \setminus \mathcal{R}_{k-1}} g(\mathbf{x}_n; \mathcal{R}_{k-1})$, $\mathcal{R}_k \leftarrow \mathcal{R}_{k-1} \cup \{\hat{\mathbf{r}}_k\}$ during $k \leq K$ together with increment $k \leftarrow k + 1$. From the obtained K medoids $\mathcal{R} = \{\mathbf{r}_1, \cdots, \mathbf{r}_K\}$, we can calculate each cluster as $\mathcal{X}^{(k)} = \{\mathbf{x}_m \in \mathcal{X}; \mathbf{r}_k = \arg\min_{\mathbf{r} \in \mathcal{R}} d(\mathbf{x}_m, \mathbf{r})\}$.

4 Pivot Generation Method: PGM2

As mentioned earlier, we need a large number of distance calculations for obtaining the resultant set of K medoids. To avoid unnecessary distance calculations, we employ an approach based on the pivot pruning technique, where each pivot is defined as a vector in the S-dimensional object space. More specifically, we propose a new pivot generation method, called PGM2 (Pivot Generation on Manhattan distance using 2 pivots), which enhances the efficiency of original PGM proposed by Kobayashi et al. [4]. Here, the objective function with respect to the pivots $\mathcal{P} = \{\mathbf{p}_1, \mathbf{p}_2\}$ is defined by

$$J(\mathcal{P}) = \sum_{n=1}^{N-1} \sum_{m=n+1}^{N} \max_{1 \leq h \leq H} |d(\mathbf{x}_n, \mathbf{p}_h) - d(\mathbf{x}_m, \mathbf{p}_h)|, \tag{3}$$

where H stands for the number of pivots and is set to $H = 2$ for the PGM2 method. Note that conventional pivot selection techniques such as a method proposed by Bustos et al. [11] select each pivot from the set of objects ($\mathbf{p}_h \in \mathcal{X}$), while the PGM2 method generates each pivot as an arbitrary point in the S-dimensional Euclidean space ($\mathbf{p}_h \in \mathbb{R}^S$). Below we overview the PGM2 algorithm that produces a pair of pivots $\mathcal{P}$ from a set of objects $\mathcal{X}$.

G1: Initialize $\mathcal{P}$ as randomly selected objects in $\mathcal{X}$.
G2: Iterate the following three steps until convergence.
G2-1: Calculate distance $d(\mathbf{x}, \mathbf{p})$ for each pair of $\mathbf{x}$ and $\mathbf{p}$.
G2-2: Compute the objective function $J(\mathcal{P})$.
G2-3: Update pivots $\mathcal{P}$ so as to maximize $J(\mathcal{P})$.
G3: Output $\mathcal{P}$ and terminate.

By using our new technique described below, we can efficiently compute the objective function in Step **G2-2** with computational cost of $O(TN \log N)$, instead of $O(TN^2)$ required by original PGM [4], where T means the number of iterations until PGM2 converges. Here note that even in a general metric space, we can straightforwardly utilize this technique for pivot selection based on Eq. (3) by setting $\mathbf{p}_h \in \mathcal{X}$. Thus, since computational cost in Step **G2-1** amounts to $O(TSN)$, and $\max\{O(TSN), O(SN \log N)\}$ in Step **G2-3** [4], the computational complexity of PGM2 becomes $\max\{O(TSN), O(TN \log N), O(SN \log N)\}$.

Below we describe our new technique for efficiently calculating the objective function in case of $H = 2$. Let $\phi_1(n)$ and $\phi_2(n)$ be the descending order ranks of object $\mathbf{x}_n$ with respect to $d(\mathbf{x}_n, \mathbf{p}_1) + d(\mathbf{x}_n, \mathbf{p}_2)$ and $d(\mathbf{x}_n, \mathbf{p}_1) - d(\mathbf{x}_n, \mathbf{p}_2)$, respectively. Then, by using a set of object numbers defined by $\mathcal{N}_h(n) = \{m \mid \phi_h(n) < \phi_h(m)\}$ for $h \in \{1, 2\}$. We can eliminate each absolute operation as follows:

$$\max_{1 \le h \le H} |d(\mathbf{x}_n, \mathbf{p}_h) - d(\mathbf{x}_m, \mathbf{p}_h)| = \begin{cases} d(\mathbf{x}_m, \mathbf{p}_1) - d(\mathbf{x}_n, \mathbf{p}_1) & m \in \mathcal{N}_1(n) \cap \mathcal{N}_2(n), \\ d(\mathbf{x}_m, \mathbf{p}_2) - d(\mathbf{x}_n, \mathbf{p}_2) & m \in \mathcal{N}_1(n) \setminus \mathcal{N}_2(n), \\ d(\mathbf{x}_n, \mathbf{p}_2) - d(\mathbf{x}_m, \mathbf{p}_2) & m \in \mathcal{N}_2(n) \setminus \mathcal{N}_1(n), \\ d(\mathbf{x}_n, \mathbf{p}_1) - d(\mathbf{x}_m, \mathbf{p}_1) & m \notin \mathcal{N}_1(n) \cup \mathcal{N}_2(n). \end{cases}$$

Thus, in the calculation of $J(\mathcal{P})$ defined in Eq. (3), the frequencies with which $-d(\mathbf{x}_n, \mathbf{p}_1)$ and $d(\mathbf{x}_n, \mathbf{p}_1)$ appear are $|\mathcal{N}_1(n) \cap \mathcal{N}_2(n)|$ times and $N - 1 - |\mathcal{N}_1(n) \cup \mathcal{N}_2(n)|$ times, respectively, while the frequencies with which $-d(\mathbf{x}_n, \mathbf{p}_2)$ and $d(\mathbf{x}_n, \mathbf{p}_2)$ appear are $|\mathcal{N}_1(n) \setminus \mathcal{N}_2(n)|$ times and $|\mathcal{N}_2(n) \setminus \mathcal{N}_1(n)|$ times, respectively. Therefore, by using the relations, $|\mathcal{N}_1(n) \cup \mathcal{N}_2(n)| + |\mathcal{N}_1(n) \cap \mathcal{N}_2(n)| = |\mathcal{N}_1(n)| + |\mathcal{N}_2(n)|$, $|\mathcal{N}_2(n) \setminus \mathcal{N}_1(n)| - |\mathcal{N}_1(n) \setminus \mathcal{N}_2(n)| = |\mathcal{N}_2(n)| - |\mathcal{N}_1(n)|$, and $|\mathcal{N}_h(n)| + 1 = \phi_h(n)$, we can calculate the objective function as follows:

$$J(\mathcal{P}) = \sum_{n=1}^{N} \sum_{h=1}^{H} c_h(n) \, d(\mathbf{x}_n, \mathbf{p}_h). \tag{4}$$

where the coefficient $c_h(n)$ is defined as $c_1(n) = N - (\phi_1(n) + \phi_2(n)) + 1$ and $c_2(n) = \phi_2(n) - \phi_1(n)$. Thus, after sorting the distance terms described above with cost of $O(N \log N)$, we can calculate the objective function shown in Eq. (4) with cost of $O(N)$, where recall $H = 2$. Therefore, the complexity required in Step **G1-2** is $O(N \log N)$.

5 Proposed Algorithm

Algorithm 1 shows the detail of our proposed algorithm, where mg means a temporal variable for marginal gain $g(\mathbf{x}_n; \mathcal{R}_{k-1})$, and $LB(\mathbf{x}_n, \mathbf{x}_m; \mathcal{P})$ denotes the lower bound distance of $d(\mathbf{x}_n, \mathbf{x}_m)$ as shown later. We first select the first medoid by using the fast first medoid selection technique described below. After that, as mentioned earlier, in case that the number N of objects is too large and memory space is not enough, we need to re-calculate most of the $N(N-1)/2$ distances for each of K (> 2) greedy steps of the K-medoids algorithm, which amounts to computational cost of $O(KN^2S)$. In order to overcome this problem, we propose two techniques which are referred to as the lazy evaluation and pivot pruning techniques. Recall that the efficiency of the pivot construction is enhanced by PGM2.

In order to describe our fast first medoid selection technique, we introduce a partial objective function defined by $F_s(n) = \sum_{m=1}^{N} |x_{n,s} - x_{m,s}|$, and assume $x_{1,s} \le \cdots \le x_{N,s}$ without loss of generality. Note that $F(\{\mathbf{x}_n\}) = \sum_{s=1}^{S} F_s(n)$. Then, by using coefficients $b(n) = N - 2n + 1$ and $\beta(n) = \sum_{m=n+1}^{N} x_{m,s} - \sum_{m=1}^{n-1} x_{m,s}$, we can compute the partial objective function as $F_s(n) = -b(n)x_{n,s} + \beta(n)$. Thus, after initializing $\beta(1) = \sum_{m=2}^{N} x_{m,s}$, by noting $\beta(n+1) = \beta(n) - x_{n,s} - x_{n+1,s}$, we can successively compute $F_s(n)$ from $n = 1$ to N by $F_s(n) = -b(n)x_{n,s} + \beta(n)$ with cost of $O(N)$. Thus, after sorting the feature values of objects so that $x_{1,s} \le \cdots \le x_{N,s}$ for every s with cost of $O(SN \log N)$, we can obtain the objective function value $F(\{\mathbf{x}_n\})$ for every n with cost of $O(SN)$. Therefore, computational cost of fast first medoid selection technique amounts to $O(SN \log N)$, which can be a significantly smaller cost in comparison to the straightforward computation with cost of $O(SN^2)$.

As for the lazy evaluation technique [2] applied at the k-th medoid selection step, we utilize an upper bound value $UB(\mathbf{x}_n)$ of the marginal gain $g(\mathbf{x}_n; \mathcal{R})$ for each candidate object $\mathbf{x}_n \in \mathcal{X}$. More specifically, after initializing as $UB(\mathbf{x}_n) \leftarrow F(\{\mathbf{x}_n\})$ by the values obtained by our fast first medoid selection technique, we update $UB(\mathbf{x}_n) \leftarrow g(\mathbf{x}_n; \mathcal{R}_h)$ in case that $g(\mathbf{x}_n; \mathcal{R}_h)$ is actually calculated at the h-th medoid selection step. Evidently, due to the submodular property, it is guaranteed that $g(\mathbf{x}_n; \mathcal{R}_k) \le UB(\mathbf{x}_n)$ for $k > h$. Thus, let g_k^* be the current best marginal gain at the selection step for obtaining the k-th medoid; then we can skip to calculate $g(\mathbf{x}_n; \mathcal{R}_k)$ when $UB(\mathbf{x}_n) \le g_k^*$. In order to obtain better g_k^* at an early stage, by sorting the candidates objects in descending order with respect to $UB(\mathbf{x}_n)$ we evaluate these candidates from the top of the sorted list.

As for the pivot pruning technique [12] applied before actually calculating $g(\mathbf{x}_n; \mathcal{R}_k)$, we utilize a lower bound distance $LB(\mathbf{x}_n, \mathbf{x}_m; \mathcal{P})$ of the distance $d(\mathbf{x}_n, \mathbf{x}_m)$ for examining the pruning condition $d(\mathbf{x}_n, \mathbf{x}_m) \ge \mu(\mathbf{x}_n; \mathcal{R})$, where $\mathcal{P} = \{\mathbf{p}_1, \mathbf{p}_2\}$ is a set of pivots described in Sect. 4. Here note that from Eq. (2), we do not add the gain when the pruning condition $d(\mathbf{x}_n, \mathbf{x}_m) \ge \mu(\mathbf{x}_n; \mathcal{R})$ holds. More specifically, from the triangle inequality, we can utilize the following lower bound distance $LB(\mathbf{x}_n, \mathbf{x}_m; \mathcal{P}) = \max_{\mathbf{p} \in \mathcal{P}} |d(\mathbf{x}_n, \mathbf{p}) - d(\mathbf{x}_m, \mathbf{p})| \le d(\mathbf{x}_n, \mathbf{x}_m)$.

Algorithm 1. Proposed algorithm

1: **Input:** $\mathcal{X} = \{\mathbf{x}_1, \ldots, \mathbf{x}_N\}, K$
2: **Output:** $\mathcal{R}_K = \{\mathbf{r}_1, \ldots, \mathbf{r}_K\} \subset \mathcal{X}$
3: **Initialize:** $\mathcal{R}_0 \leftarrow \emptyset, \quad g(\mathbf{x}; \mathcal{R}_0) \leftarrow 0$ for each object $\mathbf{x}$
4: $\mathbf{r}_1 \leftarrow$ FastSelection$(\mathcal{X}), \quad \mathcal{R}_1 \leftarrow \mathcal{R}_0 \cup \{\mathbf{r}_1\}$ $\triangleright\ O(SN \log N)$
5: $\mathcal{P} \leftarrow$ PivotGeneration$(\mathcal{X})$ $\triangleright\ O(TN \log N)$
6: **for** $k = 2$ to K **do**
7:      **for** $\rho = 1$ to N **do**
8:          $n = rank(\rho)$
9:          **if** $g(\mathbf{x}_n; \mathcal{R}_{k-2}) < g_k^*$ **then**
10:             **break** $\triangleright$ (Lazy evaluation)
11:          **end if**
12:          **Initialize:** $mg \leftarrow 0$
13:          **for** $m = 1$ to N **do**
14:             **if** $LB(\mathbf{x}_n, \mathbf{x}_m; \mathcal{P}) \geq \mu(\mathbf{x}_m; \mathcal{R}_{k-1})$ **then**
15:                 **continue** $\triangleright$ (Pivot pruning)
16:             **end if**
17:             $d(\mathbf{x}_n, \mathbf{x}_m) \leftarrow \sum_{s=1}^{S} |x_{n,s} - x_{m,s}|$
18:             **if** $d(\mathbf{x}_n, \mathbf{x}_m) < \mu(\mathbf{x}_m; \mathcal{R}_{k-1})$ **then**
19:                 $mg\ += \mu(\mathbf{x}_m; \mathcal{R}_{k-1}) - d(\mathbf{x}_n, \mathbf{x}_m)$
20:             **end if**
21:          **end for**
22:          $g(\mathbf{x}_n; \mathcal{R}_{k-1}) \leftarrow mg$
23:          **if** $g_k^* < mg$ **then**
24:             $g_k^* \leftarrow mg$
25:          **end if**
26:      **end for**
27:      $\mathbf{r}_k \leftarrow \underset{\mathbf{x}_n \in \mathcal{X} \setminus \mathcal{R}_{k-1}}{\arg\max}\ g(\mathbf{x}_n; \mathcal{R}_{k-1}), \quad \mathcal{R}_k \leftarrow \mathcal{R}_{k-1} \cup \{\mathbf{r}_k\}$
28:      **for** $n = 1$ to N **do**
29:          $d(\mathbf{x}_n, \mathbf{r}_k) \leftarrow \sum_{s=1}^{S} |x_{n,s} - r_{k,s}|, \quad \mu(\mathbf{x}_n; \mathcal{R}_k) \leftarrow \min\{d(\mathbf{x}_n, \mathbf{r}_k), \mu(\mathbf{x}_n; \mathcal{R}_{k-1})\}$
30:      **end for**
31:      **rank** $= Sort(\{g(\mathbf{x}_1; \mathcal{R}_{k-1}), \ldots, g(\mathbf{x}_N; \mathcal{R}_{k-1})\})$
32: **end for**

Hereafter, the three techniques of fast first medoid selection, lazy evaluation, and pruning by generated pivots are referred to as the FS, LE, and GP techniques, respectively. In our proposed method, the FS, LE, and GP techniques are applied in this order. This is because when the marginal gain calculation of $g(\mathbf{x}_n; \mathcal{R})$ is skipped by the LE technique, we can simultaneously prune all the distance calculation of $d(\mathbf{x}_n, \mathbf{x}_m)$ for any $\mathbf{x}_m \in \mathcal{X}$.

6 Experiments

We employed the CoPhIR (Content-based Photo Image Retrieval) test-collection [6]. From this collection, we used the four MPEG-7 global descriptors of scalable color (SC for short), color structure (CS), edge histogram (EH), and homogeneous texture (HT), whose numbers of dimensions are 64, 64, 12, 80, and 62, respectively. Each of their recommended dissimilarity measures is the L_1 distance. In addition, we used a concatenated descriptor of these descriptors referred to as MIX, whose dimensionality is 340. For each descriptor, we prepared several datasets consisting of up to $N = 1{,}000{,}000$ objects by random selection from the original dataset. In our experiments, our programs implemented in C were executed on a computer system equipped with Xeon processor E5-2697 2.7 GHz CPU and 256 GB main memory.

For comparison, we consider a baseline method consisting only of the LE technique referred to as LE, and another method consisting of the LE and FS techniques referred to as FS. In order to evaluate the performance of the GP technique in our method, we consider three comparison methods equipped with standard pivot selection techniques referred to as OP, RP, and MD, where the pivots are selected as outlier objects for OP, as random objects for RP, and as medoids for MD. More specifically, after obtaining the first medoid $\mathbf{r}_1$, the first and the successive h-th outlier pivots are selected by $\hat{\mathbf{q}}_1 = \arg\max_{v \in \mathcal{X}} d(v, \mathbf{r}_1), P \leftarrow P \cup \{\hat{\mathbf{q}}_1\}$ and $\hat{\mathbf{q}}_h = \arg\max_{v \in \mathcal{X}} \min_{\mathbf{p} \in P} d(\mathbf{v}, \mathbf{p}), P \leftarrow P \cup \{\hat{\mathbf{q}}_h\}$, respectively; All of the random pivots are selected in advance; the h-th medoid pivot is added immediately after the h-th medoid is determined. Hereafter, for the sake of comparison purpose, the proposed method is also referred to as GP. Here note that all of LE, FS, OP, RP, MD, and GP can produce completely the same clustering results because they are different only at the techniques of skipping unnecessary distance calculations.

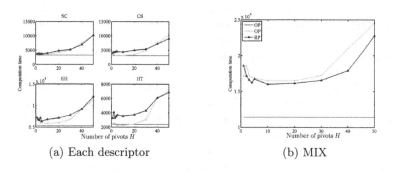

(a) Each descriptor   (b) MIX

Fig. 1. Computation time with respect to the number of pivots

First, we evaluated the computation time of the three pivot-based methods, GP, OP, and RP, by changing the numbers of pivots in the range of $1 \leq H \leq 50$, where the numbers of objects and medoids were fixed at $N = 100{,}000$ and $K = 100$, respectively. From our experimental results shown in Fig. 1, we can see that for all the descriptors, GP worked faster than OP and RP for any number of pivots, and the performances of OP and RP significantly degraded by increasing the number of pivots. In general, OP and RP might work reasonably by selecting

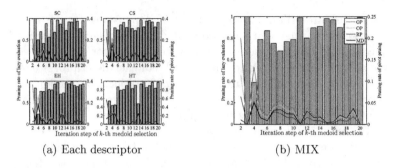

(a) Each descriptor   (b) MIX

Fig. 2. Pruning rate of each iteration step k (Color figure online)

adequate numbers of pivots, but we cannot know such numbers in advance. On the other hand, the number of pivots used by GP is 2 in all of the experiments, i.e., GP is free from such parameter selection. These experimental results suggest that our approach based on two generated pivots is useful and promising.

Next, we evaluated the pruning performance by changing the number of medoids k. Namely, let $LE(k)$, $GP(k)$, $OP(k)$, $RP(k)$, and $MD(k)$ be the sets of object pairs such that their actual distance calculations are skipped by LE, GP, OP, RP, and MD, respectively. Then, we can compute the actual pruning rate $\alpha(\cdot)$ by LE, GP, OP, RP, and MD as $\alpha(LE(k)) = |LE(k)|/N^2$, $\alpha(GP(k)) = (|LE(k) \cup GP(k)| - |LE(k)|)/N^2$, $\alpha(OP(k)) = (|LE(k) \cup OP(k)| - |LE(k)|)/N^2$, $\alpha(RP(k)) = (|LE(k) \cup RP(k)| - |LE(k)|)/N^2$, and $\alpha(MD(k)) = (|LE(k) \cup MD(k)| - |LE(k)|)/N^2$, respectively. Here, recall that the LE technique is applied prior to each pruning technique employed by GP, OP, RP, and MD. Figure 2 shows the experimental results in the range of $2 \le k \le 20$, where the number of objects was set to $N = 100,000$ and the numbers of pivots for OP and RP were set to $H = 10$ according to our experimental results shown in Fig. 1. Here, the gray bar (indicated by the left vertical axis) stands for the pruning rates of $\alpha(LE(k))$, and the red, lime, blue, and black lines (indicated by the right vertical axis) stand for $\alpha(GP(k))$, $\alpha(OP(k))$, $\alpha(RP(k))$, and $\alpha(MD(k))$, respectively. From Fig. 2, we can see that for all the descriptors, the LE technique could not skip any marginal gain calculation at the step of $k = 2$ and also worked quite poorly at the step of $k = 4$. This result indicates that each upper bound $UB(\mathbf{w})$ was a quite rough approximation to the actual marginal gain $g(\mathbf{w}; R)$ at these steps. We can also see that MD showed relatively poor pruning rates at the step of $k = 2$ because just only one pivot was used. Therefore, we can see that GP, OP, and RP could achieve reasonably high pruning rates stably even after pruning by lazy evaluation, which indicates that for all the descriptors, our pivot pruning techniques worked in a complementary manner for the LE technique.

Figure 3 compares the computation times by changing the number of medoids from $k = 10$ to 100, where the horizontal and vertical axes stand for the number of medoids k and the number of computation time, respectively. We can confirm that for all the descriptors, GP worked substantially faster than the other methods. Especially, in our experimental results, we should emphasize that our proposed GP method achieved from two to three times better performance than LE, which is one of state-of-the-art baseline methods. Here, MD required relatively large computation time in case that the number of medoids is large.

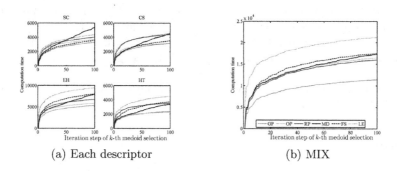

(a) Each descriptor   (b) MIX

Fig. 3. Computation time of each iteration step k

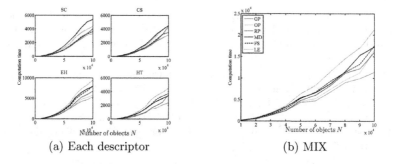

(a) Each descriptor   (b) MIX

Fig. 4. Computation time with respect to the number of objects

Finally, we evaluated the computation times by changing the numbers of objects from $N = 10,000$ to $100,000$, where these objects were selected randomly from the original dataset of each descriptor. Figure 4 shows the experimental results, where the number of medoids was set to $K = 100$ for all the methods, and the number of pivots was set to $H = 10$ for OP and RP. From Fig. 4, we can confirm that for all the descriptors, our proposed method GP achieved two times better performance than LE, which is one of state-of-the-art baseline methods. Moreover, we can see that the difference of computation times of each method from GP increase as the number of objects becomes large. Although not shown due to space limitations, for the dataset with $N = 1,000,000$ objects, we can confirm that GP worked substantially faster than the other methods in these case of larger numbers of objects. Here we note that the actual computation times of LE and GP for MIX were about 21 and 11 days, respectively, which will bring about practically significant differences for further experiments.

7 Conclusion

By combining the fast first medoid selection, lazy evaluation, and pivot pruning techniques, we proposed a new method for accelerating the greedy K-medoids algorithm with L_1 distance, where the efficiency of the pivot construction was enhanced by our new pivot generation method called PGM2. In our experiments using the CoPhIR datasets with L_1 distance, we confirmed that our proposed method achieved a reasonably high acceleration performance. In future, we plan to evaluate our method using various types of datasets.

Acknowledgments. This work was supported by JSPS Grant-in-Aid for Scientific Research (No. 16K16154).

References

1. Zhu, X., He, W., Li, Y., Yang, Y., Zhang, S., Hu, R., Zhu, Y.: One-step spectral clustering via dynamically learning affinity matrix and subspace. In: Proceedings of the 31st AAAI Conference on Artificial Intelligence (AAAI 2017), pp. 2963–2969 (2017)
2. Leskovec, J., Krause, A., Guestrin, C., Faloutsos, C., VanBriesen, J., Glance, N.: Cost-effective outbreak detection in networks. In: Proceedings of the 13th ACM SIGKDD International Conference on Knowledge Discovery and Data Mining, pp. 420–429. ACM, New York (2007)
3. Hjaltason, G.R., Samet, H.: Index-driven similarity search in metric spaces (survey article). ACM Trans. Database Syst. **28**(4), 517–580 (2003)
4. Kobayashi, E., Fushimi, T., Saito, K., Ikeda, T.: Similarity search by generating pivots based on Manhattan distance. In: Pham, D.-N., Park, S.-B. (eds.) PRICAI 2014. LNCS (LNAI), vol. 8862, pp. 435–446. Springer, Cham (2014). doi:10.1007/978-3-319-13560-1_35
5. Rubner, Y., Tomasi, C., Guibas, L.J.: The earth mover's distance as a metric for image retrieval. Int. J. Comput. Vis. **40**(2), 99–121 (2000)
6. Bolettieri, P., Esuli, A., Falchi, F., Lucchese, C., Perego, R., Piccioli, T., Rabitti, F.: CoPhIR: a test collection for content-based image retrieval. CoRR abs/0905.4627v2 (2009)
7. Kaufman, L., Rousseeuw, P.: Clustering large data sets (with discussion), pp. 425–437. Elsevier/North Holland (1986)
8. Park, H.S., Jun, C.H.: A simple and fast algorithm for k-medoids clustering. Expert Syst. Appl. **36**(2, Part 2), 3336–3341 (2009)
9. Jiang, C., Li, Y., Shao, M., Jia, P.: Accelerating clustering methods through fractal based analysis. In: The 1st Workshop on Application of Self-similarity and Fractals in Data Mining (KDD 2002 Workshop) (2002)
10. Paterlini, A.A., Nascimento, M.A., Traina, C.J.: Using pivots to speed-up k-medoids clustering. J. Inf. Data Manag. **2**(2), 221–236 (2011)
11. Bustos, B., Navarro, G., Chavez, E.: Pivot selection techniques for proximity searching in metric spaces. Pattern Recogn. Lett. **24**(14), 2357–2366 (2003)
12. Zezula, P., Amato, G., Dohnal, V., Batko, M.: Similarity Search: The Metric Space Approach. Advances in Database Systems, vol. 32, 1st edn. Springer, New York (2006)

On the Existence of Kernel Function for Kernel-Trick of k-Means

Mieczysław A. Kłopotek[(⊠)]

Institute of Computer Science of the Polish Academy of Sciences,
ul. Jana Kazimierza 5, 01-248 Warszawa, Poland
`klopotek@ipipan.waw.pl`

Abstract. This paper corrects the proof of the Theorem 2 from the Gower's paper [1, p. 5]. The correction is needed in order to establish the existence of the kernel function used commonly in the kernel trick e.g. for k-means clustering algorithm, on the grounds of distance matrix. The correction encompasses the missing if-part proof and dropping unnecessary conditions.

1 The Problem

Kernel based k-means[1] clustering algorithm (clustering objects $1, \ldots, m$ into k clusters $1, \ldots, k$) consists in switching to a multidimensional feature space $\mathcal{F}$ and searching therein for prototypes $\boldsymbol{\mu}_j^{\Phi}$ minimizing the error $\sum_{i=1}^{m} \min_{1 \leq j \leq k} \|\Phi(i) - \boldsymbol{\mu}_j^{\Phi}\|^2$ where $\Phi \colon \{1, \ldots, m\} \to \mathcal{F}$ is a (usually non-linear) mapping of the space of objects into the feature space. The so-called "kernel trick" means the possibility to apply k-means clustering without knowing explicitly the $\Phi(i)$ function and using so-called kernel matrix with elements $k_{ij} = \Phi(i)^T \Phi(j) = K(i, j)$ instead.

For algorithms like k-means, instead of the kernel matrix the distance matrix D between the objects in the feature space may be available, being the Euclidean distance matrix. We will call D Euclidean matrix. The question is now:

(1) can we obtain the matrix K from such data?
(2) can we obtain from the matrix K the function $\Phi()$ such that the distances in the feature space are exactly the same as given by the D matrix?
(3) if we derived the matrix K from D and K turns out to yield $\Phi()$, can we know then that D was really an Euclidean distance matrix?

A number of transformations yielding the kernel matrix have been proposed [3,4], the most general by Gower [1, Theorem 2, p. 5]. The proof of the validity of the latter is, however, incomplete as explained below.

Let us recall that a matrix $D \in \mathbb{R}^{m \times m}$ is an Euclidean distance matrix between points $1, \ldots, m$ if and only if there exists a matrix $X \in \mathbb{R}^{m \times n}$ rows of which $(\mathbf{x_1}^T, \ldots, \mathbf{x_m}^T)$ are coordinate vectors of these points in an n-dimensional Euclidean space and $d_{ij} = \sqrt{(\mathbf{x_i} - \mathbf{x_j})^T (\mathbf{x_i} - \mathbf{x_j})}$. Gower in [1] claims that

---

[1] For an overview of kernel k-means algorithm see e.g. [2].

© Springer International Publishing AG 2017
M. Kryszkiewicz et al. (Eds.): ISMIS 2017, LNAI 10352, pp. 97–104, 2017.
DOI: 10.1007/978-3-319-60438-1_10

Theorem 1. *D is Euclidean iff the matrix $F = \left(\mathbf{I} - \mathbf{1s}^T\right)\left(-0.5\right)D_{sq}\left(\mathbf{I} - \mathbf{s1}^T\right)$ is positive semidefinite for any vector $\mathbf{s}$ such that $\mathbf{s}^T\mathbf{1} = 1$ and $D_{sq}\mathbf{s} \neq \mathbf{0}$.*

We claim here is that the Gower's theorem has the following deficiencies

- requirement $D_{sq}\mathbf{s} \neq \mathbf{0}$ is not needed in Theorem 1.
- the if-part of Theorem 1 nor of his theorem correction in [5] was demonstrated.

We provide a correction, completing Gower's proof in Sect. 2. In Sect. 3 we draw some conclusions from the corrective proof.

2   Correction of Gower's Result

In this section we shall correct the Gower's result from [1].

For construction purposes we need still another formulation of the theorem, which is slightly more elaborate:

Theorem 2. *1. If D is Euclidean then for each vector $\mathbf{s}$ such that $\mathbf{s}^T\mathbf{1} = 1$ the matrix*

$$F = \left(\mathbf{I} - \mathbf{1s}^T\right)\left(-0.5\right)D_{sq}\left(\mathbf{I} - \mathbf{s1}^T\right) \tag{1}$$

is positive semidefinite.
2. If D is a symmetric matrix with zero diagonal and for a vector $\mathbf{s}$ such that $\mathbf{s}^T\mathbf{1} = 1$ the matrix $F = \left(\mathbf{I} - \mathbf{1s}^T\right)\left(-0.5\right)D_{sq}\left(\mathbf{I} - \mathbf{s1}^T\right)$ is positive semidefinite then D is Euclidean.
3. If D is Euclidean then for each vector $\mathbf{s}$ such that $\mathbf{s}^T\mathbf{1} = 1$ the matrix D can be derived from matrix $F = \left(\mathbf{I} - \mathbf{1s}^T\right)\left(-0.5\right)D_{sq}\left(\mathbf{I} - \mathbf{s1}^T\right)$ as $d_{ij}^2 = f_{ii} + f_{jj} - 2f_{ij}$.
4. If D is Euclidean then for each vector $\mathbf{s}$ such that $\mathbf{s}^T\mathbf{1} = 1$ the matrix $F = \left(\mathbf{I} - \mathbf{1s}^T\right)\left(-0.5\right)D_{sq}\left(\mathbf{I} - \mathbf{s1}^T\right)$ can be expressed as $F = YY^T$ where Y is a real-valued matrix, and the rows of Y can be considered as coordinates of data points the distances between which are those from the matrix D.

Let $D \in \mathbb{R}^{m \times m}$ be a matrix of Euclidean distances between objects. Let D_{sq} be a matrix of squared Euclidean distances d_{ij}^2 between objects with identifiers $1, \ldots, m$. This means that there must exist a matrix $X \in \mathbb{R}^{m \times n}$ for some n, rows of which represent coordinates of these objects in an n-dimensional space. This real-valued matrix X represents an embedding of the Euclidean distance matrix D into $\mathbb{R}^{m \times n}$. A distance matrix can be called Euclidean if and only if an embedding exists. If $E = XX^T$ (E with dimensions $m \times m$), then $d_{ij}^2 = e_{ii} + e_{jj} - 2e_{ij}$.

As a rigid set of points in Euclidean space can be moved (shifted, rotated, flipped symmetrically[2]) without changing their relative distances, there may exist many other matrices Y rows of which represent coordinates of these same objects in the same n-dimensional space after some isomorphic transformation. Let us denote the set of all such embeddings $\mathcal{E}(D)$. And if a matrix

---

[2] Gower does not consider flipping.

$Y \in \mathcal{E}(D)$, then for the product $F = YY^T$ we have $d_{ij}^2 = f_{ii} + f_{jj} - 2f_{ij}$. We will say that $F \in \mathcal{E}_{dp}(D)$.

For an $F \in \mathcal{E}_{dp}(D)$ define a matrix $G = F + 0.5D_{sq}$. Hence $F = G - 0.5D_{sq}$. Obviously then

$$d_{ij}^2 = f_{ii} + f_{jj} - 2f_{ij} \tag{2}$$

$$= (g_{ii} - 0.5d_{ii}^2) + (g_{jj} - 0.5d_{jj}^2) - 2(g_{ij} - 0.5d_{ij}^2) \tag{3}$$

$$= g_{ii} + g_{jj} - 2g_{ij} + d_{ij}^2 \tag{4}$$

(as $d_{jj} = 0$ for all j). This implies that $0 = g_{ii} + g_{jj} - 2g_{ij}$ i.e. $g_{ij} = \frac{g_{ii}+g_{jj}}{2}$. So G is of the form $G = \mathbf{g}1^T + 1\mathbf{g}^T$ with components of $\mathbf{g} \in \mathbb{R}^m$ equal $g_i = 0.5g_{ii}$.

Therefore, to find $F \in \mathcal{E}_{dp}(D)$ for an Euclidean matrix D we need only to consider matrices deviating from $-0.5D_{sq}$ by $\mathbf{g}1^T + 1\mathbf{g}^T$ for some $\mathbf{g}$. Let us denote with $\mathcal{G}(D)$ the set of all matrices F such that $F = \mathbf{g}1^T + 1\mathbf{g}^T - 0.5D_{sq}$. So for each matrix F if $F \in \mathcal{E}_{dp}(D)$ then $F \in \mathcal{G}(D)$, but not vice versa. We stress that we work with an Euclidean matrix D. So we would like to find an F such that F is decomposable into real-valued matrices Y such that $F = YY^T$ so that Y would represent an embedding of an Euclidean distance matrix. But first of all even if D is not Euclidean, or even not metric, such an embedding may be found. (see Gower and Legendre [5]). So even if D is actually an Euclidean distance matrix, and $F = -0.5D_{sq} + \mathbf{g}1^T + 1\mathbf{g}^T$, there is no warranty, that the distance matrix induced by corresponding Y is identical with D.

For an $F \in \mathcal{G}(D)$ consider the matrix $F^* = (\mathbf{I} - 1\mathbf{s}^T) F (\mathbf{I} - 1\mathbf{s}^T)^T$. We obtain

$$F^* = (\mathbf{I} - 1\mathbf{s}^T) F (\mathbf{I} - 1\mathbf{s}^T)^T \tag{5}$$

$$= (\mathbf{I} - 1\mathbf{s}^T) (1\mathbf{g}^T + \mathbf{g}1^T - 0.5D_{sq}) (\mathbf{I} - 1\mathbf{s}^T)^T \tag{6}$$

$$= (\mathbf{I} - 1\mathbf{s}^T) 1\mathbf{g}^T (\mathbf{I} - 1\mathbf{s}^T)^T + (\mathbf{I} - 1\mathbf{s}^T) \mathbf{g}1^T (\mathbf{I} - 1\mathbf{s}^T)^T$$
$$- 0.5 (\mathbf{I} - 1\mathbf{s}^T) D_{sq} (\mathbf{I} - 1\mathbf{s}^T)^T \tag{7}$$

Let us investigate $(\mathbf{I} - 1\mathbf{s}^T) 1\mathbf{g}^T (\mathbf{I} - 1\mathbf{s}^T)^T$:

$$(\mathbf{I} - 1\mathbf{s}^T) 1\mathbf{g}^T (\mathbf{I} - \mathbf{s}1^T) = 1\mathbf{g}^T - 1\mathbf{g}^T \mathbf{s}1^T - 1\mathbf{s}^T 1\mathbf{g}^T + 1\mathbf{s}^T 1\mathbf{g}^T \mathbf{s}1^T \tag{8}$$

Let us make the following choice (always possible) of $\mathbf{s}$ with respect to $\mathbf{g}$: $\mathbf{s}^T 1 = 1$, $\mathbf{s}^T \mathbf{g} = 0$.

Then we obtain from the above equation

$$(\mathbf{I} - 1\mathbf{s}^T) 1\mathbf{g}^T (\mathbf{I} - \mathbf{s}1^T) = 1\mathbf{g}^T - 101^T - 1\mathbf{g}^T + 1\mathbf{s}^T 1 \cdot 0 \cdot 1^T = 00^T \tag{9}$$

By analogy

$$(\mathbf{I} - 1\mathbf{s}^T) \mathbf{g}1^T (\mathbf{I} - 1\mathbf{s}^T)^T = ((\mathbf{I} - 1\mathbf{s}^T) 1\mathbf{g}^T (\mathbf{I} - \mathbf{s}1^T))^T = 00^T \tag{10}$$

By substituting (9) and (10) into (7) we obtain

$$F^* = \left(\mathbf{I} - \mathbf{1s}^T\right) F \left(\mathbf{I} - \mathbf{1s}^T\right)^T = -0.5 \left(\mathbf{I} - \mathbf{1s}^T\right) D_{sq} \left(\mathbf{I} - \mathbf{1s}^T\right)^T \qquad (11)$$

So for any $\mathbf{g}$, hence an $F \in \mathcal{G}(D)$ we can find an $\mathbf{s}$ such that:

$$\left(\mathbf{I} - \mathbf{1s}^T\right) F \left(\mathbf{I} - \mathbf{1s}^T\right)^T = -0.5 \left(\mathbf{I} - \mathbf{1s}^T\right) D_{sq} \left(\mathbf{I} - \mathbf{1s}^T\right)^T$$

For any matrix $F = -0.5 \left(\mathbf{I} - \mathbf{1s}^T\right) D_{sq} \left(\mathbf{I} - \mathbf{1s}^T\right)^T$ for some $\mathbf{s}$ with $\mathbf{1}^T\mathbf{s} = 1$ we say that F is in multiplicative form or $F \in \mathcal{M}(D)$.

If $F = YY^T$, i.e. F is decomposable, then also

$$F^* = \left(\mathbf{I} - \mathbf{1s}^T\right) YY^T \left(\mathbf{I} - \mathbf{1s}^T\right)^T = ((\mathbf{I} - \mathbf{1s}^T) Y)((\mathbf{I} - \mathbf{1s}^T) Y)^T = Y^* Y^{*T}$$

is decomposable. But $Y^* = \left(\mathbf{I} - \mathbf{1s}^T\right) Y = Y - \mathbf{1s}^T Y = Y - \mathbf{1v}^T$, where $\mathbf{v} = Y^T\mathbf{s}$ is a shift vector by which the whole matrix Y is shifted to a new location in the Euclidean space. So the distances between objects computed from Y^* are the same as those from Y, hence if $F \in \mathcal{E}_{dp}(D)$, then $Y^* \in \mathcal{E}(D)$.

Therefore, to find a matrix $F \in \mathcal{E}_{dp}(D)$, yielding an embedding of D in the Euclidean n dimensional space we need only to consider matrices of the form -0.5 $\left(\mathbf{I} - \mathbf{1s}^T\right) D_{sq} \left(\mathbf{I} - \mathbf{1s}^T\right)^T$, subject to the already stated constraint $\mathbf{s}^T\mathbf{1} = 1$, i.e. ones from $\mathcal{M}(D)$.

So we can conclude: If D is a matrix of Euclidean distances, then there must exist a positive semidefinite matrix $F = -0.5 \left(\mathbf{I} - \mathbf{1s}^T\right) D_{sq} \left(\mathbf{I} - \mathbf{s1}^T\right)$ for some vector $\mathbf{s}$ such that $\mathbf{s}^T\mathbf{1} = 1$, $\det((\mathbf{I} - \mathbf{1s}^T)) = 0$ and $D_{sq}\mathbf{s} \neq \mathbf{0}$. These last two conditions are implied by the following fact: D_{sq} is known to be not negative semidefinite, so that F would not be positive semidefinite in at least the following cases: $\det((\mathbf{I} - \mathbf{1s}^T)) \neq 0$ (see reasoning prior to formula (19)) or $D_{sq}\mathbf{s} = \mathbf{0}$ (see reasoning prior to formula (20)). So if D is an Euclidean distance matrix, then there exists an $F \in \mathcal{M}(D) \cap \mathcal{E}_{dp}(D)$.

Let us investigate other vectors $\mathbf{t}$ such that $\mathbf{t}^T\mathbf{1} = 1$. Note that

$$(\mathbf{I} - \mathbf{1t}^T)(\mathbf{I} - \mathbf{1s}^T) = \mathbf{I} - \mathbf{1t}^T - \mathbf{1s}^T + \mathbf{1t}^T \mathbf{1s}^T \qquad (12)$$

$$= \mathbf{I} - \mathbf{1t}^T - \mathbf{1s}^T + \mathbf{1s}^T = \mathbf{I} - \mathbf{1t}^T \qquad (13)$$

Therefore, for a matrix $F \in \mathcal{M}(D)$

$$(\mathbf{I} - \mathbf{1t}^T)F(\mathbf{I} - \mathbf{1t}^T)^T = -0.5(\mathbf{I} - \mathbf{1t}^T)\left(\mathbf{I} - \mathbf{1s}^T\right) D_{sq} \left(\mathbf{I} - \mathbf{1s}^T\right)^T (\mathbf{I} - \mathbf{1t}^T)^T$$

$$= -0.5(\mathbf{I} - \mathbf{1t}^T)D_{sq}(\mathbf{I} - \mathbf{1t}^T)^T \qquad (14)$$

But if $F = YY^T \in \mathcal{E}_{dp}(D)$, then

$$F' = (\mathbf{I} - \mathbf{1t}^T)F(\mathbf{I} - \mathbf{1t}^T)^T \qquad (15)$$

$$= (\mathbf{I} - \mathbf{1t}^T)YY^T(\mathbf{I} - \mathbf{1t}^T)^T \qquad (16)$$

$$= (Y - \mathbf{1}(\mathbf{t}^T)Y)(Y - \mathbf{1}(\mathbf{t}^T)Y)^T \qquad (17)$$

and hence each $-0.5(\mathbf{I} - \mathbf{1t}^T)D_{sq}(\mathbf{I} - \mathbf{1t}^T)^T$ is also in $\mathcal{E}_{dp}(D)$, though with a different placement (by a shift) in the coordinate systems of the embedded data points. So if one element of $\mathcal{M}(D)$ is in $\mathcal{E}_{dp}(D)$, then all of them are.

So we have established that: if D is an Euclidean distance matrix[3], then there exists a decomposable matrix $F = YY^T \in \mathcal{E}_{dp}(D)$ which is in $\mathcal{G}(D)$, hence $\mathcal{E}_{dp}(D) \subset \mathcal{G}(D)$. For each matrix in $\mathcal{G}(D) \cap \mathcal{E}_{dp}(D)$ there exists a multiplicative form matrix in $\mathcal{M}(D) \cap \mathcal{E}_{dp}(D)$. But if it exists, all multiplicative forms are there: $\mathcal{M}(D) \subset \mathcal{E}_{dp}(D)$.

In this way we have proven points 1, 3 and 4 of the Theorem 2. And also the only-if-part of Gower's theorem correction in [5].

However, two things remain to be clarified and are not addressed in [1] nor in [5]: the if-part of [5] theorem correction (given a matrix D such that -0.5 $(\mathbf{I} - \mathbf{1s}^T)D_{sq}(\mathbf{I} - \mathbf{1s}^T)^T$ is positive semidefinite, is D an Euclidean distance matrix? – see point 2 of the Theorem 2) and the status of the additional condition $D_{sq}\mathbf{s} \neq \mathbf{0}$ in Theorem 1.

Gower [1] makes the following remark: $F = (\mathbf{I} - \mathbf{1s}^T)(-0.5D_{sq})(\mathbf{I} - \mathbf{s1}^T)$ is to be positive semidefinite for Euclidean D. However, for non-zero vectors $\mathbf{u}$

$$\mathbf{u}^T F\mathbf{u} = -0.5\mathbf{u}^T(\mathbf{I} - \mathbf{1s}^T)D_{sq}(\mathbf{I} - \mathbf{1s}^T)^T\mathbf{u}$$
$$= -0.5((\mathbf{I} - \mathbf{1s}^T)^T\mathbf{u})^T D_{sq}((\mathbf{I} - \mathbf{1s}^T)^T\mathbf{u}) \tag{18}$$

But D_{sq} is known to be not negative semidefinite, so that F would not be positive semidefinite in at least the following cases: $\det((\mathbf{I} - \mathbf{1s}^T)) \neq 0$ and $D_{sq}\mathbf{s} = \mathbf{0}$. Let us have a brief look at these conditions and why they are neither welcome nor actually existent:

1. Situation $\det((\mathbf{I} - \mathbf{1s}^T)) \neq 0$ *is not welcome*, because there exists a vector $\mathbf{u}'$ such that $\mathbf{u}'^T D_{sq}\mathbf{u}' > 0$ and under $\det((\mathbf{I} - \mathbf{1s}^T)) \neq 0$ we could solve the equation $(\mathbf{I} - \mathbf{1s}^T)^T\mathbf{u} = \mathbf{u}'$ and thus demonstrate that for some $\mathbf{u}$

$$\mathbf{u}^T F\mathbf{u} < 0 \tag{19}$$

However this situation *is impossible*, because for $F \in \mathcal{M}(D)$

$$(\mathbf{I} - \mathbf{1s}^T)\mathbf{1} = \mathbf{1} - \mathbf{1} = \mathbf{0}$$

which means that the rows are linearly dependent, hence $\det((\mathbf{I} - \mathbf{1s}^T)) = 0$ is guaranteed by earlier assumption about $\mathbf{s}$; so this concern by Gower needs to be dismissed as pointless.

2. Situation $D_{sq}\mathbf{s} = \mathbf{0}$ *is not welcome*, because then

$$\mathbf{u}^T(\mathbf{I} - \mathbf{1s}^T)D_{sq}(\mathbf{I} - \mathbf{1s}^T)^T\mathbf{u} = \mathbf{u}^T D_{sq}(\mathbf{I} - \mathbf{1s}^T)^T\mathbf{u} = \mathbf{u}^T\mathbf{u} > 0$$

---

[3] This means that there exists a matrix X such that rows are coordinates of objects in an Euclidean space with distances as in D.

and thus

$$\mathbf{u}^T F \mathbf{u} < 0 \tag{20}$$

denying positive semidefiniteness of F. Gower does not consider this further, but such a situation *is impossible*. Recall that because D is Euclidean, there must exist a vector $\mathbf{r}$ such that $\mathbf{r}^T \mathbf{1} = 1$ and

$$F^{(r)} = YY^T = -0.5 \left(\mathbf{I} - \mathbf{1}\mathbf{r}^T\right) D_{sq} \left(\mathbf{I} - \mathbf{r}\mathbf{1}^T\right)$$

is in $\mathcal{E}_{dp}(D)$. Hence for any $\mathbf{s}$ such that $\mathbf{s}^T \mathbf{1} = 1$

$$\begin{aligned}
\left(\mathbf{I} - \mathbf{1}\mathbf{s}^T\right) F^{(r)} \left(\mathbf{I} - \mathbf{s}\mathbf{1}^T\right) &= ((\mathbf{I} - \mathbf{1}\mathbf{s}^T)\, Y)((\mathbf{I} - \mathbf{1}\mathbf{s}^T)\, Y)^T \\
&= -0.5 \left(\mathbf{I} - \mathbf{1}\mathbf{s}^T\right) D_{sq} \left(\mathbf{I} - \mathbf{s}\mathbf{1}^T\right) \tag{21}
\end{aligned}$$

is positive semidefinite. This allows us to conclude that for such $\mathbf{s}$ $D\mathbf{s} \neq \mathbf{0}$. Therefore if $D\mathbf{s} = \mathbf{0}$ then $\mathbf{s}^T \mathbf{1} = 0$. What is more, if $det(D) \neq 0$ then $D_{sq}\mathbf{s} = \mathbf{0}$ implies $\mathbf{s} = \mathbf{0}$, for which of course $\mathbf{s}^1 \mathbf{1} = 0$.

Hence the last assumption of if-part of Theorem 1 needs to be dropped as unnecessary which simplifies it to corrected theorem in [5].

As we can see from the first point above, F, given by

$$F = -0.5 \left(\mathbf{I} - \mathbf{1}\mathbf{s}^T\right) D_{sq} \left(\mathbf{I} - \mathbf{1}\mathbf{s}^T\right)^T$$

does not need to identify uniquely a matrix D, as $\left(\mathbf{I} - \mathbf{1}\mathbf{s}^T\right)$ is not invertible. Though of course it identifies *an* Euclidean distance matrix.

Let us now demonstrate the missing part of Gower's proof that D is uniquely defined given a decomposable F.

So assume that for some D (of which we do not know if it is Euclidean, but is symmetric and with zero diagonal), $F = -0.5 \left(\mathbf{I} - \mathbf{1}\mathbf{s}^T\right) D_{sq} \left(\mathbf{I} - \mathbf{1}\mathbf{s}^T\right)^T$ and F is decomposable, i.e. $F = YY^T$. Let $\mathcal{D}(Y)$ be the distance matrix derived from Y (i.e. the distance matrix for which Y is an embedding). That means F is decomposable into properly distanced points with respect to $\mathcal{D}(Y)$. And F is in additive form with respect to it, i.e. $F \in \mathcal{G}(\mathcal{D}(Y))$ Therefore there must exist some $\mathbf{s}'$ such that the $F' = -0.5 \left(\mathbf{I} - \mathbf{1}\mathbf{s}'^T\right) \mathcal{D}(Y)_{sq} \left(\mathbf{I} - \mathbf{s}'\mathbf{1}^T\right)$ as valid multiplicative form with respect to $\mathcal{D}(Y)$, and it holds that $F' = \left(\mathbf{I} - \mathbf{1}\mathbf{s}'^T\right) F \left(\mathbf{I} - \mathbf{s}'\mathbf{1}^T\right)$. But recall that $\left(\mathbf{I} - \mathbf{1}\mathbf{s}'^T\right) F \left(\mathbf{I} - \mathbf{s}'\mathbf{1}^T\right) =$

$$\begin{aligned}
&= \left(\mathbf{I} - \mathbf{1}\mathbf{s}'^T\right) (-0.5 \left(\mathbf{I} - \mathbf{1}\mathbf{s}^T\right) D_{sq} \left(\mathbf{I} - \mathbf{s}\mathbf{1}^T\right)) \left(\mathbf{I} - \mathbf{s}'\mathbf{1}^T\right) \\
&= -0.5(\left(\mathbf{I} - \mathbf{1}\mathbf{s}'^T\right) \left(\mathbf{I} - \mathbf{1}\mathbf{s}^T\right)) D_{sq}(\left(\mathbf{I} - \mathbf{1}\mathbf{s}'^T\right) \left(\mathbf{I} - \mathbf{1}\mathbf{s}^T\right))^T \\
&= -0.5 \left(\mathbf{I} - \mathbf{1}\mathbf{s}'^T\right) D_{sq} \left(\mathbf{I} - \mathbf{s}'\mathbf{1}^T\right) \tag{22}
\end{aligned}$$

Hence

$$-0.5\left(\mathbf{I} - \mathbf{1s'}^{T}\right) D_{sq}\left(\mathbf{I} - \mathbf{s'1}^{T}\right) = -0.5\left(\mathbf{I} - \mathbf{1s'}^{T}\right) \mathcal{D}(Y)_{sq}\left(\mathbf{I} - \mathbf{s'1}^{T}\right)$$

So we need to demonstrate that for two symmetric matrices with zero diagonals D, D' such that $-0.5\left(\mathbf{I} - \mathbf{1s}^{T}\right) D_{sq}\left(\mathbf{I} - \mathbf{s1}^{T}\right) = -0.5\left(\mathbf{I} - \mathbf{1s}^{T}\right) D'_{sq}\left(\mathbf{I} - \mathbf{s1}^{T}\right)$ the equation $D = D'$ holds.

It is easy to see that $-0.5\left(\mathbf{I} - \mathbf{1s}^{T}\right)\left(D_{sq} - D'_{sq}\right)\left(\mathbf{I} - \mathbf{s1}^{T}\right) = \mathbf{00}^{T}$. Denote $\Delta = D_{sq} - D'_{sq}$.

$$\left(\mathbf{I} - \mathbf{1s}^{T}\right) \Delta \left(\mathbf{I} - \mathbf{s1}^{T}\right) = \mathbf{00}^{T}$$

$$\Delta - \mathbf{1s}^{T}\Delta - \Delta \mathbf{s1}^{T} + \mathbf{1s}^{T}\Delta \mathbf{s1}^{T} = \mathbf{00}^{T}$$

With $\overline{\Delta}$ denote the vector $\Delta \mathbf{s}$ and with c the scaler $\mathbf{s}^{T}\Delta \mathbf{s}$. So we have

$$\Delta - \mathbf{1}\overline{\Delta}^{T} - \overline{\Delta}\mathbf{1}^{T} + c\mathbf{1}\mathbf{1}^{T} = \mathbf{00}^{T}$$

So in the row i, column j of the above equation we have: $\delta_{ij} + c - \overline{\delta}_i - \overline{\delta}_j = 0$. Let us add cells ii and jj and subtract from them cells ij and ji. $\delta_{ii} + c - \overline{\delta}_i - \overline{\delta}_i + \delta_{jj} + c - \overline{\delta}_j - \overline{\delta}_j - \delta_{ij} - c + \overline{\delta}_i + \overline{\delta}_j - \delta_{ji} - c + \overline{\delta}_j + \overline{\delta}_i = \delta_{ii} + \delta_{jj} - \delta_{ij} - \delta_{ji} = 0$. But as the diagonals of D and D' are zeros, hence $\delta_{ii} = \delta_{jj} = 0$. So $-\delta_{ij} - \delta_{ji} = 0$. But $\delta_{ij} = \delta_{ji}$ because D, D' are symmetric. Hence $-2\delta_{ji} = 0$ so $\delta_{ji} = 0$. This means that $D = D'$.

This means that D and $\mathcal{D}(Y)$ are identical. Hence decomposition of $F = -0.5\left(\mathbf{I} - \mathbf{1s}^{T}\right) D_{sq}\left(\mathbf{I} - \mathbf{1s}^{T}\right)^{T}$ is sufficient to prove Euclidean space embedding of D and yields this embedding. This proves the if-part of Gower's Theorem 1 and of the corrected theorem in [5] and point 2 of Theorem 2.

3 Concluding Remarks

The question that was left open by Gower was: do there exist special cases where two different $\Phi()$ functions, complying with a given kernel matrix, generate different distance matrices in the feature space, maybe in some special, "sublimated" cases? The answer given to this open question in this paper is definitely NO. We closed all the conceivable gaps in this respect. So usage of (linear and non-linear) kernel matrices that are semipositive definite, is safe in this respect. For more detailed presentation see https://arxiv.org/abs/1701.05335.

References

1. Gower, J.C.: Euclidean distance geometry. Math. Sci. **7**, 1–14 (1982)
2. Dhillon, I.S., Guan, Y., Kulis, B.: Kernel k-means: spectral clustering and normalized cuts. In: Proceedings of the Tenth ACM SIGKDD International Conference on Knowledge Discovery and Data Mining, KDD 2004, pp. 551–556. ACM, New York (2004)
3. Balaji, R., Bapat, R.B.: On Euclidean distance matrices. Linear Algebra Appl. **424**(1), 108–117 (2007)

4. Schoenberg, I.J.: Remarks to Maurice Fréchet's article "Sur la définition axioma-
   tique d'une classe d'espace distanciés vectoriellement applicable sur l' espace de
   Hilbert". Ann. Math. **36**(3), 724–732 (1935)
5. Gower, J.C., Legendre, P.: Metric and Euclidean properties of dissimilarity coefficients. J. Classif. **3**(1), 5–48 (1986). (Here Gower: 1982 is cited in theorem 4, but
   with a different form of condditions for D and s)

Using Network Analysis to Improve Nearest Neighbor Classification of Non-network Data

Maciej Piernik$^{(\boxtimes)}$, Dariusz Brzezinski, Tadeusz Morzy, and Mikolaj Morzy

Institute of Computing Science, Poznan University of Technology, ul. Piotrowo 2,
60-965 Poznan, Poland
maciej.piernik@cs.put.poznan.pl

Abstract. The nearest neighbor classifier is a powerful, straightforward, and very popular approach to solving many classification problems. It also enables users to easily incorporate weights of training instances into its model, allowing users to highlight more promising examples. Instance weighting schemes proposed to date were based either on attribute values or external knowledge. In this paper, we propose a new way of weighting instances based on network analysis and centrality measures. Our method relies on transforming the training dataset into a weighted signed network and evaluating the importance of each node using a selected centrality measure. This information is then transferred back to the training dataset in the form of instance weights, which are later used during nearest neighbor classification. We consider four centrality measures appropriate for our problem and empirically evaluate our proposal on 30 popular, publicly available datasets. The results show that the proposed instance weighting enhances the predictive performance of the nearest neighbor algorithm.

Keywords: Classification · Instance weighting · Nearest neighbors · Network analysis · Centrality measures

1 Introduction

Instance weighted classification enables data analysts to specify the importance of each training example in order to steer the learning process towards more promising instances. The weights can either be assigned manually by an expert or automatically as a result of some additional process. For these weights to be used by a classifier, it has to possess a certain structure, as not all learning schemes are designed to incorporate weights. One type of learners which has a very straightforward way of using weights is the nearest neighbor classifier [1]. In this method, the class prediction for each new example is based on the classes of the most similar training instances, so instance weights can directly influence the model's predictions.

A domain in which instance weights also play a significant role is network analysis, where the importance of nodes is assessed by centrality measures [2]. As network analysis focuses on relations between nodes rather than their individual

© Springer International Publishing AG 2017
M. Kryszkiewicz et al. (Eds.): ISMIS 2017, LNAI 10352, pp. 105–115, 2017.
DOI: 10.1007/978-3-319-60438-1_11

characteristics, centrality measures carry the information about the importance of each node based on its position in the network. Clearly, such information is absent in non-network data where the information about the topology of relationships between nodes is missing.

In this paper, we show how to incorporate network centrality measures into instance-weighted nearest neighbor classification of non-network data. This process consists of three main steps. The first step is the transformation of data into a network using a distance measure and a nearest neighbor approach, so that each node in the network corresponds to one training example and the differences in classes between nodes are expressed as signs of their connections (positive—if from the same class, negative—otherwise). The second step is the assessment of importance of each node in the network. This goal is achieved using one of several centrality measures capable of processing negative ties which we discuss in this paper. The final step is incorporating the centrality values as instance weights into the nearest neighbor classification algorithm. The experiments conducted on 30 datasets show that our weighted nearest neighbor algorithm surpasses the non-weighted version in terms of predictive performance.

In particular, the main contributions of this paper are as follows.

- We propose an automatic instance-weighting algorithm for the nearest neighbor classifier based on network centrality measures.
- We illustrate how to transform a non-network dataset into a network format.
- We discuss several centrality measures and select those which are suitable for instance weighted classification.
- We evaluate our algorithm by performing an experiment involving classification of 30 popular, publicly available datasets.

2   Related Work

The k-nearest neighbors classifier was first proposed by Fix and Hodges in 1951 [1], and has been gaining popularity ever since. In 1976, Dudani [3] proposed a version which weighted the votes of the neighbors according to their distance to the classified example. This method was further refined by Gou et al. [4], where the authors address the problem of sensitivity w.r.t. parameter k. Recently, Samworth [5] proposed an optimal (under certain assumptions) weighting function for nearest neighbor classification of examples with continuous attributes. This approach is based on the ranked similarity of neighbors and is able to automatically select the value of k. These proposals compute the weights based either on attributes or some distance function, but do so dynamically adjusting neighbor weights for each testing example. In contrast, to the best of our knowledge, our proposal is a first attempt to calculate static weights for each training instance based on network analysis.

Network analysis has been a hot research topic for several years now. The area of particular interest to scientists in this field has been the assessment of the importance of nodes in the network. This goal can be achieved using one of many centrality measures proposed to date [6]. Standard measures, like degree [2], betweenness [2], closeness [2], clustering coefficient [7], or eigenvector [8] centrality, or measures of influence, like Katz [9] or Bonacich [10] centrality, are widely adopted in general network analysis. However, they are designed for unimodal networks (i.e., networks in which all nodes belong to the same category), whereas in this paper we consider multi-modal networks with nodes belonging to multiple categories (where each category corresponds to a single class). Considering the fact that we are expressing the differences in classes between nodes as signs of the weights of their connections, we also have to discard measures designed for networks with exclusively negative ties, like negative degree [11] or h^* centrality [11]. Consequently, we are focusing on measures designed for signed and weighted networks: modified degree, modified clustering coefficient [12], status [13], and PN [11], which will be described in Sect. 3.

3   Centrality-Weighted Nearest Neighbors

Given a dataset of training examples $\mathcal{D} = \{(\boldsymbol{x}_1, y_1), (\boldsymbol{x}_2, y_2), ..., (\boldsymbol{x}_n, y_n)\}$, where $\boldsymbol{x}_i = (x_{i1}, x_{i2}, ..., x_{im})$ is a vector of attribute values and $y_i \in \mathcal{Y}$ is a class label, the task of a classifier is to assign each new unlabeled example $\hat{\boldsymbol{x}}$ to one of the predefined classes $\hat{y} \in \mathcal{Y}$. In case of the nearest neighbor classifier, each new example $\hat{\boldsymbol{x}}$ is classified according to the majority voting of a subset of training examples $\mathcal{N}_{\hat{x}} \subseteq \mathcal{D}$ that are closest to $\hat{\boldsymbol{x}}$, called nearest neighbors of $\hat{\boldsymbol{x}}$:

$$\hat{y} = \underset{y \in \mathcal{Y}}{\mathrm{argmax}} \sum_{(\boldsymbol{x}_i, y_i) \in \mathcal{N}_{\hat{x}}} I(y_i = y), \tag{1}$$

where $I(p)$ equals 1 if predicate p is true and 0, otherwise. The instances can also be weighted, in which case instead of counting examples in each class we simply sum their weights, so the formula from Eq. (1) becomes:

$$\hat{y} = \underset{y \in \mathcal{Y}}{\mathrm{argmax}} \sum_{(\boldsymbol{x}_i, y_i) \in \mathcal{N}_{\hat{x}}} I(y_i = y) w_i, \tag{2}$$

where w_i is the weight of the i-th training example. Notice that if $w_i = 1$ for all training examples $(\boldsymbol{x}_i, y_i) \in \mathcal{D}$, the formulas for weighted and unweighted classifiers are equivalent.

The nearest neighbors of $\hat{\boldsymbol{x}}$ are selected according to some distance function Δ defined on the examples (e.g., euclidean distance). The selection can be carried out in several ways. Common approaches include: choosing k nearest examples (where k is a user-defined value), known as the k-nearest neighbors approach (knn); using only a single closest example (a variant of knn with $k = 1$), known as 1-nearest neighbor (1nn); selecting all examples within a user-defined distance range. In this paper, we will use the most popular knn approach with instance weighting as presented in Eq. (2). We will refer to this algorithm as the weighted k-nearest neighbors classifier (wknn).

Let us now define the distance between two examples $\boldsymbol{x}_i$ and $\boldsymbol{x}_j$, which we will use to select the nearest neighbors. For this purpose, we will use a measure that is capable of comparing instances on both numerical and categorical attributes [14]. Given that x_{il} is the value of the l-th attribute in example $\boldsymbol{x}_i$, the distance δ between two corresponding attribute values x_{il}, x_{jl} of examples $\boldsymbol{x}_i$ and $\boldsymbol{x}_j$ is defined as follows:

$$\delta(x_{il}, x_{jl}) = \begin{cases} |x_{il} - x_{jl}| & \text{attribute } l \text{ is numerical} \\ I(x_{il} \neq x_{jl}) & \text{attribute } l \text{ is categorical.} \end{cases}$$

Assuming that δ_N gives normalized (rescaled to $\langle 0, 1 \rangle$) values of δ, we define the distance Δ between two examples $\boldsymbol{x}_i$, $\boldsymbol{x}_j$ as an average normalized distance over all attributes:

$$\Delta(\boldsymbol{x}_i, \boldsymbol{x}_j) = \frac{1}{m} \sum_{l=1}^{m} \delta_N(x_{il}, x_{jl}), \tag{3}$$

where m is the number of attributes.

With all parts of the weighted k-nearest neighbors classifier introduced, let us now describe the proposed process of calculating weights based on centrality measures. The process takes a training dataset $\mathcal{D}$ as input and outputs a vector of weights $\boldsymbol{w}$ where each weight w_i corresponds to one training example $(\boldsymbol{x}_i, y_i) \in \mathcal{D}$. Conceptually, this process consists of three steps: (1) transformation of the input dataset into a network so that each training example is represented by a single node, (2) calculating a selected network centrality measure for each node, (3) assigning the calculated centrality values as instance weights to training examples corresponding to the nodes. This procedure is illustrated in Fig. 1.

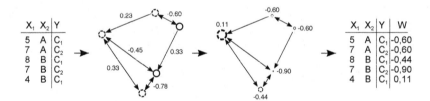

Fig. 1. The process of transforming a dataset into a weighted, signed network (using knn graph with $k_g = 2$), calculating the degree measure, and using the values as weights of instances. Dashed circles represent class C_1, solid circles C_2.

In order to transform the dataset $\mathcal{D}$ into a network, we use an approach inspired by the k-nearest neighbors classifier—the knn graph. First, we calculate the distances between all training examples according to Eq. (3). Next, each training example $(\boldsymbol{x}_i, y_i) \in \mathcal{D}$ becomes a node in the network and is connected with k_g other examples which are closest to it. These connections form directed edges in the network, starting at a given node and ending at its nearest neighbors. As the distances between examples carry potentially valuable information for further processing, we want to include this information in the network in the form of edge weights. However, because in network analysis edges represent the strength of connections, first, we need to convert the distances into similarities. To achieve this goal, we use a common conversion method [14] and rescale values back to $\langle 0, 1 \rangle$, defining the similarity between two examples x_i, x_j as:

$$s_N(x_i, x_j) = \frac{1 - \Delta(x_i, x_j)}{1 + \Delta(x_i, x_j)}.$$

Finally, as each node in the network corresponds to a training example of a certain class, we express this information in the network by adding a sign for each edge. The weight of an edge is positive if the adjacent nodes correspond to training examples from the same class, and negative, otherwise. This gives us the final edge weight between two training examples as:

$$\omega(x_i, x_j) = s_N(x_i, x_j)(2I(y_i = y_j) - 1). \tag{4}$$

After this process is complete, we obtain a network expressed as a graph $G = \langle V, \mathcal{E} \rangle$ in which V is a set of nodes, where each node $v_i \in V$ corresponds to one training example x_i, and $\mathcal{E}$ is a set of edges, where each edge $(v_i, v_j, w_{ij}) \in \mathcal{E}$ represents a connection (tie) from node v_i to node v_j with weight w_{ij}. The graph is represented with two matrices: adjacency matrix $\mathbf{A}$, where each element A_{ij} denotes an edge directed from node v_i to v_j, and weight matrix $\mathbf{W}$, where each element W_{ij} denotes the weight of edge A_{ij}. Figure 1 presents an example of a complete transformation from a training dataset to a network with directed, weighted, and signed ties.

After constructing the network, a selected centrality measure is calculated for every node. Since centrality values will be used as instance weights in wknn classification, the measure has to take into account the signs of the edges. Consequently, one can choose between four different centrality measures: degree, clustering coefficient [12], status [13], and PN [11].

Degree centrality is a classical measure which can be easily adapted to weighted, signed networks. Originally, for a given node v_i, degree centrality d_i is calculated as the number of its connections. In weighted and signed networks it simply becomes a sum of all connection weights of a given node:

$$d_i = \sum_{j=1}^{|V|} W_{ji}. \tag{5}$$

Because we are interested in the importance of an instance from the perspective of other instances, we only take into account the in-degree of each node.

Clustering coefficient is another classical centrality measure which calculates the level of connectedness of a node's nearest neighborhood. For node $v_i \in V$, clustering coefficient is the ratio of the number of edges between the nodes which are connected with v_i to the number of all possible edges between them. In unweighted and unsigned networks, clustering coefficient c_i of node v_i is usually calculated according to the Watts and Strogatz definition [7]:

$$c_i = \frac{\sum_{j,k=1}^{|V|} (A_{ij} A_{ik} A_{jk})}{k_i(k_i - 1)},$$

where k_i is the degree of node v_i. Several versions have been proposed for weighted and signed networks [12], however, given the special meaning of edge signs in our problem, we propose the following definition:

$$c_i = \frac{\sum_{j,k=1}^{|\mathcal{V}|} [(W_{ij} + W_{ik} + W_{jk})(A_{ij}A_{ik}A_{jk})]}{3k_i(k_i - 1)}. \tag{6}$$

The measure produces values between -1 (if all nodes in the neighborhood are fully connected and belong to different classes) and 1 (if all nodes in the neighborhood are fully connected and belong to the same class as node v_i).

Status [13] is a modification of the eigenvector centrality, which measures the relative importance of a node w.r.t. the importance of its adjacent nodes using a recursive definition of importance (a node is important if it is adjacent to important nodes). In order to approximate the value of the status measure we rely on the power iteration method defined as follows:

$$s^l = \frac{\mathbf{W}s^{l-1}}{max(\mathbf{W}s^{l-1})}, \tag{7}$$

where s^l is a vector of statuses for all nodes in iteration $l > 0$ and $s^0 = \mathbf{1}$.

Finally, PN [11] is a measure designed specifically for networks with signed ties and is defined as follows:

$$pn = \left(\mathbf{I} - \frac{1}{2|\mathcal{V}| - 1}\mathbf{M}\right)^{-1} \mathbf{1}. \tag{8}$$

In the equation, pn is a vector of PN values for all nodes and $\mathbf{M} = \mathbf{P} - 2\mathbf{N}$, where $\mathbf{P}$ and $\mathbf{N}$ denote two matrices containing positive and negative ties, respectively.

In the following section, we experimentally verify the predictive performance of wknn with instances weighted according to the described centrality measures.

4 Experiments

4.1 Experimental Setup

The goal of this paper is to assess the effect of centrality instance weighting on the predictive performance of the nearest neighbor classifier. For this purpose, we compare the knn classifier with instances weighted:

- identically (equivalent of unweighted knn; uniform),
- randomly according to a uniform distribution between 0 and 1 (random),
- using clustering coefficient (cluster),
- based on the in-degree (degree),
- based on PN (PN),
- using status (status),
- using the centrality measure with the highest validation score (bestCV).

The first two approaches serve as baselines of traditional (unweighted) knn and random instance weighting; the remaining approaches test the usefulness of the proposed data transformation and centrality measures.

To test the proposed solution we first divided each dataset into a training set and a holdout test set consisting of 50% of the original data. Next, we performed 5×2-fold cross-validation [15] on the training set to select the best k for knn. This way, we tune knn on each dataset to have the best possible performance without instance weighting and make the comparison more challenging. After setting k for each dataset, we performed 5×2-fold cross-validation on the training set once again, but this time to select the k_g parameter used to create the knn graph for each centrality measure. During parameter tuning, we additionally highlight the centrality measure that achieved the best mean cross-validation score as bestCV. Finally, each model was evaluated on a holdout test set.[1] Figure 2 depicts the experimental procedure.

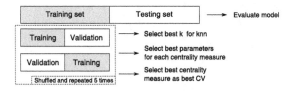

Fig. 2. Experimental procedure.

Model selection as well as evaluation on the test set were performed using the κ statistic [15]. In the context of classification, the κ statistic (or Cohen's κ) compares the accuracy of the tested classifier with that of a chance classifier: $\kappa = \frac{p_0 - p_c}{1 - p_c}$, where p_0 is the accuracy of the tested classifier and p_c is the accuracy of a chance classifier [15]. The κ statistic can achieve values from $(-\infty; 1]$, where zero means that the tested model is no better than a chance classifier and values above/below zero indicate how much better/worse the model performs compared to chance. We selected κ as the evaluation measure in this study, because it is suitable for datasets that suffer from class-imbalance and can be applied to binary as well as multi-class problems.

4.2 Datasets

In our experiments, we used 30 datasets with various numbers of classes, imbalance ratios, and containing nominal as well as numeric attributes. All of the used datasets are publicly available through the UCI machine learning repository [16]. Table 1 presents the main characteristics of each dataset.

The datasets were selected based on their availability and popularity among other studies. As can be seen in Table 1, the testbed consists of binary as well as multiclass problems with a wide range of instance and attribute counts.

---

[1] Source code in R and reproducible test scripts available at: http://www.cs.put.poznan.pl/dbrzezinski/software.php.

Table 1. Characteristic of datasets

Dataset	Inst.	Attr.	Classes	Dataset	Inst.	Attr.	Classes
balance-scale	625	4	3	monks-1	556	6	2
breast-cancer	699	9	2	monks-2	601	6	2
car-evaluation	1,728	6	4	monks-3	554	6	2
cmc	1,473	9	3	pima	768	8	2
credit-screening	690	15	2	promoters	106	57	2
dermatology	366	34	6	sonar	208	60	2
ecoli	336	7	8	spect	267	22	2
glass	214	9	6	statlog-Australian	690	14	2
haberman	306	3	2	statlog-heart	270	13	2
heart-disease	303	13	5	tae	151	5	3
hepatitis	155	19	2	tic-tac-toe	958	9	2
house-votes-84	435	16	2	vowel-context	990	10	11
image-segmentation	2,310	19	7	wine	178	13	3
ionosphere	351	34	2	yeast	1,484	8	10
iris	150	4	3	zoo	101	16	7

4.3 Results

Table 2 presents evaluation results for the analyzed instance weighting schemes. Values marked in bold represent the best result on a given dataset.

We can notice that there is no clear winner in this comparison. Additionally, it is worth highlighting that whenever knn without instance weighting (uniform) achieves best results, usually one of the centrality measures is equally accurate.

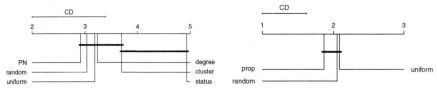

(a) Centrality measures separately   (b) Measure chosen during validation

Fig. 3. Performance ranking of instance weighting schemes. Measures that are not significantly different according to the Nemenyi test (at $\alpha = 0.05$) are connected.

To verify the significance of the observed differences, we performed the nonparametric Friedman test [15]. The null-hypothesis of the Friedman test (that there is no difference between the performance of all the tested instance weighting schemes) was rejected. To verify which instance weighting performs better than the other, we computed the critical difference (CD) chosen by the Nemenyi

Table 2. Kappa statistic for knn with different instance weighting schemes.

Dataset	Uniform	Random	Degree	Cluster	Status	PN	BestCV
balance-scale	0.83	0.80	0.79	0.83	0.47	**0.83**	**0.83**
breast-cancer	0.92	**0.93**	0.92	0.92	0.92	0.92	0.92
car-evaluation	**0.80**	0.74	0.45	**0.80**	0.34	0.76	**0.80**
cmc	0.14	0.11	0.15	0.14	0.12	**0.16**	**0.16**
credit-screening	0.69	0.67	**0.72**	0.69	0.63	0.69	0.69
dermatology	**0.95**	**0.95**	0.94	0.88	0.86	0.94	0.88
ecoli	**0.76**	0.74	0.73	0.73	0.70	**0.76**	**0.76**
glass	0.53	0.52	**0.60**	0.52	0.54	0.55	0.55
haberman	0.03	0.04	**0.16**	0.11	0.00	0.09	0.09
heart-disease	0.22	**0.26**	0.23	0.19	0.21	0.22	0.21
hepatitis	0.37	0.39	0.30	0.37	0.00	**0.48**	0.37
house-votes-84	0.81	**0.83**	**0.83**	0.81	0.80	0.81	0.81
image-segmentation	**0.96**	**0.96**	**0.96**	**0.96**	**0.96**	**0.96**	**0.96**
ionosphere	**0.81**	0.78	0.77	0.70	0.77	0.70	0.70
iris	0.92	0.92	**0.94**	0.92	0.90	0.92	**0.94**
monks-1	0.82	0.81	0.66	0.82	0.40	**0.86**	0.82
monks-2	−0.02	**0.06**	0.04	0.04	0.00	−0.01	0.04
monks-3	**0.96**	0.93	0.85	**0.96**	0.46	0.95	**0.96**
pima	0.34	0.35	0.34	0.34	0.24	**0.37**	0.34
promoters	0.46	**0.50**	0.42	0.46	0.42	0.42	0.42
sonar.all	**0.71**	**0.71**	**0.71**	**0.71**	**0.71**	**0.71**	**0.71**
spect	**0.44**	0.39	**0.44**	0.36	0.43	0.40	**0.44**
statlog-Australian	**0.68**	0.66	0.67	**0.68**	0.59	**0.68**	**0.68**
statlog-heart	0.56	0.58	**0.59**	0.35	0.52	0.56	0.56
tae	0.28	**0.30**	0.28	0.28	**0.30**	0.28	**0.30**
tic-tac-toe	**0.97**	0.92	0.61	**0.97**	0.29	0.92	**0.97**
vowel-context	**0.91**	**0.91**	**0.91**	**0.91**	**0.91**	**0.91**	**0.91**
wine	**0.93**	0.91	**0.93**	**0.93**	0.83	**0.93**	**0.93**
yeast	0.46	**0.46**	0.42	0.46	0.35	0.46	0.46
zoo	0.95	**0.97**	**0.97**	0.95	**0.97**	**0.97**	0.95

post-hoc test [15] at $\alpha = 0.05$. Figure 3 depicts the results of the test by connecting the groups of measures that are not significantly different.

We can notice that the status and cluster measures perform poorly compared to the remaining weighting schemes. PN and degree, on the other hand, present promising results, with PN achieving the best average rank in the Friedman test. Interestingly, if we select the best centrality measure during model selection, the result is not better than when using PN alone. This may suggest that some of the analyzed measures may be prone to overfitting or generalize poorly with more data. This in turn may be an interesting guideline when attempting to propose new network-based measures for instance weighting.

5   Conclusions

In this paper, we presented a new method of weighting instances in the nearest neighbor classifier based on network analysis. The approach relies on transforming the dataset into a weighted and signed network, calculating the centrality of each node, and later using this information during classification. We discussed the transformation process as well as several centrality measures capable of dealing with weighted and signed connections. The experiments performed on 30 popular datasets show that our weighted approach can perform favorably to the unweighted version.

As future research, we plan on experimenting with alternative ways of constructing networks from non-network data, as well as proposing new centrality measures dedicated for the stated problem. Regarding centrality measures, an extensive discussion of their properties w.r.t. classification would also constitute a very interesting line of future research. Furthermore, as evidenced by our experiments, the performance of measures varies with different datasets, therefore, it would be interesting to verify which measures suit which data characteristics better. Finally, generalizing our solution to other classification algorithms would be of high interest as it would make our proposal a generic approach to enhancing classifier performance.

Acknowledgments. This research is partly funded by the Polish National Science Center under Grant No. 2015/19/B/ST6/02637. D. Brzezinski acknowledges the support of an FNP START scholarship.

References

1. Fix, E., Hodges, J.L.: Discriminatory analysis, nonparametric discrimination: consistency properties. US Air Force Sch. Aviat. Med. **Technical Report 4**(3), 477+ (1951)
2. Freeman, L.C.: Centrality in social networks conceptual clarification. Soc. Netw. **1**(3), 215–239 (1978)
3. Dudani, S.A.: The distance-weighted k-nearest-neighbor rule. IEEE Trans. Syst. Man Cybern. **SMC-6**(3), 325–327 (1976)
4. Gou, J., Du, L., Zhang, Y., Xiong, T.: A new distance-weighted k-nearest neighbor classifier. J. Inf. Comput. Sci. **9**(6), 1429–1436 (2012)
5. Samworth, R.J.: Optimal weighted nearest neighbour classifiers. Ann. Stat. **40**(5), 2733–2763 (2012)
6. Kaur, M., Singh, S.: Analyzing negative ties in social networks: a survey. Egypt. Inform. J. **17**(1), 21–43 (2016)

7. Watts, D.J., Strogatz, S.H.: Collective dynamics of 'small-world' networks. Nature **393**(6684), 409–410 (1998)
8. Bonacich, P.: Factoring and weighting approaches to status scores and clique identification. J. Math. Sociol. **2**(1), 113–120 (1972)
9. Katz, L.: A new status index derived from sociometric analysis. Psychometrika **18**(1), 39–43 (1953)
10. Bonacich, P.: Power and centrality: a family of measures. Am. J. Sociol. **92**(5), 1170–1182 (1987)
11. Everett, M., Borgatti, S.: Networks containing negative ties. Soc. Netw. **38**, 111–120 (2014)
12. Costantini, G., Perugini, M.: Generalization of clustering coefficients to signed correlation networks. PLoS ONE **9**(2), 1–10 (2014)
13. Bonacich, P., Lloyd, P.: Calculating status with negative relations. Soc. Netw. **26**(4), 331–338 (2004)
14. Tan, P.N., Steinbach, M., Kumar, V.: Introduction to Data Mining. Addison Wesley, Boston (2005)
15. Japkowicz, N., Shah, M.: Evaluating Learning Algorithms: A Classification Perspective. Cambridge University Press, Cambridge (2011)
16. Lichman, M.: UCI machine learning repository (2013)

An Accurate and Efficient Method to Detect Critical Links to Maintain Information Flow in Network

Kazumi Saito[1(✉)], Kouzou Ohara[2], Masahiro Kimura[3], and Hiroshi Motoda[4,5]

[1] School of Administration and Informatics, University of Shizuoka, Shizuoka, Japan
k-saito@u-shizuoka-ken.ac.jp
[2] Department of Integrated Information Technology, Aoyama Gakuin University,
Sagamihara, Japan
ohara@it.aoyama.ac.jp
[3] Department of Electronics and Informatics, Ryukoku University, Kyoto, Japan
kimura@rins.ryukoku.ac.jp
[4] Institute of Scientific and Industrial Research, Osaka University, Suita, Japan
motoda@ar.sanken.osaka-u.ac.jp
[5] School of Computing and Information Systems,
University of Tasmania, Hobart, Australia

Abstract. We address the problem of efficiently detecting critical links in a large network. Critical links are such links that their deletion exerts substantial effects on the network performance such as the average node reachability. We tackle this problem by proposing a new method which consists of one existing and two new acceleration techniques: redundant-link skipping (RLS), marginal-node pruning (MNP) and burn-out following (BOF). All of them are designed to avoid unnecessary computation and work both in combination and in isolation. We tested the effectiveness of the proposed method using two real-world large networks and two synthetic large networks. In particular, we showed that the new method can compute the performance degradation by link removal without introducing any approximation within a comparable computation time needed by the bottom-k sketch which is a summary of dataset and can efficiently process approximate queries, *i.e.*, reachable nodes, on the original dataset, *i.e.*, the given network.

1 Introduction

Studies of the structure and functions of large networks have attracted a great deal of attention in many different fields of science and engineering [9]. Developing new methods/tools that enable us to quantify the importance of each individual node and link in a network is crucially important in pursuing fundamental network analysis. Networks mediate the spread of information, and it sometimes happens that a small initial seed cascades to affect large portions of networks [11]. Such information cascade phenomena are observed in many situations: for example, cascading failures can occur in power grids (*e.g.*, the August

© Springer International Publishing AG 2017
M. Kryszkiewicz et al. (Eds.): ISMIS 2017, LNAI 10352, pp. 116–126, 2017.
DOI: 10.1007/978-3-319-60438-1_12

10, 1996 accident in the western US power grid), diseases can spread over networks of contacts between individuals, innovations and rumors can propagate through social networks, and large grass-roots social movements can begin in the absence of centralized control (*e.g.*, the Arab Spring). These problems have mostly been studied from the view point of identifying influential nodes under some assumed information diffusion model. There are other studies on identifying influential links to prevent the spread of undesirable things. See Sect. 2 for related work.

We have studied this problem from a slightly different angle in a more general setting [10], that is to answer "Which links are most critical in maintaining a desired network performance?". For example, when the desired performance is to minimize contamination, the problem is reduced to detecting critical links to remove or block. If the desired performance is to maximize evacuation or minimize isolation, the problem is to detect critical links that reduce the overall performance if these links do not function. This problem is mathematically formulated as an optimization problem when a network structure is given and a performance measure is defined. In this paper, we define the performance to be the average node reachability with respect to a link deletion, *i.e.* average number of nodes that are reachable from every single node when a particular link is deleted/blocked. The problem is to rank the links in accordance with the performance and identify the most critical link(s).

Since the core of the computation is to estimate reachability, an efficient method of counting reachable nodes is needed. In our previous work we borrowed the idea of bottom-k sketch [4] which can estimate the approximate number of reachable nodes efficiently by sampling a small number of nodes. Our focus was how to implement bottom-k sketch algorithm to make it run fast and devised two acceleration techniques, called *redundant-link skipping* (*RLS*) and marginal-link updating (MLU). The difference from the previous paper is that, in this paper, we compute reachability accurately without resorting to approximation method, evaluate the accuracy of the approximation method and reduce the computation time to a comparable level by generalizing MLU which was tailored to bottom-k sketch, now called *marginal-node pruning* (*MNP*), and further devising the third technique called *burn-out following* (*BOF*).

We have tested our method using two real-world benchmark networks taken from Stanford Network Analysis Project and two synthetic networks which we designed to control the structural properties. We confirmed that the new method can compute the performance degradation by link removal without introducing any approximation within a computation time comparable to that needed by bottom-k sketch. We showed that depending on the network structure bottom-k sketch needs a larger k (than those used in the experiments) to obtain a result close to the correct one and in this case it needs much larger computation time even using all three techniques. We also analyzed which acceleration technique works better. The results depend on the network structure, but using all three together always works best.

The paper is organized as follows. Section 2 briefly explains studies related to this paper. Section 3 explains the proposed method with the three acceleration techniques. Section 4 reports four datasets used and the experimental results: computational efficiency and comparison with the results obtained by bottom-k sketch. Section 5 summarizes the main achievement and future plans.

2   Related Work

The problem of finding critical links in a network is related to the *influence maximization problem*, which has recently attracted much attention in the field of social network mining [3,6]. This is the problem of finding a limited number of influential nodes that are effective for the spread of information under an appropriate diffusion model. This problem is motivated by *viral marketing*, where in order to market a new product, we target a small number of influential individuals and trigger a cascade of influence by which their friends will recommend the product. Kempe et al. [6] introduced basic probabilistic diffusion models called the *independent cascade* and *linear threshold* models, and formulated this problem as a combinatorial optimization problem under these diffusion models. However, their algorithm was inefficient since it used the Monte-Carlo simulation to evaluate the influence function. Thereafter, a large number of studies have been made to reduce the running time to solve this problem, and several techniques [3,6] that efficiently prune unnecessary evaluations of the influence function have been proposed to speed up the greedy algorithm.

Recently, Borgs et al. [2] proposed an algorithm based on reverse reachability search, which can be regarded as a kind of *sketch-based* method, and proved that it runs in near liner time and provides theoretical guarantees on the approximation quality. Cohen et al. [5] presented another bottom-k sketch-based method of the greedy algorithm for influence maximization, which is called the greedy Sketch-based Influence Maximization (SKIM). Bottom-k sketch is a summary of dataset and can efficiently process approximate queries on the original dataset [4]. It is obtained by assigning a random value independently drawn from some probability distribution to each node. The bottom-k estimator requires the set of nodes having the k smallest values, and the kth smallest value is used for the estimation. This estimate has such a Coefficient of Variation (CV) that is never greater than $1/\sqrt{k-2}$ and well concentrated [4], where CV is defined by the ratio of the standard deviation to the mean. Moreover, for any $c > 0$, it is enough to set $k = (2 + c)\epsilon^{-2} \log N$ to have a probability of having relative error larger than ϵ bounded by N^{-c} [4], where N is the number of nodes. The bottom-k sketch in the network can be efficiently calculated by reversely following links over the network. In Sect. 4 we compare the proposed method with the bottom-k sketch method.

The problem we pose is more closely related to efficiently preventing the spread of undesirable things such as contamination and malicious rumors. Many studies have been made on finding effective strategies for reducing the spread of infection by removing nodes in the network [1]. Moreover, there exists a study of contamination minimization [7] that is converse to the influence maximization problem, where an effective method of minimizing the spread of contamination by blocking a small number of links was explored under a probabilistic information diffusion model. In this paper, we deal with the problem of efficiently finding critical links in terms of reachability degradation.

3   Proposed Method

Let $G = (\mathcal{V}, \mathcal{E})$ be a given simple directed network without self-loops, where $\mathcal{V} = \{u, v, w, \cdots\}$ and $\mathcal{E} = \{e, f, g, \cdots\}$ are sets of nodes and directed links, respectively. Each link e is also expressed as a pair of nodes, *i.e.*, $e = (u, v)$. Below we denote the numbers of nodes and links by $N = |\mathcal{V}|$ and $M = |\mathcal{E}|$, respectively. Let $\mathcal{R}(v; G)$ and $\mathcal{Q}(v; G)$ be the sets of reachable nodes by forwardly and reversely following links from a node v over G, respectively, where note that $v \in \mathcal{R}(v; G)$ and $v \in \mathcal{Q}(v; G)$. Also, let $\mathcal{R}_1(v; G)$ and $\mathcal{Q}_1(v; G)$ be the sets of those nodes adjacent to v, *i.e.*, $\mathcal{R}_1(v; G) = \{w \in \mathcal{R}(v; G) \mid (v, w) \in \mathcal{E}\}$ and $\mathcal{Q}_1(v; G) = \{u \in \mathcal{Q}(v; G) \mid (u, v) \in \mathcal{E}\}$, respectively. Now, let $G_e = (\mathcal{V}, \mathcal{E} \setminus \{e\})$ be the network obtained after removing a link $e = (v, w)$, then we can define the reachability degradation value with respect to $e \in \mathcal{E}$ as follows:

$$F(e; G) = \sum_{x \in \mathcal{V}} (|\mathcal{R}(x; G)| - |\mathcal{R}(x; G_e)|)/N. \tag{1}$$

In this paper, we focus on the problem of accurately and efficiently calculating $F(e; G)$ for every $e \in \mathcal{E}$. Of course, network performance measure is not unique. It varies from problem to problem, but computing $\mathcal{R}(v; G_e)$ for every node $v \in \mathcal{V}$ can be a fundamental task. Note that our proposed method and techniques can directly contribute to this task.

A simple method would be to straightforwardly compute the reachability size, *i.e.*, $|\mathcal{R}(x; G_e)|$, for every node $x \in \mathcal{V}$ after removing every link $e \in \mathcal{E}$. Let $R(G)$ be the average number of reachable nodes by forwardly following links over G, *i.e.*, $R(G) = \sum_{x \in \mathcal{V}} |\mathcal{R}(x; G)|/N$. Then, the computational complexity of this simple method is approximately $O(M \times N \times R(G))$ under the situation that $R(G) \approx R(G_e)$ for most links $e \in \mathcal{E}$, and it generally requires a large amount of computation for large-scale networks. In fact, we obtain $M \times N \times R(G) \in [2.6 \times 10^{13}, 2.0 \times 10^{14}]$ for our networks used in our experiment. In order to overcome this problem, we propose a new method by borrowing and extending the basic ideas of pruning techniques proposed in [8,10]. Below we revisit an existing technique called *redundant-link skipping* (*RLS*) [10] for the sake of readers' convenience. After that, we describe a revised technique called *marginal-node pruning* (*MNP*) which shares a basic idea of the marginal component pruning (MCP) technique proposed in [8] and the marginal link updating (MLU) tailored to bottom-k sketch in [10], and then propose a new acceleration technique called *burn-out following* (*BOF*).

3.1 RLS: Redundant-Link Skipping

The RLS technique selects each link $e \in \mathcal{E}$ for which $F(e; G) = 0$ and prune some subset of such links. Here, we say that a link $e = (v, w) \in \mathcal{E}$ is a *skippable link* if there exists some node $x \in \mathcal{V}$ such that $f = (v, x) \in \mathcal{E}$ and $g = (x, w) \in \mathcal{E}$, i.e., $x \in \mathcal{R}_1(v; G) \cap \mathcal{Q}_1(w; G)$, which means $|\mathcal{R}_1(v; G) \cap \mathcal{Q}_1(w; G)| \geq 1$. Namely, we can skip removing the link e for the purpose of solving our problem due to $F(e; G) = 0$. Moreover, we say that a link $e = (v, w) \in \mathcal{E}$ is a *prunable link* if $|\mathcal{R}_1(v; G) \cap \mathcal{Q}_1(w; G)| \geq 2$. Namely, we can prune such a link e for our problem by setting $G \leftarrow G_e$ due to $F(f; G_e) = F(f; G)$ for any link $f \in \mathcal{E}$.

For each node $v \in \mathcal{V}$, let $\mathcal{S}(v)$ and $\mathcal{P}(v)$ be sets of skippable and prunable links from v. We can compute $\mathcal{S}(v)$ and $\mathcal{P}(v)$ as follows: for each child node $w \in \mathcal{R}_1(v; G)$, we first initialize $c(v, w; G) \leftarrow 0$, $\mathcal{S}(v) \leftarrow \emptyset$ and $\mathcal{P}(v) \leftarrow \emptyset$. Then, for each node $x \in \mathcal{R}_1(v; G)$, we repeatedly set $c(v, w; G) \leftarrow c(v, w; G) + 1$ and $\mathcal{S}(v) \leftarrow \mathcal{S}(v) \cup \{(v, w)\}$ if $w \in \mathcal{R}_1(x; G)$, and set $\mathcal{P}(v) \leftarrow \mathcal{P}(v) \cup \{(v, w)\}$ and $G \leftarrow G_{(v,w)}$ if $c(v, w; G) \geq 2$.

3.2 MNP: Marginal-Node Pruning

The MNP technique recursively performs pruning every node with degree 1 such that its in- and out-degrees are 1 and 0 (or 0 and 1), respectively. Let v be such a node with degree 1, i.e., $|\mathcal{Q}_1(v; G)| = 1$ and $|\mathcal{R}(v; G)| = |\{v\}| = 1$ ($|\mathcal{R}_1(v; G)| = 0$). Then, after removing a link $e = (u, v)$, we can compute the reachability degradation value as $F(e; G) = |\mathcal{Q}(u; G)||\mathcal{R}(v; G)|/N$ where note that $|\mathcal{R}(x; G_e)| = |\mathcal{R}(x; G)| - 1$ if $x \in \mathcal{Q}(u; G)$; $|\mathcal{R}(x; G_e)| = |\mathcal{R}(x; G)|$ otherwise. Now, let $\eta(x)$ be the number of the pruned nodes which are reachable from node x, i.e., after initializing $\eta(w) \leftarrow 1$ for each $w \in \mathcal{V}$, we count the number of forwardly reachable nodes $|\mathcal{R}(x; G_e)|$ by adding $\eta(w)$ when the node w is followed. Then, by updating $\eta(u)$ as $\eta(u) \leftarrow \eta(u) + \eta(v)$, we can recursively prune each node whose in- and out-degrees are 1 and 0 with keeping the accurate calculation of the reachability size for each node.

Clearly, we can apply the similar arguments for each node v such that in- and out-degrees are 0 and 1. Namely, after removing a link $e = (v, w)$, we can also compute the reachability degradation value as $F(e; G) = |\mathcal{Q}(v; G)||\mathcal{R}(w; G)|/N$. By introducing $\mu(x)$ for counting the number of reversely reachable nodes $|\mathcal{Q}(x; G_e)|$, just like $\eta(x)$, and updating $\mu(w)$ as $\eta(w) \leftarrow \eta(w) + \eta(v)$, we can recursively prune each node whose in- and out-degrees are 0 and 1 with keeping the accurate calculation of the reachability size for each node. Namely, we can see that the MNP technique can recursively perform pruning every node with degree 1.

3.3 BOF: Burn-Out Following

For a removed link $e = (v, w)$, we can state that $|\mathcal{R}(x; G)| = |\mathcal{R}(x; G_e)|$ if $x \notin \mathcal{Q}(v; G)$ or $x \in \mathcal{Q}(w; G_e)$. Namely, the reachable size of a node $x \in \mathcal{V}$ changes when x is reachable to v, i.e., $x \in \mathcal{Q}(v; G)$, but becomes not reachable to w after removing a link e, *i.e.*, $x \notin \mathcal{Q}(w; G_e)$. Thus, we can obtain a baseline method which computes the reachability size $|\mathcal{R}(x; G_e)|$ for every node $x \in \mathcal{Q}(v; G) \setminus \mathcal{Q}(w; G_e)$ after removing every link $e \in \mathcal{E}$. Then, by noting that $Q(G) = R(G)$ where $Q(G)$ means the average number of reachable nodes by reversely following links over G, we can see that the computational complexity of this baseline method can be bounded by $O(M \cdot R(G)^2)$. Thus, the baseline method also requires a large amount of computation for large-scale networks. In fact, we still obtain $M \times R(G)^2 \in [1.6 \times 10^{11}, 6.8 \times 10^{12}]$ for our networks used later.

The BOF technique further reduces the computation time needed to follow the same links repeatedly. More specifically, for each node $x \in \mathcal{Q}(v; G) \setminus \mathcal{Q}(w; G_e)$ that is utilized by the baseline method when a link $e = (v, w)$ is removed, we propose to compute the reachable size of x by $|\mathcal{R}(x; G_e)| = |\mathcal{R}(v; G_e)| + |\mathcal{R}(x; G_e) \setminus \mathcal{R}(v; G_e)|$. Namely, after calculating (burning out) the set of reachable nodes from v, *i.e.*, $\mathcal{R}(v; G_e)$, we can compute the reachable size $|\mathcal{R}(x; G_e)|$ by only following the nodes uniquely reachable from x, *i.e.*, $\mathcal{R}(x; G_e) \setminus \mathcal{R}(v; G_e)$. Below we summarize the BOF technique that computes $F(e; G)$ from a network G and its removal link $e = (v, w) \in \mathcal{E}$.

B1: Compute $\mathcal{R}(v; G_e)$ by forwardly following links from v over G_e, and if it happens that $w \in \mathcal{R}(v; G_e)$, output $F(e; G) \leftarrow 0$ and terminate.

B2: Compute $\mathcal{Q}(w; G_e)$ by backwardly following links from w over G_e, and then compute $\mathcal{Q}(v; G) \setminus \mathcal{Q}(w; G_e)$ by backwardly following each link x from v over G unless $x \in \mathcal{Q}(w; G_e)$.

B3: After initializing $F(e; G) \leftarrow 0$, for each node $x \in \mathcal{Q}(v; G) \setminus \mathcal{Q}(w; G_e)$, compute $|\mathcal{R}(x; G_e) \setminus \mathcal{R}(v; G_e)|$ forwardly following each link y from x over G_e unless $y \in \mathcal{R}(v; G_e)$, and then set $F(e; G) \leftarrow F(e; G) + |\mathcal{R}(v; G_e)| + |\mathcal{R}(x; G) \setminus \mathcal{R}(v; G_e)|$.

B4: Output $F(e; G) \leftarrow F(e; G)/N$ and terminate.

Here we should note that the above step **B1:** can be regarded as a generalized version of skippable link calculation by the *RLS* technique.

3.4   Summary of Proposed Method

In our proposed method referred to as PM, we apply the RLS, MNP and BOF techniques to the baseline method in this order, since it is naturally conceivable that the RLS and MNP techniques decrease the numbers of links and nodes in our network G. Clearly we can individually incorporate these techniques into the baseline method. Hereafter, we refer to the proposed method without the RLS technique as the \RLS method, the method without the MNP technique as the \MNP method, and the method without the BOF technique as the \BOF method. Since it is difficult to analytically examine the effectiveness of these techniques, we empirically evaluate the computational efficiency of the proposed method in comparison to these three other methods in which only two techniques are used and the remaining one not used.

4   Experiments

We evaluated the effectiveness of the proposed method using two benchmark and two synthetic networks as we did in [10]. Namely, we employed two benchmark networks obtained from SNAP (Stanford Network Analysis Project)[1]. The first one is a high-energy physics citation network from the e-print arXiv[2], which covers all the citations within a dataset of 34,546 papers (nodes) with 421,578 citations (links). If a paper u cites paper v, the network contains a directed link from u to v. The second one is a sequence of snapshots of the Gnutella peer-to-peer file sharing network from August 2002[3]. There are total of 9 snapshots of Gnutella network collected in August 2002. The network consists of 36,682 nodes and 88,328 directed links, where nodes represent hosts in the Gnutella network topology and links represent connections between the Gnutella hosts. In addition, we utilized two synthetic networks (around 35,000 nodes and 350,000 links) with a DAG (Directed Acyclic Graph) property, which were generated by using the DCNN and DBA methods described in [8], respectively. Here, networks generated by DCNN have both the small-world and scale-free properties, while those by DBA have only the scale-free property.

We refer to these two benchmark networks of citation and pear-to-pear and those generated by the DCNN an DBA methods as CIT, P2P, DCN and DBA networks. Table 1 summarizes the basic statistics of these networks, consisting of the numbers of nodes and links, N and M, the average number of reachable nodes $R(G)$, the number of nodes with in-degree 1 and out-degree 0, $|\mathcal{D}_{1,0}|$, the number of node with in-degree 0 and out-degree 1, $|\mathcal{D}_{0,1}|$, and the numbers of skippable and prunable links, $|\mathcal{S}|$ and $|\mathcal{P}|$. From this table, we can conjecture that the \BOF method will be comparable to the PM method for the DCN network because $R(G)$ is relatively small. On the other hand, the \MNU method may work poorly for the P2P network because $|\mathcal{D}_{1,0}|$ is relatively large, and the \RLS method may also work poorly for the CIT and DCN networks because $|\mathcal{S}|$ and $|\mathcal{P}|$ are relatively large. Here note that the numbers of skippable and prunable links in the DCN network inevitably become larger than the DBA network because the DCNN method has a link creation mechanism between potential pairs.

Table 1. Basic statistics of networks

Name	N	M	$R(G)$	$|\mathcal{D}_{1,0}|$	$|\mathcal{D}_{0,1}|$	$|\mathcal{S}|$	$|\mathcal{P}|$
CIT	34,546	421,578	14,059.0	469	858	302,248	176,224
DBA	35,000	351,317	12,225.1	1,651	1,649	85,815	24,690
DCN	35,000	350,807	2,137.6	2,943	2.839	289,398	175,211
P2P	36,682	88,328	8,482.6	16,409	24	1,502	29

---

[1] https://snap.stanford.edu/.

[2] https://snap.stanford.edu/data/cit-HepPh.html.

[3] https://snap.stanford.edu/data/p2p-Gnutella30.html.

4.1 Evaluation of Acceleration Techniques

First, we evaluated the efficiency of the proposed acceleration techniques by comparing the computation times of the \BOF, \MNP, \RLS, and the proposed (PM) methods. Figure 1 shows our experimental results which compares the actual processing times of these methods, where our programs implemented in C were executed on a computer system equipped with two Xeon X5690 3.47 GHz CPUs and a 192 GB main memory with a single thread within the memory capacity. From Fig. 1, we can clearly see that except for the DCN network, the \BOF method required much computation times compared with the other three methods. As described earlier, these experimental results can be naturally explained from our conjecture that the \BOF method would work well for the DCN network because $R(G)$ is relatively small. We can also see that the \MNU method exhibited the worst performance for the P2P network, while the \RLS for the DCN network. These experimental results are to be expected and explained from the $|\mathcal{D}_{1,0}|$ value and the pair of the $|\mathcal{S}|$ and $|\mathcal{P}|$ values in Table 1.

Which technique works best depends on the network characteristics. Overall BOF which was newly introduced in this paper works the best. MNP and RLS are similar and work less. The proposed method PM combining all the three techniques BOF, MNP and RLS is most reliable and produces the best performance, but the actual reduction of computation time depends on network structure. These results demonstrate the effectiveness of the proposed method.

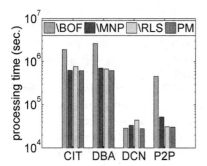

Fig. 1. Evaluation of acceleration techniques

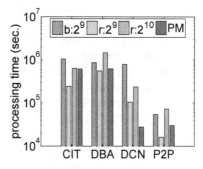

Fig. 2. Comparison with approximation method

4.2 Comparison with Approximation Methods

Next, we evaluated the efficiency of the proposed method in comparison to the approximation method based on bottom-k sketch [4]. The merit of bottom-k sketch is mentioned in Sect. 2. Here, we briefly revisit the implementation algorithm and describe the way to estimate the number of the reachable nodes from each node $v \in \mathcal{V}$, i.e., $|\mathcal{R}(v;G)|$. First, we assign to each node $v \in \mathcal{V}$ a value $r(v)$ uniformly at random in $[0,1]$. When $|\mathcal{R}(v;G)| \geq k$, let $\mathcal{B}_k(v;G)$ be the subset of the k smallest elements in $\{r(w) \mid w \in \mathcal{R}(v;G)\}$, and $b_k(v;G) = \max \mathcal{B}_k(v;G)$ be the k-th smallest element. Here, $\mathcal{B}_k(v;G)$ is set to $\mathcal{R}(v;G)$ when $|\mathcal{R}(v;G)| < k$. Then, we can unbiasedly estimate the number of the reachable nodes from v by $H(v;G) = |\mathcal{B}_k(v;G)|$ if $|\mathcal{B}_k(v;G)| < k$; otherwise $H(v;G) = (k-1)/b_k(v;G)$. We can efficiently calculate the bottom-k sketch $\mathcal{B}_k(v;G)$ for each node $v \in \mathcal{V}$ by reversely following links $k|\mathcal{E}|$ times. Namely, we first initialize $\mathcal{B}_k(v;G) \leftarrow \emptyset$ and sort the random values as $(r(v_1), \cdots, r(v_i), \cdots, r(v_{|\mathcal{V}|}))$ in ascending order, i.e., $r(v_i) \leq r(v_{i+1})$. Then, from $i = 1$ to $|\mathcal{V}|$, for $w \in \mathcal{Q}(v_i;G)$, we repeatedly insert $r(v_i)$ into $\mathcal{B}_k(w;G)$ by reversely following links from v_i if $|\mathcal{B}_k(w;G)| < k$.

Based on the bottom-k sketches described above, we can estimate our reachability degradation value $F(e;G)$ as $J(e;G) = N^{-1} \sum_{v \in \mathcal{V}} (H(v;G) - H(v;G_e))$. Then, we can straightforwardly obtain a baseline approximation method which re-calculates the bottom-k sketches, $\mathcal{B}_k(v;G_e)$ with respect to G_e for all nodes each time from scratch. We can accelerate this baseline approximation method by introducing two acceleration techniques: RLS and MLU as mentioned in Sect. 1. RLS is the same as the first acceleration technique explained in Sect. 3. MLU locally updates the bottom-k sketches of some nodes when removing links incident to a node with in-degree 0 or out-degree 0 in the network G. Hereafter, the baseline approximation method is referred to as baseline BKS, while the revised approximation method that uses these two acceleration techniques as revised BKS.

Figure 2 shows our experimental results by setting the parameter k of the baseline BKS method to 2^9 and the revised BKS method to 2^9 and 2^{10}, denoted as b:2^9, r:2^9, and r:2^{10}, respectively. From these results, we can see that the proposed method substantially outperforms both the baseline BKS and the revised BKS methods for the DCN network. For the other networks, it is better than the baseline BKS method for $k = 2^9$ and the revised BKS method for 2^{10}. Below we will see that setting k at 2^{10} is not large enough to attain a good accuracy especially for the P2P network. We can say that the proposed method is competitive to the approximation method in terms of computation efficiency and has a merit of computing the correct values for reachability degradation.

The exact solutions obtained by our method can be used as the ground-truth for evaluating the approximation method. Let $E(m)$ be the set of the top-m links according to $F(e;G)$. Figure 3 shows the average relative error of the estimated value $J(e;G)$ over $E(5)$, i.e., $\sum_{e \in E(5)} |1 - J(e;G)/F(e;G)|/5$, where we set k to one of $\{2^7, 2^8.2^9, 2^{10}\}$ for each network. From these experimental results, we observe that quite accurate estimation results were obtained for the CIT network, and the relative errors decreased monotonically by using a larger k. If we request the relative error to be less than 0.01, we need the parameter settings greater than $k = 2^8$. For other networks we observe that the results of the DBA and DCN networks are somewhat accurate around 0.1 when $k = 2^{10}$, but the results of the P2P network were quite inaccurate. We need much larger k and the computation time for BKS will overly exceed that of the present method.

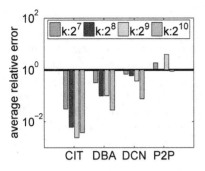

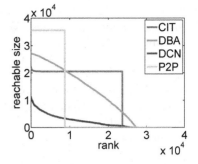

Fig. 3. Relative errors of approximation method

Fig. 4. Reachability distributions of networks

We discuss below why the BKS method worked very poorly for the P2P network. As a typical situation for a given removed link $e = (u,v) \in \mathcal{E}$, assume that $\mathcal{R}(w; G_e) \cap \mathcal{R}(v; G_e) = \emptyset$ for any $w \in \mathcal{Q}(u; G_e)$, then, we obtain $\mathcal{Q}(u; G) = \mathcal{Q}(u; G_e)$, $\mathcal{R}(v; G) = \mathcal{R}(v; G_e)$, and the reachability degradation value $F(e; G) = |\mathcal{Q}(u; G)| \times |\mathcal{R}(v; G)|/N$. However, when $|\mathcal{R}(u; G)| \geq k$, the BKS method returns its estimation as $J(e; G) = 0$ if $b_k(u; G) < \min_{w \in \mathcal{R}(v;G)} r(w)$, where recall that $r(w)$ and $b_k(u; G)$ mean a random value assigned to the node w and the k-th smallest element in $\{r(w) \mid w \in \mathcal{R}(u; G)\}$. This situation is likely to occur when $|\mathcal{R}(u; G)| \approx N$ and $|\mathcal{R}(v; G)|$ is quite small. On the other hand, when $|\mathcal{R}(u; G)| \approx N$, the BKS method may widely underestimate $J(e; G)$ if $b_k(u; G) > \min_{w \in \mathcal{R}(v;G)} r(w)$. This is because for a large number $|\mathcal{R}(u; G)|$ of nodes, say $x \in \mathcal{R}(u; G)$, $H(x; G_e)$ substantially decreases due to the removal of $\min_{w \in \mathcal{R}(v;G)} r(w)$ from $\mathcal{B}_k(x; G)$. We confirm that such situations are likely to occur on the P2P network. Figure 4 shows distributions of reachability size $|\mathcal{R}(v; G)|$ with respect to its rank. From this figure, we can clearly see that there exist two groups of nodes in the P2P network, those reachable to almost all of the other nodes, just like $|\mathcal{R}(u; G)| \approx N$ discussed above, and those reachable to almost only themselves, which includes the nodes in $\mathcal{D}_{1,0}$. These results clearly support our above explanation.

5   Conclusion

In this paper we have proposed a novel computational method that can detect critical links efficiently without introducing any approximation for a large network. The problem is reduced to finding a link that reduces the network performance substantially with respect to its removal. Such a link is considered critical in maintaining the good performance. There are many problems that can be mapped to this critical link detection problem, e.g. contamination minimization be it physical or virtual, evacuation trouble minimization, road maintenance prioritization, etc.

There are many things to do. Reachability computation is a basic operation and is a basis for many applications. We continue to explore techniques to further reduce computation time, to elaborate other useful network performance measures, and to clarify the difference in characteristics of extracted critical links from those of the links chosen by the existing measures such as edge betweenness that has no notion of reachability. Our immediate future plan is to apply our method to a real world application and show that it can solve a difficult problem efficiently, e.g. identifying important hot spots in transportation network or evacuation network.

Acknowledgments. This material is based upon work supported by the Air Force Office of Scientific Research, Asian Office of Aerospace Research and Development (AOARD) under award number FA2386-16-1-4032, and JSPS Grant-in-Aid for Scientific Research (C) (No. 17K00314).

References

1. Albert, R., Jeong, H., Barabási, A.L.: Error and attack tolerance of complex networks. Nature **406**, 378–382 (2000)
2. Borgs, C., Brautbar, M., Chayes, J., Lucier, B.: Maximizing social influence in nearly optimal time. In: Proceedings of the 25th Annual ACM-SIAM Symposium on Discrete Algorithms (SODA 2014), pp. 946–957 (2014)
3. Chen, W., Lakshmanan, L., Castillo, C.: Information and influence propagation in social networks. Synth. Lect. Data Manag. **5**(4), 1–177 (2013)
4. Cohen, E.: Size-estimation framework with applications to transitive closure and reachability. J. Comput. Syst. Sci. **55**, 441–453 (1997)
5. Cohen, E., Delling, D., Pajor, T., Werneck, R.F.: Sketch-based influence maximization and computation: scaling up with guarantees. In: Proceedings of the 23rd ACM International Conference on Information and Knowledge Management, pp. 629–638 (2014)
6. Kempe, D., Kleinberg, J., Tardos, E.: Maximizing the spread of influence through a social network. Theory Comput. **11**, 105–147 (2015)
7. Kimura, M., Saito, K., Motoda, H.: Blocking links to minimize contamination spread in a social network. ACM Trans. Knowl. Discov. Data **3**, 9:1–9:23 (2009)
8. Kimura, M., Saito, K., Ohara, K., Motoda, H.: Speeding-up node influence computation for huge social networks. Int. J. Data Sci. Anal. **1**, 1–14 (2016)
9. Newman, M.: The structure and function of complex networks. SIAM Rev. **45**, 167–256 (2003)
10. Saito, K., Kimura, M., Ohara, K., Motoda, H.: Detecting critical links in complex network to maintain information flow/reachability. In: Proceedings of the 14th Pacific Rim International Conference on Artificial Intelligence, pp. 419–432 (2016)
11. Watts, D.: A simple model of global cascades on random networks. Proc. Natl. Acad. Sci. U. S. A. **99**, 5766–5771 (2002)

Deep Learning

Continuous Embedding Spaces for Bank Transaction Data

Ali Batuhan Dayioglugil[1]([⊠]) and Yusuf Sinan Akgul[2]

[1] Cybersoft R&D Center, İstanbul, Turkey
alibatuhandayioglugil1@gmail.com
[2] Gebze Technical University, 41400 Gebze, Kocaeli, Turkey
akgul@gtu.edu.tr

Abstract. In the finance world, customer behavior prediction is an important concern that requires discovering hidden patterns in large amounts of registered customer transactions. The purpose of this paper is to utilize this customer transaction data for the sake of customer behavior prediction without any manual labeling of the data. To achieve this goal, elements of the banking transaction data are automatically represented in a high dimensional embedding space as continuous vectors. In this new space, the distances between the vector positions are smaller for the elements with similar financial meaning. Likewise, the distances between the unrelated elements are larger, which is very useful in automatically capturing the relationships between the financial transaction elements without any manual intervention.

Although similar embedding space work has been used in the other fields such as natural language processing, our work introduces novel ideas in the application of continuous word representations technology for the financial sector. Overall, we find the initial results very encouraging and, as the future work, we plan to apply the introduced ideas for the abnormal financial customer behavior detection, fraud detection, new banking product design, and making relevant product offers to the bank customers.

Keywords: Feature embedding space · Word representation · Bank customer segmentation · Deep learning · Fraud detection

1 Introduction

Financial institutions are required by law to keep their data (customer, account, credit, etc.) in a structured form. This form is sufficient for basic daily operations like transactions and report generations. However, when the company needs to extract extra information from the data, domain knowledge becomes essential. Currently, human experts provide most of the domain knowledge into the information extraction process in the financial data, which makes this process time consuming, subjective, difficult to keep updated, and prone to human errors. Hence, automatically discovering hidden patterns in the observed customer data is essential for several financial tasks such as fraud detection, new product offers, and customer behavior analysis.

© Springer International Publishing AG 2017
M. Kryszkiewicz et al. (Eds.): ISMIS 2017, LNAI 10352, pp. 129–135, 2017.
DOI: 10.1007/978-3-319-60438-1_13

The availability of raw customer transaction data made it very suitable for the application of machine learning techniques for this purpose. Many of these methods need numeric inputs and numeric output labels. Although financial data includes mostly non-numeric data, clustering, ranking and dummification [9] techniques can be used to produce digitized input and output data. However, obtaining the numeric or non-numeric output data is not trivial because it mostly requires hand labeling of the data by the experts, which is very expensive and time consuming for big scaled finance data.

The most popular techniques for financial data extraction are rule based approaches and other supervised machine learning methods. When rule based approaches are considered, keeping all the rules up to date and regularly adding the new ones (due to new customers, new scenarios or changing habits) are inevitable and also a constraint for a system's performance. Supervised machine learning methods are also employed for detecting patterns in financial data and they are widely used in modelling customer behavior and fraud detection fields [1, 3, 5, 10], but they remain incapable of detecting deep relations and they require hand labeled data.

Recently, continuous word representations in high dimensional spaces brought a great impact in Natural Language Processing (NLP) community by their ability to unsupervisedly capture syntactic and semantic relations between words, phrases [6, 7] and even complete documents [10]. Employment of these representations produced very promising results with the help of available large text bases in the fields of language modeling and language translation. Similar ideas were applied to parallel but on different modality corpora: Weston et al. [4] used word embeddings along with image embeddings for an image captioning system.

Motivated from the success of continuous word representations in the NLP world, this work proposes to represent the financial transaction data in a continuous embedding space to take advantage of the large unlabeled financial data. To the best of our knowledge, our system is the first to propose the continuous embedding space methods for the finance data. The resulting vector representations of the transactions are similar for the semantically similar financial concepts. We argue that, by employing these vector representations one can automatically make information extraction from the raw financial data. We performed experiments to show the benefits of these representations.

This paper is organized as follows; in Sect. 2 data properties are given, accordingly parameters and details to create word vectors are explained. Section 3 gives the experiments with the proposed model. Section 4 discusses the obtained results and argues possible financial applications of this method as the future work.

2   Embedding Spaces for Financial Data

Raw transaction data consists of timewise ordered individual transactions. We argue that this data is very similar to natural language sentences. Researchers of NLP domain recently achieved very favorable language modeling results by projecting the words and sentences to a continuous embedding space. We propose to use a similar idea for the raw transaction data. Representation of the transaction data in a new space can lead us to demonstrate semantic and syntactic relations among transaction elements.

2.1   Transaction Data

Throughout this work, a darkened transaction data on a medium-sized Turkish bank is used. The transaction data $TRX = \{T_i\}$, $i = 1..n$, is an ordered set of individual transactions T_i, where n is the total number of transactions that can reach up to hundreds of millions. Each transaction $T_i = \{t_{i,j}\}$, $j = 1..m$, is a set of m transaction elements. In our case, we took $m = 10$ after eliminating irrelevant elements (such as date and voucher id) and elements including private (Name, security number etc.) or missing data. Each transaction element is a variable that can take either numerical or categorical values. The two numerical transaction elements are *age* and *amount*. Categorical elements with their corresponding possible number of values are *gender* (3), *housing status* (6), *marital status* (8), *education level* (9), *business segment* (15), *customer type* (19), *profession* (10), and *transaction process group code* (56)[1]. We eliminated the transactions that were initiated by the bank because these types of transactions do not provide any significant information about the customer behavior. Thus, our dataset contains only customer based transactions. Since the numbers of transaction element values are much bigger for the numerical elements, we coarsely clustered age and amount transaction element values as shown in Table 1. We have 126 different transaction element values for only categorical elements and 11 different clustered elements values, which make a total of 137 possible element values.

Table 1. Categorical value assignments for numeric values

Transaction element name	Assigned cluster values	Number of values
Transaction amount (Turkish Liras)	Amnt1(Amount < 1), Amnt2(1 < Amount < 100), Amnt3 (100 < Amount < 200), Amnt4(200 < Amount < 1000), Amnt5(1000 < Amount < 3000), Amnt6 (3000 < Amount < 10000), Amnt7(10000 < Amount)	7
Age	Young(age <=25), Young_Adult(25 < age <=40), Middle_Age(40 < age <=55), Old(55 < age), No_Age (age is Null)	4

As mentioned before, we propose to use continuous vector representations for the data *TRX* as it is done for the natural language words. If we like to draw more concrete parallels between transaction data and the natural languages; our transaction words can be considered as natural language words and individual transactions can be seen as natural language sentences. The total number of possible element values can be compared to the number of unique words in a natural language. As expected, since the possible numbers of transaction element values are much smaller than the natural language vocabulary sizes, our embedding space dimensions will be much smaller than the frequently used dimension sizes in the NLP domain.

---

[1] All feature names are translated from Turkish to English.

2.2   Transaction Elements Vectors

Generating word embeddings from context dependent corpora using neural networks is successfully applied by feed-forward networks [2] and later by Recurrent Neural Networks (RNN) with higher accuracies [8, 9]. RNN is a special form of ANN which has self-recurrent connections on hidden nodes that can keep the short time memory of word relations with respect to a predefined window size. Other types of methods, such as [6, 7], were proposed with varying efficiencies and precision. In this work, we adopted word2vec library [6] to create our transaction element value vectors (Fig. 1).

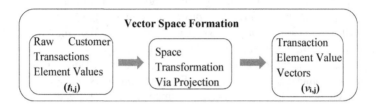

Fig. 1. Proposed vector space formation model

We selected window size as the total number of transaction elements in a transaction ($w = 10$). This is similar to choosing the window size as the average sentence length in NLP applications. Different dimensions of vectors (100, 50 and 20) were tested. For each of these dimensions, we measured the distances between semantically similar transaction element values such as "small farmer" and "medium farmer". The tested dimensions produced almost the same similarity values for the same element values. In order to reduce the complexity of the network with a small vocabulary of 137 values, we chose an embedding space with 20 dimensions.

For the vector similarities, we used the cosine similarity (Eq. 1) which is one of the most popular metrics used for this type of comparison.

$$sim(A, B) = \frac{A.B}{\|A\|\|B\|} = \frac{\sum_{i=1}^{d} A_i B_i}{\sqrt{\sum_{i=1}^{d} A_i^2}\sqrt{\sum_{i=1}^{d} B_i^2}}, \tag{1}$$

where A and B are the vectors of dimension d, A_i and B_i are the elements of these vectors, respectively.

We observed that the skip-gram method captures stronger relations between the transaction element value vectors with similar financial meanings within the embedding space. These relations are given statistically in Table 2. We use cosine similarity to measure the distance between the transformed transaction elements. Note that cosine similarity can be at most 1.0. A quick analysis of Table 2 reveals that most similar transaction elements for a given transaction element value are very similar in terms of financial semantics. For example, the most similar transaction element value to "Young Adult" is the "Middle Age" value. Note that these vector assignments were obtained fully automatically without any human supervision.

Table 2. Closest neighbours of three transaction element values (age, business segment and profession respectively) with respect to cosine similarities

Young adult		Small farmer		Student	
Middle age	0.958798	Medium farmer	0.995632	Unemployed	0.868379
Old	0.879234	Big farmer	0.944026	Housewife	0.833956
Young	0.833032	Small business owner	0.829680	Public sector employee	0.766121
TGC32[a]	0.490597	Micro business	0.774104	Private sector employee	0.737213
No age	0.460723	Agricultural enterprise	0.768568	Retired	0.692351
		Private	0.748798	No profession	0.652012
		SME[b]	0.707125	Retired working	0.589541
		Corporate	0.652397	Self-employment	0.581086

[a]Transaction Process Group Code 32
[b]Small and Medium Sized Enterprise

To visualize all the segment vectors in a reduced 2 dimensional space, Principal Component Analysis (PCA) is applied on the obtained embedded vectors and the calculated relations are shown graphically in Fig. 2(a) for the *business segment* element values. As can be seen in this figure, semantically close values are assigned closer vector positions such as all 3 farmer types (small, medium, and large farmers).

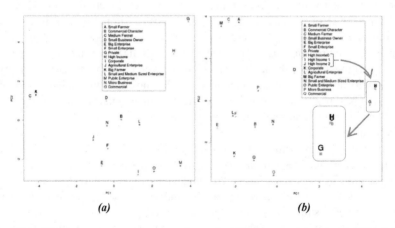

(a)   *(b)*

Fig. 2. PCA of b*usiness segment* element value embedding vectors *(a)* and embedding vectors of same element values with artificially divided 'High-income' value *(b)*

2.3   Representing Transactions as Vectors

After the transaction element value vectors are estimated, a complete transaction T_i can be represented in the vector form by concatenating the transaction element value vectors of $v_{i,j}$ into the vector V_i. Since we use 20 dimensional transaction element vectors and we have 10 transaction elements in a transaction, the resulting transaction vector V_i would be 200 dimensional.

3   Experiments

We performed two types of experiments for the validation of the estimated embedding vectors of financial data. For the first experiment, we took all the transactions that contain the "high income" element value as the business segment. We artificially replaced the original business segment values with 3 different dummy values (High_Income_0, High_Income_1, and High_Income_2) for these transactions in a random manner. We then estimated the vectors for this modified transaction data as described in Sect. 2. The obtained element value vectors are shown in Fig. 2(b) in reduced dimensions. As the figure shows, the artificially modified element value vectors turn out to be very close in this space, which shows that the proposed model is able to capture the semantic relations in the finance data.

For the second experiment, we like to explore the idea of using semantically closer representations of feature labels may successfully lead to recognition of hidden patterns in data. For this purpose, the transaction vectors ($\bar{V}_i$) without the *business segment* element vectors are used as inputs of an ANN. The *business segment* vectors (B_i) is used as the output variable. In other words, the input to the ANN is a 180 dimensional vector that does not include any *business segment* information. The output is a 20 dimensional vector that we expect to produce vectors positions close to *business segment* element value vectors.

Our dataset contains around 1.8 Million transactions (A four week period). In order to have exact scalar values for output nodes, an activation function is not implemented in the output layer of the ANN. The ANN uses cosine similarity as it is statistically more robust, to calculate the differences (error) between model output vectors and ground truth vectors. Stochastic Gradient Descent (SGD) optimizer is adopted with different learning rates and for different hidden layer node counts using hyperbolic tangent (*tanh*) as the activation function in hidden layer. In the training part of the network, the best results are obtained by an ANN which contains a single hidden layer with 60 nodes with learning rate of 0,018 where batch size was 100 for error backpropagation.

In order to avoid overfitting, 4-fold cross validation method is used during training and testing. After cross-validation process, we have predicted output vectors (B_i) for each transaction in input data. These 20 dimensional predicted vectors are compared with true element value vectors using cosine similarities. Each possible element value vector is assigned a similarity score, which are sorted to produce the top-1, top-3, and top-5 most likely predictions. Table 3 lists the final scores obtained from this experiment. Although around 70% business segment prediction accuracy seems not very high, we should note that these results are obtained without any supervision. Furthermore, considering only top 5 matching results, one can easily conclude that these unsupervised predictions can be used as reliable cues for a fraud detection system.

Table 3. Proposed model accuracies

	Top 1 match	Top 3 matches	Top 5 matches
Accuracy (%)	72,4	90,4	93,8

4   Discussion and Future Work

In the financial world, decision making processes need reliable arguments which must always be up-to-date in order to keep up with dynamic nature of the domain. Institutions register and process higher volumes of data day by day. While having such big data brings many risks to manage, it also gives many opportunities to create relevant products and also ensure customer content by analyzing customer behavior. We believe that the interpretation of large scaled structured data by traditional systems is not efficient anymore and the application of more recent methods, such as embedding vectors, can produce strong results. As the future work, we plan to extend the proposed method to detect fraud attacks, define abnormal customer activities, and provide valuable knowledge to create new products.

Acknowledgments. We would like to thank Tuğba Halıcı and Sait Şimşek for their valuable technical contribution to this work.

References

1. Ando, Y., Hidehito, G., Hidehiko, T.: Detecting Fraudulent Behavior Using Recurrent Neural Networks (2016)
2. Bengio, Y., Ducharme, R., Vincent, P., Jauvin, C.: A neural probabilistic language model. J. Mach. Learn. Res. **3**, 1137–1155 (2003)
3. Busta, B., Weinberg, R.: Using Benford's law and neural networks as a review procedure. Manag. Audit. J. **13**(6), 356–366 (1998)
4. Kirkos, E., Spathis, C., Manolopoulos, Y.: Data mining techniques for the detection of fraudulent financial statement. Expert Syst. Appl. **32**, 995–1003 (2007)
5. Mikolov, T., Sutskever, I., Chen, K., Corrado, G.S., Dean, J.: Distributed representations of words and phrases and their compositionality. In: Advances in Neural Information Processing Systems, pp. 3111–3119 (2013)
6. Mikolov, T., Zweig, G.: Context dependent recurrent neural network language model. SLT **12**, 234–239 (2012). doi:10.1109/slt.2012.6424228
7. Mikolov, T., Karafiát, M., Burget, L., Cernocký, J., Khudanpur, S.: Recurrent neural network based language model. In: Interspeech, vol. 2, p. 3 (2010)
8. Sohl, J.E., Venkatachalam, A.R.: A neural network approach to forecasting model selection. Inf. Manag. **29**(6), 297–303 (1995)
9. Wiese, B.J.: Credit card transactions, fraud detection, and machine learning: modelling time with LSTM recurrent neural networks. Dissertation, Department of Computer Science, University of the Western Cape (2007). doi:10.1007/978-3-642-04003-0_10
10. Zou, W.Y., Socher, R., Cer, D.M., Manning, C.D.: Bilingual word embeddings for phrase-based machine translation. In: EMNLP, pp. 1393–1398 (2013)

Shallow Reading with Deep Learning: Predicting Popularity of Online Content Using only Its Title

Wojciech Stokowiec[1,3]([✉]), Tomasz Trzciński[2,3], Krzysztof Wołk[1], Krzysztof Marasek[1], and Przemysław Rokita[2]

[1] Polish-Japanese Academy of Information Technology, Warsaw, Poland
wojciech.stokowiec@pjwstk.edu.pl
[2] Warsaw University of Technology, Warsaw, Poland
{t.trzcinski,pro}@ii.pw.edu.pl
[3] Tooploox, Wrocław, Poland

Abstract. With the ever decreasing attention span of contemporary Internet users, the title of online content (such as a news article or video) can be a major factor in determining its popularity. To take advantage of this phenomenon, we propose a new method based on a bidirectional Long Short-Term Memory (LSTM) neural network designed to predict the popularity of online content using only its title. We evaluate the proposed architecture on two distinct datasets of news articles and news videos distributed in social media that contain over 40,000 samples in total. On those datasets, our approach improves the performance over traditional shallow approaches by a margin of 15%. Additionally, we show that using pre-trained word vectors in the embedding layer improves the results of LSTM models, especially when the training set is small. To our knowledge, this is the first attempt of applying popularity prediction using only textual information from the title.

1 Introduction

The distribution of textual content is typically very fast and catches user attention for only a short period of time [2]. For this reason, proper wording of the article title may play a significant role in determining the future popularity of the article. The reflection of this phenomenon is the proliferation of *click-baits* - short snippets of text whose main purpose is to encourage viewers to click on the link embedded in the snippet. Although detection of click-baits is a separate research topic [3], in this paper we address a more general problem of predicting popularity of online content based solely on its title.

Predicting popularity in the Internet is a challenging and non-trivial task due to a multitude of factors impacting the distribution of the information: external context, social network of the publishing party, relevance of the video to the final user, etc. This topic has therefore attracted a lot of attention from the research community [2,13,14,17].

© Springer International Publishing AG 2017
M. Kryszkiewicz et al. (Eds.): ISMIS 2017, LNAI 10352, pp. 136–145, 2017.
DOI: 10.1007/978-3-319-60438-1_14

In this paper we propose a method for online content popularity prediction based on a bidirectional recurrent neural network called BiLSTM. This work is inspired by recent successful applications of deep neural networks in many natural language processing problems [6,21]. Our method attempts to model complex relationships between the title of an article and its popularity using novel deep network architecture that, in contrast to the previous approaches, gives highly interpretable results. Last but not least, the proposed BiLSTM method provides a significant performance boost in terms of prediction accuracy over the standard shallow approach, while outperforming the current state-of-the-art on two distinct datasets with over 40,000 samples.

To summarize, the contributions presented in this paper are the following:

- Firstly, we propose title-based method for popularity prediction of news articles based on a deep bidirectional LSTM network.
- Secondly, we show that using pre-trained word vectors in the embedding layer improves the results of LSTM models.
- Lastly, we evaluate our method on two distinct datasets and show that it outperforms the traditional shallow approaches by a large margin of 15%.

The remainder of this paper is organized in the following manner: first, we review the relevant literature and compare our approach to existing work. Next, we formulate the problem of popularity prediction and propose a model that takes advantage of BiLSTM architecture to address it. Then, we evaluate our model on two datasets using several pre-trained word embeddings and compare it to benchmark models. We conclude this work with discussion on future research paths.

2   Related Work

The ever increasing popularity of the Internet as a virtual space to share content inspired research community to analyze different aspects of online information distribution. Various types of content were analyzed, ranging from textual data, such as Twitter posts [2] or Digg stories [17] to images [9] to videos [5,13,18]. Although several similarities were observed across content domains, e.g. log-normal distribution of data popularity [19], in this work we focus only on textual content and, more precisely, on the popularity of news articles and its relation to the article's title.

Forecasting popularity of news articles was especially well studied in the context of Twitter - a social media platform designed specifically for sharing textual data [1,8]. Not only did the previous works focus on the prediction part, but also on modeling message propagation within the network [11]. However, most of the works were focused on analyzing the social interactions between the users and the characteristics of so-called social graph of users' connections, rather than on the textual features. Contrary to those approaches, in this paper we base our predictions using only textual features of the article title. We also validate our proposed method on one dataset collected using a different social media platform, namely Facebook, and another one created from various news articles [14].

Recently, several works have touched on the topic of popularity prediction of news article from a multimodal perspective [4,14]. Although in [14] the authors analyze news articles on a per-modality basis, they do not approach the problem of popularity prediction in a holistic way. To address this shortcoming, [4] have proposed a multimodal approach to predicting popularity of short videos shares in social media platform Vine[1] using a model that fuses features related to different modalities. In our work, we focus only on textual features of the article title for the purpose of popularity prediction, as our goal is to empower the journalists to quantitatively assess the quality of the headlines they create before the publication. Nevertheless, we believe that in future research we will extend our method towards multimodal popularity prediction.

3   Method

In this section we present the bidirectional LSTM model for popularity prediction. We start by formulating the problem and follow up with the description of word embeddings used in our approach. We then present the Long Short-Term Memory network that serves as a backbone for our bidirectional LSTM architecture. We conclude this section with our interpretation of hidden bidirectional states and describe how they can be employed for title introspection.

3.1   Problem Formulation

We cast the problem of popularity prediction as a binary classification task. We assume our data points contain a string of characters representing article title and a popularity metric, such as number of comments or views. The input of our classification is the character string, while the output is the binary label corresponding to popular or unpopular class. To enable the comparison of the methods on datasets containing content published on different websites and with different audience sizes, we determine that a video is popular if its popularity metric exceeds the median value of the corresponding metric for other points in the set, otherwise - it is labeled as unpopular. The details of the labeling procedure are discussed separately in the Datasets section.

3.2   Text Representation

Since the input of our method is textual data, we follow the approach of [15] and map the text into a fixed-size vector representation. To this end, we use word embeddings that were successfully applied in other domains. We follow [6] and use pre-trained GloVe word vectors [12] to initialize the embedding layer (also known as look-up table). Section 4.3 discusses the embedding layer in more details.

---

[1] https://vine.co/.

3.3   Bidirectional Long Short-Term Memory Network

Our method for popularity prediction using article's title is inspired by a bidirectional LSTM architecture. The overview of the model can be seen in Fig. 1.

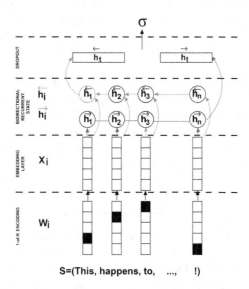

Fig. 1. A bidirectional LSTM architecture with 1-of-K word encoding and embedding layer proposed in this paper.

Let $x_i \in \mathbb{R}^d$ be d-dimensional word vector corresponding to the i-the word in the headline, then a variable length sequence: $\mathbf{x} = (x_1, x_2, \ldots, x_n)$ represents a headline. A recurrent neural network (RNN) processes this sequence by recursively applying a transformation function to the current element of sequence x_t and its previous hidden internal state h_{t-1} (optionally outputting y_t). At each time step t, the hidden state is updated by:

$$h_t = \sigma(W_h \cdot [h_{t-1}, x_t] + b_h), \tag{1}$$

where σ is a non-linear activation function. LSTM network [7] updates its internal state differently, at each step t it calculates:

$$\begin{aligned}
i_t &= \sigma(W_i \cdot [h_{t-1}, x_t] + b_i), \\
f_t &= \sigma(W_f \cdot [h_{t-1}, x_t] + b_f), \\
o_t &= \sigma(W_o \cdot [h_{t-1}, x_t] + b_o), \\
\widetilde{c}_t &= \tanh(W_{\widetilde{c}} \cdot [h_{t-1}, x_t] + b_{\widetilde{c}}), \\
c_t &= f_t \odot c_{t-1} + i_t \odot \widetilde{c}_t, \\
h_t &= o_t \odot \tanh(c_t),
\end{aligned} \tag{2}$$

where σ is the *sigmoid* activation function, *tanh* is the hyperbolic tangent function and $\odot$ denotes component-wise multiplication. In our experiments we used

128, 256 for the dimensionality of hidden layer in both LSTM and BiLSTM. The term in Eq. 2 i_t, is called the input gate and it uses the input word and the past hidden state to determine whether the input is worth remembering or not. The amount of information that is being discarded is controlled by forget gate f_t, while o_t is the output gate that controls the amount of information that leaks from memory cell c_t to the hidden state h_t. In the context of classification, we typically treat the output of the hidden state at the last time step of LSTM as the document representation and feed it to sigmoid layer to perform classification [22].

Due to its sequential nature, a recurrent neural network puts more emphasis on the recent elements. To circumvent this problem [16] introduced a bidirectional RNN in which each training sequence is presented forwards and backwards to two separate recurrent nets, both of which are connected to the same output layer. Therefore, at any time-step we have the whole information about the sequence. This is shown by the following equation:

$$\overrightarrow{h_t} = \sigma(\overrightarrow{W_h} \cdot [h_{t-1}, x_t] + \overrightarrow{b_h}),$$
$$\overleftarrow{h_t} = \sigma(\overleftarrow{W_h} \cdot [h_{t+1}, x_t] + \overleftarrow{b_h}). \tag{3}$$

In our method, we use the bidirectional LSTM architecture for content popularity prediction using only textual cues. We have to therefore map the neural network outputs from a set of hidden states $(\overrightarrow{h_i}, \overleftarrow{h_i})_{i \in 1...n}$ to classification labels. We evaluated several approaches to this problem, such as max or mean pooling. The initial experiments showed that the highest performance was achieved using late fusion approach, that is by concatenating the last hidden state in forward and backward sequence. The intuition behind this design choice is that the importance of the first few words of the headline is relatively high, as the information contained in $\overleftarrow{h_1}$, i.e. the last item in the backward sequence, is mostly taken from the first word.

3.4   Hidden State Interpretation

One interesting property of bidirectional RNNs is the fact, that the concatenation of hidden states $\overrightarrow{h_t}$ and $\overleftarrow{h_t}$ can be interpreted as a context-dependent vector representation of word w_t. This allows us to introspect a given title and approximate the contribution of each word to the estimated popularity. To that end one can process the headline representation $\mathbf{x} = (x_1, x_2, \ldots, x_n)$ through the bidirectional recurrent network and then retrieve pairs of forward and backwards hidden state $[\overrightarrow{h_t}, \overleftarrow{h_t}]$ for each word x_t. Then, the output of the last fully-connected layer $\sigma([\overrightarrow{h_t}, \overleftarrow{h_t}])$ could be interpreted as context-depended popularity of a word x_t.

3.5   Training

In our experiments we minimize the binary cross-entropy loss using Stochastic Gradient Descent on randomly shuffled mini-batches with the Adam optimization algorithm [10]. We reduce the learning rate by a factor of 0.2 once learning plateaus. We also employ early stopping strategy, i.e. stopping the training algorithm before convergence based on the values of loss function on the validation set.

4 Evaluation

In this section, we evaluate our method and compare its performance against the competitive approaches. We use k-fold evaluation protocol with $k = 5$ with random dataset split. We measure the performance using standard accuracy metric which we define as a ratio between correctly classified data samples from test dataset and all test samples.

4.1 Datasets

In this section we present two datasets used in our experiments: The NowThisNews dataset, collected for the purpose of this paper, and The BreakingNews dataset [14], publicly available dataset of news articles.

The NowThisNews Dataset contains 4090 posts with associated videos from NowThisNews Facebook page[2] collected between 07/2015 and 07/2016. For each post we collected its title and the number of views of the corresponding video, which we consider our popularity metric. Due to a fairly lengthy data collection process, we decided to normalize our data by first grouping posts according to their publication month and then labeling the posts for which the popularity metric exceeds the median monthly value as popular, the remaining part as unpopular.

The Breaking News Dataset [14] contains a variety of news-related information such as images, captions, geo-location information and comments which could be used as a proxy for article popularity. The articles in this dataset were collected between January and December 2014. Although we tried to retrieve the entire dataset, we were able to download only 38,182 articles due to the dead links published in the dataset. The retrieved articles were published in main news channels, such as Yahoo News, The Guardian or The Washington Post. Similarly, to The NowThisNews dataset we normalize the data by grouping articles per publisher, and classifying them as popular, when the number of comments exceeds the median value for given publisher.

4.2 Baselines

As a first baseline we use Bag-of-Words, a well-known and robust text representations used in various domains [20], combined with a standard shallow classifier,

---

[2] https://www.facebook.com/NowThisNews.

namely, a Support Vector Machine with linear kernel. We used LIBSVM[3] implementation of SVM.

Our second baseline is a deep Convectional Neural Network applied on word embeddings. This baseline represents state-of-the-art method presented in [14] with minor adjustments to the binary classification task. The architecture of the CNN benchmark we use is the following: the embedding layer transforms one-hot encoded words to their dense vector representations, followed by the convolution layer of 256 filters with width equal to 5 followed by max pooling layer (repeated three times), fully-connected layer with dropout and l_2 regularization and finally, *sigmoid* activation layer. For fair comparison, both baselines were trained using the same training procedure as our method.

Table 1. Popularity prediction results on NowThisNews dataset. Our proposed BiLSTM method provides higher performances than the competitors in terms of classification accuracy.

Model	Word embeddings	Fine-tuned	Dim	Accuracy
BoW + SVM				0.5832
CNN	GloVe (W + G5)	No	100	0.6320
	GloVe (W + G5)	Yes	100	0.6454
	GloVe (W + G5)	No	200	0.6308
	GloVe (W + G5)	Yes	200	0.6479
	GloVe (W + G5)	No	300	0.6247
	GloVe (W + G5)	Yes	300	0.6295
	GloVe (CC)	No	300	0.6528
	GloVe (CC)	Yes	300	0.6653
LSTM 128	Glove (W + G5)	No	300	0.63081
	Glove (W + G5)	Yes	300	0.64792
LSTM 256	Glove (W + G5)	No	300	0.64914
	Glove (W + G5)	Yes	300	0.66504
BiLSTM 128	Glove (W + G5)	No	300	0.6552
	Glove (W + G5)	Yes	300	0.6479
BiLSTM 256	Glove (W + G5)	No	300	0.6564
	Glove (W + G5)	Yes	300	**0.6711**

4.3   Embeddings

As a text embedding in our experiments, we use publicly available GloVe word vectors [12] pre-trained on two datasets: Wikipedia 2014 with Gigaword5 (W + G5) and Common Crawl (CC)[4]. Since their output dimensionality can be

---

[3] https://www.csie.ntu.edu.tw/ cjlin/libsvm/.
[4] http://nlp.stanford.edu/projects/glove/.

modified, we show the results for varying dimensionality sizes. On top of that, we evaluate two training approaches: using static word vectors and fine-tuning them during training phase.

4.4 Results

The results of our experiments can be seen in Tables 1 and 2. Our proposed BiLSTM approach outperforms the competing methods consistently across both datasets. The performance improvement is especially visible for The NowThis-News dataset and reaches over 15% with respect to the shallow architecture in terms of accuracy. Although the improvement with respect to the other methods based on deep neural network is less evident, the recurrent nature of our method provides much more intuitive interpretation of the results and allow for parsing the contribution of each single word to the overall score.

Table 2. Popularity prediction results on BreakingNews dataset. Our BiLSTM method outperforms the competitors - the performance gain is especially visible with respect to the shallow architecture of BoW + SVM.

Model	Word embeddings	Fine-tuned	Dim	Accuracy
BoW + SVM				0.7300
CNN	GloVe (W + G5)	No	100	0.7353
	GloVe (W + G5)	Yes	100	0.7412
	GloVe (W + G5)	No	200	0.7391
	GloVe (W + G5)	Yes	200	0.7379
	GloVe (W + G5)	No	300	0.7319
	GloVe (W + G5)	Yes	300	0.7416
	GloVe (CC)	No	300	0.7355
	GloVe (CC)	Yes	300	0.7394
LSTM 128	Glove (W + G5)	Yes	300	0.6694
	Glove (W + G5)	No	300	0.6663
LSTM 256	Glove (W + G5)	Yes	300	0.6619
	Glove (W + G5)	No	300	0.6624
BiLSTM 128	Glove (W + G5)	Yes	300	0.7167
	Glove (W + G5)	No	300	0.7406
BiLSTM 256	Glove (W + G5)	Yes	300	0.7149
	Glove (W + G5)	No	300	**0.7450**

To present how our model works in practice, we show in Table 3 a list of 3 headlines from NowThisNews dataset that are scored with the highest probability of belonging to a popular class, as well as 3 headlines with the lowest score. As can be seen, our model correctly detected videos that become viral at the

Table 3. Top and bottom 3 headlines from the NowThisNews dataset as predicted by our model and their views 168 hours after publication.

Top 3 headlines	Views
This teen crossed a dangerous highway to play Pokmon Go and then was hit by a car	20'836'692
This dancer dropped her phone in the water but a dolphin had her back	1'887'482
A man shoved a bag of sh*t down this womans pants and was caught on camera	784'588
Bottom 3 headlines	Views
We're recapping some of the biggest stories from last night and this morning	47'803
We're recapping some of the big stories you might have missed	64'740
Violent clashes between protesters and police broke out in Hong Kong	256'357

same time assigning low score to content that underperformed. We believe that BiLSTM could be successfully applied in real-life scenarios.

5   Conclusions

In this paper we present a novel approach to the problem of online article popularity prediction. To our knowledge, this is the first attempt of predicting the performance of content on social media using only textual information from its title. We show that our method consistently outperforms benchmark models. Additionally, the proposed method could not only be used to compare competing titles with regard to their estimated probability, but also to gain insights about what constitutes a good title. Future work includes modeling popularity prediction problem with multiple data modalities, such as images or videos. Furthermore, all of the evaluated models function at the word level, which could be problematic due to idiosyncratic nature of social media and Internet content. It is, therefore, worth investigating, whether combining models that operate at the character level to learn and generate vector representation of titles with visual features could improve the overall performance.

Acknowledgment. The authors would like to thank NowThisMedia Inc. for enabling this research by providing access to data and hardware.

References

1. Bandari, R., Asur, S., Huberman, B.A.: The pulse of news in social media: forecasting popularity. CoRR, abs/1202.0332 (2012)
2. Castillo, C., El-Haddad, M., Pfeffer, J., Stempeck, M.: Characterizing the life cycle of online news stories using social media reactions. In: CSCW (2014)

3. Chakraborty, A., Paranjape, B., Kakarla, S., Ganguly, N.: Stop clickbait: detecting and preventing clickbaits in online news media. CoRR, abs/1610.09786 (2016)
4. Chen, J., Song, X., Nie, L., Wang, X., Zhang, H., Chua, T.: Micro tells macro: predicting the popularity of micro-videos via a transductive model. In: ACMMM (2016)
5. Chesire, M., Wolman, A., Voelker, G., Levy, H.M.: Measurement and analysis of a streaming-media workload. In: USITS (2001)
6. Collobert, R., Weston, J., Bottou, L., Karlen, M., Kavukcuoglu, K., Kuksa, P.P.: Natural language processing (almost) from scratch. CoRR, abs/1103.0398 (2011)
7. Hochreiter, S., Schmidhuber, J.: Long short-term memory. Neural Comput. **9**(8), 1735–1780 (1997)
8. Hong, L., Dan, O., Davison, B.: Predicting popular messages in Twitter. In: Proceedings of International Conference Companion on World Wide Web (2011)
9. Khosla, A., Sarma, A., Hamid, R.: What makes an image popular? In: WWW (2014)
10. Kingma, D.P., Ba, J.: Adam: a method for stochastic optimization. CoRR, abs/1412.6980 (2014)
11. Osborne, M., Lavrenko, V.: RT to win! predicting message propagation in Twitter. In: ICWSM (2011)
12. Pennington, J., Socher, R., Manning, C.D.: Glove: global vectors for word representation. In: Empirical Methods in Natural Language Processing (EMNLP) (2014)
13. Pinto, H., Almeida, J., Gonçalves, M.: Using early view patterns to predict the popularity of Youtube videos. In: WSDM (2013)
14. Ramisa, A., Yan, F., Moreno-Noguer, F., Mikolajczyk, K.: Breakingnews: article annotation by image and text processing. CoRR, abs/1603.07141 (2016)
15. Salton, G., Wong, A., Yang, C.S.: A vector space model for automatic indexing. Commun. ACM **18**(11), 613–620 (1975)
16. Schuster, M., Paliwal, K.K.: Bidirectional recurrent neural networks. IEEE Trans. Signal Process. **45**, 2673–2681 (1997)
17. Szabo, G., Huberman, B.: Predicting the popularity of online content. Commun. ACM **53**(8), 80–88 (2010)
18. Trzcinski, T., Rokita, P.: Predicting popularity of online videos using support vector regression. CoRR, abs/1510.06223 (2015)
19. Tsagkias, M., Weerkamp, W., de Rijke, M.: News comments: exploring, modeling, and online prediction. In: ECIR (2010)
20. Wang, S., Manning, C.: Baselines and bigrams: simple, good sentiment and topic classification. In: ACL (2012)
21. Zhang, X., LeCun, Y.: Text understanding from scratch. CoRR, abs/1502.01710 (2015)
22. Zhou, C., Sun, C., Liu, Z., Lau, F.C.M.: A C-LSTM neural network for text classification. CoRR, abs/1511.08630 (2015)

Recurrent Neural Networks for Online Video Popularity Prediction

Tomasz Trzciński[1,2]([✉]), Paweł Andruszkiewicz[1], Tomasz Bocheński[1], and Przemysław Rokita[1]

[1] Warsaw University of Technology, Warsaw, Poland
{t.trzcinski,pro}@ii.pw.edu.pl, {pandrus1,tbochens}@mion.elka.pw.edu.pl
[2] Tooploox, Wrocław, Poland

Abstract. In this paper, we address the problem of popularity prediction of online videos shared in social media. We prove that this challenging task can be approached using recently proposed deep neural network architectures. We cast the popularity prediction problem as a classification task and we aim to solve it using only visual cues extracted from videos. To that end, we propose a new method based on a Long-term Recurrent Convolutional Network (LRCN) that incorporates the sequentiality of the information in the model. Results obtained on a dataset of over 37'000 videos published on Facebook show that using our method leads to over 30% improvement in prediction performance over the traditional shallow approaches and can provide valuable insights for content creators.

1 Introduction

The problem of online content predicting popularity is extremely challenging since the popularity patterns are driven by many factors that are complex to model, such as social graph of the users, propagation patterns and content itself. The problem has therefore gained a lot of attention from the research community and various types of content were analyzed in this context, including but not limited to news articles [1,2], Twitter messages [3], images [4] and videos [5–7].

In this paper, we focus on the problem of popularity prediction for videos distributed in social media. We postulate that the popularity of the video can be predicted with a higher accuracy if a prediction model incorporates the sequential character of the information presented in the video. To that end, we propose a method based on a deep recurrent neural network architecture called Long-term Recurrent Convolutional Network (LRCN) [8], since in LRCN, the information about the sequentiality of the inputs is stored as parameters of the network. We are inspired by the successful application of this architecture to other video-related tasks, such as video activity recognition or video captioning. We call our method Popularity-LRCN and evaluate it on a dataset of over 37'000 videos collected from Facebook. Our results show that the Popularity-LRCN method leads to over 30% improvement in prediction performance with respect to the state of the art.

© Springer International Publishing AG 2017
M. Kryszkiewicz et al. (Eds.): ISMIS 2017, LNAI 10352, pp. 146–153, 2017.
DOI: 10.1007/978-3-319-60438-1_15

The application of this method can be used by video content creators to optimize the chances that their work will become popular. In the context of everyday work, they often produce multiple videos on the same subject and choose the one that has the highest popularity potential based on their subjective opinion. Since our method predicts the popularity of the content based solely on the visual features, it can be applied on multiple videos before their publication to select the one that has the highest probability to become popular.

The contributions of the paper are the following:

– an online video popularity prediction method that relies only on visual cues extracted from the video
– application of a deep recurrent neural network architecture for the purpose of popularity prediction

The remainder of this paper is organized as follows: after a brief overview of the related work, we define the problem and describe the popularity prediction method we propose to address it. We then evaluate our method against the state of the art. Lastly, we conclude the paper with final word and future research.

2 Related Work

Popularity prediction of online content has gained a lot of attention within the research community due to the ubiquity of Internet and a stunning increase in the number of its users [5,9,10]. None of the above mentioned works, however, considers using visual information for popularity prediction, while the method proposed in this paper relies solely on the visual cues.

The first attempt to use visual cues in the context of popularity prediction is the work of [4], where they address the problem of popularity prediction for online images available on Flickr. Using a dataset of over 2 million images, the authors train a set of Support Vector Machines using visual features such as image color histogram or deep neural network outputs and apply them in the context of popularity prediction. Following this methodology, we use recently proposed neural network architectures for the purpose of video popularity prediction based on visual cues only.

The deep neural network architecture we use in this paper is known as Long-term Recurrent Convolutional Network (LRCN) [8] and is a combination of convolutional networks, initially used for image classification purposes, and Long-Short Term Memory (LSTM) [11] units, mostly used in the context of natural language processing and other text related tasks, where sequence of inputs plays an important role. Thanks to the combination of those two architectures LRCN was proven to be a good fit for several video-related tasks, such as activity recognition or video annotation. In this paper, we extend the list of potential applications of the LRCN-related architectures to the task of online video popularity prediction.

3   Method

In this section, we describe our method for popularity prediction of online videos that uses recurrent neural network architecture. We first define the popularity prediction problem as a classification task and then present the details of the deep neural network model that attempts to solve it.

3.1   Problem

We cast the problem of video popularity prediction as a binary classification task, similarly to [12] where they analyze the popularity of fashion images using a classification approach. Our objective is to classify a video as popular or unpopular before its publication using only visual information from its frames.

More formally, let us a define a set of N samples $\{x_i, l_i\}_{i=1}^{N}$, where x_i is a video and $l_i \in \{0, 1\}$ is a corresponding popularity label equal to 1, if a video is going to be popular after publication and 0, if not. The label of the video is determined based on the number of times it has been watched by social network users after its publication, which we refer to as a *viewcount*. In other words, we aim to train a classifier C that given a set of images assigns the sample with a predicted label $\hat{l} = f_C(x, \theta)$, where θ is a set of parameters of a classifier C. Our goal is to find the solution of the corresponding objective function:

$$\min_{\theta} \sum_{i=1}^{N} \left(\hat{l}_i - l_i \right)^2 = \min_{\theta} \sum_{i=1}^{N} \left(f_C(x, \theta) - l_i \right)^2 \qquad (1)$$

which is a differentiable function that can be solved within the frames of deep neural network architectures, as we show next.

3.2   Popularity-LRCN

Since the problem definition presented in the previous section is quite general, a plethora of classifiers can be used to minimize it. In this section we propose an approach based on the recently proposed Long-term Recurrent Convolutional Network (LRCN) architecture [8]. We call this method Popularity-LRCN and show that its recurrent character, which incorporates the sequential character of the information displayed in the video frames, outperforms the methods that consider only individual frames.

The architecture of the recurrent neural network used in Popularity-LRCN is a combination of a convolutional network with Long Short-Term Memory units, and is inspired by the architecture used in [8] for the task of activity recognition in videos. After initial experiments, we set the size of our network's input to 18 frames of size $227 \times 227 \times 3$ that represent the video. Our network consists of eight layers with learnable parameters. The first five layers are convolutional filters, followed by a fully-connected layer with 4096 neurons, LSTM and a final fully-connected layer with 2 neurons. We use soft-max as a classification layer.

Since our method is trained to perform a binary classification, the output of the network is a 2-dimensional vector of probabilities linked to each of the popularity labels. The output is computed by propagating each of the video frames through the above architecture in a sequential manner and averaging the probability vector across all the frames, as done in [8].

To increase transitional invariance of our network, we use max pooling layers after the first, second and fifth convolutional layer. Local Response Normalization layers [13] follow the first and the second max pooling layer. We use ReLU as a non-linear activation function and apply it to the output of every convolutional and fully-connected layer. To avoid overfitting, dropout layers are placed after the first fully-connected layer (with dropout ratio of 0.9) and after the LSTM layer (with dropout ratio of 0.5).

When training, we input video frames of size $320 \times 240 \times 3$, crop it at random to $227 \times 227 \times 3$ and perform data augmentation with mirroring. We train our network for 12 epochs (30'000 iterations each) with batch size equal to 12. At prediction time, we also use data augmentation by cropping all frames from their four corners and around the center, as well as mirroring all of them. This way we generate 10 synthetical video representations from one test video representation. The final prediction output is the result of averaging probabilities across the frames and then across all generated representations.

4 Evaluation

In this section, we present the evaluation procedure used to analyze the performance of the proposed Popularity-LRCN method and compare it against the state of the art. We first present the dataset of over 37'000 videos published on Facebook, that is used in our experiments, and we describe how the popularity labels of the samples are assigned. We then discuss the evaluation criteria and baseline approaches. Lastly, we present the performance evaluation results.

4.1 Dataset

Since our work is focused on predicting popularity of video content in social media, we collect the data from the social network with the highest number of users – Facebook – with reported 1.18 billion active everyday users worldwide[1]. To collect the dataset of the videos along with their corresponding viewcounts, we implemented a set of crawlers that use Facebook Graph API[2] and ran it on the Facebook pages of over 160 video creators listed on a social media statistics website, TubularLabs[3]. To avoid crawling videos whose viewcounts are still changing, we restricted our crawlers to collect the data about videos that were online for at least 2 weeks. On top of the videos and their viewcounts, we also collected first and preferred thumbnails of the videos, as well as the number of

---

[1] http://newsroom.fb.com/company-info/.
[2] https://developers.facebook.com/docs/graph-api.
[3] http://tubularlabs.com/.

publishers' followers. The Facebook videos missing any of this information were discarded. The resulting dataset consists of 37'042 videos published between June 1st 2016 and September 31st 2016.

4.2 Labeling

The distribution of the viewcounts across the videos from our dataset exhibits a large variation, as our dataset contains not only popular videos that were watched millions of times, but also those that were watched less than a few hundreds times. Similarly to [4], we deal with this variation using log transform. Additionally, to reduce the bias linked to the fact that popular publishers receive more views on their videos irrespectably of their quality, we also include in our normalization procedure the number of followers of publishers' page. Our normalized popularity score is therefore computed according to the following formula:

$$\text{normalized popularity score} = \log_2 \left(\frac{\text{viewcount} + 1}{\text{number of publisher's followers}} \right) \quad (2)$$

The additional increment in the numerator prevents from computing logarithm of zero, in case of the videos that did not receive any views.

After the normalization, we divide the dataset into two categories: popular videos and unpopular videos. To obtain an equal distribution of two classes in the training dataset, we split the dataset using median value of the normalized popularity score, following the approach of [12].

4.3 Evaluation Criteria

For the evaluation purposes, we used a k-fold train-test evaluation protocol with 5 folds. The Facebook dataset described above was split randomly and in each k-fold iteration 29'633 videos were used for training and 7'409 for testing. The performance metrics related to the classification accuracy and prediction quality were therefore averaged across 5 splits.

To evaluate the performances of the methods, we use classification accuracy, computed simply as a ratio of correctly classified samples on the test set. As a complementary performance metric we use Spearman correlation between video probabilities of belonging to the popular class and their normalized popularity score. This is because one of the goals of the proposed model is to help the creators to make an data-driven decision on which video should be published based on its predicted popularity. Since our model is trained to classify videos only as popular or unpopular, it provides a very granular evaluation. Therefore, we decided to use the probabilities of the popular class generated by the evaluated models as popularity scores and, following the evaluation protocol of [4] report their Spearman rank correlation with the normalized popularity score of the videos. We report the values of those metrics averaged across $k = 5$ folds of our dataset along with the standard deviation values.

4.4   Baselines

Following the methodology presented in [4,10], we used as baselines traditional shallow classifiers, i.e. logistic regression classifier and SVM with RBF kernel. As input for these classifiers we used the following visual features that proven to be successful in other computer vision related tasks:

- HOG: a 8100-dimensional Histogram of Oriented Gradients descriptor [14].
- GIST: a 960-dimensional GIST descriptor, typically employed in image retrieval tasks [15].
- CaffeNet: we use as 1000-dimensional feature vectors activations of the last fully-connected layer of BVLC CaffeNet model[4] (pre-trained on ILSVRC-2012[5] object classification set).
- ResNet: similarly to the above, we use 1000 activations of the last fully-connected layer of the ResNet-152 [16] neural network.

4.5   Results

We evaluate the performances of all methods using the dataset of Facebook videos. Methods are implemented in Python using scikit-learn[6] and caffe[7] libraries. It is worth noticing that Facebook increments a viewcount of a video after at least three seconds of watching it and by default turns the autoplay mode on. Therefore, in our experiments, we focus on the beginning of the video and take as input of the evaluated methods representative frames coming from the first six seconds of a video. In case of baseline features, we use early fusion approach and concatenate them before classification stage. For our Popularity-LRCN, we input raw RGB frames.

Table 1 shows the results of the experiments. Popularity-LRCN clearly outperforms the competing shallow architectures, both in terms of classification accuracy and Spearman correlation by a large margin of 7% and 34%, respectively. We explain that by the fact that our Popularity-LRCN incorporates the sequentiality of the information, that is relevant in terms of popularity prediction, within the model. What is worth noticing, out of the competing methods, those using the activations of recently proposed convolutional neural networks perform better than those based on traditional features (such as GIST or HOG).

To visualize the results of the classification performed by Popularity-LRCN, in Fig. 1 we show 100 representative frames from the videos with the highest probability of belonging to the popular class and 100 frames from the videos with the lowest probability. As can be seen in Fig. 1a, the videos whose frames contain pictures of food are classified as more probable to become popular, according to the Popularity-LRCN method. On the contrary, the lowest probability of becoming popular is linked to the videos with the interview-alike scenes and dull colors present in the opening scene.

---

[4] https://github.com/BVLC/caffe/wiki/Model-Zoo.
[5] http://www.image-net.org/challenges/LSVRC/.
[6] http://scikit-learn.org/.
[7] http://caffe.berkeleyvision.org/.

152   T. Trzciński et al.

Table 1. Popularity prediction results. Classification accuracy is defined as a proportion of the videos correctly classified as popular or unpopular. Spearman correlation serves as a complementary quality evaluation metric and it is computed between the probability of a video belonging to a popular class and true normalized popularity score. We also report standard deviation values.

Model	Features	Classification accuracy	Spearman correlation
Logistic regression	HOG	0.587 ± 0.006	0.229 ± 0.014
	GIST	0.609 ± 0.007	0.321 ± 0.008
	CaffeNet	0.622 ± 0.007	0.340 ± 0.007
	ResNet	0.645 ± 0.005	0.393 ± 0.010
SVM	HOG	0.616 ± 0.004	0.359 ± 0.008
	GIST	0.609 ± 0.006	0.294 ± 0.012
	CaffeNet	0.653 ± 0.003	0.395 ± 0.007
	ResNet	0.650 ± 0.007	0.387 ± 0.015
Popularity-LRCN	Raw video frames	$\mathbf{0.7 \pm 0.003}$	$\mathbf{0.521 \pm 0.009}$

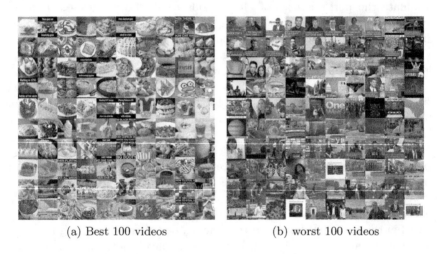

(a) Best 100 videos   (b) worst 100 videos

Fig. 1. Results of Popularity-LRCN classification. A set of 100 thumbnails with the highest (a) and lowest (b) probability of popular class.

5 Conclusions

In this paper, we proposed a new approach to the task of online video content popularity prediction called Popularity-LRCN that relies on a deep recurrent neural network architecture. It uses only visual cues present in the representative frames of the video and outputs the predicted popularity class along with the corresponding probability. This method can therefore be used to compare videos in terms of their future popularity before their publication. To our best knowledge, this is the first attempt to address the video popularity prediction problem in this manner. Future research includes casting visual-based popularity prediction problem as a regression problem in the domain of video viewcounts and verifying the effectiveness of deep neural network architectures in this context.

Acknowledgment. The authors would like to thank NowThisMedia Inc. for enabling this research with their hardware resources.

References

1. Bandari, R., Asur, S., Huberman, B.A.: The pulse of news in social media: forecasting popularity. CoRR abs/1202.0332 (2012)
2. Castillo, C., El-Haddad, M., Pfeffer, J., Stempeck, M.: Characterizing the life cycle of online news stories using social media reactions. In: Proceedings of ACM Conference on Computer Supported Cooperative Work and Social Computing (2014)
3. Hong, L., Dan, O., Davison, B.D.: Predicting popular messages in Twitter. In: Proceedings of International Conference Companion on World Wide Web (2011)
4. Khosla, A., Sarma, A.D., Hamid, R.: What makes an image popular? In: Proceedings of International World Wide Web Conference (WWW) (2014)
5. Pinto, H., Almeida, J.M., Gonçalves, M.A.: Using early view patterns to predict the popularity of Youtube videos. In: Proceedings of ACM International Conference on Web Search and Data Mining, pp. 365–374 (2013)
6. Tatar, A., de Amorim, M.D., Fdida, S., Antoniadis, P.: A survey on predicting the popularity of web content. J. Internet Serv. Appl. **5**, 8 (2014)
7. Chen, J., Song, X., Nie, L., Wang, X., Zhang, H., Chua, T.S.: Micro tells macro: predicting the popularity of micro-videos via a transductive model. In: Proceedings of ACM on Multimedia Conference (2016)
8. Donahue, J., Hendricks, L.A., Guadarrama, S., Rohrbach, M., Venugopalan, S., Saenko, K., Darrell, T.: Long-term recurrent convolutional networks for visual recognition and description. In: CVPR (2015)
9. Szabo, G., Huberman, B.A.: Predicting the popularity of online content. Commun. ACM **53**(8), 80–88 (2010)
10. Ramisa, A., Yan, F., Moreno-Noguer, F., Mikolajczyk, K.: Breakingnews: article annotation by image and text processing. CoRR abs/1603.07141 (2016)
11. Hochreiter, S., Schmidhuber, J.: Long short-term memory. Neural Comput. **9**(8), 1735–1780 (1997)
12. Simo-Serra, E., Fidler, S., Moreno-Noguer, F., Urtasun, R.: Neuroaesthetics in fashion: modeling the perception of fashionability. In: CVPR (2015)
13. Krizhevsky, A., Sutskever, I., Hinton, G.E.: ImageNet classification with deep convolutional neural networks. In: NIPS (2012)
14. Dalal, N., Triggs, B.: Histograms of oriented gradients for human detection. In: CVPR (2005)
15. Douze, M., Jégou, H., Sandhawalia, H., Amsaleg, L., Schmid, C.: Evaluation of GIST descriptors for web-scale image search. In: Proceedings of International Conference on Image and Video Retrieval (2009)
16. He, K., Zhang, X., Ren, S., Sun, J.: Deep residual learning for image recognition (2015). arXiv preprint arXiv:1512.03385

Intelligent Information Systems

User-Based Context Modeling for Music Recommender Systems

Imen Ben Sassi[1]([✉]), Sadok Ben Yahia[1], and Sehl Mellouli[2]

[1] Faculté des Sciences de Tunis, LIPAH-LR 11ES14,
Université de Tunis El Manar, 2092 Tunis, Tunisie
`imen.bsassi@gmail.com, sadok.benyahia@fst.rnu.tn`
[2] Department of Information Systems, Laval University,
Quebec, PQ, Canada
`sehl.mellouli@sio.ulaval.ca`

Abstract. One of the main issues that have to be considered before the conception of context-aware recommender systems is the estimation of the relevance of contextual information. Indeed, not all user interests are the same in all contextual situations, especially for the case of a mobile environment. In this paper, we introduces a multi-dimensional context model for music recommender systems that solicits users' perceptions to define the relationship between their judgment of items relevance and contextual dimensions. We have started by the acquisition of explicit items rating from a population in various possible contextual situations. Next, we have applied the Multi Linear Regression technique on users' perceived ratings, to define an order of importance between contextual dimensions and generate the multi-dimensional context model. We summarized key results and discussed findings that can be used to build an effective mobile context-aware music recommender system.

Keywords: Recommender systems · Context model · Multi Linear Regression

1 Introduction

Recommender systems are systems that produce individualized recommendations as output or those that guide a user through a personalized process of interesting or useful objects in a large space of possible options [6]. Recently, more and more industrial recommender systems have been developed in a variety of domains such as books in Amazon.com, movies in MOVIELENS, and so on. Music recommendation also represents a fascinating area which requires a special focus. With the technological progress and the spread of smartphones, a large volume of online music and digital music channels are accessible to people, for streaming and downloading, like YOUTUBE, DEEZER and SPOTIFY. SOURCE-TONE categorizes its list of songs in three classes: mood, activity and health to help users to select songs they want to listen regarding their emotional state, current activity, and health state. In the same context, LAST.FM makes use of

© Springer International Publishing AG 2017
M. Kryszkiewicz et al. (Eds.): ISMIS 2017, LNAI 10352, pp. 157–167, 2017.
DOI: 10.1007/978-3-319-60438-1_16

the user's location (respectively time) to offer him the top songs in his country (respectively upcoming events). These recommender systems have gone beyond the idea that considers the user's musical preferences as a fixed recommendation parameter and assumed that these preferences change dynamically according to his/her context. We currently know that recommender systems become more powerful as far as they integrate contextual information [3]. However, the factors that can be used, in each application domain, are not well identified. One of the main issues that have to be solved before the design of a context-aware recommender system is the estimation of the relevance of contextual information before collecting data from mobile environment, in order to minimize real data acquisition cost and enhance the recommendations quality [1,4]. Roughly speaking, it is necessary to study the dependencies between the user's musical preferences in various scenarios to adapt the recommended music to his/her context and understand how can the users's ratings change as far as his/her surrounding situation is shifted. To do so, we have collected explicit ratings from a population in various perceived contextual situations. The main objective of this paper is the definition of a context model for music context-aware recommender systems. Our methodology consists of: *(i)* selecting the music genres to represent users' musical preferences; *(ii)* identifying the contextual dimensions used to generate the context model; and *(iii)* collecting users' explicit ratings towards the selected music genres, in various perceived context situations, to define a multi-dimensional context model based on a priority order between the contextual dimensions. The remainder of the paper is organized as follows. We firstly detail the methodology that we have adapted to acquire the data describing the dependency between musical preferences and context dimensions in Sect. 2. In Sect. 3, we present the Multi Linear Regression (MLR) background. Next, in Sect. 4, we present our multi-dimensional contextual model by presenting our experimental evaluation and discuss our obtained results. Finally, we summarize our work and outline future directions in Sect. 5.

2   Methodology

The main objective of context-aware recommender systems is about adapting the recommended item to the user's contextual situation. So, they start by the acquisition of item rating in various possible contextual situations. In this paper, we rely on users' perceptions to express the role of context into their decisions. We have opted for the "perceived rating" rather than the "actual rating" because it is very difficult if not impossible to have all users in the actual context for rating. Thus, we proposed a user-based methodology aiming to assess the relationship between contextual factors and musical genres through two questions: *(i)* can a contextual factor influence users' judgment; and *(ii)* how can they rate an item in a particular perceived contextual situation.

2.1 Contextual Factors

Context Concept. Given the complexity and the broadness of the context, many definitions of this concept have been proposed in the literature. According to WORDNET SEARCH 3.1, a context is *the set of facts or circumstances that surround a situation or event.* Dey suggests the following definition of a context: *Context is any information that can be used to characterize the situation of an entity* [7]. Other approaches have defined a context by examples and properties. The authors in [13] define an entity context through five categories: individuality, activity, location, time, and relations that the entity has set up with others. However, several works have defined a context based on the application area particularities. For example, when recommending a movie, [2] have explained the user's context based on the following questions: when the movie was seen? where? and with whom?

Context Dimensions. In order to identify contextual factors that can influence mobile users' listening preferences, we have surveyed former works on recommender systems and context-aware systems literature to extract the most used contextual factors (c.f., Table 1).

Table 1. Context's dimensions

Dimension	Attribute	Dimension possible values
Temporal information	Part of the day [2]	Morning, afternoon, night
	Day of the week [2]	work day, weekend or day off
Location information	Type of location [10]	Home, work or school, eating, entertainment, recreation, shopping
Physical information	Weather [5]	Sunny, cloudy, rainy, thunderstorm, clear sky, snowing
Activity information	Activity of daily living [11]	Housework, reflection, sports, transportation, shopping, entertainment, relaxation
Emotional information	Emotions [9]	Joy, sadness, anger, fear, disgust, surprise
Social information	Companion [2]	Alone, with friends/colleagues, with children, with girlfriend/boyfriend, with family

We have modeled the context as a set of contextual dimensions as shown by Eq. 1, where $\mathcal{C}_T$ (respectively $\mathcal{C}_L$, $\mathcal{C}_P$, $\mathcal{C}_A$, $\mathcal{C}_E$, and $\mathcal{C}_S$) refers to the temporal (respectively location, physical, activity, emotional, and social) information.

$$\mathcal{C} = (\mathcal{C}_T, \mathcal{C}_L, \mathcal{C}_P, \mathcal{C}_A, \mathcal{C}_E, \mathcal{C}_S) \tag{1}$$

2.2 Music Preferences

Represent genres, artists or pieces of music liked by people. Thus, many ways can be employed to express people musical preferences using various levels of abstraction. For example, a person can express its preferences to a special music through a given song, e.g., "Simply Falling", an artist, e.g., "Iyeoka", a genre, e.g., "Jazz", a sub-genre, e.g., "Soul Jazz", or even some music attributes, e.g., "Vocal", "Instrumental" or "Afro-American". Thereby, studies have to identify the level of abstraction that will be used to categorize music. The simplest idea is to adopt the level that individuals naturally use to express their musical preferences. Music genres are considered as the optimal level of abstraction to assess people musical preferences. However, expressing musical preferences with genres assume that listeners have an acceptable knowledge about all music genres. This hypothesis makes raise a problem especially when we talk about different ages, i.e., old generation listeners are unfamiliar with new styles listened by young people. This limitation was discarded in our study as far as we have targeted a "young" population. We have also noted that there is no unique categorization of music genres. To solve this second problematic, we have chosen to start with ITUNES store[1] musical genres and validate this set through the focus group technique. Thus, we have retained 22 music genres (c.f., Table 2).

2.3 User-Based Study

In order to achieve our main goal and define an exhaustive contextual model for a music recommender system, we have been faced with the need to express the relationship between the user's context and the type of music (s)he is listening to. For example, when we are sad do we prefer to listen to sad music or to happy songs to get out of our mood? Or do some of us prefer the first category while others prefer the second style? However, it is not easy to define and describe the above relationship between music preferences and the context in which they appear. Thus, recommender systems designers need a lot of human efforts to ensure a reliable ground truth. Hence, we have initiated a survey study which asks participants to express their musical needs.

Participants. When we talk about mobile computing, it gives us the picture of young adults quite keen to adopt new technologies. Indeed, this particular population behave differently with respect to technology compared to those over 40 years old who suffer from some technological cognitive shortcomings. Many studies[2,3] were based on young adults like students and have proved that this population produces generally positive outcomes. Thus, we have invited a total of

---

[1] http://www.apple.com/itunes/.

[2] http://www.slideshare.net/digitalamysw/wearable-techineducationschmitzweiss.

[3] http://www.ipsos-na.com/news-polls/pressrelease.aspx?id=3124.

109 academics to respond to the online questionnaire designed to investigate the correlation between users' current contexts and their musical preferences. These participants included 59 women (54.1%) and 50 man (45.9%). Their average age is ranged from 17 to 36 (age:17–19: 15; age:20–29: 86; age:30–36: 8) with different educational backgrounds (college student: 69; engineer: 21; Master student: 6; Ph.D. student: 13).

Procedure. Online surveys represent an efficient and low cost way to collect data rapidly especially as we have addressed a population that have good computer skills and can easily access to Internet. Then, we have chosen the online survey and developed an online questionnaire[4]. In order to evaluate the questionnaire quality, we have considered two criteria during its design, i.e., validity and reliability. We have shared our questionnaire via an online research group. Hence, to allow the interviewees to expresses their degrees of agreement or disagreement versus a given question, we have used the Likert Scale. This latter has the advantage that it do not expect a yes/no response, but enables people to precise their degree of opinion, even when they have no opinion at all, in order to collect quantitative and subjective data.

3   Problem Representation: Multi Linear Regression

Let us consider a vector of contextual information $\mathcal{C} = (\mathcal{C}_1, \mathcal{C}_2, \ldots, \mathcal{C}_n)$ describing the user's environment, e.g., profile, task, etc. The objective of the context modeling problem is to represent these information as context dimensions and define an order of priority between them regarding the application field. Indeed, former works have combined many contextual dimensions into their recommender systems definition independently of users' judgment over these dimensions [12]. An interesting idea is to solicit the users' contribution to define the relation between their judgment of relevance and these dimensions through a multi-dimensional model. Usually, the problem of MLR [8] is used to express relationships between multiple criteria. It has been proved that MLR is very powerful when it comes to extrapolate or to generalize beyond the range of an experimental values, even with a relatively small data set. In this step, our objective is to define the relationship between musical genres and contextual factors. Let us consider the following components:

- $\mathcal{C} = (\mathcal{C}_1, \mathcal{C}_2, \ldots, \mathcal{C}_n)$ is a multi-dimensional vector of contextual dimension $\mathcal{C}_i$, where $\mathcal{C}_i$ is the ith dimension of the context $\mathcal{C}$; $i = 1, \ldots, n$; and n is the number of linear terms. In this paper, we have used 6 linear terms to describe the user's context as $\mathcal{C} = (\mathcal{C}_T, \mathcal{C}_L, \mathcal{C}_P, \mathcal{C}_A, \mathcal{C}_E, \mathcal{C}_S)$.
- $\mathcal{G} = (\mathcal{G}_1, \mathcal{G}_2, \ldots, \mathcal{G}_m)$ is the set of musical genres, where m is the number of musical genres. In our case, $\mathcal{G} = (Blues, Children's\_music, \ldots, Easy\_Listening)$.

---

[4] An English version is available on http://goo.gl/forms/xroRPBH5qs.

- $Coef_{i,j}$ is the regression coefficient to be computed representing the importance of the context dimension C_i regarding the musical genre G_j; $i = 1, \ldots, n$ and $j = 1, \ldots, m$.
- $\mathcal{P}_{Gj}$ is the user's preference of the genre G_j; $j = 1, \ldots, m$ and $\mathcal{P}_{Gj} \in [1,5]$.
- p is the number of experimental observations, i.e., 109 participants.

The regression problem is presented through Eq. 2 and allows to estimate the parameters $Coef_{i,j}$ by exploiting the totality of the p experimental observations.

$$\begin{cases} \mathcal{P}_{G1} = Coef_{1,1}C_1 + Coef_{2,1}C_2 + \ldots + Coef_{n,1}C_n \\ \mathcal{P}_{G2} = Coef_{1,2}C_1 + Coef_{2,2}C_2 + \ldots + Coef_{n,2}C_n \\ \ldots \\ \mathcal{P}_{Gm} = Coef_{1,m}C_1 + Coef_{2,m}C_2 + \ldots + Coef_{n,m}C_n \end{cases} \tag{2}$$

More specifically, we have defined the relationship between the preference of a musical genre $\mathcal{P}_{Gj}$ and the contextual dimension C_i through Eq. 3.

$$\mathcal{P}_{Gj} = Coef_T C_T + Coef_L C_L + Coef_P C_P + Coef_A C_A + Coef_E C_E + Coef_S C_S \tag{3}$$

where $\mathcal{P}_{Gj}$ is the participant's preference of the jth musical genre G_j and C_i is a contextual dimension where $i \in \{$ "Time", "Location", "Physical", "Activity", "Emotion", "Social"$\}$. Our objective is to compute the optimal values of $Coef_T$, $Coef_L$, $Coef_P$, $Coef_A$, $Coef_E$, and $Coef_S$, that will serve to attribute weights for each contextual dimension while defining the context representation and minimize the impact of experimental errors. The mathematical representation of the MLR problem is expressed through Eq. 4.

$$MLR_{\mathcal{G}} = \sum_{j=1}^{p=109} \left(y_j - \sum_{i=1}^{n=6} Coef_i \ x_{i,j} \right)^2 \tag{4}$$

where $\mathcal{G}$ stands for the 22 musical genres; p is the number of data points (i.e., 109 participants); n is the number of linear terms (i.e., 6 contextual dimensions); y_j is the jth dependant variable value; $x_{i,j}$ represents the jth measured independent variable value for the ith variable; and $Coef_i$ is the regression coefficient. We have formulated the regression problem as a Least Squares minimization problem (c.f., Eq. 5) that aims to compute the minimum values of $Coef_T$, $Coef_L$, $Coef_P$, $Coef_A$, $Coef_E$, and $Coef_S$, with respect to all the coefficients.

$$\sum_{j=1}^{p=109} x_{k,j} y_j = \sum_{i=1}^{n=6} C_i \sum_{j=1}^{p=109} x_{i,j} x_{k,j}; \quad k = 1, \ldots, n \tag{5}$$

where y_j stands for the jth dependant variable value; $x_{i,j}$ represents the jth measured independent variable value for the ith variable.

4   Results and Discussion

Understanding mobile users' musical needs is of a paramount importance task to improve the design of mobile music recommender systems. In this paper, individual music preferences are expressed by participants via the online questionnaire. More precisely, for each contextual dimension, the participant is asked to evaluate the list of music genres using a Likert preference scale, i.e., No value = "I do not know or I do not want to say", 1 = "I do not like very much", 2 = "I like a little", 3 = "I like", 4 = "I often like", and 5 = "I really like". We have analyzed the participants' responses to unveil their experiences with music and concretize the effect of contextual situations on the type of listened music.

4.1   Contextual Factors Influence

In order to identify the influence of contextual factors on people's preferred musical genres, we have asked them to express their opinions through Likert scale. Our objective is to evaluate whether a given contextual dimension has a positive influence, a negative one or have no influence on participants' judgment of music genres (c.f., Eq. 6). Hence, we have computed the difference between the participants' judgment to a musical genre, in a perceived contextual situation, denoted $pref^+$, and their judgment without taking any contextual factor into account denoted $pref^-$, to determine the type of the influence denoted I.

$$I = \begin{cases} positive\ influence & if\ pref^+ - pref^- > 0; \\ negative\ influence & if\ pref^+ - pref^- < 0; \\ no\ influence & otherwise. \end{cases} \qquad (6)$$

In Table 2, we classify for each musical genre, the most influential contextual factor C_{inf+} (respectively C_{inf-}) that have influenced positively (respectively negatively) participants' preferences, and its corresponding normalized degree of influence. These degrees of influence are computed using the arithmetic mean that represent the sum of the degrees related to all participants (c.f., Eq. 6) divided by the number of participants. The computed values were finally normalized within the unit interval using the Min-Max Feature scaling normalization. Table 2 screens out that the emotion dimension is the most influential contextual factor on people preferred genres. We have found that emotions have increased the preference of participants' and produced a positive influence to 19 musical genres. It is also worth to mention that people are likely to listen to religious music when they are in a sad mood, i.e., their preferences to religious music have increased by 0.573. However, they prefer listening to rock music over anger situation with an average growth of 0.484. For the remainder of music genres, "joy" was the first influential condition having enhanced users' preferences, i.e., this contextual condition have increased participants' judgement about 17 musical genres. We also have noticed that a variety of social encounters can differently influence people musical choices. Indeed, people tend to listen to Children's music while surrounded by kids with a normalized average value of 0.511.

Table 2. Normalized influence of contextual factors on participants' preferences

Genre	$\mathcal{C}_{inf+}$	Value	$\mathcal{C}_{inf-}$	Value
Blues	Emotion	0.523	Weather	0.322
Children's music	Companion	0.511	Weather & location	0.383
Classical	Activity	0.646	Activity	0.300
Country	Emotion	0.412	Location	0.376
Electronic	Emotion	0.425	Companion & time	0.241
Holiday	Emotion	0.464	Weather	0.303
Singer/song writer	Emotion	0.501	Location	0.367
Jazz	Emotion	0.592	Weather	0.367
Latino	Emotion	0.534	Weather & location	0.286
New age	Activity	0.622	Location & activity	0.382
Pop	Emotion	0.515	Companion	0.300
R&B/Urban	Emotion	0.544	Companion	0.294
Soundtracks	Emotion	0.571	Location	0.312
Dance	Emotion	0.543	Location	0.322
Hip Hop/Rap	Emotion	0.562	Companion & activity	0.300
Word	Emotion	0.525	Location	0.347
Alternative	Emotion	0.533	Location	0.345
Rock	Emotion	0.484	Location & companion	0.343
Religious	Emotion	0.573	location	0.278
Vocal	Emotion	0.525	Weather	0.303
Reggae	Emotion	0.601	Weather	0.386
Easy listening	Emotion	0.522	Companion	0.361

In the case of relaxation and rest situations, participants are keen to listen to Classical and New Age music with the respective improved values 0.646 and 0.622. However, many other contextual dimensions would negatively influence participants' listened music. Location followed by weather conditions are the most influential factors that have decreased participants' judgment to music genres. In fact, participants have mentioned that some types of locations may badly affect their judgments to 50% of music genres. For instance, in shopping places, people avoid listening to Children's music, Dance, and Rock music with the respective normalized lowered values 0.383, 0.322, and 0.343. Recreation places have also decreased participants preferences to Soundtracks music (normalized average value decreased with 0.312). Entertainment places also play a negative role in people listening to some music genres, i.e., New age (normalized average value decreased with 0.382) and Religious music (normalized average value decreased with 0.278). Although many musical genres represent a good motivation to accomplish tasks and activities, we have found that other genres

Table 3. Normalized relevance judgment of contextual factors

Genre	Time	Location	Physical	Activity	Emotion	Social
Blues	0.370	0.520	0.105	0.030	0.685	0.110
Children's music	0.020	0.540	0.170	0.200	0.725	0.435
Classical	0.225	0.110	0.165	0.165	0.660	0.385
Country	0.095	0.420	0.105	0.065	0.605	0.155
Electronic	0.245	0.420	0.425	0.360	0.600	0.175
Holiday	0.210	0.795	0.425	0.480	0.465	0.195
Singer/song writer	0.550	0.695	0.115	0.390	0.780	0.325
Jazz	0.055	0.160	0.020	0.225	0.645	0.180
Latino	0.140	0.255	0.005	0.015	0.735	0.010
New age	0.210	0.580	0.040	0.305	0.470	0.350
Pop	0.545	0.495	0.255	0.770	0.400	0.425
R&B/Urban	0.180	0.290	0.115	0.290	0.505	0.385
Soundtracks	0.440	0.110	0.120	0.715	0.540	0.180
Dance	0.165	0.565	0.200	0.190	0.785	0.300
Hip Hop/Rap	0.105	0.060	0.135	0.120	0.500	0.270
Word	0.010	0.780	0.005	0.370	0.750	0.405
Alternative	0.280	0.430	0.020	0.060	0.610	0.215
Rock	0.250	0.590	0.195	0.195	0.555	0.295
Religious	0.140	0.185	0.190	0.155	0.795	0.000
Vocal	0.610	0.875	0.115	0.155	0.750	0.115
Reggae	0.405	0.550	0.105	0.095	0.660	0.075
Easy listening	0.015	0.335	0.010	0.390	0.605	0.195

are not preferred while performing some activities. For example, people judgment have decreased with 0.3 to classical music while playing sport activities and to Hip Hop/Rap music while eating. In 31.82% of cases, weather conditions have played a remarkable negative influence in the choice of users to musical genres. In other situations, the temporal context has played a negative influence in participant's preferred music. In the morning, people dislike electronic music, i.e., their judgment have decreased by 0.241. Sometimes, the same contextual dimension can have both positive and negative influence. For example, in the case of Classical music, activities were the most influential contextual factors, i.e., relaxation tends to improve the participants' preferences with 0.646. However, sports activities have decreased their judgments to Classical music with 0.3. We discovered that as far as they are looking for something to listen to, participants have no specific idea about the music track or even the genre of music to choose. However, they have a particular priority scale, e.g., they want to have fun, to be entertaining, to relax, etc., that seems absolutely subjective. We noticed that these priorities are context dependent, such as time, past events, mood, people around or habits, and are expressed identically in the same situations, e.g., after a workday people are usually searching for soft music to relax when they are alone or for emotional music whenever they are with their companions.

4.2   Contextual Factors Relevance

In order to unveil the underlying relationship between each musical genre G and the different contextual dimensions, we have used the MLR technique [8].

The normalized results of the application of this measure are detailed in Table 3. For each musical genre, it lists the contextual factors and their importance on the variation of participant preferences. For example, in the case of Jazz music, we have found that the six contextual dimensions are sorted, using the importance order ($\succ$), with respect to their computed degrees of importance as: $C_E \succ C_A \succ C_S \succ C_L \succ C_T \succ C_P$. In this case, the emotional context has the most important influence (i.e., 0.645), followed by the activity-based context (i.e., 0.225). However, the physical context came in the last position (i.e., 0.020). The general relevance order of different contextual dimensions is made by combining their relevance, regarding all musical genres, and is defined as: $\mathcal{C}_E \succ \mathcal{C}_L \succ \mathcal{C}_A \succ \mathcal{C}_T \succ \mathcal{C}_S \succ \mathcal{C}_P$. The generalization of our results, detailed in Table 3, leads us to propose a multi-dimensional representation of the context without the specification of the musical genre (c.f., Eq. 7).

$$\mathcal{C} = 0.123 * \mathcal{C}_T + 0.228 * \mathcal{C}_L + 0.071 * \mathcal{C}_P + 0.134 * \mathcal{C}_A + 0.323 * \mathcal{C}_E + 0.121 * \mathcal{C}_S \tag{7}$$

5   Conclusion and Perspectives

In summary, most of the approaches that integrate contextual information in their music recommendation process are data-driven. Roughly speaking, they use contextual data without understanding their relation with music. In this paper, we introduced a knowledge-driven approach for a subjective evaluation to define a new context model for music recommender systems based on users' perceptions that express the role of context into their judgments. As a result, our study gives a sharp interest to the definition of the relationship between musical needs and contextual information that motivate these needs. The collected data can be used to generate a training model for rating prediction that associates music genres to contextual situations. In addition, the proposed model should be of great assistance to deal with the cold start problem, which occurs when a new user is registered to the system.

References

1. Adomavicius, G., Jannach, D.: Preface to the special issue on context-aware recommender systems. User Model. User-Adap. Inter. **24**, 1–5 (2014)

 2. Adomavicius, G., Sankaranarayanan, R., Sen, S., Tuzhilin, A.: Incorporating contextual information in recommender systems using a multidimensional approach. ACM Trans. Inf. Syst. **23**(1), 103–145 (2005)
 3. Adomavicius, G., Tuzhilin, A.: Context-aware recommender systems. In: Ricci, F., Rokach, L., Shapira, B., Kantor, P.B. (eds.) Recommender Systems Handbook, pp. 217–253. Springer, New York (2011). doi:10.1007/978-0-387-85820-3_7
 4. Baltrunas, L., Ludwig, B., Peer, S., Ricci, F.: Context relevance assessment and exploitation in mobile recommender systems. Pers. Ubiquitous Comput. **16**(5), 507–526 (2012)
 5. Braunhofer, M., Elahi, M., Ge, M., Ricci, F., Schievenin, T.: STS: design of weather-aware mobile recommender systems in tourism. In: Proceedings of AI*IA International Workshop on Intelligent User Interfaces, Turin, Italy, pp. 40–46 (2013)
 6. Burke, R.: Hybrid recommender systems: survey and experiments. User Model. User-Adap. Inter. **12**(4), 331–370 (2002)
 7. Dey, A.K.: Providing architectural support for building context-aware applications. Ph.D. thesis, Georgia Institute of Technology, Atlanta, GA, USA (2000)
 8. Draper, N.R., Smith, H.: Applied Regression Analysis. Wiley, New York (1998)
 9. Ekman, P., Friesen, W.V.: The repertoire of nonverbal behavior: categories, origins, usage and coding. Semiotica **1**(1), 49–98 (1969)
10. Hasan, S., Zhan, X., Ukkusuri, S.V.: Understanding urban human activity and mobility patterns using large-scale location-based data from online social media. In: Proceedings of International Workshop on Urban Computing, New York, NY, USA, pp. 1–8 (2013)
11. Jiang, S., Ferreira, J., Gonzlez, M.C.: Clustering daily patterns of human activities in the city. Data Min. Knowl. Discov. **25**(3), 478–510 (2012)
12. Schedl, M., Knees, P., McFee, B., Bogdanov, D., Kaminskas, M.: Music recommender systems. In: Ricci, F., Rokach, L., Shapira, B. (eds.) Recommender Systems Handbook, pp. 453–492. Springer, New York (2015). doi:10.1007/978-1-4899-7637-6_13
13. Zimmermann, A., Lorenz, A., Oppermann, R.: An operational definition of context. In: Kokinov, B., Richardson, D.C., Roth-Berghofer, T.R., Vieu, L. (eds.) CONTEXT 2007. LNCS, vol. 4635, pp. 558–571. Springer, Heidelberg (2007). doi:10.1007/978-3-540-74255-5_42

Uncalibrated Visual Servo for the Remotely Operated Vehicle

Chi-Cheng Cheng$^{(\boxtimes)}$ and Tsan-Chu Lu

Department of Mechanical and Electro-Mechanical Engineering,
National Sun Yat-Sen University, Kaohsiung, Taiwan, Republic of China
chengcc@mail.nsysu.edu.tw

Abstract. In this paper, an image-based uncalibrated visual servo scheme is proposed for image tracking tasks in highly disturbed environment, such as a remotely operated vehicle performing observation or investigation influenced by severe underwater current. Under the conditions of unknown target model and camera parameters, the control framework applies the scale invariant feature transform (SIFT) technique to extract image features. Furthermore, a robust adaptive control law is implemented to overcome the effect caused by camera parameters. Then by using three different types of camera's motion: pan, tilt, and zoom to achieve the visual servo mission by maintaining the target always at the central position on the image plane.

Keywords: Adaptive control · Remotely operated vehicles · Scale invariant feature transform · Visual servo

1 Introduction

About two thirds of the surface of the earth is covered by the water, where is filled with abundant nature resources. Therefore, exploration of ocean has never been stopped since human history started long time ago. In order to prevent humans from dangerous and unexpected underwater environment, the remotely operated vehicle (ROV) has been an important device for underwater explorations and inspections. While an ROV is conducting a specific mission, its positioning usually needs to be maintained to have a steady platform for effective observation and operation. The GPS technique, commonly used for our daily life, cannot be employed beneath the water because the energy of the electro-magnetic wave will be absorbed easily. Furthermore, conventional acoustic positioning methods for marine vehicles, such as the long baseline (LBL) approach, require installation of acoustic devices with significantly high cost. Fortunately, the monitoring camera on the ROV for observation and inspection can be directly applied for relative positioning purpose using the visual servo technique. Due to advantages of low lost, high flexibility, small volume, and easy for visual servo implementation, many research achievements, such as station keeping and feature tracking, have been reported [1].

All presented visual servo techniques can be categorized into two different groups, the position-based approaches and the image-based approaches. The position-based visual servo approaches estimate motion of a moving object by directly solving its

© Springer International Publishing AG 2017
M. Kryszkiewicz et al. (Eds.): ISMIS 2017, LNAI 10352, pp. 168–174, 2017.
DOI: 10.1007/978-3-319-60438-1_17

position relative to the camera [2]. As a result, a prior knowledge about the 3D camera model is usually required and feature matching plays an important role in the position-based approaches. Nevertheless, the image-based visual servo approaches extract motion information according to the difference between consecutive image frames regardless of the 3D camera model [3]. A popular method in the image-based category appears to be the optical flow technique, which can be formulated by the brightness constancy constraint stemming from the reasonable assumption of constant brightness of a particular point in an image pattern for steady lighting condition and small sampling increment. To adapt to variation of environmental illumination, a multiplier and an offset field to modify the brightness constancy equation was proposed [4]. Most visual servo schemes rely on accurate camera parameters to achieve the goal of motion estimation. Moreover, parameters calibration usually requires certain instruments and devices with extensive computation and time. Therefore, the calibration-free visual servo scheme has been attracting attention of researchers [5, 6]. This paper focuses on developing an adaptive visual servo algorithm without priori knowledge about camera's parameters to maintain a given target always to be located at the image center under the influence of unknown disturbance from ocean current.

2 The Adaptive Visual Servo Controller

Figure 1 illustrates the block diagram of the presented uncalibrated visual servo control system. The desired image is the first image taken by the camera installed on the vehicle when the mission starts. Then features matching and motion estimation by comparing the desired and the current images are performed. (x_d, y_d) and (x, y) are average positions of image features of the target image and the current image, respectively. The visual controller with an adaptive scheme is developed to control the driving motors for the camera. R_X and R_Y indicate angular velocities along tilting and panning directions. The block of zoom in/out stands for the zooming function of the camera.

The scale-invariant feature transform (SIFT) approach was chosen to implement the proposed visual servo scheme because of its reliable tracking performance [7]. The SIFT method combining techniques of the difference of Gaussian and the image pyramid is able to overcome the feature matching problems due to not only translation, rotation, and intensity variation, but also the scaling.

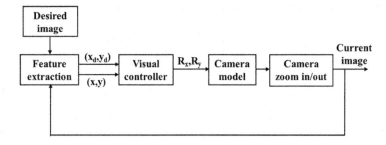

Fig. 1. Block diagram of the presented uncalibrated visual servo system

2.1   Camera's Dynamic Model

Assume a camera with focal length f moves with a linear velocity (T_X, T_Y, T_Z) and an angular velocity (R_X, R_Y, R_Z) in three-dimensional space as illustrated in Fig. 2. A fixed spatial point $P(X, Y, Z)$ is projected on the image plane at $p(x, y)$. Since the camera is installed at the bottom of an underwater vehicle, three modes of motion, which include pan, tilt, and zoom in/out, for the camera and negligible R_Z are assumed. The dynamic behavior of the camera in terms of $(\dot{x}, \dot{y})$ on the image plane can be described by the following equations [8]:

$$\dot{x} = \frac{fT_X - xT_Z}{Z} - y(\frac{x}{f}R_X + R_Z) + (f + \frac{x^2}{f})[R_Y - \tan^{-1}(\frac{x}{f})R_{zoom}] \tag{1}$$

$$\dot{y} = \frac{-fT_Y - yT_Z}{Z} + x(\frac{y}{f}R_Y + R_Z) - (f + \frac{y^2}{f})[R_X + \tan^{-1}(\frac{y}{f})R_{zoom}], \tag{2}$$

where R_{zoom} denotes the zoom factor of the camera.

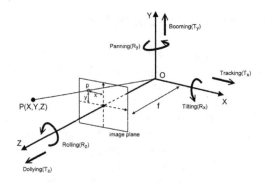

Fig. 2. Illustrative motion of the camera reference frame

2.2   Controller Design

An ideal controller using the feedback linearization approach is started first. The adaptive scheme will then be included to overcome the problem of uncertain parameters. Based on the camera's dynamic model, a matrix form can be written as follow:

$$\begin{bmatrix} \dot{x} \\ \dot{y} \end{bmatrix} = \begin{bmatrix} A_{11} & A_{12} \\ A_{21} & A_{22} \end{bmatrix} \begin{bmatrix} x \\ y \end{bmatrix} + \begin{bmatrix} B_{11} \\ B_{21} \end{bmatrix} + \begin{bmatrix} C_{11}R_y \\ C_{21}R_x \end{bmatrix}, \tag{3}$$

where

$$A_{11} = -\frac{y}{f}R_X - \frac{x}{f}R_{zoom}\tan^{-1}(\frac{x}{f}), \quad A_{22} = \frac{x}{f}R_Y - \frac{y}{f}R_{zoom}\tan^{-1}(\frac{y}{f}) \tag{4}$$

$$A_{12} = A_{21} = 0, \ C_{11} = f + \frac{x^2}{f}, \ C_{21} = -f - \frac{y^2}{f} \tag{5}$$

$$B_{11} = \frac{fT_X - xT_Z}{Z} - fR_{zoom} \tan^{-1}\left(\frac{x}{f}\right), \ B_{21} = \frac{-fT_Y - yT_Z}{Z} - fR_{zoom} \tan^{-1}\left(\frac{y}{f}\right). \tag{6}$$

Under the definitions of $e_x = x - x_d$ and $e_y = y - y_d$, the following feedback linearization controller can be derived by allowing that $\dot{e}_x + fK_1 e_x = 0$ and $\dot{e}_y + fK_2 e_y = 0$:

$$\begin{bmatrix} C_{11}R_Y \\ C_{21}R_X \end{bmatrix} = -\begin{bmatrix} A_{11} & A_{12} \\ A_{21} & A_{22} \end{bmatrix}\begin{bmatrix} e_x + x_d \\ e_y + y_d \end{bmatrix} + \begin{bmatrix} \dot{x}_d \\ \dot{y}_d \end{bmatrix} - \begin{bmatrix} B_{11} \\ B_{21} \end{bmatrix} - \begin{bmatrix} K_1 e_x \\ K_2 e_y \end{bmatrix}, \tag{7}$$

where K_1 and K_2 are positive constants to assure stability of the control system and tracking errors will diminish to zero when time goes to infinity.

The terms with T_X, T_Y and T_Z can be treated as the displacement of the underwater vehicle caused by unknown ocean current and will be modeled as unknown disturbances D_1 and D_2. Besides, the camera's focal length f and zoom factor R_{zoom} are also considered as unknown system parameters. As a result,

$$R_Y = -R_{XY}(\hat{\theta}_{11}e_x + \hat{\theta}_{16}x_d) - \tan^{-1}\left(\frac{x}{f}\right)(\hat{\theta}_{12}xe_x + \hat{\theta}_{13}xx_d + \hat{\theta}_{14}) - \hat{\theta}_{15}\dot{x}_d - \hat{\theta}_{17}R_Y x^2 - K_1 e_x - D_1 \tag{8}$$

$$R_X = R_{YX}(\hat{\theta}_{21}e_y + \hat{\theta}_{26}y_d) + \tan^{-1}\left(\frac{y}{f}\right)(\hat{\theta}_{22}ye_y + \hat{\theta}_{23}yy_d + \hat{\theta}_{24}) + \hat{\theta}_{25}\dot{y}_d + \hat{\theta}_{27}R_X y^2 + K_2 e_y + D_2, \tag{9}$$

where $\hat{\theta}_{ij}$ stands for the estimate of the parameter θ_{ij}. Although some θ_{ij} s are related to one another, all θ_{ij} s are treated as individual parameters to take into account possibly unmodeled dynamics. Since there are limitations for both pan and tilt angles, $\hat{\theta}_{1j}\tan^{-1}(x/f)$ and $\hat{\theta}_{2j}\tan^{-1}(y/f)$, $j = 2, 3, 4$ can be replaced by $1.5\text{sgn}(e_x)$ and $1.5\text{sgn}(e_y)$, respectively. In addition, unknown disturbances D_1 and D_2 are estimated by $K_3\text{sgn}(e_x)$ and $K_4\text{sgn}(e_y)$. Finally, the adaptive controllers can be obtained as

$$(\hat{\theta}_{11}e_x + \hat{\theta}_{16}x_d)yR_X + (1 + \hat{\theta}_{17}x^2)R_Y = -\hat{\theta}_{15}\dot{x}_d - 1.5\text{sgn}(e_x)(1 + |xe_x| + |xx_d| + K_3/1.5) - K_1 e_x \tag{10}$$

$$(1 - \hat{\theta}_{27}y^2)R_X - (\hat{\theta}_{21}e_y + \hat{\theta}_{26}y_d)xR_Y = \hat{\theta}_{25}\dot{y}_d + 1.5\text{sgn}(e_y)(1 + |ye_y| + |yy_d| + K_4/1.5) + K_2 e_y \tag{11}$$

and the corresponding adaptation laws are

$$\dot{\hat{\theta}}_{11} = R_X y e_x, \ \dot{\hat{\theta}}_{15} = \dot{x}_d, \ \dot{\hat{\theta}}_{16} = R_X y x_d, \ \dot{\hat{\theta}}_{17} = x^2 R_Y \tag{12}$$

$$\dot{\hat{\theta}}_{21} = R_Y x e_y, \ \dot{\hat{\theta}}_{25} = \dot{y}_d, \ \dot{\hat{\theta}}_{26} = R_Y x y_d, \ \dot{\hat{\theta}}_{27} = y^2 R_X. \tag{13}$$

It can be shown that the overall adaptive control system is asymptotically stable using the Lyapunov theorem by allowing the Lyapunov function to be defined as

$$V = \frac{1}{2f}(e_x^2 + e_y^2) + \sum_{i=1,5,6,7} (\theta_{1i} - \hat{\theta}_{1i})^2 + \sum_{i=1,5,6,7} (\theta_{2i} - \hat{\theta}_{2i})^2 \qquad (14)$$

and $K_3 > \|D_1\|$ and $K_4 > \|D_2\|$.

After the control of pan-tilt of the camera has been done, the camera's zooming control can be accomplished by examining distance variation between image features. The calculation of the zoom factor is based on the ratio of the sum of distances between features of the target image to the current image, i.e.,

$$\text{Zoom factor} = \sum_{\text{target image}} \overline{P_i P_{i-1}} \bigg/ \sum_{\text{current image}} \overline{P_i' P_{i-1}'}, \qquad (15)$$

where P_i and P_i' represent the i-th matched image features for the target image and the current image, respectively.

3   Simulation Studies

In order to verify the effectiveness of the proposed visual servo algorithm, simulation studies on underwater pipeline following were conducted by incorporating 3D virtual underwater environment generated by Autodesk 3D Studio Max into the Virtual Reality Toolbox of Matlab/Simulink.

The simulations were performed assuming that the underwater vehicle was at the height of 3 m above the seafloor initially with its camera pointing downwards. The camera owns three operational modes, which are pan, tilt and zoom in/out. A constant magnitude external velocity 0.5 m/s in all directions was considered to simulate a possible ocean current in the water. Besides, a Gaussian white noise, with 10 dB of the signal to noise ratio, was intentionally added to reflect actual undersea situation.

Control performance will be demonstrated by average position of image features as depicted in Fig. 3. Solid and dashed curves represent average locations of all feature

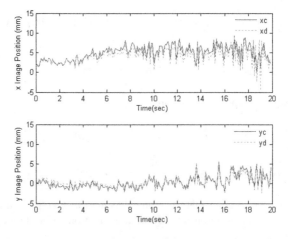

Fig. 3. Image tracking performance of the presented control scheme

points in current image and the target image, respectively. It appears that the image position was successfully kept at the center of the image plane and those two curves are close to each other, which verify the robustness of the proposed control scheme. Variations of the pan-tilt angle and the zoom factor of the camera during the pipeline following task are presented in Fig. 4. In order to maintain the camera always aims at the pipeline, both the pan and tilt angles of the camera need to be adaptively adjusted. Furthermore, to achieve visual servo for the desired image command, the zoom factor of the camera has to be enlarged when the underwater vehicle moves away from the pipeline. Apparently, the number of matched features declines when the zoom factor increases.

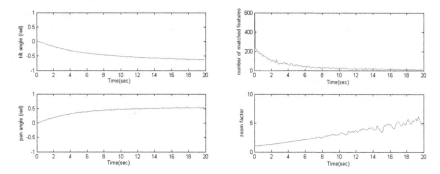

Fig. 4. Adjustments of the camera and number of matched features

Figure 5 exhibits the sampled images taken by the camera without and with the proposed visual servo control scheme for the case of constant disturbance. It appears that the underwater pipeline was located outside the image plane without control at 20th

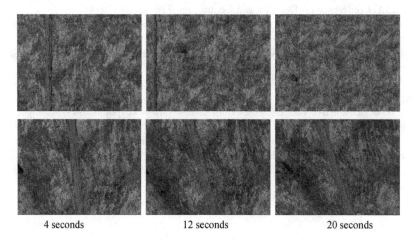

| 4 seconds | 12 seconds | 20 seconds |

Fig. 5. Images taken by the camera without (top) and with (bottom) the proposed visual servo control scheme

second. Significant control performance using the proposed visual servo control framework is clearly demonstrated.

4   Conclusions

Satisfactory visual servo performance usually cannot be expected without camera calibration. The camera calibration process, identifying both intrinsic and extrinsic parameters for the camera reference system, reduces simplicity and applicability of most existing visual servo control schemes. Therefore, a uncalibrated adaptive visual servo framework was developed to release the constraint of camera calibration. The scenario of station-keeping of a underwater pipeline for a remotely operated vehicle was chosen for performance demonstration. In order to accomplish inspection or observation, the remotely operated vehicle needs to follow a prescribed target, i.e., the underwater pipeline in this study. The camera on the underwater vehicle was assumed to own three types of adjustment, i.e., pan, tilt, and zoom, to follow the pipeline lying on the sea floor. A constant external disturbance with additive Gaussian noises was considered to justify the robustness of the proposed visual servo algorithm. Simulation results exhibit excellent visual servo performance on pipeline following tasks.

References

1. Chaumette, F., Hutchinson, S.: Visual servo control. IEEE Robot. Autom. Mag. **13**, 82–90 (2006)
2. Thuilot, B., Martinet, P., Cordesses, L., Gallice, J.: Position based visual servoing: keeping the object in the field of vision. In: IEEE International Conference on Robotics and Automation, pp. 1624–1629 (2002)
3. Chaumette, F.: Visual servoing using image features defined upon geometrical primitives. In: 33rd IEEE Conference on Decision and Control, pp. 3782–3787 (1994)
4. Negahdaripour, S., Xu, X., Jin, L.: Direct estimation of motion from sea floor images for automatic station-keeping of submersible platforms. IEEE J. Ocean. Eng. **24**, 370–382 (1999)
5. Piepmeier, J.A., McMurray, G.V., Lipkin, H.: Uncalibrated dynamic visual servoing. IEEE Trans. Robot. Autom. **20**, 143–147 (2004)
6. Liu, Y.H., Wang, H.S., Wang, C.Y., Lam, K.K.: Uncalibrated visual servoing of robots using a depth-independent interaction matrix. IEEE Trans. Robot. **22**, 804–817 (2006)
7. Lowe, D.G.: Distinctive image features from scale-invariant keypoints. Int. J. Comput. Vis. **60**, 91–110 (2004)
8. Xiong, W., Lee, J.C.-M.: Efficient scene change detection and camera motion annotation for video classification. Comput. Vis. Image Underst. **71**, 166–181 (1998)

Comparative Analysis of Musical Performances by Using Emotion Tracking

Jacek Grekow[(⊠)]

Faculty of Computer Science, Bialystok University of Technology,
Wiejska 45A, 15-351 Bialystok, Poland
j.grekow@pb.edu.pl

Abstract. Systems searching musical compositions on Internet databases more and more often add an option of selecting emotions to the basic search parameters, such as title, composer, genre, etc. Finding pieces with a similar emotional distribution throughout the same composition is an option that further extends the capabilities of search systems. In this study, we presented a comparative analysis of musical performances by using emotion tracking. A dimensional approach of dynamic music emotion recognition was used in the analysis. Values of arousal and valence, predicted by regressors, were used to compare performances. We analyzed the emotional content of performances of 6 musical works. The obtained results confirm the validity of the assumption that tracking and analyzing the values of arousal and valence over time in different performances of the same composition can be used to indicate their similarities.

Keywords: Emotion tracking · Musical performances · Similarity

1 Introduction

Systems searching musical compositions on Internet databases more and more often add an option of selecting emotions to the basic search parameters, such as title, composer, genre, etc. Finding pieces with a similar emotional distribution throughout the same composition is an option that further extends the capabilities of search systems. In this paper, we present a computer system that enables finding which performances of the same composition are closer to each other and which are quite distant in terms of shaping emotions over time. We analyzed 6 musical works, of which there were 5 different versions.

Comparisons of multiple performances of the same piece often focused on piano performances [5,9]. Tempo and loudness information were the most popular characteristics used for performance analysis. They were used to calculate correlations between performances in [9,10]. In the study [5] tempo and loudness derived from audio recordings were segmented into musical phrases, and then clustering was used to find individual features of the pianists' performances. Four selected computational models of expressive music performance were reviewed in [12]. In addition, research on formal characterization of individual performance

© Springer International Publishing AG 2017
M. Kryszkiewicz et al. (Eds.): ISMIS 2017, LNAI 10352, pp. 175–184, 2017.
DOI: 10.1007/978-3-319-60438-1_18

style, like performance trajectories and performance alphabets, was presented. A method to compare orchestra performances by examining a visual spectrogram characteristic was proposed in [7]. Principal component analysis on synchronized performance fragments was applied to localize areas of cross-performance variation in time and frequency. A connection between music performances and emotion was presented in [3], where a computer program was used to produce performances with different emotional expression. The program used a set of rules characteristic for each emotion (fear, anger, happiness, sadness, solemnity, tenderness), which were used to modify such parameters of MIDI files as tempo, sound level, articulation, tone onsets and delays.

2   Music Data for Regressor Training

In our approach, emotion recognition was treated as a regression problem. The data set for regressor training consisted of 324 6-second fragments of different genres of music: classical, jazz, blues, country, disco, hip-hop, metal, pop, reggae, and rock. The tracks were all 22050 Hz mono 16-bit audio files in .wav format.

Data annotation of perceived emotion was done by five music experts with a university musical education. During annotation of music samples, we used the two-dimensional arousal-valence (A-V) model to measure emotions in music [8]. The model consists of two independent dimensions of valence (horizontal axis) and arousal (vertical axis).

Each person making annotations, after listening to a music sample, had to specify values on the arousal and valence axes in a range from -10 to 10 with step 1. On the arousal axis, a value of -10 meant low while 10 high arousal. On the valence axis, -10 meant negative while 10 positive valence. The data collected from the five music experts was averaged.

The amount of examples in the quarters on the A-V emotion plane w: Q1 (A:high, V:high): 93, Q2 (A:high, V:low): 70, Q3 (A:low, V:low): 80; Q4 (A:low, V:high): 81. Pearson correlation coefficient was calculated to check if valence and arousal dimensions are correlated in our music data. The obtained value $r = -0.03$ indicates that arousal and valence values are not correlated, and the music data are a good spread in the quarters on the A-V emotion plane. This is an important element according to the conclusions formulated in [1]. Considering the internal reliability of the collected data, arousal and valence both showed high mutual consistency, with a Cronbach's α of 0.98 and 0.90, respectively.

The previously prepared, labeled by A-V values, music data set served as input data for the tool used for feature extraction. For feature extraction, we used a tool for audio analysis Essentia [2]. The obtained by Essentia length of feature vector was 530 features.

3   Regressor Training

We built regressors for predicting arousal and valence. For training and testing, the following regression algorithms were used: SMOreg, REPTree, M5P. Before constructing regressors arousal and valence annotations were scaled between $[-0.5, 0.5]$. We evaluated the performance of regression using the tenfold cross validation technique (CV-10).

The highest values for determination coefficient (R^2) were obtained using SMOreg (implementation of the support vector machine for regression). After applying attribute selection (attribute evaluator: WrapperSubsetEval, search method: BestFirst), we obtained $R^2 = 0.79$, for arousal and $R^2 = 0.58$ for valence. Mean absolute error reached values $MEA = 0.09$ for arousal and $MEA = 0.10$ for valence. Predicting arousal is a much easier task for regressors than valence and the values predicted for arousal are more precise.

4   Analyzed Performances

The collection of analyzed performances consisted of the following compositions by Frédéric Chopin (1810–1849): Prelude in C major, Op.28, No.1; Prelude in D major, Op.28, No.5; Prelude in F minor, Op.28, No.18; Prelude in C minor, Op.28, No.20 (the first 8 bars). All the analyzed Chopin performances were audio recordings played by 5 famous pianists: Artur Rubinstein recorded in 1946; Emil Gilels recorded in 1953; Grigory Sokolov recorded in 1990; Martha Argerich recorded in 1997; Rafał Blechacz recorded in 2007. Additionally, we analyzed approximately one-minute long beginnings of two symphonies by Ludwig van Beethoven (1770–1827): Symphony No.5 in C minor, Op.67 (the first 58 bars) and Symphony No.3 in E-flat major, Op.55 'Eroica' (the first 46 bars) performed by 5 different orchestras.

We analyzed 6 musical works, with 5 different performances of each. Detailed results are available on the web[1].

5   Results

We used the best obtained regression models for predicting arousal and valence of the musical performances. The performances were divided into 6-second segments with a 3/4 overlap. For each segment, features were extracted and models for arousal and valence were used. As a result, we obtained arousal and valence values for every 1.5 seconds of a musical piece [6].

During the comparison of different performances of the same composition, it is very important to compare the same musical fragment of these performances. Thus, we implemented a module for the alignment of audio recordings in our system. We used MATCH [4], a toolkit for accurate automatic alignment of different renditions of the same piece of music.

---

[1] http://aragorn.pb.bialystok.pl/~grekowj/HomePage/Performances.

5.1   Performances and Arousal, Valence over Time

Due to the limited nature of our research, we have restricted the analysis and presentation of results to 5 different performances of Prelude in C major, Op.28, No.1 by Frédéric Chopin.

Observation of the course of arousal in the performances (Fig. 1a) shows that the performance by G. Sokolov had significantly lower values than the remaining pieces. The reason for this is that the performance was played at a slower pace and lower sound intensity. Between samples 15 and 20, there was a clear rise in arousal for all performances, but the performance by E. Gilels achieved maximum (Arousal = 0.3). Also, the performance by E. Gilels is the most dynamically aroused in this fragment. It is not always easy to detect on the graph which performances are similar. However, we can notice a convergence of lines between A. Rubinstein and G. Sokolov (samples 30–50), and between R. Blechacz and E. Gilels (samples 37–50).

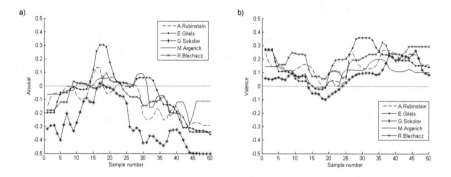

Fig. 1. Arousal (a) and valence (b) over time for 5 performances of Prelude in C major, Op.28, No.1 by Frédéric Chopin

Observation of the course of valence in the performances (Fig. 1b) shows that there was a similar decrease in this value for samples 15–20 in all performances. There is a similar line shape, i.e. good correlation, between R. Blechacz and G. Sokolov (samples 5–20), and between M. Argerich and E. Gilels (samples 35–50).

Another possibility of comparing performances is to take arousal and valence into account simultaneously. To compare performances described by two sequences of values (arousal and valence), these sequences should be joined. In order to join the two sequences of arousal and valence values of one performance, standard deviation and the mean of the two sequences of data should be equivalent. We decided to leave arousal values without change and convert valence values; although we could have converted arousal and left valence without a change and this would not have affected the correlation results.

Table 1 presents the Pearson correlation coefficients r for joined arousal and valence, calculated for each pair of performances. We see that the most similar

Table 1. Correlation coefficient r for joined arousal and valence calculated for each pair of performances of Prelude in C major, Op.28, No.1 by F. Chopin

	A. Rubinstein	E. Gilels	G. Sokolov	M. Argerich	R. Blechacz
A. Rubinstein	1.00	0.58	**0.81**	0.67	0.77
E. Gilels	0.58	1.00	0.60	0.72	0.64
G. Sokolov	**0.81**	0.60	1.00	0.64	**0.82**
M. Argerich	0.67	0.72	0.64	1.00	0.72
R. Blechacz	0.77	0.64	**0.82**	0.72	1.00

performances are by R. Blechacz and G. Sokolov ($r = 0.82$) and the most different by A. Rubinstein end E. Gilels ($r = 0.58$). It can be stated that in terms of arousal and valence we have two groups of performances. The first group consists of performances by R. Blechacz, G. Sokolov and A. Rubinstein, and the second group by E. Gilels and M. Argerich.

5.2 Arousalscape, Valencescape and AVscape

Comparing musical performances by using correlation coefficient r calculated for the whole length of a composition is only a general analysis of similarities between recordings. In order to analyze similarities between the performances in greater detail, we used scape plot.

Scape plot is a plotting method that allows presenting analysis results for segments of varying lengths on one image. Their advantage is that they enable seeing the entire structure of a composition. The scape plotting method was designed by Craig Sapp for structural analysis of harmony in musical scores [11].

In this paper, a scape plot is used to present calculated correlations between analogous segments of the examined recordings. A comparison of sequences of emotional features (arousal and valence) in different performances of the same composition is a novel application for scape plots.

Figure 2a presents a method for creating a scape plot to analyze a sample composition consisting of 5 elements: a, b, c, d, e. These elements are first examined separately, and then grouped by sequential pairs: ab, bc, cd, de. Next, three-element sequences are created: abc, bcd, cde; followed by four-element sequences: $abcd, bcde$; and finally one sequence consisting of the entire composition: $abcde$. The obtained sequences are arranged on a plane in the form of a triangle, where at the base are the analysis results of examining the shortest sequences and at the top of the triangle are the results of analyzing the entire length of the composition. In a scape plot, the horizontal axis represents time in the composition, while the vertical axis represents the length of the analyzed sequence.

Figure 2b shows the Arousalscape, a scape plot generated for the arousal value sequence for 5 different performances of Prelude in C major, Op.28, No.1 by Frédéric Chopin. The performance by A. Rubinstein was selected as the reference performance. Arousalscape illustrates which performances are the most correlated to A. Rubinstein, by different lengths of the examined sequences. In the lower levels of the triangle, it is difficult to choose the winner; but in the higher levels of the triangle, the winner is unequivocal: the performance by G. Sokolov. His area in the Arousalscape is the biggest and reaches a value of 59%. It is interesting that on the first half of Prelude, the performance by E. Gilels is the most similar to A. Rubinstein. The remaining two performances (M. Argerich, R. Blechacz) were covered by the first two winners during comparison with the reference performance (A. Rubinstein).

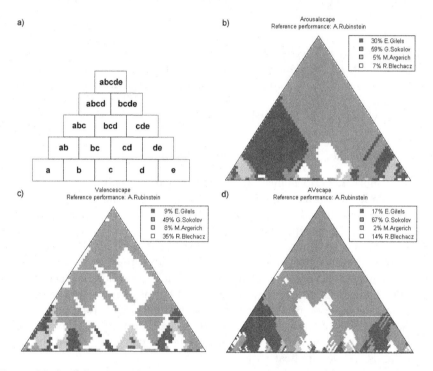

Fig. 2. Method for creating a scape plot from 5 elements (a); Arousalscape (b), Valencescape (c) and AVscape (d) for Prelude in C major, Op.28, No.1, reference performance: A. Rubinstein

Figure 2c shows the Valencescape. The situation here is not as unequivocal as for Arousalscape for the same composition (Fig. 2b). Once again the winner was the performance by G. Sokolov, but its area on the Valencescape is smaller (49%). The triangle area is ruptured by the color white, which represents the performance by R. Blechacz (35% of the area). This means that this performance was also well correlated with the reference performance at many moments.

Figure 2d shows the AVscape, a scape plot generated for sequences of joined arousal and valence values. The definitive winner of the comparisons of correlation values with various lengths of sequences is the performance by G.

Sokolov (67% of the area), with E. Gilels in second place (17% of the area), and R. Blechacz in third place (14% of the area). At the beginning of the composition, the reference performance by A. Rubinstein is similar to E. Gilels (dark colored left lower part of the triangle), and in the middle of the composition to R. Blechacz (white space in the middle lower part of the triangle).

When comparing 5 different performances of one compositions, we can create 5 AVscapes, 5 Arousalscapes, and 5 Valencescapes. Each rendition is consecutively selected as the reference performance and the degree of similarity is visualized in detail in comparison with the other 4 performances.

6   Evaluation

6.1   Parameters Describing the Most Similar Performances

To find the most similar performance to the reference performance, you can use several indicators that result from the construction of the obtained scape plots [10]:

- Score S_0 - the most general result indicating the most similar performance. The winner is the one with the best correlation for the entire sequence, entire length of time of the composition. On the scape plot, it is the top element of the triangle.
- Score S_1 - indicates the performance with the biggest area in the scape plot.
- Score S_2 - the next best similar performance from the scape plot, calculated after removing the winner S_1. If 2 performances are very similar, then one will always win and cover the wins of the second. To calculate the S_2 score, a new scape plot is generated without the winner as indicated by S_1. Shows the performance with the biggest area in the newly created scape plot.

6.2   Ground Truth for Performing Similarity Assessments Between Performances

Three music experts with a university musical education participated in the experiment. We used a similarity matrix form [5 × 5] to collect expert opinions, which is a symmetric matrix, with the main diagonal values equal to 10. The experts' task was to determine which renditions of a given performance are the most similar and which are the least. Each expert had to determine these values in the similarity matrix form for each composition. Each value described the degree of similarity between performances. The filled values were subsequent natural numbers in the range of [1, 9], where 9 meant a very similar performance, and 1 very different. A value of 10 on the diagonal meant that the given performance is maximally similar. Information on the performer of a given rendition was kept from the evaluators.

After collecting data from the experts, the data was ranked. This way we eliminated different individual opinions on the similarity scale and maintained the sequence of degree of similarity. Finally, the obtained values were averaged and rescaled to the range of $[1, 9]$. Thus we built a similarity matrix obtained from experts. It constituted the ground truth for similarities for a given composition and was used to compare with the matrix of similarity between performances obtained by a computer system.

6.3   Evaluation Parameters

Spearman's Rank Correlation Coefficient. The first evaluation parameter we used was Spearman's rank correlation coefficient (Spearman's ρ), which is the Pearson correlation coefficient between ranked variables. Before calculating correlations, variables are converted to ranks. The rank correlation coefficient ranges from 1 to -1. A positive ρ indicates a positive relationship between the two variables, while a negative ρ expresses a negative relationship.

In our case, Spearman's ρ measures how much the similarity values provided by the computer system and experts have a similar rank. While calculating Spearman's ρ between the similarity matrix obtained from experts and the similarity matrix obtained from the computer system, only elements below the main diagonal from the matrix were taken into account. Matrixes are symmetric; diagonal and upper diagonal elements are irrelevant. The greater the obtained Spearman's rank correlation coefficient, the closer the system's results were to the experts' opinions.

Maximal Similar Number of Hits. The next parameters evaluated the concordance of the indicators on the similarity matrix obtained from the experts and the similarity matrix obtained from the system. We compared indicators of the most similar performers according to the experts and the system. First, from among the experts' opinions we found the most similar performance to the reference performance, and then checked if it was confirmed by the system. If the indicators from both sides were convergent, we had a hit. The comparisons were performed for all reference performances, and the result was a percentage of hits - MSH (maximal similar hits) defined in Eqs. 1 and 2.

$$MSH = \frac{\sum_{i=1}^{n} H_i}{n} \times 100\% \qquad (1)$$

$$H_i = \begin{cases} 1 & \text{if } MS_i(EX) = MS_i(CS) \\ 0 & \text{if } MS_i(EX) \neq MS_i(CS) \end{cases} \qquad (2)$$

where EX is the similarity matrix obtained from experts, CS is the similarity matrix obtained from the computer system, $MS_i()$ is the most similar performance to the reference performance i, and n is the number of performances.

Calculating MSH, we can compare the similarity matrix obtained from the experts to the similarity matrix obtained from the system on the basis of different indicators: S_0 or S_1 (Sect. 6.1).

To check if the searched most similar performance indicated by the experts is in the top indications by the computer system, we introduced a variant of the previous parameter - $MSH2F$ (maximal similar hits 2 first). The $MSH2F$ calculation checks if the most similar performance according to the experts is among the top 2 indicated by the system. In the case of comparison with the results obtained on the scape plot, the first 2 most similar performances are indicated by S_1 and S_2 ($MSH2F_{S_1 S_2}$).

Evaluation Results. The obtained results of the evaluation are presented in Table 2. The first columns present Spearman's ρ calculated for the results between the experts' opinions and three similarity matrices obtained from the system: arousal-valence similarity matrix ρ_{AV}, arousal similarity matrix ρ_A, and valence similarity matrix ρ_V. The positive Spearman's ρ values (avg. $\rho_{AV} = 0.53$) indicate a clear relation and accordance with the experts' opinions and the computer system's calculations. Spearman's ρ taking into account arousal and valence ρ_{AV} obtained better results on average than Spearman's ρ for arousal and valence separately ($\rho_A = 0.46$, $\rho_V = 0.47$).

Table 2. Evaluation parameters for the analyzed compositions

Composition	ρ_{AV}	ρ_A	ρ_V	MSH_{S_0} %	$MSH2F_{S_0}$ %	MSH_{S_1} %	$MSH2F_{S_1 S_2}$ %
Prelude No.1	0.52	0.60	0.52	40	80	40	80
Prelude No.5	0.72	0.30	0.61	40	80	40	60
Prelude No.18	0.50	0.69	0.30	80	80	80	100
Prelude No.20	0.55	0.88	0.43	40	100	60	100
Symphony No.3	0.58	−0.12	0.74	20	80	20	80
Symphony No.5	0.34	0.41	0.22	20	60	40	40
Averages	0.53	0.46	0.47	40	80	47	77

MSH and $MSH2F$ were calculated between the experts' opinions and the arousal-valence similarity matrix. Analyzing the indicators for the most similar performance according to the experts as well as the system, the average accuracy of the applied method was 40% when using score S_0, and 47% score S_1. However, the higher values of avg. $MSH2F_{S_0}$ and avg. $MSH2F_{S_1 S_2}$ (80% and 77%) indicate that the results provided by the experts are in the top results obtained from the system.

7   Conclusions

In this study, we presented a comparative analysis of musical performances by using emotion tracking. Use of emotions for comparisons is a novel approach, not found in other hitherto published papers. Values of arousal and valence, predicted by regressors, were used to compare performances. We analyzed the emotional content of performances of 6 musical works. We found which performances of the same composition are closer to each other and which are quite distant in terms of the shaping of arousal and valence over time. We evaluated the applied approach comparing the obtained results with the opinions of music experts. The obtained results confirm the validity of the assumption that tracking and analyzing the values of arousal and valence over time in different performances of the same composition can be used to indicate their similarities. The presented method gives access to knowledge on the shaping of emotions by a performer, which had previously been available only to music professionals.

Acknowledgments. This research was realized as part of study no. S/WI/3/2013 and financed from Ministry of Science and Higher Education funds.

References

1. Aljanaki, A., Yang, Y.H., Soleymani, M.: Emotion in music task: lessons learned. In: Working Notes Proceedings of the MediaEval 2016 Workshop, Hilversum, The Netherlands (2016)
2. Bogdanov, D., Wack, N., Gómez, E., Gulati, S., Herrera, P., Mayor, O., Roma, G., Salamon, J., Zapata, J., Serra, X.: ESSENTIA: an audio analysis library for music information retrieval. In: Proceedings of the 14th International Society for Music Information Retrieval Conference, pp. 493–498, Curitiba, Brazil (2013)
3. Bresin, R., Friberg, A.: Emotional coloring of computer-controlled music performances. Comput. Music J. **24**(4), 44–63 (2000). Winter
4. Dixon, S., Widmer, G.: MATCH: a music alignment tool chest. In: Proceedings of 6th International Conference on Music Information Retrieval, ISMIR 2005, London, UK, 11–15 September 2005, pp. 492–497 (2005)
5. Goebl, W., Pampalk, E., Widmer, G.: Exploring expressive performance trajectories: six famous pianists play six chopin pieces. In: Proceedings of the 8th International Conference on Music Perception and Cognition (ICMPC 8), Evanston, IL, USA, pp. 505–509 (2004)
6. Grekow, J.: Music emotion maps in arousal-valence space. In: Saeed, K., Homenda, W. (eds.) CISIM 2016. LNCS, vol. 9842, pp. 697–706. Springer, Cham (2016). doi:10.1007/978-3-319-45378-1_60
7. Liem, C.C.S., Hanjalic, A.: Comparative analysis of orchestral performance recordings: an image-based approach. In: Proceedings of the 16th International Society for Music Information Retrieval Conference, ISMIR 2015, Málaga, Spain, pp. 302–308 (2015)
8. Russell, J.A.: A circumplex model of affect. J. Pers. Soc. Psychol. **39**(6), 1161–1178 (1980)
9. Sapp, C.S.: Comparative analysis of multiple musical performances. In: Proceedings of the 8th International Conference on Music Information Retrieval, ISMIR 2007, Vienna, Austria, pp. 497–500 (2007)
10. Sapp, C.S.: Hybrid numeric/rank similarity metrics for musical performance analysis. In: 9th International Conference on Music Information Retrieval, ISMIR 2008, Drexel University, Philadelphia, PA, USA, 14–18 September 2008, pp. 501–506 (2008)
11. Sapp, C.S.: Harmonic visualizations of tonal music. In: Proceedings of the 2001 International Computer Music Conference, ICMC 2001, Havana, Cuba (2001)
12. Widmer, G., Goebl, W.: Computational models of expressive music performance: the state of the art. J. New Music Res. **33**(3), 203–216 (2004)

Automated Web Services Composition with Iterated Services

Alfredo Milani[1] and Rajdeep Niyogi[2(✉)]

[1] Department of Mathematics and Computer Science,
University of Perugia, 06123 Perugia, Italy
milani@unipg.it
[2] Department of Computer Science and Engineering,
Indian Institute of Technology Roorkee, 247667 Roorkee, India
rajdpfec@iitr.ac.in

Abstract. In the last decade there has been a proliferation of web services based application systems. In some applications (e.g., e-commerce, weather forecast) a web service is invoked many times with different actual parameters to obtain a composed service. In this paper we introduce the notion of iterated services that are obtained from given atomic services by iteration. The iterated services provide compact and elegant solutions to such complex composition problems that are unsolvable using the existing approaches. We define a new service dependency graph model to capture web services with sets of objects as input/output. We give a translation of the web services composition problem to a planning problem. Finally, we transform a plan to a composed web service. We have implemented our approach using the BlackBox planner.

Keywords: Web services composition · Iterated services · Automated planning

1 Introduction

Let us consider a typical e-commerce scenario that is used as the running example in this paper. Bob would like to buy a new smartphone. He looks for shops that sell such smartphones. He then makes inquiry from the shops regarding the price and the delivery time. He goes for the shop for which the price and the delivery time look suitable for him. A complete description of this domain is given in Fig. 1. We assume that the following atomic web services are available.

ws1: search-item input: an item; output: a set of shops that sell the item.

ws2: inquiry-item-shop input: an item and a shop; output: the price and delivery time of the item.

ws3: best-buy-item input: an item, a set of shops that sells the item, and the corresponding prices and delivery times; output: a shop for which price and delivery time is within a specified limit and the corresponding price.

ws4: buy-item-shop: input: an item, a shop, a price of item, a valid credit card; output: gives a receipt.

© Springer International Publishing AG 2017
M. Kryszkiewicz et al. (Eds.): ISMIS 2017, LNAI 10352, pp. 185–194, 2017.
DOI: 10.1007/978-3-319-60438-1_19

The objective is to obtain a composed web service that finds a suitable shop using the given services *ws1, ws2, ws3, ws4*.

Let us see why it is not easy to build a composed service, if one exists, from the given services, for the example scenario. Let *ws1* returns as output a set of shops. The service *ws2* returns as output the price and delivery time corresponding to a shop and an item. Now *ws3* takes as input a set of shops that sells the item and the corresponding prices (a set of prices) and delivery times (a set of delivery times). The task of *ws3* is to return as output a suitable shop and the corresponding price by comparing the price and delivery time of the different shops. The outputs of *ws2* are two objects (price and delivery time) whereas the inputs of *ws3* are sets of objects (a set of prices and a set of delivery times). Thus the output of *ws2* cannot be matched with the input of *ws3*.

Several automated planning based intricate and powerful techniques [1–6] as well as graph search based techniques [7,8] have been suggested for solving complex service composition problems. However, none of these approaches are able to solve the above composition problem.

So we introduce the notion of iterated services to handle service composition problems as in the example scenario. An iterated service is fundamentally different from repetition of the same action or service several times as in [5,9,10]. The composed web service for the example is *ws1;ws5;ws3;ws4*, where *ws5* is an iterated service obtained from *ws2* (see Fig. 4).

The contributions of our paper are: (i) develop the notion of iterated services (Sect. 3), (ii) develop a new service dependency graph model to accommodate iterated services (Sect. 2), and (iii) suggest a technique that translates such web services composition problem to a planning problem which is then solved using a classical planner (Sect. 4).

The rest of the paper is organized as follows. In Sect. 5 we give a PDDL description of an e-commerce domain and the implementation details. We present the related work and the concluding remarks in Sect. 6.

2   Proposed Service Dependency Graph Model

A Service Dependency Graph (SDG) is a directed graph that shows dependencies among Inputs and Outputs of the web services. SDGs modeled as AND/OR graphs have been used for web service composition [7,8]. We extend the classical SDGs [7,8] to handle iterated services.

2.1   Service Dependency Graph for Iterated Web Services (SDG-Iter)

Definition 1. An SDG is a directed bipartite graph defined by a tuple $\langle W, V, E \rangle$ where W is a set of nodes representing the set of atomic web services, V is a set of nodes representing the set of input and output types of all the atomic web services, E is a set of oriented edges between nodes of W and V.

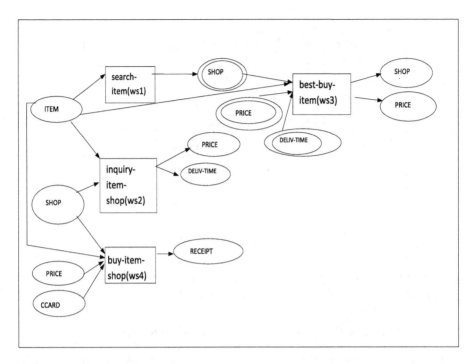

Fig. 1. Service dependency graph of an e-commerce domain (double circle: set of objects of the labelled type; single circle: an object of the labelled type)

Definition 2. An SDG-Iter is defined by a tuple $\langle W \cup W_{iter}, V, E, \mu, L_{in}, L_{out} \rangle$ where W, V, E are as in an SDG, W_{iter} is the set of iterated web services obtained by iterating a service from the set W, μ is a boolean function on $V \times V$, such that $\mu(v', v'')$ returns true *iff* the type v' is a set of objects where each object is of type v'', L_{in}, L_{out} are the set of input and output logical formulas of a web service $w \in W \cup W_{iter}$.

The dependency graph is built by adding to E an edge $e = (v_{in}, w)$ for each input parameter v_{in} and an edge $e = (w, v_{out})$ for each output parameter v_{out} of web service $w \in W$. We discuss how iterated services are constructed from atomic services in the following section.

The semantics of edges and nodes is analogous to classical service dependency AND/OR graphs, where service nodes W correspond to AND nodes (shown in boxes, see Fig. 2), since all the inputs in the incoming edges should be available to run the services, and V nodes correspond to OR node (shown in circles), since at least one service on the starting edge of incoming links should be executed in order to produce the needed data type. A set of objects has type T^* when each object of the set has an atomic type T (e.g., integer, string, float). We represent nodes with type T^* by a double circle. In Fig. 2 a dotted arch point out the potential connection between the output parameter of $w4$ of type M and the

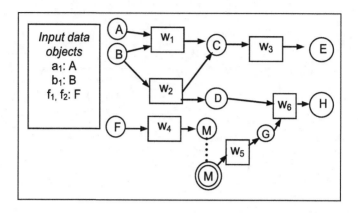

Fig. 2. A potential iterated web service ($w4$)

input parameter of $w5$ of type M^*, since $\mu(M^*,\ M) = true$. This indicates that $w4$ may be iterated to obtain $w4\text{-}iter$ which may then be composed with $w5$.

2.2   Web Services Composition Problem for SDG-Iter

A composition problem for SDG-Iter is described by input/output pair (In, Out), $In = \{in_1 : T_{in_1}, \ldots, in_m : T_{in_m}; P\}$ and $Out = \{out_1 : T_{out_1}, \ldots, out_n : T_{out_n}; Q\}$ where $in_i : T_{in_i}$ ($out_i : T_{out_i}$) denotes the type of an input (output) object, and P, Q are logical formulas (predicates) specifying the properties of the input and output respectively.

3   Iterated Web Service

In general there are many different ways in which a service, with an output of atomic type T, can be iterated in order to produce an output of type T^*. For instance, consider a 2-input service w_1 with inputs of atomic types $A,\ B$ and output of atomic type C, represented as $w_1(a, b, c)$. There are three possible iterated versions (iteration on A, iteration on B, iteration on both A and B) and for each a corresponding composition program is shown in Fig. 3.

Iteration Lemma. A web service w with input $In = \{in_1 : T_{in_1}, \ldots, in_k : T_{in_k}, \ldots, in_m : T_{in_m}; P\}$, output $Out = \{out_1 : T_{out_1}, \ldots, out_n : T_{out_n}; Q\}$ where the types are atomic, and logical specifications P, Q, iterated on any input variable in_k will result in a new service w_{iter} with input $In = \{in_1 : T_{in_1}, \ldots, in_k : T^*_{in_k}, \ldots, in_m : T_{in_m}; P\}$ and output $Out = \{out_1 : T^*_{out_1}, \ldots, out_n : T^*_{out_n}; Q\}$, and the resulting service can be implemented by a program in a service composition language which admits iteration and creation of set of objects.

Proof. It can be constructively shown the existence of the program generalizing the example of Fig. 3. We omit the details due to space constraints.   □

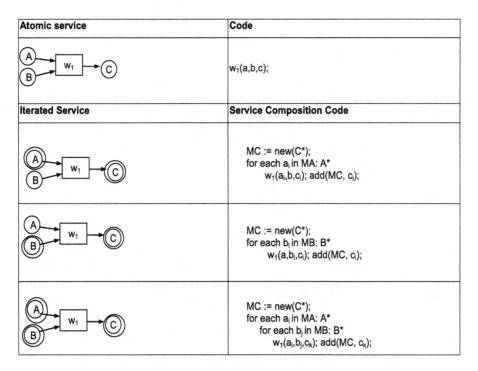

Fig. 3. Iterated Web Service: an example

The iteration lemma can be easily extended to any number of input variables. In general, any web service w with input types $T_{in_1}, \ldots, T_{in_n}$ and output types $T_{out_1}, \ldots, T_{out_n}$ can be iterated with $2^n - 1$ distinct inputs in order to obtain output type $T^*_{out_1}, \ldots, T^*_{out_n}$. Although, in theory, the number of iterated services would be exponential, in practical cases the number of iterated services increases by a small bounded factor since web services are usually characterized by a limited number of parameters. Since the inputs and outputs of a service w are atomic types, so we cannot obtain set of sets as an output type in the corresponding iterated service w_{iter} (see Fig. 3).

The SDG-Iter for the e-commerce domain is shown in Fig. 4 where each service contains input/output logical formulas. Whereas, $have(s, i)$ is a predicate to denote that a shop s has an item i, $have(S, i)$ is a predicate to denote that a set of shops S has an item i. In Fig. 4, $ws5$ is an iterated service obtained from $ws2$ using the construction shown in Fig. 3.

SDG-Iter contains iterated services that are obtained by iteration of atomic services in a given SDG. The aim is to allow solutions to problems which were previously unsolvable in the original service dependency graph without any iteration. The following theorem holds.

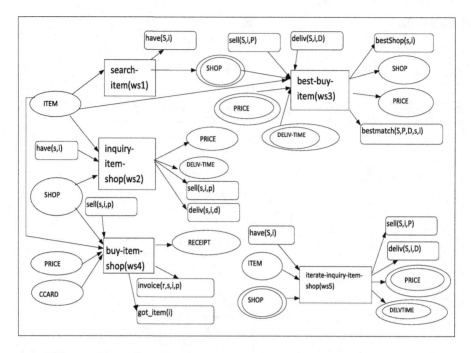

Fig. 4. SDG-Iter for the e-commerce domain (S, P, D denote set of objects; s, p, i, d, r denote an object)

Theorem 1. Let γ_{iter} denote an SDG-Iter derived from a given SDG γ. A service composition solution obtained in γ_{iter} can be implemented using only services in γ.

Proof. It follows from the Iteration Lemma.   □

An important consequence of the theorem is that solutions for γ_{iter} (obtained by a classical planner, say BlackBox [11]) can be easily mapped to solutions for γ by simple code substitution.

3.1 Advantage of Iterated Web Service

Let us consider a web service *wf(x)* which returns the weather forecast of a given city *x*, and if we want to obtain the weather forecast of three cities, say Boston, Madrid, and Tokyo then two possible compositions are:

(1) *wf(Boston); wf(Madrid); wf(Tokyo)* and
(2) *for each x in S wf(x)* where $S = \{Boston, Madrid, Tokyo\}$

Although (1) and (2) are equivalent from the execution point of view, composition (2) is more compact and general. In fact if the number of cities change, the instruction in composition (1) would change while that in (2) remains unchanged. Moreover, solution (2) only requires that the set S to be finite but the objects in the set need not be known *a priori*.

4   Translation: Web Services Composition Problem for SDG-Iter to a Planning Problem

Given an SDG-Iter, a services composition problem (In, Out) can be encoded as an equivalent planning problem (D, I, G) by iteratively applying the following domain and problem encoding rules, where D is a planning domain, I is an initial state, and G is a goal state. A solution plan P is defined as a totally ordered sequence of actions that if executed in the initial state I will produce a state where all the predicates in G are true.

Planning Domain Encoding Rules

r1. Data types. For each v in V introduce an unary predicate p_v in D.

r2. Logical specifications. For each n-ary logical label p introduce an n-ary predicate p in D.

r3. Service specification. For each w in W with $In = \{\ldots; P\}$ and $Out = \{\ldots; Q\}$ introduce an operator w with preconditions P and effects Q where P and Q are the logical specifications.

r4. Service input parameters. For each $e = (v, w)$ in E add a new parameter o to w and a precondition $v(o)$. If $\mu(v, v') = true$, add $v'(o) \wedge set(o)$ to the preconditions.

r5. Service output parameters. For each $e = (w, v)$ in E add to w a parameter o, a precondition $output(o)$ and effects $v(o) \wedge \neg output(o)$. If $\mu(v, v') = true$, add $v'(o) \wedge set(o) \wedge \neg output(o)$ to the effects of w. The predicate $output(o)$ is used to model output variables in planning.

r6. Set manipulation. For each type T, include dummy operators new_T, add_T and a predicate $element$. new_T has precondition $output(o)$ and effects $v(o) \wedge set(o) \wedge \neg output(o)$; add_T has preconditions $set(o) \wedge v(o)$ and effect $element(o)$. Actions new and add allow to build sets starting from elementary elements.

For instance, service w_1 in Fig. 4 will be encoded as an action w_1 with precondition $A(a) \wedge B(b) \wedge P(a, b) \wedge output(c)$ and effect $C(c) \wedge Q(a, b, c) \wedge \neg output(c)$.

In the rules we used the predicate set, as in [12], to explicitly state that there is a set of objects. Thus $v(o) \wedge set(o)$ means that there is a set of objects whose type is v. For instance, the effect of the action corresponding to the web service *search-item* would be specified in PDDL (Fig. 5) as $(and\ (Set\ ?s)(shop\ ?s))$.

Problem Encoding Rules

r7. Problem input. The problem input $In = o_1 : T_1, \ldots, o_m : T_m; P$ is encoded by creating an initial state I which contains a predicate $T_i(o_i)$ for each input object of type T_i, and if $\mu(T_i, T_i') = true$, add $T_i'(o_i) \wedge set(o_i)$ to the effects. P is also added to the initial state I.

r8. Problem output. The problem output $Out = o_1 : T_1, , o_m : T_m, Q$ is encoded by creating a dummy goal g and a dummy action *dummy-out-goal* which contains as preconditions: the logical specification Q, predicates $T_i(o_i)$ as well as $output(o_i)$ for each object of type T_i in Out. The effects of *dummy-out-goal* will contain $\neg output(o_i)$ for each output object o_i of type T_i, and if $\mu(T_i, T_i') = true$, add $T_i'(o_i) \wedge set(o_i)$ and the dummy goal g.

Note that a sufficiently large number of facts $output(o)$ with distinct symbols o should be defined in the initial state I in order to implement planning with output variables as in [2].

5   Implementation and PDDL Specification

We give the implementation details of the web services composition for the *e-commerce* domain. The SDG as well as SDG-Iter are also provided. The composition problem has been encoded as a planning problem in Planning Domain Definition Language (PDDL) (Fig. 5) by repeatedly applying the rules r1-r8 given in Sect. 4. Experiments were performed on a personal computer, with Intel Core2 Duo CPU, 1.8 GHz, 1 GB RAM, Windows XP, by using the Blackbox planner [11].

The SDG and SDG-iter for the e-commerce domain are shown in Fig. 1 and Fig. 4 respectively. The SDG-Iter composition problem for the *e-commerce* domain is given by $In = \{i1 : item,\ c1 : card\}$, $Out = \{s : shop,\ bestShop\ (s,\ i1),\ got\_item(i1)\}$, i.e., starting with an item description $i1$ and a credit card $c1$, we want to end up with having the item bought from the best shop.

```
e-commerce Domain
(define (domain purchase)
  (:action search-item    :parameters (?i ?s)
     :precondition (and  (item ?i)  (output ?s))
     :effect (and (Set ?s) (shop ?s ) (have ?s ?i) (not (output ?s)) ))

  (:action inquiry-item-shop :parameters (?i ?s ?p ?d)
     :precondition (and (item ?i) (shop  ?s) (have ?s ?i) (output ?p) (output ?d)
     :effect (and (sell ?s ?i ?p) )(deliver ?s ?i ?d) (price ?p)(delivery-time ?d)
            (not (output ?p)) (not (output ?d)) ))

 (:action buy-item-shop :parameters (?i ?s ?p ?c ?r )
     :precondition (and (item ?i) (shop ?s) (card ?c) (sell ?s ?i ?p) (output ?r) )
     :effect ( and  (invoice ?r ?s ?i ?p) (got_item ?i)  (not (output ?r)) ))

 (:action best-buy-item  :parameters (?i ?s1 ?p1 ?d1 ?s ?p)
     :precondition (and (item ?i) (Set ?s1) (shop ?s1) (Set ?p1) (price ?p1) (Set ?d1)
           (delivery-time ?d1) (sell ?s1 ?i ?p1) (deliver ?s1 ?i ?d1) (output ?s)(output ?p)
     :effect ( and (best-match  ?s1 ?p1 ?d1 ?s ?p) (shop ?s)(price ?p) (bestShop ?s ?i)
            (not (output ?s))(not (output ?p)) ))

 (:action iterate-inquiry-item-shop :parameters (?i  ?s ?p ?d )
     :precondition (and (item ?i) (Set ?s) (shop ?s) (have ?s ?i) (output ?p) (output ?d))
     :effect (and (sell ?s ?i ?p) (deliver ?p ?i ?d) (Set ?p) (price ?p)
            (Set ?d)(delivery-time ?d)  (not (output ?p)) (not (output ?d))))

 (:action dummy-out-goal :parameters (?s)
     :precondition (and(got_item i1) (bestShop ?s i1) (output ?s))
     :effect  (and (dummy_goal ) (not (output ?s))
------------------------------------------------------------------
Initial State:  (item i1) (card c1)
------------------------------------------------------------------
Goal State:  (got_item i1) (bestShop ?s i1) // see dummy action dummy-out-goal
```

Fig. 5. PDDL specification of the e-commerce domain

The solution plan for the e-commerce domain generated by BlackBox planner is $P = search\text{-}item(i1, s1); iterate\text{-}inquiry\text{-}item\text{-}shop(i1, s1, p1, d1); best\text{-}buy\text{-}item(i1, s1, p1, d1, s2, p2); buy\text{-}item\text{-}shop(i1,s2,p2,c1,r1;dummy\text{-}out\text{-}goal(s2)$ where $s1, s2, p1, p2, d1, r1$ are the symbols used to bind output objects to input parameters of services. Finally, the transformation of plan (sequence of actions) to a composed web service (sequence of web services), is obtained by substituting an action with its corresponding code as given in Fig. 3. For example, the action *iterate-inquiry-item-shop* will be expanded by the following code fragment:

$p1 := new(Price*); d1 := new(Deliv\_time*)$
*for each s_temp in s1:Shop**
    {*inquiry-item-shop(i1, s_temp, p_temp, d_temp)*;
    add(p_temp, p1); add(d_temp, d1);}

Some additional dummy actions and predicates have been introduced in the planning domains in order to overcome the limitation of available PDDL planners. Output variables have been implemented by the special predicate (*output ?x*) which denotes that the symbol x is an output variable [2]. Moreover every action using a new symbol s negates (*output ?s*) in its effects, thus avoiding that further actions use the same symbol to denote a different output. A dummy action *dummy-out-goal* is used to represent output objects in goals, where the expression containing the output variable is used in the preconditions.

It has been observed that even in the case of problems which were solvable in an SDG with classical graph extraction techniques, the proposed method gives more compact solutions in the corresponding SDG-Iter. For example, a long sequence of service calls (say, $w(o_1), \ldots, w(o_n)$) is equivalently replaced by a single iterated service over a set, which is not only compact but it also reflects better the logic underlying the solution.

6   Concluding Remarks

It is worth noticing that an iterated web service in our work is very different from repetition of the same action or service several times as in [5,9,10]. In [9], plan synthesis with repeating actions is studied, whereas we considered synthesis of plans with iterative actions—that can be instantiated to sets of objects. In order to analyze the dynamic behavior of web services, a Petri net based algebra has been suggested in [10]. A major limitation of the repetition operator in [10] is that the operator merely consists of repetition of the same service call on the same parameters, and no form of dynamic instantiation is provided. Repetition of a same action has also been addressed in [5] in the context of temporally extended goals. However such actions cannot be instantiated on sets of objects.

On the other hand the ability of operating on sets and fulfilling goals expressed on sets allows to manage a more compact and expressive representation as well as to manage and to solve more complex problem descriptions.

We considered a web services composition problem where some web services are invoked many times with different actual parameters. In order to model such services, we introduced the notion of iterated services that are obtained by iteration of atomic services. Such services provide compact and elegant solutions to composition problem, that would be unsolvable using the existing approaches [1–8]. Moreover, an iterated service only requires that a set be finite but the objects in the set need not be known *a priori*. We have solved such composition problems using a translation to a planning problem. The proposed technique has been implemented by plan compilation with output variables. To the best of our knowledge, no other proposal of this kind for iterated services has been investigated. We did not consider conditional constructs in the service composition problem; this would be part of our future work.

Acknowledgements. The authors thank the anonymous reviewers of ISMIS2017 for their valuable suggestions.

References

1. Bertoli, P., Pistore, M., Traverso, P.: Automated composition of web services via planning in asynchronous domains. Artif. Intell. **174**(3), 316–361 (2010)
2. Milani, A., Rossi, F., Pallottelli, S.: Planning based integration of web services. In: Proceedings of the IEEE/WIC/ACM international conference on Web Intelligence and Intelligent Agent Technology, pp. 125–128. IEEE Computer Society (2006)
3. Sirin, E., Parsia, B., Wu, D., Hendler, J., Nau, D.: HTN planning for web service composition using shop2. Web Semant.: Sci. Serv. Agents World Wide Web **1**(4), 377–396 (2004)
4. Sohrabi, S., McIlraith, S.A.: Preference-based web service composition: a middle ground between execution and search. In: Patel-Schneider, P.F., Pan, Y., Hitzler, P., Mika, P., Zhang, L., Pan, J.Z., Horrocks, I., Glimm, B. (eds.) ISWC 2010. LNCS, vol. 6496, pp. 713–729. Springer, Heidelberg (2010). doi:10.1007/978-3-642-17746-0_45
5. Traverso, P., Pistore, M.: Automated composition of semantic web services into executable processes. In: McIlraith, S.A., Plexousakis, D., Harmelen, F. (eds.) ISWC 2004. LNCS, vol. 3298, pp. 380–394. Springer, Heidelberg (2004). doi:10.1007/978-3-540-30475-3_27
6. Zou, G., Lu, Q., Chen, Y., Huang, R., Xu, Y., Xiang, Y.: Qos-aware dynamic composition of web services using numerical temporal planning. IEEE Trans. Serv. Comput. **7**(1), 18–31 (2014)
7. Liang, Q.A., Su, S.Y.: AND/OR graph and search algorithm for discovering composite web services. Int. J. Web Serv. Res. **2**(4), 48 (2005)
8. Yan, Y., Xu, B., Gu, Z.: Automatic service composition using AND/OR graph. In: 10th IEEE Conference on E-Commerce Technology and the Fifth IEEE Conference on Enterprise Computing, E-Commerce and E-Services, pp. 335–338. IEEE (2008)
9. Levesque, H.J.: Planning with loops. In: IJCAI, pp. 509–515 (2005)
10. Hamadi, R., Benatallah, B.: A petri net-based model for web service composition. In: Proceedings of the 14th Australasian Database Conference-Volume 17, pp. 191–200. Australian Computer Society Inc. (2003)
11. Blackbox. http://www.cs.rochester.edu/~kautz/satplan/blackbox
12. Niyogi, R., Milani, A.: Planning with sets. In: Esposito, F., Pivert, O., Hacid, M.-S., Raś, Z.W., Ferilli, S. (eds.) ISMIS 2015. LNCS (LNAI), vol. 9384, pp. 399–409. Springer, Cham (2015). doi:10.1007/978-3-319-25252-0_43

On the Gradual Acceptability of Arguments in Bipolar Weighted Argumentation Frameworks with Degrees of Trust

Andrea Pazienza[✉], Stefano Ferilli, and Floriana Esposito

Dipartimento di Informatica, Università di Bari, Bari, Italy
{andrea.pazienza,stefano.ferilli,floriana.esposito}@uniba.it

Abstract. Computational models of argument aim at engaging argumentation-related activities with human users. In the present work we propose a new generalized version of abstract argument system, called Trust-affected Bipolar Weighted Argumentation Framework (T-BWAF). In this framework, two mainly interacting components are exploited to reason about the acceptability of arguments. The former is the BWAF, which combines and extends the theoretical models and properties of bipolar and weighted Argumentation Frameworks. The latter is the Trust Users Graph, which allow us to quantify gradual pieces of information regarding the source (who is the origin) of an argument. The synergy between them allow us to consider further gradual information which lead to a definition of intrinsic strength of an argument. For this reason, the evaluation of arguments for T-BWAF is defined under a *ranking-based semantics*, i.e. by assigning a numerical acceptability degree to each argument.

1 Introduction

Argumentation is a major component of our everyday lives, in that people try to convince each other by using arguments to sustain their respective point of views. Many strategies can be found in the literature for the identification of the successful arguments in an argumentation dispute context [2,5–7,10]. One of the most influential computational models of argument was presented by Dung's Argumentation Frameworks [6] (in short, AF), which is roughly a directed graph where the vertices are the abstract arguments and the directed edges correspond to attacks between them. As there are no restrictions on the attack relation, cycles, self–attackers, and so on, are all allowed. Abstract arguments are intended as arguments whose internal structure or specific interpretation is ignored. The abstract nature of the arguments, and the relationship with non-monotonic reasoning formalisms, yield a very simple and quite general model that allows to easily understand which sets of arguments are mutually compatible. The precise conditions for their acceptance are defined by the semantics. Semantics produce acceptable subsets of the arguments, called *extensions*, that correspond to various positions one may take based on the available arguments.

© Springer International Publishing AG 2017
M. Kryszkiewicz et al. (Eds.): ISMIS 2017, LNAI 10352, pp. 195–204, 2017.
DOI: 10.1007/978-3-319-60438-1_20

Actually, another family of acceptability semantics has been defined recently, namely ranking-based semantics [1], in which argument cannot only be accepted or rejected, but using a membership criterion, a qualitative acceptability degree is assigned to each argument.

As a counterpart of this generality, abstract argument systems feature a limited expressiveness and can hardly be adopted as a modeling formalism directly usable in application contexts. Therefore, basic AFs may not necessarily be the best target systems for the representation of a real discussion. In order to address this problem, a research direction is to extend AFs by equipping relations with more expressive concepts such as support relations, giving rise to Bipolar Argumentation Frameworks (BAFs) [5], and weighted attacks, giving rise to Weighted Argumentation Frameworks (WAFs) [7].

During an argumentation process, whether an argument is finally accepted or not depends, among other things, on the truth behind the knowledge used to build the argument. More than often, the truth of the knowledge that is used to build the arguments may depend on the source of that knowledge. The stream of information received from an informant together with its reputation are the only elements a human user (or an agent) can use to decide whether an argument, built from that information, can be accepted or not. The use of trust models for this purpose is straightforward. Extending this kind of evaluation to an argument, we could evaluate the truthfulness of it.

Therefore, our purpose is to explore definitions of acceptability of arguments using a combination of standard dialectical acceptability and trustworthiness/reputation of the source of the argument. Moreover, considering social relations in the acceptance of arguments is important too. For instance, arguments produced by someone holding the authority usually produce a bigger change in the beliefs of the hearer. Exploring the social dimension seems an interesting line of work that has not been studied in depth yet.

This paper is organized as follows. Section 2 briefly recalls the background on abstract argumentation and subsequent generalizations of Dung's Framework. Section 3 introduces our proposal, i.e. the T-BWAF with its two major components BWAF and Trust Users Graph and their useful properties, and a new ranking-based semantics dealing with it. Finally, Sect. 4 concludes the paper.

2   Background and Related Work

Let us start by providing the basics of Abstract Argumentation.

Definition 1. *An* Argumentation Framework *(**AF**) is a pair* $F = \langle \mathcal{A}, \mathcal{R} \rangle$*, where* $\mathcal{A}$ *is a finite set of arguments and* $\mathcal{R} \subseteq \mathcal{A} \times \mathcal{A}$*. Given two arguments* $a, b \in \mathcal{A}$*, the relation* $a\mathcal{R}b$ *means that* a *attacks* b*.*

There are a few central concepts when evaluating the justification of an argument:

Definition 2. *Given an AF $F = \langle \mathcal{A}, \mathcal{R} \rangle$, and $S \subseteq \mathcal{A}$:*

- *S is conflict-free if $\nexists a, b \in S$ s.t. $a\mathcal{R}b$;*
- *$a \in \mathcal{A}$ is defended by S if $\forall b \in \mathcal{A}: b\mathcal{R}a \Rightarrow \exists c \in S$ s.t. $c\mathcal{R}b$;*
- *$f_F: 2^{\mathcal{A}} \mapsto 2^{\mathcal{A}}$ s.t. $f_F(S) = \{a \mid a \text{ is defended by } S\}$ is called the* characteristic function *of F;*
- *S is admissible if S is conflict-free and S is defended by itself, i.e. $\forall a \in S, \forall b \in \mathcal{A}: b\mathcal{R}a \Rightarrow \exists c \in S$ s.t. $c\mathcal{R}b$;*
- *S is a complete extension iff S is admissible and $S = f_F(S)$;*
- *S is a grounded extension iff S is the $\subseteq$-minimal complete extension;*
- *S is a preferred extension iff S is a $\subseteq$-maximal complete extension;*
- *S is a stable extension iff $\forall a \in \mathcal{A}, a \notin S, \exists b \in S$ s.t. $b\mathcal{R}a$.*

While Dung's framework is both natural and powerful, the conventional solutions considered wrt this framework have obvious limitations: there can be multiple solutions or, conversely, it may result that the only solution is the empty set. There have been a number of proposals for extending Dung's framework in order to allow for more sophisticated modelling and analysis of conflicting information, in order to overcome these limitations.

A Bipolar AF (*BAF*) [5] is an extension of Dung's AF in which two kinds of interactions between arguments are possible: the attack relation and the support relation. These two relations are independent and lead to a bipolar representation of the interaction between arguments. A BAF can be represented by a directed graph in which two kinds of edges are used, in order to differentiate between the two relations. In BAFs, new kinds of attack emerge from the interaction between the direct attacks and the supports: there is a *supported attack* for an argument b by an argument a iff there is a sequence of supports followed by one attack, while, there is an *indirect attack* for an argument b by an argument a iff there is an attack followed by a sequence of supports. In particular, we assume to say that a supports b if there is a sequence of direct supports from a to b.

A Weighted AF (*WAF*) [7] is another extension of Dung's AF in which attacks between arguments are associated with a weight, indicating the relative strength of the attack. Note that allowing 0-weight attacks is counter-intuitive since it can be interpreted as absence of attack relation. In this framework, some inconsistencies are tolerated in subsets S of arguments, provided that the sum of the weights of attacks between arguments of S does not exceed a given inconsistency budget $\beta \in \mathbb{R}_*^+$. The meaning is that attacks up to a total weight of β are neglected. Dung's argument systems assume an inconsistency budget of 0, while, by relaxing this constraint, WAFs can achieve more solutions.

A Social Abstract Argumentation Framework (*SAF*) [8] is a step towards using argumentation in social network by assigning a strength to each argument, taking into account both the opinion expressed through the votes, but also the structure of the argumentation graph composed of the arguments and attacks.

In Multi-Agent Systems, trust plays an important role. The degree to which agents trust one another will inform what they believe, and, as a result the reasoning that they perform and the conclusions that they come to when that involves information from other agents. In such interactions, agents will have to reason about the degree to which they should trust those other entities, whether they are trusting those entities to carry out some task, or whether they are trusting those entities to not misuse crucial information. In [9] agents build arguments by maintaining a *trust network* of their acquaintances, which includes ratings of how much those acquaintances are trusted, and how much those acquaintances trust their acquaintances, and so on. As well, [10] associates arguments with weights that express their source's *authority degree* and defines a strategy to combine them in order to determine which arguments withstand in a dispute concerning a given domain.

For applications involving a large number of arguments, it can be problematic to have only two levels of evaluations (arguments are either accepted or rejected). For instance, such a limitation can be questionable when using argumentation for debate platforms on the Web for a discussion. In order to fix these problems, a solution consists in using semantics that distinguish arguments not with the classical accepted/rejected evaluations, but with a larger number of levels of acceptability. Another way to select a set of acceptable arguments is to rank arguments from the most to the least acceptable ones. *Ranking-based semantics* [1] aim at determining such a ranking among arguments.

Definition 3. *A* Ranking-based semantics S *associates to any argumentation framework* $F = \langle \mathcal{A}, \mathcal{R} \rangle$ *a ranking* $\succeq_F^S$ *on* $\mathcal{A}$*, where* $\succeq_F^S$ *is a preorder (a reflexive and transitive relation) on* $\mathcal{A}$*. Given two arguments* $a, b \in \mathcal{A}$*,* $a \succeq_F^S b$ *means that* a *is at least as acceptable as* b*.*

Hence, the purpose of ranking-based semantics is to sort arguments from the most to the least acceptable ones.

3   Contribution

3.1   Bipolar Weighted Argumentation Framework

A first ensemble of some models of argument is required in order to capture all the useful information from a discussion, including the strength of positive (i.e., supports) and negative (i.e., attacks) relations between arguments. The combination of BAF and WAF results suitable for this purpose.

Definition 4. *A* Bipolar Weighted Argumentation Framework *(BWAF) is a triplet* $G = \langle \mathcal{A}, \hat{\mathcal{R}}, w_{\hat{\mathcal{R}}} \rangle$*, where* $\mathcal{A}$ *is a finite set of arguments,* $\hat{\mathcal{R}} \subseteq \mathcal{A} \times \mathcal{A}$ *and* $w_{\hat{\mathcal{R}}} \colon \hat{\mathcal{R}} \mapsto [-1, 0[\,\cup\,]0, 1]$ *is a function assigning a weight to each relation.*
We define attack relations as $\hat{\mathcal{R}}_{att} = \{ \langle a, b \rangle \in \hat{\mathcal{R}} \mid w_{\hat{\mathcal{R}}}(\langle a, b \rangle) \in [-1, 0[\,\} \subseteq \hat{\mathcal{R}}$ *and support relations as* $\hat{\mathcal{R}}_{sup} = \{ \langle a, b \rangle \in \hat{\mathcal{R}} \mid w_{\hat{\mathcal{R}}}(\langle a, b \rangle) \in]0, 1] \,\} \subseteq \hat{\mathcal{R}}$*.*

The main feature of BWAF regards the assignment of negative weights for attack relations and of positive weights for support relations. Nevertheless, all BAF and WAF notions and properties are still valid in BWAF. In order to tolerate inconsistencies also for weighted support relations, we present a new extended definition of inconsistency budget.

Definition 5. *Let $G = \langle \mathcal{A}, \hat{\mathcal{R}}, w_{\hat{\mathcal{R}}} \rangle$ be a BWAF, $\hat{\mathcal{R}} = \hat{\mathcal{R}}_{att} \cup \hat{\mathcal{R}}_{sup}$, $X \subseteq \hat{\mathcal{R}}$ a relation and $\alpha \in \mathbb{R}_{<0}$, $\beta \in \mathbb{R}_{>0}$ two inconsistency budgets. We define:*

$$subatt(\hat{\mathcal{R}}_{att}, w_{\hat{\mathcal{R}}}, \alpha) = \{Y \mid Y \subseteq \hat{\mathcal{R}}_{att} \wedge \sum_{\langle a,b \rangle \in Y} w_{\hat{\mathcal{R}}}(\langle a,b \rangle) \geq \alpha\}$$

$$subsup(\hat{\mathcal{R}}_{sup}, w_{\hat{\mathcal{R}}}, \beta) = \{Z \mid Z \subseteq \hat{\mathcal{R}}_{sup} \wedge \sum_{\langle a,b \rangle \in Z} w_{\hat{\mathcal{R}}}(\langle a,b \rangle) \leq \beta\}$$

$$\mathcal{E}_\sigma^G(\langle \mathcal{A}, \hat{\mathcal{R}}, w_{\hat{\mathcal{R}}} \rangle, \alpha, \beta) =$$

$$\{S \subseteq \mathcal{A} \mid \exists R \in \{subatt(\hat{\mathcal{R}}_{att}, w_{\hat{\mathcal{R}}}, \alpha) \vee subsup(\hat{\mathcal{R}}_{sup}, w_{\hat{\mathcal{R}}}, \beta)\} \wedge S \in \mathcal{E}_\sigma(\langle \mathcal{A}, \hat{\mathcal{R}} \setminus R \rangle)\}$$

where $\mathcal{E}_\sigma(\langle \mathcal{A}, \hat{\mathcal{R}} \rangle) = \{S \subseteq \mathcal{A} \mid \sigma(S)\}$ returns the set of subsets of $\mathcal{A}$ for acceptability semantics $\sigma = \{$complete, grounded, preferred, stable$\}$.

The function $subatt$ returns the set of subsets S of $\mathcal{A}$ whose total weight does not (negatively) exceed α. While, the function $subsup$ returns the set of subsets S of $\mathcal{A}$ whose total weight does not (positively) exceed β. So, $\mathcal{E}_\sigma^G(\langle \mathcal{A}, \hat{\mathcal{R}}, w_{\hat{\mathcal{R}}} \rangle, \alpha, \beta)$ yields a subset of the power set of $\mathcal{A}$ whose elements contain only those arguments that are in relation with a weight less than α and greater than β. Therefore, a set $S \in \mathcal{E}_\sigma^G(\langle \mathcal{A}, \hat{\mathcal{R}}, w_{\hat{\mathcal{R}}} \rangle, \alpha, \beta)$ can be an extension like the standard ones in Definition 2 (or better, like in [5]) with the change in the parameters α and β.

One may wonder how the weight of attack and support relations affect the overall strength of a path between two arguments. A *path* from a to b is a sequence of nodes $s = \langle a_0, \ldots, a_n \rangle$ such that from each node there is an edge to the next node in the sequence: $a_0 = a, a_n = b$ and $\forall i < n, (a_i, a_{i+1}) \in \hat{\mathcal{R}}$. Also, a path from a to b is a *branch* if a is not attacked nor supported. In particular, it is a *support branch* (resp. *attack branch*) if a is a supporter (resp. attacker) of b.

Hence, many paths involving both attack and support relations (which hold, respectively, a negative and positive weight) may exist between two arguments. Additionally, an argument can be involved many cycles, each of which may contain, in turn, arguments involved in other cycles, and so on. For this reason, we define an operator to assess the strength propagation of weighted relations that is able to deal with cycles.

Definition 6 (Strength Propagation). *Let $G = \langle \mathcal{A}, \hat{\mathcal{R}}, w_{\hat{\mathcal{R}}} \rangle$ be a BWAF. The strength propagation (in short, sp) of a path from a starting argument $a \in \mathcal{A}$ towards an argument $b \in \mathcal{A}$ is defined as:*

$$sp(a, b) = \sum_{\langle a, \ldots, b \rangle} path\_weight(\langle a, \ldots, b \rangle) \cdot \prod_{c \in \langle a, \ldots, b \rangle} influence(c)$$

Function $path\_weight(\langle v_1, \ldots, v_n \rangle)$ computes the strength of a path $\langle v_1, \ldots, v_n \rangle$, as $\prod_{i=2}^{n} w_{\hat{\mathcal{R}}}(\langle v_{i-1}, v_i \rangle)$. Function $influence(c)$ computes the influence of an argument $c \in \mathcal{A}$ within the path on the basis of cycles to which it belongs. If it belongs to a cycle, it is computed as $\prod_{\langle c, \ldots, c \rangle} path\_weight(\langle c, \ldots, c \rangle)$, as 1 otherwise. In this way, cycles and possibly involved sub-cycles are traversed exactly once. Hence, function $sp(a, b)$ returns a positive value if there exists a defense or a support from a to b. Otherwise, it returns a negative value if there exists an attack between them. Such a function gives a measure of *overall strength* of the relationship between two arguments as part of a whole discussion.

An interesting point of view of strength propagation function is that it generalizes the notions of BAF's indirect attack and supported attack in a single definition, suitably extended in a weighted context.

Our purpose is to model an argument system that accounts for all those valuable information which can be extracted from a debate. The BWAF is a very useful argument system since it ensembles all the notions and advantages of various single extended Argumentation Frameworks. Also, it is already mappable to an online debate, in the way that it can represent an argument graph with worthwhile quantitative information. Nevertheless, in the argumentation process, a variety of further impact factors are considered in human reasoning in order to evaluate the justification of an argument. In this context, the BWAF is not able to model all those external information, but it represents a encouraging starting point to define a new wider revised argument system.

3.2   Trust-Affected BWAF

The acceptability of an argument is often affected by how reliable is the source of that argument. Quantify this information is a non-trivial task. The first step in doing this is to develop a system of argumentation that can make use of trust information. Trust is a mechanism for managing the uncertainty about autonomous entities and the information they deal with. Therefore, we start with presenting a model of trust for users involved in a debate.

Definition 7 (Trust Users Graph). *A* Trust Users Graph *is complete directed weighted graph* $T = \langle \mathcal{U}, \mathcal{E}, w_{\mathcal{U}}, w_{\mathcal{E}} \rangle$ *where* $\mathcal{U}$ *is a set of users involved in a debate,* $\mathcal{E} = \mathcal{U} \times \mathcal{U}$, $w_{\mathcal{U}} \colon \mathcal{U} \mapsto [0,1]$ *is a function assigning a weight to each user,* $w_{\mathcal{E}} \colon \mathcal{E} \mapsto [0,1]$ *is a function assigning a weight to each relation of user pairs.*

In a situation in which trust needs to be considered, it seems to be a natural step to think about how argumentation-based interactions might be extended with a model of trust. To this extent, we design the Trust Users Graph as a complete directed weighted graph. In this fashion, it allows us to retrieve two kind of quantitative information that actually take part in human argumentative reasoning, namely:

- $w_{\mathcal{U}}$ represents the *Internal Confidence* a user hold about a certain domain;
- given $u_1, u_2 \in \mathcal{U}, w_{\mathcal{E}}(u_1, u_2)$ represents the **Trust** that user u_1 has for user u_2 using a real value trust within $[0,1]$, where 0 means fully distrust, 0.5 means fully ignorant, and 1 means fully trust.

In particular, the trust values for a user are given by the edges' weight of all its adjacents. The value of trust reflects the confidence over the trust knowledge. Having collected the trust from peer users, we need to aggregate these trust values to derive a common global value for each user. This issue could be addressed by choosing an appropriate aggregation function, which gives us a notion of authority.

Definition 8 (Authority Degree). *Let* $T = \langle \mathcal{U}, \mathcal{E}, w_{\mathcal{U}}, w_{\mathcal{E}} \rangle$ *be a Trust Users Graph,* $u_i, u_j \in \mathcal{U}$, *the* Authority Degree *can be defined as:*

$$\text{authority}(u_i) = \max_i (\min_j \{w_{\mathcal{E}}(u_j, u_i), w_{\mathcal{U}}(u_i)\}) \text{ , with } i \neq j$$

The Authority Degree is a T-conorm operator which infers, starting from the internal confidence of a user in a certain topic and the trust that other users have placed in him, a degree of authority for a user in that given topic. Thus, holding the authority of each user yields a social factor which could affect the acceptability of arguments and hence it must be considered in the process of arguments evaluation.

Another information to be accounted is the relation between a user and the arguments that he carries out.

Definition 9 (User Argument Confidence). *Let* $\mathcal{A}$ *be a set of arguments,* $\mathcal{U}$ *a set of users,* $a \in \mathcal{A}, u \in \mathcal{U}$, *then the* User Argument Confidence *is defined as* conf : $\mathcal{U} \times \mathcal{A} \mapsto [0, 1]$.

It is important to point out the difference between the user's Internal Confidence and Argument Confidence: the former is related to the degree of confidence a user holds about a given topic of interest (e.g., soccer knowledge), the latter refers to the degree of confidence that a user holds putting forward an argument (e.g., the argument "I am *almost sure* that Inter Milan football club is stronger than Juventus football club" shows the defeasibility of the statement "almost sure" which gives a hint of user's confidence degree on that argument of about 0.9). Finally, we can define a revised version of the BWAF that is endowed with all the new notions we have explicited above.

Definition 10. *A Trust-affected BWAF* (***T-BWAF***) *is a tuple* $F = \langle \mathcal{A}, \hat{\mathcal{R}}, w_{\mathcal{A}}, w_{\hat{\mathcal{R}}}, \mathcal{K}, \text{conf} \rangle$, *where* $\mathcal{A}$ *is a finite set of arguments,* $\hat{\mathcal{R}} \subseteq \mathcal{A} \times \mathcal{A}$, $w_{\mathcal{A}} : \mathcal{A} \mapsto [0, 1]$ *is a function assigning a weight to each argument,* $w_{\hat{\mathcal{R}}} : \mathcal{A} \mapsto [-1, 0[\cup]0, 1]$ *is a function assigning a weight to each relation,* $\mathcal{K} = \{T_i\}_{i \in \mathcal{T}}$ *is a set of complete directed graphs, where* $\mathcal{T}$ *is a set of abstract topics,* $T_i = \langle \mathcal{U}, \mathcal{E}, w_{\mathcal{U},i}, w_{\mathcal{E},i} \rangle$ *is an instantiation of a Trust Users Graph for the topic* i *and* conf *is the User Argument Confidence function.*

Here, $\mathcal{A}, \hat{\mathcal{R}}, w_{\hat{\mathcal{R}}}$ are the same as in the BWAF. We assume that our argument system to deal with online debate platforms (web forums, blogs, social networks, etc.), thus we envision users to get some outcome for their effort. The first refinement in this scenario is to endow this framework with the social outcomes of a debate. A community of users may understand and follow a debate, so that small changes in the underlying argumentation framework and its social feedback (i.e., votes) should result in small changes to the formal outcome of the debate. Therefore, we represent this information on the arguments in the same way that [8] does, with the Simple Vote Aggregation function. So, our function $w_{\mathcal{A}}$ represents the *Crowd's Agreement* of a community. Given an argument $a \in \mathcal{A}$, the function $V^+(a)$ (resp. $V^-(a)$) denotes the number of positive (resp. negative) votes for argument a, therefore:

$$w_{\mathcal{A}}(a) = \begin{cases} 0 & V^+(a) = V^-(a) = 0 \\ \frac{V^+(a)}{V^+(a)+V^-(a)} & \text{otherwise} \end{cases}$$

The second refinement regards the distinction of arguments made by domain experts of a particular topic from those made by novices or by outsiders of the topic's domain in the argumentation. The topic represents a context in which a person shows an expertise. More confident is a person within a topic of discussion, higher will be his internal confidence in that given topic. Also, users involved in the debate may convey a trust degree for every other user. Hence, for each abstract topic $i \in \mathcal{T}$, we can instantiate a specific Trust Users Graph $T_i = \langle \mathcal{U}, \mathcal{E}, w_{\mathcal{U},i}, w_{\mathcal{E},i} \rangle$ in which we are able to map each user Internal Confidence $w_{\mathcal{U},i}$, and the Trust on each pair of users $w_{\mathcal{E},i}$. At this point, we can compute the Authority Degree for each user depending on the topic i. Authoritative evidence shows that not all opinions are equal. The opinions of experts are more convincing that are those of individuals with no specialized knowledge. In the end, what is important is not just the quality of evidence but also the credibility of the person offering it.

3.3 Intrinsic Strength of an Argument

As stated in [2], an argument has an intrinsic strength which may come from different sources: the certainty degree of its reason [3], the importance of the value it promotes if any [4], the reliability of its source [9].

From the T-BWAF, we hold all the quantitative information to address the notion of intrinsic strength for an argument: in this context, we can define an aggregation function which combines the notion of Authority Degree, User Argument Confidence and Crowd's Agreement.

Definition 11 (Argument Strength). *Let* $F = \langle \mathcal{A}, \hat{\mathcal{R}}, w_{\mathcal{A}}, w_{\hat{\mathcal{R}}}, \mathcal{K}, \text{conf} \rangle$ *be a T-BWAF,* $a \in \mathcal{A}$ *an argument,* $u \in \mathcal{U}$ *a user,* $i \in \mathcal{T}$ *a topic,* $\text{authority}_i(u)$ *the Authority Degree of user* u *for the topic* i, $\text{conf}(u, a)$ *the Argument Confidence of user* u *for argument* a, $w_{\mathcal{A}}(a)$ *the Crowd's Agreement for argument* a, *then the intrinsic Argument Strength is defined as*

$$\text{strength}(a) = \alpha \cdot \text{conf}(u, a) \cdot \text{authority}_i(u) + (1 - \alpha) \cdot w_{\mathcal{A}}(a), \text{with } \alpha \in [0, 1].$$

The parameter α allow us to balance the trade-off between Argument Confidence, Authority Degree and the appropriate feedback of the community so that we can easily assess the strength of each argument, taking into account not only the logical consequences of the debate, but also the popular opinion and all its subjectiveness.

3.4   T-Ranking Semantics

Whatever its intrinsic strength (strong or weak), an argument may be attacked or supported by other arguments. An attack amounts to undermining one of the components of an argument, and has thus a negative impact on its target. Vice versa, a support amounts to stand up for an argument, and has thus a positive impact on its target. An evaluation of the overall strength (or overall acceptability) of an argument becomes mandatory, namely for judging whether or not its conclusion is reliable.

In the following, we propose a new semantics, i.e. T-BWAF Ranking Semantics, which compares arguments by computing a ranking on them by assigning a numerical acceptability degree to each argument. Such a ranking will deal with both the intrinsic strength of arguments and the strength propagation of support and attack branches ending to it.

Definition 12. *Let* $F = \langle \mathcal{A}, \hat{\mathcal{R}}, w_{\mathcal{A}}, w_{\hat{\mathcal{R}}}, \mathcal{K}, \text{conf} \rangle$ *be a T-BWAF,* $a \in \mathcal{A}$ *an argument,* strength(a) *the intrinsic Argument Strength of* a, $sp(\cdot, a)$ *the strength propagation function for support (resp. attack) branch ending to* a *and let* $P = \{p_1, \ldots, p_n\}$ *the set of all directed paths for F. The* rank *function* $rank \colon \mathcal{A} \mapsto [0, 2]$ *is defined as:*

$$rank(a) = \begin{cases} \text{strength}(a) & \text{if } \forall x \in \mathcal{A} \colon \langle x, a \rangle \notin \hat{\mathcal{R}} \\ \frac{\text{strength}(a)}{n} \sum_{p_i = \langle \alpha_{i1}, \ldots, \alpha_{im_i} \rangle \in P} 1 + sp(\alpha_{i1}, a) & \text{otherwise} \end{cases}$$

Our approach is based two principles: the impact of unattacked argument, which play a key role in the (extension-based) acceptability of an argument, and the strength propagation of its attackers and of its supporters. It is important to note that, in considering the strength propagation function $sp(\cdot, x)$, we select only the support (resp. attack) branches (i.e., the *longest directed paths*) ending to $x \in \mathcal{A}$, in order to facing with the entire chains of arguments.

Definition 13. *The* ranking-based semantics T-Ranking *associates to any T-BWAF* $F = \langle \mathcal{A}, \hat{\mathcal{R}}, w_{\mathcal{A}}, w_{\hat{\mathcal{R}}}, \mathcal{K}, \text{conf} \rangle$ *a ranking* $\succeq_F^{rank}$ *on* $\mathcal{A}$ *such that* $\forall a, b \in \mathcal{A}, a \succeq_F^{rank} b$ *iff* $rank(a) \geq rank(b)$.

Assuming that all arguments hold an Argument Strength of 1, the rank function can be also applicable to simple BWAFs. Therefore, the T-Ranking Semantics can give gradual acceptability in both argument systems.

4   Conclusions and Future Work

This work proposed a new generalization of Dung's AF. We presented the T-BWAF, which is made up of two interacting components: the BWAF, which are able to express weighted attack/support relations, and the Trust Users Graph, a new model of Trust graph for users associated with the notions of Internal Confidence and their global level of reliability, summarized as their Authority

Degree. We defined a new characterization of inconsistency budget dealing with weighted supports. We considered also the User Argument Confidence and the Crowd's Agreement, two similar but different measure of reliability wrt an argument, depending on the source (i.e., the origin) of that information. We leverage all these gradual pieces of knowledge to a novel definition of Intrinsic Strength of an argument and exploit it together with the propagated strength of indirect relations to propose a new ranking-based semantics.

Our work opens some issues for further research. We consider only the direct trust relations between pairs of users. An important line of inquiry in this context could be what inference is reasonable in such graphs, and the propagation of trust and provenance. Another interesting frontier of the evaluation process of arguments would be to assess a ranking-degree of acceptability to extension-based semantics, dealing with a mixed approach of both ranking-based and collective justification of arguments.

Acknowledgments. This work was partially funded by the Italian PON 2007–2013 project PON02_00563_3489339 'Puglia@Service'.

References

1. Amgoud, L., Ben-Naim, J.: Ranking-based semantics for argumentation frameworks. In: Liu, W., Subrahmanian, V.S., Wijsen, J. (eds.) SUM 2013. LNCS (LNAI), vol. 8078, pp. 134–147. Springer, Heidelberg (2013). doi:10.1007/978-3-642-40381-1_11
2. Amgoud, L., Ben-Naim, J.: Axiomatic foundations of acceptability semantics. In: Proceedings of the International Conference on Principles of Knowledge Representation and Reasoning, KR, vol. 16 (2016)
3. Amgoud, L., Cayrol, C.: A reasoning model based on the production of acceptable arguments. Ann. Math. Artif. Intell. **34**(1–3), 197–215 (2002)
4. Bench-Capon, T.: Persuasion in practical argument using value-based argumentation frameworks. J. Logic Comput. **13**(3), 429–448 (2003)
5. Cayrol, C., Lagasquie-Schiex, M.C.: On the acceptability of arguments in bipolar argumentation frameworks. In: Godo, L. (ed.) ECSQARU 2005. LNCS (LNAI), vol. 3571, pp. 378–389. Springer, Heidelberg (2005). doi:10.1007/11518655_33
6. Dung, P.M.: On the acceptability of arguments and its fundamental role in nonmonotonic reasoning, logic programming and n-person games. Artif. Intell. **77**(2), 321–357 (1995)
7. Dunne, P.E., Hunter, A., McBurney, P., Parsons, S., Wooldridge, M.: Weighted argument systems: basic definitions, algorithms, and complexity results. Artif. Intell. **175**(2), 457–486 (2011)
8. Leite, J., Martins, J.: Social abstract argumentation. In: IJCAI, vol. 11, pp. 2287–2292 (2011)
9. Parsons, S., Tang, Y., Sklar, E., McBurney, P., Cai, K.: Argumentation-based reasoning in agents with varying degrees of trust. In: AAMAS-Volume, vol. 2, pp. 879–886 (2011)
10. Pazienza, A., Esposito, F., Ferilli, S.: An authority degree-based evaluation strategy for abstract argumentation frameworks. In: Proceedings of the 30th Italian Conference on Computational Logic, pp. 181–196 (2015)

Automatic Defect Detection by Classifying Aggregated Vehicular Behavior

Felix Richter[1]([✉]), Oliver Hartkopp[1], and Dirk C. Mattfeld[2]

[1] Volkswagen AG, 38440 Wolfsburg, Germany
{felix.richter,oliver.hartkopp}@volkswagen.de
[2] Technische Universität Braunschweig, Mühlenpfordtstrasse 23, 38106
Braunschweig, Germany
d.mattfeld@tu-braunschweig.de

Abstract. Detecting defects is a major task for all complex products, as automobiles. Current symptoms are the failure codes a vehicle produces and the complaints of a customer. An important part on the defect detection is the vehicular behavior. This paper highlights the analysis of vehicular data as a new symptom in the customer service process. The proposed concept combines the necessary preprocessing of vehicular data, especially the feature-based aggregation of this data, with the analysis on different sets of features for detecting a defect. In the modeling part a Support Vector Machine classifier is trained on single observed situations in the vehicular behavior and a Decision Tree is used to abstract the model output to a trip decision. The evaluation states a detection quality of 0.9418 as the F1-score.

Keywords: Support Vector Machine · Failure analysis · Data aggregation

1 Introduction

In current vehicular systems two different types of failure situations can occur. The first situation is characterized by the generation of a failure code, in technical term a diagnostic trouble code (DTC). If an electronic control unit (ECU) identifies the malfunction of the system, for example the exceeding of a parameter, it reacts with a DTC. This means a DTC is generated by a rule defined in advance. But not all failure situations can be described by a DTC, for example systematical deviations in the damper.

In the other situation the defect is characterized by a significant deviation in the vehicular behavior, but no generated DTC. The only way to analyze these situations is the modeling of the deviation and the analysis of the verbalized customer complaint. A deviation can occur within the history of the vehicle itself. In this situation a trend of deviations is detected to identify a systematical malfunction. Another possibility is to compare the history of the observed vehicle with others in a fleet. Either by comparing the behavior over a long time or by

© Springer International Publishing AG 2017
M. Kryszkiewicz et al. (Eds.): ISMIS 2017, LNAI 10352, pp. 205–214, 2017.
DOI: 10.1007/978-3-319-60438-1_21

applying a model trained on historical known situations. This paper focuses on the latter case.

The paper is structured as followed. Section 2 shows current approaches in the area of defect detection by vehicular behavior and highlights their limitations. In a next step the proposed concept is described in Sect. 3, including the problem statement, data preprocessing and modeling. For the evaluation the test setup and the results are shown in Sect. 4. The paper ends with the description of further research, especially the integration of the modeling results with existing customer service symptoms. This is highlighted in Sect. 5.

2   State of the Art

Vehicular data is produced by sensors, sensing the environmental conditions of the vehicle. Sensors are connected to ECUs and read in discrete moments of the time. Though, the continuous physical behavior of a sensor is transformed to a time series with values in the frequency of the readings. The ECU has a time-discrete representation of the sensor values. Additionally, the data is also value-discrete as the sensor values are mapped to the digital variable space of the ECU.

A related work using vehicular data as a time series for anomaly detection is given by [5]. Neural networks are trained on the relationship between the input parameters and the temperature for normal operation. Then, a dataset with anomalous temperatures is tested. For this approach data has to be available in a high frequency. In the automotive context this means the data has to be transmitted continuously to the back end or the training has to take place in the vehicle itself. Transmitting this amount of data via on-line connections is challenging. On the other side a generalized training of normal operation conditions needs more than the data of one vehicle.

In order to overcome the limitations of raw sensor data, different techniques to extract features exist. These are from the research area of data streams. An overview of summarization methods is given in [3]:

1. Sampling: Probabilistic approach to determine whether a data element will be processed or not
2. Load Shedding: Dropping of data stream sequences
3. Sketching: Random projection of a subset of features, though vertical sampling of the stream
4. Synopsis Data Structures: Describes the application of summarization techniques as wavelet analysis, histograms, quantiles and frequency moment.
5. Aggregation: Calculation of statistical measures as the mean and variance

As aggregating is a valid technique to generate features of vehicular data there are different works dealing with this concept. In [8] Support Vector Machines (SVM) are used to classify bearing faults. This is done on the Fast-Fourier Transformation, as spectral features, and a set of statistical features describing the data. In [2] the authors are using histograms, though synopsis data structures, for predicting battery lifetime. These histograms represent the relative amount a bin occurs in the observed time interval. An example is the relative amount of time the first gear is operating in a rpm interval from 500–1000 rpms. Based on the histograms a cumulative distribution and statistical features, for example the mean and standard deviation, are derived. For the prediction a Random Survival Forest is used. Another work using histograms for deviation detection is [1]. The histograms are clustered hierarchically and trend detection is done by the cluster changes over the time. The core assumption is that healthy and faulty situations are significantly differing.

The limitation of the given works is their scope. All authors take use of features extracted by vehicle data, or the high frequent vehicle data itself, in order to detect defects. But none of the works is highlighting the possible integration of their result in the current after sales data. This should be shown in this paper by two contributions. On one side the classification of defect situations is evaluated and on the other side it will be shown, as a future perspective, how to integrate this information with existing customer service data.

3   Using Vehicle Behavior for Automatic Detection of Defects

Today, defects of a vehicle can just be detected in one way. A customer identifies a malfunction of his vehicle by an indicator lamp or a different behavior of the vehicle. The only available data on these situations is the verbalized complaint of the customer and a set of DTCs. This set of information is used to detect the failures root cause in the workshop. An alternative approach is proposed in this paper. The vehicular behavior is generated while driving and analyzed for defect patterns in advance.

In order to systematically analyze the vehicular behavior, the *Cross Industry Standard Process of Data Mining* (CRISP-DM) can be used [9]. This process consists of the six phases Business Understanding, Data Understanding, Data Preparation, Modeling, Evaluation and Deployment. These steps are visualized in Fig. 1. The modeling phase has two parts, the modeling of single situations and the decision on trips. In order to describe the proposed concept in detail a description of the focused steps is given in the following.

3.1   Problem Description

This paper addresses a concept on detecting defects by analyzing the vehicular behavior. The vehicle collects data on its behavior and transmits it to the back end. In the back end a model is applied to an incoming trip in order to detect its characteristic, though whether it is normal or defect. A trip is the collection of observed situations, for example all occurred breaking procedures in a trip. The models used in the proposed approach are trained on single situations, in order to decide on their characteristic.

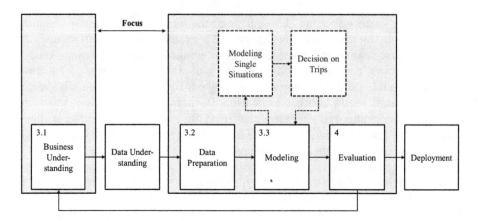

Fig. 1. CRISP-DM with the highlighted focus of this paper, the related sections and the two step approach in the modeling

The proposed concept will be evaluated on a data set for transmission problems. An unhealthy situation is characterized by high frequent amplitude changes of the rotation speed value. A trip is characterized as unhealthy with a number u of observed unhealthy situations. The reason for using multiple situation detections as the criteria is the statistic. An unhealthy situation can be reasoned by a defect on the transmission system or by bad road conditions. In both situations the sensor identifies high frequent changes in the rotation. Though, from an after sales perspective the decision on single situations is not an interesting one and an abstraction to a trip decision has to be built.

3.2   Preprocessing of Vehicular Data

From a data structure perspective vehicle data is a time series, as the data consists of time-ordered values. These data streams will be aggregated in order to compare trips over different vehicles in the back end without transmitting the whole time series. Then, these aggregated time series represent the vehicular behavior.

At the beginning of the preprocessing features will be derived from the time series in the vehicle. As stated in [2] and [1] histograms will be built to get a characteristic representation of the observed situation. The result is a distribution of the counting objective (i.e. the time a signal is active or the count of a signal occurrence) in a relevant situation, though a trip consists of s histograms with s as the number of observed situations. A bin i, with $i \in 1, 2, 3, \ldots, b$ and b as the number of bins, of the histogram is then considered as a single feature observed. The bin consists of the observed value o_i and the centric value x_i of the bins interval. The value p_i is the relative amount of the bin i in the observed situation o_i. To get the distribution of the signal, and with it the histogram comparable, the bins are normalized with the result

$$\sum_{i=1}^{b} p_i = 1. \tag{1}$$

Based on the relative distribution, a cumulative one, with the features

$$c_i = \sum_{k=1}^{i} p_k \tag{2}$$

is built. For a detailed description of the underlying data distribution, the *mean*, *variance*, *standard deviation* and *10th*, *50th* and *90th percentiles* and the *tail values* are built, as mentioned in [2]. In case of the tail values two situations are generated. The first tail-value describes the proportional amount of the observed situation within the lower and upper tail of the distribution, named the *Ptail*. For the second tail-value the difference of the lower and upper tail is calculated, though the tails tendency of the distribution. This value is named as the *Mtail*. In contrast to [2] the tail values are built on the 0.15 tails of the distribution.

3.3 Modeling the Behavior of Defect and Non-Defect Vehicles

The proposed concept will use a classifier to model single situations and a decision tree as the abstraction to the trip level. In Fig. 2 the steps of the concept and their relation is shown. A Support Vector Machine classifier is trained on the preprocessed data. SVM is chosen as it is an applicable technique regarding [8]. In order to access the separability of the healthy and defect situations different kernel methods $K(x,y)$ are addressed, with y as the unlabeled input and x the labeled one. The SVM is trained with a polynomial, radial basis and linear function (Eq. (3) to (5)).

$$K(x,y) = x^T y \quad \text{Linear kernel function} \tag{3}$$

$$K(x,y) = (x^T y)^d \quad \text{Polynomial kernel function} \tag{4}$$

$$K(x,y) = \exp(-\gamma |x - y|^2) \quad \text{Radial basis kernel function (RBF)} \tag{5}$$

As stated before the data consists of each observed situation occurring in a trip. The trained models then represent the separation of healthy and defect situations. For the detection of whole trips a decision system, representing the characteristics of them, is necessary. This is done by aggregating the predictions of single situations on trip level. The first aggregation parameter is the average prediction per trip. This value can be in the interval $[0, 1]$, where zero is indicating just defect predictions and one just healthy ones. The second parameter is the average probability of the prediction for both situations healthy and defect. Finally, the average probability for all healthy predictions and on the other side all defect is built.

Finally, a decision tree is trained to differentiate the aggregated characteristics of healthy and defect trips. The SVM models are applied on the training data and the described aggregations are built on the results. Based on this data a decision tree is trained as a trivial human readable model.

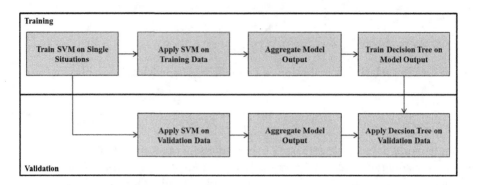

Fig. 2. Training and validation of the proposed concept

4 Evaluation of the Proposed Concept

In the following section the proposed concept is put in practice. The first part describes the test setup, though the used dataset and the performed analysis in detail. This is followed by a description and discussion of the analysis results.

4.1 Test Setup

For evaluating the proposed concept data on defect and healthy vehicle exists. The data is in the form of sensor value time series. In order to evaluate the proposed concept the data needs to be preprocessed. This is done by identifying driveways, as the observed situation, and generating the described features. For evaluating the proposed concept an overall dataset of 574 trips consisting of 3,825 driveaways exists. Within this data 326 trips are healthy ones and consist of 1,809 driveaways. On the other side 248 trips with a total amount of 2,016 driveaways are defect. Not all of this data is useful for an analysis. Before the modeling a filtering is performed. All trips with more than three driveaways will be selected. The resulting dataset has 287 trips with 1,872 driveaways. This dataset is split in a training and validation set. 70% of the data is used for training and 30% for independently validating the models.

A characteristic copy of the vehicular behavior is given by the relative histogram. Each bin represents the proportion of amplitude deltas of a single driveaway. The cumulative histogram highlights the distribution of the amplitudes along the driveaway. In addition, statistical values are built. In order to deal with the descriptive importance of the features the modeling is applied on three different trainings' data sets:

- Use of all variables: All generated variables are used to model healthy and defect vehicular behavior without considering the importance of the variables.
- Feature Selection by Information Gain (IG): The features are selected based on the importance calculated by the information gain.

– Feature Selection by Chi-Square statistic (CHI): The features are selected based on the importance calculated by the Chi-Square statistic.

The aim of using all generated variables is to get a baseline on detecting defects. Applying feature selection techniques should highlight the detection quality dealing with the important features. For all the feature selection methods the top 20 important features are selected, in order to have a reduction of more than 50%. There are two reasons for selecting a subset of features. On the one side it is important to focus on the features describing the different situations at best. From an application perspective, on the other side, it is necessary to reduce the amount of transmitted data from the vehicle, too.

Another important aspect in modeling is the selection of train and test data, to prevent effects reasoned by the data selection. Based on the overall train data mentioned, 70% of the overall data, a 10-fold cross-validation is performed for each set of parameters. In each run the model is trained on nine folds (90% of the dataset) and tested on one fold. The *F1-score* is used to address the quality of the cross-validation, because it is the harmonic mean of precision and recall. For identifying the best parameters a weighted F1-measure is used and is built by the F1-score for each class weighted by the class ratio.

The quality of SVM is dependent on the kernel functions parameters. Based on the train data the best set of parameters for the selected models is determined. This is done by a grid search. Each combination of parameters is evaluated. Table 1 shows the parameters of each kernel function and the interval that is evaluated by a cross-validation.

Table 1. Parameters evaluated for each of the kernel functions

Kernel function	Parameter	Optimization interval
Linear	C	Start: 0.001; End: 1000; Steps: 15
RBF	C	Start: 0.001; End: 1000; Steps: 15
RBF	γ	Start: 0; End: 0.5; Steps: 25
Polynomial	C	Start: 0.001; End: 1000; Steps: 15
Polynomial	Degree	2, 3

After determining the best model parameters, a decision tree is trained. The learned SVM models are applied on the train data. Then, the model output is aggregated on a trip level as stated in the proposed concept. This data is used to train a decision tree with gain ratio as criterion for attribute splits and a maximal depth of 20.

The modeling part of the evaluation, the grid search and cross-validation is implemented with the *Scikit-learn* Python framework [7]. All other analysis parts are performed with *RapidMiner Studio*.

4.2   Results on Defect Detection by the Vehicular Behavior

Before building models the feature selection techniques IG and CHI has been applied and the result are three datasets. The dataset with all 42 features, one with the top 20 features of the IG and the last one with the top 20 features of the CHI technique. Both feature selection sets differ in one feature. Based on the three different datasets a grid search is performed in order to detect the optimal parameters. This optimum is determined for the combination algorithm and feature set. The results are shown in Table 2 including the calculated F1-score of the parameter set and the deviation along the cross-validation. These models then are evaluated on the validation dataset.

Table 2. Optimal parameters selected by the grid search based on a weighted F1-score

Algorithm	Features	C	γ	Degree	F1-score	F1-score deviation
C-SVM Linear	All	1	-	-	0.8281	0.0259
C-SVM Polynomial	All	0.0518	-	2	0.8462	0.0274
C-SVM RBF	All	7.1969	0.0208	-	0.8781	0.0278
C-SVM Linear	IG	0.3728	-	-	0.8249	0.0241
C-SVM Polynomial	IG	0.0518	-	3	0.8284	0.0208
C-SVM RBF	IG	7.1969	0.1458	-	0.8553	0.0205
C-SVM Linear	CHI	0.3728	-	-	0.8197	0.0156
C-SVM Polynomial	CHI	0.0027	-	3	0.8203	0.0237
C-SVM RBF	CHI	7.1969	0.1667	-	0.8578	0.0243

Table 3 highlights the results of applying the models on the validation data and aggregating to a trip decision by a decision tree. The metrics shown are the accuracy and the F1-score for the classes healthy and unhealthy. It can be seen the best detection rate for the all feature case is given by *C-SVM RBF* with a mean F1-score of 0.9418 and an accuracy of 0.9419. The result of the trainings step is confirmed as it stated this technique as the best one too. Another important result is the similar quality of the dataset with all features and the one reduced by information gain.

The resulting decision tree has one node. Trips with an average probability for the healthy class with more than 0.557 is healthy and with less equal defect. All other aggregated variables are not relevant for the distinction.

5   Further Research - Use of Detected Patterns as New Customer Service Symptom

The proposed concept shows the qualities on analyzing the aggregated vehicular behavior for defect detection. A defect can be identified with the F1-score of 0.9418 with a trained classifier. The different qualities with a reduced feature

Table 3. Results of applying the trained models on the validation data

Algortihm	Features	Accuracy	F1-score healthy	F1-score defect
C-SVM Linear	All	0.9302	0.9348	0.9250
C-SVM Polynomial	All	0.9302	0.9318	0.9286
C-SVM RBF	All	0.9419	0.9438	0.9398
C-SVM Linear	IG	0.9302	0.9362	0.9231
C-SVM Polynomial	IG	0.9070	0.9091	0.9048
C-SVM RBF	IG	0.9419	0.9425	0.9412
C-SVM Linear	CHI	0.8953	0.9072	0.8800
C-SVM Polynomial	CHI	0.8721	0.8764	0.8675
C-SVM RBF	CHI	0.9186	0.9231	0.9136

set shows the possibility to reduce the amount of data by keeping the quality. Another important result is the combination of SVMs and decision trees. It is possible to train the model on single observed situations and generate a trip decision by applying using a decision tree.

But, not each failure situation is known a priori to train a supervised model. In a conclusion further research will address the unsupervised detection of deviations. An analysis of the vehicular behavior along its own history and of a vehicle and a whole fleet will highlight the characteristics of a trip, though if it deviates significant or not. One possibility on detecting deviations is to train a model on the normal vehicular behavior as stated in [4].

A detected deviation or defect, as shown in this paper, can be seen as a new symptom in the customer service data. One possibility to integrate all customer service data is given in [6]. Figure 3 shows the integration of anomalies or failures detected by a trained model as an enrichment of the current set of symptoms. This data structure can be used to automatically derive the necessary repair actions and spare parts by detecting patterns in the symptoms.

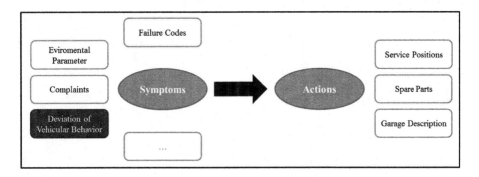

Fig. 3. Integration of detected vehicular deviations with current symptoms

References

1. Fan, Y., Nowaczyk, S., Rögnvaldsson, T.: Using histograms to find compressor deviations in bus fleet data. In: The SAIS Workshop 2014 Proceedings, Swedish Artificial Intelligence Society (SAIS), pp. 123–132 (2014)
2. Frisk, E., Krysander, M., Larsson, E.: Data-driven lead-acid battery prognostics using random survival forests. In: Proceedings of the 2nd European Conference of the PHM Society (2014)
3. Gaber, M.M., Zaslavsky, A., Krishnaswamy, S.: Mining data streams: a review. ACM SIGMOD Rec. **34**(2), 18–26 (2005)
4. Hawkins, S., He, H., Williams, G., Baxter, R.: Outlier detection using replicator neural networks. In: Kambayashi, Y., Winiwarter, W., Arikawa, M. (eds.) DaWaK 2002. LNCS, vol. 2454, pp. 170–180. Springer, Heidelberg (2002). doi:10.1007/3-540-46145-0_17
5. Kusiak, A., Verma, A.: Analyzing bearing faults in wind turbines: a data-mining approach. Renew. Energy **48**, 110–116 (2012)
6. Müller, T.C.: Neuronale Modelle zur Offboard-Diagnostik in komplexen Fahrzeugsystemen. Ph.D. thesis, Technische Universität Braunschweig (2011)
7. Pedregosa, F., Varoquaux, G., Gramfort, A., Michel, V., Thirion, B., Grisel, O., Blondel, M., Prettenhofer, P., Weiss, R., Dubourg, V., Vanderplas, J., Passos, A., Cournapeau, D., Brucher, M., Perrot, M., Duchesnay, E.: Scikit-learn: machine learning in Python. J. Mach. Learn. Res. **12**, 2825–2830 (2011)
8. Rojas, A., Nandi, A.K.: Practical scheme for fast detection and classification of rolling-element bearing faults using support vector machines. Mech. Syst. Signal Process. **20**(7), 1523–1536 (2006)
9. Shearer, C.: The CRISP-DM model: the new blueprint for data mining. J. Data Warehouse. **5**(4), 13–22 (2000)

Automatic Speech Recognition Adaptation to the IoT Domain Dialogue System

Maciej Zembrzuski[1], Heesik Jeon[2], Joanna Marhula[1], Katarzyna Beksa[1],
Szymon Sikorski[1], Tomasz Latkowski[1], and Paweł Bujnowski[1(✉)]

[1] Samsung R&D Institute Poland,
Warsaw Spire, Pl. Europejski 1, 00-844 Warsaw, Poland
{m.zembrzuski,heesik.jeon,j.marhula,k.beksa,s.sikorski,t.latkowski,
p.bujnowski}@samsung.com
[2] Samsung Electronics, Seoul R&D Campus,
33 Seongchon-gil, Seocho-gu, Seoul 06765, Korea

Abstract. The paper addresses the issue of error correction of the domain-specific output from the cloud Automatic Speech Recognition (ASR) system. The research and the solution were built for the Internet of Things (IoT) data collected in the process of applying voice control over home appliances. We describe an ASR post-processing module that reduces the word error rate (WER) of ASR hypotheses and consequently improves prediction of users' intentions by the Natural Language Understanding (NLU) module. The study also compares three various English proficiency level groups of speakers and makes observations about differences of recognition accuracy ratio of users' speech by a dialogue system.

Keywords: Automatic Speech Recognition · Natural Language Understanding · Dialogue systems · Machine learning

1 Introduction

The research concentrates on the improvement of Automatic Speech Recognition (ASR) cloud service results for a domain-specific dialogue system. We investigate linguistic data collected from voice control over home appliances integrated within the Internet of Things (IoT) system. In the preliminary research, we studied the Natural Language Understanding (NLU) module for the IoT data in English. Our internal laboratory experiments showed that using the output from ASR systems decreases the correct detection within the NLU module in comparison with the transcribed data (original input text) by 14% to 30% depending on the task and the speaker's language level. For example, speech utterance *Turn down sound* ***a little bit*** was recognized by the ASR cloud service as *Turn down sound* ***Elizabeth***, *Change lamp colour to* ***red*** as *Change lamp colour to* ***read***, or ***Lights off*** as ***White sauce***. The ASR cloud services are mostly of high quality. For example, Google ASR technology has been revealed to have only 8% word error rate (WER) for general language [13]. ASR web services are supposed to support utterances from various domains, which significantly limits the

© Springer International Publishing AG 2017
M. Kryszkiewicz et al. (Eds.): ISMIS 2017, LNAI 10352, pp. 215–226, 2017.
DOI: 10.1007/978-3-319-60438-1_22

adaptation of such systems to a desired domain. The basic adaptation method depends on providing lists of domain-specific words. However, since the cloud-based services do not allow for training a domain-specific language model, the simplest adaptation method is infeasible. In view of this fact, building the ASR post-processing adaptation component is essential for the success of the whole domain-specific dialogue system.

Apart from the study of the ASR adaptation to the IoT domain, we address the issue of speech recognition and understanding in the case of interlocutors at different language proficiency levels, especially non-native English speakers. In this respect our research develops technology that seems to be applicable to future well-suited personal assistant solutions.

The article is built as follows. In Sect. 2, related works are recalled. In Sect. 3, we describe the IoT domain. Section 4 presents the architecture of our solution. In Sect. 5, we explain the data collection process, the methods used and we show a number of evaluation tests. In Sect. 6, we discuss the implications of the study and the plans for future work.

2   Related Work

There are several approaches in the literature that focus on the adaptation of ASR services to a specific domain. The solution described in [4] applies open source Sphinx-based language models to reduce WER of the Google speech recognizer results. Traum et al. [3] combine independent outputs of Google and Sphinx ASR services to receive general and domain-specific outcomes. Besides, Wang et al. [10] merge n-grams of tokens and phonemes into a mixture model in order to improve the statistical natural language understanding (NLU) module. In order to enhance NLU, [9] explore the reranking of the list of n results (hypotheses) received from one or more ASR systems.

In our solution, we explore some ideas of [1] and [11], but we extend them by applying other techniques, for example, using Conditional Random Fields (CRFs) for error detection. For this purpose we also use various syntactic methods which were broadly researched for speech recognition, particularly by [6]. Additionally, within the area of recognition and understanding of various proficiency-level speakers by the dialogue system, our approach does not exploit acoustic modeling methods [5], but it concentrates on contextual language modeling enhancement.

The implementation of CRF model for error detection and error correction was proposed by [14]. We adopt the idea of using CRF model for error detection. Our approach, however, focuses on grammatical features and reranking of ASR hypotheses, whereas [14] present the usage of n-grams, confusion network and Normalized Web Distance – a measure of words' co-occurrence on web pages for both error detection and error correction. In contrast, the proposed error correction approach focuses on phoneme similarity. Moreover, we discuss an approach limited only to reranking, which proved to be significantly more time-efficient than the error correction approach as described in Sect. 5. In our solution we do not apply the lattices from the ASR output, as they are not used by some services, e.g. famous Google web ASR.

Table 1. Examples of different dialogue act utterances with their optional parameters

Dialogue act	Utterance	Parameters
Statement	*Light*	*in the bathroom*
Command	*Set the AC*	*at twenty degrees*
Question	*Is the TV on?*	*at the moment*

3   Language Domain

The idea behind the Internet of Things refers to the interconnection of every-day objects into a larger intelligent network which allows for exchanging data between different elements, as well as controlling the objects remotely by a network infrastructure. Our study concentrates on the IoT within a smart home solution which allows the user to interact with a dialogue agent and control a range of home appliances – a system similar to the one described by [2]. Using an NLU module, the dialogue agent is able to detect the intention of the user and manage the relevant devices in line with the user's goal. Each user's utterance addressed to the personal assistant is categorized as a Command, Statement, Answer or Question, and relevant parameters such as Time, Date and Location are identified. For example, the utterance *Turn on the TV in the kitchen at 5 pm* is interpreted as a command referring to the TV set in the kitchen (parameter: Location) and ordering *power.on* action to be performed at 5 pm (parameter: Time). Table 1 presents sample utterances for three different dialogue act categories that we primarily focused on in our study.

4   Proposed System

The developed system (Fig. 1) consists of three main components: (1) the normalization and preprocessing of ASR results, (2) error detection (which tests each token's correctness) and (3) error correction containing tools to substitute error words with correct domain-specific words. The details of these components are described in this section. The system receives all ASR hypotheses as its input and returns a single output hypothesis which fits the domain best. The output can be handled by the NLU module in the next step.

4.1   Normalization and Preprocessing

During normalization, the input text is split into sentences and tokenized with methods from the Stanford CoreNLP library [7] version 3.4.1. Stanford toolkit

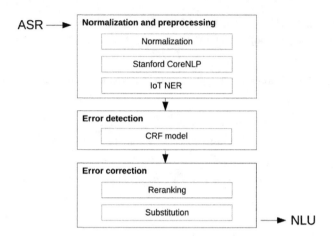

Fig. 1. The domain-specific ASR corrector

performs lemmatization, Part-Of-Speech (POS) tagging, Named Entity (NE) recognition and dependency parsing. Moreover, NE recognizer is extended by the dictionary-based NE tagger for labeling IoT domain-specific instances of locations, devices, their modes, programmes, etc. (more than 20 categories). The tokens are standardized with the use of regex and dictionary-based transformations to unify the spelling.

Next, the tokens are transformed to retrieve the most probable sequence of phonemes provided as the ASR input. For example, digits are replaced by their text representation (with some exceptions for time, date and negative numbers). Furthermore, we apply grapheme-to-phoneme representation of words. For this purpose we use the CMU dictionary [12] extended by out-of-vocabulary entities from the IoT domain. Sometimes a few phoneme representations are used for one token. Finally, to cover the unknown words, we employ the FSA-based phoneme guesser from the tool FreeTTS [15].

4.2   ASR Error Detection

The ASR error detection task consists in assigning *correct* or *incorrect* labels to tokens. To that end the tokens are assigned with:

1. original word,
2. lemma,
3. Part-Of-Speech (POS) tag,
4. Named Entity (NE),
5. syntactic dependence, i.e. the sentence role from the parser dependency tree, e.g.: root, dobj, nsubj, det, etc.,
6. distance between the parent and the child from the dependency tree counted as the number of tokens between them in the sentence,

7. correctness of labels (*true, false*) determined by the comparison of the ASR hypotheses with the original input (used only for training and evaluation purposes).

Error detector is based on the CRF model. The CRF labels the input tokens with *correct* or *incorrect* tags and assigns empirical probability of the selected labeling to each token. This approach is related to the methods described by [1]; however, we combined the proposed error detectors into one model and added syntactic features (i.e. (5) and (6) from the list above).

The CRF model is trained by using combinations of the above features for each token. Additionally, these combinations are enhanced by the respective features of the neighboring tokens. The token window contains up to two preceding and two following tokens. Selecting the most successful combination of variables from possible features (the list above) is done experimentally.

Domain Independent Error Detection. The solution parallel to the described ASR error detection depends on training CRF models with universal NLP data only. Such data include a limited number of features compared to the presented ones (see the list above): only POS tags (3) and universal Stanford NEs for seven classes [7], without dictionary-based entities as in (4), syntactic dependencies (5) and distances between dependent tokens (6). We believe this solution is domain-independent (DI) as it is related neither to the lexical forms of tokens, nor to domain-specific NEs.

4.3   Error Correction

Reranking. For each ASR hypothesis, the average of CRF support probabilities is computed in consideration of all the sentence tokens (see Sect. 4.2). If a token is labeled as erroneous, the probability is multiplied by -1 before calculating the average. The final sentence score is calculated with the formula:

$$Reranking\_score = CRF\_support + init\_position/a + sent\_length/b, \quad (1)$$

where *init_position* is the initial position of the hypothesis in the ASR hypotheses list, *sent_length* is the number of words in the sentence and a and b are experimentally chosen constants (set to 4 and 10 respectively). The hypothesis with the highest *Reranking_score* is considered to be the most reliable and is selected for further processing.

Substitution. The goal of the substitution component is to find the best candidate from the domain-specific vocabulary that is pronounced similarly to the phrases that are discovered to be incorrect. The detailed procedure is presented in Algorithm 1. The system builds a ranking of substitution candidates for each token labeled as erroneous at the previous stage. The candidates are generated on the basis of skip-grams dependent on the tokens preceding the erroneous word and separately on the tokens following it (the skip-grams' matched parts). The skip-grams are produced beforehand from sentences of the training set (the same data are used in the error detection process). We limited the experiments to 0-skip-2-grams and 1-skip-2-grams. The remaining (non-matched) part of the skip-gram (candidate token, marked as c in Algorithm 1) is transformed into phoneme representation (see Sect. 4.1) and compared with a corresponding phoneme-transformed erroneous token (e in Algorithm 1) based on Modified Levenshtein distance (MLD) value (see the description below). Finally, for each candidate, the system computes the fitting score using logistic regression model (LSR). The LSR model applies MLD results and other features (see below LRS subsection). The candidate with the highest LRS support value that exceeds the threshold replaces the erroneous token. Further skip-gram candidate generation for any following erroneous tokens will depend on the recently corrected candidate.

Algorithm 1. Substitution algorithm

1: **function** SUBSTITUTION(*sentence*, *errors*)
2:    $bestC \leftarrow null$
3:    $maxScore \leftarrow 0.5$
4:           ▷ 0.5 or above denotes a correct token
5:    $referenceCRFScore \leftarrow avgCRF(\text{sentence})$
6:        ▷ average CRF support for the sentence
7:    **for all** $e \in errors$ **do**
8:       $C \leftarrow forwardSGramCandidates(e) \cup backwardSGramCandidates(e)$
9:       ▷ combining candidates from forward and backward skip-gram models
10:       **for all** $c \in C$ **do**
11:          $\alpha \leftarrow phonemeDistance(e, c)$
12:          ▷ modified Levenshtein distance
13:          $\beta \leftarrow CRFModel(\text{sentence}, e, c)$
14:          ▷ CRF support for candidate correctness
15:          $\bar{\gamma} \leftarrow avgCRF(subst(\text{sentence}, e, c))$
16:          ▷ average CRF support for new sentence
17:          $\Delta \leftarrow \bar{\gamma} - referenceCRFScore$
18:          $lD \leftarrow lengthDifference(e, c)$
19:          ▷ normalized token lengths difference
20:          $r \leftarrow logisticModel(\alpha, \beta, \bar{\gamma}, \Delta, lD)$
21:          **if** $r > maxScore$ **then**
22:             $maxScore \leftarrow r$
23:             $bestC \leftarrow c$
24:       **if** $bestC \neq null$ **then**
25:          $sentence \leftarrow subst(\text{sentence}, e, bestC)$
26:          ▷ substitute error e with the best candidate
27:    **return** *sentence*

Modified Levenshtein Distance (MLD). We propose modified Levenshtein distance as a modification of the original Levenshtein distance. We substitute the indicator function (0 or 1 for equality or inequality, respectively) with a membership function calculated by multiplying the inequality cost (originally 1) by the position factor defined as $(20 - p)/20$, where p stands for the phoneme position in the erroneous word. If the factor is below 0.01, it is normalized to 0.01. The position factor affects substitution, deletion and insertion costs.

The inequality cost is a phoneme similarity measure. The measure is supposed to penalize differing phonemes (like /k/ and /z/) and reward similar ones (like /z/ and /s/). The calculation is based on the alignment between the phonemes of the ASR hypotheses and the phonemes of the original utterances. A similar approach was adopted by Twiefel et al. [4] and Ziolko et al. [8]. In our method, the similarity measure equals 0, if the phoneme confusion probability is below threshold 0.85; otherwise, its value is 1. The threshold was established experimentally.

Logistic Regression Model for Substitution Scoring (LRS). The LRS is used as the final score for the substitution process. It depends on the normalized values of:

1. Modified Levenshtein distance (MLD) – see the subsection above,
2. CRF support for the candidate word (Sect. 4.2),
3. normalized word length difference between the original and candidate words,
4. the average CRF support for the sentence with the erroneous token replaced with the candidate (subsection Reranking in 4.3),
4. the difference between the average CRF support of the original sentence and the CRF support for the new sentence (with the erroneous word replaced by the candidate word).

5   Experiments

5.1   Data Collection

For the purposes of our study we collected voice data for the IoT domain in English. We invited 91 speakers representing thirty different countries and six continents. All participants were assessed in terms of their English proficiency and classified as intermediate (B1–B2), advanced (C1–C2) and native speaker (NATIVE) users of English. Language proficiency assessment was carried out on the basis of language certificates held by study participants as well as the evaluation of their English language skills during an interview with a professional English instructor.

All study participants were invited to an approximately two-hour long speech recording session during which they were given two different tasks: the first one consisted in producing spontaneous utterances (Task 1); the second one was devoted to reading out sentences or phrases displayed on a prompter (Task 2). Within Task 1, the speakers were asked to give commands to the IoT system concerning eight different home appliances and approximately two hundred related actions. Sample actions with their optional parameters for selected devices are given in Table 2. Speakers also produced questions concerning the status of home equipment (*Is the TV set in the bedroom turned on?*) or the weather forecast (*Is it going to rain tomorrow?*). In Task 2, the participants read out IoT related sentences or phrases displayed on the screen: they included commands and questions addressed to the system as well as greetings and names of different household appliances and locations. In this task we could strictly study differences of ASR recognition accuracy on the same utterances of 3 various language proficiency groups of speakers. Additionally, we were able to indicate the most problematic phrases. In total, we recorded over 35 h of speech data and collected about 7,700 spontaneous (Task 1) and nearly 32,000 read out voice utterances (Task 2).

Table 2. Sample actions for selected devices with their optional parameters

Device	Action	Parameters
TV	Power on/off	In the kitchen at 5 o'clock by 3°
	Wi-fi connect/Disconnect	
Air conditioner	Cooling level	
	Up/down	
Robot cleaner	Cleaning start/stop	

5.2   Method

The numeric experiment for the ASR post-processor was conducted using the WER comparison and the accuracy test of the NLU module (part of the IoT dialogue system). The NLU module was implemented separately for the IoT domain and was trained on text data only.

Error Detection. For the purpose of error detection, we used the logistic regression model trained on n-gram probabilities as the baseline. As an alternative, we evaluated CRF models with various features. All models were evaluated using a 10-fold cross-validation. For Task 1 (spontaneous speeches) we processed about 7,700 utterances that produced over 35,000 ASR hypotheses. For Task 2 (the reading exercise) we used nearly 32,000 with over 171,000 ASR hypotheses.

Error Correction. Error substitution was evaluated separately in each language proficiency group by using a 10-fold cross-validation of commands gathered in Task 1. Each set consisted of approximately 2,150 original user utterances. The error substitution mechanism in our simplified solution can substitute one token with only one best candidate. Moreover, since n-grams for candidate generation contain NE classes, they need to be expanded into actual candidates for phoneme comparison. The expansion does not cover dates, numbers and time expressions.

5.3   Results

Error Detection. Table 3 presents the comparison of error detection accuracy for two datasets. The baseline model, *n-gram LR*, is a logistic regression model trained on n-gram probabilities. In comparison, instead of n-grams, the *CRF model* used a list of features described in Sect. 4.2. The last row reports on an additional experiment. The *CRF model (DI)* presents the domain-independent model separated from word surface representations (see Sect. 4.2). Despite relying only on universal NLP data, it still has potential for WER improvement (though, the recall is significantly lower). The results for both tasks proved to be quite similar, with Task 2 data gaining slight advantage over Task 1 data. The output may be influenced by the fact that the reading part contains more data; however, the sentences in the reading part are longer and more complex than the spontaneously created ones. For further experiments the *CRF model* was used as the better one.

Table 3. Comparison of error detection models for spontaneous (1) and read (2) data

Model	Task 1 Prec	Recall	Acc	Task 2 Prec	Recall	Acc
n-gram LR	85.18%	64.64%	87.98%	79.23%	60.05%	88.10%
CRF model	**85.21%**	**79.62%**	**91.10%**	**88.94%**	**82.73%**	**92.07%**
CRF model (DI)	58.06%	31.08%	76.21%	65.12%	39.58%	76.51%

Error Correction in Different Proficiency Groups. The reduction of WER has been observed in all proficiency groups. Detailed results of error correction based on the logistic regression model (see subsection *Substitution* in 4.3) are presented in Table 4 for Task 1 data. Applying the ASR adaptation component results in improving WER by 22%, 23% and 27% for intermediate (B1–B2), advanced (C1–C2) and native English users, respectively. The WER reduction is more significant when the language proficiency level is higher.

In our experiment, we have additionally noticed an interesting time performance result: the mere reranking procedure worked 17–27 times faster than the combined approach, reranking and substitution, with a very slight accuracy decrease. The average reranking time was around 0.3 sec per sentence and between 4.4 sec and 7.5 sec when we added the substitution process (based on the logistic regression score). This means that the simple reranking approach may be preferable for real-time human-computer dialogue systems.

Natural Language Understanding Results of ASR Error-Correction. We have verified the impact of the ASR post-processor on the NLU system using IoT voice control data. The NLU detects user's intentions described by more that 20 variables: dialogue acts, user's goals, actions, objects, devices, locations, negations, colors, numbers, time and date expressions. In the evaluated dialogue

Table 4. Comparison of final WER improvements for spontaneous data

Group	Initial WER	Resulting WER	WER reduction in %
B1–B2	19.03%	14.82%	22.12%
C1–C2	13.62%	10.46%	23.20%
NATIVE	8.79%	6.38%	27.42%

system, the NLU performance is measured by using all of the above parameters: in order to classify an utterance as correct all of them must be recognized properly.

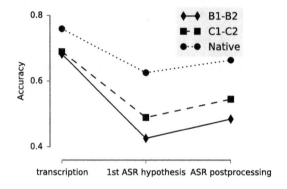

Fig. 2. Accuracy of the Natural Language Understanding module for spontaneous speech data sets (from the left: transcribed utterance, 1st ASR hypothesis, ASR domain post-processing) regarding three language levels: B1–B2, C1–C2 and NATIVE

Figure 2 presents NLU evaluation results for three different language proficiency groups for the spontaneous commands (Task 1) containing about 2,150 utterances each. We confront three various sets of the same data: (1) original transcription, (2) the first ASR hypothesis and (3) the post-processed data of the ASR results. Here, our method (Sect. 4) was tested in a 10-fold cross-validation experiment and the results confirm its effectiveness. We observe an improvement of approximately 6 percentage points in NLU accuracy in both non-native groups compared to 4 percentage points in the native group (still the detection accuracy of the NLU module for native speakers remains higher than for the non-native English users). In effect, the raised accuracy level of the NLU should enhance the quality of natural human-computer conversation and boost the performance of the IoT domain dialogue system.

6   Conclusion and Future Work

The study brings out the necessity for the ASR domain-specific post-processing module as a low cost element of a dialogue system. Our experiment also points at the need for improvements in the area of understanding interlocutors at different levels of linguistic proficiency. We have demonstrated that the application of ASR post-processor significantly reduces the WER and increases NLU accuracy. An intermediate speaker after ASR post-processing can be understood at a success rate similar to that of an advanced language user without the use of such a system.

The CRF models we used and evaluated here seem to suit the task of error detection and reranking even when used with a limited set of features or small datasets. The mere ASR hypotheses reranking approach turned out to be more practical due to the significant time advantage over the combined approach (which contains the substitution process), with only a minor drop in WER improvement. The proposed domain-independent reranking approach may be useful for universal ASRs (with low recall trade-off), but needs more testing with different data.

We believe that the presented solution may be applied to refine the dialogue manager and properly handle language imprecisions of a wide range of users.

In future work we plan to focus on detecting user's proficiency level and implementing level-specific adaptations to the correction mechanism. We also plan to compare a deep learning approach, particularly LSTM, with the proposed CRF approach and other authors' solutions.

References

1. Choi, J., Lee, D., Ryu, S., Lee, K., Kim, K., Noh, H., Geunbae Lee, G.: Engine-independent ASR Error Management for Dialog Systems Situated Dialogue in Speech-Based Human-Computer Interaction. Signals and Communication Technology. Springer, Switzerland (2016)
2. Jeon, H., Oh, H.R., Hwang, I., Kim, J.: An intelligent dialogue agent for the IoT home. In: The Workshops of the Thirtieth AAAI Conference on Artificial Intelligence. Artificial Intelligence Applied to Assistive Technologies and Smart Environments: Technical report WS-16-01 (2016)
3. Traum, D., Georgila, K., Artstein, R., Leuski, A.: Evaluating spoken dialogue processing for time-offset interaction. In: Proceedings of the 16th SIGDIAL Conference, Praha, Czech Republic (2015)
4. Twiefel, J., Baumann, T., Heinrich, S., Wermter, S.: Improving domain-independent cloud-based speech recognition with domain-dependent phonetic post-processing. In: Proceedings of the 28th AAAI Conference on Artificial Intelligence, CA (2014)
5. Razavi, M., Magimai Doss, M.: On recognition of non-native speech using probabilistic lexical model. In: Proceedings of the 15th Annual Conference of the International Speech Communication Association. Interspeech 2014 (2014)
6. Lambert, B., Raj, B., Singh, R.: Discriminatively trained dependency language modeling for conversational speech recognition. In: Proceedings of the 14th Annual Conference of the International Speech Communication Association. Interspeech 2013 (2013)
7. Manning, C.D., Surdeanu, M., Bauer, J., Finkel, J., Bethard, S.J., McClosky, D.: The stanford CoreNLP natural language processing toolkit. In: ACL 2014 Demo Session (2014)

8. Ziolko, B., Galka, J., Skurzok, D., Jadczyk, T.: Modified weighted Levenshtein distance in automatic speech recognition. In: Proceedings of KKZMBM (2010)
9. Morbini, F., Audhkhasi, K., Artstein, R., Van Segbroeck, M., Sagae, K., Georgiou, P., Traum, D.R., Narayanan, S.: A reranking approach for recognition and classification of speech input in conversational dialogue systems. In: Proceedings of the 2012 IEEE Spoken Language Technology Workshop (SLT), pp. 49–54. IEEE (2012)
10. Wang, W.Y., Artstein, R., Leuski, A., Traum, D.: Improving spoken dialogue understanding using phonetic mixture models. In: Proceedings of the 24th International Florida Artificial Intelligence Research Society Conference, pp. 329–334 (2011)
11. Sarma, A., Palmer, D.D.: Context-based speech recognition error detection and correction. In: Proceedings of the Human Language Technology Conference of the North American Chapter of the Association for Computational Linguistics, pp. 85–88 (2004)
12. CMUdict: Carnegie Mellon University open-source grapheme-to-phoneme dictionary 1993–2015 (2015). http://svn.code.sf.net/p/cmusphinx/code/trunk/cmudict/cmudict-0.7b
13. Novet, J.: Google says its speech recognition technology now has only an 8% word error rate (2015). http://venturebeat.com/2015/05/28/google-says-its-speech-recognition-technology-now-has-only-an-8-word-error-rate
14. Byambakhishig, E., Tanaka, K., Aihara, R., Nakashika, T., Takiguchi, T., Ariki, Y.: Error correction of automatic speech recognition based on normalized web distance. In: Proceedings of the INTERSPEECH 2014, pp. 2852–2856 (2014)
15. Speech Integration Group of Sun Microsystems Laboratories: Phoneme guesser from the tool FreeTTS v1.2 (2005). http://freetts.sourceforge.net/docs/index.php

Knowledge-Based Systems

Rule-Based Reasoning with Belief Structures

Łukasz Białek[1], Barbara Dunin-Kęplicz[1], and Andrzej Szałas[1,2(✉)]

[1] Institute of Informatics, University of Warsaw, Warsaw, Poland
{bialek,keplicz,andrzej.szalas}@mimuw.edu.pl
[2] Department of Computer and Information Science, Linköping University,
Linköping, Sweden

Abstract. This paper introduces $4QL^{Bel}$, a four-valued rule language designed for reasoning with paraconsistent and paracomplete belief bases as well as belief structures. Belief bases consist of finite sets of ground literals providing (partial and possibly inconsistent) complementary or alternative views of the world. As introduced earlier, belief structures consist of constituents, epistemic profiles and consequents. Constituents and consequents are belief bases playing different roles. Agents perceive the world forming their constituents, which are further transformed into consequents via the agents' or groups' epistemic profile.

In order to construct $4QL^{Bel}$, we extend 4QL, a four-valued rule language permitting for many forms of reasoning, including doxastic reasoning. Despite the expressiveness of $4QL^{Bel}$, we show that its tractability is retained.

Keywords: Rule languages · Doxastic reasoning · Paraconsistency · Paracompleteness

1 Reasoning About Beliefs

Contemporary intelligent systems, especially autonomous, require a variety of reasoning techniques. The rule-based approach leading to executable logic-based specifications is of special importance as it provides adequate knowledge/belief bases equipped with a specialized reasoning machinery. Successful implementation of an autonomous system relies on the quality of the underlying model of an environment it is situated in. Realistically, sensor-based perception together with heterogeneity and diverse credibility of information sources typically results in inconsistent and incomplete information.

The essential problem of belief modeling and formation when information is partly missing and inconsistent, is addressed, e.g., in [4–6,9–11,19,21,22,24]. To adequately model beliefs and reasoning about them, *epistemic profiles* and *belief structures* have been introduced in [9–11]. In short, epistemic profiles encapsulate agents' or groups' reasoning, perception and communication capabilities, including context-specific methods of both information disambiguation and completion. To make the approach as general as possible, epistemic profiles are mappings transforming one belief base (*constituents*) into another belief base

Supported by the Polish National Science Centre grant 2015/19/B/ST6/02589.

M. Kryszkiewicz et al. (Eds.): ISMIS 2017, LNAI 10352, pp. 229–239, 2017.
DOI: 10.1007/978-3-319-60438-1_23

(*consequents*). While constituents contain sets of beliefs acquired by perception, expert-supplied knowledge, communication and other ways, consequents contain final, "mature" beliefs or beliefs capturing an updated view on the environment. Together with epistemic profiles, constituents and consequents constitute belief structures. Importantly, belief structures ensure a unified approach to both individual and group reasoning [12]. They have been already applied in argumentation [8] and modeling complex dialogues [7], where their use led to a simplified and tractable framework.

In [9–11], as an implementation tool we have suggested 4QL, a four-valued rule-based query language designed in [17, 19, 23].[1] 4QL allows for negation both in premises and in conclusions of rules. It delivers simple, yet powerful constructs (modules and external literals) for expressing nonmonotonic rules. 4QL enjoys tractable query computation and captures all tractable queries [18]. Apart from providing firm foundations for paraconsistent knowledge bases and nonmonotonic reasoning, 4QL opens the space for a diversity of applications. For surveys of closely related areas see, e.g., [3, 16].

The semantics of 4QL is defined by *well-supported models*, i.e., models consisting of (positive or negative) ground literals, where each literal is a conclusion of a derivation starting from facts. Generally, 4QL programs consist of modules. For any module, its corresponding well-supported model is uniquely determined [18, 19]. Namely, each module can be treated as a finite set of ground literals and this set can be computed in deterministic polynomial time [17, 19].

There is a rich literature on beliefs and belief bases (see [2, 13–15, 20, 21] and references there). In our approach belief bases consist of finite sets of ground literals what creates a natural correspondence between 4QL modules and belief bases, thus also between 4QL and belief structures [9–11]. However, 4QL does not include belief operators nor other constructs useful in doxastic reasoning. Therefore we aim here to:

– define the syntax and semantics of $4QL^{Bel}$, extending 4QL with constructs for reasoning about beliefs, belief bases and belief structures;
– show tractability of computing well-supported models of $4QL^{Bel}$ programs;
– show tractability of querying well-supported models of $4QL^{Bel}$ programs.

In order to offer a more flexible formalism, we relax the definition of semantics of formulas given in [11]. Roughly speaking, in addition to referring to belief structures, we allow to refer to their belief bases (that is their constituents and consequents).

The paper is structured as follows. In Sect. 2 we define the base logic and recall our approach to belief bases and indeterministic belief structures [11]. Next, in Sect. 3, we introduce the $4QL^{Bel}$ rule language and discuss its complexity. In Sect. 4 we outline the methodology of specifying belief structures using $4QL^{Bel}$. Finally, Sect. 5 concludes the paper.

---

[1] For an open-source implementations of 4QL see http://4ql.org. The INTER4QL 3.0 interpreter includes some features for querying belief bases, implemented by A. Bułanowski.

2   Base Logic and Belief Bases

In order to construct paraconsistent and/or paracomplete belief bases, we need to extend classical first-order language over a given vocabulary without function symbols. We assume that *Const* is a fixed finite set of constants, *Var* is a fixed set of variables and *Rel* is a fixed set of relation symbols. Following the convention, we begin variable names with capital letters and constants with small letters.

In addition to classical truth values t (true) and f (false), we also have i (inconsistent) and u (unknown). By *truth ordering* and *information ordering* we understand respectively the (reflexive and transitive closure of) orderings on truth values shown in Fig. 1. Truth ordering is used to evaluate formulas.

Information ordering reflects the process of gathering and fusing information. It is used to merge results concerning the same (perhaps negated) ground literal concluded from different rules, possibly originating from different information sources. Starting from the lack of information, in the course of belief acquisition, evidence supporting or denying hypotheses is collected, leading finally to a decision about its truth value.

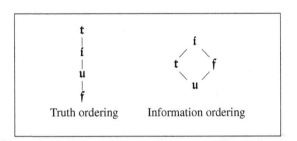

Fig. 1. Orderings on truth values.

Definition 1. A *literal* is an expression of the form $R(\bar{\tau})$ or $\neg R(\bar{\tau})$, with τ being a sequence of arguments, $\bar{\tau} \in (Const \cup Var)^k$, where k is the arity of R. *Ground literals over Const*, denoted by $\mathcal{G}(Const)$, are literals without variables, with all constants in *Const*. If $\ell = \neg R(\bar{\tau})$ then $\neg\ell \stackrel{\text{def}}{=} R(\bar{\tau})$. For an *assignment* $v : Var \longrightarrow Const$ and a literal ℓ, by $\ell(v)$ we understand the ground literal obtained from ℓ by substituting each variable X occurring in ℓ by constant $v(X)$.   ◁

If S is a set then $\mathrm{FIN}(S)$ denotes the set of all finite subsets of S. By $\mathbb{C} \stackrel{\text{def}}{=} \mathrm{FIN}(\mathcal{G}(Const))$ we denote the set of all finite sets of ground literals over the set of constants *Const*.

Definition 2. By a *belief base over a set of constants Const* we understand any finite set Δ of finite sets of ground literals over *Const*, i.e. any finite set $\Delta \subseteq \mathbb{C}$.   ◁

The intuition behind belief bases is that every set in a belief base represents a possible perception of the world. Sets constituting belief bases can be seen as a four-valued generalization of Kripke-like possible worlds or impossible worlds [21].

Definition 3. The *truth value* of a literal ℓ wrt a set of ground literals L and an assignment v, denoted by $\ell(L, v)$, is defined as follows:

$$\ell(L, v) \stackrel{\text{def}}{=} \begin{cases} \mathsf{t} \text{ if } \ell(v) \in L \text{ and } (\neg \ell(v)) \notin L; \\ \mathsf{i} \text{ if } \ell(v) \in L \text{ and } (\neg \ell(v)) \in L; \\ \mathsf{u} \text{ if } \ell(v) \notin L \text{ and } (\neg \ell(v)) \notin L; \\ \mathsf{f} \text{ if } \ell(v) \notin L \text{ and } (\neg \ell(v)) \in L. \end{cases} \qquad \triangleleft$$

Syntax of the base logic is given in Table 1, where among others, we allow formulas:

- $\mathrm{Bel}_\Delta (\langle Formula \rangle)$, expressing beliefs rooted in belief bases (indicated by Δ);
- $f(\Delta).\alpha$, allowing one to evaluate $\mathrm{Bel}()$-free formulas in belief bases: here f is a mapping transforming a belief base into a single set of ground literals, e.g., $f(\Delta)$ may be $\bigcup_{D \in \Delta} D$ or $\bigcap_{D \in \Delta} D$ (further denoted by $\bigcup \Delta$, $\bigcap \Delta$, respectively).

Table 1. Syntax of the base logic.

$$\begin{aligned} \langle Formula \rangle ::= &\ \langle Literal \rangle \mid \neg \langle Formula \rangle \mid \langle Formula \rangle \wedge \langle Formula \rangle \mid \\ &\ \langle Formula \rangle \vee \langle Formula \rangle \mid \langle Formula \rangle \rightarrow \langle Formula \rangle \mid \\ &\ \forall X \langle Formula \rangle \mid \exists X \langle Formula \rangle \mid \\ &\ \langle Formula \rangle \in \langle TruthValues \rangle \mid \\ &\ \mathrm{Bel}_\Delta (\langle Formula \rangle) \mid f(\Delta). \langle Formula \rangle \end{aligned}$$

where:
- $\langle Literal \rangle$ represents the set of literals;
- $\langle TruthValues \rangle$ represents nonempty subsets of $\{\mathsf{t}, \mathsf{f}, \mathsf{i}, \mathsf{u}\}$;
- Δ is a belief base,
- f is a mapping transforming a belief base into a (single) finite set of ground literals; if f is not specified, by default $f(\Delta) \stackrel{\text{def}}{=} \bigcup \Delta$, i.e., $\Delta.\alpha \stackrel{\text{def}}{=} (\bigcup \Delta).\alpha$.

The semantics of propositional connectives, quantifiers and doxastic operators is given in Tables 2, 3 and 4. Observe that definitions of $\wedge$ and $\vee$ reflect minimum and maximum wrt the truth ordering (see Fig. 1), as advocated, e.g., in [1, 17, 23]. Such a semantics appears to be natural. It also reflects intuitions of classical two-valued logic.

Table 2. Truth tables for $\wedge$, $\vee$, $\rightarrow$ and $\neg$ (see [17, 19]).

$\wedge$	f	u	i	t		$\vee$	f	u	i	t		$\rightarrow$	f	u	i	t		$\neg$	
f	f	f	f	f		f	f	u	i	t		f	t	t	t	t		f	t
u	f	u	u	u		u	u	u	i	t		u	t	t	t	t		u	u
i	f	u	i	i		i	i	i	i	t		i	f	f	t	f		i	i
t	f	u	i	t		t	t	t	t	t		t	f	f	t	t		t	f

Table 3. Semantics of first-order formulas (with $\in$).

- if α is a literal then $\alpha(L, v)$ is defined in Definition 3;
- $(\neg\alpha)(L, v) \stackrel{\text{def}}{=} \neg(\alpha(L, v))$;
- $(\alpha \circ \beta)(L, v) \stackrel{\text{def}}{=} \alpha(L, v) \circ \beta(L, v)$, where $\circ \in \{\vee, \wedge, \rightarrow\}$;
- $(\forall X \alpha(X))(L, v) \stackrel{\text{def}}{=} \min_{a \in Const} \{(\alpha(X/a)(L, v)\}$;
- $(\exists X \alpha(X))(L, v) \stackrel{\text{def}}{=} \max_{a \in Const} \{(\alpha(X/a)(L, v)\}$;
- $(\alpha \in T)(L, v) \stackrel{\text{def}}{=} \begin{cases} \mathbf{t} \text{ when } \alpha(L, v) \in T \\ \mathbf{f} \text{ otherwise,} \end{cases}$

where:
- L is a set of ground literals;
- min, max are respectively minimum and maximum wrt truth ordering;
- $\alpha(X/a)$ denotes the formula obtained from α by substituting all free occurrences of variable X by constant a.

For a belief base Δ, we define $\alpha(\Delta, v) \stackrel{\text{def}}{=} \alpha(\bigcup \Delta, v)$.

Table 4. Semantics of the Bel$()$ and $f()$. operators.

- $(\mathrm{Bel}_\Delta(t))(v) \stackrel{\text{def}}{=} t$, for $t \in \{\mathbf{t}, \mathbf{f}, \mathbf{i}, \mathbf{u}\}$;
- $(\mathrm{Bel}_\Delta(\alpha))(v) \stackrel{\text{def}}{=} \mathrm{LUB}\{\alpha(D, v) \mid D \in \Delta\}$;
- $(f(\Delta).\alpha)(v) \stackrel{\text{def}}{=} \alpha(f(\Delta), v)$,

where:
- Δ is a belief base;
- $v: Var \longrightarrow Const$ is an assignment of constants to variables;
- α is a first-order formula (for nested Bel$()$ s, one starts with the innermost one.)
- LUB denotes the least upper bound wrt the information ordering (see Figure 1).

3   The 4QL^{Bel} Language

3.1   Syntax of 4QL^{Bel}

The 4QL^{Bel} language inherits many features of 4QL (in particular its extended version 4QL^{+} of [23]). Its basic components are *modules*. Their syntax is shown in Table 5,[2] where *rules* are of the form:

$$\langle Literal \rangle :- \langle Formula \rangle . \qquad (1)$$

Facts are rules with the $\langle Formula \rangle$ part being **t**. In such cases we write $\langle Literal \rangle$ rather than $\langle Literal \rangle :- \langle Formula \rangle$.

Note that formulas at the righthand side of :− can, among others, have the form $f(\Delta).\alpha$. In cases when Δ consists of a single set of ground literals represented by a module named m, we often write m.α rather than $\{m\}.\alpha$. However, these references cannot include cycles. This assumption is formalized in Definitions 4 and 5.

Table 5. Syntax of modules of 4QL^{Bel}.

```
1  module module_name:
2      domains ... :
3      relations ... :
4      facts ... :
5      rules ... :
6  end.
```

Definition 4. Let $M = \{\mathtt{m}_1, \ldots, \mathtt{m}_k\}$ $(k \geq 1)$ be a set of modules. By a *reference graph* of M we understand a graph $\langle M, E \rangle$ such that for $m_i, m_j \in M$: ·

$(m_i, m_j) \in E$ iff rules *in* $\mathtt{m}_i$ contain a subformula of the form $\mathtt{m}_j.\alpha$.

If $(m_i, m_j) \in E$ then we say that module m_i *refers to* module m_j.   ◁

Definition 5. By a 4QL^{Bel} *program* we understand a set of 4QL^{Bel} modules whose reference graph is acyclic.   ◁

Acyclicity of a reference graph of any 4QL^{Bel} program can be determined in deterministic polynomial time wrt the size of the program. We also assume *strong typing* in the following sense:

- when a value occurs as an argument of a relation, it has to belong to the domain associated with that argument;
- when a variable occurs in a rule as an argument in more than one place, all such arguments have to be specified as belonging to the same domain;
- a domain consists of all values of all modules occurring as relations' arguments specified as belonging to that domain.

---

[2] Here we adjust the syntax used in implementations of 4QL (see http://4ql.org).

3.2   Semantics of 4QL^{Bel}

The semantics of 4QL^{Bel} modules is defined via well-supported models. Such models consist of ground literals. To simplify the presentation, in the current section we assume that all literals are ground and all universal (existential) quantifiers are substituted by conjunctions (disjunctions) of formulas. In particular, rather than allowing rules with variables, we consider their ground and quantifier-free instances. The definition provided here is adopted from [19], where also related intuitions are discussed.

The difficulty in generating well-supported models for 4QL programs depends on deriving conclusions on the basis of facts being temporarily true or false and later becoming inconsistent. We iterate the following method until no new conclusions are generated:

- generate the least set of conclusions by Datalog-like reasoning;
- retract conclusions based on defeated premises;
- correct (minimally) the obtained set of literals to make all facts and rules true.

The following definitions realize this method,where by $Pos(S)$ (respectively, $Pos(L)$) we understand the 4QL^{Bel} program (respectively, the set of literals) obtained from S (respectively, from L) by replacing each negative literal $\neg\ell$ by ℓ', where ℓ' is obtained from ℓ by changing relation symbol of ℓ by a fresh relation symbol (for simplicity denoted by its primed symbol).

Let us start with the case of modules not referring to any other modules.

Definition 6. Let S be a set of (ground) rules. Then gen^S is obtained from the least model of $Pos(S)$ by substituting literals of the form ℓ' with $\neg\ell$. For and X, Y being sets of literals, we define:

$$\delta_X^S(Y) \overset{\text{def}}{=} Y \cup \{\ell, \neg\ell \mid \text{ there is a rule } `\ell :\!-\ \beta.` \in S \text{ such that } \beta(X \cup Y) = \mathfrak{i}$$
$$\text{and } \ell(X \cup Y) \neq \mathfrak{i}\},$$

$$\Delta^S(X) \overset{\text{def}}{=} \text{the least fixpoint of } \delta_X^S,$$
$$correct^S(X) \overset{\text{def}}{=} X \cup \Delta^S(X). \qquad \triangleleft$$

For X being a set of ground literals, let $incons(X) \overset{\text{def}}{=} \{\ell, \neg\ell \mid X(\ell) = \mathfrak{i}\}$.

Definition 7. Let S be a set of (ground) rules. By the set of *pre-consequences* of a set of literals X of S, denoted by $Pre^S(X)$, we understand the set of literals defined by:

$$Pre^S(X) \overset{\text{def}}{=} correct^S\big(incons(X) \cup gen^{T(X)}\big),$$

where $T(X) \overset{\text{def}}{=} S - \{\varrho \in S \mid concl(\varrho) \in incons(X)\}$. $\qquad \triangleleft$

Definition 8. For any set S of (ground) rules, the well supported model of S, denoted by W^S, is defined by $W^S = \bigcup_{i>0}(Pre^S)^i(gen^S)$, where:

$$(Pre^S)^i(X) \overset{\text{def}}{=} \begin{cases} Pre^S(X) & \text{when } i = 1; \\ Pre^S((Pre^S)^{i-1}(X)) & \text{when } i > 1. \end{cases} \qquad \triangleleft$$

Let us now consider the case when modules of a 4QL^{Bel} program refer to other modules. Since the reference graphs of 4QL^{Bel} programs are acyclic, the computation of well-supported models can be organized by performing the following steps until well-supported models of all modules are computed:

1. compute the well-supported models of modules not referring to other modules;
2. replace formulas involving '$\in$' by truth values in all modules referring to those with the already computed well-supported models.[3]

3.3   Complexity of 4QL^{Bel}

As regards complexity issues, the following theorems can be proved similarly to analogous results for 4QL [17–19]. When considering a 4QL^{Bel} program P, by $\#D$ we denote the sum of the sizes of all domains of P and by $\#P$ we denote the number of modules in P.

Theorem 1. *For every 4QL^{Bel} program P, well-supported models of its modules can be computed in deterministic polynomial time in* $\max\{\#D, \#P\}$.   ◁

Theorem 2. *The querying problem for 4QL^{Bel} has deterministic polynomial time complexity, i.e., for every 4QL^{Bel} program P and formula α, the set of all tuples satisfying α in the well-supported model of P can be computed in deterministic polynomial time in* $\max\{\#D, \#P\}$.   ◁

Theorem 3. *4QL^{Bel} captures deterministic polynomial time over linearly ordered domains. That is, every polynomially computable query to a belief base (therefore, to a belief structure, too) can be expressed in 4QL^{Bel}.*   ◁

4   Specifying Belief Structures in 4QL^{Bel}

Let us first recall belief structures and epistemic profiles as introduced in [11]. Further on we fix a finite set of constants *Const*. Recall that $\mathbb{C} = \text{Fin}(\mathcal{G}(\text{Const}))$ is the set of all finite sets of ground literals over *Const*.

Definition 9

- By a *constituent* we understand any set $C \in \mathbb{C}$;
- by an *indeterministic epistemic profile* we understand any function $\mathcal{E}$ of the sort $\text{Fin}(\mathbb{C}) \longrightarrow \text{Fin}(\mathbb{C})$;
- by an *indeterministic belief structure over an indeterministic epistemic profile* $\mathcal{E}$ we mean $\mathcal{B}^{\mathcal{E}} = \langle \mathcal{C}, \mathcal{F} \rangle$, where:
  - $\mathcal{C} \subseteq \mathbb{C}$ is a nonempty set of constituents;
  - $\mathcal{F} \stackrel{\text{def}}{=} \mathcal{E}(\mathcal{C})$ is the set of *consequents* of $\mathcal{B}^{\mathcal{E}}$.   ◁

---

[3] Note that due to the acyclicity of the reference graph, new modules not referring to other modules are obtained in this step.

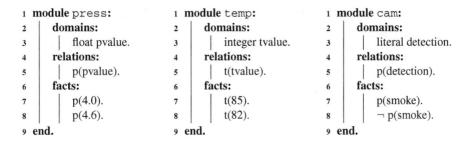

Fig. 2. Sample constituents.

To implement belief structures using $4QL^{Bel}$ we simply encode constituents and consequents by $4QL^{Bel}$ modules and query them using, among others, the $Bel()$ operator indexed by constituents and consequents, respectively. Epistemic profiles can be defined by auxiliary $4QL^{Bel}$ modules or by rules directly included in modules specifying consequents.

In order to evaluate $Bel()$-free formulas in belief structures one can use formulas indexed by constituents or consequents, as needed. To illustrate the idea let us consider robots equipped with temperature and pressure sensors together with a camera. In this case beliefs can be formed on the basis of robots' perception resulting from sensor measurements and interpretations of camera images. A natural belief structure for this simple scenario is $\mathcal{B}$ with constituents represented by modules {press, temp, cam} shown in Fig. 2, where (identifying modules with their well-supported models):

- press gathers measurements of pressure sensors, press = {p(4.0), p(4.6)};
- temp gathers measurements of temperature sensors, temp = {t(85), t(82)};
- cam gathers information extracted from camera, cam = {p(smoke), ¬p(smoke)}.

The $\mathcal{B}$'s epistemic profile is encoded by rules in modules f1 and f2 shown in Fig. 3. The well-supported models of these modules are consequents of

```
1 module f1:
2 |   relations:
3 |   |   danger().
4 |   rules:
5 |   |   danger() :–
6 |   |   ∃ X(press.p(X)∧ X≥ 4.3)
7 |   |   ¬danger() :–
8 |   |   ∀ X(temp.t(X)→ X≤ 60).
9 end.
```

```
1 module f2:
2 |   relations:
3 |   |   danger().
4 |   rules:
5 |   |   danger() :–
6 |   |   cam.p(fire)∨cam.p(smoke).
7 end.
```

Fig. 3. Sample consequents and rules defining an epistemic profile.

238   Ł. Białek et al.

$\mathcal{B}$. They represent two alternative views on the world: $\mathtt{f1} = \{\mathrm{danger}()\}$ and $\mathtt{f2} = \{\mathrm{danger}(), \neg \mathrm{danger}()\}$.

Now, for example $\mathrm{Bel}_{\{\mathtt{f1},\mathtt{f2}\}}(danger) = \mathtt{i}$ and $\{\mathtt{press}, \mathtt{temp}\}.(t(10) \wedge \neg p(4.6)) = \mathtt{f}$ (the latter because $\bigcup\{\mathtt{press}, \mathtt{temp}\} = \{p(4.0), p(4.6), t(85), t(82)\}$).

5 Conclusions

In the current paper we have defined a rule-based four-valued language $4\mathrm{QL}^{\mathrm{Bel}}$ allowing one for paraconsistent and paracomplete doxastic reasoning with belief bases and belief structures. We have also shown the tractability of the approach. Moreover, all tractable queries to belief bases and belief structures can be expressed in our language.

Due to the fact that epistemic profiles encapsulate reasoning processes, including non-deductive ones possibly rooted in argumentation or computational social choice theory, we can abstract from the way in which individual agents and groups form their consequents. This essential property allows us to address individual and group reasoning in a uniform way. As soon as these profiles are computable in deterministic polynomial time, they can be expressed within the $4\mathrm{QL}^{\mathrm{Bel}}$ language.

To the best of our knowledge, no other rule language allows for combining paraconsistent and paracomplete doxastic reasoning with nonmonotonic rules for completing missing knowledge and disambiguating inconsistencies, still retaining tractability of computing well supported models and queries.

References

1. de Amo, S., Pais, M.S.: A paraconsistent logic approach for querying inconsistent databases. Int. J. Approx. Reason. **46**, 366–386 (2007)
2. Balbiani, P., Pearce, D., Uridia, L.: On logics of group belief in structured coalitions. In: Michael, L., Kakas, A. (eds.) JELIA 2016. LNCS, vol. 10021, pp. 97–111. Springer, Cham (2016). doi:10.1007/978-3-319-48758-8_7
3. Béziau, J.Y., Carnielli, W., Gabbay, D. (eds.): Handbook of Paraconsistency. College Publications, Norcross (2007)
4. Cholvy, L., Hunter, A.: Information fusion in logic: a brief overview. In: Gabbay, D.M., Kruse, R., Nonnengart, A., Ohlbach, H.J. (eds.) ECSQARU/FAPR-1997. LNCS, vol. 1244, pp. 86–95. Springer, Heidelberg (1997). doi:10.1007/BFb0035614
5. Cholvy, L., Hunter, A.: Merging requirements from a set of ranked agents. Knowl.-Based Syst. **16**(2), 113–126 (2003)
6. da Costa, N., Bueno, O.: Belief change and inconsistency. Log. Anal. **41**(161–163), 31–56 (1998)
7. Dunin-Kęplicz, B., Strachocka, A.: Tractable inquiry in information-rich environments. In: Proceedings of the 24th IJCAI, pp. 53–60 (2015)
8. Dunin-Kęplicz, B., Strachocka, A.: Paraconsistent argumentation schemes. Web Intell. **14**, 43–65 (2016)

9. Dunin-Kęplicz, B., Szałas, A.: Epistemic profiles and belief structures. In: Jezic, G., Kusek, M., Nguyen, N.-T., Howlett, R.J., Jain, L.C. (eds.) KES-AMSTA 2012. LNCS, vol. 7327, pp. 360–369. Springer, Heidelberg (2012). doi:10.1007/978-3-642-30947-2_40

10. Dunin-Kęplicz, B., Szałas, A.: Taming complex beliefs. In: Nguyen, N.T. (ed.) Transactions on Computational Collective Intelligence XI. LNCS, vol. 8065, pp. 1–21. Springer, Heidelberg (2013). doi:10.1007/978-3-642-41776-4_1

11. Dunin-Kęplicz, B., Szałas, A.: Indeterministic belief structures. In: Jezic, G., Kusek, M., Lovrek, I., J. Howlett, R., Jain, L.C. (eds.) Agent and Multi-agent Systems: Technologies and Applications. AISC, vol. 296, pp. 57–66. Springer, Cham (2014). doi:10.1007/978-3-319-07650-8_7

12. Dunin-Kęplicz, B., Szałas, A., Verbrugge, R.: Tractable reasoning about group beliefs. In: Dalpiaz, F., Dix, J., Riemsdijk, M.B. (eds.) EMAS 2014. LNCS, vol. 8758, pp. 328–350. Springer, Cham (2014). doi:10.1007/978-3-319-14484-9_17

13. Dunin-Kęplicz, B., Verbrugge, R.: Teamwork in Multi-Agent Systems. A Formal Approach. Wiley, Hoboken (2010)

14. Fagin, R., Halpern, J.: Belief, awareness, and limited reasoning. Artif. Intell. **34**(1), 39–76 (1988)

15. Fagin, R., Halpern, J., Moses, Y., Vardi, M.: Reasoning About Knowledge. MIT Press, Cambridge (2003)

16. van Harmelen, F., Lifschitz, V., Porter, B.: Handbook of Knowledge Representation. Elsevier, Amsterdam (2007)

17. Małuszyński, J., Szałas, A.: Living with inconsistency and taming nonmonotonicity. In: Moor, O., Gottlob, G., Furche, T., Sellers, A. (eds.) Datalog 2.0 2010. LNCS, vol. 6702, pp. 384–398. Springer, Heidelberg (2011). doi:10.1007/978-3-642-24206-9_22

18. Małuszyński, J., Szałas, A.: Logical foundations and complexity of 4QL, a query language with unrestricted negation. J. Appl. Non-Class. Logics **21**(2), 211–232 (2011)

19. Małuszyński, J., Szałas, A.: Partiality and inconsistency in agents' belief bases. In: Barbucha, D., Le, M., Howlett, R., Jain, L. (eds.) Frontiers in Artificial Intelligence and Applications, vol. 252, pp. 3–17. IOS Press, Amsterdam (2013)

20. Meyer, J.J.C., van der Hoek, W.: Epistemic Logic for Computer Science and Artificial Intelligence. Cambridge University Press, Cambridge (1995)

21. Priest, G.: Special issue on impossible worlds. Notre Dame J. Formal Logic **38**(4), 481–660 (1997)

22. Priest, G.: Paraconsistent belief revision. Theoria **67**(3), 214–228 (2001)

23. Szałas, A.: How an agent might think. Logic J. IGPL **21**(3), 515–535 (2013)

24. Tanaka, K.: The AGM theory and inconsistent belief change. Logique & Analyse **48**(189–192), 113–150 (2005)

Combining Machine Learning and Knowledge-Based Systems for Summarizing Interviews

Angel Luis Garrido[1]([⊠]), Oscar Cardiel[2], Andrea Aleyxendri[2],
and Ruben Quilez[2]

[1] IIS Department, University of Zaragoza, Zaragoza, Spain
garrido@unizar.es
[2] Computer Department, Grupo Heraldo, Zaragoza, Spain
{ocardiel,aaleyxendri,rquilez}@heraldo.es

Abstract. Achieving optimal results of an automatic summarization process is frequently conditioned by the knowledge of the domain. The performance of general methods is always lower than what can be achieved by introducing custom modifications taking into account the context. Nevertheless, these type of custom adjustments represents a hard work by experts and developers, which is not always possible to achieve due to the high costs. In this work we aim to leverage the features of the documents in order to classify them by using machine learning methods. Once the typology is identified, the application of improvements is done by a knowledge-based system that allows users to easily customize both the summarization process, and the presentation to the final user. The proposed method has been applied with promising results to interviews in a real environment of a major Spanish media group.

Keywords: Summarization · Machine learning · Automatic classification · Knowledge-based systems · Interviews

1 Introduction

Text Summarization is defined as the act of condensing a text into a shorter version preserving its most relevant information. For a computer, or even for a human, summarizing is a challenging task which involves the utilization of several different techniques, all them related with Artificial Intelligence. Although summarization algorithms have became increasingly used both in research and industry, the emergence of the World Wide Web triggered a blowup in the amount of textual information in all areas, and thereupon automatic summarization gained prominence thanks to its different applications.

Practically in any domain we can find that text documents do not always have the same format or structure. A clear example would be the legal documents, where very different formats can be found for judicial sentences, laws, notarial documents, etc. Attempting to make summaries of these documents with the

© Springer International Publishing AG 2017
M. Kryszkiewicz et al. (Eds.): ISMIS 2017, LNAI 10352, pp. 240–250, 2017.
DOI: 10.1007/978-3-319-60438-1_24

same approach and with a similar presentation format probably does not lead to an optimal outcome, and it seems logical to think that the presentation format of the summary will not be the most appropriate.

From a viewpoint of a documentalist, the summaries of each type of text in a domain could need a specific format with the aim of being more readable. For instance, in the news domain, if we pay attention to interviews, on the one hand is interesting to have a short summary of the interviewed character and, on the other hand, the topics covered in the questions and the opinions extracted from the answers should appear separately in the summary. Hence, in this work, we hypothesize that a good summary should keep the most relevant information taking into account the typology on these three aspects: linguistic content, structure and presentation of the summary.

In this work, we describe a new methodology for the design of specialized summarizer systems. This methodology has been applied on a software called "NESSY" (NEws Summarization SYstem), which summarizes the daily news production of several Spanish newspapers, and with a minimal effort, it has been possible to handle the summarization process of certain typologies for obtaining much better results. The main contribution of this work is to make use of the automatic categorization of news, to then facilitate the summary generation process, by using a knowledge-based system (KBS). That KBS contains the specialized information both to perform the summarization process, and to present the results in an optimized way. We have selected an appropriate dataset for testing the methodology, specifically one related to news, and we have selected a particular type of news: the interview. It is an outstanding example of the type of document that by its structure and content is very different from the rest of the texts in its context. The experiments show a good outcome on a real working environment, provided by *Grupo Heraldo*[1], a major Spanish media group.

In the remainder of this paper, we study the state of the art in Sect. 2, and the methodology is detailed in Sect. 3. Thereafter, we present in Sect. 4 the results of the empirical study conducted to assess the our proposal. Finally, we offer some concluding remarks and directions for future work in Sect. 5.

2 Related Work

First works related with automatic summarization are from almost seventy years ago: in [1] is proposed to weight the sentences of a document as a function of high frequency words and disregarding the very high frequency common words. Other approaches [2] leveraged the use of certain words, the titles, and the structure for assessing the weight of the sentences in the text. Machine learning techniques started to be used in the 90's. Some systems used naive-Bayes methods [3], and others leveraged on learning algorithms [4]. On the other hand, hidden Markov models were used in several works [5].

---

[1] http://www.grupoheraldo.com.

Our work deals with the single-document generic extractive summaries, taking into account an important fact: the aforementioned approaches always works with a homogeneous corpus. We think that is not the most realistic view, since specific typologies of documents can easily be found in any domain, as for example the interviews in the news domain, which own a very special structure and characteristics.

Regarding the use of knowledge bases, they are widely used as resources to allow exchange knowledge between humans and computers. They are based on the principles of *ontologies*. An ontology [6] is a formal and explicit specification of a shared conceptualization that can be used to model human knowledge and to implement intelligent systems. Ontologies are widely used nowadays [7]; they can be used to model data repositories [8], or they can serve for purposes of classification [9,10]. They also can be used to guide extraction data processes from heterogeneous sources [11–14]. When the extraction methodology is based on the use of ontologies in an information extraction system, it belongs to the OBIE (Ontology Based Information Extraction) system group [15].

News represent an important field of application for automatic summarization. News summarizer systems achieves acceptable results in the traditional format of journalistic articles, but the summarization of more complex formats, such as reviews, interviews or opinion pieces, is currently beyond the state of the art, due to the large differences in structure and style. If we analyze, for example, interviews, we realize that is a very special type of news. The interviews disclose the ideas and opinions of a character through a dialog between the interviewee and the interviewer. When writing an interview it is important to make sure of having a great title that makes readers want to read this piece. Pictures, captions, and quotes are also extremely important. The beginning paragraph should describe the interviewed. The answers also include parenthetical notes, which helps the reader see what happened during the interview and give the interviewee emotion, which can be difficult with the written word. All these characteristics make the interviews especially difficult to summarize without a customized work, and there are no much research on this topic.

Regarding other works related to news, we can cite Newsblaster [16], a system to summarize texts that can help users find the most important news. Another sample is SweSum [17], whose basic architecture is the keyword search based on frequency, the weights of the sentences according to their position in the text, and the weights of text elements (bold, numbers, etc.), user's keywords, etc. An example of news aggregator and summarizer is Google News [18], a computer-generated news site that aggregates headlines from news sources worldwide, groups similar stories together and displays them according to each reader's personalized interests. All the above mentioned works, do not consider applying previous efforts to categorize the news in order to be able to optimize the summaries that are made afterwards. Besides, as far as we have been able to study, there are also no scientific papers aimed at studying the automatic construction good summaries of a type of news as widespread as the interview.

3 Methodology

This section explains the working methodology applicable to any system dedicated to producing single summaries from a single-source document, in a domain where there is at least one document typology that, due to the specificity of its content or its structure, can not be optimally summarized by a generic summarizer.

The key of this automatic summarization process is to guess the specific typology of the source document. Then we should define in an ontology both the customization of the process for this type and the presentation to the final user. Hence, the advantage of our modular development is we can interchange these methods easily in order to work with other typologies, or in other domains, only by modifying the ontology.

As described in Fig. 1, the system's input is a document that will go through a set of specialized treatments until the production of the summary. Our document summary generation process is divided into four stages: categorizing, preprocessing, ranking, and showing. Knowing the type of text allows the process to exploit its features by using the knowledge stored in the ontology. The size of the summary is set by a *compression rate*, given by a specific number of words, or by a percentage of total text size.

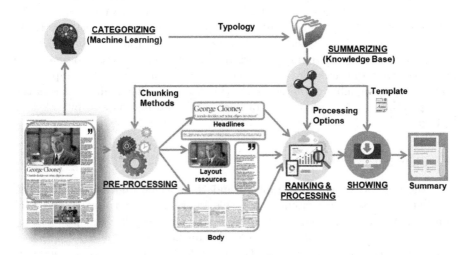

Fig. 1. Overview of the methodology

3.1 Categorizing

Supervised learning algorithms are used in this stage to train the categorizer in recognizing the specific typologies whose summaries we want to improve. Each type has unique features that can be used for classification purposes: writing style, predominant vocabulary, linguistic resources, etc.

3.2   Pre-processing

This stage consists of breaking up the original text in order to process each of the parts separately. This task is customized according to an external ontology which guides the process using a set of different chunking extraction methods. These methods are designed apart from the main process, and it is the ontology that serves to link with these methods. Therefore, according to the information provided by the ontology, the most adequate methods are used for analyzing the text and fragmenting into groups of sentences, taking into account the typology. These methods can base their design on any technique, such as pattern rules or machine learning techniques.

3.3   Ranking and Processing

Depending on the structure (obtained in the pre-processing stage) and the typology (obtained in the categorizing stage) of the input text, different ranking and processing options apply. The ontology indicates what information should be extracted, where it is, and how can be obtained. The ranking stage scores the sentences by following a list of criteria or characteristics defined by the typology. The standard ranking is made by generating a catalog with the frequencies of keywords and named entities (persons, locations, organizations, products, etc.) in the original text. By using TF-IDF method [19] each word receives a weight according to their frequency. To get the score of each sentence, the way to do it is by adding a weight value to each word, and then the total sum of these weights is divided by the number of words in each sentence. The sentences are sorted according to their score, and considering the aforementioned compression rate to filter the number of sentences. At this stage, specific modifications of the scoring method can be adopted according to the information also provided by the ontology.

3.4   Showing

After the sentences have been ranked, and the specific information has been obtained, the system concatenates all this information and creates the final summary. At this stage the ontology indicates what specific and professional template should be used for each type of document. The template is composed by a set of presentation rules related to the information given in the previous stages.

4   Evaluation

In order to validate our methodology, we have applied it on a real system called NESSY, devoted to summarize news. We have chosen the news domain because it is as an appropriate use case: news present a number of common structural elements [20]: headlines, captions (from photographs or graphics), date, author, body, and layout resources (quotes, numbers, tables, etc.). Summaries of each type of news must leverage these different attributes to achieve good results. We have performed the experiments in collaboration with *Grupo Heraldo*[2], a major Spanish media group.

4.1   Datasets

In our experiments, we have used three datasets, all them containing news taken from *Heraldo de Aragón*[3], a major Spanish newspaper:

- *DSHA-1* is a corpus composed by 14,000 news. These news had been previously categorized by the documentation department of the company.
- *DSHA-2* is another corpus of 400 standard news. Each of the news has an associated summary made by professional documentalists. This dataset is used to empirically obtain parameters of the summarizer (see Subsect. 4.3).
- *DSHA-2i* is a smaller corpus of 40 interviews. Each of the document has an associated summary made by professional documentalists. This dataset is used to test the precision and the recall of the summarization task.

These datasets are available upon request to the authors exclusively for research purposes, subject to confidentiality agreements due to copyright issues.

4.2   Automatic Classification

We have categorized the news with Support Vector Machines (SVM), an avowed supervising learning method [21]. We have chosen this technique because SVMs are able of extracting an optimal solution with a very small training set size, it works properly when data is randomly scattered, and also for its simplicity and its speed [22]. Besides, our previous experience with SVM has provided us with very good results [23–25].

4.3   Standard Ranking and Processing

As a general methodology for any document other than an interview, NESSY gives an extra score to the first sentence of the body, the first sentence of each paragraph, and those sentences containing keywords or named entities located in the title. The score of these aforementioned sentences is multiplied by an α value. The α value has been empirically obtained with experimental tests over the whole *DSHA-2* dataset, setting this value to 10. Therefore, the total score of each sentence is the sum of the weights of their keywords, named entities multiplied by this extra score during the ranking process. Eventually, the final summary is a concatenation of the highest scoring sentences sorted by their position in the original document.

---

[2] http://www.grupoheraldo.com.
[3] http://www.heraldo.es.

4.4   Using the Knowledge Base to Customize Interview's Summaries

We have designed a specific knowledge base system (KBS) containing detailed information in order to guide the different stages of the summarization process when an interview is detected. A fragment of the ontology can be appreciated in Fig. 2. The ontology includes references to the different methods used during the process. One advantage of this architecture is that new typologies can be added dynamically in a quick way.

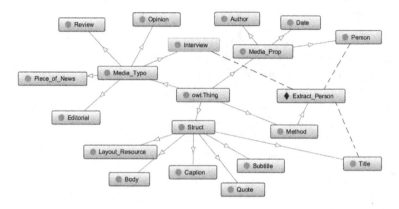

Fig. 2. A partial sample of the ontology model, showing one of the extraction methods

The first task is to identify the aforementioned parts of the text by using a set of simple rules stored in the own KBS. The body of the interview is divided into an optional block of introduction (usually found before the first question) and several blocks of questions and answers ("Q-A"). Otherwise, the body of the interview begins directly with blocks Q-A. Moreover, the system is able to detect layout resources like the interviewee's quotes or biographical notes. Making use of the ontology, the interviewee's name, and his/her main quote are extracted from the headline. Typically, the profession of the interviewee is also located in the headline. We can obtain this information with simple pattern rules and specialized databases of surnames[4].

To punctuate the words, NESSY use a catalog with the frequencies of keywords and named entities in the original text. Applying TF-IDF (Term Frequency-Inverse Document Frequency) each word receives a weight according to its frequency. The blocks of questions are syntactically analyzed to obtain a list of noun phrases. The topics covered will be the highest scored noun phrases containing the keywords and named entities that appear in the blocks of questions, (i.e. if one of the main keyword is "film", one of the topics could be "the film that makes you feel proud"). Opinions of the interviewee are the sentences with the highest score literally taken from the response block. Specific heuristics are used to score blocks and sentences of the text. The features that have been considered for the ranking are:

---

[4] E.g., http://dbpedia.org/page/Category:Spanish-language_surnames.

a. Introduction text: The paragraphs of the introduction of the interview contain keywords and named entities, each of them multiply their score by α.
b. Location of sentences: Answer sentences which appear on the top position of each block Q-A usually represent the views of the interviewee. The value of these phrases multiplies also their score by α.
c. Clustering Q-A blocks by topics: The distribution of certain keywords in several Q-A blocks define a set of blocks on the same topic. Each cluster of answers is considered as another single document, so the process covers first all the topics with at least one sentence of each cluster, before introducing more sentences of the same cluster. This approach is inspired in multi-document summarizers like [26].

Finally, the score of each sentence is the sum of the weights of their keywords and named entities. Regarding the output template stored in the ontology, it has been made with the help of a Documentation Department. For example, in interviews, the final summary is divided into a header and a body with two blocks: main topics, and opinions of the interviewee. The header is formed from the name of the interviewee, date, author, and a characteristic phrase. In the body we apply two rules: (1) no more than two sentences from the same Q-A block will be extracted, (2) sentences must have a minimum of ten words to appear in the final summary.

4.5   Comparison

In order to compare results, we have used a set of on-line summarizers: SWE-SUM[5] (Sw), Tools4noobs[6] (T4n), Autosummarizer[7] (AS), and the Mashape Tools[8] (MT). We have configured all these summarizers for achieving a compression rate of 20%. Finally, we have used the ROUGE-L[9] method to compare the automatic summaries obtained with the examples of handmade summaries existing in our *DSHA-2i* dataset.

4.6   Experiments

For the first experiment 5,000 news from the dataset *DSHA-1* have been selected, and the most outstanding 5 typologies have been considered: Editorial, Interview, Opinion, Piece of News, and Review. So, we have used 1,000 news of each type: 800 texts are used to train the model and 200 are used to test it. In this experiment we have used SVM Multiclass with radial basis function (RBF) kernel, achieving a 92.15% of accuracy for 5-fold validation categorization. We select

---

[5] http://swesum.nada.kth.se/index-eng.html.
[6] https://www.tools4noobs.com/summarize/.
[7] http://autosummarizer.com/.
[8] http://textsummarization.net/.
[9] ROUGE-L is one of the five evaluation metrics avaliable in ROUGE (a recall-based metric for fixed-length summaries), and it is based on founding the longest common subsequence.

a random set of 320 news from the *DSHA-1* dataset, corresponding to all the types except Interviews (80 of each).

Then, we have introduced in this set the 40 interviews of the *DSHA-2i* dataset. We used the aforementioned classifier to classify the whole 400 news, and we have applied the enhanced summarization methodology to the subset classified as "Interviews" by the SVM tool. Finally, we have compared the obtained summaries with the adhoc summaries prepared by a professional documentation department by using ROUGE-L. The results of NESSY compared with the rest of summarizers are shown in Table 1.

As it can be seen, we have obtained satisfactory results, although it is noteworthy to point out that the mistakes in the classification stage negatively affect the generation of the summaries, since a piece of news classified in a wrong way will be summarized in the second stage using an inadequate procedure.

Table 1. F-measure results regarding the single-source summarization task over a subset of 40 interviews within a dataset of 400 news

	Sw	T4n	AS	MT	Nessy
Interview	0.51	0.46	0.45	0.52	0.66
Editorial	0.49	0.21	0.29	0.48	0.50
Opinion	0.41	0.35	0.36	0.39	0.43
Review	0.36	0.37	0.33	0.26	0.39
Piece of news	0.42	0.33	0.34	0.33	0.44

Even so, the improvement that is obtained in the summary process of the interviews is quite significant (30% higher), so we can conclude that the applied methodology optimizes the process.

5   Conclusions and Future Work

In this paper, we have introduced a supervised learning methodology for building automatic extractive summaries in a specific context with presence of specialization of certain type of documents. The methodology is based on performing an automatic categorization of the input texts, and then, with the help of a knowledge base system, customize the process of generating summaries for each typology. It can be applied in multiple working environments, with the advantage that the system, from a sample, is able to optimize a specific type of document summarization with little effort and without affecting the rest of typologies. For assessing this approach, we have applied it with the specific case of the interviews in the news domain, performing an evaluation over a real dataset. As future work we will test it with different contexts and typologies of documents to verify its correct behavior.

Acknowledgments. This research work has been supported by the CICYT project TIN2013-46238-C4-4-R, TIN2016-78011-C4-3-R (AEI/FEDER, UE), and DGA/FEDER.

References

1. Luhn, H.P.: The automatic creation of literature abstracts. IBM J. Res. Dev. **2**(2), 159–165 (1958)
2. Edmundson, H.P.: New methods in automatic extracting. J. ACM (JACM) **16**(2), 264–285 (1969)
3. Kupiec, J., Pedersen, J., Chen, F.: A trainable document summarizer. In: Proceedings of the 18th International ACM SIGIR Conference on Research and Development in Information Retrieval (SIGIR 1995), pp. 68–73. ACM (1995)
4. Lin, C.Y.: Training a selection function for extraction. In: Proceedings of the 8th International Conference on Information and Knowledge Management (CIKM 1999), pp. 55–62. ACM (1999)
5. Conroy, J.M., O'leary, D.P.: Text summarization via hidden Markov models. In: Proceedings of the 24th International ACM SIGIR Conference on Research and Development in Information Retrieval (SIGIR 2001), pp. 406–407. ACM (2001)
6. Gruber, T.R.: A translation approach to portable ontology specifications. Knowl. Acquis. **5**(2), 199–220 (1993)
7. Bobed, C., Yus, R., Bobillo, F., Ilarri, S., Bernad, J., Mena, E., Trillo-Lado, R., Garrido, Á.L.: Emerging semantic-based applications. In: Workman, M. (ed.) Semantic Web, pp. 39–83. Springer, Cham (2016). doi:10.1007/978-3-319-16658-2_4
8. Barbau, R., Krima, S., Rachuri, S., Narayanan, A., Fiorentini, X., Foufou, S., Sriram, R.D.: Ontostep: enriching product model data using ontologies. Comput.-Aided Des. **44**(6), 575–590 (2012)
9. Vogrinčič, S., Bosnić, Z.: Ontology-based multi-label classification of economic articles. Comput. Sci. Inf. Syst. **8**, 101–119 (2011)
10. Garrido, A.L., Gómez, O., Ilarri, S., Mena, E.: An experience developing a semantic annotation system in a media group. In: Bouma, G., Ittoo, A., Métais, E., Wortmann, H. (eds.) NLDB 2012. LNCS, vol. 7337, pp. 333–338. Springer, Heidelberg (2012). doi:10.1007/978-3-642-31178-9_43
11. Kara, S., Alan, Ö., Sabuncu, O., Akpınar, S., Cicekli, N.K., Alpaslan, F.N.: An ontology-based retrieval system using semantic indexing. Inf. Syst. **37**(4), 294–305 (2012)
12. Borobia, J.R., Bobed, C., Garrido, A.L., Mena, E.: SIWAM: using social data to semantically assess the difficulties in mountain activities. In: 10th International Conference on Web Information Systems and Technologies (WEBIST 2014), pp. 41–48 (2014)
13. Buey, M.G., Garrido, A.L., Bobed, C., Ilarri, S.: The AIS project: boosting information extraction from legal documents by using ontologies. In: Proceedings of the 8th International Conference on Agents and Artificial Intelligence (ICAART 2016), Rome, Italy, pp. 438–445. SCITEPRESS (2016)
14. Garrido, A.L., Buey, M.G., Muñoz, G., Casado-Rubio, J.-L.: Information extraction on weather forecasts with semantic technologies. In: Métais, E., Meziane, F., Saraee, M., Sugumaran, V., Vadera, S. (eds.) NLDB 2016. LNCS, vol. 9612, pp. 140–151. Springer, Cham (2016). doi:10.1007/978-3-319-41754-7_12
15. Wimalasuriya, D.C., Dou, D.: Ontology-based information extraction: an introduction and a survey of current approaches. J. Inf. Sci. **36**(3), 306–323 (2010)

16. Evans, D.K., Klavans, J.L., McKeown, K.R.: Columbia newsblaster: multilingual news summarization on the web. In: Demonstration Papers at HLT-NAACL 2004, pp. 1–4. Association for Computational Linguistics (2004)
17. Dalianis, H.: Swesum: a text summarizer for Swedish. KTH (2000)
18. Das, A.S., Datar, M., Garg, A., Rajaram, S.: Google news personalization: scalable online collaborative filtering. In: Proceedings of the 16th International Conference on World Wide Web, pp. 271–280. ACM (2007)
19. Salton, G., Buckley, C.: Term-weighting approaches in automatic text retrieval. Inf. Process. Manag. **24**(5), 513–523 (1988)
20. Bell, A.: The discourse structure of news stories. In: Approaches to Media Discourse, pp. 64–104 (1998)
21. Joachims, T.: Text categorization with support vector machines: learning with many relevant features. In: Nédellec, C., Rouveirol, C. (eds.) ECML 1998. LNCS, vol. 1398, pp. 137–142. Springer, Heidelberg (1998). doi:10.1007/BFb0026683
22. Shin, K.S., Lee, T.S., Kim, H.J.: An application of support vector machines in bankruptcy prediction model. Expert Syst. Appl. **28**(1), 127–135 (2005)
23. Garrido, A.L., Gomez, O., Ilarri, S., Mena, E.: NASS: News annotation semantic system. In: Proceedings of the 23rd IEEE International Conference on Tools with Artificial Intelligence (ICTAI 2011), pp. 904–905. IEEE (2011)
24. Garrido, A.L., Buey, M.G., Ilarri, S., Mena, E.: GEO-NASS: a semantic tagging experience from geographical data on the media. In: Catania, B., Guerrini, G., Pokorný, J. (eds.) ADBIS 2013. LNCS, vol. 8133, pp. 56–69. Springer, Heidelberg (2013). doi:10.1007/978-3-642-40683-6_5
25. Garrido, A.L., Buey, M.G., Escudero, S., Peiro, A., Ilarri, S., Mena, E.: The GENIE project-a semantic pipeline for automatic document categorisation. In: Proceedings of the 10th International Conference on Web Information Systems and Technologies (WEBIST 2014), pp. 161–171. SCITEPRESS (2014)
26. Silveira, S.B., Branco, A.: Extracting multi-document summaries with a double clustering approach. In: Bouma, G., Ittoo, A., Métais, E., Wortmann, H. (eds.) NLDB 2012. LNCS, vol. 7337, pp. 70–81. Springer, Heidelberg (2012). doi:10.1007/978-3-642-31178-9_7

Validity of Automated Inferences in Mapping of Anatomical Ontologies

Milko Krachunov$^{(\boxtimes)}$, Peter Petrov, Maria Nisheva, and Dimitar Vassilev

Faculty of Mathematics and Informatics, Sofia University "St. Kliment Ohridski",
5 James Bourchier Blvd., 1164 Sofia, Bulgaria
milkok@fmi.uni-sofia.bg

Abstract. A system for automated prediction and inference of cross-ontology links is presented. External knowledge sources are used to create a primary body of predictions. The structure of the projected super-ontology is then used to automatically infer additional predictions. Probabilistic scores are attached to all of these predictions, allowing them to be filtered using a statistically-selected threshold. Three anatomical ontologies were mapped in pairs, and all the predicted mapping links were individually checked by a manual curator, allowing a closer look at the quality of the chosen prediction procedures, and the validity of the resulting mappings.

Keywords: Anatomical ontologies · Ontology mappings · Probabilities · Automated inference · Knowledge representation

1 Introduction

Ontologies are a powerful tool for knowledge representation that can be used to model the complex interrelationships of the entities in given problem domain. They provide a high expressiveness, in a standardised structure that allows knowledge sharing and interoperability, even across different domains of discourse. They are a popular instrument in Biology and Bioinformatics. The most significant undertaking is the Gene Ontology project [1], which aims to create a unified cross-species controlled vocabulary and annotation of genes and gene products. It is part of the larger OBO project [14], which is an effort to create a large body of controlled vocabularies that can be used across biological and medical domains.

This paper is aimed at creating software for automated and semi-automated ontology mapping [3]. This is the process of establishing semantic links between the concepts and terms of two or more input ontologies. Even with a common representation, and the existence of projects aiming at interoperability, this is usually not a straightforward goal. The mapping may be hampered by the use of different terminology, or because the ontologies aren't complete in their coverage of the problem domains. Conversely, the same terms may on occasion be

© Springer International Publishing AG 2017
M. Kryszkiewicz et al. (Eds.): ISMIS 2017, LNAI 10352, pp. 251–260, 2017.
DOI: 10.1007/978-3-319-60438-1_25

used to refer to different concepts. This is particularly an issue when specialised ontologies are mixed with general-purpose ones.

We focus specifically on the task of mapping the anatomical ontologies of different biological species, for the purpose of facilitating the transfer of acquired knowledge from one species to another. This task poses additional difficulties, because corresponding anatomical terms across species do not carry an equivalence relation. Instead, they have a fuzzy relation defined by a shared function and shared origin, with a varying degree of correspondence.

This paper presents a composite computational approach for discovering and scoring predicted mappings between pairs of ontologies, in particular anatomical ontologies taken from the OBO Foundry. In addition to the direct terminology matching between such compatible ontologies, the approach uses external vocabularies and structural matching to generate more potential predictions. The resulting predictions were checked by an expert in the field of animal anatomies for validation of the procedures and scoring schemes, as well as for evaluation of the use of automated structural inferences that attempt to discover links between terms of the projected super-ontology.

In terms of popular mapping technqiues, the presented approach utilises aggregated matching combining string-based mapping, linguistic resource mapping and taxonomic mapping [8], in which the linguistic resources include specialised ontologies, and the aggregated technique has been extended with probabilistic scores. It falls within the categories of terminological, semantic and structural matching [6]. Our taxonomic structural procedure CMP uses a similar strategy to the contextual similarity used in [5], but is limited to structure that is unexpected and is extended with predicted taxonomy and probabilistic scores as well.

2   The Problem of Merging Ontologies

The main goal is to predict the semantic links between the terms across a pair of anatomical ontologies of two different species, for example the ontology of the *mouse* anatomy and the ontology of the *zebrafish* anatomy. We are interested in predicting semantic links that correspond to one of the following relations $\mathcal{R}$: R_1—synonymy, R_2—hypernymy, R_3—hyponymy, R_4—holonymy, R_5—meronymy, which cover the full spectrum of relevant relations found in the utilised external knowledge sources, including the linguistic thesauri.

The predicted semantic links are slightly different from the relationships that comprise the input anatomical ontologies. Each predicted link would carry a *confidence score*, which would be a number between 0 and 1. In this work, it is interpreted as the *degree of certainty*, or a naïve estimate of the probability, that the relationship in question exists and should be added to the super-ontology. However, in a refined model this can be extended to the discovery of fuzzy relationships, indicating, for example, the strength of the correspondence between two terms.

Another special feature is the addition of the R_1 synonym relation, which doesn't exist in the structure of the input ontologies. That is because completely synonymous terms would be merged in a single term, but due to the cross-species nature of the mapping, the matches are more likely to represent different terms that are only functionally or ancestrally equivalent, in which case merging them in the same node may be incorrect, as well as reduce the knowledge represented in the mapped pair of ontologies.

All the input ontologies are taken from the Open Biomedical Ontologies (OBO) initiative, and their library—the OBO Foundry [14]. They are all in the OBO file format, as described in [4].

As external knowledge sources (ontologies and vocabularies), we utilise:

1. Unified Medical Language System (UMLS) [2]—a comprehensive thesaurus of biomedical concepts.
2. The Foundational Model of Anatomy (FMA) ontology [13]—a reference ontology in the domain of anatomy.
3. WordNet [7,9]—an unspecialised thesaurus of the English language.

2.1 The Task of Mapping

To discover candidate mapping links, we would only be interested in the $\mathcal{R}$ relations, of which only $R_2 \ldots R_5$ are present in the input ontologies. This allows us to discard any more complex relationships during the mapping procedure, and represent each input ontology as a directed acyclic graph, with an edge colouring function assigning one of the following relations to the edges:

- **is-a**: A parent-child relation between a superclass and subclass, where the parent is the hypernym of the child, and the child is the hyponym of the parent. For example, 'arm' and 'leg' are children (subclasses, hyponyms) of 'limb', which can be expressed with arm $\rightarrow_{\text{is-a}}$ limb.
- **part-of**: A transitive relation between a whole and its part, which will be also considered further as parent-child relation where the parent is the holonym, and the child is the meronym. For example, 'eye' and 'nose' are children (meronyms) of 'face', which can be expressed with eye $\rightarrow_{\text{part-of}}$ face.

We thus use the following graph model:

$$
\begin{aligned}
O_1 : G_1 &= (V_1, E_1); \\
F_1 : E_1 &\rightarrow C = \{\text{is-a}, \text{part-of}\} \\
O_2 : G_2 &= (V_2, E_2); \\
F_2 : E_2 &\rightarrow C = \{\text{is-a}, \text{part-of}\}
\end{aligned}
\tag{1}
$$

The two anatomical ontologies O_i ($i \in \{1, 2\}$) are represented by the directed acyclic graphs G_i, with the vertices V_i corresponding to the two sets of ontology terms, and edges E_i corresponding to the aforementioned *is-a* and *part-of* relations. The types of relations are assigned by the colouring functions F_i. For example, if $v_1 = $ 'arm' $\in V_1$ and $v_2 = $ 'limb' $\in V_1$, then $(v_1, v_2) \in E_1$ and $F(v_1, v_2) = $ is-a.

During the mapping, we would use the external knowledge sources $\mathcal{T}$: T_1—UMLS, T_2—FMA and T_3—WordNet. From each source $T_s \in \mathcal{T}$, we would get the set of **terms** $M_s = \{t_{s1}, \ldots, t_{sm_s}\}$ defined in it, as well as the set of **relations** $Q_s^{\text{is-a}}, Q_s^{\text{part-of}} \subseteq M_s \times M_s$.

Each source $T_s \in \mathcal{T}$ was assigned a fixed score $f(T_s)$, denoting the expected accuracy of the predictions acquired from it. These accuracies were empirically evaluated in [10].

The main goal of this work is to generate a set of predictions:

$$D = \{(v_{1k}, v_{2k}, r_k, s_k) | k = 1, 2, \ldots, |D|\} \tag{2}$$

Here, $v_{1k} \in V_1$ is a term from the ontology O_1, $v_{2k} \in V_2$ is a term from the ontology O_2, and they are predicted to have a relation $r_k \in \mathcal{R}$, with a confidence score $s_k \in (0, 1]$.

3   Matching and Predicting Procedures

Three procedures are used for the building of the set of predictions D. They are described in higher detail in [11].

The first procedure is *direct matching* (DM) between the terms of the two ontologies. The ontologies from the OBO initiative are constructed for interoperability, guaranteeing a certain level of terminological consistency. For this reason, any pair of terms $t_1 \in V_1$ and $t_2 \in V_2$ which have the same name are marked as synonyms, $r_k = R_1$, with a score s_k of 1.0.

The second procedure is *source matching predictions* (SMP), which extracts relationship predictions from the external sources $\mathcal{T}$ by textually matching the terms V_1 and V_2 from the ontologies and the terms M_s from a given source $T_s \in \mathcal{T}$. Two rules are used to create these predictions:

- *Rule (A)*. If the terms $t_1 \in V_1$ and $t_2 \in V_2$ are textual matches for $t \in T_s$, then t_1 and t_2 are marked as synonyms, $r_k = R_1$, with a score of $s_k = f(T_s)$.
- *Rule (B)*. If $t_j \in V_j$ is a textual match for $t' \in T_s$, and if $t_{3-j} \in V_{3-j}$ is a textual match for $t'' \in T_s$, and t'' is a *is-a*/**part-of** child/parent of t', then t_j is marked as predicted a cross-ontology *is-a*/**part-of** child/parent of t_{3-j} ($j \in \{1, 2\}$).

The final procedure is *child matching predictions* (CMP) which uses the structure of the projected super-ontology (or super-graph) that would be a result of a naïve union of the two ontologies or their graphs G_1 and G_2. The procedure works on terms $t_1 \in V_2$ and $t_2 \in V_2$ from a different ontology, and their children t_{ch1}, t_{ch2}. It tries to recognise one of the following three patterns:

(i) $t_1 \in V_1 \leftarrow t_{ch1} \in V_1 \sim t_{ch2} \in V_2 \rightarrow t_2 \in V_2$—the **U-pattern**
(ii) $t_1 \in V_1 \leftarrow t_{ch2} \in V_2 \sim t_{ch1} \in V_1 \rightarrow t_2 \in V_2$—**X-pattern**
(iii) $t_1 \in V_1 \leftarrow t_{ch1} \in V_1 \rightarrow t_2 \in V_2$ or
      $t_1 \in V_1 \leftarrow t_{ch2} \in V_2 \rightarrow t_2 \in V_2$ —the **V-pattern**

Here, $\rightarrow$ and $\leftarrow$ denote non-CMP parent-child links, either from the relations described by the edges E_1 and E_2 (as in the U-pattern), or new relations from the growing set of predictions D generated by DM or SMP (as in the X-pattern). The $\sim$ symbol denotes synonyms, which are symmetrical links. All asymmetrical links in a pattern must be of the same type—either both of them have to be *is-a*, or both of them have to *part-of*. Each such pattern is called *pattern instance*, as is utilised during the scoring.

If one of the three patterns is found between two terms t_1 and t_2, a synonym prediction between those two terms is added to the prediction set D. To calculate the score of that CMP prediction, however, we need a more complex probabilistic model.

4   Scoring

The scoring of the individual DM and SMP predictions is straightforward. To merge these predictions, or to score the CMP predictions, however, requires a collection of functions. The merging of predictions from multiple sources requires the use of the disjunctive function $Disj$ defined below. The calculation of the CMP link requires the use of the aggregate function F_{aggr} defined below, which is applied on scores calculated for the individual CMP pattern instances. Different functions can be chosen for these tasks depending on the chosen scoring scheme.

4.1   Scoring Definitions

Conjunctive Function. $Conj : [0,1]^N \rightarrow [0,1]$ is an N-ary function which accumulates scores in conjunctive associations of events that are *all* necessary to occur.

Disjunctive Function. $Disj : [0,1]^N \rightarrow [0,1]$ is an N-ary function which accumulates scores in disjunctive associations of events that reaffirm each other.

Score of a Set of Non-CMP Links. Given an evidence set $\overline{S_k}$ for a given link between v_{1k} and v_{2k} of type r_k, with scores of s_{kj}, the merged score s_k is computed using the disjunctive function, because each piece of evidence s_{kj} reaffirms the existence of the link.

$$s_k = score(\overline{S_k}) = Disj_{j=1}^m(s_{kj}) = Disj(s_{k1}, \ldots, s_{km}) \qquad (3)$$

Score of a CMP Pattern Instance. If a pattern e is defined by a sequence of links with individual scores s_i, the score of the CMP pattern instance is computed using the conjunctive function, because all links making the pattern must exist for the pattern to be confirmed, and is multiplied by a penalization constant $p \in [0,1]$, which corresponds to the probability that a CMP pattern in which all links are certain identifies a new synonym.

$$score(e) = p \cdot Conj_{i=1}^n(s_i) = Conj_{i=1}^n(score(\overline{S_i})), \qquad (4)$$

Aggregate Function. Given K different CMP pattern instances e_l for the same nodes $t_1 \in V_1$ and $t_2 \in V_2$, an aggregation function $F_{aggr} : [0,1]^K \to [0,1]$ is a K-ary function that, given the scores of all the patterns, aggregates them in a single score for the final CMP prediction.

5   Scoring Schemes

Three different scoring schemes have been proposed and used, defining different functions $Conj$, $Disj$ and F_{aggr}. They are loosely based on the notion of probability of a union and intersection of events.

Scheme #1 ("simple"). The first scoring scheme is directly using the formulas for multiplication and addition of probabilities as conjunctive and disjunctive function. Making the naïve assumption that the different predictions are independent, the disjunction of scores would be equal to the probability that any of the corresponding predictions was true, and the conjunction of scores would be equal to the probability that all the corresponding predictions were true.

(1a) $Conj(s_1, s_2) = s_1 \cdot s_2$
    $Conj(s_1, s_2, \ldots, s_N) = Conj(Conj(s_1, s_2, \ldots, s_{N-1}), s_N)$
(1b) $Disj(s_1, s_2) = s_1 + s_2 - s_1 \cdot s_2$
    $Disj(s_1, s_2, \ldots, s_N) = Disj(Disj(s_1, s_2, \ldots, s_{N-1}), s_N)$
(1c) $F_{aggr}(s_1, s_2, \ldots, s_N) = \max(s_1, s_2, \ldots, s_N)$

Since the CMP patterns are assumed to not be independent, the F_{aggr} function uses the maximum value instead of addition of probabilities, which corresponds to a significant dependence between the events leading to the same CMP prediction.

After the validation detailed in the results Sect. 6.2, it was observed that this simple score produces good enough results. The more complex scoring schemes are, nevertheless, also detailed here.

Scheme #2 ("staircase"). One obvious modification of the first scoring scheme is to assume that the different CMP patterns correctly predicting a given inter-ontology link are *also* independent events, and thus to test what the result would be if we used addition of probabilities for F_{aggr} as well.

(2a) $Conj(s_1, s_2) = s_1 \cdot s_2$
    $Conj(s_1, s_2, \ldots, s_N) = Conj(Conj(s_1, s_2, \ldots, s_{N-1}), s_N)$
(2b) $Disj(s_1, s_2) = s_1 + s_2 - s_1 \cdot s_2$
    $Disj(s_1, s_2, \ldots, s_N) = Disj(Disj(s_1, s_2, \ldots, s_{N-1}), s_N)$
(2c) $F_{aggr}(s_1, s_2, \ldots, s_N) = Disj(s_1, s_2, \ldots, s_N)$

Such a modification risks an increase in the false positive rate, because of the CMP scores growing too large.

Scheme #3 ("hybrid"). A big shortcoming of these two scoring schemes is that they make naïve assumptions about the prediction dependence. For that reason, we also tested a more complex hybrid scoring scheme, which introduces an α parameter which is related to the expected correlation between a couple of sources/predictions.

(3a) $Conj(s_1, s_2) = s_1 \cdot s_2$
$Conj(s_1, s_2, \ldots, s_N) = Conj(Conj(s_1, s_2, \ldots, s_{N-1}), s_N)$
(3b) $Disj(s_1, s_2) = \alpha(s_1 + s_2 - s_1 \cdot s_2) + (1 - \alpha)\max(s_1, s_2)$
$Disj(s_1, s_2, \ldots, s_N) = Disj(Disj(s_1, s_2, \ldots, s_{N-1}), s_N)$
(3c) $F_{aggr}(s_1, s_2, \ldots, s_N) = Disj(s_1, s_2, \ldots, s_N)$

Tuning this parameter in the interval $[0, 1]$ for each pair of sources, or for predictions within the same source or procedure, can produce any of the other scoring schemes, as well as model much more complex relationships with varying degrees of dependance.

Once the final score has been calculated for a merged prediction, a threshold is needed to determine if it is a valid or an invalid one. Because all the scores ended in two easily identifiable groups, they were clustered, and the average of their means shifted by the standard deviations was chosen, thus getting a threshold equal to $(\mu_c - v_c + \mu_e + v_e)/2$

6   Results and Discussion

To validate the predictions generated using the various sources and the automated inferences, and to validate the scoring procedure, the anatomical ontologies of three species were selected from the OBO Foundry—mouse, zebrafish (danio rerio) and xenopus (clawed frog). Three pairs of mappings were constructed, and an expert in animal anatomy was asked to curate the predictions by placing them in three sets—correct, incorrect, and semi-correct. The last category refers to terms that were correctly predicted to be related, but the prediction doesn't correctly describe the actual relationship between them, and for which the expert's opinion was that they fitted in neither the correct nor incorrect categories.

In addition to validating the cumulative set of predictions, a focus was placed on determined the validity of the links automatically inferred from the projected super-ontology structure using the CMP procedure, both raw and after scoring.

6.1   Validity of the Predictions and Automated Inferences—Raw Numbers

Since the predictions were checked by an expert before scoring functions were applied, we were able to judge the raw quality of the sets of unscored predictions. The merging of mouse and xenopus had the most correct predictions—100% using DM, as well as 97.60% synonyms and 93.95% parent-child relationships

Table 1. Raw accuracy of CMP predictions

Pair	Source	Count	Correct	Semi-correct	Incorrect
Mouse–Zebrafish	CMP	802	125 (15.59%)	521 (64.96%)	156 (19.45%)
Mouse–Zebrafish	CMP Only	693	21 (3.03%)	517 (74.60%)	155 (22.37%)
Mouse–Zebrafish	CMP+Other	109	104 (95.41%)	4 (3.67%)	1 (0.92%)
Mouse–Xenopus	CMP	720	146 (20.28%)	507 (70.42%)	67 (9.30%)
Mouse–Xenopus	CMP Only	595	26 (4.37%)	503 (84.54%)	66 (11.09%)
Mouse–Xenopus	CMP+Other	125	120 (96.00%)	4 (3.20%)	1 (0.80%)
Zebrafish–Xenopus	CMP	712	160 (22.47%)	434 (60.96%)	118 (16.57%)
Zebrafish–Xenopus	CMP Only	566	23 (4.06%)	427 (75.44%)	116 (20.50%)
Zebrafish–Xenopus	CMP+Other	146	137 (93.84%)	7 (4.79%)	2 (1.37%)

using SMP, while zebrafish–xenopus was consistently the worst. Direct matching had the highest accuracy among predictors, as it relied on the consistent terminology used by the OBO Foundry, although there were still 1.23% incorrect predictions between zebrafish and xenopus, as well as some (2–4%) semi-correct predictions. The accuracy of the DM and SMP predictions was good enough to justify their use even without the scoring functions.

The most interesting question remained the raw quality of the predictions inferred by the structural CMP procedure, which is heuristic and does not rely on any thesaurus data. In Table 1, the verdicts for the CMP method are shown, as well as how the quantity changes if we only include the predictions exclusive to the CMP method, or only the CMP predictions confirmed by either SMP or DM. These were estimated with the complete, untrimmed set of predictions, without excluding any of the sources.

As it can be seen in the table, the majority of the CMP predictions are semi-correct. This implies that while CMP adds new valuable information about relationships between pairs of nodes, the type of that relationship isn't accurately predicted. Among the accurate predictions, some are done exclusively by CMP, and are missed by either DM or SMP. As a heuristic, CMP had the lowest overall accuracy of all the procedures, which is correctly reflected by the scoring functions.

6.2   Validity of the Predictions Selected by Score

After the complete set of predictions had been curated by an expert, the verdicts produced by the expert were used to evaluate the results from different sources and procedures, and threshold filtering using the different scoring schemes. Table 2 shows the accuracy of prediction selection using the three scoring schemes, and the results both with and without CMP. The hybrid scoring scheme (#3) is placed in the middle, as it uses a liner combination of the aggregation function of the other two scoring schemes. For this evaluation, the semi-correct predictions are considered incorrect.

Table 2. Prediction selection accuracy of the three scoring schemes

	Without CMP				With CMP			
	Accepted		Rejected		Accepted		Rejected	
	Correctly	Incorr.	Correctly	Incorr.	Correctly	Incorr.	Correctly	Incorr.
Mouse–Zebrafish								
#1	231	**5**	6	28	**257**	7	346	14
#3	223	**5**	6	36	**257**	7	347	14
#2	231	**5**	6	28	256	31	323	15
Xenopus–Mouse								
#1	196	**3**	9	22	216	9	309	14
#3	196	**3**	9	22	216	9	309	14
#2	196	**3**	9	22	**219**	32	287	11
Xenopus–Zebrafish								
#1	393	**23**	21	16	405	36	311	14
#3	393	**23**	21	16	405	36	311	14
#2	393	**23**	21	16	**406**	61	285	13

The staircase scoring scheme (#2) leads to more false positives, without any improvement in the true positives. This confirms the expectation from Sect. 5 that the assumption of independence is incorrect for CMP patterns. At the same time, the hybrid and the simple scoring schemes have a better and nearly identical performance. This confirms that the more complex scheme is unnecessary, and the assumptions of the simple scheme provides enough accuracy.

The structure-based inferences of the CMP procedure lead to more correct predictions, just as it did for the number of correct raw predictions in Sect. 6.1. The addition of a scoring function significantly reduces the number of incorrect/semi-correct CMP predictions by rejecting most of them.

7   Conclusion

This paper presented an algorithmic approach for predicting and scoring links between terms across pairs of anatomical ontologies. It uses direct matches, relying on terminological consistency, multiple external knowledge sources, as well as the structure of the ontologies themselves to automatically infer new links. All of the predictions between the anatomies of three biological species were individually checked by a curator.

It was determined that the direct predictions (DM) and the predictions from external knowledge sourced (SMP) had a high accuracy even prior to the application of scoring functions to filter them. It was also determined that the CMP procedure to automatically infer links produced new valid predictions, but that was at the cost of a lot of incorrect or mislabelled links. The scoring procedure was able to filter most of those undesired links, but the use of a human curator to individually validate the predicted links would still be preferable.

The procedures discussed here are used in the software program AnatOM [12]. F/OSS implementation is available at https://launchpad.net/anatom.

Acknowledgements. This work has been supported by the National Science Fund of Bulgaria within the "Methods for Data Analysis and Knowledge Discovery in Big Sequencing Datasets" Project, Contract I02/7/2014.

References

1. Ashburner, M., et al.: Gene ontology: tool for the unification of biology. Nat. Genet. **25**(1), 20–29 (2000)
2. Bodenreider, O.: The unified medical language system (UMLS): integrating biomedical terminology. Nucleic Acids Res. **32**, 267–270 (2004)
3. de Bruijn, J., et al.: Ontology mediation, merging, and aligning. In: Davies, J., Studer, R., Warren, P. (eds.) Semantic Web Technologies, pp. 95–113. Wiley, Hoboken (2006)
4. Day-Richter, J.: OBO flat file format specification, version 1.2 (2006)
5. Diallo, G.: An effective method of large scale ontology matching. J Biomed. Seman. **5**, 44 (2014). 189[PII]
6. Euzenat, J., Shvaiko, P.: Classifications of ontology matching techniques. In: Euzenat, J., Shvaiko, P. (eds.) Ontology Matching, pp. 73–84. Springer, Heidelberg (2013). doi:10.1007/978-3-642-38721-0_4
7. Fellbaum, C.: WordNet: An Electronic Lexical Database. MIT Press, Cambridge (1998)
8. Hooi, Y.K., Hassan, M.F., Shariff, A.M.: A survey on ontology mapping techniques. In: Jeong, H.Y., S. Obaidat, M., Yen, N.Y., Park, J.J.J.H. (eds.) CSA 2013. LNEE, vol. 279, pp. 829–836. Springer, Heidelberg (2014). doi:10.1007/978-3-642-41674-3_118
9. Miller, G.: A lexical database for English. Commun. ACM **38**(11), 39–41 (1995)
10. van Ophuizen, E., Leunissen, J.: An evaluation of the performance of three semantic background knowledge sources in comparative anatomy. J. Integr. Bioinform. **7**, 124–130 (2010)
11. Petrov, P., Krachunov, M., van Ophuizen, E., Vassilev, D.: An algorithmic approach to inferring cross-ontology links while mapping anatomical ontologies. Serdica J. Comput. **6**, 309–332 (2012)
12. Petrov, P., Krachunov, M., Todorovska, E., Vassilev, D.: An intelligent system approach for integrating anatomical ontologies. Biotechnol. Biotechnol. Equipment **26**(4), 3173–3181 (2012)
13. Rosse, C., Mejino, J.: A reference ontology for biomedical informatics: the foundational model of anatomy. J. Biomed. Inform. **36**(6), 478–500 (2003)
14. Smith, B., et al.: The OBO foundry: coordinated evolution of ontologies to support biomedical data integration. Nat. Biotechnol. **25**, 1251–1255 (2007)

An Experiment in Causal Structure Discovery. A Constraint Programming Approach

Antoni Ligęza$^{(\boxtimes)}$ (iD)

AGH University of Science and Technology, 30-059 Kraków, Poland
ligeza@agh.edu.pl

Abstract. The problem of Causal Structure Discovery is defined by given inputs, outputs, auxiliary knowledge of components and possible internal connections. Constraints Programming is employed to discover admissible system models. Existence of internal connections and predefined functionality of components is handled through reification.

Keywords: Causal structure discovery · Constraint programming

1 Introduction

Discovering the causal structure of a given system is a challenging task of numerous potential applications. In fact, it seems to be a prevailing activity in many areas of investigation of natural systems and technology-oriented research. For example, mathematical techniques are used in modeling cyber-physical systems [8]. Such models are the core of automated diagnosis. In social sciences, economy, biology and medicine Bayesian Network modeling is often applied [7].

Typical causal discovery requires knowledge of variables and definition of causal dependencies among them. Such dependencies are typically represented with a Directed Acyclic Graph (DAG). The graph is built from a universal skeleton on the base of statistical data and using statistical tests of conditional variable independence; as a result Bayesian Networks are produced [7]. A very thorough review of the current state-of-the-art in algorithms for probabilistic causality discovery is presented in [9].

It is symptomatic that vast majority of research on discovering some *regularities* in data (including various notions related to *causality*) employ methods based on syntax, relative frequencies, probabilities, fuzzy sets or rough sets; for a review of such methods see, e.g. the Handbook [2]. The obtained models are useful but apparently they cover *shallow knowledge* of the investigated phenomena. Following this line, a recent study of causal rule mining is presented in [5].

Recently, an interest was paid to using Constraint Programming technology to Data Mining [1]; in fact, many of the well-known data mining models can be defined in terms of constraints. An extensive survey on such approaches is provided in the recent seminal report paper [3]. In [4] an application of Answer Set Programming technology to causal discovery is reported.

© Springer International Publishing AG 2017
M. Kryszkiewicz et al. (Eds.): ISMIS 2017, LNAI 10352, pp. 261–268, 2017.
DOI: 10.1007/978-3-319-60438-1_26

In this paper a preliminary study of an experiment with causal structure discovery for deterministic systems is presented. In contrast to the mentioned above papers, the focus is on discovery of existing connections and functional dependencies. Exact, numerical models are searched for and only the deterministic ones are considered. As the underlying tool constraint programming methods are introduced. The main focus is on discovering a causal structure with functional dependencies such that the observed behavior can be explained. A priori knowledge about potential components, their functional behavior and existing connections can be incorporated in the constraints. The main technique in use is the *reification*: existence of potential connections is conditioned with Boolean variables. This alternative approach incorporating Model-Based Reasoning seems to constitute an interesting complement to the available methodologies.

2   Problem Formulation and Methodology

The main focus of this work is on the problem of internal structure discovery of given systems. Any system is assumed to have some inputs, specified by a vector of variables $\mathbf{U} = [U_1, U_2, \ldots, U_k]$ and some outputs, given by a vector $\mathbf{Y} = [Y_1, Y_2, \ldots, Y_m]$. Only static, feed-forward, and deterministic systems of discrete values are considered for simplicity. At a specific instant of time the values of $\mathbf{Y}$ can be calculated as

$$\mathbf{Y} = f(\mathbf{U}, \mathbf{P}) \qquad (1)$$

where f defines the behavior of the system (its internal structure and components behavior) and $\mathbf{P}$ is a vector of internal system parameters.

An example of such system is presented in Fig. 1. The input variables vector is $\mathbf{U} = [A, B, C, D, E]$ and its output is $\mathbf{Y} = [F, G]$. There are also some auxiliary, internal variables X, Y, Z. The operation f is defined as $F = A * C + B * D$ and $G = B * D + C * E$. In a network like that there are three issues of interest: (i) definition of causal dependencies (i.e. which variables defines values of other variables), (ii) the functional dependencies (how the values of the dependent variables are determined), and (iii) the overall structure of the network (including potential internal variables).

Causal Structure Discovery Problem. Being given a set of training examples $\{(\mathbf{U}_i, \mathbf{Y}_i) | i = 1, 2, \ldots, n\}$, set of legal components *Comps* and perhaps some additional knowledge $\boldsymbol{\Delta}$ (e.g. concerning existing internal connections) discover the internal structure of the system satisfying all the examples.

A solution to this problem is a complete network specification. It is admissible if it satisfies all the examples, i.e. for a given input it produces the expected output. Obviously, since there can be infinitely many solutions we intend to restrict our interest to some reasonable number of simplest models. This can be achieved with use of additional knowledge $\boldsymbol{\Delta}$ used as auxiliary constraints. The principal technique proposed to solve this problem is to use constraint programming in a specific way. There are two levels of constraints to be used:

(i) basic level constraints concerning variables and functional operation, and
(ii) meta-level constraints concerning existence (or not) of basic level constraints.

The specification of basic level constraints is straightforward. Below we show some intuitive examples using the SWI-Prolog notation of the constraint programming library over finite domains. The typical relation symbols are used but they are preceded with the # sign. For example

```
A #= B,   Y #< Z,   X #= A*C,
```

define the requirements that A must be equal to B, Y must be less than Z and X must be equal to the result of multiplication of A and C.

The way to specify the meta-level constrains – aimed in fact in encoding connections within the network structure – is by *reification*. Below we have some three examples:

```
P #==> A #= B,    P #==> X #= A*C,    P #==> G #= Y+Z
```

The meaning is that if P is True (or 1) then the constraint of the right-hand side must also hold. Instead of P its negation in the form #\P can be used.

2.1   Types of Causal Constraints

The simplest logical constraints may be defined with use of logical connectives, such as conjunction, disjunction, negation, implication, etc. In SWI-Prolog clpfd library[1] these are: #/\, #\/, #\, #==>. Moreover, relational constraints over numbers can be defines in a usual way, e.g. #=, #>, #<, #>=, #=<, #\=.

Logical OR Constraints: They are defined by the logical formula $V_1 \lor V_2 \lor \ldots V_k \Longrightarrow V$; the interpretation is as follows: it is enough for a single symptom V_i on the left to be true to cause the V on the right to be true (OR-causality).

Logical AND Constraints: They are defined by the logical formula $V_1 \land V_2 \land \ldots V_k \Longrightarrow V$; the interpretation is also straightforward: all the symptoms V_i on the left must be true to cause the V on the right to be true (AND-causality).

Functional Constraints: They are defined with the functional (arithmetical) specification of the form $Y = f(X_1, X_2, \ldots, X_k)$. It is interpreted as follows: if all the X_i on the left take specific values then Y is forced to take value given by the definition of f (functional causality).

2.2   Modeling Causal and Functional Structure

The potential structure of the system can be modeled with the constraints introduced in Sect. 2.1. In order to define existence of a potential constraint the reification technique is used. Further, in order to model a potential connection, a simple equality constraint of some two variables can be defined.

For example, consider the following definition:

---

[1] The library clpfd for constraint programming with finite domains was incorporated; see http://www.pathwayslms.com/swipltuts/clpfd/clpfd.html for details.

```
A_X #==> A #= X,
B_Y #==> B #= Y,
MXY #==> Z #= X*Y,
AXY #==> Z #= X+Y,
#\(MXY /\ AXY),
```

The specification can be read as follows: (i) there is a possible connection of signal A to input X (provided that A_X = 1), (ii) there is a possible connection of signal B to input Y (provided that B_Y = 1), (iii) there can be a multiplying component producing Z as X multiplied by Y (provided that MXY = 1), (iv) there can be an adding component producing Z as X plus Y (provided that AXY = 1), and finally (v) it is impossible to have both a multiplier and an adder.

3   An Example Case Study

3.1   Example Network

Let us briefly recall a classical diagnostic example of a feed-forward arithmetic circuit. This is the multiplier-adder example presented in the seminal paper by R. Reiter [8]. Here we shall focus on its *causal structure* and *functional components*.

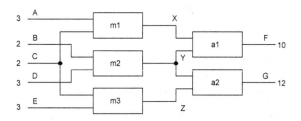

Fig. 1. The multiplier-adder system: a faulty behavior (on F) is observed

The basic, intuitive schema of the system is presented in Fig. 1. It can be observed that the current value of F is incorrect, namely F = 10; the system is faulty. The final *diagnoses* for the considered case are: $D_1 = \{m1\}$, $D_2 = \{a1\}$, $D_3 = \{a2, m2\}$ and $D_4 = \{m2, m3\}$; for details see [6,8].

3.2   Simple Case: Connection Discovery

The first experiment reported here concerns the search for admissible connections between the outputs of multipliers $m1$, $m2$ and $m3$, i.e. X, Y and Z, and the inputs of the two adders, i.e. $a1$ and $a2$. In order to enable the search, names of input variables are introduced as XA1, and YA1 for the adder $a1$, and YA2 and ZA2 for the adder $a2$, respectively. A simple mnemonically convention is kept, so that in the original structure we have the following four connections: X_XA1,

Y_YA1, Y_YA2, and Z_ZA2. In this experiment it is assumed that both inputs of each adder should be receiving some signal, but not necessarily any output of a multiplier must be connected. This opens space for as many as 81 (i.e. 3^4) potential structures. The following constraints have been defined in SWI-Prolog clpfd library.

```
Vars = [XA1,YA1,YA2,ZA2],
Vars ins 0..100,   Conns =
[X_XA1,X_YA1,X_YA2,X_ZA2,Y_XA1,Y_YA1,Y_YA2,Y_ZA2,Z_XA1,Z_YA1,Z_YA2,Z_ZA2],
Conns ins 0..1,
```

The first two lines define the inputs of the adders; in line 3 and 4 we have binary variables activating potential connections. The reification constraints define existence of a connection if the appropriate variable takes value 1; they are as follows:

```
X_XA1 #==> X #= XA1, Y_XA1 #==> Y #= XA1, Z_XA1 #==> Z #= XA1,
X_YA1 #==> X #= YA1, Y_YA1 #==> Y #= YA1, Z_YA1 #==> Z #= YA1,
X_YA2 #==> X #= YA2, Y_YA2 #==> Y #= YA2, Z_YA2 #==> Z #= YA2,
X_ZA2 #==> X #= ZA2, Y_ZA2 #==> Y #= ZA2, Z_ZA2 #==> Z #= ZA2,
```

As mentioned above, each adder input must be connected; these constraints take the disjunctive form:

```
X_XA1 #\/ Y_XA1 #\/ Z_XA1,
X_YA1 #\/ Y_YA1 #\/ Z_YA1,
X_YA2 #\/ Y_YA2 #\/ Z_YA2,
X_ZA2 #\/ Y_ZA2 #\/ Z_ZA2,
```

Finally, each adder input can be connected to at most one multiplier output:

```
#\(X_XA1 #/\ Y_XA1), #\(X_XA1 #/\ Z_XA1), #\(Y_XA1 #/\ Z_XA1),
#\(X_YA1 #/\ Y_YA1), #\(X_YA1 #/\ Z_YA1), #\(Y_YA1 #/\ Z_YA1),
#\(X_YA2 #/\ Y_YA2), #\(X_YA2 #/\ Z_YA2), #\(Y_YA2 #/\ Z_YA2),
#\(X_ZA2 #/\ Y_ZA2), #\(X_ZA2 #/\ Z_ZA2), #\(Y_ZA2 #/\ Z_ZA2).
```

The complete specification covers further details, such as specifications of the multipliers and adders, and some technical details. Since adders are symmetrical w.r.t. their inputs, it may be reasonable to use the symmetry breaking constraints:

```
XA1 #=< YA1,
YA2 #=< ZA2,
```

The program was run for several specifications of the inputs variables; some symptomatic results are summarized in Table 1. The first two cases are simple to explain: there are 81 (3^4) possible connections. Introducing the symmetry breaking does not help; the values of X, Y and Z are equal to 6. In the two following cases the input values of the two adders are pairwise different, what leads to a unique model with symmetry breaking. Removing that constraint leads to 4 models since the inputs of every adder can be interchanged.

Table 1. Example results of internal connections discovery

(A,B,C,D,E)	(F,G)	Symmetry breaking	No. of models
(3,2,2,3,3)	(12,12)	No	81
(3,2,2,3,3)	(12,12)	Yes	81
(1,3,5,7,11)	(26,76)	No	4
(1,3,5,7,11)	(26,76)	Yes	1
(1,2,3,4,5)	(11,23)	No	4
(1,2,3,4,5)	(11,23)	Yes	1

3.3   Extended Case: Function Selection

The second experiment reported in this paper is conceptually more complex. This time we shall try to discover the functionality of the components and – as before – the internal connections. Two types of functional components are allowed: the multipliers and the adders.

In order to model the constraints on components reification will be used. Five new binary variables are introduced: M1M, M2M, M3M, A1A, A2A. Each of them taking the value 1 (**True**) defines the expected mode of work (multiplication at first level of components and addition in the second level), while the value 0 (**False**) means the other mode. The constraints are defined as:

```
Funs = [M1M,M2M,M3M,A1A,A2A],
Funs ins 0..1,
M1M #==> X #= A*C,    % If M1M=1 (M2M=1, M3M=1) then the operational
M2M #==> Y #= B*D,    % mode is multiplication
M3M #==> Z #= C*E,
#\M1M #==> X #= A+C,  % If M1M=0 (M2M=0, M3M=0) then the operational
#\M2M #==> Y #= B+D,  % mode is addition
#\M3M #==> Z #= C+E,
A1A #==> F #= XA1 + YA1,  % If A1A=1 (A2A=1) then the operational mode
A2A #==> G #= YA2 + ZA2,  % is addition
#\A1A #==> F #= XA1*YA1,  % If A1A=0 (A2A=0) then the operational mode
#\A2A #==> G #= YA2*ZA2,  % is multiplication
```

The other constraints (see Sect. 3.2), including symmetry breaking are in place. Some example results are summarized in the table below. The result of the first line is caused by both a number of similar results produced by different operations for these specific data (e.g. $2 * 3 = 3 + 3$) as well as equal values of variables in the intermediate level. On the other hand, the results shown in line 2 and 3 demonstrate that the input sequences are highly distinctive – in both cases there is only one (correct) model (Table 2).

Table 2. Example results of functionality and internal connections discovery

(A,B,C,D,E)	(F,G)	No. of models
(3,2,2,3,3)	(12,12)	132
(1,3,5,7,11)	(26,76)	1
(1,2,3,4,5)	(11,23)	1

4 Concluding Remarks

This paper presents selected issues concerning causal and functional structure discovery. The proposed approach seems to be different from the classical algorithms applied mostly for causal graph generation from experimental data with statistical methods – as in the case of Bayesian Networks. First, the methodological background is based on Constraint Programming rather than statistical tests. In consequence, the discovered models are *strict* – exact w.r.t representation of the causal and functional behavior. Second, functional aspects can be covered, as shown with the multiplier-adder example. Third, several different types of causal connections and constraints are distinguished; all of them are of strict, deterministic character. Last but not least, the phenomena of *causality* and its understanding is much more technical and defined in terms of logic, signal flow, and numerical operation models.

The proposed approach can – and should be used with as much additional knowledge as possible. This include coverage of preliminary knowledge of partial structure/connections, accessible components and their operational specification, and all kinds of variables: input, output and the internal ones, as well as their domains and constraints over them. The more knowledge we have, the more precise resulting models can be expected.

References

1. Bessiere, C., De Raedt, L., Kotthoff, L., Nijssen, S., O'Sullivan, B., Pedreschi, D. (eds.): Data Mining and Constraint Programming. Foundations of a Cross-Disciplinary Approach. LNCS (LNAI), vol. 10101. Springer, Cham (2016)
2. Cios, K.J., Pedrycz, W., Swinarski, R.W., Kurgan, L.A.: Data Mining. A Knowledge Discovery Approach. Springer, New York (2007)
3. Grossi, V., Romei, A., Turini, F.: Survey on using constraints in data mining. Data Mining Knowl. Discov. **31**, 424–464 (2017)
4. Hyttinen, A., Eberhardt, F., Järvisalo, M.: Constraint-based causal discovery: conflict resolution with answer set programming. In: Proceedings of Uncertainty in Artificial Intelligence, Quebec, Canada, pp. 340–349 (2014)
5. Li, J., Le, T.D., Liu, L., Liu, J.: From observational studies to causal rule mining. ACM Trans. Intell. Syst. Technol. **7**(2), 14:1–14:27 (2015)
6. Ligęza, A., Kościelny, J.M.: A new approach to multiple fault diagnosis. combination of diagnostic matrices, graphs, algebraic and rule-based models. the case of two-layer models. Int. J. Appl. Math. Comput. Sci. **18**(4), 465–476 (2008)

7. Pearl, J.: Causality. Models, Reasoning and Inference, 2nd edn. Cambridge University Press, New York (2009)
8. Reiter, R.: A theory of diagnosis from first principles. Artif. Intell. **32**, 57–95 (1987)
9. Yu, K., Li, J., Liu, L.: A review on algorithms for constraint-based causal discovery. University of South Australia. arXiv:1611.03977v1 (2016)

Machine Learning

Actively Balanced Bagging for Imbalanced Data

Jerzy Błaszczyński$^{(\boxtimes)}$ and Jerzy Stefanowski

Institute of Computing Science, Poznań University of Technology,
Piotrowo 2, 60-965 Poznań, Poland
{jerzy.blaszczynski,jerzy.stefanowski}@cs.put.poznan.pl

Abstract. Under-sampling extensions of bagging are currently the most accurate ensembles specialized for class imbalanced data. Nevertheless, since improvements of recognition of the minority class, in this type of ensembles, are usually associated with a decrease of recognition of majority classes, we introduce a new, two phase, ensemble called Actively Balanced Bagging. The proposal is to first learn a bagging classifier and then iteratively improve it by updating its bootstraps with a limited number learning examples. The examples are selected according to an active learning strategy, which takes into account: decision margin of votes, example class distribution in the training set and/or in its neighbourhood, and prediction errors of component classifiers. Experiments with synthetic and real-world data confirm usefulness of this proposal.

Keywords: Class imbalance · Active learning · Bagging · Ensembles of classifiers · Neighbourhood Balanced Bagging

1 Introduction

The problem of learning from class-imbalanced data has been a topic of intensive research in recent years, motivated by an observation that standard learning methods are inherently biased to focus on the majority classes and fail to sufficiently recognize the minority class which is often of particular interest in the given application. For reviews of new methods see, e.g., [6,10].

Ensembles are among the most effective methods for improving recognition of the minority class. Most of them extend strategies from bagging or boosting. They either employ pre-processing methods before learning component classifiers or embed a cost-sensitive framework in the ensemble learning process; see review in [9,10]. Experimental comparisons [3,11,12] have shown that extensions of bagging work better than those of boosting and other solutions.

The basic idea behind extending bagging for imbalanced data is to modify the distribution of examples from minority and majority classes in bootstraps [9]. This may be achieved in many ways, which usually balance the numbers of examples from both classes. Experimental studies show that under-sampling, i.e., reduction of examples from the majority class, performs better than over-sampling (i.e., multiplication of examples from the minority class) [3,4,11]. Improvements are observed even for simplest extensions, like Exactly

© Springer International Publishing AG 2017
M. Kryszkiewicz et al. (Eds.): ISMIS 2017, LNAI 10352, pp. 271–281, 2017.
DOI: 10.1007/978-3-319-60438-1_27

Balanced Bagging (EBBag) [7,9]. Other studies [3,11] demonstrate that Roughly Balanced Bagging (RBBag), which applies a more sophisticated random under-sampling, achieves the best G-mean among compared ensembles. We have proposed Neighbourhood Balanced Bagging (NBBag) being competitive to RBBag. NBBag modifies bootstrap sampling by weighting examples [4]. The weight of an minority example depends on quantity of examples in the whole training set, and in the example neighbourhood, which belong to the majority class.

However, these improvements may come with a decrease of recognition majority class examples [4]. We claim that there is still a possibility to better learn the trade-off between performance in both classes. This leads us to a research question how to achieve it. To best of our knowledge the current research focuses on various modifications of bootstrap samples in under-sampling bagging extensions. In this paper, we want to consider another hypothesis: given an already good technique of constructing a bagging extension, is it possible to perform additional steps of updating its bootstraps by selecting a limited number of learning examples, which are important to improve performance in both classes.

This hypothesis directs our attention toward *active learning methods* [15]. Recall that, in active learning, a learning method may achieve a better classification performance when it is allowed to choose examples on which it learns. Usually it is performed on partially labeled examples, i.e., after the initial step of full supervised learning, the active method should select a limited number of examples to be labeled and further used to update the classifier. However, the active learning could also be used to filter examples in the complete labeled examples [2]. In this way active strategies have been already applied to imbalanced learning - although these attempts are still quite limited (see Sect. 2).

We introduce a new approach, called Active Balancing Bagging (ABBag), to extend bagging for imbalanced data by an active learning modification of bootstrap samples of previously constructed under-sampling bagging extension. In this proposal, after the initial training, the ensemble (here we consider two algorithms: EBBag and NBBag) is re-trained on bootstraps enlarged by batches, i.e., small portions of learning examples, selected according to active selection strategies. Batch querying, although being very practical by nature, is rarely used by existing active learning approaches [5]. According to our best knowledge no similar extension has been considered in the literature so far.

The paper is organized as follows. The next section summarizes related works on bagging extensions proposed for imbalanced data and active learning strategies. The following Sect. 3, describes the proposed active balancing bagging extension. Experimental validation of the proposed methods is carried out in Sect. 4. The paper is concluded in Sect. 5.

2   Related Works

Besides working on new algorithms, some researchers also studied the characteristics of imbalanced data and sources of difficulties for learning classifiers. It has been noticed that the global imbalance ratio is not the only or even not the most important factor which makes learning difficult. Other data difficulty factors such as class overlapping, small disjuncts or lack of representativeness significantly deteriorate the quality of an induced model even on exactly balanced data [13, 14]. They also inspire development of new methods, see e.g. [4].

Several extensions of bagging approach for imbalanced data have been proposed [4, 9, 16]. In the simplest one, called Exactly Balanced Bagging (EBBag), while constructing bootstrap sample, the entire minority class is copied and combined with the randomly chosen subset of the majority class to exactly balance cardinalities of examples in both classes. The first bagging extension which uses knowledge of data difficulty factors is Neighbourhood Balanced Bagging (NBBag) [4]. NBBag focuses bootstrap sampling toward difficult minority examples by using certain type of weights. The weight of a minority example depends on the analysis of class labels among its k nearest neighbours. A minority example is considered the more unsafe, the more it has majority examples in its neighbourhood. Thus, this part of the weight reflects a local balancing factor. Moreover, this weight is also aggregated with a global balancing factor, which takes into account the imbalance ratio between classes. Hence, the formula for minority example weight is the following: $w = 0.5 \times \left(\frac{(N'_-)^\psi}{k} + 1 \right)$ where N'_- is the number of majority examples among k nearest neighbours of the example and ψ is a scaling factor. Setting $\psi = 1$ causes a linear amplification of the example weight with an increase of unsafeness and setting ψ to values greater then 1 effects in an exponential amplification. Each majority example is assigned a constant weight $w = 0.5 \times \frac{N_+}{N_-}$.

Active learning approaches for imbalanced data are rather limited. The active strategy proposed by Ertkin et al. [8] relies on iterations of selecting single uncertain example and rebuilding the model. However, selecting more that one instance at time, and rebuilding model on a batch, often can greatly reduce the computation time [5]. This is especially true when the cost of constructing classifier is higher than for online SVM [8]. A simple active extension of bagging has been proposed for unlabeled data in imbalanced medical problems [16].

3   Active Selection of Examples in Balanced Bagging

In this section we describe Actively Balanced Bagging (ABBag), which is composed of two phases. The first phase consists in learning an ensemble classifier by one of approaches to construct under-sampling extensions of bagging. Although one can choose any extension, we will further consider one simple, yet effective one: Exactly Balanced Bagging (EBBag) [9], and better performing, yet more complex one: Neighbourhood Balanced Bagging NBBag [4]. For more information on constructing these ensemble refer to Sect. 2. Here, we focus on an explanation of the second phase, which is specific to ABBag, and which we call *active selection of examples*. It consists in: (1) an iterative modification of bootstrap samples, constructed in the first phase, by adding selected examples from the training set; (2) re-training of component classifiers on modified bootstraps. The examples in (1) are added to bootstraps in small portions called batches.

The proposed active selection of examples can be seen as a variant of Query-by-committee (QBC) approach [1]. In QBC, one constructs an ensemble of diverse component classifiers and then ranks all classified examples with respect to a selected committee disagreement measure (a decision margin). Originally, this margin is defined as a difference between number of votes for a class, or probabilities of a class, between the two most supported classes. Note that QBC has been already successfully applied in active learning of ensemble of diverse component classifiers in [5]. Ensembles constructed in the first phase of ABBag are attractive to combine with QBC, because, as we have previously shown in [4], they promote component classifiers diversity. However, QBC does not take into account global (i.e., concerning the whole training set) properties of examples distribution, and in result, it can focus on selecting outliers and sparse regions [5].

Recall that single example selection is a typical strategy in active learning [2, 15]. In ABBag we promote to select, in each iteration, a small batch instead of one example. It is motivated by a potential reduction of computation time as well as increasing diversity of examples in the batch. As it was observed in [5], a greedy selection of example with respect to a single criterion, e.g., typical for active strategy: highest utility/uncertainty measure [2, 15], does not provide desired diversity. In our view, giving chance to also randomly select some slightly sub-optimal examples besides the best ones may result in higher diversity of new bootstraps and increased diversity of re-trained component classifiers.

We address the above mentioned issues twofold. First, and foremost, the proposed active selection of examples considers multiple factors. Among them, we consider: uncertainty of example prediction by component classifiers, and a prediction error of a single component classifier. In addition, we consider factors specific to imbalanced data, which reflect more global (i.e., concerning the whole training set) and/or local (i.e., concerning example neighbourhood) class distribution of examples. Second, we use a specific variant of the rejection sampling to enforce diversity within the batch through randomization.

The pseudo-code of the algorithm for learning ABBag ensemble is presented as Algorithm 1. It starts with training set LS, and m_{bag} bootstrap samples $\boldsymbol{S}$, and it results, in the first phase (lines 1–3), in an under-sampling extension of bagging. It makes use of initial balancing weights $\boldsymbol{w}$, which are calculated in accordance with the under-sampling bagging extension. These initial *balancing weights* $\boldsymbol{w}$ allow us to direct sampling toward more difficult to learn examples in comparison to uniform sampling typical for bagging. In case of EBBag, $\boldsymbol{w}$ reflects the global imbalance of example in the training set. For NBBag, $\boldsymbol{w}$ expresses both global and local imbalance of example in the training set [4].

In the second phase, the active selection of examples is performed between lines 4, and 10. All bootstraps from $\boldsymbol{S}$ are iteratively (m_{al} times) enlarged by adding batches, and new component classifiers are re-learned. In each iteration, new weights $\boldsymbol{w'}$ of examples are calculated according to weights update method um (which is described in the next paragraph), and then they are sorted (lines 6–7). Eachbootstrap is enhanced by n_{al} examples selected ran-

---

Algorithm 1. Actively Balanced Bagging Algorithm

Input : LS training set; TS testing set; CLA component classifier learning
algorithm; m_{bag} number of bootstrap samples; S bootstrap samples; w
initial weights of examples in LS; um weights update method; m_{al}
number of active learning iterations; n_{al} size of active learning batch

Output: C ensemble classifier

1 *Learning phase*;
2 **for** $i := 1$ *to* m_{bag} **do**
3 $\quad \lfloor \; C_i := $ CLA (S_i) {generate a component classifier} ;

4 **for** $l := 1$ *to* m_{al} **do**
5 $\quad$ **for** $i := 1$ *to* m_{bag} **do**
6 $\quad\quad w' := $ updateWeights(w, C, um) {update weights used in sampling} ;
7 $\quad\quad$ sort all x with respect to $w'(x)$, so that $w'(x_1) \geq w'(x_2) \geq \ldots \geq w'(x_n)$;
8 $\quad\quad S'_i := $ random sample from $x_1, x_2, \ldots, x_{n_{al}}$ according to w' {rejection
sampling from top n_{al} x sorted according to w'; $\alpha = w'(x_{n_{al}})$ } ;
9 $\quad\quad S_i := S_i \cup S'_i$;
10 $\quad\quad C_i := $ CLA (S_i) {re-train a new component classifier} ;

11 *Classification phase*;
12 **foreach** x *in* TS **do**
13 $\quad C(x) := $ majority vote of $C_i(x)$, **where** $i = 1, \ldots, m$ {the prediction for
example x is a combination of predictions of component classifiers C_i} ;

---

domly with rejection sampling according to $\alpha = w'(x_{n_{al}}) + \epsilon$, i.e., n_{al} random
examples with weights w' higher than α are selected (lines 8–9). Parameter ϵ
introduces an additional (after α) level of randomness into the sampling. Finally,
after each bootstrap sample is enlarged, new component classifier C_i is trained
resulting in new ensemble classifier C (line 10).

We consider four different weights update methods. The simplest method,
called *margin* (m), is substituting the initial weights of examples with a decision
margin between component classifiers in C. For a given example it is defined as:
$m = 1 - \left| \frac{V_{maj} - V_{min}}{m_{bag}} \right|$, where V_{maj} is number of votes for majority class, V_{min}
is number of votes for minority class, and m_{bag} is the number of component
classifiers. Since the margin may not be directly reflecting the characteristic
of imbalanced data (indeed under-sampling somehow should reduce bias of the
classifiers) we consider aggregating it with additional factors. As a result, three
variants of weights update methods are proposed. In the first variant, called,
margin and weight (mw), new weight w' is a product of m and initial balancing
weight w. Additionally we reduce the influence of w in subsequent iterations
of active example selection, as active selection iteration index l is increasing.
The reason for this reduction of influence is that we expect that margin m is
improving with subsequent iterations, and thus initial weights w are becoming
less important. More precisely, $mw = m \times w^{\left(\frac{m_{al} - l}{m_{al}} \right)}$.

Recall that both considered so far weights update methods produce bootstrap samples which, in the same iteration l, differ only according to randomization introduced by rejection sampling, i.e., weights $\boldsymbol{w}'$ are the same for each i. That is why, we consider yet another modification of m and mw, which makes $\boldsymbol{w}'$, and, consequently, each bootstrap dependent on performance of the corresponding component classifier. These two new update methods: *margin and component error* (*mce*), and *margin, weight and component error* (*mwce*) are defined, respectfully, as follows. $mce = m + 1_e \times w$, and $mwce = mw + 1_e \times w$. In this notation, 1_e is an indicator function defined so that $1_e = 1$ when a component classifier is making a prediction error on the example, and $1_e = 0$ otherwise.

4   Experiments

We consider two aims of the experiments. First, we check whether the predictive performance of Actively Balanced Bagging is improved in comparison to known well performing under-sampling extensions of bagging. To examine this part more generally we have decided to choose two efficient extensions based on different principles: Exactly Balanced Bagging (EBBag) [7] and Neighbourhood Balanced Bagging (NBBag) [4]. Our second aim is to compare different variants of active selection methods, which result in different versions of ABBag.

In order to examine both aims we use several synthetic and real-world imbalanced data sets representing various difficulty factors in the distribution of the minority class. All considered synthetic data sets, i.e., `flower`, and `paw`, were constructed, as described in [17]. They have a controlled level of difficulty, i.e., decomposition of the minority class into sub-concepts, overlapping area, presence of rare examples and outliers. The percentage of minority examples representing safe, borderline, rare and outlier examples are given in the suffix of the data set name. The examples from the minority class were generated randomly inside predefined spheres and the majority class examples were randomly distributed in an area surrounding them. Following similar motivations and analysis from works on difficulty imbalanced data [4,14], we selected UCI data sets, and additional `scrotal-pain` data set[1]. Due to page limits we redirect the reader to the description and characteristics of these data sets in [4]. Note that some of data sets contain multiple majority classes, which were aggregated into one class.

The experiments were carried out in two variants with respect to the first phase of the active selection. In the first variant, standard EBBag or under-sampling NBBag was considered. In the other variant, the size of each of the classes in bootstrap samples was reduced to 50% of the size of the minority class in the training set. Active selection parameters, used in the second phase, m_{al}, and n_{al} were chosen in a way, which enables the bootstrap samples constructed in ABBag to excess the size of standard under-sampling bootstrap by a factor not higher than two. The size of ensembles m_{bag}, in accordance with previous experiments [4], was always fixed to 50 J4.8 decision trees. NBBag parameters were

---

[1] We are grateful to prof. W. Michalowski and the MET Research Group from the University of Ottawa for allowing us to use `scrotal-pain` data set.

set the same as in [4]. All presented results are estimated by a stratified 10-fold cross-validation repeated three times to improve reproducibility. The number of repetitions is the same for all considered data sets for better comparability.

Due to page limits, we present results of two main experiments, which are good representatives of tendencies observed in other analyzed settings.[2] In Tables 1 and 2, we present values of G-mean measure, since we want to find a good trade-off between recognition in both classes [4]. More precisely, we show the best G-mean value, which can be achieved with a limited number of active

Table 1. G-mean of actively balanced 50% under-sampling EBBag

Data set	EBBag	m-EBBag	mce-EBBag	mw-EBBag	$mwce$-EBBag
abalone	79.486	79.486	79.486	79.486	79.574
breast-cancer	57.144	59.817	58.355	59.793	59.827
car	96.513	97.312	98.202	97.208	98.192
cleveland	70.818	74.823	72.347	75.027	71.940
cmc	64.203	65.229	64.900	65.435	64.786
ecoli	87.836	88.398	88.667	88.287	88.639
flower5-3d-10-20-35-35	0.000	53.445	52.992	52.768	53.809
flower5-3d-100-0-0-0	92.315	93.272	94.094	93.402	94.398
flower5-3d-30-40-15-15	77.248	77.867	78.591	77.780	78.543
flower5-3d-30-70-0-0	91.105	91.947	92.971	91.876	93.039
flower5-3d-50-50-0-0	91.966	92.311	93.414	92.381	93.558
haberman	62.908	65.551	65.423	65.909	65.378
hepatitis	78.561	79.302	80.169	79.427	81.044
paw3-3d-10-20-35-35	0.000	51.245	51.047	51.984	50.788
paw3-3d-100-0-0-0	90.857	93.004	94.193	92.926	94.312
paw3-3d-30-40-15-15	74.872	76.707	78.429	77.971	78.091
paw3-3d-30-70-0-0	88.545	91.159	91.535	90.827	91.438
paw3-3d-50-50-0-0	91.424	91.427	92.249	91.427	92.232
scrotal-pain	72.838	73.915	73.581	73.572	73.344
solar-flare	82.048	82.834	82.771	82.464	82.674
transfusion	66.812	66.847	66.812	67.607	66.812
vehicle	95.506	96.275	96.663	96.409	96.573
yeast	82.658	85.225	85.305	84.702	85.084
average rank	4.891	2.739	2.109	3.000	2.261

---

[2] We have published detailed results for specific values of parameters, on-line, in the appendix, http://www.cs.put.poznan.pl/jblaszczynski/ISMIS17/resABBag.pdf.

Table 2. G-mean of actively balanced under-sampling NBBag

Data set	NBBag	m-NBBag	mce-NBBag	mw-NBBag	mwce-NBBag
abalone	78.714	79.426	79.585	79.621	80.066
breast-cancer	58.691	62.037	62.130	62.223	62.201
car	96.200	97.094	97.414	97.126	97.724
cleveland	73.004	73.004	73.718	74.197	75.318
cmc	65.128	65.184	65.128	65.184	65.128
ecoli	88.581	88.867	88.581	88.867	88.581
flower5-3d-10-20-35-35	0.000	52.425	52.685	52.601	51.340
flower5-3d-100-0-0-0	92.373	93.244	93.185	93.048	93.998
flower5-3d-30-40-15-15	76.914	78.520	78.282	77.998	78.412
flower5-3d-30-70-0-0	91.120	92.508	93.019	92.044	92.997
flower5-3d-50-50-0-0	92.003	93.154	92.870	92.726	93.321
haberman	64.128	65.404	65.126	66.098	66.199
hepatitis	78.017	79.323	79.648	79.080	79.551
paw3-3d-10-20-35-35	0.000	52.294	51.545	50.043	50.825
paw3-3d-100-0-0-0	90.122	92.915	93.420	93.101	94.346
paw3-3d-30-40-15-15	63.966	76.500	76.423	76.987	77.921
paw3-3d-30-70-0-0	87.208	90.707	90.438	90.296	90.730
paw3-3d-50-50-0-0	91.317	91.603	91.803	91.527	91.973
scrotal-pain	73.205	74.836	76.003	75.515	75.336
solar-flare	83.435	83.738	83.929	83.435	83.605
transfusion	65.226	66.612	65.391	66.612	65.375
vehicle	95.339	96.102	96.922	96.491	97.437
yeast	84.226	85.368	85.444	85.359	84.605
average rank	4.870	2.739	2.435	2.870	2.087

learning iterations (we considered $m_{al} \leq 10$) for relatively small batches. The size of batch in experiments was set as percentage of the size minority class ($n_{al} = \{5\%, 10\%\}$). We show two variants of the first phase of the active selection. Each of them is appropriate for the analyzed type of ABBag. Table 1, presents results of active balancing of 50% under-sampling EBBag. While in Table 2, we present results of active balancing of standard under-sampling NBBag. The last row of Tables 1 and 2 contains average ranks calculated as in the Friedman test – the lower average rank, the better the classifier.

The first, general conclusion resulting from our experiments is that ABBag performs better than under-sampling extensions of bagging, both: EBBag, and NBBag. Let us treat EBBag, and NBBag as baselines in Tables 1 and 2, respectively. The observed improvements of G-mean are statistically significant regardless of the version of ABBag. More precisely, following Friedman statistical test, each actively balanced EBBag has lower average rank than the baseline EBBag, and, similarly, each actively balanced NBBag has lower average rank than the baseline NBBag. Moreover, Friedman tests results in p-values $\ll$ 0.00001. According to Nemenyi post-hoc test, critical difference CD between average ranks is around 1.272. CD is thus higher than the difference between average ranks of each actively balanced EBBag and the baseline EBBag. An analogous observation holds for each actively balanced NBBag and the baseline NBBag. Some more detailed observations concerning results presented in Tables 1 and 2 may be given. The most striking difference in performance between baseline bagging extensions and ABBag is visible for synthetic data sets with the most difficult to learn distribution of safe, borderline, rare and outlier examples (i.e., data sets with suffix 10-20-35-35). In that case, both baseline EBBag and NBBag are not able to learn the minority class at all. ABBag shows significantly better performance for these data sets, regardless of the weights update strategy.

Moving to the next research question on the role of the proposed active modifications, i.e., weights update methods in the active selection, we make the following observations. First, if we consider actively balanced EBBag, *margin and component error* weights update method, thus (*mce*-EBBag), has the best average rank. Nevertheless, *margin, weight and component error* weights update method, thus (*mwce*-EBBag), has the best value of median calculated for all G-mean in Table 1. These observations are, not statistically significant according to the critical difference and results of Wilcoxon test for a selected pair of classifiers. On the other hand, in all of experiments with actively balanced EBBag we note that one of the two weights update methods: *mce* or *mwce* yields the best results.

Similar observations are valid for actively balanced NBBag. In this case, the best weights update method according to average rank is *margin, weight and component error*, and thus (*mwce*-NBBag). On the other hand, *margin and component error* weights update method, and thus (*mce*-NBBag), has the best value of median. Again these observations are not statistically significant. However, Wilcoxon paired test between *mwce*-NBBag and *mw*-NBBag resulted in p-value close to 0.045, and *mwce*-NBBag has the higher value of median. We can take it as another indication that inclusion of *component error* inside the weights update method yields better G-mean results.

To sum up, we can interpret results obtained with all four different weights update methods applied in the active selection in the following way. First, they show that all considered elements of weights update strategy: margin of classifiers in ensemble, weight of example, and component error are important for improving ABBag performance. Second, two combinations of considered elements tend to give better results than the others. These are: *margin and component error* (*mce*), and *margin, weight and component error* (*mwce*).

5   Conclusions

In this paper we examined the following research question: is it possible to improve classification performance of an under-sampling bagging ensemble with an active learning strategy? To address this question we introduced a new approach, called Actively Balanced Bagging (ABBag). The proposed active selection of examples involves iterative updating of bootstraps with batches composed of examples selected from the training set. These examples were sampled according to the distribution of weights, which expressed: a decision margin of ensemble votes, balancing of example class distribution in the training set and/or in its neighbourhood, and prediction errors of component classifiers.

The results of experiments have clearly shown that the active selection of examples has improved G-mean performance of considered under-sampling bagging extensions, which are known to be very good classifiers for imbalanced data. Another important observation resulting from experiments is that the active selection strategy performs best when it integrates the ensemble disagreement factor (typical for active learning) with information on class distribution in imbalanced data and prediction error of component classifiers.

We hope that this study may open future research lines on adapting active learning strategies to improve ensemble classifiers.

Acknowledgment. The research was funded by the Polish National Science Center, grant no. DEC-2013/11/B/ST6/00963.

References

1. Abe, N., Mamitsuka, H.: Query learning strategies using boosting and bagging. In: Proceedings of 15th International Conference on Machine Learning, pp. 1–10 (2004)
2. Aggarwal, C., Kong, X., Gu, Q., Han, J., Yu, P.: Active learning: a survey. In: Data Classification: Algorithms and Applications, pp. 571–606. CRC Press (2015)
3. Błaszczyński, J., Stefanowski, J., Idkowiak, L.: Extending bagging for imbalanced data. In: Burduk, R., Jackowski, K., Kurzynski, M., Wozniak, M., Zolnierek, A. (eds.) Proceedings of the 8th CORES 2013. AISC, vol. 226, pp. 269–278. Springer, Heidelberg (2013)
4. Błaszczyński, J., Stefanowski, J.: Neighbourhood sampling in bagging for imbalanced data. Neurocomputing **150A**, 184–203 (2015)
5. Borisov, A., Tuv, E., Runger, G.: Active batch learning with Stochastic Query-by-Forest (SQBF). In: JMLR Workshop on Active Learning and Experimental Design 2011, vol. 16, pp. 59–69 (2011)
6. Branco, P., Torgo, L., Ribeiro, R.: A survey of predictive modeling on imbalanced domains. ACM Comput. Surv. (CSUR) **49**(2), 31 (2016)
7. Chang, E.: Statistical learning for effective visual information retrieval. In: Proceedings of ICIP 2003, pp. 609–612 (2003)
8. Ertekin, S., Huang, J., Bottou, L., Giles, L.: Learning on the border: active learning in imbalanced data classification. In: Proceedings of the 16th ACM Conference on Information and Knowledge Management, pp. 127–136 (2007)

9. Galar, M., Fernandez, A., Barrenechea, E., Bustince, H., Herrera, F.: A review on ensembles for the class imbalance problem: bagging-, boosting-, and hybrid-based approaches. IEEE Trans. Syst. Man Cybern. Part C Appl. Rev. **99**, 1–22 (2011)
10. He, H., Yungian, M. (eds.): Imbalanced Learning. Foundations, Algorithms and Applications. IEEE - Wiley, Hoboken (2013)
11. Hido, S., Kashima, H.: Roughly balanced bagging for imbalance data. In: Proceedings of the SIAM International Conference on Data Mining, pp. 143–152 (2008) - An extended version in Stat. Anal. Data Mining **2**(5–6), 412–426 (2009)
12. Khoshgoftaar, T., Van Hulse, J., Napolitano, A.: Comparing boosting and bagging techniques with noisy and imbalanced data. IEEE Trans. Syst. Man Cybern.-Part A **41**(3), 552–568 (2011)
13. Napierala, K., Stefanowski, J.: The influence of minority class distribution on learning from imbalance data. In: Corchado, E., Snášel, V., Abraham, A., Woźniak, M., Graña, M., Cho, S.-B. (eds.) HAIS 2012. LNCS, vol. 7209, pp. 139–150. Springer, Heidelberg (2012)
14. Napierala, K., Stefanowski, J.: Types of minority class examples and their influence on learning classifiers from imbalanced data. J. Intell. Inf. Syst. **46**(3), 563–597 (2016)
15. Settles, B.: Active learning literature survey. Technical report, Computer Sciences Technical Report (2009)
16. Yang, Y., Ma, G.: Ensemble-based active learning for class imbalance problem. J. Biomed. Sci. Eng. **3**(10), 1022–1029 (2010)
17. Wojciechowski, S., Wilk, S.: The generator of synthetic multi-dimensional data. Poznan University of Technology Report RB-16/14 (2014)

A Comparison of Four Classification Systems Using Rule Sets Induced from Incomplete Data Sets by Local Probabilistic Approximations

Patrick G. Clark[1], Cheng Gao[1], and Jerzy W. Grzymala-Busse[1,2(✉)]

[1] Department of Electrical Engineering and Computer Science,
University of Kansas, Lawrence, KS 66045, USA
{cheng.gao,jerzy}@ku.edu, patrick.g.clark@gmail.com
[2] Department of Expert Systems and Artificial Intelligence,
University of Information Technology and Management, 35-225 Rzeszow, Poland

Abstract. This paper is a continuation of our previous research in which we compared four classification strategies using rule sets induced from incomplete data sets and global probabilistic approximations. In our current research we use local probabilistic approximations. In our incomplete data sets, missing attribute values are interpreted as lost values and "do not care" conditions. Our current results are that for symbolic data and numerical data with a few attributes, the best strategy is *strength with support*, while for data sets with many numerical attributes the best strategy is *probability only*. Our results for incomplete data sets with many numerical attributes are supported by only three data sets so further research is required.

Keywords: Incomplete data · Lost values · "do not care" conditions · MLEM2 rule induction algorithm · Local probabilistic approximations

1 Introduction

In this paper we compare four strategies used in classification systems. In such a system rule sets, induced from training incomplete data sets, are used for classification of unseen testing data. Quality of a classification strategy is determined by an error rate computed by ten-fold cross validation.

This paper presents a continuation of our previous research in which we compared four classification strategies using rule sets induced from incomplete data sets and global probabilistic approximations [1]. Such approximations, for complete data sets, were discussed in [2–6]. Probabilistic approximations were extended to incomplete data in [7].

In incomplete data sets used for our experiments, missing attribute values are interpreted as lost values and "do not care" conditions. A lost value, denoted by "?" is interpreted as a value that is currently not accessible, e.g., it was erased. A "do not care" condition, denoted by "*", is interpreted as any possible

M. Kryszkiewicz et al. (Eds.): ISMIS 2017, LNAI 10352, pp. 282–291, 2017.
DOI: 10.1007/978-3-319-60438-1_28

value of the attribute. Other methods of handling missing attribute values were presented, e.g., in [8,9]. A comparison of methods of handling missing attribute values, based on imputation, such as *most common values* and *concept most common values* for symbolic attributes and *average values* and *concept average values* for numeric attributes, *closest fit* and *concept closest fit* and CART approach with methods based on rough-set interpretations of missing attribute values was summarized in [10]. There is no universally best approach to missing attribute values.

Description of all four classification strategies is included in Sect. 5. These four classification strategies were compared using global probabilistic approximations [1]. As follows from [7], for incomplete data sets we may use singleton, subset or concept probabilistic approximations, all three approaches are based on global probabilistic approximations, constructed from characteristic sets [11,12]. Our conclusions in [1] were dependent on a kind of interpretation of missing attribute values, and were relatively weak, for both interpretations one of the classification strategies was better than another strategy.

In this paper we use local probabilistic approximations, more precise than global probabilistic approximations, since global probabilistic approximations are constructed from characteristic sets, while local probabilistic approximations are constructed from more elementary granules than characteristic sets, i.e., from blocks of attribute-value pairs. Search for such blocks is conducted on the set of all blocks of attribute-value pairs, while characteristic sets are constructed from selected blocks of attribute-value pairs.

In general, local approximations were introduced in [13] and were generalized to probabilistic local approximations in [14,15]. In the approach for computation of local probabilistic approximations used in this paper, both local approximations and rule sets are computed concurrently. Such rules, as we show in Sect. 4, are more informative than rules induced from global approximations. Additionally, as follows from our current research, our conclusions are more decisive and precise and are dependent on a type of attributes rather than on interpretation of missing attribute values.

2   Incomplete Data Sets

An example of incomplete data set is presented in Table 1. A *concept* is a set of all cases with the same decision value. In Table 1 there are two concepts, e.g., the set of all cases with flu is the set $\{1, 2, 3, 4\}$.

We use notation $a(x) = v$ if an attribute a has the value v for the case x. The set of all cases will be denoted by U. In Table 1, $U = \{1, 2, 3, 4, 5, 6, 7, 8\}$.

For complete data sets, for an attribute-value pair (a, v), a *block* of (a, v), denoted by $[(a, v)]$, is the following set

$$[(a, v)] = \{x | x \in U, a(x) = v\}.$$

For incomplete decision tables the definition of a block of an attribute-value pair must be modified in the following way [11,16]:

Table 1. An incomplete data set

	Attributes			Decision
Case	Temperature	Headache	Cough	Flu
1	High	Yes	Yes	Yes
2	High	?	No	Yes
3	Normal	No	*	Yes
4	Normal	*	Yes	Yes
5	?	?	No	No
6	*	No	*	No
7	High	?	?	No
8	*	No	No	No

- If for an attribute a and a case x, $a(x) =?$, the case x should not be included in any blocks $[(a, v)]$, for all values v of attribute a,
- If for an attribute a and a case x, $a(x) = *$, the case x should be included in blocks $[(a, v)]$, for all specified values v of attribute a.

For a case $x \in U$ the *characteristic set* $K_B(x)$ is defined as the intersection of the sets $K(x, a)$, for all $a \in B$, where B is a subset of the set A of all attributes and the set $K(x, a)$ is defined in the following way:

- If $a(x)$ is specified, then $K(x, a)$ is the block $[(a, a(x))]$ of attribute a and its value $a(x)$,
- If $a(x) =?$ or $a(x) = *$ then the set $K(x, a) = U$.

For the data set from Table 1, the set of blocks of attribute-value pairs is

$[(Temperature, normal)] = \{3, 4, 6, 8\}$,
$[(Temperature, high)] = \{1, 2, 6, 7, 8\}$,
$[(Headache, no)] = \{3, 4, 6, 8\}$,
$[(Headache, yes)] = \{1, 4\}$,
$[(Cough, no)] = \{2, 3, 5, 6, 8\}$,
$[(Cough, yes)] = \{1, 3, 4, 6\}$.

For Table 1, the characteristic sets are

$K_A(1) = \{1\}$,
$K_A(2) = \{2, 6, 8\}$,
$K_A(3) = \{3, 4, 6, 8\}$,
$K_A(4) = \{3, 4, 6\}$,
$K_A(5) = \{2, 3, 5, 6, 8\}$,
$K_A(6) = \{3, 4, 6, 8\}$,
$K_A(7) = \{1, 2, 6, 7, 8\}$,
$K_A(8) = \{3, 6, 8\}$.

3 Global Probabilistic Approximations

For incomplete data sets three different types of global approximations may be used: singleton, subset and concept. In this paper we will use only *concept* approximations.

The global B-*lower approximation* of X, denoted by $\underline{appr}(X)$, is defined as follows

$$\cup \{K_B(x) \mid x \in X, K_B(x) \subseteq X\}.$$

Such global lower approximations were introduced in [16].

The global B-*upper approximation* of X, denoted by $\overline{appr}(X)$, is defined as follows

$$\cup \{K_B(x) \mid x \in X, K_B(x) \cap X \neq \emptyset\} = \cup \{K_B(x) \mid x \in X\}.$$

These approximations were studied in [16,17].

A global B-probabilistic approximation of the set X with the threshold α, $0 < \alpha \leq 1$, denoted by $B\text{-}appr_\alpha(X)$, is defined as follows

$$\cup\{K_B(x) \mid x \in X, \ Pr(X|K_B(x)) \geq \alpha\},$$

where $Pr(X|K_B(x)) = \frac{|X \cap K_B(x)|}{|K_B(x)|}$ is the conditional probability of X given $K_B(x)$. Global A-probabilistic approximations of X with the threshold α will be denoted by $appr_\alpha(X)$.

For Table 1, we have

$$appr_1(\{1, 2, 3, 4\}) = \underline{appr}(\{1, 2, 3, 4\}) = \{1\}$$

and

$$appr_1(\{5, 6, 7, 8\}) = \underline{appr}(\{5, 6, 7, 8\}) = \emptyset.$$

The following single rule may be induced from such approximations
(Headache, yes) & (Temperature, high) → (Flu, yes).

4 Local Probabilistic Approximations

An extensive discussion on local probabilistic approximations is presented in [14,15]. Let X be any subset of the set U of all cases. Let $B \subseteq A$. A set T of attribute-value pairs, where all attributes belong to set B and are distinct, will be called a *B-complex*. Any A-complex will be called—for simplicity—a *complex*.

The most general definition of a local probabilistic approximation assumes only an existence of a family $\mathcal{T}$ of B-complexes T with the conditional probability $Pr(X|[T])$ of $X \geq \alpha$, where $Pr(X|[T]) = \frac{|X \cap [T]|}{|[T]|}$.

A *B-local probabilistic approximation* of the set X with the parameter α, $0 < \alpha \leq 1$, denoted by $appr_\alpha^{local}(X)$, is defined as follows

$$\cup\{[T] \mid \exists \ a \ family \ \mathcal{T} \ of \ B\text{-}complexes \ T \ of \ X \ with \ \forall \ T \in \mathcal{T}, \ Pr(X|[T]) \geq \alpha\}.$$

In general, for given set X and α, there exists more than one A-local probabilistic approximation.

In our experiments we used a heuristic version of the local probabilistic approximation, denoted by $appr_\alpha^{mlem2}(X)$, since it is inspired by the MLEM2 rule induction algorithm [18]. In this approach, $appr_\alpha^{mlem2}(X)$ is constructed from complexes Y that are the most relevant to X, i.e., with $|X \cap Y|$ as large as possible, if there is more than one complex that satisfies this criterion, the largest conditional probability of X given Y is the next criterion to select a complex. An algorithm to compute the local probabilistic approximation $appr_\alpha^{mlem2}(X)$ is presented in [15]. The same algorithm may be used for rule induction, we applied it in our experiments.

For Table 1, we have

$$appr_1^{mlem2}(\{1,2,3,4\}) = \{1,4\}$$

and

$$appr_1^{mlem2}(\{5,6,7,8\}) = \{6,8\}.$$

From these approximations we may induce the following rules
(Headache, yes) $\rightarrow$ (Flu, yes), and
(Headache, no) & (Temperature, high) $\rightarrow$ (Flu, no).
It is clear that the rule set induced from local probabilistic approximations is more informative than the rule set induced from global probabilistic approximations.

5   Classification

Rule sets are usually applied for classification of new, unseen cases. A *classification system* has two inputs: a rule set and a data set containing unseen cases. The classification system classifies every case as a member of some concept. The LERS classification system used in our experiments is a modification of the well-known bucket brigade algorithm [19–21].

In general, two parameters are used to decide to which concept a case belongs: *strength* and *support*. *Strength* is the total number of cases correctly classified by a rule during rule induction. The second parameter, *support*, is defined as the sum of strengths for all matching rules indicating the same concept. The concept C for which the support, i.e., the following expression

$$\sum_{\text{matching rules } r \text{ describing } C} Strength(r)$$

is the largest is the winner and the case is classified as being a member of C. This strategy is called *strength with support*. There exist three additional strategies. We may decide to which concept a case belongs on the basis of the strongest rule matching the case. This strategy will be called *strength only*. Another strategy is based on computing ratios of rule strength to the total number of cases matching

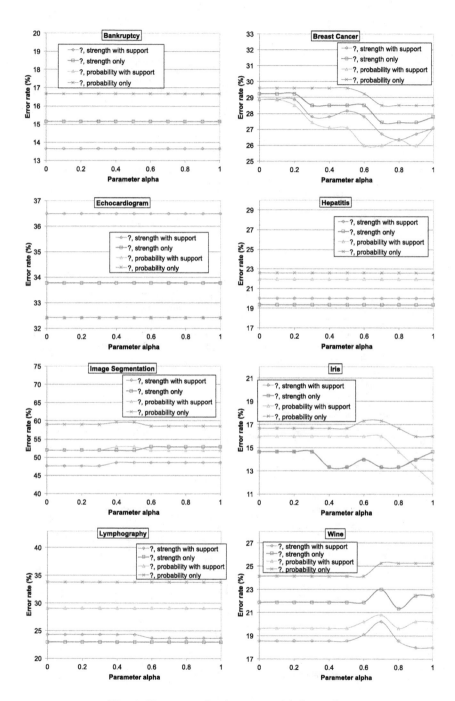

Fig. 1. Error rate for data sets with lost values

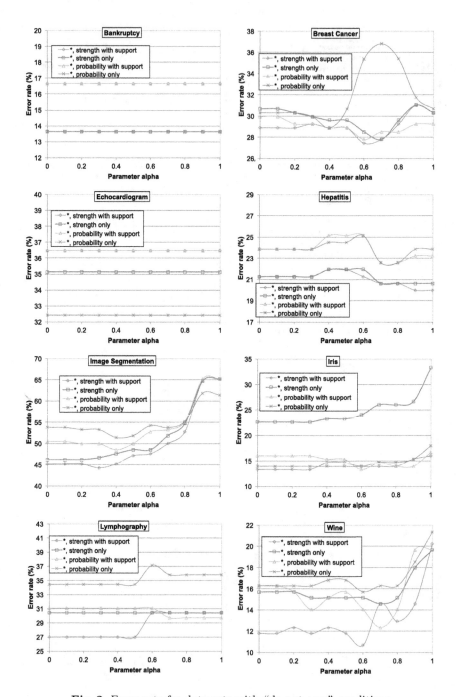

Fig. 2. Error rate for data sets with "do not care" conditions

the left-hand side of the rule. This ratio is a conditional probability of the concept given rule domain. A rule with the largest probability decides to which concept a case belongs. This strategy is called *probability only*. The fourth strategy, highly heuristic, in which all probabilities for rules indicating the same concept are added up, is called *probability with support*.

The problem is how to classify unseen cases with missing attribute values. In the LERS classification system, when an unseen case x is classified by a rule r, case x is considered to be not matched by r if for an attribute a, $a(x) = ?$ and the rule r contained a condition of the type (a, v), where v was a value of a. If for an attribute a, $a(x) = *$ and if the rule r contained a condition of the type (a, v), then case x is considered to be matched by r, does not matter what v is.

6   Experiments

Our experiments were conducted on eight data sets available from the University of California at Irvine *Machine Learning Repository*. For every data set, an incomplete data set was created. First, we used "?"s (lost values) for a random replacement of 35% specified values. Then an additional incomplete data set was created by global editing, all "?"s were replaced by "*"s ("do not care" conditions).

Our objective was to compare four classification systems in terms of an error rate of the rule sets induced from the local probabilistic approximations. Results of our experiments, presented in Figs. 1 and 2, show that there are significant differences between the four classification strategies. Our experiments were conducted on 16 incomplete data sets, eight with lost values and eight with "do not care" conditions. For all 16 data sets the Friedman Rank Sums test shows strong evidence to reject the null hypothesis that the four strategies are equivalent. Additionally, distribution-free multiple comparisons based on Friedman rank sum show that there is a difference between two classes of data: symbolic data and numeric data with a few attributes in one class and numeric data with many attributes in the other class. For the former class, consisting with 13 data sets, the best strategy is *strength with support* (it is the best for ten data sets, for remaining three data sets there is no better strategy than *strength with support*. For these 13 data sets the worst strategy is based on *probability only* (for 11 data sets, for one data set there is a tie between *probability only* and *probability with support* and in one case the worst strategy is *strength only*.

The latter class, i.e., data sets with numeric attributes and many attributes, the best strategy is *probability only* for one data set (*echocardiogram* with "do not care" conditions), for one data set (*echocardiogram* with lost values) there is a tie between *probability only* and *probability with support*, and for one data set (*bankruptcy* with "do not care" conditions) there is a tie between *strength only* and *probability only*. For this class *strength with support* is either the worst strategy or - for two data sets -there is a tie between *strength with support* and *probability with support*. More experiments are required to come up with conclusions about numeric data with a few attributes since our current experiments are restricted to three data sets only.

7 Conclusions

We compare four classification strategies: *strength with support, strength only, probability with support* and *probability only* applied for rule sets induced from incomplete data sets using local probabilistic approximations. Our results show that for symbolic data and numerical data with a few attributes, the best strategy is *strength with support*, while for data sets with many numerical attributes the best strategy is *probability only*. Since our results for incomplete data sets with many numerical attributes are supported by only three data sets, further research is required.

References

1. Clark, P.G., Grzymala-Busse, J.W.: A comparison of classification systems for rule sets induced from incomplete data by probabilistic approximations. In: Proceedings of the ALLDATA 1-st International Conference on Big Data, Small Data, Linked Data and Open Data, pp. 46–51 (2015)
2. Grzymala-Busse, J.W., Ziarko, W.: Data mining based on rough sets. In: Wang, J. (ed.) Data Mining: Opportunities and Challenges, pp. 142–173. Idea Group Publishing, Hershey (2003)
3. Pawlak, Z., Wong, S.K.M., Ziarko, W.: Rough sets: probabilistic versus deterministic approach. Int. J. Man-Mach. Stud. **29**, 81–95 (1988)
4. Wong, S.K.M., Ziarko, W.: INFER–an adaptive decision support system based on the probabilistic approximate classification. In: Proceedings of the 6-th International Workshop on Expert Systems and their Applications, pp. 713–726 (1986)
5. Yao, Y.Y.: Probabilistic rough set approximations. Int. J. Approximate Reasoning **49**, 255–271 (2008)
6. Ziarko, W.: Variable precision rough set model. J. Comput. Syst. Sci. **46**, 39–59 (1993)
7. Clark, P.G., Grzymala-Busse, J.W.: Experiments on probabilistic approximations. In: Proceedings of the 2011 IEEE International Conference on Granular Computing, pp. 144–149 (2011)
8. Bruha, I.: Missing attribute values. In: Sammut, C., Webb, G.I. (eds.) Encyclopedia of Machine Learning, pp. 674–680. Springer, Heidelberg (2010)
9. Grzymala-Busse, J.W., Grzymala-Busse, W.J.: Handling missing attribute values. In: Maimon, O., Rokach, L. (eds.) Data Mining and Knowledge Discovery Handbook, 2nd edn, pp. 33–51. Springer, Heidelberg (2010)
10. Grzymala-Busse, J.W.: A rough set approach to incomplete data. In: Ciucci, D., Wang, G., Mitra, S., Wu, W.-Z. (eds.) RSKT 2015. LNCS, vol. 9436, pp. 3–14. Springer, Cham (2015). doi:10.1007/978-3-319-25754-9_1
11. Grzymala-Busse, J.W.: Three approaches to missing attribute values–a rough set perspective. In: Proceedings of the Workshop on Foundation of Data Mining, in Conjunction with the Fourth IEEE International Conference on Data Mining, pp. 55–62 (2004)
12. Clark, P.G., Grzymala-Busse, J.W.: Experiments on rule induction from incomplete data using three probabilistic approximations. In: Proceedings of the 2012 IEEE International Conference on Granular Computing, pp. 90–95 (2012)

13. Grzymala-Busse, J.W., Rzasa, W.: Definability of approximations for a generalization of the indiscernibility relation. In: Proceedings of the 2007 IEEE Symposium on Foundations of Computational Intelligence, pp. 65–72 (2007)

14. Clark, P.G., Grzymala-Busse, J.W., Kuehnhausen, M.: Local probabilistic approximations for incomplete data. In: Proceedings of the 20-th International Symposium on Methodologies for Intelligent Systems, pp. 93–98 (2012)

15. Grzymala-Busse, J.W., Clark, P.G., Kuehnhausen, M.: Generalized probabilistic approximations of incomplete data. Int. J. Approximate Reasoning **132**, 180–196 (2014)

16. Grzymala-Busse, J.W.: Rough set strategies to data with missing attribute values. In: Notes of the Workshop on Foundations and New Directions of Data Mining, in Conjunction with the Third International Conference on Data Mining, pp. 56–63 (2003)

17. Lin, T.Y.: Topological and fuzzy rough sets. In: Slowinski, R. (ed.) Intelligent Decision Support. Handbook of Applications and Advances of the Rough Sets Theory, pp. 287–304. Kluwer Academic Publishers, Dordrecht (1992)

18. Grzymala-Busse, J.W., Rzasa, W.: A local version of the MLEM2 algorithm for rule induction. Fundamenta Informaticae **100**, 99–116 (2010)

19. Booker, L.B., Goldberg, D.E., Holland, J.H.: Classifier Systems and Genetic Algorithms. MIT Press, Boston (1990)

20. Holland, J.H., Holyoak, K.J., Nisbett, R.E.: Induction: Processes of Inference, Learning, and Discovery. MIT Press, Boston (1986)

21. Stefanowski, J.: Algorithms of Decision Rule Induction in Data Mining. Poznan University of Technology Press, Poznan (2001)

Robust Learning in Expert Networks: A Comparative Analysis

Ashiqur R. KhudaBukhsh[(✉)], Jaime G. Carbonell, and Peter J. Jansen

Carnegie Mellon University, Pittsburgh, USA
{akhudabu,jgc,pjj}@cs.cmu.edu

Abstract. Learning how to refer effectively in an expert-referral network is an emerging challenge at the intersection of Active Learning and Multi-Agent Reinforcement Learning. Distributed interval estimation learning (`DIEL`) was previously found to be promising for learning appropriate referral choices, compared to greedy and Q-learning methods. This paper extends these results in several directions: First, learning methods with several multi-armed bandit (MAB) algorithms are compared along with greedy variants, each optimized individually. Second, `DIEL`'s rapid performance gain in the early phase of learning proved equally convincing in the case of multi-hop referral, a condition not heretofore explored. Third, a robustness analysis across the learning algorithms, with an emphasis on capacity constraints and evolving networks (experts dropping out and new experts of unknown performance entering) shows rapid recovery. Fourth, the referral paradigm is successfully extended to teams of Stochastic Local Search (SLS) SAT solvers with different capabilities.

Keywords: Active learning · Referral networks · SLS SAT solvers

1 Introduction

Human experts may refer to other experts when given problems outside their area of expertise. In a clinical network, a physician may diagnose and treat a patient or refer the patient to another physician whom she believes may have more appropriate knowledge, conditioned on the presenting symptoms. Referral networks are common across other professions as well, such as members of large consultancy firms. In addition to human professional networks, we can envision referral networks of automated agents, such as the SAT solver network introduced in this paper, or heterogeneous networks of human and machine agents. In all cases, unless the experts are omniscient they often need to consult or refer problems to their colleagues. This paper addresses learning to whom an expert should refer given a problem and topic, attempting to optimize performance based on accumulated experience of the results of previous referrals. We address the distributed learning-to-refer setting, without a "boss agent" telling all the others when to try and solve a problem or when to refer and if so to whom.

© Springer International Publishing AG 2017
M. Kryszkiewicz et al. (Eds.): ISMIS 2017, LNAI 10352, pp. 292–301, 2017.
DOI: 10.1007/978-3-319-60438-1_29

Much of this work is based on DIEL, a simple but efficient algorithm balancing exploration with exploitation, which has been proposed in [1]. Previous work is summarized in Sect. 2. Section 3 presents our assumptions and the structure of referral networks and expertise. Sections 4, 5 and 6 describe the distributed learning algorithms we used in the comparison, our experimental setup, and the results.

2 Related Work

Our primary predecessor for this work was the referral learning framework proposed in [1], although that work did not extend to comparison among competing algorithms, nor address capacity constraints, nor robustness to unexpected dropouts and additions to the network, which is needed for modeling actual networks of human experts. This paper also addresses fully-autonomous SAT-solver agents, vs just simulated experts. A completely different approach was taken in [2], which extends the same referral framework by augmenting the learning setting with an advertising mechanism, where experts can post estimates of their skill level in different tasks, encouraging truth-in-advertisement.

The problem of learning appropriate referrals can be cast in various ways. Taking it as a multi-armed bandit (MAB) selection problem, we accordingly enlisted several MAB algorithms [3–5] with known finite-time regret bound for performance comparison, none of which have as far as we know been studied in the context of referral learning before. An analysis of referral networks also exhibits similarities with the study of task allocation [6,7], where minimizing turn-around time corresponds to the maximizing the probability of a correct answer. The FAL algorithm described in [7] uses a variant of ϵ-greedy Q-learning similar to one we compared favorably against in the current work.

Finally, in part of this work, we used SATenstein [8], a highly parameterized Stochastic Local Search (SLS) SAT solver to generate non-synthetic experts and expertise data. This was done previously [9] in the context of the augmented setting discussed in [2]. Here we systematically apply a set of referral algorithms (as opposed to two), and use continuous rewards (as opposed to binary) to introduce the notion of *solution quality*.

3 Referral Networks

A *referral network* can be represented by a graph (V, E) of size k in which each vertex v_i corresponds to an expert e_i $(1 \leq k)$ and each bidirectional edge $\langle v_i, v_j \rangle$ indicates a *referral link*. We call the set of experts linked to an expert e_i by a referral link, the *subnetwork* of expert e_i. In a referral *scenario*, a set of m instances $(q_1, \ldots, q_m)$ belonging to n topics $(t_1, \ldots, t_n)$ are to be addressed by the k experts $(e_1, \ldots, e_k)$.

Assuming a query budget of $Q = 2$, the following steps are executed for each instance q_j.

1. A user issues an *initial query* to an expert e_i (*initial expert*) chosen uniformly at random.
2. Expert e_i examines the instance and solves it if able, depending on the *expertise* of e_i wrt. q_j, defined as the probability that e_i can solve q_j correctly.
3. If not, she passes a *referral query* to a *referred expert* within her subnetwork. *Learning-to-refer* means improving the estimate of who is most likely to solve the problem.
4. If the referred expert succeeds, she communicates the solution to the initial expert, who in turn, communicates it to the user.

We also considered in our experiments the case $Q > 2$, when the recipient of a referral can herself re-refer to another expert (with reduced budget).

For the simulations, we follow (initially) the *assumptions* made in [1], notably that: network connectivity depends on (cosine) similarity between the topical expertise, expertise is stationary, and its distribution can be characterized by a mixture of Gaussian distributions (for further details, see [1,2]).

4   Distributed Referral Learning

Essentially, from the point of view of a single expert, learning appropriate referral choices for a given topic is an action selection problem. Action selection using Interval Estimation Learning (IEL) works in the following way [2,10]. First, for each action a, the upper confidence interval for the mean reward $(UI(a))$ is estimated by

$$UI(a) = m(a) + \frac{s(a)}{\sqrt{n}} \tag{1}$$

where $m(a)$ is the mean observed reward for a, $s(a)$ is the sample standard deviation of the reward, n is the number of observed samples from a. Next, IEL selects the action with the highest upper confidence interval. The intuition behind selecting the action with the highest mean plus upper confidence interval is that high mean indicates good expected performance and high variance indicates we lack knowledge about said performance. This naturally trades off exploitation (selecting high mean) and exploration (resulting in variance reduction). An earlier version of DIEL [1] used an additional Student's t-distribution parameter, here we are using the version reported in [2], which is parameterless and outperformed the earlier version.

In a distributed setting, each expert is running a thread of action selection for each topic in parallel. So basically, DIEL consists of multiple IELs for each topic/expert pair. Categorized into three broad categories: Q-Learning variants, UCB-variants and greedy-variants, our choice of referral algorithms is presented in Table 1. The distributed versions of these algorithms function the same way as DIEL – only their action selection procedure is different. Ideally, the distributed version of UCB1 should be called DUCB1; but since we have both Q-Learning [11] and DQ-Learning [12] in our pool of algorithms, we slightly abuse the notation in order to avoid confusion.

Table 1. Referral algorithms

Category	Algorithm	Parameters
IEL	DIEL [2]	None
Greedy	DMT [1]	None
Greedy	ϵ-Greedy [13]	c
Greedy	ϵ-Greedy1	α
UCB	UCB1 [3]	None
UCB	UCB2 [13]	None
UCB	UCBNormal [4]	None
UCB	UCBV [5]	θ
Q-learning	Q-learning [11]	α, γ, ϵ
Q-learning	DQ-learning [12]	α, γ, ϵ

A primary challenge in the distributed setting is that there is no global visibility of rewards, i.e., $reward(e_i, topic_p, e_j)$ (a function of the initial expert e_i, instance topic $topic_p$, and referred expert e_j) is only visible to expert e_i. Also, because of the scale, in a practical setting, we cannot afford a large number of referrals for finding suitable referral choices. For this reason, a high performance in the early phase of learning is crucial.

In the following description of the action selection procedures, $m(a)$ is the mean observed reward for action a, n_a is the number of observations of a, and N is the total number for all actions.

DMT: Unlike DIEL, DMT only considers the mean observed reward and always greedily picks the action with the highest mean reward.

ϵ-Greedy: DMT, being purely greedy, can easily get stuck with a sub-optimal referral choice. ϵ-Greedy performs a diversification step with a probability ϵ. i.e., with probability ϵ, it randomly chooses one of the connected experts for referral.

ϵ-Greedy1: ϵ-Greedy1 differs from ϵ-Greedy only in its way of setting the diversification probability parameter (set to $\frac{\alpha * K}{N}$ where K is the subnetwork size, i.e., the total number of referral choices).

UCB1: UCB1 selects the action with highest $m(a) + \sqrt{\frac{2lnN}{n_a}}$. This implies among two actions with equal mean reward, UCB1 will favor the least sampled one.

UCB2: UCB2 executes in an episodic fashion. Once an action is selected, it is executed for an episode. For each action a, it first initializes r_a to 0 where r_a denotes the episode length and each action is executed once in the beginning. If the last selected action j has been played for r_j times in a row, the new action is selected by maximizing $m(a) + \sqrt{\frac{(1+\alpha).ln(eN\tau(r_a))}{2\tau(r_a)}}$ where $\tau(r_a) = (1 + \alpha)^{r_a}$ and α is a configurable parameter.

UCB-normal: UCB-normal performs any action that has been executed less than $\lceil logN \rceil$. Otherwise, the action with highest $m(a) + \sqrt{16.\frac{sq(a)-n_a.m(a)^2}{n_a-1}.\frac{ln(N-1)}{n_a}}$ is chosen ($sq(a)$ is the sum of squared rewards obtained from action a).

UCBV: Similar to DIEL, UCBV also uses variance to compute expected reward. However, it uses a different exploration function, $\frac{logN}{n_a}$. UCBV selects the action with highest $m(a) + s(a).\sqrt{\frac{2\theta log(N)}{n_a}} + \frac{3\theta log(N)}{n_a}$. [5] reported a value of 1.2 for the parameter θ to guarantee logarithmic convergence.

DQ-Learning: Double Q-Learning, or DQ-learning, consists of two standard Q-Learning algorithms running in tandem (with Q functions: Q_A and Q_B, say). Whenever an action is chosen based on Q_A, the observed reward is used to update Q_B and vice versa. In practice, DQ-Learning tends to converge faster than Q-Learning (for further details, see [12]). For Q-Learning, we considered ϵ-Greedy-Q-Learning.

5   Experimental Setup

We compared the DIEL referral-learning algorithm against the nine other algorithms in Table 1), a topical upper bound, and a random (expertise-blind) baseline. Each parameterized referral algorithm was tuned on a separate training set constructed using the same parameter distribution described in [1]. For each algorithm, we ran 100 random instantiations of the algorithm on the training data set and selected the configuration that performed best on this set. The ϵ-Greedy algorithm, as presented in [13], requires prior knowledge about the reward distribution in order to set the value of the hyper-parameter d. However, we found that estimating d from the observations created sub-par performance. Setting instead ϵ to $\frac{\alpha*K}{N}$ (where K is the subnetwork size and N is the number of total observations) gave rise to a good performance when appropriately configured. We followed a similar procedure to set ϵ for ϵ-Greedy Q-learning.

Our test set for performance evaluation is the same data set used in [1,2]. It consisted of 1000 scenarios, each with 100 experts (average connection density 16.05 ± 4.99), 10 topics and a referral network. Our measure of performance is the overall task accuracy of our multi-expert system. For the sake of comparability, for a given simulation across all algorithms, we chose the same sequence of initial expert and topic pairs. For our per-instance query budget, Q, we chose the values 2, 3, and 4, corresponding to single-hop, two-hop and three-hop referrals, respectively. Following [1], our upper bound for single-hop referral is the performance of a network where every expert has access to an oracle that knows the true topic-mean (i.e., $mean(Expertise(e_i, q) : q \in topic_p) \forall i, p$) of every expert-topic pair. For two-hop referrals, we use an upper bound based on calculating optimal referral choices up to depth 2. Finally, the baseline is an Expertise-Blind algorithm where the initial expert randomly chooses a connected expert for referral.

The 100 `SATenstein` solvers we used are obtained by configuring `SATenstein2.0` on six well-known SAT distributions (distribution and solver details can be found in [8]). We used the test sets of the SAT distributions as our pool of tasks. Our experiments were carried out on a cluster of dual-core 2.4 GHz machines with 3 MB cache and 32 GB RAM running Linux 2.6.

6   Results

6.1   Performance Comparison on Synthetic Data

Single Hop Referral: For single referral, Table 2 presents the mean task accuracy across the entire data set at specific points of the horizon (samples per subnetwork). For a given horizon, the best performance is highlighted in bold. Our results show that except during the very early stages of the simulation, DIEL dominated all the other referral algorithms with a performance approaching the optimal upper bound. A paired t-test reveals that beyond the crossover point (1000 samples per subnetwork), DIEL is better than all other referral algorithms with p-value less than 0.0001. Algorithms with provable performance guarantees may catch up with DIEL given a sufficiently large horizon, but from a practical standpoint, DIEL is an effective referral algorithm to handle real-world scenarios. We extended a random subset of 200 scenarios up to a horizon of 20,000 samples per subnetwork, at which time none of the top performing referral-algorithms from each category had caught up with DIEL.

All referral learning algorithms performed better than our baseline, the expertise-blind referral. Although `DQ-learning` and `ε-Greedy1`, the best algorithms in the Q-learning and greedy category respectively, obtained a performance close to DIEL, this was conditional to tuning on training data of similar

Table 2. Performance comparison of referral algorithms with query budget $Q = 2$

	500	1000	1500	2000	3000	4000	5000
Upper Bound	79.47	79.31	79.27	79.42	79.38	79.41	79.47
DIEL	67.73	**73.35**	**75.35**	**76.33**	**77.33**	**77.76**	**77.96**
DMT	**70.63**	73.06	73.83	74.20	74.54	74.71	74.69
ε-Greedy	55.33	56.63	57.80	58.95	60.57	61.95	62.97
ε-Greedy1	70.22	72.91	73.97	74.48	74.92	75.19	75.32
UCB1	57.78	59.60	60.80	61.61	63.28	64.49	65.49
UCB2	63.71	64.16	64.19	64.21	64.18	64.19	64.28
UCB-normal	54.38	54.47	54.71	54.97	56.43	58.91	61.39
UCBV	54.99	55.92	56.44	56.87	57.83	58.60	59.15
Q-Learning	65.46	69.19	70.98	72.08	73.46	74.27	74.75
DQ-learning	70.23	72.68	73.74	74.37	75.14	75.60	75.91
Expertise-Blind	54.48	54.46	54.45	54.41	54.48	54.44	54.60

distributional properties. In contrast, DIEL is parameterless and thus does not require additional configuration.

For the remaining results, we retained only the best-performing algorithms in each category, as follows: DIEL (IEL category), ε-Greedy1 (Greedy category), UCB1 (UCB category) and DQ-learning (Q-learning category). For comparison, we included additionally, DMT, a horizon-free algorithm.

Multi-hop Referral: In a multi-hop setting, a referred expert can continue referring an instance to another expert as long as the budget permits (excluding cyclic referrals). Figure 1 compares our top-performing referral algorithms with query budget 3 and 4. In Fig. 1(a), we compare the performance with an upper bound that calculates optimal choice to depth 2 (optimal choice to depth 1 but the same query budget 3 achieved a task accuracy of 93.05%). Understandably, with a higher query budget, the overall task accuracy of every learning algorithm increases. However, DIEL's rapid performance gain in the early phase of learning still enables it to obtain a superior performance. The practical benefit of DIEL against algorithms with theoretical convergence guarantees is particularly evident when compared against UCB1. In fact, DIEL with a lower query budget ($Q = 3$) achieves a better performance than UCB1 with a higher budget ($Q = 4$).

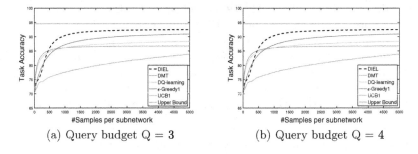

(a) Query budget Q = 3   (b) Query budget Q = 4

Fig. 1. Multi-hop referrals

6.2   Robustness of Performance

Dynamic Network Behavior: In practice, referral networks are not static; they evolve over time with new links are forged, experts drop out or new ones join, experts gain or degrade expertise, etc. Here we focused primarily on addition/deletion of experts to the network, both as a one time event with 20% of the experts in the network replaced at iteration 100, and a *distributed change* (modelling more closely a real-world gradual change: 5% of the network changes every 50 iterations). We also ran experiments where the network changes are distributed across time-steps and found qualitatively similar performance.

Figure 2 compares the performance of DIEL on a static network with that on a dynamic network. Our results show that DIEL coped fairly well with a distributed change, and in spite of multiple changes in the network at a regular interval, the final DIEL performance on a dynamic network (task accuracy

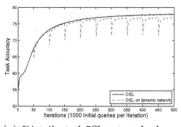

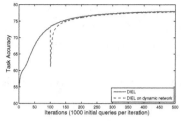

(a) Distributed 5% network change   (b) Single point 20% network change

Fig. 2. Performance of DIEL with network changes

76.91%) is slightly worse than DIEL on a static network, but still better than any other referral learning algorithm presented in Table 2. In addition, we ran experiments where no experts leave or join, but new referral links get created. Then too, the performance of DIEL proved robust, exhibiting qualitatively similar characteristics. We also found that DIEL could easily cope with a large one time network change (see, Fig. 2(b)).

Capacity Constraints: Capacity constraints on experts are rarely taken into account in Active Learning (though *Proactive Learning* [14] considers similar aspects). In reality, of course, experts can handle only a limited number of tasks at any given time, and the capacity of the best experts can easily be exceeded. This was borne out by our simulations – for our DIEL simulations, expertise and load were correlated with correlation coefficient $r = 0.69$.

We simulated transient (bursty) overloading. Let $load(e_i, m)$ denote the number of tasks expert e_i received among the last m tasks (*initial* or *referred*) the network received. In a network of k experts, a fair load for every expert is $\frac{m}{k}$. An expert is overloaded if $load(e_i, m) \geq c * \frac{m}{k}$, where the load-factor $c > 1$. In our experiments we assumed that an expert reaching her load limit becomes unavailable until completing one or more current tasks. Even with a tight value of $c = 1.5$, we find that the performance of the referral-algorithms degrades gracefully, and surprisingly, sometimes causing a performance improvement because of forced exploration as tasks are sent to other experts. For example, as shown in Fig. 3(b), the load-balanced version of DMT with a load-factor of 2 slightly

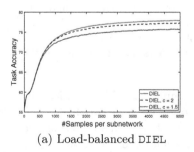

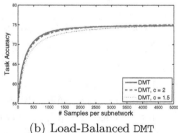

(a) Load-balanced DIEL   (b) Load-Balanced DMT

Fig. 3. Performance of DMT and DIEL for different values of the load-factor c

outperforms DMT without any capacity constraint. That we observed a graceful performance degradation with all the referral algorithms leads us to conjecture that load balancing is facilitated by the distributed nature of the learning setting.

6.3 SATenstein SLS Solvers as Experts

So far, we have presented our results on synthetic data and binary rewards. Here, we describe our results where experts are Stochastic Local Search (SLS) solvers and the task is to solve a SAT problem instance. In addition to the attractive properties of SAT solvers listed in [9] (e.g., easy availability of a large number of experts with differential expertise (Fig. 4(a)), and the straight-forward verifiability of solutions), they allow us to easily express solution quality as a function of run time, allowing us to test the referral algorithms under continuous rewards.

In these experiments, in order to save computational cycles, we solely focus on the referral learning behavior of the network; i.e., we assume that the initial expert always refers a task to a connected expert. We set the budget C for solving each instance to 1 CPU second, which is the maximum time in which, on a similar computing architecture, configured high-performance SATenstein solvers were found to solve a majority of the instances in their expertise area [8] (This was corroborated in our experiments). The reward is computed as $(C - r_t)$ where r_t is the run time (when a solver fails to solve an instance, $r_t = 1$). With C set to 1 in our experiments, the reward is bounded by $[0, 1)$ with a failed task fetching a reward of 0 and higher rewards implying faster solutions. So in this setting, through continuous reward, we have incorporated solution quality (in this case, run time) in our experiments.

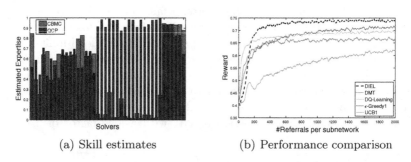

(a) Skill estimates (b) Performance comparison

Fig. 4. Expertise estimates of a subset of solvers on background data of two SAT distributions and performance comparison with SATenstein solvers as experts

Figure 4(b) presents the performance comparison of referral-learning algorithms where experts are SAT solvers and topics are SAT problem distributions on 10 randomly chosen referral networks. We found that DIEL outperformed all other algorithms, with DMT, DQ-learning, and ϵ-Greedy1 achieving a performance close to DIEL (even when we extended the runs to 4000 referrals per subnetwork for ϵ-Greedy1, it had not yet caught up with DIEL). Similar to the results obtained on our synthetic data, we found that UCB had the slowest rate of improvement in the initial stage of learning. These results highlight the following. First, even with real experts, a well-defined task and very few distributional assumptions on expertise, it is possible to learn effective referral choices. Second, DIEL's superiority over other referral-learning algorithms is not just restricted to synthetic data, nor dependent on binary rewards.

Acknowledgements. This research is partially funded by the National Science Foundation grant EAGER-1649225.

References

1. KhudaBukhsh, A.R., Jansen, P.J., Carbonell, J.G.: Distributed learning in expert referral networks. In: European Conference on Artificial Intelligence (ECAI), vol. 2016, pp. 1620–1621 (2016)
2. KhudaBukhsh, A.R., Carbonell, J.G., Jansen, P.J.: Proactive skill posting in referral networks. In: Kang, B.H., Bai, Q. (eds.) AI 2016. LNCS, vol. 9992, pp. 585–596. Springer, Cham (2016). doi:10.1007/978-3-319-50127-7_52
3. Agrawal, R.: Sample mean based index policies with o (log n) regret for the multi-armed bandit problem. Adv. Appl. Probab. **27**(4), 1054–1078 (1995)
4. Lai, T.L., Robbins, H.: Asymptotically efficient adaptive allocation rules. Adv. Appl. Math. **6**(1), 4–22 (1985)
5. Audibert, J.-Y., Munos, R., Szepesvári, C.: Tuning bandit algorithms in stochastic environments. In: Hutter, M., Servedio, R.A., Takimoto, E. (eds.) ALT 2007. LNCS, vol. 4754, pp. 150–165. Springer, Heidelberg (2007). doi:10.1007/978-3-540-75225-7_15
6. Abdallah, S., Lesser, V.: Learning the task allocation game. In: Proceedings of AAMAS 2006, pp. 850–857. ACM (2006)
7. Zhang, C., Lesser, V., Shenoy, P.: A multi-agent learning approach to online distributed resource allocation. In: Proceedings of IJCAI 2009, Pasadena, CA, vol. 1, pp. 361–366 (2009)
8. KhudaBukhsh, A.R., Xu, L., Hoos, H.H., Leyton-Brown, K.: SATenstein: automatically building local search SAT solvers from components. Artif. Intell. **232**, 20–42 (2016)
9. KhudaBukhsh, A.R., Carbonell, J.G., Jansen, P.J.: Proactive-DIEL in evolving referral networks. In: European Conference on Multi-Agent Systems, Springer, Heidelberg (2016)
10. Donmez, P., Carbonell, J.G., Bennett, P.N.: Dual strategy active learning. In: Kok, J.N., Koronacki, J., Mantaras, R.L., Matwin, S., Mladenič, D., Skowron, A. (eds.) ECML 2007. LNCS (LNAI), vol. 4701, pp. 116–127. Springer, Heidelberg (2007). doi:10.1007/978-3-540-74958-5_14
11. Watkins, C.J., Dayan, P.: Q-learning. Mach. Learn. **8**(3–4), 279–292 (1992)
12. Hasselt, H.V.: Double Q-learning. In: Advances in Neural Information Processing Systems, pp. 2613–2621 (2010)
13. Auer, P., Cesa-Bianchi, N., Fischer, P.: Finite-time analysis of the multiarmed bandit problem. Mach. Learn. **47**(2–3), 235–256 (2002)
14. Donmez, P., Carbonell, J.G.: Proactive learning: cost-sensitive active learning with multiple imperfect oracles. In: Proceedings of CIKM 2008, vol. 08, pp. 619–628 (2008)

Efficient All Relevant Feature Selection with Random Ferns

Miron Bartosz Kursa[✉]

Interdisciplinary Centre for Mathematical and Computational Modelling,
University of Warsaw, Pawińskiego 5A, 02-106 Warsaw, Poland
M.Kursa@icm.edu.pl

Abstract. Many machine learning methods can produce variable importance scores expressing the usability of each feature in context of the produced model; those scores on their own are yet not sufficient to generate feature selection, especially when an all relevant selection is required. There are wrapper methods aiming to solve this problem, mostly focused around estimating the expected distribution of irrelevant feature importance. However, such estimation often requires a substantial computational effort.

In this paper I propose a method of incorporating such estimation within the training process of a random ferns classifier and evaluate it as an all relevant feature selector, both directly and as a part of a dedicated wrapper approach. The obtained results prove its effectiveness and computational efficiency.

Keywords: Feature importance · Feature selection · Random Forest · Random ferns

1 Introduction

The crucial part of any machine learning application is to find a good representation of the observed data; its ability to express the contained information in a way that is well exposed for the used modelling method is often the most important component of the final performance. On the other hand the robustness of this process is also critical, as it is very easy to introduce over-fitting this way; either by leaking information which model should not be given or by amplifying false, random associations which are only present in training data due to its finite size.

There are also cases when the original data is already in a 'tabular' form, i.e. as a series of independently measured features, and it is strongly desired to retain this structure because of known links between features and certain physical aspects of the investigated phenomenon. Then, the preferred form of representation altering is *feature selection*, that is reducing the original set of attributes to its strict subset.

Most prominent practical examples here are data sets obtained via high throughput biological experiments: they can simultaneously capture the activity

© Springer International Publishing AG 2017
M. Kryszkiewicz et al. (Eds.): ISMIS 2017, LNAI 10352, pp. 302–311, 2017.
DOI: 10.1007/978-3-319-60438-1_30

of thousands or even millions of agents representing a substantial fraction of a full state of a given system, though only a handful is expected to be connected to the investigated state or behaviour. Finding important yet previously unknown agents in such case can lead, through targeted studies, to a discovery of novel mechanisms, consequently become even more important than the original task of building a black-box model.

While such data may have even few orders of magnitude more features than cases, it often contains aforementioned false associations of a predictive potential matching the true interactions. This not only increases the risk of over-fitting, but may also cause masking of true interactions, consequently hindering the explanatory role of feature selection, especially when the analysis methods applied aim at reducing the redundancy of the result.

Furthermore, many of such measurements are done blindly (or re-used in other studies), thus an effective analysis pipeline can not relay on the assumption that some significant relations must be present, consequently be able to reveal inconclusive problems—this is another important challenge for feature selection in such context.

2 Background

Fundamentally, feature selection methods can be divided into two classes, *minimal optimal* and *all relevant* [12]—methods of the first group attempt to find a smallest subset of features on which certain model achieves optimal performance; second group collects methods which attempt to remove features irrelevant to the problem, consequently retaining those features which may be useful for modelling. The first aim is straightforward to implement as an accuracy optimisation problem, although it is expected miss important features that carry redundant information. Moreover, it is possible that the analysed problem contains random, false associations between features and the decision which may be indistinguishable or even more informative than the true ones, especially if the true mechanism is complex and the set is of a $p \gg n$ class. In that case a minimal optimal approach is likely to return mostly noise and still produce a deceivingly well performing model [11]. Consequently, only all relevant selection is applicable for deciphering mechanisms behind the analysed data.

There is also a technical taxonomy due to how the selection is coupled with modelling [17]: there are *filters*, algorithms which are independent from the modelling method, *embedded methods* which integrate selection and modelling into a single algorithm, finally *wrappers* which relay on some modelling method, but only as an oracle of an efficiency of a given subset of features.

All relevant wrappers [7, 8, 19] generally relay on an external model to return feature importance (also called variable importance measure, VIM), yet try to establish a VIM threshold separating relevant and irrelevant features. While it is mostly impossible to analytically obtain the distribution of VIM for irrelevant variables, aforementioned wrappers use a kind of permutation test approach to estimate it—utilising either permutation of the decision attribute or features irrelevant by design (called *shadows* or *contrasts*) injected within original ones. Such an approach obviously requires many repetitions to stabilise the approximation; some methods also try to progressively eliminate irrelevant features, as both stability and the quality of VIM often decrease with the dimensionality of a set.

The aim of this work is to propose a methodology to combine the estimation of irrelevant VIM distribution with the estimation of VIM itself, consequently integrating both within the model training procedure. In particular, I will follow the heuristics behind the Boruta method [9], as it is generic and proved effective in a demanding assessment [11], and random ferns, an efficient, stochastic ensemble classifier. Furthermore, I will assess usability and efficiency of the obtained *shadow importance* in all relevant feature selection, both directly and as a core of a simple wrapper approach.

3   Methodology

3.1   Boruta

Boruta [8,9] is an all relevant feature selection wrapper, originally developed for Random Forest [2], but in a present form capable of using arbitrary classifier that outputs VIM. The algorithm works in a following way. First, VIM is calculated on a modified data set extended with explicitly generated shadows, i.e. nonsense features created from the original ones by permuting the order of values within them (hence wiping out information but preserving distribution). Then, all original variables which importance was greater than the maximal importance of a shadow are assigned a *hit*.

This procedure is repeated (with shadows re-generated each time), while the proportion of hits for each feature is observed for being significantly lower or higher than half; in the first case the feature is claimed rejected and removed from further consideration, in the latter claimed confirmed and added to final result. The algorithm stops either when the status of all original features is decided or when a previously given maximal number of iterations is exhausted, in which case some variables may be left undecided.

3.2   Random Ferns

Random ferns are classifiers introduced in [15] and named as such in [14]. They were developed to serve as an computationally efficient alternative to Random Forests in demanding computational vision tasks [1,13]. I have previously shown, however, that with certain modifications they can perform well in a generic machine learning context [10], as well as produce variable importance measure (VIM) of a similar quality as Random Forest, though in a much shorter time [11]. Here I will briefly present the method in the aforementioned generalised version, called *rFerns*.

Let's assume a set of N training objects $\mathcal{T} \subset \mathcal{X}$, where $\mathcal{X}$ is composed of M attributes, i.e., $\mathcal{X} = \prod_{j=1}^{M} \mathcal{X}_j$, where $\mathcal{X}_j$ is a domain of the j-th attribute. Each object $\mathbf{x} \in \mathcal{T}$ is assigned $Y(\mathbf{x})$, one of C disjoint classes.

rFerns is an ensemble of K ferns $\mathbf{F}^k(\mathcal{T}') : \mathcal{X} \to \mathbb{R}^C$ returning a vector of class scores indicating confidence how a given object fits within each class. The ensemble is built using bagging, thus k-th fern is built on a separate random multiset of training objects called bag, $\mathcal{B}_k := B(\mathcal{T}, k)$, where $B(\mathcal{X}, \lambda)$ denotes sampling with replacement the same number of elements as in the input set, and λ functions as a random seed of sampling procedure, i.e., samples with different λ would be different and statistically independent. The prediction of the ensemble is the class which gets maximal sum of scores over all ferns, i.e.,

$$\hat{Y}(\mathbf{x}) := \arg\max_y \sum_k F_y^k(\mathbf{x}; \mathcal{B}_k). \tag{1}$$

Each fern is practically a Bayes classifier over a random partition of $\mathcal{X}$. The base of $\mathbf{F}$ is a $trunk$ function $T : \mathcal{X} \to 1..2^D$ classifying given object into its corresponding $leaf$, which is then associated with a vector of classes' scores estimated using $\mathcal{T}$. D is a hyper-parameter of the classifier and is called $fern$ $depth$; it allows one to control the profoundness of the classification, as it is a maximal dimension of interaction random ferns can catch. Trunk is defined by a sequence of D split attribute indexes $j_i \in 1..M$ and subsets $\Xi_i \subset \mathcal{X}_{j_i}$, so that an element $\mathbf{x} = (x_1, x_2, \ldots, x_M) \in \mathcal{X}$ belongs to a leaf

$$T(\mathbf{x}) := \sum_{i=0}^{D} 2^i \cdot I(x_{j_i} \in \Xi_i), \tag{2}$$

where I is an indicator function. As rFerns is a highly stochastic method, trunks are generated randomly: D split attributes are drawn uniformly with replacement from $1..M$, and subsets Ξ_i are either random subsets of $\mathcal{X}_{j_i}$ (for finite, unordered domains) or generated from some random threshold value θ_i, that is $\Xi_i = \{x \in \mathcal{X}_{j_i} : x \geq \theta_i\}$.

On the other hand, scores are based on the distribution of training objects' classes over leaves, adjusted for the class imbalance and with add-one smoothing. For convenience, let's first define

$$\mathcal{L}_{\mathcal{B},T}(\mathbf{x}) = \{\xi \in \mathcal{B} : T(\xi) = T(\mathbf{x})\}, \tag{3}$$

a subset of bag objects in the same leaf as $\mathbf{x}$ and

$$\mathcal{Y}_{\mathcal{B}}(y) = \{\xi \in \mathcal{B} : Y(\xi) = y\}, \tag{4}$$

a subset of bag objects of a class y. Full fern is then defined as

$$\exp(F_y(\mathbf{x}; \mathcal{B})) := \frac{1 + \#(\mathcal{L}_{\mathcal{B}}(\mathbf{x}) \cap \mathcal{Y}_{\mathcal{B}}(y))}{C + \#\mathcal{L}_{\mathcal{B}}(\mathbf{x})} \cdot \frac{C + \#\mathcal{B}}{1 + \#\mathcal{Y}_{\mathcal{B}}(y)}, \tag{5}$$

where $\#$ denotes multiset cardinality. Note that the score expresses the deviation of class proportion within leaf with respect to what is expected from random assignment; under-represented classes have negative scores while over-represented positive. For balanced classes and $\mathcal{L}_{\mathcal{B}}(\mathbf{x}) = \varnothing$, the score is $\mathbf{0}$.

3.3 Shadow Importance

Similarly to Random Forest, the regular rFerns VIM is defined as a loss of model accuracy due to a random permutation of a given attribute; the crucial difference is that rFerns uses a loss of score for a true class rather than strict accuracy difference [10]. It is estimated using the *out-of-bag* or OOB objects, i.e., those which are not in the bag of a given fern; I denote this set here as $\mathcal{B}^* := \mathcal{T} - \mathcal{B}$. Precisely, rFerns VIM for an attribute a is defined as

$$I_a = \sum_{k \in \mathcal{A}(a)} \frac{1}{\#\mathcal{A}(a) \cdot \#\mathcal{B}_k^*} \left(\sum_{\xi \in \mathcal{B}_k^*} F_{Y(\xi)}^k(\xi; \mathcal{B}_k) - \sum_{\xi^\circ \in \mathcal{B}_{k,a}^{*\circ}} F_{Y(\xi^\circ)}^k(\xi^\circ; \mathcal{B}_k) \right), \quad (6)$$

where $\mathcal{A}(a)$ is a set of ferns that incorporate feature a, and $\mathcal{B}_{k,a}^{*\circ}$ is $\mathcal{B}_k^*$ in which values within attribute a has been randomly shuffled, thus decoupled from other attributes' values and object classes, consequently making a irrelevant. The involved permutation is different for every (k, a) pair.

In the Boruta method, distribution of VIM of irrelevant attributes is estimated by calculating VIM on the training set augmented with shadows, randomly permuted copies of all true features. This method is straightforward but inefficient—augmented set occupies twice as much memory, also the search space of potential interactions becomes significantly larger. To this end the shadow attributes for rFerns shadow VIM are created implicitly during the importance calculation loop, utilising fern trunks built only on actual features. This yields *shadow importance*, which is defined as

$$
\begin{aligned}
J_a = &\sum_{k \in \mathcal{A}(a)} \frac{1}{\#\mathcal{A}(a) \cdot \#\mathcal{B}_k^*} \sum_{\xi \in \mathcal{B}_k^*} F_{Y(\xi)}^k(\xi; B(\mathcal{T}_a^\circ, k)) + \\
&- \sum_{k \in \mathcal{A}(a)} \frac{1}{\#\mathcal{A}(a) \cdot \#\mathcal{B}_k^*} \sum_{\xi^\circ \in \mathcal{B}_{k,a}^{*\circ}} F_{Y(\xi^\circ)}^k(\xi^\circ; B(\mathcal{T}_a^\circ, k)),
\end{aligned}
\quad (7)
$$

where $\mathcal{T}_a^\circ$ is $\mathcal{T}$ in which values of an attribute a have been randomly shuffled. Note that while $\mathcal{B}_{k,a}^{*\circ}$ is different for each (k, a) pair, $\mathcal{T}_a^\circ$ is the same for all ferns; this is crucial because implicit shadows are required to mimic the behaviour of original attributes, which obviously do not change due to being used by different ferns.

3.4 Feature Selection with Shadow Importance

In order to further mimic the Boruta method, the final selection should be defined as a subset of attributes which VIM was higher than the maximal value of shadow VIM for all attributes, i.e.

$$\mathcal{S} = \{a \in 1..M : I_a > \max_{a'} J_{a'}\}. \quad (8)$$

I will later refer to this heuristic as *shadow cutoff* method and use it as a baseline shadow VIM approach. One should note that it is an embedded method, as it can be realised within a single run of rFerns training.

Algorithm 1. Naïve wrapper over shadow VIM

Given: training set $\mathcal{T}$, sub-model count Q and parameters K and D, weight–hit
relation parameter λ, subspace size S
$\mathbf{w} \leftarrow 1$
for all $1..Q$ **do**
    $S \leftarrow$ weighted random sample of S features using weights $\mathbf{w}$
    $\mathbf{F}^k \leftarrow$ rFerns ensemble of K ferns of depth D on $\mathcal{T}_\mathbf{S}$
    $\mathcal{S} \leftarrow$ subset of attributes selected by $\mathbf{F}^k$ according to Eq. 8
    for all $a \in \mathcal{S}$ **do**
        $w_a \leftarrow w_a + \lambda$
    end for
    for all $a \in \mathbf{S} - \mathcal{S}$ **do**
        $w_a \leftarrow \max(w_a - 2\lambda, \frac{1}{10})$
    end for
end for
return $\{a : w_a > \max(\max(\mathbf{w})/2, 3\lambda + 1)\}$

Still, the fully stochastic nature of rFerns has a disadvantage of a lack of
feedback between the current knowledge about usability of certain attributes
model has and their prevalence in its structure. Let's assume a problem with
a nonlinear, multivariate mechanism involving m relevant attributes; in a most
pessimistic scenario a fern would need to cover all of them to properly assess
their importance, but it is obvious that the probability of constructing such
fern during training may easily become impractically small. To this end, I will
also consider a simple wrapper approach implementing such a feedback loop. It
employs a super-ensemble of Q small rFerns forests, each built on a subspace of S
attributes, which is selected randomly, but with a higher selection probability for
attributes claimed important by previous sub-models. This relation is controlled
through a parameter λ; initial weight of all features is one, and it grows by
λ on each confirmation, as well as drops by 2λ on each rejection (but never
below 0.1). In this study $\lambda = 5$ will be used. The final selection is a group of
attributes which hold a substantial weight at the end of the procedure. The
precise implementation is presented as the Algorithm 1.

4   Assessment

The proposed feature selection approaches are assessed on a series of 4 synthetic
problems built based on established benchmark datasets. All of them contain *a
priori* known set of relevant attributes; in case of one of them this set is empty,
as it was derived through a total randomisation of a benchmark data. Precisely:

- IRI (150 cases × 5 000 features)—a derivative of the Fisher's iris data [4],
  expanded by adding 4 996 irrelevant features generated by shuffling original
  ones.
- MAD (2 000 × 500)—the Madelon dataset from the NIPS 2003 feature selection challenge [6], a 5-dimensional XOR problem extended with 15 random

linear combinations of 5 main attributes (also considered relevant in this work) and 480 irrelevant features containing random values.
- RND (102 × 12 533)—a derivative of the Singh et al. microarray data [18], made nonsense by randomly shuffling values within decision (so that all inter-attribute relation present in the original data are retained). It is used as a model of an inconclusive experiment, for which no attributes shall be selected regardless of its rich internal structure.
- SGW (2 000 × 6 000)—simulated genome-wide association study (GWAS) data, with a realistic inter-attribute relations (linkage disequilibrium), derived from simuCC data from the R package genMOSS [5] by converting original number-coded features back into categorical form. Original simulation was performed with the simuPOP software [16]. Decision was generated based on two out of 6 000 attributes.

These problems were designed to comprehensively test all crucial traits of an all relevant feature selector: ability to reject attributes involved only in spurious interactions present only due to $p \gg n$ nature of the data (IRI, RND and SGW), ability to detect complex interactions (MAD), ability to retain full set of relevant features, regardless of their redundancy (MAD and SGW), finally ability to provide negative answer for an inconclusive data (RND).

For each of the benchmark problems, a series of shadow cutoff feature selections has been performed: over 10 repetitions with different random seeds and over a comprehensive subspace of hyper-parameters. The investigated values of the depth parameter D were 1, 3, 5, 7, 10 and 12. For sake of comparability between cases for different D and sizes of the data, the ensemble size parameter K was set so that each feature would be considered, at average, 1 000 times; such number was chosen experimentally, as conservatively enough for the selection to stabilise. The wrapper approach is assessed in a similar manner, for $K = 100$, $S = 30$ and 1 000 iterations. As a base for comparison, a similar analysis has been performed with Boruta using, as in its original implementation, Random Forest VIM; it was also stabilised by taking 10 repetitions and applied for 500, 5 000 and 50 000 trees in the ensemble.

Moreover, VIM-based methods are also compared with an information-based filter approach. Unfortunately, most of such methods only produce ranking of features, which disallows fair comparison with methods that output strict selection. To this end, a greedy optimiser of conditional mutual information was used; this method works in a following way. First, the set of selected attributes is initialised empty. Then, during each iteration, this set is expanded with a previously unselected attribute of a highest, positive conditional mutual information with the decision given values of all already selected features. The algorithm stops when no such feature can be found. For the experiments presented in this paper, the CondMI implementation from the FEAST library [3] was used; while it can only work on categorical data, attributes in IRI, MAD and RND were discretised into 10 equal bins before applying it.

A comprehensive comparison of accuracy of all investigated methods is shown as Table 1, while a summary of their computational demands as Table 2. The

Table 1. The accuracy of investigated feature selection methods, shown as an average number of false positive (FP) and false negative (FN) attribute selections. Results for shadow selection are shown for approximately 1 000 attribute scans.

Method	IRI		MAD		RND		SGW	
	FP	FN	FP	FN	FP	FN	FP	FN
Naïve wrapper, depth 5	0	0	0	0	0.8	0	5.0	0
Naïve wrapper, depth 7	0	0	0	0	0.1	0	3.8	0
Naïve wrapper, depth 10	0	0	0	0	0	0	1.9	0
Shadow selection, depth 5	1.2	0	0.4	7.0	0	0	10.2	0
Shadow selection, depth 7	1.0	0	0.2	6.8	0	0	9.7	0
Shadow selection, depth 10	0.8	0	1.3	6.5	0	0	8.0	0
Boruta, 500 trees	2.7	0	0	0.1	2.2	0	1.7	0
Boruta, 5 000 trees	3.8	0	1.0	0	1.3	0	5.0	0
Boruta, 50 000 trees	2.7	0	1.8	0	0.3	0	6.5	0
CondMI filter	2.0	3.0	3.0	16.0	4.0	0	11.0	0
Feature count	5 000		500		12 533		6 000	

Table 2. The average single-core computational time taken by the analysed feature selection methods. Results for shadow selection are shown for approximately 1 000 attribute scans.

Method	IRI	MAD	RND	SGW
Naïve wrapper, depth 5	48 s	5 s	33 s	5 s
Naïve wrapper, depth 7	58 s	7 s	44 s	7 s
Naïve wrapper, depth 10	37 s	3 s	29 s	3 s
Shadow selection, depth 5	3 min	4 min	6 min	48 min
Shadow selection, depth 7	4 min	4 min	7 min	49 min
Shadow selection, depth 10	6 min	4 min	10 min	51 min
Boruta, 500 trees	10.4 h	27.4 h	17.8 h	6.4 h
Boruta, 5 000 trees	1.8 h	7.1 h	2.5 h	2.7 h
Boruta, 50 000 trees	12.4 h	2.5 h	14.0 h	2.2 h
CondMI filter	11 s	13 s	20 s	2 min

naïve wrapper approach clearly improves the selection accuracy of shadow VIM; it produced no false negatives, and even a perfect selection for IRI and MAD. This shows that although the shadow cutoff selection fails for MAD, this is an easy to circumvent consequence of stochastic attributes sampling rather than some hypothetical sensitivity deficit inherent to the shadow VIM.

False positive counts for the naïve wrapper were non-zero only for SGW and RND sets, still were not substantial and decreasing with D, which shows that this approach can effectively leverage detection of spurious associations with sensitivity to even complex interactions.

The Boruta method yields similar results like the naïve wrapper, although at a drastically higher computational cost—execution time for Boruta was in an order of hours, while the wrapper never took more than a minute. As the naïve wrapper, for more than 5 000 trees it never produced a false negative; though it allowed a visible amount of false positives, which shows that it is more prone to spurious correlations.

On the other hand, the CondMI method was as fast as the naïve wrapper, yet showed significantly inferior accuracy: it failed to detect inconclusiveness of RND, and missed relevant features not only for MAD but also for IRI. It has also produced false positives in case of all problems, although not substantially more than Boruta.

These results are certainly not the upper bound of filter method capabilities, however they prove that the raw interaction of attribute (or attribute group) and the decision can be deeply deceiving when stripped from the context of the whole data.

5   Conclusions

The all relevant feature selection, despite its advantages of being more robust and superior for explanatory machine learning, is a very hard task in practice, requiring utilisation of targeted algorithms with strong assumptions or involving very time-consuming computational methods. Latter approaches are mostly built around some heuristic approximation of how plausible it is that a certain, seemingly important feature is actually irrelevant, and its usability for modelling only comes from spurious correlations arisen at random. In this paper, I propose a computationally efficient implementation of this idea, based on an extension of random ferns VIM, called shadow importance. While it can be used either directly or as a part of a broader algorithm, I also propose a naïve wrapper method utilising it.

The results of numerical assessment of the proposed approach show that it is highly specific, robust and computationally efficient; its quality is at least on par with the quality of Boruta, a random forest wrapper which inspired it, yet leads to significant speed-ups ranging from 100 to even 1 000×. Such computational demands are similar to those required by filter methods, which proves that certain wrappers can be directly applied to larger datasets even without utilising HPC resources. Moreover, both proposed approaches are capable of producing negative results for inconclusive problems, which is crucial for reliability of potential pipelines involving them. Overall, it is clear that shadow VIM is a viable attribute relevance criterion and a promising building block for a powerful feature selection methods.

Acknowledgements. This work has been financed by the National Science Centre, grant 2011/01/N/ST6/07035, as well as with the support of the OCEAN—*Open Centre for Data and Data Analysis* Project, co-financed by the European Regional Development Fund under the Innovative Economy Operational Programme. Computations were performed at ICM, grant G48-6.

References

1. Bosch, A., Zisserman, A., Munoz, X.: Image classification using random forests and ferns. In: 2007 IEEE 11th International Conference on Computer Vision, pp. 1–8. IEEE (2007)
2. Breiman, L.: Random forests. Mach. Learn. **45**, 5–32 (2001)
3. Brown, G., Pocock, A., Zhao, M., Luján, M.: Conditional likelihood maximisation: a unifying framework for information theoretic feature selection. J. Mach. Learn. Res. **13**, 27–66 (2012)
4. Fisher, R.A.: The use of multiple measurements in taxonomic problems. Ann. Eugenics **7**(2), 179–188 (1936)
5. Friedlander, M., Dobra, A., Massam, H., Briollais, L.: genMOSS: Functions for the Bayesian Analysis of GWAS Data, rpackageversion 1.2 (2014). https://CRAN. R-project.org/package=genMOSS
6. Guyon, I., Gunn, S., Ben-Hur, A., Dror, G.: Result analysis of the NIPS 2003 feature selection challenge. Adv. Neural Inf. Process. Syst. **17**, 545–552 (2005)
7. Huynh-Thu, V.A., Wehenkel, L., Geurts, P.: Exploiting tree-based variable importances to selectively identify relevant variables. In: JMLR: Workshop and Conference Proceedings, pp. 60–73 (2008)
8. Kursa, M.B., Jankowski, A., Rudnicki, W.R.: Boruta – a system for feature selection. Fundamenta Informaticae **101**(4), 271–285 (2010)
9. Kursa, M.B., Rudnicki, W.R.: Feature selection with the Boruta package. J. Stat. Softw. **36**(11), 1–13 (2010)
10. Kursa, M.B.: rFerns: an implementation of the random ferns method for general-purpose machine learning. J. Stat. Softw. **61**(10), 1–13 (2014)
11. Kursa, M.B.: Robustness of random forest-based gene selection methods. BMC Bioinform. **15**(1), 8 (2014)
12. Nilsson, R., Peña, J., Björkegren, J., Tegnér, J.: Consistent feature selection for pattern recognition in polynomial time. J. Mach. Learn. Res. **8**, 612 (2007)
13. Oshin, O., Gilbert, A., Illingworth, J., Bowden, R.: Action recognition using randomised ferns. In: 2009 IEEE 12th International Conference Computer Vision Workshops (ICCV Workshops), pp. 530–537. IEEE (2009)
14. Özuysal, M., Calonder, M., Lepetit, V., Fua, P.: Fast keypoint recognition using random ferns. Image Process. (2008)
15. Özuysal, M., Fua, P., Lepetit, V.: Fast keypoint recognition in ten lines of code. In: 2007 IEEE Conference on Computer Vision and Pattern Recognition, pp. 1–8, June 2007
16. Peng, B., Amos, C.I.: Forward-time simulation of realistic samples for genome-wide association studies. BMC Bioinform. **11**(1), 1–12 (2010)
17. Saeys, Y., Inza, I.N., Larrañaga, P.: A review of feature selection techniques in bioinformatics. Bioinformatics **23**(19), 2507–2517 (2007)
18. Singh, D., Febbo, P.G., Ross, K., Jackson, D.G., Manola, J., Ladd, C., Tamayo, P., Renshaw, A.A., D'Amico, A.V., Richie, J.P., Lander, E.S., Loda, M., Kantoff, P.W., Golub, T.R., Sellers, W.R.: Gene expression correlates of clinical prostate cancer behavior. Cancer Cell **1**(2), 203–209 (2002)
19. Tuv, E., Borisov, A., Torkkola, K.: Feature selection using ensemble based ranking against artificial contrasts. In: The 2006 IEEE International Joint Conference on Neural Network Proceedings, pp. 2181–2186. IEEE (2006)

Evaluating Difficulty of Multi-class Imbalanced Data

Mateusz Lango$^{(\boxtimes)}$, Krystyna Napierala, and Jerzy Stefanowski

Institute of Computing Science, Poznan University of Technology, Poznań, Poland
{mateusz.lango,krystyna.napierala,jerzy.stefanowski}@cs.put.poznan.pl

Abstract. Multi-class imbalanced classification is more difficult than its binary counterpart. Besides typical data difficulty factors, one should also consider the complexity of relations among classes. This paper introduces a new method for examining the characteristics of multi-class data. It is based on analyzing the neighbourhood of the minority class examples and on additional information about similarities between classes. The experimental study has shown that this method is able to identify the difficulty of class distribution and that the estimated minority example safe levels are related with prediction errors of standard classifiers.

Keywords: Imbalanced data · Multiple classes · Supervised classification

1 Introduction

Learning from class-imbalanced data has been a topic of intensive research in recent years. On one hand, several new specialized algorithms as well as data pre-processing methods have been developed; see their reviews in [3,5]. On the other hand, a growing research interest has also been put into better understanding the imbalanced data characteristics which cause the learning difficulties [11].

Most of these works concentrate on binary imbalanced problems with a single minority class and a single majority class. This formulation is justified by focusing the interest on the most important class. If there are multiple classes, the original problem is transformed into binary one, e.g., by selecting a minority class and aggregating the remaining classes into a single one.

Nevertheless, in some situations it may be reasonable to distinguish more classes with low cardinalities. In such cases the aforementioned binarization becomes questionable. Consider for instance the medical problem of diagnosing two types of asthma (minority classes) and discerning them from healthy patients (majority class). Selecting one type of asthma as a minority class and aggregating the other one with the majority class leads to an unacceptable situation of considering ill patients as healthy. Aggregating all asthmatic patients into one minority class could be a better choice, but it still leads to the undesired loss of information about the asthma type.

© Springer International Publishing AG 2017
M. Kryszkiewicz et al. (Eds.): ISMIS 2017, LNAI 10352, pp. 312–322, 2017.
DOI: 10.1007/978-3-319-60438-1_31

Handling multiple minority classes makes the learning task more difficult as relations between classes become more complex [7,12]. The current approaches to it are adaptations of the *one-against-all* or *one-against-one* decomposition into several binary subproblems [4]. Although the selected minority classes are preserved in these approaches, the information about internal data distributions or decision boundaries is lost, while in the original problem one class influences several neighboring classes at the same time.

Moreover, these decompositions do not consider the *mutual relations* between classes that are different for majority and minority classes. Consider for instance the aforementioned asthma learning problem. The two asthma classes are more closely related to each other, while their similarity to the majority class (healthy patients) is smaller and it should be taken into account while constructing a new approach to multiple imbalanced classes.

In our opinion, modeling the relations between classes is particularly useful for studying *data difficulty factors* in imbalanced data. Previous research on binary imbalanced data showed that local factors such as small sub-concepts, overlapping or rare case are more influential than the global imbalance ratio [6,11]. They have an impact on performance of learning methods as some classifiers and preprocessing methods are more sensitive to given data types than the others [8]. Therefore, prior to designing and applying new learning methods, it is important to analyse the data characteristics. It is even more important for multi-class problems, where such approaches do not exist yet.

In [8] we have introduced a new approach to model several types of data difficulty, based on analysing the local neighbourhood of minority examples, which was successfully used to differentiate types of examples in binary problems [9,11]. The results of that work are useful for constructing pre-processing methods or ensembles specialized for imbalanced data [11].

The main aim of this study is to introduce a new method to identify different types of minority examples in multi-class imbalanced data, which refer to data difficulty. When analysing the local neighbourhood of the given example to determine its difficulty, we take into account the class of each neighbour as in [8,9]. However, we also exploit additional information about relations between classes of the analysed example and its neighbor. It is based on the priorly defined degree of *similarity* between these classes. To the best of our knowledge, this kind of handling similarity of classes has not been proposed yet for imbalanced data.

Summarizing our contribution, this work introduces a concept of class similarity and uses it to extend the method of identifying types of minority examples to a multi-class setting. It is then applied to analyze the difficulty of several artificial and real-world multi-class imbalanced datasets. Finally, the impact of data difficulty on learning abilities of several classifiers is experimentally studied.

2   Related Works

Multi-class imbalanced problems are not so intensively studied as its binary counterpart. There exist only few approaches; for their recent review see [10]. For instance, the new re-sampling techniques include static-SMOTE or Mahalanobis distance-based over-sampling [2]. Nearly all other approaches follow the idea of the decomposition of the multi-class imbalanced problem to a set of binary sub-problems. Usually either one-against-all or one-against-one class binarization is integrated with appropriate balancing of binary samples or with specialized ensembles, see e.g. [4]. Few other algorithmic modifications are designed for specific learning algorithms, like SVM or neural networks. However, as it is pointed out in a review in [7], none of these methods takes into account both individual properties of classes and their mutual relations.

Research on data characteristics of multi-class imbalanced datasets is limited to one paper [10] only. The authors considered the categorization of minority examples into four types (safe, borderline, rare, outlier), proposed for binary problems in [9]. To adapt it to multi-class problems, they decomposed data into several binary problems using one-vs-all technique, i.e. all examples from different classes are treated equally when analysing the local neighbourhood of the minority examples to determine their difficulty. Again, no notion of mutual relations between the classes is taken into account.

3 Identifying Difficulty Factors in Multi-class Imbalanced Data

The relations between multiple imbalanced classes are more complex than in the binary versions [7,12]. When dealing with multiple classes, one may easily lose performance on one minority class while attempting to improve it at another class [10]. Moreover, the mutual relations between classes show that some minority classes can be treated as more closely related to each other than to the majority class. Current decomposition approaches, which treat all pairs of classes equally, do not reflect well these issues [10,12].

Furthermore, data difficulty factors may appear only in some subsets of classes. For instance, the degree of overlapping between various classes may be different. The type of examples present in the given class distribution also strongly depends on their relations to other classes. For instance, a given example may be of a borderline type [8,9] for certain classes and at the same time a safe example for the remaining classes. Using existing binary class approaches to estimate data difficulty is not straightforward in case of multiple class imbalance. There is a need for a deeper insight into these complex relations and for a new and more flexible approach to analyse multi-class data difficulty factors.

3.1 Handling Multiple Class Relations with Similarity Information

Modeling relations between multiple imbalanced classes can be realized by means of additional information acquired from users. Following the motivations described in Sect. 1, we will exploit information about *similarity* between pairs of classes. More precisely, given a certain class we need information about similarities of other classes to it. An intuition behind it is the following: if example

x from a given class has some neighbors from other classes, then neighbors with higher similarity are more preferred. For instance, consider the asthma learning problem, in which two asthma classes are defined as more similar to each other than to the no-asthma class. If an example from asthma-type-1 is not surrounded only by examples from its class (which is the most preferred situation), then we would prefer it to have neighbors from asthma-type-2 class rather than from the no-asthma class. Such neighborhood would let us consider the analysed example to be safer – easier recognized as a member of its class (as it will be less prone to suffer from the algorithm bias toward the majority classes)[1].

We assume that for each pair of classes C_i, C_j the degree of their similarity will be defined as a real valued number $\mu_{ij} \in [0; 1]$. Let us discuss its main properties. Similarity of a class to itself is defined as $\mu_{ii} = 1$. The degree of similarity does not have to be symmetric, i.e. for some classes C_i, C_j it may happen that $\mu_{ij} \neq \mu_{ji}$.

Although the values of μ_{ij} are defined individually for each dataset, we claim that for the given minority class C_i its similarity to other minority classes should be relatively higher than to the majority classes.

Degrees of similarity should be provided by the expert or can come from the domain knowledge. If neither is available, we recommend for other minority classes C_g $\mu_{ig} \to 1$, while similarities to majority classes C_h should be $\mu_{ih} \approx 0$.

3.2 Data Difficulty with Respect to a Safe Level of Minority Examples

In our earlier research [8, 9, 11] we claimed that (1) imbalanced data difficulty factors correspond to *local data characteristics*, occurring in some sub-regions of the minority class distribution and (2) the mutual position of an example with respect to examples from other classes of both minority and majority classes influences learning classifiers. We linked these difficulty factors to different types of examples – *safe* and *unsafe* (difficult) for recognizing the minority class. Safe examples are located in the homogeneous sub-regions belonging to one class while unsafe examples are categorized into borderline, rare cases or outliers. In [8] we introduced the method of assessing the type of example by analyzing class labels of its surrounding examples. The neighborhood was constructed based either on k–nearest neighbors or on kernel functions.

In this study we consider the k–nearest neighbors variant[2]. Determining the number of examples from the majority class in the neighborhood of the minority example allows to assess how safe the example is, and then to establish its type. Below we adapt this idea to the multiple imbalanced class framework.

---

[1] Note that in our proposal of similarity between classes, we do not model directly misclassifications between minority classes, which alternatively could be handled by yet another approach with costs of misclassifications between classes.

[2] Refer to [9] for details of the neighborhood construction, recommended distance functions and neighborhood size tuning.

Considering a given example x belonging to the minority class C_i its safe level is defined with respect to l classes of examples in its neighborhood as:

$$safe(x_{C_i}) = \frac{\sum_{j=1}^{l} n_{C_j} \mu_{ij}}{n}$$

where μ_{ij} is a degree of similarity, n_{C_j} is a number of examples from class C_j inside the considered neighborhood of x and n is a total number of neighbors.

Given the safe levels calculated for all learning examples, one can analyse them in two ways: either the numeric distribution of safe levels in the learning set for each class, or transform the continuous safe levels into discrete intervals corresponding to types of example (as done in [8,9]). Here we follow the first option and then aggregate the distributions for each class, e.g., by the average. They should be interpreted in the following way: the lower the average value, the more unsafe (difficult) is the minority class. The statistics for each minority class can be analysed independently or can be further aggregated into a single criterion describing the difficulty of the whole learning set. Alternatively, the histograms of safe levels in each class can be presented to the user.

4   An Experimental Evaluation

In the experiments we want to examine three aims: (1) verify whether the new approach to evaluate the safe level (see Sect. 3.2) sufficiently reflects the difficulty of multi-class datasets; (2) compare this approach against its binary predecessor; (3) check whether values of safe levels relate to classification performance of standard algorithms. In order to check these aims, we will use several synthetic and real-world multi-class imbalanced datasets.

The artificial datasets were constructed to control their level of difficulty [13]. They are two-dimensional with two minority classes, having elliptic shapes, surrounded by the examples from the majority class. Each data set contains 1200 examples with the class ratio 1:2:7. In the first dataset (A1), two minority classes are well separated from each other and also from the majority class (see Fig. 1). Then, it is modified to (A1b) version by introducing an overlapping border with the majority class. In the third dataset (A2) minority class ellipses are additionally overlapping (see Fig. 1). The most difficult dataset (A3) additionally contains rare cases, outliers and more borderline examples.

Following similar motivations and the previous research on difficulty of binary imbalanced data [9] we chose three UCI datasets: new thyroid (NT) is a safe (easy) dataset, ecoli (EC) is a borderline dataset, and cleveland (CL) is a rare/outlier dataset. Characteristics of these datasets are presented in Table 1 and their visualisations after the reduction to two dimensions using the Multidimensional Scaling (MDS) are presented on Fig. 1. If some of these datasets contain more majority classes, we aggregated them to be consistent with the artificial data setup.

We chose three different configurations of similarity values μ_{ij} - see Table 2 to examine various potential relations between classes. In the version called

Table 1. Characteristics of real-world datasets

Dataset	Abbrev.	Size	Min1 name	Min1 size	Min2 name	Min2 size	Min3 name	Min3 size
new-thyroid	NT	215	2	35	3	30		
ecoli	EC	336	imU	35	om	20	pp	52
cleveland	CL	303	2	36	3	35	4	13

`Safety1`, we set the similarity between minority classes to a high value (0.8), following the recommendation from Sect. 3.1. In `Safety2` we also assume high similarity between minority classes (0.7), but we assign a small similarity between majority and minority classes as 0.2. The last configuration `Safety3`, models the situation when there is no prior information about classes relation (quite small similarity between minority classes and no similarity with the majority class). The column called `Safety` refers to the previous binary class version [9].

Table 2. Different configurations of similarity degrees

	Safety	Safety1	Safety2	Safety3
$\mu_{min1\ min2}$	0	0.8	0.7	0.5
$\mu_{min\ maj}$	0	0	0.2	0

All experiments were performed in scikit-learn or WEKA frameworks. Classification performance was evaluated in 5-fold cross validation. Following earlier related studies, we selected CART decision tree, PART rules, Naive Bayes and 3-nearest neighbors classifier for the experiment. All classifiers were used with default parameter values. Values of average safe levels are presented in Table 3, while sensitivity (true-positive-rate) of minority classes is given in Table 4. Note that sensitivity is reported for each class separately and no additional aggregation over classes is used[3].

4.1   Analysing Results for Artificial Datasets

Let us consider the results for each dataset in the order of their increasing difficulty (from A1 to A3).

In dataset A1, classes are easily separable, so the average safe level of both minority classes is close to 1 (Table 3). It is diminished by the safe level of some minority examples on the border of the class, which are surrounded by majority neighbors. As in `Safety2` the degree of similarity $\mu_{min\ maj}$ is higher (0.2) than

---

[3] Artificial datasets and more detailed results with additional evaluation measures are available at www.cs.put.poznan.pl/mlango/publications/multi-typology.html.

Table 3. Average safe levels for real and artificial datasets.

	Safety			Safety1			Safety2			Safety3		
	Min1	Min2	Min3	Min1	Min2	Min3	Min1	Min2	Min3	Min1	Min2	Min3
NT	0.77	0.78		0.77	0.78		0.82	0.82		0.77	0.78	
EC	0.57	0.74	0.82	0.57	0.91	0.86	0.66	0.90	0.88	0.57	0.85	0.84
CL	0.14	0.13	0.08	0.29	0.32	0.34	0.41	0.42	0.42	0.23	0.25	0.24
A1	0.91	0.96		0.91	0.96		0.93	0.97		0.91	0.96	
A1b	0.68	0.80		0.68	0.80		0.75	0.84		0.68	0.80	
A2	0.53	0.70		0.71	0.79		0.74	0.82		0.64	0.76	
A3	0.32	0.47		0.55	0.59		0.60	0.65		0.46	0.54	

Table 4. Sensitivity of minority classes for studied classifiers.

	CART			NB			3NN			PART		
	Min1	Min2	Min3	Min1	Min2	Min3	Min1	Min2	Min3	Min1	Min2	Min3
NT	0.94	0.83		0.94	0.86		0.71	0.80		0.94	0.83	
EC	0.60	0.85	0.78	0.68	0.30	0.90	0.48	0.75	0.84	0.46	0.80	0.79
CL	0.28	0.11	0.07	0.14	0.25	0.15	0.08	0.00	0.00	0.20	0.11	0.08
A1	0.93	0.93		0.47	0.84		0.94	0.97		0.88	0.97	
A1b	0.74	0.82		0.60	0.78		0.77	0.84		0.67	0.70	
A2	0.56	0.73		0.32	0.56		0.54	0.79		0.50	0.73	
A3	0.25	0.42		0.00	0.02		0.20	0.39		0.14	0.57	

for the remaining configurations, the average safety for this configuration is also slightly increased.

A1b is similar to A1, but has higher overlapping with the majority class. Therefore, its average safe level is smaller than for A1. Similarly to A1, values for `Safety2` are higher than for other configurations (where $\mu_{min\ maj} = 0$).

Dataset A2 is a more difficult modification of A1b dataset, with additional overlapping between minority classes. It is reflected in the lower values of the average safe level compared to A1 and A1b. Let us observe, however, that in case of `Safety1`, where we defined a very high similarity between minority classes (0.8), this additional overlapping is mostly neglected, which reduces the dataset A2 to dataset A1b. It can be noticed by almost identical average safe levels of both datasets for `Safety1`.

Finally, dataset A3 is the most difficult, with additional rare cases and outliers. It has also the lowest average safe levels from all datasets, independent of the similarity degrees configuration.

For all artificial datasets, class Min1 has lower safe levels than Min2 because it is smaller, so fewer examples are placed in homogeneous safe regions.

Looking at predictions of classifiers, one can notice that the values of safe levels are related to the sensitivity of minority classes (Table 4) – dataset A1 is best recognized by all classifiers, while dataset A3 is the most difficult. Class Min1 is always recognized worse than Min2. Majority class recognition was always at approximately 0.9.

To sum up, we have shown that the proposed approach is related to data difficulty – by rating the datasets from the safest (A1) to the most difficult (A3). The difficulty is also strongly related to the recognition of the minority classes by different classifiers. Moreover, it has been shown that analyzing our enhanced safeness (related to similarity degree) allows the user to differentiate overlapping of different classes, giving a better insight into the structure of the imbalanced dataset.

4.2 Analysing Results for Real-World Datasets

Looking at the MDS visualisation of NT dataset (Fig. 1), notice that all the classes are clearly separated. It is also reflected in average safe levels (Table 3). Its values of Safety, Safety1, and Safety3 are similar, while its Safety2 is slightly higher. This suggests that analogously to datasets A1 and A1b, the overlapping occurs only between minority and majority classes (minority classes do not overlap each other) – it is confirmed by the MDS visualisation. The results of classifiers on this dataset also show that it is of a safe type.

EC dataset is more difficult. On MDS visualisation, class Min2 partially overlaps with Min3, then Min3 overlaps with Min2 and Maj, while Min1 overlaps with Maj. It could be devised also from values of the safe levels. For Min1 the safe level is stable for Safety, Safety1 and Safety3 configurations and increases slightly for Safety2 – which confirms that it overlaps only with majority class. The safe level of Min2, on the other hand, is the smallest for Safety ($\mu_{min1\ min2}$ = 0) and the highest for Safety3 ($\mu_{min1\ min2}$ = 0.8) which suggests that the overlapping is mostly between minority classes. From the classification point of view, this dataset is more difficult than NT, and Min1 is the most difficult class for all classifiers (except NB). It is due to a fact that this class is surrounded by the majority class, towards which standard classifiers have a strong bias. The latter observation supports our intuition expressed in Sect. 3.1, that the minority neighborhood should be considered as safer (easier) than majority neighborhood, and that it should be taken into account when estimating the difficulty of the multi-class imbalanced dataset.

CL dataset is the most difficult. The MDS visualisation clearly shows that this dataset consists mostly of mixed rare and outlier examples for all classes. Its average safe level is also very low, and the classes are hardly recognized by any of the classifiers.

To sum up, the analysis on real-world datasets also shows that the proposed approach can sufficiently well estimate the difficulty of the dataset, which is consistent with both MDS visualisations and with performance of classifiers.

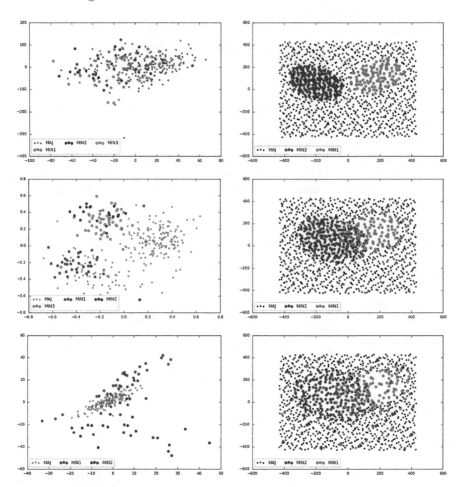

Fig. 1. MDS visualisation of studied datasets. In the first column from the top: CL, EC, NT real world datasets are presented, while the second column shows A1, A2, A3 artificial datasets, respectively.

5   Concluding Remarks

Learning from imbalanced multi-class data is particularly challenging and requires more extensive research on its nature and sources of its difficulty. In our opinion, it is necessary to analyze types of examples (safe vs. unsafe). As such research is still not well-developed, we have introduced a new method. It is based both on analyzing the neighborhood of the minority class example and on the additional information about similarity of neighboring classes to the class of this example. To the best of our knowledge, similar approaches to handle complex relations among classes have not been considered yet – they were put in the main open research points of [7].

The results of experiments show that this method sufficiently identifies difficulties of minority class distributions in various artificial and real-world datasets – which is expressed by values of safe levels for appropriate minority examples. Furthermore, these values are well related to predictions of standard classifiers.

Although our method requires defining values of similarities among classes, we claim that by using them we are able to identify data difficulty factors, e.g. we could evaluate which classes overlap. Note that considering various sets of class similarities has led us to observe that the class surrounded by the majority examples is more difficult to recognize than overlapped minority classes (see an analysis of `ecoli` in Sect. 4.2).

Our proposal could also be used to construct new preprocessing methods, e.g. by exploiting safe levels to adaptively tune re-sampling. Furthermore, they could be used inside new algorithms, similarly to earlier attempts of using the local neighborhood in generalizations of under-bagging [1].

Acknowledgment. The research was funded by the Polish National Science Center, grant no. DEC-2013/11/B/ST6/00963. The work of the last author was also, partially supported by DS internal grant of PUT.

References

1. Błaszczyński, J., Stefanowski, J.: Neighbourhood sampling in bagging for imbalanced data. Neurocomputing **150A**, 184–203 (2015)
2. Abdi, L., Hashemi, S.: To combat multi-class imbalanced problems by means of over-sampling techniques. IEEE Trans. Knowl. Data Eng. **28**(1), 238–251 (2016)
3. Branco, P., Torgo, L., Ribeiro, R.: A survey of predictive modeling under imbalanced distributions. ACM Comput. Surv. (CSUR) **49**(2), 31 (2016)
4. Fernandez, A., Lopez, V., Galar, M., Jesus, M., Herrera, F.: Analysis the classification of imbalanced data sets with multiple classes, binarization techniques and ad-hoc approaches. Knowl. Based Syst. **42**, 97–110 (2013)
5. He, H., Yungian, Ma. (eds.): Imbalanced Learning. Foundations, Algorithms and Applications. IEEE - Wiley, Hoboken (2013)
6. Japkowicz, N., Stephen, S.: Class imbalance problem: a systematic study. Intell. Data Anal. J. **6**(5), 429–450 (2002)
7. Krawczyk, B.: Learning from imbalanced data: open challenges and future directions. Progress Artif. Intell. **5**, 221–232 (2016)
8. Napierala, K., Stefanowski, J.: Identification of different types of minority class examples in imbalanced data. In: Corchado, E., Snášel, V., Abraham, A., Woźniak, M., Graña, M., Cho, S.B. (eds.) HAIS 2012. LNCS, vol. 7209. Springer, Heidelberg (2012). doi:10.1007/978-3-642-28931-6_14
9. Napierala, K., Stefanowski, J.: Types of minority class examples and their influence on learning classifiers from imbalanced data. J. Intell. Inf. Syst. **46**(3), 563–597 (2016)
10. Seaz, J., Krawczyk, B., Wozniak, M.: Analyzing the oversampling of different classes and types in multi-class imbalanced data. Pattern Recogn. **57**, 164–178 (2016)
11. Stefanowski, J.: Dealing with data difficulty factors while learning from imbalanced data. In: Mielniczuk, J., Matwin, S. (eds.) Challenges in Computational Statistics and Data Mining, pp. 333–363. Springer, Heidelberg (2016)

12. Wang, S., Yao, X.: Mutliclass imbalance problems: analysis and potential solutions. IEEE Trans. Syst. Man Cybern. Part B **42**(4), 1119–1130 (2012)
13. Wojciechowski, S., Wilk, S.: The generator of synthetic multi-dimensional data. Poznan University of Technology Report RB-16/14 (2014)

Extending Logistic Regression Models with Factorization Machines

Mark Pijnenburg[1,2](✉) and Wojtek Kowalczyk[1]

[1] Leiden Institute of Advanced Computer Science, Leiden, The Netherlands
mgf.pijnenburg@belastingdienst.nl, w.j.kowalczyk@liacs.leidenuniv.nl
[2] Netherlands Tax and Customs Administration, Utrecht, The Netherlands

Abstract. Including categorical variables with many levels in a logistic regression model easily leads to a sparse design matrix. This can result in a big, ill-conditioned optimization problem causing overfitting, extreme coefficient values and long run times. Inspired by recent developments in matrix factorization, we propose four new strategies of overcoming this problem. Each strategy uses a Factorization Machine that transforms the categorical variables with many levels into a few numeric variables that are subsequently used in the logistic regression model. The application of Factorization Machines also allows for including interactions between the categorical variables with many levels, often substantially increasing model accuracy. The four strategies have been tested on four data sets, demonstrating superiority of our approach over other methods of handling categorical variables with many levels. In particular, our approach has been successfully used for developing high quality risk models at the Netherlands Tax and Customs Administration.

Keywords: Categorical variables with many levels · Logistic regression · Interactions · Matrix factorization · Value grouping · Tax administration

1 Introduction

Logistic regression is a well-known classification algorithm that is frequently used by businesses and governments to model risks. However, logistic regression will run into problems when one tries to include categorical variables with many levels. The standard approach will transform each level of each categorical variable into a binary ('dummy') variable (see, e.g., [7]), resulting in a large, sparse design matrix. The sparsity usually leads to an ill-conditioned optimization problem, resulting in overfitting, extremely large values of model coefficients and long run times or even lack of convergence. The size and sparsity of the design matrix will increase even more if interactions are included between categorical variables with many levels or interactions between these categorical variables and some numeric variables. Finally, a large design matrix leads to a model with many coefficients, making it difficult to interpret.

© Springer International Publishing AG 2017
M. Kryszkiewicz et al. (Eds.): ISMIS 2017, LNAI 10352, pp. 323–332, 2017.
DOI: 10.1007/978-3-319-60438-1_32

Existing approaches for incorporating multi-level categorical variables into logistic regression reduce the problem of sparsity at the price of losing some information from data and consequently leading to models of inferior quality, see Sect. 2.1 for an overview. This became clear to the authors when working on a risk model for selecting risky VAT tax returns for the Netherlands Tax and Customs Administration (NTCA). Although the risk model performed pretty well, one experienced auditor was able to outperform the model by manually extracting information from categorical variables with many levels such as industry sector code and zip code. This information was clearly not picked up by the model, where standard approaches had been followed to include these categorical variables. The real-life example of the NTCA will be referred to in this paper several times.

Logistic regression is traditionally used at the NTCA since it is a standard tool in the industry and the role of various features can be easily interpreted. Moreover, it performs well and its output can be interpreted as probabilities. This latter fact is important since it allows a decoupling of two key components of a tax return risk model: the probability of an erroneous tax return and the size of the financial loss connected to the error in the tax return, see [2].

In this paper we propose four strategies for including categorical variables with many levels into a logistic regression model, by making use of Factorization Machines, [16]. Factorization Machines transform the categorical variables into vectors of numerical ones, taking interactions of the categorical variables into account. Factorization Machines can be viewed as an extension of matrix factorization methods, [11], that in turn have been developed in the context of recommendation systems, stimulated by the Netflix Challenge.

The paper is organized as follows. Section 2 reviews the related literature. Section 3 introduces the four strategies of combining Factorization Machines and Logistic Regression. Section 4 describes the experiments on four data sets. Section 5 contains conclusions and a discussion of the results.

2 Related Research

2.1 Existing Methods for Many Levels

A number of approaches have been suggested to deal with categorical variables with many levels. Many approaches focus on grouping levels of categorical variables into a smaller number of levels. This grouping can be done in a supervised manner (i.e. involving the target) or in an unsupervised way.

A well-known supervised way of grouping levels comes from the decision tree algorithm CART [4]. Here, levels are ordered by the percentage of cases in the target class. Subsequently, all levels with the percentage above a certain threshold value are grouped into one new level, and the remaining levels into another one. Breiman et al. [4] proved that this approach will find the optimal partitioning of a train set under the conditions that the target is binary, the new categorical variable has two levels, and a convex criterion (like Gini) is used to measure the quality of the partition. Several extensions of this approach have been developed, e.g. [5,6], but they either lack the guarantee of finding the optimal partitioning, or have a substantially higher order of complexity.

Other, frequently employed, supervised ways of grouping levels include search methods, like forward or stepwise search [3]. Typically one starts with each level forming a group of its own. Then groups are merged one at a time based on various criteria, leading to fewer groups. This approach is used, for instance, in the CHAID algorithm [10].

A simple, but often effective, unsupervised way of grouping levels has been proposed by Hosmer and Lemeshow [9]. They suggest to group levels that occur infrequently. This approach gave the best results on our data sets from all supervised and unsupervised groupings tried. Another frequently used unsupervised approach is to let experts group the levels.

Other methods than grouping levels have been put forward. For example, in a data pre-processing step the categorical variable with many levels can be transformed into a numeric variable with help of an Empirical Bayes criterion, [15]. Another approach, see [1], is to find additional numeric predictors. For instance, if the categorical variable is "city", the number of inhabitants of the city could be added to the data set as a predictor.

2.2   Factorization Machines

Factorization Machines are introduced in a seminal paper of Rendle [16]. This class of models is often employed for recommendation systems, where categorical variables with many levels occur frequently. The model equation of a Factorization Machine is given by:

$$\hat{y} = w_0 + \sum_{j=1}^{s} w_j x_j + \sum_{j=1}^{s} \sum_{k=j+1}^{s} \langle \mathbf{v}_j, \mathbf{v}_k \rangle x_j x_k \tag{1}$$

where $\hat{y}$ is the predicted value for an observation with variables $x_1, \ldots, x_s$. The w's are numeric coefficients to be fitted and the $\mathbf{v}$'s are vectors to be fitted (one for each variable). All vectors $\mathbf{v}$ have the same (usually small) length r, which is an input parameter. These vectors can be interpreted as low dimensional numeric representations of levels.

The interesting part of Eq. (1) are the interaction terms. Instead of assigning a new coefficient w_{jk} to each interaction term, a Factorization Machine models the interaction coefficients as an inner product between the vectors $\mathbf{v}_j$ and $\mathbf{v}_k$. The introduction of such a vector for each variable reduces the number of interactions from $O(s^2)$ to $O(rs)$, so from quadratic to linear in the number of variables s. Typically, variables x_j in a Factorization Machine are binary variables resulting from transforming a categorical variable with many levels in dummy variables. In this case the number of coefficients is thus not quadratic in the number of levels, but linear. Note that when s is small (e.g., $s \leq 2r+1$) there is no reduction in the number of coefficients.

The loss function that is used to find the optimal values of parameters in (1) usually involves a regularization term which controls the L^2 norm of model parameters. To find an optimum of the loss function several techniques can be used, among them Markov Chain Monte Carlo, Alternating Least Squares and Stochastic Gradient Descent [17].

3   Combining Logistic Regression with Factorization Machines

In this section we propose four strategies for extending Logistic Regression with Factorization Machines. The key idea is the usage of Factorization Machines for squeezing relevant information from many-level categorical variables and their interactions into numeric variables and incorporating these latter in a logistic regression model. In this way, potential problems with large and sparse design matrix are handled by Factorization Machines, while Logistic Regression takes care of combining "non-sparse" variables in a standard way.

Notation. We model a binary target y with help of p numeric variables $x_1, \ldots, x_p$, and q categorical variables $d_1, \ldots, d_q$ with $l_1, \ldots, l_q$ levels.

We will compare the performance of our four strategies with two benchmarks: (1) a logistic regression model without categorical variables, and (2) a logistic regression model where infrequent levels have been grouped as suggested by [9], see Sect. 2.1. We start by introducing these latter methods in more detail.

3.1   Plain Logistic Regression (PLM)

The PLM model consists of a standard logistic regression model of the numeric variables $x_1, \ldots, x_p$. The model equation is:

$$\hat{y} = \frac{1}{1 + e^{-z}}, \text{ where } z = \alpha_0 + \alpha_1 x_1 + \ldots + \alpha_p x_p. \tag{2}$$

The coefficients α_i are estimated by finding the unique maximum of the log-likelihood function over the train set.

3.2   Logistic Regression with Grouping (LRG)

This model groups infrequent levels of the categorical variables with many levels $d_1, \ldots, d_q$ into a default level for each d_j. The actual threshold for calling a level 'infrequent' depends on the size of the data set and can be found in Table 1. The grouping of infrequent levels leads to a new set of categorical variables $\tilde{d}_1, \ldots, \tilde{d}_q$ with less levels. Next, a standard approach is followed to replace categorical variables by dummy variables, i.e. each $\tilde{d}_j$ is transformed into $l_j - 1$ binary variables. Subsequently, PLM is applied on these binary variables and the numeric predictors.

In mathematical terms, the LRG model is given by:

$$\hat{y} = \frac{1}{1 + e^{-z}}, \quad \text{where } z = \alpha_0 + \alpha_1 x_1 + \ldots \alpha_p x_p +$$

$$\alpha_{11} b_{11} + \ldots + \alpha_{1l_1-1} b_{1l_1-1} + \ldots + \alpha_{ql_q-1} b_{ql_q-1}, \quad (3)$$

where b's are binary variables. Note that in order to limit notation, we denote the coefficients of the logistic regression in all model Eqs. (2), (3), (4), (5), (6), and (7) with $\alpha_0, \alpha_1, \ldots$, despite the fact that the values of these coefficients differ.

3.3   LRFM1

The model LRFM1 (Logistic Regression with Factorization Machines 1) is the first model showing our new approach. The categorical variables with many levels $d_1, \ldots, d_q$ are first put into a Factorization Machine f_0 whose coefficients are estimated from a train set with the target variable y. The output of f_0 — denoted by g_0 — is then added to the model equation of the logistic regression. Therefore,

$$g_0 = f_0(d_1, \ldots, d_q) \quad \text{and}$$

$$\hat{y} = \frac{1}{1+e^{-z}}, \quad \text{where } z = \alpha_0 + \alpha_1 x_1 + \ldots \alpha_p x_p + \alpha_{g_0} g_0. \quad (4)$$

3.4   LRFM2

Although LRFM1 is able to model interactions between categorical variables with many levels, it does not model interactions between the variables $d_1, \ldots, d_q$ and one or more numeric variables x_j. For this reason we allow the model Eq. (4) to be extended with additional variables $g_1, \ldots g_t$, where $t \leq p$. Each g_j is a prediction from a Factorization Machine f_j that takes as input the categorical variables $d_1, \ldots, d_q$ and a variable $\bar{x}_j$. The variable $\bar{x}_j$ is a discretized version of x_j, obtained by an equal frequency binning with 5 bins.

The coefficients of the Factorization Machine f_j are learned from a train set that contains the target y. Only variables g_j that significantly improve the results on the train set (compared to the model with only g_0, significance level $\alpha = 0.05$) will enter the model equation. Therefore, the model equation for LRFM2 is:

$$g_j = f_j(\bar{x}_j, d_1, \ldots, d_q) \text{ and } \hat{y} = \frac{1}{1 + e^{-z}}, \quad \text{where}$$

$$z = \alpha_0 + \alpha_1 x_1 + \ldots \alpha_p x_p + \alpha_{g_0} g_0 + \alpha_{g_1} g_1 + \ldots + \alpha_{g_t} g_t. \quad (5)$$

3.5   LRFM3

Instead of learning the coefficients of a Factorization Machine f on a train set with known binary target y, we can do an intermediate step. We first fit a logistic regression model with the numeric variables $x_1, \ldots, x_p$ on the train set (so without $d_1, \ldots, d_q$), and then compute the deviance residuals r_i (see [9]):

$$r_i = \pm \sqrt{2 \left[y_i \log \frac{y_i}{\hat{y}_i} + (1 - y_i) \log \frac{1 - y_i}{1 - \hat{y}_i} \right]},$$

where $\hat{y}_i$ denotes the predicted probability that $y_i = 1$ and the sign is $+$ iff $y_i = 1$. The residual vector $\mathbf{r}$ can then be used to train the coefficients of the Factorization Machine instead of the original target $\mathbf{y}$. This will give a Factorization Machine $\tilde{f}$. Note that the Factorization Machine is now performing a regression task, instead of classification. LRFM3 is described by the equations:

$$h_0 = \tilde{f}(d_1, \ldots, d_q) \text{ and}$$
$$\hat{y} = \frac{1}{1 + e^{-z}}, \text{ where } z = \alpha_0 + \alpha_1 x_1 + \ldots \alpha_p x_p + \alpha_{h_0} h_0. \tag{6}$$

3.6   LRFM4

Similarly as LRFM1 was extended to LRFM2, we can extend LRFM3 to LRFM4. More specifically, we form additional variables h_j by including a discretized numeric variable $\bar{x}_j$ in the Factorization Machine that is trained on residuals. Only variables h_j that significantly ($\alpha = 0.05$) improve the result on the train set will enter the model equation. This provides our last strategy:

$$h_j = \tilde{f}(\bar{x}_j, d_1, \ldots, d_q) \text{ and } \hat{y} = \frac{1}{1 + e^{-z}}, \text{ where}$$
$$z = \alpha_0 + \alpha_1 x_1 + \ldots \alpha_p x_p + \alpha_{h_0} h_0 + \alpha_{h_1} h_1 + \ldots + \alpha_{h_r} h_r. \tag{7}$$

4   Experiments

Interest in including categorical variables with many levels was raised by a practical problem at the NTCA. Since this data set is not publicly available, we found additionally three publicly available data sets with similar characteristics in the UCI Repository [13]. See Table 1 for some key characteristics. We considered a variable to have 'many levels' if the number of levels exceeds 30. This number corresponds roughly with the situation where the design matrix becomes sparse in our four data sets. Below we will describe the data sets, the exact parameters of the experiments and the results.

4.1   Data Sets

Tax Administration. This data set consists of approximately 80.000 audited VAT tax returns. A small part of these tax returns (17.5%) were found to contain one or more erroneous statements when audited. The data set has 33 numeric variables that are the result of a stepwise feature selection process that started with over 500 variables.

Table 1. Summary of data sets

	Data set			
	Tax	kdd98	Retail	Census
# observations	86,235	95,412	532,621	199,523
# numeric variables	33	18	3	23
% target = 1	17.5%	5.0%	29.1%	6.2%
% target = 0	82.5%	95.0%	70.9%	93.8%
Threshold infrequent level (LRG method)	100	100	1000	100
Categorical features (# levels)	Zipcode	DMA (207)	InvoiceNo	Industry code (52)
	(1,027)	RFA 11 (101)	(25,900)	Occ. code (47)
	Industry	RFA 14 (95)	Description	Prev. state (51)
	Sector	RFA 23 (87)	(4148)	Househ. stat (38)
	(3,747)	OSOURCE (869)	CustomerID	Country father (43)
		ZIP (19,938)	(4,373)	Country mother (43)
			Country (38)	Country birth (43)

KDD 98 Cup. This data set comes with a binary target and a numeric target. We only use the former as we focus on classification. Additionally, we only used the 'learning' data set and not the 'validation' data set. The original data set contains 480 variables. We selected 22 variables to get a data set similar to the tax data. The variable selection has been done by keeping the variables reported in [8] (p. 147, Table 6) and adding the two variables with many levels: OSOURCE and ZIP. Missing values have been replaced by 0, except for WEALTH where the median has been inserted.

Online Retail. The following data processing steps have been performed: (1) canceled transaction have been removed, (2) InvoiceDate has been split in a date part and a time part, (3) StockCode has been removed since its values can be mapped almost one-to-one to the values of Description. The data set does not contain a target. We created a binary target by defining the target to be 1 if the variable Quantity is larger or equal to 10, and 0 otherwise. After this, Quantity has been removed from the data set.

Census. The following data processing steps have been taken: (1) Weight has been discarded as is advised in the data description, (2) variables that are aggregates of other variables have been removed. For instance, Major Industry Code aggregates the levels of Industry Code. For this reason the variables: Major Industry Code, Major Occupation Code, Previous Region, Household Summary, Migration Code Reg, Migration Sunbelt, and Live1year House have been removed. Similarly, (3) Veteran Questionnaire is removed since most relevant information is in Veteran Benefits. Finally, (4) Migration Code MSA has been removed since it is highly collinear with Previous State.

4.2 Model Quality Measures

Various performance measures can be used to assess the results of a classification algorithm (e.g. accuracy, precision, area under the curve, recall). At the NTCA, interest lies mostly in observations with the highest risk scores; these observations will be selected for audit. For this reason *precision* is a natural measure. We applied precision at a '10% cut-off level', measured on a test set (i.e., we select the 10% highest scoring observations of a test set and then compute the precision), similarly as is done at the NTCA.

For completeness, we have also provided the frequently used Area Under the Curve (AUC). For confidentiality reasons the precision and AUC could not be reported for the tax data set. Instead we have provided the *increase in precision*, measured using the Plain Logistic Regression as a baseline.

In our experiments we used 5-fold cross validation to get reliable estimates of all quality measures listed above.

4.3 Settings Factorization Machines

In our experiments Factorization Machines were constructed with libFM software [17] with the following settings: number of iterations: 25, lengths of the parameter vectors $\mathbf{v}$: 16 (see Eq. (1)), and the optimization technique is set to 'MCMC'. The standard deviation of the normal distribution that is used for initializing the parameter vectors $\mathbf{v}$ in MCMC is set to 0.1. When building the Factorization Machines in approaches LRFM1 and LRFM2 we set the 'task' to 'classification', and in the remaining two cases to 'regression'.

4.4 Results

The results of applying two benchmark methods Plain Logistic Regression (PLR) and Logistic Regression with Grouping (LRG), as well as our four strategies are summarized in Table 2. The columns *precision* and *AUC* are not filled for the tax administration data set, because of confidentiality reasons.

5   Conclusion and Discussion

Looking at Table 2, some conclusions can be drawn. First, our proposed strategies give better results than plain logistic regression for all data sets. Second, when comparing with LRG (the strategy that gave the best results from the methods of Sect. 2.1), we see a subtler picture. Our methods outperform LRG clearly on the tax data and the retail data. For the tax data we see that taking interactions of the categorical variables with numeric variables into account, while training on residuals (i.e. LRFM4), can substantially improve the result. This is in agreement

Table 2. Performance (measured in precision, increase of precision with relation to Plain Logistic Regression, and Area Under Curve) for each strategy on all data sets using five-fold cross-validation.

Approach	Precision (%)				Precision (% increase w.r.t. PLR)				AUC			
	Tax	kdd98	Retail	Census	Tax	kdd98	Retail	Census	Tax	kdd98	Retail	Census
PLR	-	6.05	69.3	42.7	0.0%	0.0%	0.0%	0.0%	-	.5348	.7845	.9358
LRG	-	7.88	80.7	44.7	2.7%	30.3%	16.4%	4.6%	-	.5748	.8442	.9439
LRFM1	-	7.92	99.6	44.5	3.9%	31.0%	43.7%	4.2%	-	.5775	.9634	.9426
LRFM2	-	7.92	99.8	44.5	5.1%	31.0%	44.1%	4.2%	-	.5775	.9689	.9426
LRFM3	-	6.93	99.5	44.3	6.1%	14.6%	43.6%	3.7%	-	.5502	.9710	.9410
LRFM4	-	6.79	99.5	44.3	9.6%	12.3%	43.6%	3.7%	-	.5553	.9713	.9410

with the experienced auditor, see Sect. 1. When looking at the data set kdd98, the approaches that train directly on the target y (LRFM1 and LRFM2) are able to give slightly better results compared to LRG. However, LRG gives a slightly better result for the census data set. The latter might be caused by the relatively small number of levels of the categorical variables (maximum 52). Finally, our four strategies LRFM1, LRFM2, LRFM3, LRFM4 lead to different results on different data sets, without one strategy being the best for all data sets. The only exception is that in some cases the result of LRFM2 equals LRFM1 or the result of LRFM4 equals LRFM3. This is the case when no interaction term exceeded the significance level for entering the modeling equation. Therefore, we suggest to explore all strategies in a practical problem setting.

The results of this paper show that Factorization Machines can be successfully combined with Logistic Regression to overcome problems with categorical variables with many levels. We think that our methods can be adjusted without much effort to allow inclusion of categorical variables with many levels in other classification algorithms that suffer from a sparse, ill-conditioned model matrix, like other Generalized Linear Models or Support Vector Machines. Also a generalization to a multinomial logistic regression is straightforward. Note that some well-known classification algorithms have problems with categorical variables with many levels. For example, the standard implementation in R of randomForest [12] accepts only categorical variables with at most 53 levels.

Further research can address the issue of explainability of a Factorization Machine. Although our strategies lead to relatively simple logistic regression models, the introduction of the Factorization Machines worsens the explainability for that part of the model. We think that this problem can be solved by applying various dimensionality reduction and visualization techniques to the matrix $\mathbf{V}$ that consists of vectors $\mathbf{v}$ that represent levels of categorical variables, [14]. One could experiment as well with using an L^1 norm as a regularization term in the Factorization Machine.

Finally, we mention that the NTCA wants to avoid ethnic profiling, i.e. the selection process of audits should not be based on any variable that is strongly related to the ethnic background of a person. Since people with the same ethnic background might be clustered in certain zip codes, the NTCA wants to investigate this issue prior to including zip codes in a risk model.

References

1. Bassi, D., Hernandez, C.: Credit risk scoring: results of different network structures, preprocessing and self-organised clustering. In: Decision Technologies for Financial Engineering. Proceedings of the Fourth International Conference on Neural Networks in the Capital Markets, pp. 151–61 (1997)
2. Basta, S., Fassetti, F., Guarascio, M., Manco, G., Giannotti, F., Pedreschi, D., Spinsanti, L., Papi, G., Pisani, S.: High quality true-positive prediction for fiscal fraud detection. In: International Conference on Data Mining Workshops, ICDMW 2009, pp. 7–12. IEEE (2009)
3. Berkman, N.C.: Value grouping for binary decision trees. Technical report, University of Massachusetts (1995)
4. Breiman, L., Friedman, J., Stone, C.J., Olshen, R.A.: Classification and Regression Trees. CRC Press, Boca Raton (1984)
5. Burshtein, D., Della Pietra, V., Kanevsky, D., Nadas, A.: Minimum impurity partitions. Ann. Stat. **20**, 1637–1646 (1992)
6. Chou, P.A., et al.: Optimal partitioning for classification and regression trees. IEEE Trans. Pattern Anal. Mach. Intell. **13**(4), 340–354 (1991)
7. Friedman, J., Hastie, T., Tibshirani, R.: The Elements of Statistical Learning: Data Mining, Inference, and Prediction. Springer Series in Statistics. Springer, New York (2009)
8. Gupta, G.: Introduction to Data Mining with Case Studies. PHI Learning Pvt. Ltd., Delhi (2014)
9. Hosmer Jr., D.W., Lemeshow, S., Sturdivant, R.X.: Applied Logistic Regression, 3rd edn. Wiley, Hoboken (2013)
10. Kass, G.V.: An exploratory technique for investigating large quantities of categorical data. Appl. Stat. **29**, 119–127 (1980)
11. Koren, Y., Bell, R., Volinsky, C.: Matrix factorization techniques for recommender systems. Computer **42**(8), 30–37 (2009). http://dx.doi.org/10.1109/MC.2009.263
12. Liaw, A., Wiener, M.: Classification and Regression by randomForest. R News **2**(3), 18–22 (2002). http://CRAN.R-project.org/doc/Rnews/
13. Lichman, M.: UCI Machine Learning Repository (2013). http://archive.ics.uci.edu/ml
14. van der Maaten, L.J.P., Postma, E.O., van den Herik, H.J.: Dimensionality reduction: a comparative review. Tilburg University Technical report, TiCC-TR 2009-005 (2009)
15. Micci-Barreca, D.: A preprocessing scheme for high-cardinality categorical attributes in classification and prediction problems. ACM SIGKDD Explor. Newsl. **3**(1), 27–32 (2001)
16. Rendle, S.: Factorization machines. In: 2010 IEEE International Conference on Data Mining, pp. 995–1000. IEEE (2010)
17. Rendle, S.: Factorization machines with libFM. ACM Trans. Intell. Syst. Technol. **3**(3), 57:1–57:22 (2012)

Filtering Decision Rules with Continuous Attributes Governed by Discretisation

Urszula Stańczyk[(✉)]

Institute of Informatics, Silesian University of Technology,
Akademicka 16, 44-100 Gliwice, Poland
urszula.stanczyk@polsl.pl

Abstract. The paper presents research on selection of decision rules with continuous condition attributes while exploiting characteristics of these attributes obtained by supervised discretisation. The considered features were split into categories corresponding to numbers of intervals required for partitioning of their values, and this information was next used to divide the sets of rules by their conditions falling into specific categories. Also to each group of variables there was assigned some weight, basing on which several rule quality measures were calculated. They enabled filtering rules meeting requirements with respect to performance.

Keywords: Filtering rules · Continuous attributes · Discretisation · DRSA

1 Introduction

Construction of a good rule classification system is a challenge with many facets. Firstly, some definition of its quality needs to be formulated. Along with changing domains of application, intended implementations, or available resources, completely different qualities can be considered as important or irrelevant. Even for already defined optimisation criteria, the search for optimal solutions takes processing time. It can be done in pre-processing, by somehow improving input data, or in post-processing, by introducing modifications and enhancements to previously constructed classifiers, which can be adapted to specific needs [6]. Such post-processing procedures were employed in the described research.

In stylometry, chosen as the application domain, writing styles are often defined by usage frequencies of linguistic markers [1], resulting in continuous features. Thus either data mining technique to be employed needs to be able to work for this type of data, or discretisation is required [14]. The considered classification task involved cases of authorship attribution with balanced classes.

For construction of rule classifiers Dominance-Based Rough Set Approach (DRSA) was selected [9]. DRSA can operate on both discrete and continuous attributes, which meant that discretisation was not necessary. However, it

© Springer International Publishing AG 2017
M. Kryszkiewicz et al. (Eds.): ISMIS 2017, LNAI 10352, pp. 333–343, 2017.
DOI: 10.1007/978-3-319-60438-1_33

was used in order to learn additional characteristics of available characteristic features.

Discretisation can be performed in unsupervised approaches with disregarding class information, or with taking it into account, as happens in supervised procedures, such as Fayyad and Irani's [4] that was exploited in the described research. In the recursive processing the whole ranges of attributes values were partitioned into some numbers of intervals, called also bins, while observing class entropy. Basing on the numbers of bins found for all features they were divided into corresponding categories, exploited next in rule filtering.

The sets of rules inferred within DRSA were divided into disjoint groups reflecting categories of included attributes. Depending on weights assigned to these categories, values of the proposed rule quality measures were calculated.

The experiments performed indicate that even though for some features only single intervals were found, which suggests that their values should be irrelevant for classification, it did not always hold true, especially for more complex and difficult problems. Taking into account all categories of variables and assigned weights, many classifiers found had reduced numbers of rules while maintaining recognition at the required level or even showing some improved performance.

The text of the paper is organised as follows. Section 2 provides the background information, while Sect. 3 presents the research framework. Section 4 details experiments and their results, and Sect. 5 concludes the paper.

2   Background

The presented research was focused on characterisation of continuous features by supervised discretisation and rule filtering, described in the following sections.

2.1   Supervised Discretisation

In discretisation, points from the input continuous space are grouped into granules corresponding to intervals (or bins), within which values are indiscernible, with the aims of simplified expression of concepts, or ability to employ methods working for nominal attributes [14]. Discretisation can result in improved performance, yet it cannot be guaranteed as it always means loss of information.

Discretising procedures are grouped into two distinct approaches: unsupervised and supervised. In unsupervised approach information about class is disregarded, and the whole ranges of observed values are divided into set numbers of intervals of equal width, or into bins with specified equal frequencies.

Fayyad and Irani's [4] supervised discretisation method, employed in the presented research, takes into account class information entropy to find cut points. Starting from one interval containing all available values for each considered attribute, there is executed its recursive partitioning up until the stopping criterion of Minimal Description Length (MDL),

$$Gain < \frac{log_2(N-1)}{N} + \frac{log_2(3^k - 2) - kE + k_1 E_1 + k_2 E_2}{N}, \tag{1}$$

is met. N stands for the number of instances, k is the total number of classes, and k_1 and k_2 numbers of classes in constructed subintervals. E denotes the entropy for all instances, whereas E_1 and E_2 give entropies of samples in subintervals. Thus in recursive calculations this non-parametric procedure establishes how many bins are required for each variable, and these numbers can vary.

2.2   Rule Filtering

Rule classifiers are often preferred to other types of inducers due to the fact that in a very clear and direct way they express knowledge learned about patters discovered in the training data. By specifying conditions on characteristic features that need to be met, decision rules point to a certain class or groups of classes. Depending on the applied approach, as a result of rule induction process, decision algorithms with significantly varying numbers of constituent rules can be obtained [15]. Some algorithms return only minimal subsets of rules providing coverage, others generate even all rules that can be found for the learning examples, while still others construct rule sets meeting some requirements.

A set of inferred rules can be tested for possible pruning, that is rejecting some elements without decreasing performance. One of filtering approaches is to impose hard constraints on rule parameters, such as length, support, or included condition attributes [13]. The process can be also driven by rankings of attributes [10], or quality measures reflecting their characteristics [11].

3   Research Framework

The first step of research was devoted to preparation of input datasets and their supervised discretisation, returning discovered numbers of bins for all attributes. Then for continuous data DRSA rules were induced. Rule sets were next filtered by attributes, and by calculated measures based on assigned attribute weighs.

3.1   Input Datasets

The application domain for the presented research was stylometry, a study of writing styles [1], with its prominent task of authorship attribution [3]. To define a style, textual markers were selected and calculated basing on available text samples. The markers referred to lexical and syntactic elements of style, such as usage of certain words and employed punctuation marks [8].

The classification was binary as two pairs of writers were chosen for comparison and recognition, female and male, and their longer works were divided into smaller samples. For these texts the frequencies of usage were calculated for 17 function words (from the list of the most frequently used words in English language), and 8 punctuation marks, giving 25 features. For evaluation of classifiers test sets were used, based on separate works than those used in the learning phase. This approach gives more reliable results than cross-validation [2], which, with such construction of training sets, tends to return falsely high recognition.

3.2 Characterisation of Continuous Attributes by Discretisation

The condition attributes were split into groups corresponding to the distinct numbers of intervals found for their values in supervised discretisation by Fayyad and Irani's method. As higher numbers of bins can be viewed as more detailed study of values needed for class recognition, it can be reasoned that these numbers reflect to some extent the importance of variables, or the degree of complexity of relationships between attributes and target concepts.

In the studied cases the numbers of bins varied from 1, when all values of some variable were assigned to a single interval, to 3, which gave three distinct groups of condition attributes with different cardinalities, as listed in Table 1.

Table 1. Categories of condition attributes established by discretisation

Number of bins found	Condition attributes	
	Female writer dataset	Male writer dataset
1	and in with of what from if . !	on of this . , : (
2	but not at this as that by for to , ? (–	but not in with as to if ? ! ; –
3	on ; :	and at that what from by for

These characteristics indicate that for both classification tasks there possibly were irrelevant features, but also suggest that recognition of male writers would be harder than for female writers, since we need to pay more attention to values of more than twice as many attributes. The findings with respect to the importance of variables can be confirmed by obtaining their rankings [10,12]. It can be observed that variables with 3 bins are among highest-ranking, while these with a single interval belong with the lowest ranking attributes.

3.3 DRSA Rules

Dominance-Based Rough Set Approach (DRSA) is a soft computing methodology based on rough set theory invented by Pawlak [7], but with substituting the original indiscernibility relation with dominance [9]. This change enables to observe ordinal properties in value sets of attributes [5], and when preferences of these values are assigned to groups of classes, or cones of classes, both nominal and ordinal classification is possible. DRSA can operate on both continuous and discrete variables, it only needs definitions of preference orders for all attributes.

A decision algorithm can be induced in its minimal form, with just providing a minimal cover of the learning samples, but very short algorithms often offer neither high performance, nor much space for improvement by rule selection. That is why for both considered datasets all rules on examples were inferred for the whole set of characteristic features. This returned 46,191 rules for male writer and 62,383 rules for female writer dataset, referred to in the paper as *M-FAlg* and *F-FAlg* respectively. With such high numbers of rules, without any additional processing (such as filtering or voting), the number of correct decisions in testing was zero, because for all test samples there were ambiguous decisions, which were always treated as wrong decisions.

These two full algorithms were next filtered by imposing hard constraints on support required of rules, in order to detect the maximal classification accuracy. This allowed for correct classification of 86.67% of test samples for female, and 76.67% for male writers, with the numbers of rules 17 (with support, given as the number of supporting samples, equal 66 or higher) and 80 (with support at least 41) for female and male datasets. These results confirmed the previous expectations for male writer dataset — it proves more difficult in recognition. The two shortened algorithms served as reference points when further rule filtering was executed, and were denoted respectively *F-BAlg17* and *M-BAlg80*.

3.4   Proposed Rule Quality Measures

Basing on division of attributes into groups reflecting the numbers of intervals found for their values, there were defined four rule quality measures, as follows.

The first discretisation-based quality measure, $QMD1$, took into account a number of conditions with each number of intervals, assuming the lowest rank for a single bin. $MaxNBin$ was the highest number of bins found, and $Ncond_r(i)$ denoted the number of conditions in r rule with i intervals. All attributes had nonzero rank, but the more conditions were included in the rule, the higher the value of this measure, hence somehow other indications of rule quality needed to be introduced, that is either length or support, which led to $QMD1_L(r) = QMD1(r)/Length(r)$, where $Length(r)$ denoted rule length, and $QMD1S(r) = QMD1(r)Support(r)$, where $Support(r)$ gave the number of supporting samples.

$$QMD1(r) = \sum_{i=1}^{MaxNBin} Ncond_r(i)10^{i-1} \qquad (2)$$

The second measure enabled to test whether attributes with single intervals were in fact irrelevant, but to such degree that all rules referring to them should be disregarded. It was obtained by assuming that these features had zero rank, thus the value of the measure was zero when at least one of included variables belonged to this category. As the measure was defined as a product of fractions, the longer the rule the lower the value of the measure, thus one factor of quality was already included in this formulation. The second, that is rule support, could be also added to the equation forming $QMD2S(r) = QMD2(r)Support(r)$.

$$QMD2(r) = \prod_{i=1}^{MaxNBin} \left(\frac{i-1}{i}\right)^{Ncond_r(i)} \qquad (3)$$

The third measure also assumed irrelevancy of attributes with single intervals, but only at the stage of measure calculation, that is the value depended only on these attributes that had more than one interval found. As a result, rules with conditions only on 1-bin features had the measure evaluated as 1, while all others, with at least one condition from other categories, assumed fractional values. With this approach rules containing these seemingly irrelevant variables were not completely excluded from considerations.

$$QMD3(r) = \prod_{i=2}^{MaxNBin} \left(\frac{i-1}{i}\right)^{Ncond_r(i)} \tag{4}$$

Again, $QMD3S(r)$ denoted the value of $QMD3(r)$ multiplied by the number of samples supporting rule r. The definition of this measure led to the same values obtained for some rules that varied by conditions on attributes with single intervals, and to some extent to disregarding rule length, since these attributes had no influence on the calculated measure. To prevent that, either division by rule length could be directly added, as for $QMD1_L$, or, to variables from 1-interval category some minimal weight Min_w could be assigned, as given in the fourth proposed measure.

$$QMD4(r) = (Min_w)^{Ncond_r(1)} \prod_{i=2}^{MaxNBin} \left(\frac{i-1}{i}\right)^{Ncond_r(i)} \tag{5}$$

The usefulness of these measures was evaluated in research on rule filtering.

4   Experiments

Selection of rules was obtained in two approaches. Firstly, two generated full algorithms, F-FAlg and M-FAlg, were divided into disjoint subsets of rules, depending on included conditions for attributes with different numbers of intervals, and these subsets were tested. Secondly, for all rules there were calculated values of the proposed quality measures, and then gradually increasing subsets of rules with highest values were filtered, as is typical for processing with a ranking.

4.1   Filtering Rules by Attributes

It was established by supervised discretisation that the minimal number of intervals required was 1, while the maximal was 3, which grouped all features into three categories: C1, C2, or C3, the number directly indicating the number of bins. This division in turn led to seven groups of decision rules as follows:

- C1, C2, or C3: all condition attributes included respectively had single bins, two bins, or three bins,
- C12: attributes included had either one or two bins, but at least one of variables belonged to each category,

- C13: attributes included had either one or three bins, but at least one of variables belonged to each category,
- C23: attributes included had either two or three bins, but at least one of variables belonged to each category,
- C123: attributes included had either one, two, or three bins, but at least one of variables belonged to each category.

The tests were performed for all categories of rules individually and for some combinations, as listed in Table 2. In the bottom part of the table there were included results for tests executed for groups of rules that originally contained conditions on 1-bin attributes, but these conditions were removed, while the other conditions were kept intact. For example C2-1 included rules with conditions on both 1- and 2-bin variables, from which the former were rejected, leaving only those on 2-bin attributes. In all cases there were imposed such constraints on minimal support required of rules that led to obtaining the highest classification accuracy possible, given as correct decisions only, with fewest rules.

The results show one case of slightly increased accuracy for female writers (for category C2), two cases of rejecting a single rule from the reference algorithm *F-BAlg17*, and some cases with the same predictions but for more rules than included in it. For male dataset the same level of accuracy as for *M-BAlg80* was obtained once for the rule subset (C2+3+23), when 4/80 rules were discarded.

Furthermore, it is clearly visible that for male writers inclusion of rules with conditions on 3-intervals attributes always resulted in better accuracy than otherwise. This observation was not true for female dataset, but then in this category only 3 variables were included. There were also some cases when removal of conditions on 1-interval variables from rules improved the recognition.

4.2 Filtering Rules by Quality Measures

Filtering decision rules by values of the proposed quality measures involved firstly calculating these measures for all rules, sorting them in descending order, and then executing filtering steps with selection of gradually increasing overlapping subsets with the highest scores. Depending on a measure and dataset in each step within a series different numbers of rules can be recalled, and different numbers of steps can be performed. Table 3 displays results, where *QMD* indicates the measure employed. Notation *-1* (such as in 1*S-1* or 3-*1*) denotes considering subsets of rules with discarding those belonging to C1 category, that is containing conditions only on 1-interval features. Only selected filtering steps are shown.

For female writer dataset application of all four proposed quality measures and their variations enabled to find at some filtering step the same recognition as for the reference algorithm *F-BAlg17*, but achieved with a reduced number of included rules. The shortest of these algorithms contained 11 decision rules, which meant rejecting $6/17 = 35.29\%$ rules. For this dataset clearly 1-interval attributes were generally not needed for good classification accuracy, since disregarding rules with conditions on them did not degrade the observed performance.

Table 2. Performance of classifiers based on rule categories. Columns denote: (a) number of rules in a set, (b) classification accuracy [%] without any constraints, (c) support imposed to achieve the best recognition, (d) number of rules meeting constraints, (e) classification accuracy [%] with imposed constraints.

Labels of rule categories	Female writers					Male writers				
	(a)	(b)	(c)	(d)	(e)	(a)	(b)	(c)	(d)	(e)
C2	3798	28.89	52	40	**87.78**	900	23.33	4	174	33.33
C3	23	34.44	16			256	70.00	32	31	75.00
C1	1489	15.56	5	129	31.11	325	30.00	4	63	31.11
C23	1888	27.78	48	42	**86.67**	5158	13.33	35	50	75.00
C12	42913	0.00	50	21	82.22	11718	3.33	15	18	41.67
C13	973	21.11	15	142	65.56	2808	10.00	40	28	73.33
C123	11299	1.11	35	44	81.11	25026	0.00	32	38	70.00
C2+12	46711	0.00	52	58	**86.67**	12618	3.33	15	21	41.67
C3+13	996	21.11	15	150	65.56	3064	10.00	41	40	73.33
C2+23	5686	17.78	66	16	**86.67**	6058	11.67	35	50	75.00
C3+23	1911	27.78	48	42	**86.67**	5414	13.33	35	76	75.00
C2+3+23	5709	17.78	66	16	**86.67**	6314	11.67	35	76	**76.00**
C2+1+12	48200	0.00	52	58	**86.67**	12943	3.33	15	21	41.67
C3+1+13	2485	7.78	15	150	65.56	3389	10.00	41	40	73.33
C2+23+12+123	59898	0.00	66	17	**86.67**	42802	0.00	41	40	71.11
C3+23+13+123	14183	1.11	48	58	**86.67**	33248	0.00	41	80	**76.67**
C2+1+23+12+123	61387	0.00	66	17	**86.67**	43127	0.00	41	40	71.11
C3+1+23+13+123	15672	1.11	48	58	**86.67**	33573	0.00	41	80	**76.67**
C2+2+12+13+23+123	60894	0.00	66	17	**86.67**	45866	0.00	41	80	**76.67**
C2-1	42913	0.00	55	13	83.33	11718	0.00	16	8	41.67
C3-1	973	1.11	27	64	73.33	2808	0.00	41	25	73.33
C23-1	11299	0.00	44	17	81.11	25026	0.00	36	18	66.67
C2+2-1	46711	0.00	52	58	85.56	12618	0.00	16	10	41.67
C3+3-1	996	1.11	27	66	73.33	3064	0.00	41	40	73.33
C2+23+2-1+23-1	59898	0.00	66	17	**86.67**	42802	0.00	41	40	71.67
C3+23+3-1+23-1	14183	0.00	48	58	85.56	33248	0.00	41	80	**76.67**
C2+3+23+2-1+3-1+23-1	60894	0.00	66	17	**86.67**	45866	0.00	41	80	**76.67**

Table 3. Performance of classifiers with rules recalled by quality measures. Columns present: (a) number of recalled rules, (b) classification accuracy [%] without any constraints, (c) support imposed to achieve the best recognition, (d) number of rules meeting constraints, (e) classification accuracy [%] with imposed constraints.

QMD and step		(a)	(b)	(c)	(d)	(e)
	04	13326	1.11	48	58	**86.67**
1_L	05	17981	0.00	66	16	**86.67**
	06	60894	0.00	66	17	**86.67**
1S-1	03	301	77.78	66	11	**86.67**
	06	337	77.78	66	16	**86.67**
2	03	908	54.44	66	16	**86.67**
2S	02	33	85.56	66	12	**86.67**
	03	47	84.44	66	16	**86.67**
3-1	04	13695	0.00	63	12	**86.67**
	05	30788	0.00	66	17	**86.67**
3S	03	95	81.11	63	12	**86.67**
	05	189	72.22	66	17	**86.67**
4	02	908	54.44	66	16	**86.67**
	10	6571	8.89	66	17	**86.67**

QMD and step		(a)	(b)	(c)	(d)	(e)
	02	756	46.67	32	40	**76.67**
	03	1271	41.67	41	30	**81.67**
1_L	04	1665	31.67	41	36	**83.33**
	05	3581	11.67	41	58	**76.67**
	08	31449	0.00	41	80	**76.67**
	01	45	**78.33**	32	41	
	02	67	73.33	41	44	**83.33**
1S	04	174	66.67	41	52	**76.67**
	06	462	50.00	41	80	**76.67**
2	03	537	60.00	10	272	75.00
2S	03	267	71.67	35	75	75.00
3-1	05	17723	1.11	41	80	**76.67**
3S	03	466	45.00	41	79	**76.67**
	04	622	40.00	41	80	**76.67**
	06	7156	10.00	35	97	**76.67**
4	07	8754	5.00	41	67	**76.67**
	12	24604	1.11	41	80	**76.67**

For male writer dataset at least some attributes from 1-interval category were required to maintain satisfactory recognition. When rules with conditions on such features were excluded from considerations, as it happened for QMD2, even recalling relatively high numbers of rules was not sufficient and the classification accuracy was slightly decreased with respect to the reference algorithm M-BAlg80. Also the statement that for rules with mixed conditions, 1-bin variables could have no influence on their perceived quality, which occurred for QMD3, did not hold true, as without filtering almost all rules from the reference algorithm the prediction ratio was lower than required. The best results, with the same or even increased recognition for significantly shorter algorithms, were obtained when attributes with single intervals influenced values of quality measures, as could be observed for QMD1 and QMD4.

These test results show dependence on the considered datasets and the degree of difficulty in recognition posed by them, but they also indicate bias of supervised discretisation, which by observing entropy found some features as not really needed for recognition of classes, while for another type of classifier employed they proved to be required for acceptable performance.

The proposed methodology has a drawback of being feasible only in cases when there are not many distinct numbers of intervals found for features by discretisation, as the respective numbers of variable categories, and then groups of rules increase exponentially. However, this limitation can be easily circumvented by reducing the number of attribute categories through their merging.

5 Conclusions

The paper presents research on filtering decision rules inferred for continuous features, while applying characteristics of these variables learned from discretisation. In supervised discretisation, statistical procedures enable to discover the number of intervals required for each attribute, while taking into account information about class entropy. Depending on a number of bins defined for each variable by Fayyad and Irani's method, they were grouped into categories, with the assumption that the significance of groups increased with the number of intervals. Basing on the inclusion in one of these groups and the values of calculated quality measures, the rules from the algorithms, comprised of all rules on examples, were filtered and their performance evaluated with test sets.

The executed experiments indicate that the same or even increased performance of decision algorithms can be obtained for reduced sets of rules. The tests also show cases when for some attributes only a single interval was assigned for the whole range of values, which meant that these variables were found as irrelevant for classification. However, this conclusion is biased by supervised discretisation method, as exclusion of these attributes from considerations for rule classifiers in several cases hurt recognition, and at least some of variables from this category had to be included in the set to keep the required prediction level.

Acknowledgments. The research presented in the paper was performed at the Silesian University of Technology, Gliwice, within the project BK/RAu2/2017. In the research there was used 4eMka Software [5,9] and WEKA [14].

References

1. Argamon, S., Burns, K., Dubnov, S. (eds.): The Structure of Style: Algorithmic Approaches to Understanding Manner and Meaning. Springer, Berlin (2010)
2. Baron, G., Harężlak, K.: On approaches to discretization of datasets used for evaluation of decision systems. In: Czarnowski, I., Caballero, A.M., Howlett, R.J., Jain, L.C. (eds.) Intelligent Decision Technologies 2016. SIST, vol. 57, pp. 149–159. Springer, Cham (2016). doi:10.1007/978-3-319-39627-9_14
3. Craig, H.: Stylistic analysis and authorship studies. In: Schreibman, S., Siemens, R., Unsworth, J. (eds.) A Companion to Digital Humanities. Blackwell, Oxford (2004)
4. Fayyad, U., Irani, K.: Multi-interval discretization of continuous valued attributes for classification learning. In: Proceedings of the 13th International Joint Conference on Artificial Intelligence, pp. 1022–1027. Morgan Kaufmann (1993)
5. Greco, S., Matarazzo, B., Słowiński, R.: Dominance-based rough set approach as a proper way of handling graduality in rough set theory. Trans. Rough Sets **7**, 36–52 (2007)
6. Michalak, M., Sikora, M., Wróbel, L.: Rule quality measures settings in a sequential covering rule induction algorithm – an empirical approach. In: Proceedings of the 2015 Federated Conference on Computer Science and Information Systems. ACSIS, vol. 5, pp. 109–118 (2015)
7. Pawlak, Z.: Rough sets and intelligent data analysis. Inf. Sci. **147**, 1–12 (2002)

8. Peng, R., Hengartner, H.: Quantitative analysis of literary styles. Am. Stat. **56**(3), 15–38 (2002)
9. Słowiński, R., Greco, S., Matarazzo, B.: Dominance-based rough set approach to reasoning about ordinal data. In: Kryszkiewicz, M., Peters, J.F., Rybinski, H., Skowron, A. (eds.) RSEISP 2007. LNCS, vol. 4585, pp. 5–11. Springer, Heidelberg (2007). doi:10.1007/978-3-540-73451-2_2
10. Stańczyk, U.: Attribute ranking driven filtering of decision rules. In: Kryszkiewicz, M., Cornelis, C., Ciucci, D., Medina-Moreno, J., Motoda, H., Raś, Z.W. (eds.) RSEISP 2014. LNCS, vol. 8537, pp. 217–224. Springer, Cham (2014). doi:10.1007/978-3-319-08729-0_21
11. Stańczyk, U.: Selection of decision rules based on attribute ranking. J. Intell. Fuzzy Syst. **29**(2), 899–915 (2015)
12. Stańczyk, U.: Measuring quality of decision rules through ranking of conditional attributes. In: Czarnowski, I., Caballero, A.M., Howlett, R.J., Jain, L.C. (eds.) Intelligent Decision Technologies 2016. SIST, vol. 56, pp. 269–279. Springer, Cham (2016). doi:10.1007/978-3-319-39630-9_22
13. Stańczyk, U.: Weighting and pruning of decision rules by attributes and attribute rankings. In: Czachórski, T., Gelenbe, E., Grochla, K., Lent, R. (eds.) ISCIS 2016. CCIS, vol. 659, pp. 106–114. Springer, Cham (2016). doi:10.1007/978-3-319-47217-1_12
14. Witten, I., Frank, E., Hall, M.: Data Mining: Practical Machine Learning Tools and Techniques, 3rd edn. Morgan Kaufmann, Burlington (2011)
15. Wróbel, L., Sikora, M., Michalak, M.: Rule quality measures settings in classification, regression and survival rule induction – an empirical approach. Fundam. Inf. **149**, 419–449 (2016)

Mining Temporal, Spatial and Spatio-Temporal Data

OptiLocator: Discovering Optimum Location for a Business Using Spatial Co-location Mining and Spatio-Temporal Data

Robert Bembenik[(✉)], Jacek Szwaj, and Grzegorz Protaziuk

Institute of Computer Science, Warsaw University of Technology,
Nowowiejska 15/19, 00-665 Warsaw, Poland
{r.bembenik, g.protaziuk}@ii.pw.edu.pl,
jacek.szwaj@gmail.com

Abstract. The issue presented in the paper concerns methods allowing to pinpoint an optimum location for a given business. As long as business is based on brick-and-mortar location and real customers (as opposed to businesses available online) the location is a crucial factor contributing to the business' success. To this end we propose a method named OptiLocator based on spatial co-location mining and measures utilizing spatio-temporal aspects of social data related to the urban space under investigation that computes the optimum location for a business of a given type. The resulting recommendation concerning business localization is based on the neighborhoods of popular places similar to the business we plan to localize in the city.

Keywords: Optimum business location · Spatial data mining spatial co-location rules · Spatio-temporal data analysis · City exploration

1 Introduction

Location is the most important factor influencing the success of a business, especially retail business involving face-to-face contact with customers, e.g. restaurants or clothing stores. While doing such type of business it is very important to be close to potential customers. Even the quality of offered products or services is secondary in comparison with the location. In the case of the traditional business (not internet-based) location changes slowly. Usually at the start of the activity we select a place and we are then bound to it. The change of the place entails large costs. If we do not select business location well it may mean bankruptcy even if we have good services or products. We will be simply too far from the target group and we will not sell the goods or services. In this paper, we consider the problem of selection of the optimal location for a given type of business.

2 Related Work

Techniques currently used to support location selection are discussed in [6]. The authors based on the questionnaire conducted in the UK in 1998. It covered over 55000 sales points in 8 sectors of the economy. The study shows that the techniques used to

© Springer International Publishing AG 2017
M. Kryszkiewicz et al. (Eds.): ISMIS 2017, LNAI 10352, pp. 347–357, 2017.
DOI: 10.1007/978-3-319-60438-1_34

determine the best location for a given business are: experience, ratios comparison, multiple regression and discriminant analysis, cluster and factor analysis, spatial interaction, expert systems. Large sales companies use multiple techniques supporting location of sales points.

Data mining methods utilizing spatio-temporal data are a possible new direction in searching for the optimal location for businesses. In [3] the authors consider the problem of finding groups of people often appearing in a given location in similar time.

In [2] the authors consider the problem of store placement using knowledge from mined location-based services. Based on the analyzed spatio-temporal Twitter data they show specificity of human movement in the city. The larger the city the stronger the local character of the changes. An important factor is the type of place where a user of the LBSN (Location-Based Social Network) service is found. Communication nodes (metro stations, airports) and tourist attractions (museums, parks, hotels) are more frequent check-in places than other types of venues. Starbucks is the favorite place where people communicate their presence (several times more popular than Dunkin' Donuts or McDonald's). Another interesting observation is the neighborhood profile for the mentioned franchises. Starbucks cafes are most frequently localized near railway stations and stadiums. Dunkin' Donuts can most often be found near hostels and gas stations. McDonald's can be found in Manhattan near offices and florist shops. Some measures are proposed to track LBSN users in the context of deciding business location: area popularity, transition density, incoming flow, and transition quality.

Because the LBSN data becomes more widely available we propose an approach to determining the optimal location for a given type of business utilizing the spatio-temporal data from LBSN services. In the proposed method we use the LBSN data to compute indicators characterizing regions of the city with respect to businesses existing there in combination with spatial data mining techniques, particularly co-location rules

The rest of the paper is organized as follows. Section 2 introduces the OptiLocator algorithm. Section 3 presents experimental evaluation of the introduced approach. Section 4 concludes the paper.

3   OptiLocator Algorithm

In this chapter we introduce a new approach we called *OptiLocator* to discover the optimal location for conducting a business of a given type. The method combines the area characteristics indicators and the spatial data exploration. The input data required for the algorithm are: a set of business places across the city (with the geographical coordinates) and the selected category of business we want to locate as well as the relevant LBSN data. The result is the recommendation of the best location with positive impact for the selected business category.

OptiLocator consists of the following steps: (1) spatio-temporal data preparation and adjustment, (2) clustering of data representing existing businesses, (3) computation of area characterization indicators, (4) computation of spatial co-location rules, and (5) determination of the optimal location. Below we provide description of the steps.

The goal of the *spatio-temporal data preparation and adjustment* step is the selection of relevant data coming from a LBSN, such as Foursquare, Facebook, Instagram, or QQ with information on users' check-ins. The data has then to be enriched, i.e. missing data have to be supplemented using other sources and adjusted to the model used in the further analyses. The *clustering of data representing existing businesses* step consists in execution of a clustering algorithm to create clusters of places. In the input data set we only have information that all places are in one large area (New York City in our case). As usually business is conducted locally, there is a need to separate these small regions. To create clusters we use DBSCAN. The result is a division of the large city area into many small regions (in our case more than 1000). In the *computation of area characterization indicators* step measures describing the regions are computed. We use the following indicators: *popularity, mobility, transitivity, competitiveness, attractiveness*. All of them are presented as numeric values. They are the base for area score calculations. A detailed description of area characterization indicators is presented in Sect. 3.2. The aim of the *computation of spatial co-location rules* step is to discover the co-location rules by applying the FARICS algorithm [4]. The rules represent relationships between different business activity categories. Thanks to it, the OptiLocator algorithm knows that e.g.: an office area is a better neighborhood for a florists than a museum area. In the last step, *determination of the optimal location*, we select the best location (a region determined during clustering) for conducting business for the selected category. For this purpose the following steps are performed: (a) calculate voting rules algorithm value for each group (chapter 3.4), (b) select the group which has the highest rank in the list if it fulfills the criterion of the minimal area score. The selected group is the best location for conducting business for the selected category.

3.1 Clustering

In the selection of a localization we are interested in as small area as possible (information that the best location is a city of the size of e.g. New York for considered kind of business do not solve the problem). Smaller area guarantees greater accuracy in the choice of a location. While looking for the optimum location it is thus desirable to consider smaller regions. Such regions may be determined by using clustering methods. Clustering in OptiLocator is performed by the well-known DBSCAN algorithm [5], which allows to discover clusters of various shapes. Thanks to the venues clustering we can calculate area scores for each cluster and choose the one with the best evaluation.

During the OptiLocator tests for the test dataset we used values: MinPts = 7, Eps = 30 m in order to reduce the number of venues marked as noise. Running the algorithm with these parameters resulted in groups of compact sizes appropriate for further processing.

3.2 Area Characterization Indicators

An area is defined as a set of features of all venues located there (in our case all venues in New York). Example features of a venue are: average customer rating, number of likes, category of the place, tags added by the users, opening hours, and customers' profiles. These properties are taken into consideration when calculating the indicators, features of the entire area.

Area metrics have been proposed in [2]. The metrics have been divided into two categories: geographic features and mobility features. Geographic features are measured using: density (the number of neighbors around the assessed place), neighbors entropy (for assessing the influence of the spatial heterogeneity of the area on the popularity of a place), competitiveness (proportion of neighboring places of the same type with respect to the total number of neighboring places), Quality by Jensen (for considering spatial interactions between different place categories). Mobility features here are measured by: area popularity (the total number of check-ins empirically observed among the neighboring places in the area), transition density (the density of transitions between the venues inside the area), incoming flow (incoming flow of external user traffic towards the area of the place in question), and transition quality (for measuring the probability of nearby places as potential sources of customers to the place under prediction).

We propose our own indicators which are intuitive and contribute to high quality location predictions in our approach. These indicators, as explained below, are popularity, mobility, transitivity, competitiveness, attractiveness and area score. For each presented indicator we provide an example explaining how to compute it as well as the interpretation. Examples refer to the optimal location of a bakery near the Astor Place Subway (Astor Place Cluster).

Popularity. The measure of venues popularity for a region. It is computed by dividing the quantity of all check-ins in the selected region by the total number of check-ins in the data set. The result usually is a very small fraction, which we then multiply by 1000 to normalize the results. The highest rate of popularity in the analyzed data set was achieved for regions containing transportation hubs. The following formula is used to calculate the value of this indicator:

$$p = (ci\_r \ / \ ci\_all) * 1000$$

where ci_r – number of check-ins in the region under consideration, ci_all – all check-ins in the area. Astor Place Cluster example: $ci\_r = 1776$, $ci\_all = 225126$, $p = (216 \ /225126) * 1000 = 7.9$.

Mobility. This indicator determines the direction of users' movement as external or internal with respect to the selected region. The first case is a situation when a user is in the region and next she or he visits a place (generates the next check-in) outside of the region. The second case - when both current and next check-in are within the same region. In OptiLocator we use external mobility because both measures are proportional and their sum is 100% of all check-ins. The following formula is used to calculate the value of this indicator:

$$m = ci\_out \ / \ (ci\_out + ci\_in)$$

where: ci_out – number of check-ins outside of the region, ci_in - number of check-ins within the same region. For the Astor Place Cluster we have: $ci\_out = 1512$, $ci\_in = 254$, $m = 1512/(1512 + 254) = 0.856$.

Transitivity. This indicator tells us if the selected region is a target region for users or just a point on their path. To calculate transitivity all venues in the group are classified as transitive or intransitive and the percentage of transitive places is computed. Transitive places are these inside a user path where the next and the previous venues (places of a user's check-ins) are outside of the region. A graphical example of a transitive place is given in Fig. 1.

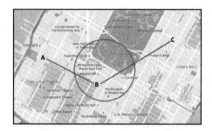

Fig. 1. Example of a transitive place (B) on the user path A → B → C

The value of the indicator is calculated according to the formula:

$$t \; = \; tv \; / \; all\_v$$

where: tv – number of transitive venues in the region, all_v – all venues in the region. For the Astor Place Cluster we have: $tv = 26$, $all\_v = 216$, $t = 26/216 = 0.12$.

Competitiveness. This indicator specifies whether there are direct competitors (for a given business category) in the region. For the indicator to be useful compact region sizes are assumed. To calculate the competitiveness we count the number of venues for the given category type in the region and divide it by the total number of venues in the region. The value of the indicator is calculated according to the formula:

$$c \; = \; c\_v \; / \; all\_v$$

where: c_v – number of venues with the same business category in region, all_v – number of all venues in the region. For the Astor Place Cluster we have: $c\_v = 4$, $all\_v = 216$, $c = 4/216 = 0.018$.

Attractiveness. The attractiveness of the region is calculated based on users' evaluations of the venues in the region. The user evaluates a venue on a scale from 0 (the worst score) to 10 (the best score). The most attractive places encourage many people, including potential customers. The value of the indicator is calculated as the average value obtained by all evaluated places in the region. During the tests of OptiLocator we collected venues evaluations from the Foursquare API[1]. The following formula is used to calculate the value of this indicator:

$$a = s\_vm \ / \ c\_v$$

where: s_vm – sum of marks of venues, c_v – number of venues in the region. For the Astor Place Cluster we have: $s\_vm = 1109.6$, $c\_v = 142$, $a = 1109.6/142 = 7.81$.

Area Score. The values of the defined indicators are contained in different intervals. In such case it is difficult to compare them and state which ones are more important and what the same value for different indicators mean. At the same time we want one value as the final mark characterizing a given region. We normalized all indicators' values to the common interval (according to the distribution of their values). The value of each indicator evaluation gets the value of 2, 3, 4 or 5. The interpretation of each of the values is as follows: 5 – highly recommended, 4 – recommended, 3 – not much recommended, 2 – not recommended.

In addition to the area score we need to evaluate the number of venues (the number of venues factor), which is also normalized to a value in the range 2–5. The quality of indicators for regions with the larger number of venues is better. An important factor is users' preferences as regards a given type of business. OptiLocator takes into account two situations. The first is whether the users inside of the region look for a selected business category (+0.4 to the region score) – the category is next on the user path. The second situation is the category is missing in the region (+0.6 to the region score). The final score for the region is in the range 2–6, where 6 denotes "excellent place", and it is calculated by the following formula:

$$as = (p \ + \ m \ + \ t \ + \ c \ + \ a \ + \ v) \ / \ 6 \ + \ x * 0.4 \ + \ y * 0.6$$

where: p – popularity score, m – mobility score, t – transitivity score, c – competitiveness score, a – attractiveness score, v – venues count score, x – the value is 0 if the selected category exists in the area or 1 if the category does not exist and is wanted, y – the value is 0 if the category is not wanted or 1 if it is wanted.

For the Astor Place Cluster we have: popularity – 4, mobility – 5, transivity – 2, competitiveness – 4, attractiveness – 4. Additionally we have the number of venues factor: $216/37772 = 0,0057$ (top 10% of all regions) – 5 after normalization, Is a bakery wanted? No, $x = 0$, lack od bakery? No, $y = 0$. So: $as = (4 + 5 + 2 + 4 + 4 + 5)/6 + 0 * 0.4 + 0 * 0.6 = 4.0$ *(recommended)*.

3.3   Spatial Co-location Rules

A very important part of OptiLocator is the determination of spatial co-location rules displaying links among different business categories. For discovering spatial co-location rules we use the FARICS algorithm [4]. In this method neighborhood is defined in terms of the Delaunay diagram (we create a Delaunay diagram for a set of

---

[1] Foursquare API, https://developer.foursquare.com.

points being businesses' locations and treat as neighbors of a given business those businesses that are linked by a direct line connection in the Delaunay diagram), so determination of a neighborhood does not require a distance parameter. The definition of a co-location rule proposed in [4] utilizes the notion of a group of types being any subset g of the set $T = \{t_1, t_2, ..., t_k\}$ of objects represented in the Delaunay diagram, where t_i represents one type of objects. To this end spatial co-location rule is a rule in the form of $g_1 \rightarrow g_2$ ($p\%$, $c\%$), where g_1 and g_2 are groups of types that do not intersect, $p\%$ is the prevalence of the rule computed in terms of participation index, $c\%$ denotes the confidence of the rule.

FARICS is performed separately for each region resulting from the clustering stage. During the tests we generated 3778 spatial co-location rules. They characterize the neighborhoods of sub-regions (parts of the Delaunay diagram) within the regions allowing for more precise location of a business with regard to the neighboring businesses.

3.4   Rules Selection

Having computed the co-location rules we now take them into consideration in the OptiLocator algorithm. In many different regions the same rules appear (e.g. Office → Coffee Shop), but with different prevalence and confidence values. The number of appearances of a given rule is treated as its "force". We believe that a rule appearing in many regions is more valuable even if it has lower values of prevalence and confidence than a rule that is present in just one region even if it has higher prevalence and confidence, as single instances of a rule may be mistakes [3]. Using our algorithm named *Rules Voting* we compute values basing on the rules to be considered in OptiLocator.

Rules Voting Algorithm

1. Select from the rules set all the rules whose consequent represents the selected category (the OptiLocator input parameter).
2. User defines the condition of rules usefulness – the minimum prevalence and confidence.
3. Combine all antecedents of the rules in one set – antecedents set.
4. Calculate the value showing the level of adjustment between the rules and the selected group of venues by means of the following formula:

$$rv\_value = (a\ /\ b) * (c\ /\ d) * 100\%$$

where a – number of unique objects in the region of types included in the set of antecedents, b – number of unique objects in the set of antecedents, c – number of occurrences of antecedents in the region, d – total number of venues in the region. It is worth to mention that the way of calculating rv_value promotes sub-regions having neighbors with many different venues (e.g. big shopping centers) as it makes probable that many antecedents are represented in a given neighbor and the values will be high.

Example 1. Let us assume we have a region with the venues: two Bakeries, one Office, one Bar, and one Subway station, and among the rules for the whole considered area we have the following rules: Bakery → Bank (0.5, 1.0), Pizza Place → Bank (0.33, 0.75). We are looking for the optimal location for the Bank. Evaluation of the region is as follows:

- >Percentage of antecedents coverage in the region. We have two antecedents: Bakery, Pizza Place. In the region only Bakery exists. First component is thus: $1/2 = 0.5$.
- Percentage of region coverage by the antecedents: the number of antecedents instances in the region is 2 ($2 \times$ Bakery), number of venues is 5. The second component is thus equal to $2/5 = 0.4$.

The final evaluation by Voting Rules is: $0.5 * 0.4 * 100\% = 20\%$. Interpretation: the analyzed region is a good location for the bank branch. The resultant value of 20% based on the experiments we conducted points to a good location. Average value for the *rv_value* parameter in the considered dataset was in the range 10–12%, 20% represents good locations.

4   Experimental Evaluation

To analyze users' activity in relation to city venues we used the New York Foursquare dataset mentioned in [1]. We complemented and enriched the data for use in OptiLocator, as the data was originally anonymized. The enrichment consisted in synthetic generation of users' profiles. For that purpose we used statistical data characterizing the population of New York as well as python scripts querying the Foursquare API and the Google Maps API. The final dataset consisted of 37772 venues (as depicted in Fig. 2), 225126 check-ins, 1083 users, 252 categories. The algorithm was implemented as a Django-based application.

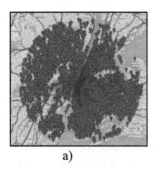
a)
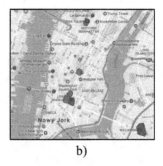
b)

Fig. 2. (a) All venues from the New York dataset (37772) marked in Google Maps, (b) Areas chosen by OptiLocator as optimum locations for: Bakery – bronze, hotel – blue, gym – orange (Color figure online)

Following the steps of OptiLocator we performed clustering on the prepared data. As a result we received 1133 groups of venues. We also calculated user paths (consecutive places visited by a given user). There are 50882 paths in the dataset, which means that on average a user has 50 paths covering 4.5 venues. User paths are used to compute some of the area characterization indicators.

The experiment was carried out for three different business categories: bakery, hotel and gym.

1. Bakery was selected for the experiments due to the nature of the business: continuous daily sales and relatively small size of business activity. The total number of check-ins for bakeries in the dataset: 1582. We discovered 21 co-location rules in which bakery is a consequent. Some rules we obtained are: Burger Joint → Bakery (0.5, 1), Drugstore/Pharmacy → Bakery (0.2, 0.5), American Restaurant → Bakery (0.25, 0.33). Based on the rules we received the following set of neighbors for bakery: Burger Joint, Bank, Electronics Store, Subway, Drugstore/Pharmacy, Scenic Lookout, Sandwich Place, Salon/Barbershop, American Restaurant, Bar, Pizza Place, Office.
2. Hotel was selected for the experiments due to tourist demand. This is the category with very high popularity. Check-ins in hotels were performed 2851 times for the test dataset. We discovered 31 co-location rules in which hotel is a consequent. Based on the rules we received the following set of neighbors for hotel: Cosmetics Shop, Convention Center, Sandwich Place, Residential Building, Electronics Store, Clothing Store, Deli/Bodega, Office Supplies, Japanese Restaurant, Office Bar, Cafe, Bank.
3. Gym was selected for the experiments due to high popularity and a lot of competition in New York City. Check-ins in gyms were performed 2851 times for the test dataset. We discovered 50 co-location rules in which gym is a consequent. Based on the rules we received the following set of neighbors for gym: Convention Center, Asian Restaurant, Sandwich Place, Bank, Coffee Shop, Cosmetics Shop Salon/Barbershop, Cafe, Bar, Spa/Massage, Clothing Store, Plaza, Vietnamese Restaurant, Neighborhood, Electronics Store, American Restaurant, Residential Building, Shop & Service, Bus Station, Event Space, Office, Pizza Place, Diner, Entertainment.

Determination of the Optimal Location. Top recommendations for localization of bakery, hotel and gym are given in Table 1. Rows having yellow shading are not recommended due to low values of area score measure. Values in the table are computed as described in the *determination of the optimal location* step of Sect. 3. Such selection procedure has the following grounds: we favor neighborhood information described by Rules Voting value over Area Score relating to the social characteristics of a given region on the grounds that for a random person the neighborhood is very important, because they tend to group their activities and neighborhood reflects their preferences in this regard in a given area/region. So the location of a new business has to be correlated with the preferences of the local community in terms of routines. The second important factor is the characteristic of the area in terms of social behaviors. The value of 4 for area score means the place is recommended (as described in Sect. 3.2). For that reason places with area scores below 4 are marked as not recommended for placing the selected business. The map presented in Fig. 2b) contains all recommended regions for the localization of bakery, hotel and gym.

Table 1. Top recommendations for localization of bakery, hotel and gym.

Recommended localization for	Region name	Neighborhood	Rules Voting	Area Score
bakery	Times Square	Subway, Bar, Office	40,08%	4,40
	Broadway-Lafayette St	Bank, Pizza Place, Subway	39,65%	4,17
	Park Ave - 32nd St	Pizza Place, Subway, Bar	37,93%	3,33
hotel	Broadway – Prince St	Cafe, Office	57,82%	4,43
	Union Square	Bar, Convention Center	36,21%	4,10
	Bryant Park	Event Space, Cafe	34,78%	3,17
gym	Grand Central Terminal	Bus Station, Coffee Shop	47,45%	3,93
	Nolita	Deli / Bodega, General Entertainment	43,67%	4,60
	Union	Bank, Plaza, Clothing Store	41,31%	4,60

5   Conclusions

In this paper we presented a new method for finding optimal business location called OptiLocator. Our solution is based on location-based social network data and spatial data mining methods. The final recommendation is done based on two different indicators: the area score and the rules voting score. The former indicator is derived from analysis of people's behavior whereas the latter reflects spatial relations between different types of businesses.

The experiments were performed for selected types of businesses and obtained results were subject to manual verification. In most cases, the results were reasonable and consistent with reality.

In the proposed approach we don't take into consideration the financial aspects such as rental prices. Here, the main problem is the lack of a reliable, up-to-date data source that could be explored automatically. Such factors can be introduced to the input data as a parameter for business categories (e.g. profitability level) and then can be taken into account in the calculation of the area score. It would raise the quality and veracity of the area characteristics.

References

1. Yang, D., et al.: Modeling user activity preference by leveraging user spatial temporal characteristics in LBSNs. IEEE Trans. Syst. Man Cybern. Syst. **45**(1), 129–142 (2015)
2. Karamshuk, D., Noulas, A., Scellato, S., Nicosia, V., Mascolo, C.: Geo-spotting: mining online location-based services for optimal retail store placement. In: Proceedings of the 19th ACM SIGKDD International Conference on Knowledge Discovery and Data Mining. ACM (2013)
3. Weiler, M., Schmid, K.A., Mamoulis, N., Renz, M.: Geo-social co-location mining. In: Second International ACM Workshop on Managing and Mining Enriched Geo-Spatial Data, pp. 19–24. ACM (2015)

4. Bembenik, R., Rybiński, H.: FARICS: a method of mining spatial association rules and collocations using clustering and Delaunay diagrams. J. Intell. Inf. Syst. **33**(1), 41–64 (2009)
5. Ester, M., Kriegel, H.P., Sander, J., Xu, X.: A density-based algorithm for discovering clusters in large spatial databases with noise. In: KDD, vol. 96, no. 34, pp. 226–231 (1996)
6. Hernandez, T., Bennison, D.: The art and science of retail location decisions. Int. J. Retail Distrib. Manag. **28**(8), 357–367 (2000)
7. Koperski, K., Han, J.: Discovery of spatial association rules in geographic information databases. In: Egenhofer, M.J., Herring, J.R. (eds.) SSD 1995. LNCS, vol. 951, pp. 47–66. Springer, Heidelberg (1995). doi:10.1007/3-540-60159-7_4

Activity Recognition Model Based on GPS Data, Points of Interest and User Profile

Igor da Penha Natal$^{(\boxtimes)}$, Rogerio de Avellar Campos Cordeiro,
and Ana Cristina Bicharra Garcia

Institute of Computing, Fluminense Federal University, Niteroi, RJ, Brazil
igorpnatal@gmail.com, {rogerio.avellar,bicharra}@ic.uff.br

Abstract. The problem of activity recognition is a topic that has been explored in the field of ubiquitous computing, the popularization of sensors on the most diverse types has been instrumental in improving the effectiveness of recognition. Smartphones offer a range of sensors (GPS, Accelerometer, Gyroscopes, etc.) that can be used to provide data for this type of problem. This work proposes a new model of activity recognition in GPS captured data and enriched with POIs (Points Of Interest) and user profile. The experiment was performed by 10 volunteers collecting data for 10 days. The model aims to recognize 13 different activities, divided into stop activities (Bank, breakfast, dining, lunch, praying, recreation, shopping, studying, waiting transport and working) and moves activities (in a car, on a bus and walking). The model was tested and compared to J48, SVM, ANN and RF algorithms, and obtained 97.4% hits.

Keywords: GPS · POIs · Activity recognition · User profile

1 Introduction

In the field of ubiquitous computing, activity recognition coupled with the popularization of sensors in smartphones (Accelerometer, Gyroscope, GPS (Global Positioning System), etc.) have become an increasingly explored subject. Several studies aim at activity recognition with data from these sensors [1–3] and other studies try to understand the data produced by Smartphone users during the continuous use of the device [4–6].

The ability to monitor people through non-invasive sensors becomes increasingly interesting, with people feeling a bit afraid of having sensors attached to their bodies or sharing their information directly. The use of sensor data from Smartphones has been applied in various types of applications: monitoring, healthcare, activity recognition, etc. [3].

The problem of activity recognition is being explored in literature and one of the main challenges is knowing which sensors and data are collected to refine the automatic recognition of activities. The most common sensors used are the GPS and the accelerometer [2].

© Springer International Publishing AG 2017
M. Kryszkiewicz et al. (Eds.): ISMIS 2017, LNAI 10352, pp. 358–367, 2017.
DOI: 10.1007/978-3-319-60438-1_35

There is no sensor capable of specifically defining what activity the user is doing. For the recognition of activities, an analysis of sensor data (GPS, Accelerometer, Gyroscope, etc.) is necessary to infer what activity the user is doing. The use of one sensor is generally not efficient when it comes to coverage of activities, due to limitations of Indoor or Outdoor activities.

This work proposes a model for the recognition of activities by enriching data from GPS of Smartphones with information from the internet (POIs - Points of Interest) and User Profile (Profession and if he has a car). This allows for greater accuracy than other algorithms in recognizing activities. It additionally maximizes coverage of activities in relation to the other works.

This paper is structured as follows. The related works to the proposed theme are in Sect. 2. In Sect. 3, the model is formulated and user in the work is described. Section 4 presents how the experiment was conducted, describing the data collection and the database used. Section 5 discusses the obtained results. Lastly, Sect. 6 presents the conclusions of the paper and some proposals for future works.

2   Related Work

The works presented in this section refers to the understanding of the data collected from the various Smartphones sensors (GPS, accelerometer and gyroscope) and also presents works that use this data for the activity recognition.

The work developed by [4] aimed to analyze and understand accelerometer data. The RF (Random Forest) model was used as the basis of work to automatically classify the data in different modes of motion. An evaluation was made with data from 12 people collected in a 6 day period. The recognized types of motion were: walking, running, cycling, in a vehicle and stopped. The results presented show good performance for both indoor and outdoor environments.

The work of [5] used the data collected from Smartphone sensors to extract characteristics for classifying of the type of transport that the user is in. The authors conducted the experiment by monitoring 15 people for one month in the city of Kobe, Japan. The algorithm used was the RF and the recognized classes were: walking, bicycle, car, bus, train and subway. The authors considered these as only part of the outdoor environment. The work presented a methodology for collecting accelerometer and GPS data taking into account different frequencies of data collection and measuring the computational cost of the algorithm.

The focus of work [6] was to detect the subjects' mode of transportation through the use of the accelerometer. The authors used concepts of segmentation for trajectories and separation between segments of different means of transport. The data was collected in Zurich (Switzerland), with 396 samples from 30 users. The work focused on detecting the following means of transportation: Walking, Car, Train and Bus. Considering these as only part of the outdoor environment. The authors concluded that the incorporation of site-type information combined with their work might impact the phenomen driven by human mobility and increase people's awareness of behavior, urban planning and agent-based modeling.

Work [1] used GPS data and POIs for activity recognition of patients with schizophrenia in outdoor environments. These, subjects had a diary to record their activities. Two methods were implemented for activity recognition, one based on time (5, 6, 8 and 10 min increments) and another based on location (radius of 50 m). Ultimately they used semantic enrichment to classify types of places and activities associated with them. The experiment was carried out with the participation of 7 people and the data collection of 5 days. The recognized activities were linked to areas of social functioning and were as follows: Work, Shopping, Sports, Social Activities, Recreational Activities and Others. The authors concluded that the use of semantic enrichment could improve refinement in activity recognition.

The work of [2] uses accelerometers and gyroscopes from Smartphones to recognize physical activities (standing, walking, running, upstairs, downstairs and lying). The authors extracted features of the two sensors and realized that the gyroscope offers more satisfactory data in dynamic activities (walking, running, upstairs and downstairs). The experiment collected data from 30 volunteers. Three algorithms were used to classify activities: SVM (Support Vector Machine), J48 and Logistic Regression.

The work of [3] dealt with the recognition of activities through GPS data, considering the trajectory of the user. The study was concerned with more accurately dealing with less frequent activities in the database. The experiment collected data from 100 volunteers in Beijing (China) for 10 days. The study environment was outdoor and the recognized activities were: at home, at work, visiting friends, shopping, dining out, going to the theater, searching for something or someone, entertaining and others. The approach presented by the authors was rooted in a time-based cost function to improve the accuracy of activities with few records. The function was implemented in a HMM (Hidden Markov Model) and with this, it was possible to balance results in relation to the number of recorded activities.

The aim of work [7] was to make the recognition of activities through GPS data and use of POIs. The model presented targets division of the users' trajectories in order to analyze the nearest POIs and subsequently recognize the activity. The experiment collected data from 10 volunteers for 30 days. The recognized activities were: Education, Sustenance, Transportation, Finance, Healthcare and Entertainment, Arts & Culture. The trajectories were analyzed and a clustered by the nearest POIs is made in order to recognize the activities.

The main difference of the work proposed here is the creation of a deterministic model for the recognition of activities independent of the environment (indoor or outdoor) with data coming from GPS of the Smartphone and enriched by period of day, type of day, average speed of the trajectory, duration of the activity, distance from the previous point, POIs, the person's profession and if the person has a vehicle. This model seeks to maximize the coverage of activities, attending to a greater number of activities in relation to the references presented.

3   Proposed Model

The proposed model seeks to recognize activities in a deterministic way through GPS data, POIs (GooglePlaces[1]) and user profile. For this to work, a significant change of velocity was used as a parameter to detect the change of activity.

Table 1. Variables and values of POI and user profile data

Variable	Value	Variable	Value
POI_Category	EDUCATION	Profession	Academic
	ENTERTAINMENT		Adm_Tech
	FINANCIAL		Professor
	FOOD	Has_Car	Yes
	HOME		No
	OTHERS		
	RELIGION		
	SHOPPING		
	TRANSPORT		

Table 2. Variables, range of values and discretization of GPS data

Variable	Range of values	Discretization
Day	Monday to Friday	Weekday
	Weekend and Holiday	Weekend
Ini_Hour/End_Hour	05:30 a.m–11:59 a.m	Morning
	12:00 p.m–05:59 p.m	Afternoon
	06:00 p.m–05:29 a.m	Night
TotalDuration (x)	$x <= 20$ (min)	Low
	$20 < x <= 40$ (min)	Medium
	$x > 40$ (min)	Long
Dist_Preview_Point (x)	$x = 0$ (m)	Null
	$x <= 250$ (m)	Low
	$250 < x <= 450$ (m)	Medium
	$x > 450$ (m)	High
AverageSpeed (x)	$x = 0$ (KM/H)	Null
	$x <= 5$ (KM/H)	Low
	$5 < x <= 30$ (KM/H)	Medium
	$x > 30$ (KM/H)	High

---

[1] https://developers.google.com/places/.

The variables used are described as follows: Day is the day of the week, Ini_Hour is the turn that the activity started, End_Hour is the turn that the activity ended, TotalDuration is the total duration of the activity, Dist_Preview_Point is The distance from the current point to the previous point, AverageSpeed is the user's average speed, POI_Category is the closest POI category of the local, Profession is the user's profession and Has_Car determines if the user has a car. All variables were discretized so that rules would be created on them. The form of discretization with the possible values for each of the variables is described in Table 1 (POI and user profile data) and in Table 2 (GPS data).

The model was generated to recognize 13 types of activities that were separated into 2 groups (Move Activities and Stop Activities), highlighted in Table 3.

Table 3. Activity groups

Move activities	Stop activities
IN_BUS	BANK
IN_CAR	BREAKFAST
WALKING	DINING
	LUNCHING
	PRAYING
	RECREATION
	SHOPPING
	STUDYING
	WAITING_TRASP
	WORKING

The condition tree of the model can be seen in Fig. 1 where the circles represent the variables, the intermediate labels are the values and the rectangles are the activities. The model contains rules using at least 1 variable (If POI_Category is FINANCIAL then activity is BANK) and at most 5 variables (If POI_Category is OTHERS and If Has_Car is Yes and If AverageSpeed is Low and TotalDuration is Medium and Profession is Academic then activity is IN_CAR).

4   Experiment

The first step of the experiment was to collect the GPS data, which was collected through an application (in Portuguese) developed by the authors for Smartphones with the Android platform. The platform was chosen based on the profile analysis of the volunteers. The application collected the GPS data (Longitude and Latitude), the system date and time, a user identifier and the activity performed (if the user filled the label), the application interface is shown in Fig. 2,

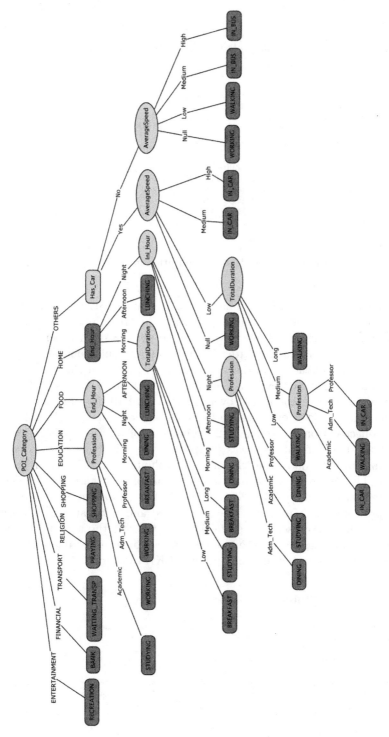

Fig. 1. Proposed model

Fig. 2. Data collection application interface

where at the top shows the user ID, underneath the button (SENSORIAMENTO DESATIVADO) to activate the sensing (activate the GPS), underneath a label for the user to register the activity, a button to save (SALVAR) and another to maintain (MANTER) in case after 30 min the user would continue with the same activity and feedback from the user's records recorded at the bottom. The app would pick up the data every 3 min, or if the user pressed the save button.

The experiment was conducted over the course of 10 days in the cities of Niteroi and Campos dos Goytacazes in the state of Rio de Janeiro (Brazil). The data was collected from a sample of 20 volunteers of both sexes between the ages 25 of 40. The data utilized was from 10 of the volunteers due to technical problems with the data from the other 10 volunteers, which were rendered incomplete. The data went through a pre-processing phase, where it was possible to infer the following data: day, start time, end time, total duration, distance from the previous point, average trajectory speed and POI category, together with profile data of the user as a profession and whether the user has car and class activity. Some examples of the treated data can be seen in Table 4. We used 1998 samples and the tests were done using WEKA software.

Table 4. Examples of data processed

Day	Ini_Hour	End_Hour	Total duration	Dist_Preview_Point	Vel_Media	POI_Category	Profession	Has_Car	ACTIVITY
Weekday	Afternoon	Afternoon	Low	Null	Null	FINANCIAL	Professor	Yes	BANK
Weekday	Morning	Morning	Low	High	Medium	OTHERS	Academic	No	IN_BUS
Weekday	Morning	Afternoon	Low	Null	Null	HOME	Academic	Yes	LUNCH
Weekend	Morning	Afternoon	Long	Null	Null	ENT	Professor	Yes	RECR

(a) weka.classifiers.trees.J48

a	b	c	d	e	f	g	h	i	j	k	l	m	<-- classified as
16	0	0	0	0	0	0	0	0	0	0	0	0	a = BANK
0	29	0	0	0	0	0	0	0	0	0	0	0	b = BREAKFAST
0	0	17	0	0	0	0	0	0	0	0	0	0	c = DINING
0	0	0	168	11	0	0	0	0	0	0	0	0	d = IN_BUS
0	0	0	0	741	0	0	0	0	0	0	0	0	e = IN_CAR
0	0	0	0	0	148	0	0	0	0	0	0	0	f = LUNCHING
0	0	0	0	0	0	10	0	0	0	0	0	0	g = PRAYING
0	0	0	0	0	0	0	10	0	0	0	0	0	h = RECREATION
0	0	0	0	0	0	0	0	23	0	0	0	0	i = SHOPPING
0	0	0	0	0	0	0	0	0	70	0	0	0	j = STUDYING
0	0	30	0	0	0	0	0	0	18	22	0	0	k = WAITING_TRANSP
0	0	0	0	0	0	0	0	0	0	0	28	0	l = WALKING
0	0	0	0	50	20	0	0	0	0	0	0	225	m = WORKING

Correctly Classified Instances   1455   83.7169 %
Incorrectly Classified Instances 283   16.2831 %

(b) weka.classifiers.trees.RandomForest

a	b	c	d	e	f	g	h	i	j	k	l	m	<-- classified as
16	0	0	0	0	0	0	0	0	0	0	0	0	a = BANK
0	29	0	0	0	0	0	0	0	0	0	0	0	b = BREAKFAST
0	39	17	0	0	0	0	0	0	0	0	0	0	c = DINING
0	0	0	168	11	0	0	0	0	0	0	0	0	d = IN_BUS
0	0	0	49	692	0	0	0	0	0	0	0	0	e = IN_CAR
0	0	0	0	0	40	0	0	0	0	0	0	0	f = LUNCHING
0	0	0	0	0	0	10	0	0	0	0	0	0	g = PRAYING
0	0	0	0	0	0	0	10	0	0	0	0	0	h = RECREATION
0	0	0	0	0	0	0	0	23	0	0	0	0	i = SHOPPING
0	0	30	0	0	0	0	0	0	37	0	0	0	j = STUDYING
0	0	0	0	0	0	0	0	0	0	22	0	0	k = WAITING_TRANSP
0	0	0	0	0	0	0	0	0	0	0	27	0	l = WALKING
0	0	0	0	50	0	0	0	0	0	0	0	225	m = WORKING

Correctly Classified Instances   1316   75.7192 %
Incorrectly Classified Instances 422   24.2808 %

(c) weka.classifiers.functions.SMO

a	b	c	d	e	f	g	h	i	j	k	l	m	<-- classified as
16	0	0	0	0	0	0	0	0	0	0	0	0	a = BANK
0	29	0	0	0	0	0	0	0	0	0	0	0	b = BREAKFAST
0	20	17	0	0	0	0	0	0	0	0	0	0	c = DINING
0	0	0	168	11	0	0	0	0	0	0	0	0	d = IN_BUS
0	0	0	0	741	0	0	0	0	0	0	0	0	e = IN_CAR
0	0	0	0	9	31	0	0	0	0	0	0	0	f = LUNCHING
0	0	0	0	0	0	10	0	0	0	0	0	0	g = PRAYING
0	0	0	0	0	0	0	10	0	0	0	0	0	h = RECREATION
0	0	10	0	0	0	0	0	23	0	0	0	0	i = SHOPPING
0	0	0	0	0	0	0	0	10	49	0	0	0	j = STUDYING
0	0	0	0	0	0	0	0	0	0	22	0	0	k = WAITING_TRANSP
0	0	0	0	0	0	0	0	0	0	0	28	0	l = WALKING
0	0	0	0	20	0	0	0	0	0	0	0	225	m = WORKING

Correctly Classified Instances   1406   80.8976 %
Incorrectly Classified Instances 332   19.1024 %

(d) weka.classifiers.functions.MultilayerPerceptron

a	b	c	d	e	f	g	h	i	j	k	l	m	<-- classified as
16	0	0	0	0	0	0	0	0	0	0	0	0	a = BANK
0	29	0	0	0	0	0	0	0	0	0	0	0	b = BREAKFAST
0	20	17	0	0	0	0	0	0	0	0	0	0	c = DINING
0	0	0	168	11	0	0	0	0	0	0	0	0	d = IN_BUS
0	0	0	0	741	0	0	0	0	0	0	0	0	e = IN_CAR
0	0	0	0	0	31	0	0	0	0	0	0	0	f = LUNCHING
0	0	0	0	0	0	10	0	0	0	0	0	0	g = PRAYING
0	0	0	0	0	0	0	10	0	0	0	0	0	h = RECREATION
0	0	10	0	0	0	0	0	23	0	0	0	0	i = SHOPPING
0	0	0	0	0	0	0	0	10	49	0	0	0	j = STUDYING
0	0	0	0	0	0	0	0	0	0	22	0	0	k = WAITING_TRANSP
0	0	0	0	0	0	0	0	0	0	0	28	0	l = WALKING
0	0	0	0	0	0	0	0	0	0	0	0	225	m = WORKING

Correctly Classified Instances   1369   78.7687 %
Incorrectly Classified Instances 369   21.2313 %

Fig. 3. Confusion matrix and accuracy of the algorithm (a) J48, (b) RF, (c) SVM and (d) ANN

The experiment was done using the ANN (Artificial Neural Network - Multilayer Perceptron), RF, J48 (Decision Tree) and SVM algorithms, with 260 training samples (20 examples from each class) and 1738 samples for testing. The presented model was used to classify the same test samples from the other algorithms. Finally, the results were compared.

5   Results

The results of the experiment show that the results achieved by the algorithms were higher than 75% hits, J48 and SVM with results higher than 80%, with 83.7% and 80.9% respectively. Figure 3 shows the confusion matrix, where the numbers of the main diagonal are the correctness of classification of the algorithm. There is also information about the number of correctly and erroneously classified instances and their percentages. It may be noted that there were a lot of mistakes confusing LUNCHING and STUDYING activities.

```
a   b   c   d   e   f   g   h   i   j   k   l   m   <-- classified as
16   0   0   0   0   0   0   0   0   0   0   0   0 |   a = BANK
 0  51   0   0   0   0   0   0   0   0   0   0   0 |   b = BREAKFAST
 0   0  56   0   0   0   0   0   0   0   0   0   0 |   c = DINING
 0   0   0 168  11   0   0   0   0   0   0   0   0 |   d = IN_BUS
 0   0   0   0 748   0   0   0   0   0   0   5   0 |   e = IN_CAR
 0   0   0   0   0 218   0   0   0   0   0   0   0 |   f = LUNCHING
 0   0   0   0   0   0  10   0   0   0   0   0   0 |   g = PRAYING
 0   0   0   0   0   0   0  10   0   0   0   0   0 |   h = RECREATION
 0   0   0   0   0   0   0   0  23   0   0   0   0 |   i = SHOPPING
 0   0   0   0   0  29   0   0   0  48   0   0   0 |   j = STUDYING
 0   0   0   0   0   0   0   0   0   0  22   0   0 |   k = WAITING_TRANSP
 0   0   0   0   0   0   0   0   0   0   0  28   0 |   l = WALKING
 0   0   0   0   0   0   0   0   0   0   0   0 295 |   m = WORKING
Correctly Classified Instances          1693              97.4108 %
Incorrectly Classified Instances          45               2.5892 %
```

Fig. 4. Confusion matrix and accuracy of the proposed model

The results using the model were above 95% (97.4%), showing the efficiency of the model, considering the experiment carried out. Figure 4 shows the results of the model in the same format as the other algorithms. In relation to J48, RF, SVM and ANN, the model obtained 238 (13.7%), 377 (21.7%), 287 (16.5%) and 324 (18.6%) hits more respectively algorithm. This promises to be better than traditional algorithms for this type of classification. It is important to note that model errors also occurred in the classification confusion of LUNCHING and STUDYING activities.

6   Conclusions and Future Works

The authors proposed a new activity recognition model based on GPS information, POIs and user profile. The model was assembled and tested in several ways until the final version was presented. The results showed the efficiency of the model in relation to the other traditional algorithms for the task of recognition of activities and also considered the maximization of the coverage of activities.

The limitation of the work was becoming very specific to the experiment in some situations of the model, but the model was easily adapted to generalize these situations in different experiments, so that the specificities happen in a very small number of situations.

As for proposals for future work, the authors intend to verify other variables that may be included in the model to join a greater coverage of activities and test the model in an adaptive way in order for databases that contain the necessary information to operate the same, verifying the efficiency in a different context.

References

1. Difrancesco, S., Fraccaro, P., Veer, S.N., Alshoumr, B., Ainsworth, J., Bellazzi, R., Peek, N.: Out-of-home activity recognition from GPS data in schizophrenic patients. In: IEEE 29th International Symposium on Computer-Based Medical Systems, pp. 324–328. IEEE (2016)
2. Hung, W.-C., Shen, F., Wu, Y.-L., Hor, M.-K., Tang, C.-Y.: Activity recognition with sensors on mobile devices. In: International Conference on Machine Learning and Cybernetics, pp. 449–454. IEEE (2014)
3. Huang, W., Li, M., Hu, W., Song, G., Xing, X., Xie, K.: Cost sensitive GPS-based activity recognition. In: 10th International Conference on Fuzzy Systems and Knowledge Discovery, pp. 962–966. IEEE (2013)
4. Zhou, X., Yu, W., Sullivan, W.C.: Making pervasive sensing possible: effective travel mode sensing based on smartphones. Comput. Environ. Urban Syst. **58**, 52–59 (2016)
5. Shafique, M.A., Hato, E.: Travel mode detection with varying smartphone data collection frequencies. Sensors **16**(5), 716 (2016)
6. Shin, D., Aliaga, D., Tunçer, B., Arisona, S.M., Kim, S., Zund, D., Schmitt, G.: Urban sensing: using smartphones for transportation mode classification. Comput. Environ. Urban Syst. **53**, 76–86 (2015)
7. Boukhechba, M., Bouzouane, A., Bouchard, B., Gouin-Vallerand, C., Giroux, S.: Online recognition of people's activities from raw GPS data: semantic trajectory data analysis. In: Proceedings of the 8th ACM International Conference on Pervasive Technologies Related to Assistive Environments. ACM (2015)

Extended Process Models for Activity Prediction

Stefano Ferilli[1]([✉]), Floriana Esposito[1], Domenico Redavid[2],
and Sergio Angelastro[1]

[1] Dipartimento di Informatica, Università di Bari, Bari, Italy
{stefano.ferilli,floriana.esposito,sergio.angelastro}@uniba.it
[2] Artificial Brain S.r.l., Bari, Italy
redavid@abrain.it

Abstract. In addition to the classical exploitation as a means for checking process enactment conformance, process models may be used to predict which activities will be carried out next. The prediction performance
may provide indirect indications on the correctness and reliability of a
process model. This paper proposes a strategy for activity prediction
using the WoMan framework for workflow management. It extends a previous approach, that has proved to be able to handle complex processes.
Experimental results on different domains show an increase in prediction
performance compared to the previous approach.

Keywords: Process mining · Activity prediction · Process model

1   Introduction and Background

A *process* consists of actions performed by agents [1,2]. A *workflow* is a formal
specification of a process. It may involve sequential, parallel, conditional, or iterative execution [11]. A process execution, compliant to a given workflow, is called
a *case*. It can be described as a list of *events* (i.e., identifiable, instantaneous
actions, including decisions upon the next activity to be performed), associated
to *steps* (time points) and collected in *traces* [12]. Relevant events are the start
and end of process executions, or of activities [2]. A *task* is a generic piece of
work, defined to be executed for many cases of the same type. An *activity* is the
actual execution of a task by a *resource* (an agent that can carry it out).

Process Management techniques are useful in domains where a production
process must be monitored (e.g. in the industry) in order to check whether the
actual behavior is compliant with a desired one. When a process model is available, new process enactments can be automatically supervised. The complexity
of some domains requires to learn automatically the process models, because
building them manually would be very complex, costly and error-prone. Process
Mining [8,12] approaches aim at solving this problem. *Declarative* process mining
approaches learn models expressed as a set of constraints, instead of a monolithic
model (usually expressed as some kind of graph) [10].

M. Kryszkiewicz et al. (Eds.): ISMIS 2017, LNAI 10352, pp. 368–377, 2017.
DOI: 10.1007/978-3-319-60438-1_36

The WoMan framework [4,5] lies at the intersection between *Declarative Process Mining* and Inductive Logic Programming (ILP) [9]. Indeed, it pervasively uses First-Order Logic as a representation formalism, that provides a great expressiveness potential and allows one to describe contextual information using relationships. Experiments proved that WoMan can handle efficiently and effectively very complex processes, thanks to its powerful representation formalism and process handling operators. Differently from all previous approaches in the literature, it is *fully incremental*: not only can it refine an existing model according to new cases whenever they become available, it can even start learning from an empty model and a single case, while others need a (large) number of cases to draw significant statistics before learning starts. This allows to carry out continuous adaptation of the learned model to the actual practice efficiently, effectively and transparently to the users [4].

A relevant issue in Process Management in general, and in Process Mining in particular, is to assess how well can a model provide hints about what is going on in the process execution, and what will happen next. Indeed, given an intermediate status of a process execution, knowing how the execution will proceed might allow the (human or automatic) supervisor to take suitable actions that facilitate the next activities. The task of activity prediction may be stated as follows: given a process model and the current (partial) status of a new process execution, guess which will be the next activity that will take place in the execution. In industrial environments, the rules that determine how the process must be carried out are quite strict; so, predicting the process evolutions is a trivial consequence of conformance checking. Other, less traditional application domains (e.g., the daily routines of people at home or at work, seen as a process), involve much more variability, and obtaining reliable predictions becomes both more difficult and more useful. Another, very relevant and interesting, application of process-related predictions is in the assessment of the quality of a model. Indeed, since models are learned automatically exactly because the correct model is not available, only an empirical validation can be run. In literature, this is typically done by applying the learned model to new process enactments.

In particular, this paper focuses on the approach adopted by WoMan [6] to carry out the activity prediction task, given its good results in various domains. We propose here two new contributions: first, we extend the Woman's process model with additional information aimed at improving the prediction performance; second, we report for the first time the detailed prediction algorithm. The next two sections present WoMan, its (extended) formalism and its approach to activity prediction. Then, Sect. 4 reports and comments about the experimental outcomes. Finally, in the last section, we draw some conclusions and outline future work issues.

2 The WoMan Formalism

WoMan representations are based on the Logic Programming formalism, and works in Datalog, where only constants or variables are allowed as terms.

Following foundational literature [1, 7], trace elements in WoMan are 6-tuples, represented in WoMan as facts $\texttt{entry}(T, E, W, P, A, O)$., that report information about relevant events for the case they refer to. T is the event timestamp, E is the type of the event, W is the name of the workflow the process refers to, P is a unique identifier for each process execution, A is the name of the activity, and O is the progressive number of occurrence of that activity in that process. An optional field, R, can be added to specify the agent that carries out activity A. In particular, E is one of {**begin|end**}_{**process|activity**}. Activity begin and end events are needed to properly handle time span and parallelism of tasks [12]. Since parallelism among activities is explicit, there is no need for inferring it by means of statistical (possibly wrong) considerations. In each case, the activities are uniquely identified by a progressive number called *step*. $E = $ **context_description** is used to describe contextual information at time T, in the form of a conjunction of FOL atoms built on domain-specific predicates.

WoMan models are expressed as sets of atoms built on different predicates. The predicates used in previous versions:

- $\texttt{task}(t, C)$: task t occurred in training cases C.
- $\texttt{transition}(I, O, p, C)$: transition[1] p, occurred in training cases C, is enabled if all input tasks in $I = [t'_1, \ldots, t'_n]$ are active; if fired, after stopping the execution of all tasks in I (in any order), the execution of all output tasks in $O = [t''_1, \ldots, t''_m]$ is started (again, in any order). If several instances of a task can be active at the same time, I and O are multisets, and application of a transition consists in closing as many instances of active tasks as specified in I and in opening as many activations of new tasks as specified in O.
- $\texttt{task\_agent}(t, A)$: an agent, matching the roles A, can carry out task t.
- $\texttt{transition\_agent}([a'_1, \ldots, a'_n], [a''_1, \ldots, a''_m], p, C, q)$: transition p, involving input tasks $I = [t'_1, \ldots, t'_n]$ and output tasks $O = [t''_1, \ldots, t''_m]$, may occur provided that each task $t'_i \in I, i = 1, \ldots, n$ is carried out by an agent matching role a'_i, and that each task $t''_j \in O, j = 1, \ldots, m$ is carried out by an agent matching role a''_j; several combinations can be allowed, numbered by progressive q, each encountered in cases C.

The core of the model is expressed by $\texttt{task/2}$ and $\texttt{transition/4}$. The latter, in particular, represent the allowed connections between activities in a very modular way. This allows WoMan to check if a new execution corresponds to at least one training case, or if it can be obtained by mixing transitions taken from different training cases. Since a task or transition t may occur many times in the same case, C is defined as a multiset. It allows WoMan to compute the probability of t by means of its relative frequency in cases (as shown in [4]). In this way, WoMan can ignore less frequent $t's$, since they could be noisy. Also, the multiplicity of a case in C allows to set a limit on the number of repetitions of t during an execution (avoiding longer loops than expected). In this work, the set of predicates was extended to express *temporal constraints* on the activities:

---

[1] Note that this is a different meaning than in Petri Nets.

- $\texttt{task\_time}(t,[b',b''],[e',e''],d)$: task t must begin at a time $i_b \in [b',b'']$ and end at a time $i_e \in [e',e'']$, and has average duration d;
- $\texttt{transition\_time}(p,[b',b''],[e',e''],g,d)$: transition p must begin at a time $i_b \in [b',b'']$ and end at a time $i_e \in [e',e'']$; it has average duration d and requires an average time gap g between the end of the last input task in I and the activation of the first output task in O;
- $\texttt{task\_in\_transition\_time}(t,p,[b',b''],[e',e''],d)$: task t, when run in transition p, must begin at a time $i_b \in [b',b'']$ and end at a time $i_e \in [e',e'']$, and has average duration d;
- $\texttt{task\_step}(t,[b',b''],[e',e''],d)$: task t must start at a step $s_b \in [b',b'']$ and end at a step $s_e \in [e',e'']$, along an average number of steps d;
- $\texttt{transition\_step}(p,[b',b''],[e',e''],g,d)$: transition t must start at a step $s_b \in [b',b'']$ and end at a step $s_e \in [e',e'']$, along an average number of steps d and requires an average gap of steps g between the end of the last input task in I and the activation of the first output task in O;
- $\texttt{task\_in\_transition\_step}(t,p,[b',b''],[e',e''],d)$: task t, when run in transition p, must start at a step $s_b \in [b',b'']$ and end at a step $s_e \in [e',e'']$, along an average number of steps d.

These temporal constraints are mined on the entire training set. Begin and end time are relative to the start of the process execution, computed as the timestamp difference between the begin of process and the event they refer to. Step information is computed on the progressive number s on which a task t is executed. Consider, for instance, task $act\_Meal\_Preparation$ and transition $p23 : \{act\_Meal\_Preparation\} \Rightarrow \{act\_Meal\_Preparation, act\_Relax\}$; an example of the new components for them might be:

$\texttt{task\_time}(act\_Meal\_Preparation,\ [5, 10],\ [10, 21], 8.62)$
$\texttt{transition\_time}(p23,\ [6, 8],\ [20, 25], 17.34, 10.12)$
$\texttt{task\_step}(act\_Meal\_Preparation,\ [s3, s5],\ [s7, s12], 2.5)$
$\texttt{transition\_step}(p23,\ [s4, s5],\ [s10, s13], 8.3, 3.3)$
$\texttt{task\_in\_trans\_step}(act\_Meal\_Preparation, p23,\ [s10, s11],\ [s11, s12], 1.4)$
$\texttt{task\_in\_trans\_time}(act\_Meal\_Preparation, p23,\ [10, 12],\ [18, 21], 8.3)$

Finally, WoMan can expresses pre/post conditions (that specify what must be true for executing a given task, transition or a task in the context of a specific transition) as FOL rules based on contextual and control flow information. Conditions are not limited to the current status of execution. They may involve the status at several steps using two predicates:

- $\texttt{activity}(s,t)$: at step s (unique identifier) t is executed;
- $\texttt{after}(s',s'',[n',n''],\ [m',m''])$: step s'' follows step s' after a number of steps ranging between n' and n'' and after a time ranging between m' and m''.

Due to concurrency, predicate $\texttt{after/3}$ induces a partial ordering on the set of steps.

3   Workflow Prediction

WoMan's supervision module, **WEST** (Workflow Enactment Supervisor and Trainer) [6], checks whether new cases are compliant with a given model, returning suitable warnings (deviations from model) or errors (case trace syntactically wrong) in case it is not. In [6], warnings concerned unexpected tasks or transitions, conditions not fulfilled, unexpected resources running a given activity. Here, additional warnings were considered, expressing deviations from the intervals specified by the new constraints on time and steps introduced in the model. Each kind of warning was associated to a numerical weight (currently determined heuristically) that quantifies the associated degree of severity. As explained in [6], in WoMan a partial process execution might be compliant to several different exploitations of the model. To handle this ambiguity, WEST maintains the set S of these exploitations, each of which is called *status* and represented as a 5-tuple of sets $\langle M, R, C, T, W \rangle$ recording the following information:

M the marking, i.e., terminated activities, not yet used to fire a transition;
R (for 'Ready') the output activities of fired transitions in the status, and that the system is waiting for in order to complete them;
C training cases that are compliant with that status;
T (hypothesized) transitions that have been fired to reach that status;
W multiset of warnings raised by the various events that led to that status.

As long as the process executions proceeds, new alternative statuses may be added to S, and statuses that are not compliant with the model may be removed. For each surviving status, its *discrepancy* from the model, $\delta(status)$, can be computed based on W, to be used in the prediction procedure. Compared to [6], here it takes into account also the execution of a task or transition being not consistent with the learned time constraints. Since each *status* may be associated with different activities to be performed next, there is also an ambiguity about which activities that will be carried out. The activity prediction module of WoMan, **SNAP** (Suggester of Next Action in Process), exploits S (maintained by WEST) to compute statistics that are useful to determine which are the expected next activities and to rank them by some sort of likelihood, according to Algorithm 1. In a preliminary phase each $status \in S$ is removed and evaluated. For each transition in the model enabled by the *Marking* component of *status*, the evolution $status'$ of *status* is added to S, and W is updated if inconsistency from observed and learned behavior is encountered. Then, the discrepancy $\delta(status)$ of each $status \in S$ is measured, and statuses that exceed a certain discrepancy tolerance threshold are removed. Based on the statuses with a low discrepancy only, the set *Nexts* of actions/tasks that may be carried out next is selected from the *Ready* component of each $status \in S$, and, finally, actions in *Nexts* are scored and ranked based on a heuristic combination of the following parameters, computed over S:

1. $\mu(a)$, multiplicity of a across the various statuses (activities that appear in more statuses are more likely to be carried out next);

Algorithm 1. Prediction of possible next actions in SNAP

Require: $\mathcal{M}$: process model
Require: *Statuses* : set of currently compliant statuses compatible with the case
Require: *Event* : current event of trace
Require: ϵ: a threshold to filter only more compliant statuses
  if $E =$ end_activity $\vee$ $E =$ begin_process **then**
    for all $S = \langle M, R, C, T, W \rangle \in Statuses$ **do**
      $Statuses \leftarrow Statuses \setminus \{S\}$
      for all $p : I \Rightarrow O \in \mathcal{M}$ **do**
        if $I \subseteq Marking$ **then**
          $C' \leftarrow C \cap C_p$ /* C_p training cases involving p */
          $W' \leftarrow W \cup W_{A,p,S}$ /* $W_{A,p,S}$ warnings raised by running A in p given S */
          $Statuses \leftarrow Statuses \cup \{\langle M \setminus I, R \cup O, C', T \& \langle t \rangle, W' \rangle\}$
    $NewStatuses \leftarrow Statuses \setminus \{S\}$
  if $E =$ begin_activity **then**
    $NewStatuses \leftarrow Statuses$
  $Nexts = \{a \mid \forall S = \langle M, R, C, T, W \rangle \in NewStatuses : discrepancy(S) > \epsilon \wedge a \in R\}$
  /* Nexts is the multiset of candidate next actions */
  $Ranking \leftarrow \{\}$
  for all $a \in Nexts$ **do**
    $StatusWithA = \{\langle M, R, C, T, W \rangle \in NewStatuses \mid a \in R\}$
    $\delta_a = \sum_{S \in StatusWithA} \delta(S)$ /* summation of status's δ involving a */
    $C_a = \bigcup_{\langle M, R, C, T, W \rangle \in StatusWithA} C$ /* union of status's cases involving a */
    $score \leftarrow (\mid C_a \mid \cdot \mid StatusWithA \mid \cdot \mu(a)) / \delta_a$
    $Ranking \leftarrow Ranking \cup \{\langle score, a \rangle\}$

---

2. C_{status}, number of cases with which each status is compliant (activities expected in the statuses supported by more training cases are more likely to be carried out next);

3. $\delta(status)$, sum of weights of warnings raised by the status in which the action is included (activities expected in statuses that raised less warnings are more likely to be carried out next).

4   Evaluation

The performance of the proposed activity prediction approach was evaluated on several datasets, concerning different kinds of processes associated with different kinds and levels of complexity. The datasets related to Ambient Intelligence concern typical user behavior. Thus, they involve much more variability and subjectivity than in industrial process, and there is no 'correct' underlying model, just some kind of 'typicality' can be expected:

Aruba from the CASAS benchmark repository[2]. It includes continuous record-ings of home activities of an elderly person, visited from time to time by

---

[2] http://ailab.wsu.edu/casas/datasets.html.

Table 1. Dataset statistics

	Cases	Events		Activities		Tasks		Transitions	
		Overall	Avg	Overall	Avg	Overall	Avg	Overall	Avg
Aruba	220	13788	62.67	6674	30.34	10	0.05	92	0.42
GPItaly	253	185844	369.47	92669	366.28	8	0.03	79	0.31
White	158	36768	232.71	18226	115.35	681	4.31	4083	25.84
Black	87	21142	243.01	10484	120.51	663	7.62	3006	34.55
Draw	155	32422	209.17	16056	103.59	658	4.25	3434	22.15

her children, in a time span of 220 days. Each day is mapped to a case of the process representing the daily routine of the elderly person. Transitions correspond to terminating some activities and starting new activities. The resources (persons) that perform activities are unknown.

GPItaly from one of the Italian use cases of the GiraffPlus project[3] [3]. It concerns the movements of an elderly person (and occasionally other people) in the various rooms of her home along 253 days. Each day is a case of the process representing the typical movements of people in the home. Tasks correspond to rooms; transitions correspond to leaving a room and entering another.

The other concerns chess playing, where again the 'correct' model is not available:

Chess from the Italian Chess Federation website[4]. 400 reports of actual top-level matches were downloaded. Each match is a case, belonging to one of 3 processes associated to the possible match outcomes: *white* wins, *black* wins, or *draw*. A task is the occupation of a square by a specific kind of piece (e.g., "black rook in a8"). Transitions correspond to moves: each move of a player terminates some activities (since it moves pieces away from the squares they currently occupy) and starts new activities (that is, the occupation by pieces of their destination squares). The involved resources are the two players: 'white' and 'black'.

Table 1 reports statistics on the experimental datasets: number of cases and number of events, activities, tasks and transitions, also on average per case. There are more cases for the Ambient Intelligence datasets than for the chess ones. However, the chess datasets involve many more different tasks and transitions, many of which are rare or even unique. The datasets are different also from a qualitative viewpoint. Aruba cases feature many short loops and some concurrency (involving up to 2 activities), optional and duplicated activities. The same holds for GPItaly, except for concurrency. The chess datasets are characterized by very high concurrency: each game starts with 32 concurrent

---

[3] http://www.giraffplus.eu.
[4] http://scacchi.qnet.it.

Table 2. Prediction statistics

New (old)	Folds in [6]	Activity prediction				
		Pred	Recall	Rank	Tasks	Quality
Aruba	3	0.88 (+0.03)	0.97 (=)	0.86 (−0.06)	6.3 (+0.24)	0.78 (=)
GPItaly	3	1.0 (+0.01)	0.99 (+0.02)	0.98 (+0.02)	8.2 (+0.28)	0.97 (+0.05)
Black	5	0.53 (+0.11)	0.98 (=)	1.0 (=)	11.09 (−0.71)	0.51 (+0.09)
White	5	0.55 (=)	0.98 (+0.01)	1.0 (=)	10.9 (−0.37)	0.5 (+0.01)
Draw	5	0.65 (+0.01)	0.98 (=)	1.0 (=)	10.6 (−0.35)	0.64 (+0.02)
Chess	5	0.58 (+0.04)	0.98 (=)	1.0 (=)	10.90 (−0.44)	0.55 (+0.02)

activities (a number which is beyond the reach of many current process mining systems [5]). This number progressively decreases (but remains still high) as long as the game proceeds. Short and nested loops, optional and duplicated tasks are present as well. The number of agent and temporal constraints is not shown, since the former is at least equal, and the latter is exactly equal, to the number of tasks and transitions.

The experimental procedure was as follows. First, each dataset was translated from its original representation to the input format of WoMan. Then, a 10-fold cross-validation procedure was run for each dataset, using the learning functionality of WoMan (see [4]) to learn models for all training sets. Finally, each model was used as a reference to call WEST and SNAP on each event in the test sets: the former checked compliance of the new event and suitably updated the set of statuses associated to the current case, while the latter used the resulting set of statuses to make a prediction about the next activity that is expected in that case (as described in the previous section).

Table 2 reports average performance for the processes on the row headings ('chess' refers to the average of the chess sub-datasets). Column *Pred* reports the ratio of cases in which SNAP returned a prediction. Indeed, when tasks or transitions not present in the model are executed in the current enactment, WoMan assumes a new kind of process is enacted, and avoids making predictions. Column *Recall* reports the ratio of cases in which the correct activity (i.e., the activity that is actually carried out next) is present in the ranking, among those in which a prediction was made. Finally, column *Rank* reports how close it is to the first element of the ranking (1.0 meaning it is the first in the ranking, and 0.0 meaning it is the last in the ranking), and *Tasks* is the average length of the ranking (the lower, the better). $Quality = Pred \cdot Recall \cdot Rank \in [0,1]$ is a global index that provides an immediate indication of the overall activity prediction performance. When it is 0, it means that predictions are completely unreliable; when it is 1, it means that WoMan always makes a prediction, and that such a prediction is correct (i.e., the correct activity is at the top of the ranking). Since the proposed approach extends the one in [6], a reference to the differences with respect to its outcomes is also reported in parentheses, for each column in Table 2 ('+' indicating an improvement, '−' a degradation, and '=' equal performance). Column *Folds* shows the number of folds used for the previous k-fold cross validation, since the current one runs the 10-fold for each dataset.

First, note that *Quality* index in the current approach outperforms the previous in [6]. This is due to the WoMan's ability to ensure more reliable predictive support, as the model includes both more constraints and cases. Specifically, *Rank* index is improved in all cases, except on Aruba. However, this decrease is balanced by improved Pred, which leaves the *Quality* unchanged. This is acceptable, because, while WoMan can make a prediction in more cases, a 6% decrease in ranking over 6.3 tasks means a loss of much less than 1 position (0.378).

WoMan is extremely reliable because the correct next activity is almost always present in the ranking (97–99% of the times), and always in the top section (first 10% items) of it, especially for the chess processes (always at the top). Compared to previous one, current *Recall* and *Rank* tend to be equal, except for GPItaly. The former is because of different fold setting, while the latter probably due to the temporal constraints. This confirm that WoMan is effective under very different conditions as regards the complexity of the models to be handled. In the Ambient Intelligence domain, this means that it may be worth spending some effort to prepare the environment in order to facilitate that activity, or to provide the user with suitable support for that activity. In the chess domain, this provides a first tool to make the machine able to play autonomously. The number of predictions is proportional to the number of tasks and transitions in the model. This was expected, because, the more variability in behaviors, the more likely it is that the test sets contain behaviors that were not present in the training sets. Compared to previous results, the number of predictions, in all domains, is increased, due to the fact that the 10-fold provide a bigger training, involving more information. WoMan is almost always able to make a prediction in the Ambient Intelligence domain, which is extremely important in order to provide continuous support to the users. While neatly lower, the percentage of predictions in the chess domain was clearly improved, and covers more than half of the match. The nice thing is that WoMan reaches this percentage by being able to distinguish cases in which it can make an extremely reliable prediction from cases in which it prefers not to make a prediction at all.

5   Conclusions

In addition to other classical exploitations, process models may be used to predict the next activities that will take place. This would allow to take suitable actions to help accomplishing those activities. This paper proposed an extended approach to make these kinds of predictions using the WoMan framework for workflow management. Experimental results on different tasks and domains suggest that the proposed approach can successfully perform such predictions.

Given the positive results, we plan to carry out further work on this topic. First of all, we plan to check the prediction performance on other domains, e.g. Industry 4.0 ones. Also, we will investigate how to improve the prediction accuracy by means of more refined strategies. Finally, we would like to embed the prediction module in other applications, in order to guide their behavior.

References

1. Agrawal, R., Gunopulos, D., Leymann, F.: Mining process models from workflow logs. In: Schek, H.-J., Alonso, G., Saltor, F., Ramos, I. (eds.) EDBT 1998. LNCS, vol. 1377, pp. 467–483. Springer, Heidelberg (1998). doi:10.1007/BFb0101003
2. Cook, J.E., Wolf, A.L.: Discovering models of software processes from event-based data. Technical Report CU-CS-819-96, Department of Computer Science, University of Colorado (1996)
3. Coradeschi, S., Cesta, A., Cortellessa, G., Coraci, L., Gonzalez, J., Karlsson, L., Furfari, F., Loutfi, A., Orlandini, A., Palumbo, F., Pecora, F., von Rump, S., Štimec, A., Ullberg, J., Ostlund, B.: Giraffplus: combining social interaction and long term monitoring for promoting independent living. In: Proceedings of the 6th International Conference on Human System Interaction (HSI), pp. 578–585. IEEE (2013)
4. Ferilli, S.: Woman: logic-based workflow learning and management. IEEE Trans. Syst. Man Cybern.: Syst. **44**, 744–756 (2014)
5. Ferilli, S., Esposito, F.: A logic framework for incremental learning of process models. Fundam. Inf. **128**, 413–443 (2013)
6. Ferilli, S., Esposito, F., Redavid, D., Angelastro, S.: Predicting process behavior in woman. In: Adorni, G., Cagnoni, S., Gori, M., Maratea, M. (eds.) AI*IA 2016. LNCS, vol. 10037, pp. 308–320. Springer, Cham (2016). doi:10.1007/978-3-319-49130-1_23
7. Herbst, J., Karagiannis, D.: An inductive approach to the acquisition and adaptation of workflow models. In Proceedings of the IJCAI 1999 Workshop on Intelligent Workflow and Process Management: The New Frontier for AI in Business, pp. 52–57 (1999)
8. van der Aalst, W., et al.: Process mining manifesto. In: Daniel, F., Barkaoui, K., Dustdar, S. (eds.) BPM 2011. LNBIP, vol. 99, pp. 169–194. Springer, Heidelberg (2012). doi:10.1007/978-3-642-28108-2_19
9. Muggleton, S.: Inductive logic programming. New Gener. Comput. **8**(4), 295–318 (1991)
10. Pesic, M., van der Aalst, W.M.P.: A declarative approach for flexible business processes management. In: Eder, J., Dustdar, S. (eds.) BPM 2006. LNCS, vol. 4103, pp. 169–180. Springer, Heidelberg (2006). doi:10.1007/11837862_18
11. van der Aalst, W.M.P.: The application of petri nets to workflow management. J. Circuits Syst. Comput. **8**, 21–66 (1998)
12. van der Aalst, W.M.P., Weijters, T., Maruster, L.: Workflow mining: Discovering process models from event logs. IEEE Trans. Knowl. Data Eng. **16**, 1128–1142 (2004)
13. Weijters, A.J.M.M., van der Aalst, W.M.P.: Rediscovering workflow models from event-based data. In: Proceedings of the 11th Dutch-Belgian Conference of Machine Learning (Benelearn 2001), pp. 93–100 (2001)

Automatic Defect Detection by One-Class Classification on Raw Vehicle Sensor Data

Julia Hofmockel[1]($\boxtimes$), Felix Richter[2], and Eric Sax[3]

[1] Audi Electronics Venture GmbH, Gaimersheim, Germany
`ju.hofmockel@googlemail.com`
[2] Volkswagen AG, Wolfsburg, Germany
[3] Institute for Information Processing Technologies,
Karlsruher Institute of Technology, KIT, Karlsruhe, Germany

Abstract. The next step in the automotive industry is the automatic detection of a defect in the vehicle behavior in addition to the current analysis of failure codes or costumer complaints. The idea of learning the normality by one-class classification is applied to the identification of an exemplary defect. Different neural network topologies for time series prediction are realized where the quality of the forecast indicates the strength of abnormality. It is compared how the detection possibilities of a concrete defect changes when the model is trained with different data extractions. A distinction is made between data from complete rides and filtered data, containing only the situations where the defect is visible. It can be shown that a generalization is possible.

Keywords: Anomaly detection · One-class classification · Time series prediction

1 Introduction

In the automotive industry, the selection of new functions based on data from a multitude of vehicles is the next step especially in the context of Advanced Driver Assistant Systems (ADAS) [1]. As shown in Fig. 1 the backend collects data from a vehicle fleet which can be used for the development of such new customer functions.

In this context, the huge amount of data is difficult to progress. Not all data are worth to analyze, instead defects should get special attention. Their early identification can reduce time and costs and avoid safety critical situations for the costumer. A defect is a rare, unusual event. That is why it can be categorized as abnormal vehicle behavior [2]. One approach is the determination of a failure. Then a classifier which distinguishes between normal and abnormal is learned. Another possibility is to define an anomaly as an *observation that deviates so much from other observations as to arouse suspicion that it was generated by a different mechanism* [7]. This approach makes no exact statements about what is causing a problem. Instead one gets a hint where to have a closer look at.

© Springer International Publishing AG 2017
M. Kryszkiewicz et al. (Eds.): ISMIS 2017, LNAI 10352, pp. 378–384, 2017.
DOI: 10.1007/978-3-319-60438-1_37

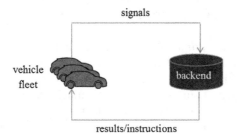

Fig. 1. Communication vehicle - backend

This paper applies the general idea of finding unusual vehicle behavior as deviation from the normal data as a one-class classification problem on a selected application: the detection of transmission problems within a ride.

The paper is structured as followed. Section 2 introduces the concepts of one-class classification and replicator neural networks for the detection of static anomalies. In Sect. 3 the idea is extended for time dependent data. As a next step, the application to the detection of transmission problems is presented. The detailed description of the problem and the data available is highlighted in Sect. 4. For the evaluation, Sect. 5 shows different set ups and results. The paper ends with a conclusion and the description of further research in Sect. 6.

2 Static Anomaly Detection

One-class classifiers are used for the identification of anomalies. They are able to distinguish between two classes while the training data are all from the same class [2]. For detecting problems in the transmission system, a binary classifier predicting defect or no defect can be trained. But with the intention to generally indicate deviations from the normal vehicle behavior, the paper analyses the possibilities of one-class classification.

The Replicator Neural Network (RNN) is one method for one-class classification based anomaly detection [8]. During the training a prediction function $f(x)$ is learned such that the difference between training point x and its reconstructed output x' is minimized for all $x \in X$, with X describing the training data. The parameters are learned with normal data and the reconstruction error of test data point x_{test} characterizes its anomaly score. The reconstruction error is calculated as the squared difference between the original value and its reconstruction:

$$||x - x'||_2^2. \tag{1}$$

Regularization methods are used to avoid overfitting or learning the identify function. One possibility is an additional sparsity term which forces the average activity of the hidden neurons to be close to a certain value. Another way is the L2-regularization. It ensures that the L2-norm of the weights are kept small [12,13].

3   Anomaly Detection in Time Series

In general, sensor data from a vehicle are in the data structure of a time series. Defects, like transmission problems, can not be determined by one single corrupted measurement in the time series. These situations occur due to the movement over time, though a simple RNN is not sufficient. One possibility to cover the time information is the extraction of features describing a time window. Accordingly, a static method as the RNN can be applied on the features instead of the raw data [4,11]. The raw data can be used as features in terms of time series prediction. A function predicts the current value x_t from a certain amount (window size ws) of previously observed data points and shifted by one time stamp [5,10]:

$$x_t' = f(x_{t-ws}, \ldots, x_{t-1}). \tag{2}$$

The idea of time series prediction for anomaly detection is comparable to the static RNN since the difference between x_t and its reconstruction x_t' can be used as anomaly score. This adapted version of the RNN is in the following declared as the description *dense layers*. Extensions are recurrent or convolutional layers. Recurrent layers have additional self-connecting weights giving them internal memory. The simplest form is a fully connected recurrent layer (*simpleRNN*), where all neurons within a layer are connected [3,6]. Long-short term memory networks (LSTMs) are a special kind of recurrent networks introduced for solving the vanishing gradient problem [9]. A Convolutional Neural Network (CNN) learns the weights as filters of a particular length which can fold the input sequence [3,14]. An additional dense layer leads from the folded sequence to the prediction of data point x_t.

4   Problem Description

The concept of anomaly detection using the reconstruction error will be evaluated on a data set recorded for the purpose of detecting transmission problems. A transmission problem can be noted by high frequent amplitude changes of the rotation speed value in the moments when the vehicle is brought up to speed, in the following named starting procedure. Since starting procedures are only one detail within one complete ride, the paper differentiates between two set ups: Set up 1 considers the possibilities of detecting transmission problems using data limited to starting procedures. Following this, the data set is increased to complete rides in set up 2. The reason for extending is the goal of identifying unusual vehicle behavior in general; the transmission problem is one possible application. The knowledge that the problem can only be recognized during starting procedures is use-case driven.

The early identification of abnormal behavior, like problems in the transmission system, is desirable from the after-sale perspective. It allows to support the costumer and avoid a total breakdown.

4.1   Data Overview

For training and evaluation 152 rides without transmission problems are given. They contain 280 faultless starting procedures. Additionally there are 8 rides with transmission problems, containing 46 faulty and 37 faultless starting procedures.

4.2   Setup 1: Training and Test on Starting Procedures

As first approach a model describing the normal progression of the rotation speed value during the starting procedure is learned on 280 of such situations. Here 70% of the data are used for training and 30% for validation. The reconstruction error of the validation set provides information about the accuracy of the forecast. At the same time, the model is evaluated as a classifier able to distinguish between normal and abnormal circumstances. Therefore, the starting procedures of the eight rides with problems, divided into 46 anomalous starting procedures and 37 normal ones, are investigated.

Under consideration are dense models processing different time windows from $ws = 0$ to $ws = 25$ and various architectures including dense, recurrent and convolutional layers. All of them are making use of a sparsity factor and L2-regularization (see Sect. 2). This upper bound for the window size is chosen based on the shortest starting procedure. The CNN uses different filter lengths from two to 24.

4.3   Setup 2: Training on Complete Rides

As already mentioned, the restriction on starting procedures within the rides derives from the application. As next step, is tested how the most promising configuration from set up 1 behaves when trained on complete rides. From the 152 rides without problems, eight are randomly excluded as additional faultless test rides. The remaining data are again divided into 70% training and 30% validation data. As in set up 1, the starting procedures from the eight rides with problems are used to evaluate the model as classifier. In addition, it is compared if there is a significant difference between the average reconstruction error of the eight test rides without and the eight rides with transmission problems.

5   Results

In the following the results based on 50 iterations are presented for set up 1 and set up 2. For evaluation, the average reconstruction error (see Eq. (1)) of the validation data is calculated. It describes the prediction power of starting procedures in set up 1 and of complete rides in set up 2. Further indicators are sensitivity and specificity resulting from the classification of positive and negative starting procedures. Hereby anomalies are declared as negatives. Sensitivity describes the true positive rate; the percentage of normal starting procedures detected as normal. Specificity is the true negative rate; the percentage of abnormal starting procedures detected as abnormal.

5.1   Setup 1: Training and Test on Starting Procedures

The models are trained on the values of the rotation speed of faultless start-ing procedures. The reconstruction error is used as anomaly score. The border between normal and abnormal data points is defined by the 97%-quantile of the anomaly scores of the training data. The choice of that value gives the model a robustness against outliers in the training data. Because of noise, one outstand-ing time stamp is not sufficient to classify the complete starting procedure as abnormal. That is why at least 5% of its time stamps needs to be above the border to be categorized as abnormal starting procedure.

A first test determines whether a positive impact of time dependencies is given. Dense models predicting the current time stamp are compared on windows with different length. Table 1 shows the 95%-confidence intervals of the rated measurements. It can be seen that specificity significantly increases when the detection of transmission problems is regarded as a time series prediction task ($ws > 0$). The layout under consideration is a model having three hidden layers with 50, 10 and 50 neurons. The best results can be found when the current time stamp is predicted based on the two previous ones. The lower bounds of sensitivity and specificity are both high while the upper bound of the validation error is satisfactory. Additional past time stamps lead to increasing sensitivity and decreasing specificity. That means more starting procedures are classified as normal. Explanation can be the worse predictive power within proper starting procedures, recognizable by the increasing validation error. It leads to a weaken border between abnormality and normality.

After that, it is checked if another architecture can improve the model. For that purpose, all models are trained with the aim to predict x_t using the past 25 time stamps. Hereby an additional layout with only one hidden layer having 50 neurons is analyzed. According to Table 2 it can be summarized that recurrence does not improve the outcomes. Especially the specificity stands out through the high standard deviation noticeable by a detection rate from 0% to more than 70%. Although the CNN with filter length equals 24 is the most promising one, it achieves worse results compared to the dense model with a window of length two ($ws = 2$) from Table 1.

Table 1. Comparison of different window sizes, dense model

	Validation error	Sensitivity	Specificity
$ws = 0$	(2.83e−05, 3.83e−05)	(81.08%, 81.08%)	(44.79%, 46.95%)
$ws = 1$	(8.35e−05, 9.87e−05)	(77.82%, 88.77%)	(74.93%, 92.46%)
$ws = 2$	(9.84e−05, 1.26e−04)	(86.09%, 92.62%)	(81.62%, 92.21%)
$ws = 5$	(1.45e−04, 2.04e−04)	(93.70%, 95.45%)	(62.93%, 79.51%)
$ws = 10$	(1.51e−04, 2.04e−04)	(90.69%, 93.96%)	(34.44%, 57.38%)
$ws = 20$	(1.59e−04, 2.25e−04)	(84.14%, 95.87%)	(45.30%, 78.18%)
$ws = 25$	(1.66e−04, 2.35e−04)	(83.66%, 96.89%)	(44.16%, 77.14%)

Table 2. Comparison of different architectures

	Validation error	Sensitivity	Specificity
Hidden layers with 50, 10 and 50 neurons			
Dense layers	(1.66e−04, 2.35e−04)	(83.66%, 96.89%)	(44.16%, 77.14%)
SimpleRNN layers	(0.00e−03, 7.56e−03)	(69.39%, 93.85%)	(12.03%, 79.45%)
LSTM layers	(0.00e−03, 1.92e−03)	(72.09%, 100.00%)	(00.00%, 71.13%)
CNN, filter length = 2	(7.57e−05, 2.47e−04)	(89.18%, 95.36%)	(50.61%, 77.13%)
CNN, filter length = 5	(1.76e−04, 2.54e−04)	(87.11%, 94.08%)	(41.47%, 62.62%)
CNN, filter length = 10	(1.58e−04, 2.42e−04)	(88.45%, 93.84%)	(43.29%, 68.36%)
CNN, filter length = 24	(7.92e−05, 1.27e−04)	(91.40%, 96.38%)	(64.86%, 76.53%)
One hidden layer with 50 neurons			
Dense layers	(8.79e−05, 1.09e−04)	(90.70%, 96.55%)	(55.93%, 74.78%)
SimpleRNN layers	(0.00e−03, 3.42e−03)	(70.64%, 94.12%)	(24.13%, 92.22%)
LSTM layers	(0.00e−03, 2.80e−03)	(73.11%, 95.43%)	(20.20%, 84.99%)
CNN, filter length = 2	(0.00e−04, 9.40e−04)	(86.82%, 97.62%)	(58.07%, 71.58%)
CNN, filter length = 5	(1.30e−04, 3.15e−04)	(87.70%, 95.87%)	(42.38%, 65.27%)
CNN, filter length = 10	(8.10e−05, 2.15e−04)	(87.57%, 97.07%)	(46.50%, 71.18%)
CNN, filter length = 24	(5.42e−05, 9.17e−05)	(83.93%, 96.07%)	(60.04%, 67.18%)

5.2 Setup 2: Training on Complete Rides

The presented results lead to the choice of a dense network which predicts the current data point x_t by the two preceding time stamps for the detection of transmission problems. As a result, it is measured if a model describing complete rides is also able to distinguish between rides with and without problems. Therefore, the model is trained with normal rides and tested on the labeled starting procedures and rides.

First of all, the accuracy of classification is calculated. It must be mentioned that the value of the border from when a data point is abnormal changes. The reason is that the extended training data leads to an adapted, decreased value of the 97%-quantile of their anomaly scores. That is why the border when a starting procedure is considered as abnormal is increased from a proportion of 5% to 40%. The neural network with dense layers and $ws = 2$ classifies in average 81.5% of the normal starting procedures and 69.6% of those having transmission problems into the correct class. When applying the model on complete rides, the average reconstruction errors of the data points between rides with and without problems are compared. With a significance level of 5% the average reconstruction error of eight normal rides lies between 4.67e−07 and 2.09e−06 while those of the eight abnormal rides lies between 3.02e−06 and 3.48e−06. Their separability indicates that the model is able to distinguish between normal and abnormal rides.

6   Conclusion and Future Work

In summary the investigations show that the idea of one-class classification, employed as time series prediction, can be used for the identification of abnormal behavior in the rotation speed value of the transmission system. Restricted to starting procedures as part of a ride, up to 92.21% of the starting procedures with transmission problems can be detected. It is demonstrated that an extension to complete rides is possible.

The next step will be the validation of the findings with an extended data set and the analysis of possible improvements by feature engineering as preprocessing step. To satisfy the broad idea of the identification of abnormal vehicle behavior, it is necessary to find the window size and its shift in a statistical way. Furthermore, the number of signals has to be extended such that more than only one signal are covered.

References

1. Bach, J., Bauer, K.-L., Holzaepfel, M., Hillenbrand, M., Sax, E.: Control based driving assistant functions' test using recorded in field data. In: 7. Tagung Fahrerassistenz (2015)
2. Chandola, V., Banerjee, A., Kumar, V.: Anomaly detection: a survey. ACM Comput. Surv. **41**, 15:1–15:58 (2009)
3. Chollet, F.: Keras. https://github.com/fchollet/keras
4. Fan, Y., Nowaczyk, S.L., Rognvaldsson, T.: Using histograms to find compressor deviations in bus fleet data. In: 28th Swedish Artificial Intelligence Society Workshop, pp. 123–132. Swedish Artificial Intelligence Society (2014)
5. Faraway, J., Chatfield, C.: Time series forecasting with neural networks: a comparative study using the airline data. J. R. Stat. Soc. Ser. C Appl. Stat. **47**, 231–250 (1998)
6. Gal, Y., Ghahramani, Z.: A theoretically grounded application of dropout in recurrent neural networks. In: arXiv preprint arXiv:1512.05287 (2015)
7. Hawkins, D.: Identification of Outliers. Springer, Netherlands (1980)
8. Hawkins, S., He, H., Williams, G., Baxter, R.: Outlier detection using replicator neural networks. In: Kambayashi, Y., Winiwarter, W., Arikawa, M. (eds.) DaWaK 2002. LNCS, vol. 2454, pp. 170–180. Springer, Heidelberg (2002). doi:10.1007/3-540-46145-0_17
9. Hochreiter, S., Schmidhuber, J.: Long short-term memory. Neural Comput. **9**, 1735–1780 (1997)
10. Ma, J., Perkins, S.: Time-series novelty detection using one-class support vector machines. In: Proceedings of the International Joint Conference on Neural Networks, pp. 1741–1745. IEEE Press, New York (2003)
11. Nanopoulos, A., Alcock, R., Manolopoulos, Y.: Feature-based classification of time-series data. In: Information Processing and Technology, pp. 49–61. Nova Science Publishers, Commack, NY, USA (2001)
12. Olshausen, B.A., Field, D.J.: Sparse coding with an overcomplete basis set: a strategy employed by V1? Vis. Res. **23**, 3311–3325 (1997)
13. Tibshirani, R.: Regression shrinkage and selection via the lasso. J. R. Stat. Soc. Ser. B Stat. Methods **58**, 267–288 (1996)
14. Van Fleet, P.: Discrete Wavelet Transformations: An Elementary Approach with Applications. Wiley, Hoboken (2011)

Visualizing Switching Regimes Based on Multinomial Distribution in Buzz Marketing Sites

Yuki Yamagishi$^{(\boxtimes)}$ and Kazumi Saito

Graduate School of Management and Information of Inovation,
University of Shizuoka,
52-1 Yada, Shizuoka, Suruga 422-8526, Japan
yamagissy@gmail.com, k-saito@u-shizuoka-ken.ac.jp

Abstract. The review scoring results in large-scale buzz marketing sites can greatly affect actual purchase activities of many users. In this paper, since the scoring tendency for an item usually changes over time due to several reasons, we propose a method for visualizing its scoring stream data as a timeline based on switching regimes. Namely, by assuming that fundamental scoring behavior of users in each regime obeys a multinomial distribution model, we first estimate the switching time steps and the model parameters by maximizing the likelihood of generating the observed scoring stream data, and then produce a timeline and its associated dendrogram as our final visualization results by calculating the probability function from the estimated switching regimes. In our experiments using not only synthetic stream data generated from a known ground truth model but also real scoring stream data collected from a Japanese buzz marketing site, we show that our proposed method can produce accurate and interpretable visualization results for such stream data.

Keywords: Stream data · Switching regimes · Multinomial distribution

1 Introduction

In recent years, reviews posted by users in buzz marketing sites are explosively increasing and those are affecting directly user's purchase decisions and after-purchase actions. This phenomenon suggests that buzz marketing sites are being an important social medium which controls sales promotion of products and services. In addition, the emergence of buzz marketing sites has provided us with the opportunity to collect a large number of user reviews for various items such as cosmetic products. Thus, researchers have conducted several types of researches using such large scale data, for instance, sentiment analysis attempt to classify contents on buzz marketing sites for certain topics [1–3], which could allow some companies to know how their products are evaluated by consumers.

Analyzing review in depth needs natural language processing. Fortunately, many buzz marketing sites offer numeric review scores, and we can use this score

© Springer International Publishing AG 2017
M. Kryszkiewicz et al. (Eds.): ISMIS 2017, LNAI 10352, pp. 385–395, 2017.
DOI: 10.1007/978-3-319-60438-1_38

instead of the review itself. A score is defined as a rating given by a user and their values vary across users and items, say, tens of thousands of users and items. Namely, we can obtain large scoring stream data for each item. In large-scale buzz marketing sites, such a scoring tendency for an item may change over time due to several reasons. Actually, if the rank of an item has dropped on a certain review site, the manufacturer that produces it would analyze the past reviews on the site in order to know when and how the consumers' evaluation of the item had changed. Thus, it must be very important in research of social media to summarize such tendency as easily interpretable visualization results.

Technically, by using such large scoring stream data, we are not only interested in knowing what is happening now and how it develops in the future, but also we are interested in knowing what happened in the past and how it caused by some changes in the distribution of the information as studied in [4,5]. For this purpose, we formulate our problem as detection of switching regimes, and propose a method for visualizing scoring stream data as a timeline and its associated dendrogram based on the estimated switching regimes, where the fundamental scoring behavior of users in each regime is assumed to obey a multinomial distribution model, and these switching time steps and model parameters are estimated by maximizing the likelihood of generating the observed scoring stream data.

In our experiments, we evaluate the performance and the characteristics of the proposed method, in comparison to Kleinberg's method [4] regarded as the state of the art for burst detection along the time axis. By generating synthetic stream data where the ground truth for the model of switching regimes is assumed, we first show that the proposed method could successfully detect the almost correct models except for the case of quite narrow switching interval values, while Kleinberg's method [4] performed very poorly for the cases of relatively large switching interval values. Then, we demonstrate that the proposed method could produce interpretable visualization results from real scoring stream data in a Japanese large-scale buzz marketing site for cosmetics. We also show that these results were very different from those by Kleinberg's method.

2   Related Work

Our research aim is in some sense the same, in the spirit, with the work by Kleinberg [4] and Swan and Allan [5]. They noted a huge volume of the stream data, tried to organize it and extract structures behind it. This is done in a retrospective framework, i.e., assuming that there is a flood of abundant data already and there is a strong need to understand it. Kleinberg's work is motivated by the fact that the appearance of a topic in a document stream is signaled by a burst of activity and identifying its nested structure manifests itself as a summarization of the activities over a period of time, making it possible to analyze the underlying content much easier. He used a hidden Markov model in which bursts appear naturally as state transitions, and successfully identified the hierarchical structure of e-mail messages. Here, such changes might involve changes in the number of reviews which posted in a certain period, i.e., changes in the posting interval and frequency, in which case existing burst detection techniques [4,6,7] would be applicable. However, if no frequency change in the time interval is involved, we would not be able to use these techniques because they do not focus on the change in the scoring tendency. In other words, they intend to detect a burst for the single topic and do not deal directly with multiple topics and the change of their distribution. Swan and Allan's work is motivated by the need to organize a huge amount of information in an efficient way. They used a statistical model of feature occurrence over time based on hypotheses testing and successfully generated clusters of named entities and noun phrases that capture the information corresponding to major topics in the corpus and designed a way to nicely display the summary on the screen (Overview Timelines). We also follow the same retrospective approach, i.e., we are not predicting the future, but we are trying to understand the phenomena that happened in the past. However, our aim is not exactly the same as theirs. We are interested in detecting changes in the scoring tendency as switching regimes.

Here we should emphasize that this kind of switching detection is substantially different from the typical research paradigm of anomaly detection or change point detection widely studied in the field of machine learning, whose techniques are more or less similar to those used in novelty detection or outlier detection [8]. For instance, statistical techniques used in anomaly detection fit a statistical model (usually for normal behavior) to the given data and then apply a statistical inference test to determine if an unseen instance belongs to this model or not. Instances that have a low probability to be generated from the learned model, based on the applied test statistic, are declared as anomalies. On the other hand, we are interested in time-varying parameter models of time series as switching regimes, rather than continuous evolutions (e.g., [9]). A conventional approach for this direction includes studies of regime-switching models in the field of economics (e.g., [10]), but these studies heavily rely on the Gaussian assumption. In the field of decision support, a number of methods have been developed for detecting and excluding unfair ratings in online reputation systems [11], but these methods are clearly categorized into a class of anomaly detection method.

In general, an average score of reviews is widely used as the social information of an item. Certainly, an average score is easy to understand for everyone as a converged result, and it is obvious that the reliability of the value is increased by much more reviews. However, if the distribution of scores has drastically changed, the reliability decreases intensely. Some possible reasons for the changing of score distribution are as follows: fake reviews by users, an influence of other media, improvement of a product or a service, changing of fashion and so on. It is certain that these events are high priority information of items, but it is hard to guess these facts from the average score and recent reviews. For this reason, a study of detecting changes of score distribution as switching regimes from a huge number of reviews is important in the field of web intelligence. Thus, from a viewpoint of retrospective in a similar way to representative researches [4,5], we propose a visualization method of these switching regimes from scoring stream data of scores in buzz marketing sites.

3   Preliminaries

Let $\mathcal{D} = \{(s_1, t_1), \cdots, (s_N, t_N)\}$ be a set of scoring stream data for a given item in a buzz marketing site, where s_n and t_n stands for the score with J-category and the submitted time of the n-th review, respectively, and $|\mathcal{D}| = N$ means the number of reviews submitted to this item from users. Here, we assume $t_1 \leq \cdots \leq t_n \leq \cdots \leq t_N$. Below the number n is referred to as time step, and let $\mathcal{N} = \{1, 2, \cdots, N\}$ be the set of the time steps. We express the starting time step of the k-th switching regime as $T_k \in \mathcal{N}$, and let $\mathcal{T}_K = \{T_0, \cdots, T_k, \cdots, T_{K+1}\}$ be the set of the switching time steps, where we set $T_0 = 1$ and $T_{K+1} = N+1$ for our convenience. Namely, $T_1, \cdots, T_K$ are the individual switching time steps to be estimated, where we assume $T_k < T_{k+1}$. Let $\mathcal{N}_k$ be the set of time steps for the k-th regime defined by $\mathcal{N}_k = \{n \in \mathcal{N}; T_k \leq n < T_{k+1}\}$ for each $k \in \{0, \cdots, K\}$. where $\mathcal{N} = \mathcal{N}_0 \cup \cdots \cup \mathcal{N}_K$.

Now, we assume that each score distribution of the regime is modeled by a J-category multinomial distribution. Let $\boldsymbol{p}_k$ be the probability vector for the k-th multinomial distribution, and $\mathcal{P}_K$ the set of the probability vectors defined by $\mathcal{P}_K = \{\boldsymbol{p}_0, \cdots, \boldsymbol{p}_K\}$. Then, given $\mathcal{T}_K$, we can define the following log-likelihood function:

$$L(\mathcal{D}; \mathcal{P}_K, \mathcal{T}_K) = \sum_{k=0}^{K} \sum_{n \in \mathcal{N}_k} \sum_{j=1}^{J} s_{n,j} \log p_{k,j}. \tag{1}$$

where $s_{n,j}$ is the dummy variable for $s_n \in \{1, \cdots, J\}$ defined by $s_{n,j} = 1$ if $s_n = j$; 0 otherwise. Then, the maximum likelihood estimators of Eq. (1) is given by $\hat{p}_{k,j} = \sum_{n \in \mathcal{N}_k} s_{n,j} / |\mathcal{N}_k|$ for $k = 0, \cdots, K$ and $j = 1, \cdots, J$. Substituting these estimators to Eq. (1) leads to

$$L(\mathcal{D}; \hat{\mathcal{P}}_K, \mathcal{T}_K) = \sum_{k=0}^{K} \sum_{n \in \mathcal{N}_k} \sum_{j=1}^{J} s_{n,j} \log \hat{p}_{k,j}. \tag{2}$$

Therefore, the problem of detecting switching time steps is reduced to the problem of finding the set $\mathcal{T}_K$ that maximizes Eq. (2).

Here note that only Eq. (2) does not allow us to evaluate directly the effect of introducing $\mathcal{T}_K$. It is important to evaluate how the log-likelihood improved over the one obtained without considering the regime switching. Thus, we reformulate the problem as the maximization problem of log-likelihood ratio. If we do not assume any changes, i.e., $\mathcal{T}_0 = \emptyset$, Eq. (2) is reduced to

$$L(\mathcal{D}; \hat{\mathcal{P}}_0, \mathcal{T}_0) = \sum_{n \in \mathcal{N}} \sum_{j=1}^{J} s_{n,j} \log \hat{p}_{0,j}, \tag{3}$$

where $\hat{p}_{0,j} = \sum_{n \in \mathcal{N}} s_{n,j} / N$. Thus, the log-likelihood ratio of the two cases, one with K time steps and the other with no switching is given by

$$LR(\mathcal{T}_K) = \mathcal{L}(\mathcal{D}; \hat{\mathcal{P}}_K, \mathcal{T}_K) - \mathcal{L}(\mathcal{D}; \hat{\mathcal{P}}_0, \mathcal{T}_0). \tag{4}$$

In summary, we consider the problem of finding the set of the switching time steps $\mathcal{T}_K$ that maximizes $LR(\mathcal{T}_K)$ defined above. If we solve the maximization problem of Eq. (4) with an exhaustive method, it is guaranteed that the optimal solution can be obtained although the time complexity becomes $O(N^K)$, and the time complexity of this naive method becomes an enormous amount when $K \geq 3$. Thus, we employ a detection method which combines a greedy search and a local search [12]. We determine an adequate number of switching time steps by introducing the significance level p from the fact that Eq. (4) asymptotically obeys a χ^2 distribution function with the degree of freedom $K(J-1)$.

4   Proposed Method

Let $\hat{\mathcal{T}}_K$ be the set of the estimated time steps by applying the above detection method. For each category j, we consider a probability function of time step n defined by $\hat{p}_j(n) = \hat{p}_{k,j}$ if $n \in \mathcal{N}_k$ where $0 \leq k \leq K$. Then, in order to visually analyze the switching regimes by using a timeline, we consider simultaneously plotting these probability functions of J-category. Here, note that we plot the probability with respect to time step n rather than actual review time t_n because we want to treat each review score with an equal weight. Actually, in case of a timeline plotted with respect to actual review time t_n, a large number of user review scores submitted during a bursty period are likely to be embedded in a relatively narrow time span.

On the other hand, we consider the following dissimilarity measure $d(i,j)$ based on a cosine similarity between these probability functions of categories i and j:

$$d(i,j) = \sqrt{1 - \frac{\sum_{n \in \mathcal{N}} \hat{p}_i(n)\hat{p}_j(n)}{\sqrt{\sum_{n \in \mathcal{N}} \hat{p}_i(n)^2}\sqrt{\sum_{n \in \mathcal{N}} \hat{p}_j(n)^2}}}. \tag{5}$$

Then, we consider constructing a dendrogram of these probability functions based on Ward's minimum variance method [13] using the above dissimilarity measure. Finally, we summarize the algorithm of our proposed method described above.

1. Compute the set of the estimated time steps $\hat{\mathcal{T}}_K$ from a given set of review scoring results $\mathcal{D}$ and significance level p.
2. Calculate probability function $\hat{p}_j(n)$ of each category j from the set of the estimated time steps $\hat{\mathcal{T}}_K$.
3. Output a timeline of probability functions and a dendrogram based on the dissimilarity of Eq. (5).

5   Experimental Results

5.1   Experiments on Synthetic Data

We show in this section that the proposed method indeed finds the true model of switching regimes at least as good as the state-of-the-art method that takes a different approach, i.e., Kleinberg's method, one of the representative methods, which uses a hidden Markov model [4]. By assuming a simple model of switching regimes associated with multinomial distributions with $(J = 3)$-category, we examine whether both the methods can accurately restore the true model from stream data generated by the model. More specifically, for the true number $K = 3$ of switching time steps, we consider the set of the true switching time steps defined by

$$\mathcal{T}_K^* = \{T_0^* = 1,\ T_1^* = 10,000,\ T_2^* = T_1^* + \delta,\ T_3^* = T_2^* + \delta,\ T_4^* = T_3^* + 10,000\},$$

where the interval value δ was set to one of five values, i.e., $100, 500, 1,000, 5,000$, or $10,000$. Note that for each value for δ, the final time step becomes $20,200$, $21,000, 22,000, 30,000$, or $40,000$. As for the multinomial distribution settings, during the 1st (initial) and 4th (final) regimes, all the probability values are set to the same one, i.e., $p_{0,1}^* = p_{0,2}^* = p_{0,3}^* = p_{3,1}^* = p_{3,2}^* = p_{3,3}^* = 1/3$, while during the 2nd and 3rd regimes, the values of the 1st category are set to $p_{2,1}^* = 1/2$ and $p_{2,1}^* = 3/4$, respectively, and the other values are set to $p_{1,2}^* = p_{1,3}^* = (1 - p_{1,1}^*)/2$ and $p_{2,2}^* = p_{2,3}^* = (1 - p_{2,1}^*)/2$. From the viewpoint of busty level detection, by setting $\alpha = 3/2$, we can see that during the 2nd and 3rd regimes, the 1st category scores appear α and α^2 times more than those during the 1st or 4th regime. Thus, we apply Kleinberg's method to a partial scoring stream data consisting only of the 1st category data (i.e., $\mathcal{D}_1 = \{(s_n, t_n) \in \mathcal{D}; s_n = 1\}$) with parameter settings, $\alpha = 1.5$ and $\gamma = 1.0$. As for our proposed method, the significance level of χ^2 test is set to $p = 0.0001$. Hereafter, Kleinberg's method and the proposed method are simply expressed as KM and PM, respectively.

First, we qualitatively evaluate the resultant probability functions by visualizing them as the timelines with respect to time step n. Figures 1 and 2 show

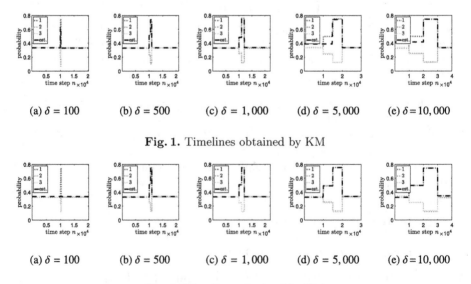

(a) $\delta = 100$ (b) $\delta = 500$ (c) $\delta = 1,000$ (d) $\delta = 5,000$ (e) $\delta = 10,000$

Fig. 1. Timelines obtained by KM

(a) $\delta = 100$ (b) $\delta = 500$ (c) $\delta = 1,000$ (d) $\delta = 5,000$ (e) $\delta = 10,000$

Fig. 2. Timelines obtained by PM

the timelines obtained by KM (Figs. 1(a) to (e)) and those by PM (Figs. 2(a) to (e)), Here, in the case of KM, by using the set of burst level change points obtained from $\mathcal{D}_1$, we estimated their probability functions for the 1st category as computed in PM and then calculated the probability functions for the other categories from the true model definitions. From these figures, we can see that in case of the middle-range interval value $\delta = 500$ or $1,000$, both the methods could estimate the timelines with reasonably high accuracy in comparison to the true ones (Figs. 1(b), (c) and 2(b), (c)), while in case of the larger interval value $\delta = 5,000$ or $10,000$, KM produced poor results whose numbers of change points are incorrect (Figs. 1(d) and (e)), and in case of the smallest interval value $\delta = 100$, PM could not find switching time step (Fig. 2(a)). These experimental results suggest that in case of not too small interval values, which is expected in our problem formulation, PM has an advantage over KM in terms of the accuracy as shown in Figs. 2(d) and (e), while KM has an advantage over our method for detecting burst activity during relatively quite small interval values as shown in Fig. 1(a).

Next, in order to quantitatively evaluate the accuracy of the obtained timelines, we introduce the following mean absolute error of the obtained probability function:

$$E(\hat{p}_1(n)) = \frac{1}{N} \sum_{k=0}^{K} \sum_{T_k^* \leq n < T_{k+1}^*} |p_{k,1}^* - \hat{p}_1(n)|. \tag{6}$$

Figure 3(a) compares the mean absolute errors produced by KM and PM, where these means were estimated with 100 trials of independently generated stream data. From this figure, we can quantitatively confirm the above discussions, i.e., both the methods showed relatively small errors for $\delta = 500$, or $1,000$, while only PM did so for $\delta = 5,000$ or $10,000$, and KM for $\delta = 100$. Figure 3(b) compares the average processing times over the 100 trials for KM

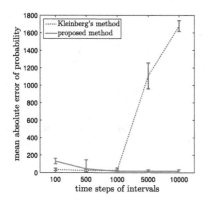
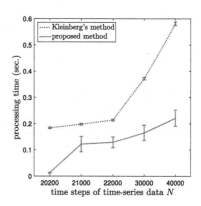

(a) Comparison of estimation errors   (b) Comparison of processing times

Fig. 3. Performance comparison of KM and PM

and PM, where the horizontal and vertical axes stand for the final time step and processing time (sec.), respectively. Here our programs were executed on a computer system equipped with two Intel(R) Xeon(R) X5690 @3.47 GHz CPUs and a 192 GB main memory with a single thread within the memory capacity. We can confirm that PM has an advantage over KM in terms of the efficiency.

5.2 Experiments on Real Data

We collected real scoring stream data from "@cosme"[1] which is a Japanese large-scale buzz marketing site for cosmetics, and utilized two items with the largest review numbers, i.e., "Oshima Tsubaki Camellia Hair Care Oil (Oshima Tsubaki)" and "Conditioner Essential (Albion)". In the experiments, we use the same parameter settings for both the methods just as before, but the range of review score is $j = 0$ to 7 in this site $((J = 8)$-category).

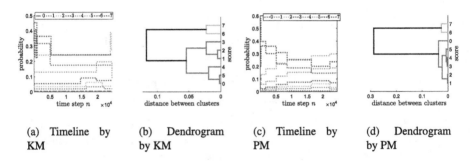

(a) Timeline by KM   (b) Dendrogram by KM   (c) Timeline by PM   (d) Dendrogram by PM

Fig. 4. Timelines and dendrograms for Oshima Tsubaki

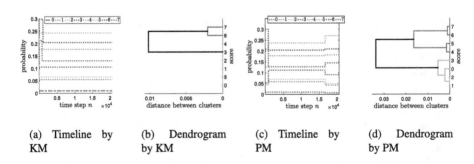

(a) Timeline by KM   (b) Dendrogram by KM   (c) Timeline by PM   (d) Dendrogram by PM

Fig. 5. Timelines and dendrograms for Albion

Figure 4 compares the obtained timelines and their associated dendrograms for Oshima Tsubaki, where those by KM are shown in Figs. 4(a) and (b), and

---

[1] http://www.cosme.net/.

those by PM in Figs. 4(c) and (d). Here note that we applied KM each category data $\mathcal{D}_j$ individually, and estimated the corresponding probability function. Thus, in general, the sum of the obtained probabilities over categories cannot be one in the case of KM. From Figs. 4(a) and (c), we can clearly see that PM produced a larger number of switching regimes than KM did. By combining the results shown in Figs. 4(c) and (d), we can naturally understand that the frequencies of the scores $j = 6$ and 7 decrease as time step proceeds, while those of the other scores are likely to increase. In contrast, it must be difficult to interpret the results obtained by KM. On the other hand, Fig. 5 compares the obtained timelines and their associated dendrograms for Albion, where those by KM are shown in Figs. 5(a) and (b), and those by PM in Figs. 5(c) and (d). From Figs. 5(a) and (c), we can clearly see that both the method produced small numbers of switching regimes. Again, from Fig. 5(d) obtained by PM, we can see that there exist two groups of scores whose probability functions behave similarly, but it must be difficult to interpret the dendrogram obtained by KM as shown in Fig. 5(b). Thus, it is expected that PM can produce more interpretable timelines and their associated dendrograms than KM do.

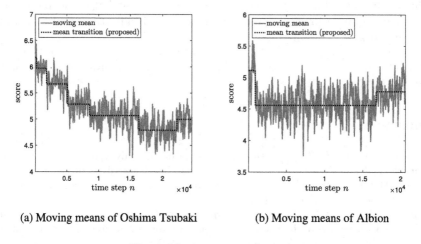

(a) Moving means of Oshima Tsubaki   (b) Moving means of Albion

Fig. 6. Moving means of scores

We analyze more closely the result obtained by PM for these two items, Oshima Tsubaki and Albion. Figures 6(a) and (b) show the moving means of scores by solid lines, and the means in the regime by dotted lines, where the moving means are calculated with the window size 100. From these figures, we can observe that these two types of means naturally coincide with each other for both the items. We consider that these experimental results also support the usefulness and the vitality of the proposed method.

6   Conclusion

In this paper, we proposed a method for visualizing scoring stream data as a timeline based on switching regimes. In this method, under the assumption that fundamental scoring behavior of users in each regime obeys a multinomial distribution model, the switching time steps and the model parameters are estimated by maximizing the likelihood of generating the observed scoring stream data, and then a timeline and its associated dendrogram are produced as our final visualization results by calculating the probability function from the estimated switching regimes. Next, we empirically evaluated the performance and the characteristics of the proposed method in comparison to Kleinberg's method. Namely, in our experiments using synthetic stream data, we showed that the proposed method could successfully detect the almost correct models except for the case of quite narrow switching interval values, while Kleinberg's method performed very poorly for the case of relatively large switching interval values. Moreover, in our experiments using real scoring stream data in a Japanese large-scale buzz marketing site for cosmetics. we demonstrated that the proposed method could produce interpretable visualization results. In future, we plan to evaluate our method using various scoring stream data and attempt to establish useful techniques for producing more interpretable visualization results by using the framework of switching regimes.

Acknowledgments. This work was supported by JSPS Grant-in-Aid for Scientific Research (C) (No. 16J11909).

References

1. Melville, P., Gryc, W., Lawrence, R.D.: Sentiment analysis of blogs by combining lexical knowledge with text classification. In: Proceedings of the 15th ACM SIGKDD International Conference on Knowledge Discovery and Data Mining (KDD 2009), pp. 1275–1284 (2009)
2. Pak, A., Paroubek, P.: Twitter as a corpus for sentiment analysis and opinion mining. In: Proceedings of the Seventh conference on International Language Resources and Evaluation (LREC 2010), pp. 1320–1326 (2010)
3. Glass, K., Colbaugh, R.: Estimating sentiment orientation in social media for business informatics. In: AAAI Spring Symposium: AI for Business Agility (2011)
4. Kleinberg, J.: Bursty and hierarchical structure in streams. In: Proceedings of the 8th ACM SIGKDD International Conference on Knowledge Discovery and Data Mining (KDD-2002), pp. 91–101 (2002)
5. Swan, R., Allan, J.: Automatic generation of overview timelines. In: Proceedings of the 23rd Annual International ACM SIGIR Conference on Research and Development in Information Retrieval (SIGIR 2000), pp. 49–56 (2000)
6. Zhu, Y., Shasha, D.: Efficient elastic burst detection in data streams. In: Proceedings of the 9th ACM SIGKDD International Conference on Knowledge Discovery and Data Mining (KDD-2003), pp. 336–345 (2003)
7. Sun, A., Zeng, D., Chen, H.: Burst detection from multiple data streams: a network-based approach. IEEE Trans. Syst. Man Cybern. Soc. Part C **40**, 258–267 (2010)

8. Chandola, V., Banerjee, A., Kumar, V.: Anomaly detection: a survey. ACM Comput. Surv. **41**, 15:1–15:58 (2009)
9. Pio, G., Lanotte, P.F., Ceci, M., Malerba, D.: Mining temporal evolution of entities in a stream of textual documents. In: Andreasen, T., Christiansen, H., Cubero, J.-C., Raś, Z.W. (eds.) ISMIS 2014. LNCS (LNAI), vol. 8502, pp. 50–60. Springer, Cham (2014). doi:10.1007/978-3-319-08326-1_6
10. Kim, C.J., Piger, J., Startz, R.: Estimation of Markov regime-switching regression models with endogenous switching. J. Econom. **143**, 263–273 (2008)
11. Josang, A., Ismail, R., Boyd, C.: A survey of trust and reputation systems for online service provision. Decis. Support Syst. **43**, 618–644 (2007)
12. Yamagishi, Y., Okubo, S., Saito, K., Ohara, K., Kimura, M., Motoda, H.: A method to divide stream data of scores over review sites. In: Pham, D.-N., Park, S.-B. (eds.) PRICAI 2014. LNCS, vol. 8862, pp. 913–919. Springer, Cham (2014). doi:10.1007/978-3-319-13560-1_78
13. Ward, J.: Hierarchical grouping to optimize an objective function. J. Am. Stat. Assoc. **58**, 236–244 (1963)

"Serial" versus "Parallel": A Comparison of Spatio-Temporal Clustering Approaches

Yongli Zhang[1(✉)], Sujing Wang[2], Amar Mani Aryal[2], and Christoph F. Eick[1]

[1] Department of Computer Science, University of Houston, Houston, TX, USA
{yzhang93,ceick}@uh.edu
[2] Department of Computer Science, Lamar University, Beaumont, TX, USA
{sujing.wang,aaryal3}@lamar.edu

Abstract. Spatio-temporal clustering, which is a process of grouping objects based on their spatial and temporal similarity, is increasingly gaining more scientific attention. Research in spatio-temporal clustering mainly focuses on approaches that use time and space in parallel. In this paper, we introduce a serial spatio-temporal clustering algorithm, called ST-DPOLY, which creates spatial clusters first and then creates spatio-temporal clusters by identifying continuing relationships between the spatial clusters in consecutive time frames. We compare this serial approach with a parallel approach named ST-SNN. Both ST-DPOLY and ST-SNN are density-based clustering approaches: while ST-DPOLY employs a density-contour based approach that operates on an actual density function, ST-SNN is based on well-established generic clustering algorithm Shared Nearest Neighbor (SNN). We demonstrate the effectiveness of these two approaches in a case study involving a New York city taxi trip dataset. The experimental results show that both ST-DPOLY and ST-SNN can find interesting spatio-temporal patterns in the dataset. Moreover, in terms of time and space complexity, ST-DPOLY has advantages over ST-SNN, while ST-SNN is more superior in terms of temporal flexibility; in terms of clustering results, results of ST-DPOLY are easier to interpret, while ST-SNN obtains more clusters which overlap with each other either spatially or temporally, which makes interpreting its clustering results more complicated.

Keywords: Spatio-temporal clustering · Cluster analysis · Batch processing · Shared nearest neighbor clustering · Density-based clustering

1 Introduction

With the development of positioning and sensing technologies, huge amounts of spatio-temporal data are generated everyday. Extracting interesting spatio-temporal patterns from such datasets is very important as it has broad real-world applications, such as earthquake analysis, detection of outbreak of epidemics [1], identifying crime patterns [2], and understanding climate changes.

© Springer International Publishing AG 2017
M. Kryszkiewicz et al. (Eds.): ISMIS 2017, LNAI 10352, pp. 396–403, 2017.
DOI: 10.1007/978-3-319-60438-1_39

The objective of spatio-temporal clustering is to identify homogeneous groups of objects and to discover interesting spatio-temporal patterns associated with each group [3]. In order to identify spatio-temporal clusters, one major challenge that needs to be addressed is to determine how spatial and temporal information is combined. In almost all existing approaches, time and space are treated in a parallel fashion, which raises a limitation on the scalability. For example, ST-DBSCAN proposed by Briant and Kut [4], as an extension of DBSCAN, introduces a temporal neighborhood radius in addition to the spatial neighborhood radius, then looks for dense regions both temporally and spatially at the same time.

In this paper, we propose a serial, density-contour based spatio-temporal clustering approach, ST-DPOLY, which uses density polygons as cluster models to cluster spatio-temporal point objects. As a serial approach, it subdivides stream of spatio-temporal point objects into batches and obtains spatial clusters for each batch first; next, it identifies continuing relationships between temporarily consecutive spatial clusters. ST-DPOLY is an improved version of ST-DCONTOUR [5], in which we use non-parametric density estimation to replace the parametric Gaussian Mixture Model which significantly improves the performance. We also compare ST-DPOLY with a parallel approach called ST-SNN [6], which relies on a distance function that combines spatial and temporal distances and then modifies the well-established generic clustering algorithm-Shared Nearest Neighbor (SNN) [7] to operate on that distance function.

This paper's main contributions include:

- We introduce a serial, density-contour based spatio-temporal clustering algorithm named ST-DPOLY which generates spatial clusters first and then forms spatio-temporal clusters by identifying continuing relationships between spatial clusters in consecutive batches.
- We give a thorough comparison between the "serial" spatio-temporal clustering approach and "parallel" spatio-temporal clustering approach. Strengths and weaknesses of each approach are discussed.
- We evaluate the two approaches in a real-world case study involving NYC taxi trips data. Experimental results that obtain and interpret spatio-temporal clusters of taxi activity in New York City are presented.

The rest of the paper is organized as follows. Section 2 introduces the serial, spatio-temporal clustering approach ST-DPOLY in detail. Section 3 demonstrates ST-DPOLY and ST-SNN with NYC taxi trips data. Section 4 provides the comparison between two approaches. Section 5 concludes the paper.

2   ST-DPOLY

ST-DPOLY mainly consists of the following 3 phases:

1. Obtain a spatial density function through non-parametric density estimation for the spatial point data collected in each batch.

2. For each batch, use a density threshold to identify spatial clusters, each of which is enclosed by a polygon that is created from density contour lines of the spatial density function that was created in Phase 1.
3. Identify continuing relationships between density polygons in consecutive batches and construct spatio-temporal clusters as continuing spatial clusters.

2.1   Phase 1: Obtain Spatial Density Distribution

For phase 1, our goal is to obtain a 2-dimensional density function for the spatial point data collected at each batch. For ST-DPOLY, we use non-parametric kernel density estimation (KDE) [8] to obtain a 2-dimensional spatial density function f. For a bivariate random sample $X_1, X_2, \ldots, X_n$ drawn from an unknown density f, the kernel density estimator is defined as follows:

$$\widehat{f}(x; H) = \frac{1}{n} \sum_{i=1}^{n} K_H(x - X_i) \tag{1}$$

where $x = (x_1, x_2)^T$, $X_i = (X_{i1}, X_{i2})^T$ $(i = 1, 2, \ldots, n)$, $K(x)$ is the kernel which is a symmetric non-negative probability density function, H is bandwidth matrix which is symmetric and positive-definite, $K_H(x) = |H|^{-1/2} K H^{-1/2} x$. Our implementation uses the KernSmooth package [9] in R to estimate the spatial density distribution for given spatial points in each batch.

2.2   Phase 2: Spatial Cluster Extraction

The goal of phase 2 is to extract spatial clusters using the spatial density function obtained in the first phase. A spatial cluster in our approach is defined as a region whose scope is described by a polygon and whose probability density of data points is above a given threshold. Our approach uses the contouring algorithm CONREC [10] and post-processing to get polygon models as spatial clusters.

Overall, phase 2 of ST-DPOLY obtains spatial clusters by performing the following 6 steps:

1. Grid the data collection area.
2. Calculate a probability density for all grid intersection points using the spatial density function (given in Eq. 1), and obtain a density matrix. Create a table T to store locations of all grid intersection points and corresponding density matrix.
3. Pass T, along with a pair of density threshold θ_1, $\hat{\theta}_1$ to CONREC, which returns two sets of contour lines.
4. Close open contour lines.
5. Classify the obtained contour lines into holes and spatial clusters;
6. Construct spatial cluster polygon using the information obtained in step 5.

As far as step 4 is concerned, for all the open contours, we use boundary lines of the data collection area to connect them and make them closed contours. As far as step 5 is concerned, the identified contour lines could be either spatial clusters for which the density increases as we move to the inside of the spatial cluster or hole contours for which the reverse is true. In order to distinguish contour holes from spatial clusters, we use another threshold, which is slightly smaller than the chosen threshold. So we get two layers of contours, and contour holes are those whose inner contour has a smaller density value. At last, we get a final list of spatial clusters which is a set of polygons which might contain holes.

2.3   Phase 3: Spatio-Temporal Cluster Extraction

After spatial clusters for each batch have been obtained, we identify spatio-temporal clusters as continuing spatial clusters in consecutive batches. To identify continuing spatial clusters, we use an overlap matrix, which measures the percentage of area overlap of spatial clusters in consecutive batches. If a pair of spatial clusters has significant overlap, we consider them as continuing clusters. For more details of phase 3 including pseudocode, see [5].

3   Experimental Evaluation

3.1   Dataset Description

The TLC Trip Record Data [11] were collected by technology providers authorized under the Taxicab and Livery Passenger Enhancement Programs, which contain data for over 1.1 billion taxi trips from January 2009 to June 2016. Each individual trip record contains precise location coordinates for where the trip started and ended, timestamps for when the trip started and ended, and a few other variables including fare amount, payment method, and distance traveled.

3.2   Experimental Results of ST-DPOLY and ST-SNN

For the experiment, we use yellow taxi pick-up locations collected in 20 min interval as batches. We analyzed taxi pickups from 6 to 7 AM on January 8th, 2014, Fig. 1 shows the clusters that have been created using ST-DPOLY: 3 clusters for 6–6:20 batch, 2 clusters for the 6:20–6:40 batch, and 4 clusters for the 6:40–7 batch. According to the result, in terms of pick-up locations, east of Midtown of New York is a hotspot which is crowded with people looking for taxis early in the morning, as well as the region centered around the Grand Central Terminal. Analyzing these 3 batches, we can see some patterns: west of the Midtown area, two clusters continue for these 3 consecutive batches, one cluster we find is near the south-west of Time Square which is closer to several train and bus terminals, such as 42 St-Port Authority Bus Terminal; the other is also centered around several bus terminals and train stations, such as 34 Street Penn Station. We infer that, early in the morning, after New Yorkers get off trains or buses, a lot of people choose to look for taxi rides, which explains the presence of high-density clusters of taxi pick-ups around the train and bus stations. Similarly, cluster 3

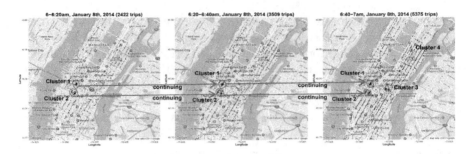

Fig. 1. Spatio-temporal clustering example of ST-DPOLY (grid size: 200 × 200, bandwidth: 0.00109, log density: 6.9)

Fig. 2. Spatio-temporal clustering example of ST-SNN (red polygons) (Color figure online)

in the 6:00–6:20 batch, and cluster 3 in the 6:40–7 batch are all centered around the Grand Central Terminal exhibiting a similar taxi pickup pattern.

We apply the ST-SNN algorithm to the same dataset. The input parameters are: $k = 100$, $MinPs = 60$, and $w = 0.5$. We use Euclidean distance to compute the spatial distance. There are 16 clusters obtained. Figure 2 visualizes clusters 2, 13 and 14, and they are centered around several bus terminals and train stations, which shows a similar pattern of the clustering results generated by ST-DPOLY. Cluster 2 and 13 are similar in the spatial domain, however, the time slots corresponding to these two clusters are different.

4   Comparison Between ST-DPOLY and ST-SNN

4.1   Time and Space Complexity

Table 1 gives the time and space complexities of ST-DPOLY and ST-SNN. With respect to ST-DPOLY, we assume that the grid size we use is $m \times m$, n is the total number of points, e is the average number of edges a spatial cluster has. For ST-DPOLY, in general, e is smaller than m, and since it is a serial approach, its overall time complexity is $O(m^2 \times n)$, in cases that the number of data points is much larger than the number of grid cells ($n >> m^2$), ST-DPOLY's

Table 1. Time/space complexities of ST-DPOLY and ST-SNN

		Time complexity		Space complexity
ST-DPOLY	Phase 1	$O(m^2 \times n)$	Phase 1	$O(n + m^2)$
	Phase 2	$O(m^2)$	Phase 2	$O(m^2 + e^2)$
	Phase 3	$O(e^2)$	Phase 3	$O(e^2)$
ST-SNN		$O(n^2)$		$O(k \times n)$

complexity becomes $O(n)$. The time complexity of ST-SNN is the same as SNN which is $O(n^2)$ without the use of an indexing structure. The space complexity is $O(k \times n)$ since only k-nearest neighbor lists need to be stored. But in cases that $m^2 << n$, the space complexity of ST-SNN is significantly worse than ST-DPOLY. Therefore, ST-DPOLY is superior to ST-SNN in terms of both time and space complexity.

4.2   Temporal Flexibility

In terms of temporal flexibility, ST-SNN is more flexible as cluster have more variation with respect to temporal mean and standard deviation. Though time intervals can be selected based on application needs, it is fixed throughout the clustering process once selected, whereas ST-SNN has the potential to detect "more optimal" time intervals. However, the clustering result of ST-DPOLY is more straightforward, and in terms of clustering data streams, ST-DPOLY as a serial approach is more appropriate.

4.3   Quality of Clusters

To compare the quality of clustering result, we measure the variation of clusters obtained, which is showed in Tables 2 and 3. The clusters generated by ST-SNN have smaller values of standard deviation and range of time, longitude, and latitude than clusters identified by ST-DPLOY.

Table 2. Variation measurements of 3 clusters of ST-DPOLY in Fig. 1

	Time	Longitude	Latitude
range	20	0.003552	0.002071
mean (μ)	9	-73.9772	40.7528
sd(σ)	7.33	0.00116	0.000509

	Time	Longitude	Latitude
range	20	0.004863	0.005282
mean (μ)	29.64	-73.9904	40.7563
sd(σ)	5.956	0.000773	0.00076

	Time	Longitude	Latitude
range	20	0.005263	0.004274
mean (μ)	50	-73.99277	40.75057
sd(σ)	5.5468	0.00147	0.000805

(a) cluster 3 from batch 1 (38 points)   (b) cluster 1 from batch 2 (216 points)   (c) cluster 2 from batch 3 (259 points)

Table 3. Variation measurements of 3 clusters of ST-SNN in Fig. 2

	Time	Longitude	Latitude			Time	Longitude	Latitude			Time	Longitude	Latitude
range	19	0.0026	0.0024		range	19	0.0025	0.0032		range	19	0.002264	0.002881
mean (μ)	9.1986	-73.9904	40.7565		mean (μ)	49.1038	-73.9904	40.7564		mean (μ)	49.7439	-73.9906	40.6862
sd(σ)	5.6204	0.0005	0.0005		sd(σ)	5.7201	0.0005	0.0006		sd(σ)	5.5195	0.0086	0.009

(a) cluster 2 (141 points)   (b) cluster 13 (212 points)   (c) cluster 14 (122 points)

5   Conclusion

The main objective of this paper is to introduce a serial spatio-temporal clustering approach and the comparison of that approach with the commonly used parallel approach (ST-SNN). In particular, we propose a serial, density contour based spatio-temporal clustering approach ST-DPOLY. We also demonstrate the effectiveness of both approaches in a case study involving the taxi pickup locations in New York city. In terms of time and space complexity, ST-DPOLY has advantages over ST-SNN, while ST-SNN is more superior in terms of temporal flexibility; in terms of clustering results, results of ST-DPOLY are easier to interpret, while ST-SNN usually obtains a significant number of clusters which overlap either spatially or temporarily, which makes interpreting its clustering results more complicated. Moreover, there is more variation with respect to the time mean value and standard deviation in ST-DPOLY results.

As far as future work is concerned, for ST-DPOLY, we will try to extend our approach to support multiple thresholds. In terms of selecting parameters for the proposed approach—in particular, batch size and density thresholds— we will investigate semi-automatic and automatic parameter selection tools to facilitate the use of ST-DPOLY.

References

1. Gaudart, J., Poudiougou, B., Dicko, A., Ranque, S., Toure, O., Sagara, I., Diallo, M., Diawara, S., Ouattara, A., Diakite, M., et al.: Space-time clustering of childhood malaria at the household level: a dynamic cohort in a mali village. BMC Public Health **6**(1), 286 (2006)
2. Grubesic, T.H., Mack, E.A.: Spatio-temporal interaction of urban crime. J. Quant. Criminol. **24**(3), 285–306 (2008)
3. Zhang, Y., Eick, C.F.: Novel clustering and analysis techniques for mining spatio-temporal data. In: Proceedings of the 1st ACM SIGSPATIAL Ph.D. Workshop, Article no. 2. ACM (2014)
4. Birant, D., Kut, A.: ST-DBSCAN: an algorithm for clustering spatial-temporal data. Data Knowl. Eng. **60**(1), 208–221 (2007)
5. Zhang, Y., Eick, C.F.: ST-DCONTOUR: a serial, density-contour based spatio-temporal clustering approach to cluster location streams. In: Proceedings of the 7th ACM SIGSPATIAL International Workshop on GeoStreaming, Article no. 5. ACM (2016)
6. Wang, S., Cai, T., Eick, C.F.: New spatiotemporal clustering algorithms and their applications to ozone pollution. In: 2013 IEEE 13th International Conference on Data Mining Workshops (ICDMW), pp. 1061–1068. IEEE (2013)

7. Ertöz, L., Steinbach, M., Kumar, V.: Finding clusters of different sizes, shapes, and densities in noisy, high dimensional data. In: Proceedings of the 2003 SIAM International Conference on Data Mining, pp. 47–58. SIAM (2003)
8. Epanechnikov, V.A.: Non-parametric estimation of a multivariate probability density. Theory Probab. Appl. **14**(1), 153–158 (1969)
9. Ripley, M.B., Suggests, M.: The kernsmooth package (2007)
10. Bourke, P.D.: A contouring subroutine. Byte **12**(6), 143–150 (1987)
11. http://www.nyc.gov/html/tlc/html/about/trip_record_data.shtml. Accessed 23 Aug 2016

Time-Frequency Representations for Speed Change Classification: A Pilot Study

Alicja Wieczorkowska[1]([⊠]), Elżbieta Kubera[2], Danijel Koržinek[1],
Tomasz Słowik[3], and Andrzej Kuranc[3]

[1] Polish-Japanese Academy of Information Technology,
Koszykowa 86, 02-008 Warsaw, Poland
alicja@poljap.edu.pl, danijel@pjwstk.edu.pl
[2] Department of Applied Mathematics and Computer Science,
University of Life Sciences in Lublin, Akademicka 13, 20-950 Lublin, Poland
elzbieta.kubera@up.lublin.pl
[3] Department of Energetics and Transportation,
University of Life Sciences in Lublin, Akademicka 13, 20-950 Lublin, Poland
{tomasz.slowik,andrzej.kuranc}@up.lublin.pl

Abstract. Speeding is an important factor influencing road traffic safety. Even though speed is monitored using radars, the drivers may increase speed after passing the radar. In this paper, we address automatic classification of speed changes (or maintaining constant speed) from audio data, as a microphone added to the radar can register the drivers' behavior both in front of and behind the radar. We propose two time-frequency based approaches to represent the audio data for speed classification purposes. These approaches have been tested in a pilot study using on-road data, and the results are presented in this paper.

Keywords: Intelligent transport system · Road traffic safety · Audio signal analysis

1 Introduction

The road traffic safety is very important to everybody, since we use the road systems as drivers or pedestrians. There are many factors that may influence the road traffic safety, including:

- technical and financial reasons: low quality of some roads and vehicles, lack of protection for pedestrians and bikers, insufficient financial support;
- human factors: reckless driving and behavior of pedestrians and bikers, low public awareness of the consequences of this recklessness; these factors include speeding, driving and walking while impaired, not using seat belts.

Although speed limits and speed cameras are often criticized by the drivers, the statistics show that limiting the speed decreases the number of fatalities in road crashes. The speed control is a way to ensure the observance of speed limits

© Springer International Publishing AG 2017
M. Kryszkiewicz et al. (Eds.): ISMIS 2017, LNAI 10352, pp. 404–413, 2017.
DOI: 10.1007/978-3-319-60438-1_40

by the drivers, but many drivers slow down in front of the speed camera, and then accelerate after passing it. The radar or the speed camera provide a single measurement of the vehicle speed, and after passing the measurement point it is not registered. If the audio data were recorded at speed measurement points and later analyzed, then the conclusions on the drivers' behavior could be drawn, and possibly new measures to improve the road traffic safety would be undertaken.

The audio automotive data have been applied already in the automatic classification tasks: in vehicle detection and classification [3,5,6], fault diagnosis [2], and noise assessment [10,13]. The speed and speed change classification was not a subject of interest, to the best of our knowledge, apart from our previous work [8]. We believe the outcomes of such research can be applied not only for speed change classification, but also for noise assessment purposes.

This paper addresses automatic classification of the speed changes (or maintaining constant speed) of vehicles, based on audio data. The audio signal, if recorded with a mic at a measurement point, represents both approaching and receding the mic. We started with experiments described in [8], with the data recorded at a dyno test bench, and then performed our first tests using on-road data, recorded for these experiments. We observed differences between data representing particular cars, but we also reported differences in results depending on how the signal is represented. Changes of audio signal with time and Doppler effect pose difficulties when parameterizing the data, so any sound representation for speed change classification must reflect changes in time. This is why we decided to investigate on time-frequency representations of the audio signal in the research presented below. We prepared a new set of on-road data, acoustically different than previously used data, but relatively uniform, to assure that the experiments on the sound representations are not affected by other factors.

2 Data Preparation

The data prepared for the purpose of this research were recorded in one 91-minute long session, at 48 kHz sampling rate, 24 bits resolution, in stereo, but only the left channel was used in further experiments. The OBD (on-board diagnostics) and GPS data were also acquired. Next, the audio data were aligned with the vehicle's GPS location and speed. Since we knew the exact GPS coordinates of the microphone, we could easily mark the points in the recording where the vehicle was closest to the microphone. The recordings were made on January 16th 2017 (winter), on a road in the outskirts of Lubartów in Lublin voivodship in Poland - *Lubartów* data. There was snow in the area and on the road itself, but the road surface just below the tires was not covered with snow. The car used in the recordings was a Renault Espace IV (2007), driven by one of the authors. The recordings included 84 examples (3*28 audio segments), representing:

- 28 drives with acceleration from 50 to 70 km/h,
- 28 drives with stable speed, 50 km/h,
- 28 drives with deceleration from 70 to 50 km/h.

The car was equipped with manual transmission, and all drives were recorded without changing gear and without applying brakes (engine braking only). Since the readings of the GPS were exact to 1 s, another procedure was used to pinpoint the time point when the vehicle position was closest to the microphone; a window of 2 s before and after the GPS derived time was applied for this purpose. For each drive we extracted a 10 s segment, centered at the maximum of the envelope of the signal amplitude for this segment. Hilbert transform [9] was applied to compute the envelope. Since the number of examples was not too big, we visually verified that all the drives were properly centered.

2.1   Data Analysis

Spectrograms show time-frequency analysis of the audio data, obtained via discrete Fourier transform (DFT) performed frame by frame on the audio data. Horizontal axis represents time, vertical axis represents frequency, and the amplitudes of the DFT spectra are represented using a selected color scale. Spectrograms are discrete graphs; each element corresponds to a particular spectral bin (a unit in the spectrum), calculated for one audio frame. Exemplary spectrograms are shown in Fig. 1 for deceleration, stable speed, and acceleration.

In the central part of each spectrogram we can observe the moment of passing the mic. As we can see, a lot of energy is concentrated in all frequencies. Also, in each spectrogram we can observe sloping lines (curves) of high energy, with the Doppler effect, i.e. frequency shift in these lines. The Doppler effect is observed when the sound source is moving wrt. the observer; the frequency is increased when the source is approaching and decreased when the source is receding.

For deceleration, the lines show the decrease of frequency with time, with the strongest decrease at the center; the lines are more visible after passing the mic. For stable speed, the Doppler shift is pronounced at the center; before and after passing the mic the line around 60 Hz only slightly decreases. For acceleration, the lines represent increasing frequency before and after passing the mic, with frequency decrease at the center. As we can see, the moment of passing the mic divides the audio data into two parts, which can be separately described.

Frequencies in the Data. The data were acquired for the drive at the fourth gear, with the engine crankshaft RPM (revolutions per minute) around 1850. This corresponds to the frequency of 61.6 Hz and explains the line at around 60 Hz in the spectrogram for our stable speed 50 km/h. When decelerating, the drive started with a higher speed, 70 km/h, so the speed when passing the mic was higher. For 60 km/h we had RPM about 2200, which corresponds to 73 Hz in the spectrogram; the further decrease of RPM to about 1620 would correspond to 55 Hz. The lines are of lower amplitude than for stable speed because of lower engine load. In the case of acceleration the lines are more intense. More fuel is used when accelerating, and therefore more energy is generated in the cylinders of the engine. This also corresponds to more noise generated in this case. Therefore, the data describing speed changes, or maintaining constant speed, also illustrate how much noise is generated by the vehicle.

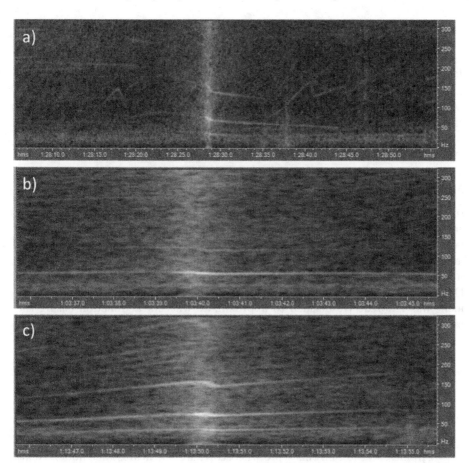

Fig. 1. Spectrogram for: (a) deceleration, (b) stable speed, (c) acceleration. A close-up of each spectrograms below 300 Hz is shown. In this setting of the color scale (grayscale), the higher the luminance level, the bigger the amplitude, and darker areas correspond to lower amplitude. Spectrograms obtained from Adobe Audition [1]

Other frequencies (also higher than 400 Hz) are also present in the audio data. Noises corresponding to tires and road are also in these data, and they can be especially useful when registering audio data for vehicles with electrical engines. However, such vehicles are still rare in Poland where we performed our experiments, so we assume the engine noise to be present in the data. The engine noise depends on the engine type, i.e. Diesel engines generate different noises than gasoline vehicles. The presented spectrograms correspond to a gasoline engine.

3 Data Representation

We decided to represent our data through features reflecting the time-frequency representation of audio data. We applied 2 representations, R_{lin} and R_{spg}, based on the spectrogram calculated via DFT. R_{lin} represents the data with 18-element feature set, with the purpose of applying classifiers typical for a relatively small feature set; we used support vector machines (SVM) and random forests (RF). R_{spg} represents the data with 5120 numbers, so the length of the representation vector is much bigger than the size of the data set (84 examples). R_{spg} was designed to be used with artificial neural networks (ANN) as classifiers; we assumed that ANN would learn the characteristics of the data from this representation, and ignore irrelevant features. In both cases, we limited our observations to 500 Hz, as this range reflects the speed-specific characteristics of audio data.

3.1 Lines

In this representation, R_{lin}, we parameterized the curves (lines) of the highest energy up to 500 Hz in the spectrogram. 8192-element spectrum is used to calculate the spectrogram, with 1/3 frame hop size, i.e. with overlap, and with Hamming windowing, i.e. multiplying signal x[n] by $w[n] = 0.54 - 0.46\cos(2\pi n/N)$, n = 0,...,8191, N = 8192. Since the center part of the spectrogram shows the Doppler effect, we decided to skip this part, so the middle 1 s of each 10 s segment was ignored. We decided to approximate with linear function (using the method of least squares) 3 lines in 2 s before and after the central 1s, as follows.

- In the 2 s when the vehicle is approaching to the mic, we look for the 3 lines of the highest energy, spanning slantwise in the spectrogram, as follows:
  - For each frame, we look for a maximum in each of the ranges: 50–100 Hz, 100–200 Hz, 200–300 Hz, 300–400 Hz, and 400–500 Hz. For each frequency bin we calculate the score as the number of times this bin was indicated as representing the maximum (in all frames within this 2 s segment). For each bin with the score greater or equal to the half of the number of frames, we sum up this score with scores for the nearest neighbors. Next, the score for the winning bin and for bins within the range of musical second is zeroed (to capture the width of the line, and avoid too close lines). This procedure is applied iteratively, to obtain 3 candidates for starting frequencies of the lines.
  - Each line is tracked through the entire 2 s segment, from the frame on the right hand side (closest to the moment of passing the mic), to the left. The starting frequency represents the maximum of energy in the range of the candidate ± musical second. In each tracking step, for a consecutive neighboring frame, the bin representing the average of the last 3 bins (if available) or 1 bin above/below is combined with the line, if it represents a local maximum of energy.
  - The obtained 3 lines (curves) are approximated with linear function. Each line is represented through the intercept, gradient, and quotient of the ending and starting frequencies, so we obtain 9 features.
- In the 2 s when the vehicle passed the mic, we similarly look for 3 lines in the same frequency ranges, starting the lines from the left hand side (again, near the moment of passing the mic). As a result, we obtain another 9 features, describing the linear functions approximating the lines in the spectrogram.

3.2 Time-Frequency Feature Set

The 2 s segments before and after passing the mic were also applied in our second approach to audio data representation, R_{spg}, but this time together with the central 1 s (5 s altogether). In this approach, the signal was first down-sampled to 1 kHz. Next, spectra were calculated for 500 ms frame, with Hamming window and 3/4 frame overlap, following the settings recommended for analysis-synthesis [11]. We added 375 ms of the signal (188 ms before and 187 ms after this 5 s segment), to obtain 40 frames. Next, a filterbank of half-overlapping triangular filters in the frequency domain was applied, to reduce the number of the analyzed frequency bins from 250 (DFT amplitude points) to 128 (filter band outputs). The filters were logarithmically spaced, in the mel-scale. For each frame, the filters' outputs were normalized, i.e. divided by the sum of these outputs. Finally, the logarithm of all these values was taken (standard approach applied to the spectral amplitude values). Thus, we obtained a 40×128 matrix. Examples of such matrices (Fig. 2) show that R_{spg} indicates characteristics typical for speed changes (or stable speed). The representing vector length is $40 \cdot 128 = 5120$. We assumed that ANN will learn the data features for our 3 classes; ANN can be even applied to high-dimensional data to reduce data dimensionality [7].

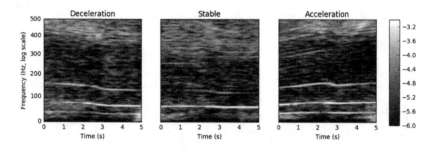

Fig. 2. R_{spg} values for deceleration, stable speed, and acceleration

We designed R_{spg} after constructing R_{lin}, with the intent to test if ANN can learn a similar concept as the lines, but using processed spectrograms from R_{spg}.

In Fig. 3 we show MDS (multi-dimensional scaling) plot from the random-Forest package in R for R_{lin} and R_{spg} for our data, to illustrate that these representation allow easy separation of the target classes.

4 Classifiers

In our previous work on speed change classification we applied ANN, SVM, and RF, as these are state of the art classifiers and yielded good results [8]. Therefore, we decided to use them for our 3 classes: deceleration, stable speed, acceleration. We applied the Keras library [4] for ANN and R [12] for other classifiers.

RF is a set of decision trees, built with minimizing bias and correlations between the trees. Each tree is built without pruning, for a different N-element bootstrap sample (i.e. obtained through drawing with replacement) of the N-element training set. For a K-element feature set, k features are randomly selected for each node of any tree. The best split on these k features is applied to split the data in the node. Gini impurity criterion is minimized to choose the split. The Gini criterion measures how often an element would be incorrectly labeled, if random labeling an object according to the distribution of labels in the subset is applied. This procedure is repeated M times, to obtain the forest of M trees. Classification using RF is performed by simple voting of all trees. Standard settings in the randomForest package in R were used (M = 500, $k = \sqrt{K}$).

SVM looks for a decision surface (hyperplane) maximizing the margin around the decision boundary between classes. The training points are called support vectors. SVM projects data into a higher dimensional space, using a kernel function; we used linear, quadratic, RBF (radial basis function) kernels. Each kernel has parameters, tuned in each classification experiment.

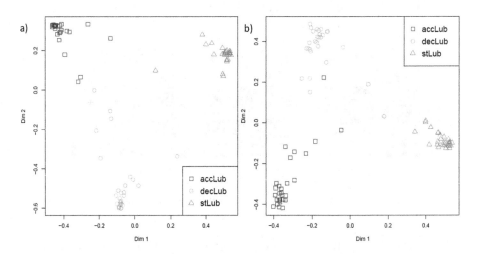

Fig. 3. MDS plot for the data described in Sect. 2 (*Lubartów*), represented with (a) R_{lin}, (b) R_{spg}, for acceleration (acc), stable speed (st), and deceleration (st)

ANN consists of the layers of interconnected neurons. We used a simple feed-forward neural net. One hidden layer was sufficient to classify our data (Sect. 2) using R_{spg}, with sigmoid function used as neuron activation function, and softmax at the output. Mean-square error was used as loss function, and stochastic gradient descent (SGD) as optimizer, with learning rate 10^{-5}, and momentum 0.9. We used the number of hidden units equal to the size of the input vector, and 3 outputs, for 3 classes. The data were normalized to the mean $\mu = 0$ and standard deviation $\sigma = 1$ before inputting to ANN.

5   Experiments

We performed the experiments on the data described in Sect. 2 through repeated runs of 3-fold crossvalidation (CV-3), i.e. with random division of the data into training and testing parts, 2/3 for training and 1/3 for testing.

The results for R_{lin}, averaged over 50 runs of each classifier, were as follows:

- RF: 98.98%,
- SVM: linear kernel 90.57%, quadratic kernel 78.28%, RBF kernel 93.62%.

For ANN and R_{spg} we performed 10 runs of CV-3, as training was much slower in this case. The average accuracy after 50 epochs training was 97.26%, for the net with 5120 input units, 5120 hidden neurons with sigmoid activation function, and 3 output units, with softmax.

Although our both representations were designed for using with particular types of classifiers, we also tested R_{lin} on ANN, and R_{spg} on SVM and RF. For R_{lin} we tested various settings of ANN, and we finally chose a 2-layer ANN. We obtained 90.42% average accuracy after 10 runs of ANN (CV-3) with the following topology: 18 input units, 18 hidden neurons in the each of the 2 hidden layers, with sigmoid activation function, and 3 output units with softmax. For the R_{spg} tested on RF and SVM (10 runs, CV-3) we obtained 96.91% accuracy for RF, 97.74% for SVM with linear kernel, 92.62% for quadratic kernel, and a low result at the level of random choice (33%) for RBF. SVM tuning was very slow for such a long input vector ($\|R_{spg}\| = 5120$), so it was performed only once. Still, the results for SVM with linear and quadratic kernels were good. Also, RF and ANN yielded good results on the data intended for other classifiers. Thus, we can apply classifiers different than intended, but then the tuning of hyper-parameters of ANN is needed, and SVM is slow on big input vectors.

These results were encouraging, so we decided to test our approach on *Ciecierzyn* data, recorded on August 2nd, 2016, in the countryside in Ciecierzyn, Lublin voivodship, Poland. These data (20 objects) differ from *Lubartów* because of different road, weather - summer, settings of the audio recorder, and vehicles. i.e. 3 different cars, including 2 with Diesel engine (2 drives of each car per class, plus 2 drives at 70 km/h for one of the cars). MDS plots in Fig. 4 show that adding these data makes discernment between our 3 classes more difficult.

In pre-tests on these data, using the classifiers (3 runs only) trained on the whole *Lubartów* dataset (Sect. 2), we obtained for R_{lin}:

- RF: 58%,
- SVM: linear kernel 75%, quadratic kernel 40%, RBF kernel 46.1%.

For R_{spg} the results were at the level of 45%. All the results are lower, but it is not surprising for the data very different than a relatively homogeneous dataset prepared for testing the time-frequency representations. Apparently, the *Lubartów* data were not representative for the *Ciecierzyn* data. Therefore, we decided to add some of the data from *Ciecierzyn* to the training, with the hope for classification improvement. Since we had 3 cars in *Ciecierzyn*, we performed

CV-3 with each of the 3 cars in a different fold, i.e. 2 cars added in training and the third one in tests. For R_{lin} we obtained:

- RF: 61%,
- SVM: linear kernel 65.5%, quadratic kernel 31.5%, RBF kernel 59.6%.

The results improved for SVM with RBF kernel, slightly improved for RF, and decreased for SVM with linear and quadratic kernel. This shows that these cars also differed one from another, so 2 cars are not representative for all 3 cars.

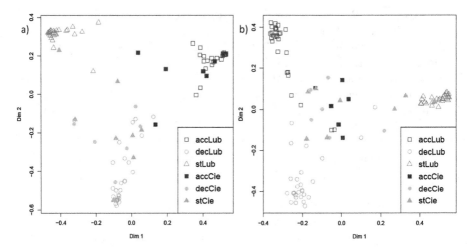

Fig. 4. MDS plot for combined *Lubartów* and *Ciecierzyn* data, represented with (a) R_{lin}, (b) R_{spg}, for acceleration (acc), stable speed (st), and deceleration (st)

We also performed tests on the *Ciecierzyn* data using ANN for R_{spg} representation. Unfortunately, the results were only 37.5% (average after 10 runs). Adding data from *Ciecierzyn* to the training did not improve the accuracy, and the decrease of the training error did not result in the decrease of the test error – we observed overfitting to the training data, and testing various ANN topologies did not help. We believe a more complicated topology or training method is required for such combined data, and more data should be acquired, to be representative for various settings: road surface, weather, engine type, etc. On the other hand, R_{lin} perform better for all classifiers, including ANN, which means that a small feature set is more representative and less sensitive to data variability and generalizes better. A huge number of details represented in the long R_{spg} vector fits more to the data, and prevents generalization.

6   Conclusions

Our experiments yielded high accuracy (99%) on our relatively homogenous data Therefore, we also decided to test this approach on another data set. The results were much lower, but still reaching 75%. We are aware that we had a limited amount of data, so this was only a pilot study, but the results are encouraging, so we are planning to acquire more data and then continue our research.

Acknowledgments. Work partially supported by the infrastructure bought within the project "Heterogenous Computation Cloud" funded by the Regional Operational Programme of Mazovia Voivodeship, and the Research Center of PJAIT, supported by the Ministry of Science and Higher Education in Poland.

References

1. Adobe. http://www.adobe.com/#
2. Ahn, I.-S., Bae, S.-G., Bae, M.-J.: Study on fault diagnosis of vehicles using the sound signal in audio signal processing. J. Eng. Technol. **3**, 89–95 (2015)
3. Averbuch, A., Rabin, N., Schclar, A., Zheludev, V.: Dimensionality reduction for detection of moving vehicles. Pattern Anal. Appl. **15**(1), 19–27 (2012)
4. Chollet, F.: Keras (2015). https://github.com/fchollet/keras
5. Duarte, M.F., Hu, Y.H.: Vehicle classification in distributed sensor networks. J. Parallel Distrib. Comput. **64**, 826–838 (2004)
6. George, J., Cyril, A., Koshy, B.I., Mary, L.: Exploring sound signature for vehicle detection and classification using ANN. Int. J. Soft Comput. **4**(2), 29–36 (2013)
7. Hinton, G.E., Salakhutdinov, R.R.: Reducing the dimensionality of data with neural networks. Science **313**, 504–507 (2006)
8. Kubera, E., Wieczorkowska, A., Słowik, T., Kuranc, A., Skrzypiec, K.: Audio-based speed change classification for vehicles. In: Appice, A., Ceci, M., Loglisci, C., Manco, G., Masciari, E., Raś, Z.W. (eds.) NFMCP 2016. LNAI, vol. 10312. Springer, Cham (to appear, 2017). http://www.springer.com/gp/book/9783319614601
9. Lyons, R.G.: Understanding Digital Signal Processing. Prentice Hall, Upper Saddle River (2012)
10. Sandberg, U.: The Multi-Coincidence Peak Around 1000 Hz in Tyre/Road Noise Spectra. Euro-Noise, Naples (2003)
11. Smith III., J.O., Serra, X.: PARSHL: an analysis/synthesis program for non-harmonic sounds based on a sinusoidal representation. In: International Computer Music Conference ICMC 1987. Computer Music Association (1987)
12. The R Foundation. http://www.R-project.org
13. Ziaran, S.: The assessment and evaluation of low-frequency noise near the region of infrasound. Noise Health **16**(68), 10–17 (2014)

Text and Web Mining

Opinion Mining on Non-English Short Text

Esra Akbas[(✉)]

Florida State University, Tallahassee, FL, USA
akbas@cs.fsu.edu

Abstract. As the type and the number of such venues increase, auto-mated analysis of sentiment on textual resources has become an essential data mining task. In this paper, we investigate the problem of mining opinions on the collection of informal short texts. Both positive and negative sentiment strength of texts are detected. We focus on a non-English language that has few resources for text mining. This approach would help enhance the sentiment analysis in languages where a list of opinion-ated words does not exist. We present a new method to automatically construct a list of words with their sentiment strengths. Then, we pro-pose a new method projects the text into dense and low dimensional fea-ture vectors according to the sentiment strength of the words. We detect the mixture of positive and negative sentiments on a multi-variant scale. Empirical evaluation of the proposed framework on Turkish tweets shows that our approach gets good results for opinion mining.

Keywords: Opinion mining · Sentiment analysis · Twitter · Text mining

1 Introduction

Users generally do an online search about a product before buying it, and online reviews about the product affect their opinion significantly. Companies can mon-itor their brand reputations, analyze how people's opinions change over time, and decide whether a marketing campaign is effective or not. As sharing opin-ions increase, getting useful information of sentiments on textual resources has become an essential data mining goal. The task of sentiment classification aims to identify the polarity of sentiment in text. The polarity is predicted on either a binary (positive, negative) or a multi-variant scale as the strength of sentiment expressed in the text. On the other hand, a text may contain a mix of posi-tive and negative sentiment; hence it is often necessary to detect both of them simultaneously [18]. An example of a text that shows this mixture is: "Carrier-X kaliteli ama çok pahalı". It is stated in the example that "Carrier-X provides quality service, but its cost is too high".

Extracting polarity from textual resources involves many challenges. Senti-ment may be expressed in a more hidden attitude. While a sentence with opinion words does not indicate sentiment, a sentence without any opinion words may

© Springer International Publishing AG 2017
M. Kryszkiewicz et al. (Eds.): ISMIS 2017, LNAI 10352, pp. 417–423, 2017.
DOI: 10.1007/978-3-319-60438-1_41

contain sentiment. The major problem we face in this paper is to classify sentiments of informal short texts in a non-English language. There have been several
studies focusing on sentiment classification on reviews, blogs, and news which
are relatively well-formed, coherent and at least paragraph-length pieces of text.
On the other hand, short-texts include only a few sentences of phrases. When
we use all words as features, corresponding feature vectors are so sparse and we
need to select those that have an important effect on the sentiment of the text.
The other challenge is that most of the available resources, such as lexicons and
labeled corpus, in the literature, are in English. Constructing these resources
manually is time-consuming.

Most of the previous studies are in the English text that has a rich resource
for sentiment analysis. Also, they use all words in the text to detect sentiment.
However, all words may not have an effect on the sentiment. To solve these
problems, we propose several methods.

Contributions of this paper are summarized as follows:

1. We focus on non-English texts in Twitter that contain informal short messages and construct our resources, lexicon and corpus for Turkish. Our work
   offers a methodology to enhance the sentiment analysis to other languages
   where such rich sources do not exist.
2. We present a new method to automatically construct a list of words with
   their sentiment strengths.
3. We propose a new representation of the text according to sentiment strength
   of the words, called *Opinion-based text representation: Grouped* as a new
   feature vector type. This method projects the text into dense and low dimensional feature vectors in contrast to BofW text representation as sparse and
   high dimensional.
4. We perform extensive evaluation of our proposed approaches by using our
   Turkish tweets dataset. According to this evaluation, our approaches give
   good accuracy results for sentiment analysis.

2   Background

2.1   Sentiment Analysis

A large collection of research on mining opinions from text has been done. The
existing works are presented as comprehensive survey in [13,15]. Most of these
works detect sentiment as positive-negative, or add a natural class to them. There
are also some studies [18,20], that detect positive-negative sentiment strength
by predicting human ratings on a scale.

Some approaches use unsupervised (lexical) methods which utilize a dictionary or a lexicon of pre-tagged (positive-negative or strength of sentiment) opinion words known as polar words and sentiment words [10,18,19]. In addition to
manually constructing this word list, it can be constructed automatically [8,14].

As a classification problem [11,16], a classifier is trained with a collection of
tagged corpus using a chosen feature vector and a supervised machine learning

method. Different kinds of feature vectors can be used to represent text in classification. In addition to words, tags, emoticons, punctuation *e.g.*, ?, !, .., negation words are also exploited as a feature type [2,6,12,15].

While most of research are for the text written in English, in recent years, some research is done for text written in Turkish [3,4,7,9,17,20].

3 Methodology

In this section, our sentiment model is described. First, we have a preprocessing step before converting text into feature vectors. Then, the sentiment of the tweets are extracted. Lexical and supervised approaches are combined to extract sentiment value of each tweet. In the lexical approach, we need a sentiment word list. Constructing a sentiment word list manually is time consuming and also, sentiment of the words depends on the dataset. Therefore, we develop an automated sentiment word list construction method. Then, we propose two methods to measure the sentiment strength of tweets. In addition to them, SentiStrength [18] is configured for Turkish.

3.1 Preprocessing

A text may contain many unnecessary information for sentiment analysis. Therefore, before converting the text into feature vectors, we remove these redundant parts of the text by applying different processes, such as punctuation and suffixes removing, spelling correction. A Turkish morphological analyzer, Zemberek [1], is used for these processes. Details of the preprocessing steps are skipped because of size limitation.

As another preprocessing step, we study on emoticons, which are also important symbols to represent the opinion in the text. There are different types of emoticons used in informal text. While some of them are used to express positive emotion, some of them are used for negative emotion. We group emoticons according to their corresponding sentiment value as "positive", "negative" and "others" and a symbol is given to each group. Then, all emoticons in the text are replaced with their symbols according to their sentiment types.

3.2 Sentiment Word List Construction

We propose a new approach to extract opinion words and their sentiment strength from a labeled corpus using relationships between words and classes which is the sentiment strength of the text. In [5], positive and negative words are extracted from documents of each class. In their algorithm, for each class C_i, all documents are examined and words are ranked based on their frequency in the document of the class. A set of top scoring positive words is labeled as good predictors of that class. So, a word w is called positive for a class c if

$$P(c|w) > 0.5 * p + 0.5 * P(c) \tag{1}$$

where $P(c|w)$ is the probability of the class c for word w and p is a parameter that is used in order to counteract the cases where the simpler relation $P(c|w) > P(c)$ leads to a trivial acceptor/rejector, for too small/large values of $P(c)$.

Similar to [5], we extract only positive words for each class. We have both 10 class values as the sentiment strength of the text and 10 sentiment strength values for words from 1 to 5 and -1 to -5. After getting positive words of each class with using Eq. 1, we give class values to the selected words of that class as sentiment strength of the words. If the word w_i is a positive word for class C_j, the value j is given to the word as its sentiment strength value. The intuition for this process is that if a word is seen more frequently in the text that has higher sentiment strength, sentiment strength of that word should be also higher and vice versa.

3.3   Sentiment Strength Detection

We propose two methods to measure the sentiment strength of tweets. As the basic method, we combine lexical and machine learning approaches, call it *Feature selection using sentiment lexicon*. As the second method, a new feature type, called *Opinion Based Text Representation: Grouped*, is created to represent the tweets. In addition to these, Sentistrength [18] is configured for Turkish as an alternative to them.

Feature Selection Using Sentiment Lexicon. We use the lexicon to eliminate the words that do not affect the sentiment of the text. Just the words in the lexicon are kept and the others are removed. Instead of looking only these words in the tweets as in the lexical approaches, we train a classifier using the words in the lexicon as the features. For this, we use the manually constructed and the automatically constructed lexicons as mentioned in the Sect. 3.2.

Opinion Based Text Representation: Grouped. BofW is the most commonly used method to represent the text as a feature vector. However, it constructs high dimensional feature vectors which include each word in the documents a feature. However, if two words have the same sentiment strength, they may have the same effect on the overall strength of the text. This means that two synonymous words have the same influence on the sentiment strength of the text, so it does not matter which one is in the text. A feature can represent the presence of both of them. Just we need to know how many of them are seen in the text, since this may have an effect on the sentiment of the text. For instance if the number of the words whose sentiment values are 5 in a text is high, sentiment value of the text should be also close to 5. Therefore, the emotion words in the lexicon are grouped according to their sentiment strength.

In our lexicon, there are five groups of positive and five groups of negative sentiment strength of the words. Also, one group for negation words such as "değil" (not), "hayır" (no) and one group for booster words such as "çok" (much), "fazla" (many) are added to them. In this new representation, each dimension of the feature vector corresponds to each group of the words. The last dimension of the feature vector is used to represent the sentiment strength (positive-negative) of the tweet as the class value.

The emotion words from groups are searched in the text. If a word from the group i is seen in the text, the value of the corresponding dimension i of the feature vector is raised by one. For instance, the word "güzel" (beautiful) has strength of 3 as a positive sentiment. If a text includes this word, the third dimension of its feature vector is increased by one. As another example, the word "işkence" (torment) has strength of -4 value as a negative sentiment. One is added to the value of the ninth $(5 + (-(-4)))$ dimension of the text's feature vector. Here is an example of the process of the converting a text into a feature vector.

A tweet;

"Carrier-Z, cok ucuzsun ama cok kalitesizsin. Evi gectim disarda bile cek-miyosun". ("Carrier-Z, your data plans are very good but your quality is zero. I cannot speak well at home but I also cannot speak well outside") p:3 n:-4.

Feature vector of the tweet; $<0, 0, 2, 1, 0, 0, 1, 0, 1, 0, 1, 2, (3, 4)>$

This shows that there are two words from group three and one word from group 4 and so on. After constructing feature vectors of tweets according to the proposed representation, a classifier with one of machine learning algorithms is trained and used to find the sentiment of test data.

4   Experimental Results

4.1   Data

A collection of tweets about three different Turkish telecommunication brands gathered over one month is used as the corpus for our experiment. Tweets are judged on a 5 point scale as follows for both positive and negative sentiment as in [18] manually. After eliminating Junk tweets, our data set includes 1420 tweets.

In addition to this, we construct a Turkish emotion word list manually as the lexicon. It includes 220 positive and negative words with a value from 1 to 5, following the format in SentiStrength [18] and booster words.

4.2   Opinion Mining

We tested our new feature type, *Opinion based text representation: Grouped*, and our proposed algorithms on the Turkish tweet data set by using a cross-validation approach. The algorithms are called; with new representation *Grouped* with the word list constructed automatically *Auto* with the word list constructed manually *Manual*. Also, the algorithms that include all words in the lexicons are called *combin*. We perform 10-fold cross-validation to train classifiers and to test our new feature type and algorithm. Different machine learning algorithms are used and best one with SVM is given in the Table 1 The results of our algorithms were compared to the result of the baseline majority class classification, the classification obtained using BofW feature vectors type and SentiStrength configured for Turkish. For SentiStrength, we change its English word list with our manually constructed Turkish Word list.

When we change p threshold values in the sentiment word list construction algorithm, different words are selected. According to selected words, the results of classifications change. First, we perform experiments for different threshold values using different machine learning algorithms mentioned above and one of threshold values and machine learning methods that give the best result is selected for other tests. The best threshold value is 0.4 for negative sentiment strength and 0.8 for positive.

Table 1. Performance of algorithms on positive and negative sentiment strength detection

Algorithms		P. Accuracy	P. Accuracy ±1	N. Accuracy	N. Accuracy ±1
Baseline		74.54%	79.60%	43.10%	77.14%
BofW		57.45%	79.25%	40.08%	73.14%
SentiStrength		56.90%	73.17%	30.77%	47.11%
Combination	Manual	74.51%	82.78%	48.42%	79.71%
	Auto	55.70%	74.54%	56.27%	81.59%
Grouped	Manual	75.50%	82.21%	49.96%	81.17%
	Auto	**78.95%**	**86.99%**	**62.94%**	**84.32%**

As we see from Table 1, using BofW feature type is not sufficient for both positive and negative sentiment learning. Since not all words affect the sentiment of the text and many features make the learning difficult and give poorer results. After selecting features using our lexicon, the results get better. However, after grouping them based on their sentiment strength, we obtain the best results. Also, the lexicon automatically constructed gives better results than one manually constructed. Since it is constructed based on the dataset and sentiment strength of words may be different for different datasets. So, sentiments of words in the lexicon constructed manually may be different from the data set and this may not give good results for sentiment extraction.

5   Conclusion

In this paper, we present a framework for mining opinions on the collection of informal short texts. We focus on Turkish and present a methodology that can be applied to other languages where rich resources do not exist.

We propose a new method to construct sentiment word list. Then, we compute sentiment strengths of the text using novel feature extraction algorithms. We apply feature selection by using sentiment lexicons to reduce the complexity and improve the accuracy of the results. Our framework includes methods to construct the sentiment lexicons, and a novel opinion based representation of text that can be applied to any language.

References

1. Akin, A.A., Akin, M.D.: Zemberek, an open source NLP library for Turkic languages (2007). http://code.google.com/p/zemberek/
2. Barbosa, L., Feng, J.: Robust sentiment detection on twitter from biased and noisy data. In: COLING 2010, pp. 36–44 (2010)
3. Cetin, M., Amasyali, F.: Active learning for Turkish sentiment analysis. In: INISTA 2013, pp. 1–4 (2013)
4. Dehkharghani, R., Saygin, Y., Yanikoglu, B., Oflazer, K.: SentiTurkNet: a Turkish polarity lexicon for sentiment analysis. Lang. Resour. Eval. **50**(3), 667–685 (2016)
5. Fragoudis, D., Meretakis, D., Likothanassis, S.: Best terms: an efficient feature-selection algorithm for text categorization. Knowl. Inf. Syst. **8**(1), 16–33 (2005)
6. Davidov, D., Tsur, O., Rappoport, A.: Enhanced sentiment learning using Twitter hashtags and smileys. In: COLING 2010, pp. 241–249 (2010)
7. Erogul, U.: Sentiment analysis in Turkish. Master's thesis, Middle East Technical University (2009)
8. Kamps, J., Marx, M., Mokken, R.J., Rijke, M.D.: Using wordnet to measure semantic orientation of adjectives. In: National Institute for, pp. 1115–1118 (2004)
9. Kaya, M., Fidan, G., Toroslu, I.: Sentiment analysis of Turkish political news. In: WI-IAT 2012, vol. 1, pp. 174–180, December 2012
10. Kennedy, A., Inkpen, D.: Sentiment classification of movie and product reviews using contextual valence shifters. Comput. Intell. **22**(2), 110–125 (2006)
11. Kim, S., Hovy, E.: Crystal: analyzing predictive opinions on the web. In: EMNLP-CoNLL 2007 (2007)
12. Liu, B.: Sentiment analysis and subjectivity. In: Handbook of Natural Language Processing (2010)
13. Liu, B.: Sentiment Analysis and Opinion Mining. Morgan and Claypool, San Rafael (2012)
14. Özsert, C.M., Özgür, A.: Word polarity detection using a multilingual approach. In: Gelbukh, A. (ed.) CICLing 2013. LNCS, vol. 7817, pp. 75–82. Springer, Heidelberg (2013). doi:10.1007/978-3-642-37256-8_7
15. Pang, B., Lee, L.: Opinion mining and sentiment analysis. Found. Trends Inf. Retrieval **2**(1–2), 1–135 (2008)
16. Pang, B., Lee, L., Vaithyanathan, S.: Thumbs up? Sentiment classification using machine learning techniques. In: Proceedings of the ACL, pp. 79–86, July 2002
17. Parlar, T., Özel, S.A.: A new feature selection method for sentiment analysis of turkish reviews. In: International Symposium on INnovations in Intelligent SysTems and Applications (INISTA), pp. 1–6. IEEE (2016)
18. Thelwall, M., Buckley, K., Paltoglou, G., Cai, D., Kappas, A.: Sentiment strength detection in short informal text. J. Am. Soc. Inf. Sci. Technol. **61**(12), 2544–2558 (2010)
19. Turney, P.D.: Thumbs up or thumbs down? Semantic orientation applied to unsupervised classification of reviews. In: Proceedings of ACL, pp. 417–424 (2002)
20. Vural, A.G., Cambazoglu, B.B., Senkul, P., Tokgoz, Z.O.: A framework for sentiment analysis in Turkish: application to polarity detection of movie reviews in Turkish. In: Gelenbe, E., Lent, R. (eds.) Computer and Information Sciences III, pp. 437–445. Springer, London (2013). doi:10.1007/978-1-4471-4594-3_45

Pathway Computation in Models Derived from Bio-Science Text Sources

Troels Andreasen[1($\boxtimes$)], Henrik Bulskov[1], Per Anker Jensen[2],
and Jørgen Fischer Nilsson[3]

[1] Computer Science, Roskilde University, Roskilde, Denmark
{troels,bulskov}@ruc.dk
[2] Management, Society and Communication, Copenhagen Business School,
Frederiksberg, Denmark
paj.msc@cbs.dk
[3] Mathematics and Computer Science,
Technical University of Denmark, Lyngby, Denmark
jfni@dtu.dk

Abstract. This paper outlines a system, ONTOSCAPE, serving to accomplish complex inference tasks on knowledge bases and bio-models derived from life-science text corpora. The system applies so-called natural logic, a form of logic which is readable for humans. This logic affords ontological representations of complex terms appearing in the text sources. Along with logical propositions, the system applies a semantic graph representation facilitating calculation of bio-pathways. More generally, the system affords means of query answering appealing to general and domain specific inference rules.

Keywords: Semantic text processing in bio-informatics · Bio-models using natural logic and semantic graphs · Querying and pathway computation

1 Introduction

This paper addresses logic-based bio-models derived from life science texts. We discuss representation languages and reasoning principles for bio-models derived from actual life science sources. In particular, we describe and exemplify the intended query answering and pathway functionality, that is, the ability to compute conceptual pathways in the stored model.

Our approach is based on the construction of a logical model for a considered bio-system and is in line with the foundational developments in [1,2]. One main challenge in the logical approach is the extraction of comprehensive bio-models from text sources and formalisation of these. This logical approach is in contrast to established and rather successful approaches to text mining based on direct references to phrases in concrete text sources and advanced information extraction techniques, cf. for example [13–16].

© Springer International Publishing AG 2017
M. Kryszkiewicz et al. (Eds.): ISMIS 2017, LNAI 10352, pp. 424–434, 2017.
DOI: 10.1007/978-3-319-60438-1_42

At first sight our approach resembles the well-known, rather simplistic entity-relationship models and RDF representations. However, our framework is unique in various respects, first of all in its generativity, that is, the ability of the models to accommodate arbitrarily complex concepts formed by composition of lexicalized classes and relationships as discussed in [3]. By way of examples, the virtually open-ended supply of concept terms such as *'cell in the liver that secretes hormone'*, *'arteria in pancreas'*, *'secrete from the exocrine pancreas'* are accomodated in the model by composing simple, given class terms into compound concept terms with an obvious resemblance to phrases in natural language, as these examples illustrate. All such encountered concepts are situated in the so-called "generative ontology" in a manner such that they can be "de-constructed" and reasoned with computationally.

The generativity and the liaison to natural language specifications is achieved, as mentioned, by adopting 'natural logic' cf. [6, 7] as the logical model language. In addition, the models come in the form of graphs with concepts as nodes and relations as edges.

The paper is organized as follows: In Sect. 2 we introduce models formalized in terms of natural logic. In Sect. 3 we derive semantic graph-based models from the natural logic specifications and in Sect. 4 we exemplify natural logic model fragments drawn from various medical text sources. Section 5 describes our prototype system and explains the pathway computation functionality, concluding finally in Sect. 6.

2   Models in Natural Logic

In the applied natural logic conception, a knowledge base or specification consists of a collection of descriptive sentences called 'propositions' in order to distinguish them from the natural language sentences from which they are derived. Propositions in the applied logic, dubbed NATURALOG, are of the following general form

$$Cterm_1 \; Relterm \; Cterm_2$$

where

- The two *Cterm*s are atomic or compound concept terms.
- The *Relterm* is a relational term, in the simplest cases corresponding to a transitive verb, e.g. 'cause', or 'secrete' or prepositions like 'in', 'via' etc.

In a logical proposition like betacell secrete insulin the two concept terms are atomic, and so is the intervening relational term. A bio-model comprising also the class inclusion ontology consists of a finite, albeit possibly huge, collection of such NATURALOG propositions. As it appears, we use *sans serif* font for propositions throughout. This model is then the basis for inferences and querying.

Propositions may contain complex structures: Compound *Cterm*s consist of a class C with attached qualifications. In a more complex proposition like

(cell that secrete insulin) is:located:in (pancreatic gland).
the first concept term consists of the atomic term cell adorned with a relative clause consisting of the relational term secrete followed by the concept term insulin. Relative clauses are indicated by the optional keyword 'that', merely to make the reading easier. Relative clauses are assumed always to act restrictively. For instance, as a matter of principle, cell that secrete insulin is recorded by the system as a sub-concept of cell in the concept inclusion structure in the ontology. Likewise, the second concept term pancreatic gland, is recognized as a sub-concept of the class gland in that all adjectives are also assumed to be interpreted restrictively. Parentheses are inserted for ease of reading and serve to ensure disambiguation. They may be omitted if there is no risk of ambiguity.

However, sub-class - and, more generally, sub-concept relationships may also be specified explicitly, namely by the relation term isa, as seen in copula sentences. Example: betacell isa cell. By contrast, the propositions (cell that secretes insulin) isa cell and (pancreatic cell) isa cell are inferred by the system according to the principles mentioned. Still, (pancreatic cell) isa (cell located-in pancreas) (and *vice versa*) has to be provided.

As it appears, the natural logic propositions are perfectly readable, if somewhat stereotypical, by domain experts by virtue of their resemblance to natural language. The converse, challenging task of automating translation from manageable parts of natural language in scientific text sources into natural logic is approached in our [3].

2.1   Quantifiers and Recursion in Concept Terms

The above propositional form $Cterm_1$ $Relterm$ $Cterm_2$ is a special case of

$$Q_1 \ Cterm_1 \ Relterm \ Q_2 \ Cterm_2$$

where the Qs are quantifiers, primarily 'all/every' or 'some'. Usually the quantifiers are absent with Q_1 then being interpreted as all and Q_2 as some by default. Accordingly, the example betacell secrete insulin is interpreted logically as the proposition all betacell secrete some insulin, where some insulin is meant to be some portion or amount of insulin. Generally speaking, classes are assumed to be non-empty (appealing to existential import), and the entities in a class of substance are taken to be arbitrary, non-empty amounts of the substance.

The propositional form all $Cterm_1$ $Relterm$ some $Cterm_2$ corresponds to the predicate logic formula $\forall x (Cterm_1[x] \rightarrow \exists y (Relterm[x,y] \wedge Cterm_2[y]))$, see further [3,4], where we also discuss the relationship to description logic. The introduced NATURALOG forms cover only those parts of binary predicate logic which are considered relevant for bio-modelling. Notable exclusions at present are logical negation and logical disjunctions.

Recall that a concept term consists of a class C followed by one or more qualifications or restrictions, where restrictions consist of a relational term followed by a concept term: $Relterm$ $Cterm$. In case of more than one restriction, these are to form a conjunction with and understood as logical conjunction proper.

By contrast, two and-aligned concept terms within the same class are conceived of as a logical disjunction (ex. beta-cell and alpha-cell produce hormone). Accordingly, concept terms have a finitely nested, recursive structure reflecting the syntax of natural language nominal phrases with possibly nested relative clauses and prepositional phrases. The handling of adjectives (ex. pancreatic gland) and compound nouns (ex. lung symptom) are both assumed to be acting restrictively. These as well as genitives will not be discussed further in this paper.

2.2   Ontologies

As mentioned above, a special case of the above propositions is class inclusion relationships corresponding to stylized copula sentences. For example, in the proposition pancreas isa (endocrine gland), isa denotes concept inclusion. The synonymy relation syn is construed as both way isa, cf. the declaration pancreas syn (pancreatic gland). Such propositions form the backbone of the ontology in our knowledge-based bio-models. Also partonomic propositions like betacell part-of (endocrine pancreas) are included in the ontology; cf. [8] for the various partonomic relations.

By contrast, a proposition like betacell secrete insulin is understood as an observational fact, an assertion, and therefore does not belong to the ontology proper. The concept of betacell would then be expected to be defined in some other way, which may or may not be part of the logical bio-model. However, the stated assertion might be replaced by the definitional proposition (cell that secrete insulin) syn betacell at the discretion of the domain expert. This proposition posits that all cells that secrete insulin (whatever their location), are to be called betacells.

3   Bio-Models as Semantic Graphs

In our framework, the natural logic propositions constituting a bio-knowledge base are parallelled by an alternative representation in the form of directed graphs as commonly used in bio-models [9–11]. The graphs come about by decomposing compound and relational concept terms into their constituents in the form of triples [6]. These triples are re-conceived of as labeled directed edges between nodes. Every concept is associated with one node and *vice versa*.

This semantic graph representation facilitates computation of relevant associations between concepts, namely by computation of connecting paths in the graph. For example, the subject concept in the proposition (cell that secrete insulin) located-in pancreas corresponding to the natural language sentence *cell that secretes insulin is located in pancreas* is internally decomposed into the two triples

(cell-that-secrete-insulin) isa cell.
(cell-that-secrete-insulin) located-in pancreas.

where the added auxiliary concept (cell-that-secrete-insulin) is conceived of as an atomic name of a node defined by the two triples. An arc symbol as in '⋪' is

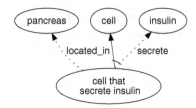

Fig. 1. The graph corresponding to the natural language sentence *cell that secretes insulin is located in pancreas.*

inserted between the defining edges in the graph rendition to express that they form the definition of the concept, in casu (cell-that-secrete-insulin).

The given proposition, which is epistemically in observational mode, then becomes represented by the triple (cell-that-secrete-insulin) located-in pancreas. So in this way a distinction is made between definitional and assertive (observational) propositions. This ensures that the original propositions, whatever their complexity, can be reconstructed modulo paraphrasation from the semantic graph as indicated with the double-headed arrow in Fig. 3. A graph representing the proposition and the decomposed subject term is shown in Fig. 1.

4   Fragments of a Bio-Model: A Case Study

To exemplify the approach, below we develop fragments of logical knowledge base representation based on excerpts from Wikipedia articles on endocrine glands and insulin. The fragments of concern are stated as propositions in an extended, relaxed form of NATURALOG, cf. [3].

Some propositions introduce sub-concepts by agglutination rather than by the use of separate words, calling for manual treatment. Conversely, some would-be compounds like *islet of Langerhans* and *Graves' disease* should not be decomposed, but should be kept as atomic class names. From the source [12] we consider the following

> **Endocrine glands are glands** of the endocrine system **that secrete hormones** directly into the blood rather than through a duct. The major **glands of the endocrine system include** the pineal gland, pituitary gland, **pancreas**, ovaries, testes, **thyroid gland, parathyroid gland, hypothalamus** and adrenal glands.

leading to the following triples

(endocrine gland) isa gland.
(endocrine gland) isa (gland that secrete hormone).
pancreas isa (endocrine gland).
hypothalamus isa (endocrine gland).
(thyroid gland) isa (endocrine gland).
(parathyroid gland) isa (endocrine gland).

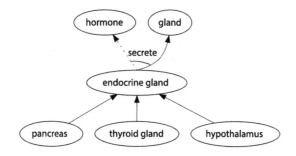

Fig. 2. Semantic graph focussing on endocrine gland.

where a few of the obvious triples corresponding to propositions about location
of the secretion and further specialisations of (endocrine gland) are omitted.
Corresponding to these propositions we derive the semantic graph shown in
Fig. 2. In addition, also from [12], we consider:

> The **pancreas**, *located in the abdomen close to the stomach*, **is both
> an exocrine and an endocrine gland**. *Calcitonin*, **produced by
> the parafollicular cells of the thyroid gland in response to rising
> blood calcium levels**, *depresses blood calcium levels by inhibiting bone
> matrix resorption and enhancing calcium deposit in bone.*
> The **parathyroid glands**, *located on the dorsal aspect of the thyroid
> gland*, **secrete parathyroid hormone**, *which* **causes an increase in
> blood calcium levels**.

from which we derive the following propositions (again omitting some to limit
the extent of the example):

pancreas isa (endocrine gland).
pancreas isa (exocrine gland).
calcitonin produced-by (parafollicular cell in the thyroid gland).
(parafollicular cell in the thyroid gland) located-in (thyroid gland).
(parafollicular cell in the thyroid gland) isa (parafollicular cell).
(parafollicular cell) isa (cell).
(rising calcium level in blood) cause (production of calcitonin).
(production of calcitonin) produce calcitonin.
(rising calcium level in blood) located-in blood.
(parathyroid glands) secrete (parathyroid hormone).
(parathyroid hormone) cause (rising calcium level in blood).

All derived propositions above, including those shown in Fig. 2, are situated in
the semantic graph shown in Fig. 4.

5   A Prototype System

As indicated above, pathway query answers are provided by first extracting
propositions from relevant texts as contributions to the semantic graph. The

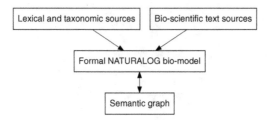

Fig. 3. Building the semantic graph.

extracted propositions are combined with knowledge from supplementary sources into a semantic graph with unified nodes. Finally, pathways are computed in a separate module based on the semantic graph. We briefly describe these tasks below.

5.1   Extracting Propositions from Text

The problem of deriving NATURALOG propositions from a natural language text remains an open issue. Our main idea is to analyse the text seen from an extended version of NATURALOG (see [3]). This extension is purely syntactical so that it captures more expression forms in the text. However, it is at present semantically conservative in the sense that propositions in the extended NATURALOG can be decomposed into the simple NATURALOG applied here. We intend to pursue this approach by stepwise extending also semantically NATURALOG to capture more meaning in the text. For instance, various forms of anaphora constructs fall outside NATURALOG, semantically. Also, for the moment, we consider only affirmative propositions, although in bio-texts one comes across negations for instance in the form of exceptions.

Furthermore, we envisage that one sentence may give rise to multiple propositions e.g. due to linguistic conjunctions, appositions, and parenthetical relative clauses. As shown in Fig. 1, one proposition in general gives rise to multiple triples in the graph rendition by a decomposition introducing nodes for compound, auxiliary terms.

5.2   Building the Semantic Graph

Given a proposition extraction module, a semantic graph is built by processing a corpus and situating the extracted triples in the graph. The graph is incrementally expanded by results derived from supplementary texts. Apart from textual input, contributions to the knowledge base may also be in the form of common knowledge lexical ressources (such as WORDNET), domain specific structured vocabularies/thesauri (such as UMLS) as well as other medical and bioscience sources that include taxonomic knowledge. Common for these sources is that they provide what can be considered concepts and relations connecting these. Therefore, the transformation into NATURALOG triples can be done by simple means. These triples, however, connect only atomic concepts. Thus, the contributions from such resources can be considered "skeleton"-ontologies to be further expanded with new atomic and compound concepts extracted from textual sources. The resource-based skeleton ontology is thus expanded into a generative ontology that grows incrementally with concepts and triples derived from new text sources. A sketch of the ontology building is shown in Fig. 3.

In Fig. 4 an example semantic graph is shown. The graph includes the example propositions derived in Sect. 4 and it includes the following additional atomic concept triples:

gland isa organ.
stomach isa organ.
grehlin isa hormone.
hormone isa protein.
insulin isa hormone.
pancreas produce insulin.

These may stem from results of other text sources or may be assumed to be part of a skeleton ontology that forms the basis for building the semantic graph.

5.3   Computational Query Answering and Pathfinding

Relationships derived by specialization of the subject and generalization of the object, know as inference by monotonicity, are identified by computational traversal of stated relationships. Concepts are connected in the semantic graph by pathways reflecting mathematical composition of the relations represented by edges in the logical bio-model, cf. [2]. It is our tenet that bio-pathways appear among the computed pathways between the given query concepts.

This computation process is supported by logical inference rules since inferred propositions may constitute shortcuts, as it were, in the graph view. For instance, the transitivity of inclusion, isa, conceptually shortens the distance from a concept to a superior concept in the ontology via intermediate concepts. Similarly for partonomic, causative and effect relations. In [4], the path finding is explained more abstractly as application of appropriate logical comprehension principles supporting the relation composition.

A miniature ontology, corresponding to a subset of the bio-model propositions listed in Sect. 4, is visualised in the graph in Fig. 4. In addition, two candidate answers to the query comprising two concepts

rising calcium level in blood ~ gland?

are indicated.

An answer is provided as a pathway connecting the two concepts and the pathway can be seen as an explanation of how the two concepts are related. Consider the graph in Fig. 4 and assume that the darkgrey nodes are not yet inserted. The reading of the (answer corresponding to) lightgrey path can be

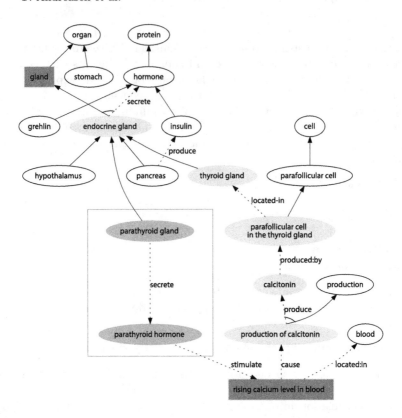

Fig. 4. A miniature ontology corresponding to a subset of the bio-model propositions listed in Sect. 4. Two pathways connecting rising calcium level in blood and gland are shown.

> Rising calcium level in the blood causes production of calcitonin in the parafollicular cells in the thyroid gland, which is an endocrine gland, which is a gland.

Suppose, at a later state, that new knowledge is being added to the base and that this include the darkgrey nodes in Fig. 4. The two query concepts are now connected by a new and shorter pathway corresponding to the following alternative answer.

> Rising calcium level in the blood is stimulated by parathyroid hormone, which is secreted by the parathyroid gland, which is an endocrine gland, which is a gland.

A pathway computation, being more than a pure inferential process, in our system is also the composition of relations guided by appropriate path computation. In our framework this computation is reduced algorithmically to search for weighted paths between concept nodes in the graph representation, utilizing standard heuristic algorithms in artificial intelligence. The intermediate propositional representations refer back to the source texts so that computed paths can be shown by highlighting excerpts in the texts.

6 Summary and Conclusion

We have described a system for querying and pathfinding in bio-models taking the form of logical knowledge bases derived from text sources. The applied logical language accommodates complex propositions, which can be queried by deductive means, and the supporting semantic graph form enables algorithmic pathfinding between concepts. A small scale prototype has been developed that translates complex propositions into a graph representation for pathfinding. This prototype is described in detail in [5]. Computational translation of text sources into the logical form is a challenging problem, which is approached in [3] by adopting enriched forms of natural logic as a specification language for bio-systems.

References

1. Schultz, S., Hahn, U.: Towards the ontological foundations of symbolic biological theories. Artif. Intell. Med. **39**, 237–250 (2007)
2. Bittner, T., Donelly, M.: Logical properties of foundational relations in bio-ontologies. Artif. Intell. Med. **39**, 197–216 (2007)
3. Andreasen, T., Bulskov, H., Fischer Nilsson, J., Jensen, P.A.: On the relationship between a computational natural logic and natural language. In: The 8th International Conference on Agents and Artificial Intelligence, ICAART, pp. 335–342 (2016)
4. Andreasen, T., Bulskov, H., Fischer Nilsson, J., Anker Jensen, P., Lassen, T.: Conceptual pathway querying of natural logic knowledge bases from text bases. In: Larsen, H.L., Martin-Bautista, M.J., Vila, M.A., Andreasen, T., Christiansen, H. (eds.) FQAS 2013. LNCS, vol. 8132, pp. 1–12. Springer, Heidelberg (2013). doi:10.1007/978-3-642-40769-7_1
5. Andreasen, T., Bulskov, H., Fischer Nilsson, J., Jensen, P.A.: A system for conceptual pathway finding and deductive querying. In: Andreasen, T., et al. (eds.) Flexible Query Answering Systems, vol. 400, pp. 461–472. Springer, Cham (2015). doi:10.1007/978-3-319-26154-6_35
6. Fischer Nilsson, J.: Diagrammatic reasoning with classes and relationships. In: Moktefi, A., Shin, S.-J. (eds.) Visual Reasoning with Diagrams. Studies in Universal Logic. Birkhäuser, Springer, Basel (2013)
7. van Benthem, J.: Essays in Logical Semantics. Studies in Linguistics and Philosophy, vol. 29. D. Reidel Publishing Company, Dordrecht (1986)
8. Smith, B., Rosse, C.: The role of foundational relations in the aligment of biomedical ontologies. In: Fieschi, M., et al. (eds.) MEDINFO 2004 (2004)
9. Vechina, A., et al.: Representation of semantic networks of biomedical terms. In: Proceedings of the International Work-Conference on Bioinformatics and Biomedical Engineering, Granada, Spain (2013)

10. Miljkovic, D., et al.: Incremental revision of biological networks from texts. In: Proceedings of the International Work-Conference on Bioinformatics and Biomedical Engineering, Granada, Spain (2013)
11. Quesada-Martinéz, M., et al.: Analysis and classification of bio-ontologies by the structure of their labels. In: Proceedings of the International Work-Conference on Bioinformatics and Biomedical Engineering, Granada, Spain (2013)
12. Endocrine Gland (2015). Wikipedia. http://en.wikipedia.org/wiki/Endocrine_gland. Accessed 6 Mar 2016
13. Li, C., Liakata, M., Rebholz-Schuhmann, D.: Biological network extraction from scientific literature: state of the art and challenges. Briefings Bioinform. **15**(5), 856–877 (2013)
14. Kaewphan, S., Kreula, S., Van Landeghem, S., Van de Peer, Y., Jones, P.R., Ginter, F.: Integrating large-scale text mining and co-expression networks: targeting NADP(H) metabolism in E. coli with event extraction. In: Proceedings of the Third Workshop on Building and Evaluating Resources for Biomedical Text Mining (BioTxtM 2012), pp. 8–15 (2012)
15. Hakala, K., Van Landeghem, S., Kaewphan, S., Salakoski, T., Van de Peer, Y., Ginter, F.: CyEVEX: literature-scale network integration and visualization through Cytoscape. In: Proceedings of SMBM 2012, pp. 91–96, Zurich, Switzerland (2012)
16. Miwa, M., Ohta, T., Rak, R., Rowley, A., Kell, D.B., Pyysalo, S., Ananiadou, S.: A method for integrating and ranking the evidence for biochemical pathways by mining reactions from text. Bioinformatics **29**(13), i44–i52 (2013)

Semantic Enriched Short Text Clustering

Marek Kozlowski$^{(\boxtimes)}$ and Henryk Rybinski

Warsaw University of Technology, Warsaw, Poland
mkozlowski@ii.pw.edu.pl
http://www.ii.pw.edu.pl

Abstract. The paper is devoted to the issue of clustering short texts, which are free answers gathered during brain storming seminars. Those answers are short, often incomplete, and highly biased toward the question, so establishing a notion of proximity between texts is a challenging task. In addition, the number of answers is counted up to hundred instances, which causes sparsity. We present three text clustering methods in order to choose the best one for this specific task, then we show how the method can be improved by a semantic enrichment, including neural-based distributional models and external knowledge resources. The algorithms have been evaluated on the unique seminar's data sets.

Keywords: Document clustering · Information retrieval · Semantic enrichment

1 Introduction

Since a decade, text clustering has become an active field of research in the machine learning community. Most of the approaches are based on the term occurrence frequency. The clustering engines group documents on the basis of their lexical similarity, and therefore suffer from lack of semantics, just to mention polysemy. Furthermore, research in text categorization has mainly focused on documents with at least the size of a paragraph. The performance of such surface-based methods can drastically decrease when the texts are too short e.g. phrase/sentence size. Actually, the difficulties with short text clustering results from the text ambiguity and a lack of significant number of features in document vectors. Solutions for it are usually based on semantics injection by expanding features with some additional external information or by assigning the meaning to the textual layer of documents.

The problem considered in the paper results from practical needs. Namely, a company runs a business, which consists in organizing brain-storming seminars for the staff of their customers. During the seminars the participants answer to a series of questions. A back-office application clusters responses, so that one can identify the company problems. The clusters of responses are presented to the audience almost in real time in order to easily identify strong ideas and weak signals. Participants can react with the results and create a dialogue. All

© Springer International Publishing AG 2017
M. Kryszkiewicz et al. (Eds.): ISMIS 2017, LNAI 10352, pp. 435–445, 2017.
DOI: 10.1007/978-3-319-60438-1_43

the participants' responses are free text answers. The answers are short, often incomplete, and highly biased toward the questions, so establishing a measure of proximity between such texts is a challenging task.

Although we start with a specific task, we would like to approach to it in a more general way. The main goal of the paper is to identify a clustering method to be used in the process of analyzing very short textual answers, so that the manual work for ordering the discussion material is minimized.

We have verified a number of well-known short document clustering approaches, including data-centric ones, description centric ones, and semantic-enhanced ones. Additionally, we have identified some improvements concerning features expansion in order to resolve the problem of sparsity. In the presented approach we combine the semantic-enhancement (distributional models and sense disambiguation) with the clustering method in order to overcome the above-mentioned flaws. The presented approach is evaluated on a custom French data set from seminars, as provided by the company.

2   Related Work

2.1   Text Clustering

The goal of text clustering in information retrieval is to discover groups of semantically related documents. At the root of all the documents clustering methods lies the van Rijsbergen's [16] hypothesis: "closely associated documents tend to be relevant to the same requests", whereas documents concerning different meanings of the input query are expected to belong to different clusters.

The approaches to short text clustering can be classified as data-centric, description-centric or semantic-enhanced. The data-centric approach focuses more on the problem of data clustering, rather than presenting the results to the user. The algorithm Scatter/Gather [1] is an example of this approach. It divides the data set into a small number of clusters and, after the selection of a group, it performs clustering again and proceeds iteratively using the Buckshot-fractionation algorithm. Other data-centric methods use bisecting k-means or hierarchical agglomerative clustering.

Description-centric approaches are focused on describing the resulting clusters. They discover diverse groups of semantically related documents associated with meaningful, comprehensible and compact text labels. Among the most popular and successful approaches are the phrase-based ones that form clusters based on recurring phrases instead of numerical frequencies of isolated terms. The Suffix Tree Clustering algorithm [18,19] (shortly STC) employs frequently recurring phrases as both document similarity feature and final cluster description. Clustering with the STC algorithm is treated as finding groups of documents sharing a high ratio of frequent phrases. Lingo algorithm [13,14] combines common phrase discovery with latent semantic indexing techniques to separate search results into meaningful groups. Lingo uses singular value decomposition of the term-document matrix to select good cluster labels among the candidates extracted from the text (frequent phrases).

The phrase-based methods usually provide good results. They reveal some problems when some topics are dominated, or the texts contain different words referring to one meaning. In [2,3,11] a novel approach for short text clustering was presented, which is based on automatic discovering word senses from raw text. The method clusters texts based on their similarity to the induced query senses. The methods of this kind are called semantic-enhanced clustering algorithms. An example of a such algorithm is also SnSRC [6,7], though it does not use any external corpora. SnSRC is well suited to cluster short texts (e.g. web snippets). It induces word senses in order to dynamically acquire an inventory of senses in a given set of retrieved documents. The number of discovered senses does not have to be predefined by the user, as it is determined solely by the corpus content.

2.2 Data Expansion

All the above mentioned methods work well with the sets containing at least hundreds of documents, each composed of at least one, even small, paragraph. Otherwise, we encounter the problem of data sparsity, which makes impossible finding relevant regularities among documents. In order to resolve the problem of insufficient text representation there are two ways of using external information resources, namely: (1) distributional semantic models (e.g. neural-based distributional models, like e.g. word2vec); and (2) well structured knowledge resources (e.g. BabelNet). Information coming from the resources can be used to expand native features.

2.3 Distributional Semantic Models

Distributional Semantic Models (DSMs) have recently received increased attention, together with the rise of neural architectures for scalable training of dense vector embeddings. One of the strongest trends in Natural Language Processing (NLP) is the use of word embeddings [5], which are vectors whose relative similarities correlate with semantic similarity. Such vectors are used both as an end in itself (for computing similarities between terms), and as a representational basis for downstream NLP tasks, like text classification, document clustering, part of speech tagging, named entity recognition, sentiment analysis, and so on.

Also, neural network based distributional semantics received a substantially growing attention. The main reason for this is a very promising approach of employing neural network language models (NNLMs) trained on large corpora to learn distributional vectors for words [15]. Recently, Mikolov et al. [9,10] introduced the Skip-gram and Continuous Bag-of-Words models, being efficient methods for learning high-quality vector representations of words from large amounts of unstructured text data. The word representations computed using neural networks are very interesting because the learned vectors explicitly encode many linguistic regularities and patterns. The most well-known tool in this field now is word2vec[1], which allows fast training on huge amounts of raw linguistic data. Word2vec takes as its input a large corpus and builds a vector space, typically of few hundred dimensions, with each unique word in the corpus represented by corresponding vector in the space. In distributional semantics, words are usually represented as vectors in a multi-dimensional space. Semantic similarity between two words is then trivially calculated as the cosine similarity between their corresponding vectors.

2.4   Babel Eco-System

The creation of very large knowledge bases has been made possible by the availability of collaboratively edited online resources such as Wikipedia and WordNet. Although these resources are only partially structured, they provide a great deal of valuable knowledge which can be harvested and transformed into structured form.

BabelNet [4,12] is a multilingual encyclopedic dictionary and semantic network, which currently covers more than 271 languages and provides both lexicographic and encyclopedic knowledge thanks to the seamless integration of WordNet, Wikipedia, Wiktionary, OmegaWiki, Wikidata and Open Multilingual WordNet. BabelNet encodes knowledge in a labeled directed graph $G = (V, E)$, where V is the set of nodes (concepts) and E is the set of edges connecting pairs of concepts. Each edge is labeled with a semantic relation. Each node contains a set of lexicalizations of the concepts for different languages. The multilingually lexicalized concepts are Babel synsets. At its core, concepts and relations in BabelNet are harvested from WordNet, and Wikipedia.

Babelfy [4,12] is a unified graph-based approach that leverages BabelNet to jointly perform word sense disambiguation and entity linking in arbitrary languages. It is based on a loose identification of candidate meanings coupled with a densest subgraph heuristic, which selects high-coherence semantic interpretations. Babelfy WSD evaluation outperforms the state-of-the-art supervised systems.

3   The Proposed Approach

In order to perform the computations we have built a dedicated tool, which processed the text data in a pipeline, starting with (1) the input data preprocessing, followed by (2) semantic enhancement of the texts, using the BabelNet resource and word2vec, respectively, and then (3) clustering by SnSRC. In addition, we have used as a baseline the results of SnSRC clustering for the texts without any semantic enhancement.

---

[1] https://code.google.com/p/word2vec/.

3.1   Preprocessing

Texts as an unstructured source of information are usually transformed into a structured representation (term-document matrix). Following the seminar scenario, the first step was to retrieve the answers for a given seminar question, and then iteratively process the answer texts into bag-of-words representations. The answers (usually short phrases or sentences) were processed by sentence segmentation, word tokenization, stop-words cleaning, spell-correction, then finally tokens were stemmed (Snowball Stemmer[2]).

In this pipeline, special importance was given to spell correction, which we applied before stemming, as the answers were very often misspelled. We customized Norvig spell corrector[3]. On the other hand, we removed from the pipeline the PoS tagging phase, because the texts were very short and any relevant word was crucial in the clustering phase.

3.2   BabelNet Enriched Clustering

In order to verify how Babelnet and Babelfy can influence the clustering we enhanced the texts with semantic features, such as synsets[4], and categories describing them[5]. We used the HTTP API provided by Babelfy and BabelNet. Babelfy was used to disambiguate the given text by finding relevant synsets for words. Then, with the disambiguated words represented by synsets we used BabelNet API in order to access the corresponding categories and semantic edges.

3.3   Word2Vec Enriched Clustering

In order to verify how the neural network based distributional model can improve the clustering quality in short textual answers, first we trained the model with negative sampling using raw textual data. We have trained two models: (1) one with the Wikipedia texts, and (2) another one with seminar data. Conceptually word2vec involves a mathematical embedding from a classical space with one dimension per word to a continuous vector space with much lower dimension usually each word is represented by several hundreds of real numbers, called word embeddings. After building the model we computed cosine similarity between a simple mean vector of the embeddings of the text's words and the vectors for each word in the distributional model. Then, the retrieved top semantically similar words were added as the additional term features to the initial bag of words text representation.

---

[2] http://snowball.tartarus.org/texts/introduction.html.
[3] http://norvig.com/spell-correct.html.
[4] https://babelfy.io/v1/disambiguate.
[5] https://babelnet.io/v3/getSynset.

4   Experiments

4.1   Data

As mentioned above, for the experiments we used data from brain-storming seminars. For each query (discussion starting point), the participants provided short answers. All the answers for a given query are then subject of automatic clustering, finally the clusters were manually cleaned, so that each cluster contained semantically closed sentences. Clearly, the manually cleaned clusters represent in our experiments gold standard. The gold standard data set used for the experiments was given as a set of XML files, each one in the form representing answer texts to a query, grouped into the final clusters. Each cluster has manually assigned a label expressing the meaning of the cluster. An example of one answer in the XML form is provided below:

```
<group id="203" lv="0">
    <answer id="1049" origin="30">Ouverture: partager une
    veille commerciale, technologique, manageriale,
    innovation, performance
    </answer>
</group>
```

The obtained gold standard set contained 60 files (each representing answers to a query). Below are some details about this data set:

1. the number of answers for each query was between 21 and 97, on average 49.28 answers per query;
2. the total number of answers in all files of the gold standard was 2957;
3. the length of each answer varies from 3 to 20 words (on average 8), after preprocessing it is reduced to 5 words on average;
4. on average, for a question there were 7.85 manually created clusters.

4.2   Processing

One of our first task was to find the best knowledge-poor clustering method for short texts provided by the company. Hence, we tested four algorithms: data centric (bisecting k-means), two description centric ones (Lingo, STC), and SnSRC. Then, we performed experiments to verify how semantic enrichment can improve short text clustering. As SnSRC has revealed to be the best one, we have limited only to verifying this algorithm. For the semantic enrichment of the texts we have used Babel eco-system (Babelnet, Babelfy), and neural-network based distributional models, for which we used the word2vec tool.

4.3   Scoring

The methods were evaluated in terms of the clustering quality. In the literature one can find many evaluation measures. In our experiments we have considered

Rand Index(RI), Adjusted Rand Index (ARI). In addition, as the company was using the entropy based measures, namely Homogeneity, Completeness, and V-measure (harmonic mean of homogeneity and completeness), we included these measures to our experiments as well.

Rand Index determines the percentage of text pairs that are in the same configuration in the evaluated clusters and the gold standard. Its main weakness is that it does not take into account the probabilistic nature of RI, the expected value of the RI of two random clusterings is not a constant value. This issue is addressed by the Adjusted Rand Index, which corrects the RI for the probability distribution [8].

The two entropy-based measures – homogeneity and completeness are introduced by Rosenberg and Hirschberg [17]. However, the two measures are not normalized with regard to random labeling. This problem can safely be ignored, when the number of samples is higher than a thousand and the number of clusters is less than 10. Let us recall, that in our experiments we worked with the sets containing from 20 to 100 answers, therefore the entropy based measures (homogeneity and completeness) should not be taken into account. However, as the company was using these measures, we have decided to assess the measures for Lingo, and demonstrate the problems, which turned out to be quite interesting for revealing 2 paradoxes. Table 1 illustrates them:

1. with increase of the number of clusters the completeness is also growing in spite of the fact that more elements should be missing in each cluster;
2. even more interesting is that both the homogeneity and completeness are getting higher in line with increasing the number of clusters, so with almost 2,5 times more clusters than in the gold standard the entropy-based parameters are both better than for clustering giving similar number of clusters as the gold standard (as a rule, the growth of precision causes loss of completeness).

To this end, in all the experiments we finally decided to use only ARI.

Table 1. The paradox of improving the entropy-based homogeneity and completeness for a small number of instances presented in 7 series of experiments producing more and more clusters, where cl. – average number of obtained clusters; and gscl. – the average number of clusters in the gold standard.

Measure	0	1	2	3	4	5	6
Homogeneity	28.79	35.18	36.33	43.72	55.74	64.28	66.48
Completeness	38.39	39.99	40.44	41.92	42.01	43.27	44.49
cl.	6.48	7.85	8.45	10.32	13.02	16.95	19.25
gscl.	7.85	7.85	7.85	7.85	7.85	7.85	7.85

5   Results

First, we have verified the four text clustering methods, i.e. SnSRC, Lingo, STC, and Bisecting k-means. The evaluations were performed without defining *a priori* the number of resulting clusters, and with limiting the number of clusters. The comparison of the algorithms is presented in Table 2 (without limiting the number of clusters), and in Table 3 (with limiting the number of cluster to the gold standard properties).

Table 2. The quality evaluation of the clustering methods measured by ARI (percentages); cl. – the average number of obtained clusters, gscl. – the average number of clusters in gold standard.

Method	ARI	cl.	gscl.
Bisecting K-means	3.71	9.45	7.85
STC	11.35	8.61	7.85
Lingo	12.62	10.32	7.85
SnSRC	**15.88**	10.65	7.85

Table 3. ARI scores (percentages) reported by clustering methods with the predefined constraint at number of clusters.

Method	ARI	cl.	gscl.
Bisecting K-means	4.61	7.85	7.85
STC	11.21	7.85	7.85
Lingo	11.11	7.85	7.85
SnSRC	**12.73**	7.85	7.85

Then, as SnSRC obtained the best results, in the consecutive experiments we have focused on experimenting with this algorithm only. Table 4 shows results of the experiment for clustering texts with semantic enrichment. The first row presents the baseline, which corresponds to SnSRC without semantic enrichment of texts. The next rows present the results obtained for the evaluated types of semantic enrichment of texts, namely:

1. Babel-synsets - the text is disambiguated using Babelfy, and the retrieved synset ids are added as additional tokens to the answer's textual data;
2. Babel-categories - the text is disambiguated using Babelfy, and the retrieved synset ids are processed with BabelNet in order to get the corresponding categories, and add them as additional tokens to the answer's textual data;
3. Babel-glosses - the text is disambiguated using Babelfy, and the retrieved synset ids are processed with BabelNet in order to get the corresponding glosses, the glosses are added as additional phrases to the texts;

4. Babel-hypernyms - the text is disambiguated using Babelfy, and the retrieved synset ids are processed with BabelNet in order to get the corresponding hypernyms, those hypernyms are added as additional tokens to the texts;
5. Word2Vec-1 presents the results of the neural network based model trained on texts taken from Wikipédia;
6. Word2Vec-2 presents the results of the neural network based model trained on the company corpus.

Table 4. ARI scores (percentages) reported with the use of BabelNet and Word2Vec enrichment.

Method	ARI	cl.	gscl.
Baseline - SnSRC	15.88	10.65	7.85
Babel-synsets	16.23	10.04	7.85
Babel-categories	15.13	10.55	7.85
Babel-glosses	14.03	9.33	7.85
Babel-hypernyms	15.21	9.56	7.85
Word2Vec-1	16.42	10.38	7.85
Word2Vec-2	**17.42**	9.88	7.85

The evaluation phase of semantic enrichment shows that the information coming from the word2vec model built on the domain oriented texts outperform others improvements.

6   Conclusions

We presented an approach to short text clustering problem, which is based on SnSRC. We evaluated possible improvements of SnSRC by means of semantic enrichment of the clustered texts. In particular, we tested the influence of enriching texts with the BabelNet/Babelfy resources, and by using neural network-based distributional models.

The first step of the evaluation was devoted to comparing with SnSRC three clustering algorithms, namely Bisecting k-Means, STC, and Lingo. Then the idea of semantic enrichment of short texts was tested on SnSRC. In particular, we tested how enhancing short texts by semantic features from BabelNet/Babelfy influences the quality of clustering, then we performed experiments with the neural network based distributional model.

The quality improvements of the semantic extensions based on the Babel-Net/Babelfy resources are modest. The expansion of answers by corresponding categories, hypernyms and glosses does not provide any quality gain against the baseline solution. The extension by synsets increases ARI by 0.35%, as compared to the baseline, which is c.a 2.2% of the relative improvement over the baseline.

The experiments with the neural network based model (implemented by means of word2vec) show better results. In particular, the model trained with the texts from the company increases ARI by 1.54%, as compared to the baseline, which is 9.6% of the relative improvement over the baseline.

In the future we plan to introduce more sophisticated improvements connected with the Babel eco-system concerning graph theories, because there must be a way to drastically improve the quality measures consuming such well-defined and organized semantic network as BabelNet.

References

1. Cutting, D., Karger, D., Pedersen, J., Tukey, J.: Scatter/gather: a cluster based approach to browsing large document collections. In: Proceedings SIGIR, Copenhagen, pp. 318–329 (1992)
2. Di Marco, A., Navigli, R.: Clustering web search results with maximum spanning trees. In: Pirrone, R., Sorbello, F. (eds.) AI*IA 2011. LNCS, vol. 6934, pp. 201–212. Springer, Heidelberg (2011). doi:10.1007/978-3-642-23954-0_20
3. Di Marco, A., Navigli, R.: Clustering and diversifying web search results with graph-based word sense induction. Comput. Linguist. **39**(3), 709–754 (2013). MIT Press
4. Flati, T., Navigli, R.: Three birds (in the LLOD cloud) with one stone: BabelNet, Babelfy and the Wikipedia Bitaxonomy. In: Proceedings of SEMANTiCS, Leipzig (2014)
5. Huang, E.H., Socher, R., Manning, C.D., Ng, A.Y.: Improving word representations via global context and multiple word prototypes. In: Proceedings of 50th Annual Meeting of the ACL (2012)
6. Kozłowski, M., Rybiński, H.: SnS: a novel word sense induction method. In: Kryszkiewicz, M., Cornelis, C., Ciucci, D., Medina-Moreno, J., Motoda, H., Raś, Z.W. (eds.) RSEISP 2014. LNCS, vol. 8537, pp. 258–268. Springer, Cham (2014). doi:10.1007/978-3-319-08729-0_25
7. Kozlowski, M., Rybinski, H.: Word sense induction with closed frequent termsets. In: Computational Intelligence (2016)
8. Manning, C., Raghavan, P., Schutze, H.: Introduction to Information Retrieval. Cambridge University Press, Cambridge (2008)
9. Mikolov, T., Chen, K., Corrado, G., Dean, J.: Efficient estimation of word representations in vector space. In: International Conference on Learning Representations (2013)
10. Mikolov, T., Le, Q.: Distributed representations of sentences and documents. In: International Conference on Machine Learning, Beijing (2014)
11. Navigli, R., Crisafulli, G.: Inducing word senses to improve web search result clustering. In: Proceedings Conference on Empirical Methods in NLP, Boston, pp. 116–126 (2010)
12. Navigli, R.: (Digital) goodies from the ERC wishing well: BabelNet, Babelfy, video games with a purpose and the Wikipedia bitaxonomy. In: Proceedings of the 2nd International Workshop on NLP and DBpedia, Italy (2014)
13. Osiński, S., Stefanowski, J., Weiss, D.: Lingo: search results clustering algorithm based on singular value decomposition. In: Kłopotek, M.A., Wierzchoń, S.T., Trojanowski, K. (eds.) IIPWM 2004, vol. 25, pp. 359–368. Springer, Heidelberg (2004)
14. Osinski, S., Weiss, D.: A concept-driven algorithm for clustering search results. IEEE Intell. Syst. **20**(3), 48–54 (2005). IEEE Press

15. Sutskever, I., Mikolov, T., Chen, K., Corrado, G.S., Dean, J.: Distributed representations of words and phrases and their compositionality. In: Advances in Neural Information Processing Systems (2013)
16. Rijsbergen, C.: Information Retrieval. Butterworths, London (1979)
17. Rosenberg, A., Hirschberg, J.: V-measure: a conditional entropy-based external cluster evaluation measure. In: Proceedings of the 2007 Joint Conference on Empirical Methods in NLP and Computational Natural Language Learning (2007)
18. Zamir, O., Etzioni, O.: Web document clustering: a feasibility demonstration. In: Proceedings of the 21st Annual International ACM SIGIR Conference on Research and Development in Information Retrieval, New York, pp. 46–54 (1998)
19. Zamir, O., Etzioni, O.: Grouper: a dynamic clustering interface to web search results. Comput. Netw. **31**(11), 1361–1374 (1999). Elsevier

Exploiting Web Sites Structural and Content Features for Web Pages Clustering

Pasqua Fabiana Lanotte[1($\boxtimes$)], Fabio Fumarola[2], Donato Malerba[1,3], and Michelangelo Ceci[1,3]

[1] University of Bari Aldo Moro, via Orabona 4, 70125 Bari, Italy
{pasqua.lanotte,donato.malerba,michelangelo.ceci}@uniba.it
[2] Unicredit Research and Development, 20100 Milan, Italy
fabio.fumarola@unicredit.eu
[3] CINI - Consorzio Interuniversitario Nazionale per l'Informatica, Bari, Italy

Abstract. Web page clustering is a focal task in Web Mining to organize the content of websites, understanding their structure and discovering interactions among web pages. It is a tricky task since web pages have multiple dimension based on textual, hyperlink and HTML formatting (i.e. HTML tags and visual) properties. Existing algorithms use this information almost independently, mainly because it is difficult to combine them. This paper makes a contribution on clustering of web pages in a website by taking into account a distributional representation that combines all these features into a single vector space. The approach first crawls the website by using web pages' HTML formatting and *web lists* in order to identify and represent the hyperlink structure by means of an adapted skip-gram model. Then, this hyperlink structure and the textual information are fused into a single vector space representation. The obtained representation is used to cluster websites using simultaneously their hyperlink structure and textual information. Experiments on real websites show that the proposed method improves clustering results.

1 Introduction

Since a web page is characterized by several dimensions (i.e. textual, structural based on HTML tags and visual/structural based on hyperlinks) the existing clustering algorithms differ in their ability of using these representations. In particular, algorithms based on textual representation typically group web pages using words distribution [6,13]. These solutions manage web pages as plain text ignoring all the other information of which a page is enriched and turn to be ineffective in at least two categories of web pages: *(i)* when there is not enough information in the text; *(ii)* when they have different content, but refer to the same semantic class. The former case refers to web pages with poor textual information, such as pages rich of structural data (e.g. from Deep Web Databases), multimedia data, or that have scripts which can be easily found also in other pages (e.g. from a CMS website). The latter case refers to pages having the same semantic type (e.g. web pages related to professors, courses, publications) but

© Springer International Publishing AG 2017
M. Kryszkiewicz et al. (Eds.): ISMIS 2017, LNAI 10352, pp. 446–456, 2017.
DOI: 10.1007/978-3-319-60438-1_44

characterized by a different distribution of terms. On the other side, clustering based on structure typically considers the HTML formatting (i.e. HTML tags and visual information rendered by a web browser) [2,3,7]. Algorithms which use these information, are based on idea that web pages are automatically generated by programs that extract data from a back-end database and embed them into an HTML template. This kind of pages show a common structure and layout, but differ in content. However, because tags are used for content displaying, it happens that most of the web pages in a website have the same structure, even if they refer to distinct semantic types. This negatively affect clusters' quality. The above described solutions exploit within-page information. Other algorithms make use of the graph defined by the hyperlink structure of a set of web pages [18,22]. Hyperlinks can be used to identify collections of web pages semantically related and relationships among these collections. In this area, DeepWalk and Line [18,22] are two embedding-based methods that exploit neural networks to generate a low-dimensional and dense vector representation of graph's nodes. DeepWalk [18] applies the skip-gram method on truncated random walks to encode long-range influences among graph's nodes. Still, this approach is not able to capture the local graph structure (i.e. nodes which can be considered similar because are strongly connected). Line [22] optimizes an objective function that incorporates both direct neighbours and neighbours of neighbours. However, both methods (DeepWalk and Line) ignore node attributes (e.g. textual content).

Most of the discussed works analyze contents, web page structure (i.e. HTML formatting) and hyperlink structure almost independently. Over the last decade, some researchers tried to combine several sources of information together. For example, [1,16] combine content and hyperlink structure for web page clustering, [4,7,19] combine web page and hyperlink structure for clustering purposes. This paper is a contribution in this direction. It combines information about content, web page structure and hyperlink structure of web pages homogeneously. It analyzes web pages' HTML formatting to extract from each page collections of links, called *web lists*, which can be used generate a compact and noise-free representation of the website's graph. Then, the extracted hyperlink structure and content information of web pages are mapped in a single vector space representation which can be used by clustering algorithms. Our approach is based on the idea that two web pages are similar if they have common terms (i.e. *Bag of words hypothesis* [23]) and they share the same reachability properties in the website's graph. In order to consider reachability, the solution we propose is inspired by the concept of *Distributional Hypothesis*, initially defined for words in natural language processing (i.e. *"You shall know a word by the company it keeps"*) [9] and recently extended to generic objects [11]. In the context of the Web we can translate that citation in *"You shall know a web page by the paths it keeps"* (i.e. two similar web pages are involved in the same paths).

2   Methodology

The proposed solution implements a four steps strategy: in the first step website crawling is performed. Crawling uses web pages' structure information and exploits web lists in order to mitigate problems coming from noisy links. The output of this phase is the website graph, where each node represents a single page and edges represent hyperlinks. In the second step, we generate a link vector by exploiting Random Walks extracted from the website's graph. In the third phase content vectors are generated. In the last one, a unified representation of pages is generated and clustering is performed on such representation.

2.1   Website Crawling

A Website can be formally described as a direct graph $G = (V, E)$, where V is the set of web pages and E is the set of hyperlinks. In most cases, the homepage h of a website represents the website's entry page and allows the website to be viewed as a rooted directed graph. As claimed in [7] *not all links are equally important to describe the website structure.* In fact, a website is rich of noisy links, which may not be relevant to clustering process, such as hyperlinks used to enforce the web page authority, short-cut hyperlinks, etc. . . . Besides, the website structure is codified in navigational systems which provide a local view of the website organization. Navigational systems (e.g. menus, navbars, product lists) are implemented as hyperlink collections having same domain name and sharing layout and presentation properties. Our novel solution is based on the usage of web lists. This has a twofold effect: from one side it guarantees that only urls useful to the clustering process are considered; on the other side, it allows the method to implicitly take into account the web page structure which is implicitly codified in the web lists available web pages [7,15,24]. Starting from the homepage h, a crawling algorithm iteratively extracts the collection of the urls having same domain of h and organized in *web lists*. Only web pages included in web lists are further explored. Following [15], a web list is:

Definition 1. *A **Web List** is a collection of two or more web elements having similar HTML structure, visually adjacent and aligned on a rendered web page. The alignment is identified on the basis of the x-axis (vertical list), the y-axis (horizontal list), or in a tiled manner (aligned vertically and horizontally).*

Figure 1 shows, in red boxes, web lists extracted from the homepage of a computer science department which will be used for website crawling. Links in box A will be excluded because their domains are different from the homepage's domain. To identify from a web page the set of web lists we implement HyLien [10]. The output of website crawling step is the sub-graph $G' = (V', E')$, where $V' \subseteq V$ and $E' \subseteq E$ will be used for link and content vectors generation steps.

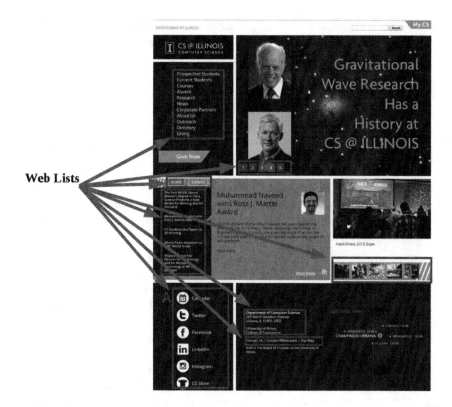

Fig. 1. Web lists extracted from a web page taken from www.cs.illinois.edu (Color figure online)

2.2 Link Vectors Generation Through Random Walks

A random walk over a linked structure is based on the idea that the connections among nodes encode information about their correlations. To codify these correlations we use the Random Walk with Restart (RWR) approach. RWR is a Markov chain describing the sequence of nodes (web pages) visited by a random walker. Starting from a random point i, with probability $(1 - \alpha)$ a walker walks to a new, connected neighbor node or, with probability α, it restarts from i.

Inspired by the field of information retrieval, we model a web page as a word, that is, a *topic indicator* and, each random walk as a document constituting the natural context of words (i.e. topical unity). Thus, we represent a collection of random walks as a document collection where topics intertwine and overlap. This enable the application of a *distributional-based* algorithm to extract new knowledge [21] from the obtained representation. In our case, we apply the skip-gram model [17], a state-of-art algorithm, to extract a vector space representations of web pages that encode the topological structure of the website. In the skip-gram model we are given a word w in a corpus of words V_W (in our case a web page w belonging to random walks) and its context $c \in V_C$ (in our case web pages in random walks which appear before and after the web page w). We consider the conditional probabilities $p(c|w)$, and given a random walks collection Rws, the goal is to set the parameters θ of $p(c|w; \theta)$ so to maximize the probability:

$$\underset{\theta}{argmax} \prod_{L \in Rws; w \in L} \left[\prod_{c \in C_L(w)} prox_L(w, c) \cdot p(c|w; \theta) \right] \tag{1}$$

where L is a random walk in Rws, w is a web page in L and $C_L(w) = \{w_{i-k}, \ldots, w_{i-1}, w_{i+1}, \ldots, w_{i+k}\}$ is the set of contexts of web page w in the list L. Moreover, $prox_L(w, c)$ represents the proximity between w and $c \in C_L(w)$. This is necessary since the skip-gram model gives more importance to the nearby context words than distant context words. One approach for parameterizing the skip-gram model follows the neural-network language models literature, and models the conditional probability $p(c|w; \theta)$ using soft-max: $p(c|w; \theta) = \frac{e^{v_c \cdot v_w}}{\sum_{c' \in V_C} e^{v_{c'} \cdot v_w}}$, where v_c, $v_{c'}$ and $v_w \in \mathbb{R}^d$ are vector space representations for c, c' and w respectively (d is defined by the user). Therefore, the optimization problem (1) leads to the identification of the web page and context matrices $W = \{v_{w_i} | w_i \in V_W\}$ and $C = \{v_{c_i} | c_i \in V_C\}$. They are dependent each other and we only use W to represent web pages (coherently with what proposed in [17] for words). The computation of $p(c|w; \theta)$ is computationally expensive due the summation $\sum_{c' \in V_C}$ and thus in [17] are presented *hierarchical softmax* and *negative-sampling approach* to make the computation more tractable. Therefore, given in input to skip-gram model a corpus data composed by the collection of random walks, it returns the matrix W which embeds each web page into a dense and low-dimensional space $\mathbb{R}^d$.

2.3 Content Vectors Generation

Here we describe the process for generating a vector representation of web pages using textual information. Differently from traditional documents, web pages are written in HTML and contain additional information, such as tags, hyperlinks and anchor text. To apply on web pages a bag-of-words representation we need a preprocessing step, in which the following operations are performed: HTML tags removal (however, we maintain terms in anchor, title and metadata since they contribute to better organize web pages [8]); unescape escaped characters; eliminate non-alphanumeric characters; eliminate too frequent (>90%) and infrequent (<5%) words. After preprocessing, each web page is converted in a plain textual document and we can apply the traditional *TF-IDF* weighting schema to obtain a content-vector representation. Due the uncontrolled and heterogeneous nature of web page contents, vector representation of web pages based on content is characterized by high-dimensional sparse data. To obtain a dense and low-dimensional space we apply Truncated SVD algorithm, a low-rank matrix approximation based on random sampling [12]. In particular, given the *TF-IDF matrix* of size $|V'| \times n$ and the desired dimensionality of content vectors m, where $m \ll n$, the algorithm returns a matrix of size $|V'| \times m$.

2.4   Content-Link Coupled Clustering

Once the content vector $v_c \in \mathbb{R}^m$ and the link vector $v_l \in \mathbb{R}^d$ of each web page in V' have been generated, the last step of the algorithm is to concatenate them in a new vector having dimension $m + d$. Before the concatenation step we normalize each vector with its Euclidean norm. In this way we ensure that components of v_l having highest weights are as important as components of v_c having highest weights. The generated matrix preserves both structural and textual information and can be used in traditional clustering algorithms based on vector space model. In this study we consider K-MEANS and H-DBSCAN [5] because they are well known and present several complementary properties.

Table 1. Description of websites

Website	#pages	#edges	#edges using web lists	#clusters
Illinois	563	9415	5330	10
Oxford	3480	44526	35148	19
Stanford	167	12372	30087	10
Princeton	3132	122493	104585	16

3   Experiments

In order to empirically evaluate our approach, we performed validation four computer science department's websites: *Illinois* (cs.illinois.edu), *Princeton* (cs.princeton.edu), *Oxford* (www.cs.ox.ac.ou), and *Stanford* (cs.stanford.edu). The motivation behind this choice is related to our competence in manually labelling pages belonging to this domain. This was necessary in order to create a ground truth for the evaluation of the clustering results. The experimental evaluation is conducted to answer the following questions: (1) which is the real contribution of combining content and hyperlink structure in a single vector space representation with respect to using only either textual content or hyperlink structure? (2) Which is the real contribution of exploiting web pages structure (i.e. HTML formatting) and, specifically, the role of using web lists to reduce noise and improve clustering results? In Table 1 the dimension of each dataset is described. In particular, to correctly analyze the contribution of web lists in the clustering process, we compare only the web pages extracted both by crawling websites using web lists and by traditional crawling (first column of Table 1). Moreover, we report the dimension of the edge set obtained with traditional crawling (second column) and crawling using web lists (third column). Finally the last column describes the number of clusters manually identified by the experts.

We evaluated the effectiveness of the approach using the following measures:

- Homogeneity [20]: each cluster should contain only data points that are members of a single class. This measure is computed by calculating the conditional entropy of the class distribution given the proposed clustering.
- Completeness [20]: all of the data points that are members of a given class should be elements of the same cluster. It is computed by the conditional entropy of the proposed cluster distribution given the real class.
- V-Measure [20]: harmonic mean between homogeneity and completeness.
- Adjusted Mutual Information (AMI): it is a variation of the Mutual Information MI. $MI = \sum_{i \in K} \sum_{j \in C} log \frac{P(i,j)}{P(i)P(j)}$ where C is the set of real classes, K is the set of learned clusters, $P(i,j)$ denotes the probability that a point belongs to both the real class i and the learned cluster j and $P(i)$ is the a priori probability that a point falls into i. MI is generally higher for two clusterings with a larger number of clusters, regardless of whether there is actually more information shared. The Adjusted Mutual Information represents an adjustment of this metric to overcome this limitation.
- Adjusted Random Index (ARI) [14]: it represents a similarity measure between two clusterings by considering all pairs of samples and counting the pairs that are assigned in the same or different clusters in the predicted and true clusterings. $RI = (a + b)/\binom{n}{2}$ where a is number of pairs of points that are in the same class and learned cluster and b is number of pairs of points that belong to different class and learned cluster.
- Silhouette: it measures how similar an object is to its own cluster (cohesion) compared to other clusters (separation). The silhouette ranges from -1 to 1, where a high value indicates that the object is well matched to its own cluster and poorly matched to neighboring clusters.

In order to respond to our research questions we ran our algorithm with different configurations:

- *Text*. We generate a vector space representation, having dimension $m = 120$, using only web pages' textual information;
- *RW-List*. We generate a vector space representation of size $d = 120$ using only hyperlink structure extracted by crawling the website using web lists. We set $\alpha = 1$, $rwrLength = 10$ and $dbLength = 100\,k$;
- *RW-NoList*. We generate a vector space representation of size 120 using only the hyperlink structure obtained with traditional crawling. We ran *rwrGeneration* with the same parameters of RW-List;
- *Comb-Lists*. We combine, as defined in Sect. 2.4, the content vector of size $m = 60$ and hyperlink structure vector of size $d = 60$ generated by crawling the website using web lists.
- *Comb-NoLists*. As in the Comb-Lists, but with a traditional crawler.

Since our goal is not that of comparing clustering algorithms, we set for K-MEANS the parameter K (i.e. total number of clusters to generate) to the number of real clusters, while we set for H-DBSCAN the *minimal cluster size* parameter to 5. Finally, since at the best of our knowledge there is no work which

Table 2. Experimental results

Configuration	Website	Clustering	Homogeneity	Completeness	V-Measure	ARI	AMI	Silhouette
Text	illinois	KMEANS	0.84	0.62	0.71	0.4	0.61	0.33
Text	illinois	H-DBSCAN	0.72	0.53	0.61	0.4	0.5	0.21
RW-Lists	illinois	KMEANS	0.72	0.53	0.61	0.27	0.51	0.42
RW-Lists	illinois	H-DBSCAN	0.81	0.47	0.6	0.18	0.43	**0.43**
RW-NoLists	illinois	KMEANS	0.71	0.52	0.6	0.25	0.5	0.42
RW-NoLists	illinois	H-DBSCAN	0.8	0.45	0.58	0.17	0.41	0.42
Comb-Lists	illinois	KMEANS	**0.9**	**0.69**	**0.78**	**0.54**	**0.68**	0.4
Comb-Lists	illinois	H-DBSCAN	0.83	0.51	0.63	0.27	0.48	0.34
Comb-NoLists	illinois	KMEANS	0.84	0.62	0.71	0.37	0.6	0.38
Comb-NoLists	illinois	H-DBSCAN	0.83	0.52	0.64	0.27	0.49	0.29
Text	Princeton	KMEANS	0.71	**0.59**	**0.64**	**0.68**	**0.58**	0.21
Text	Princeton	H-DBSCAN	0.36	0.31	0.34	0.12	0.28	-0.21
RW-Lists	Princeton	KMEANS	0.56	0.37	0.45	0.27	0.36	0.18
RW-Lists	Princeton	H-DBSCAN	0.49	0.3	0.37	0.12	0.26	-0.05
RW-NoLists	Princeton	KMEANS	0.55	0.36	0.43	0.24	0.35	0.15
RW-NoLists	Princeton	H-DBSCAN	0.48	0.3	0.37	0.1	0.26	-0.09
Comb-Lists	Princeton	KMEANS	0.76	0.54	0.63	0.55	0.53	0.14
Comb-Lists	Princeton	H-DBSCAN	0.47	0.52	0.49	0.36	0.45	**0.37**
Comb-NoLists	Princeton	KMEANS	**0.78**	0.54	**0.64**	0.49	0.53	0.13
Comb-NoLists	Princeton	H-DBSCAN	0.47	0.52	0.49	0.37	0.45	0.38
Text	Oxford	KMEANS	0.74	0.6	0.66	0.48	0.59	0.25
Text	Oxford	H-DBSCAN	0.43	0.41	0.42	0.07	0.37	-0.06
RW-Lists	Oxford	KMEANS	0.65	0.55	0.6	0.48	0.54	0.32
RW-Lists	Oxford	H-DBSCAN	0.6	0.44	0.51	0.26	0.41	0.22
RW-NoLists	Oxford	KMEANS	0.67	0.57	0.62	0.51	0.56	**0.35**
RW-NoLists	Oxford	H-DBSCAN	0.6	0.45	0.51	0.27	0.41	0.18
Comb-Lists	Oxford	KMEANS	0.79	0.67	0.73	**0.56**	0.67	0.34
Comb-Lists	Oxford	H-DBSCAN	0.58	0.49	0.53	0.15	0.47	0.08
Comb-NoLists	Oxford	KMEANS	**0.81**	**0.68**	**0.74**	0.53	**0.68**	0.28
Comb-NoLists	Oxford	H-DBSCAN	0.62	0.53	0.57	0.23	0.51	0.08
Text	Stanford	KMEANS	0.37	0.43	0.39	0.08	0.28	0.3
Text	Stanford	H-DBSCAN	0.18	0.62	0.28	0.07	0.16	0.43
RW-Lists	Stanford	KMEANS	**0.59**	0.58	**0.58**	**0.27**	**0.52**	0.31
RW-Lists	Stanford	H-DBSCAN	0.28	0.4	0.33	0.1	0.22	0.15
RW-NoLists	Stanford	KMEANS	0.47	0.54	0.5	0.14	0.39	0.53
RW-NoLists	Stanford	H-DBSCAN	0.34	0.6	0.43	0.13	0.29	**0.55**
Comb-Lists	Stanford	KMEANS	0.42	0.46	0.44	0.12	0.34	0.22
Comb-Lists	Stanford	H-DBSCAN	0.21	**0.63**	0.31	0.07	0.17	0.46
Comb-NoLists	Stanford	KMEANS	0.53	0.56	0.54	0.17	0.46	0.35
Comb-NoLists	Stanford	H-DBSCAN	0.34	0.51	0.4	0.12	0.28	0.27

uses the skip-gram model to analyze the topological structure of websites, we ran both of skip-gram versions (i.e. hierarchical softmax and SGNS) for generating link vectors. Due to space limitations, we report only results for SGNS (setting the window size to 5), which, in most cases, outperformed hierarchical softmax.

Table 2 presents the main results. In general, the experiments show that best results are obtained combining textual information with hyperlink structure. This is more evident for Illinois and Oxford websites, where content and

Table 3. Wilcoxon pairwise signed rank tests. (+) ((−)) indicates that the second (first) model wins. The results are highlighted in bold if the difference is statistically significant (at p-value = 0.05). The tests have been performed by considering the results obtained with both hierarchical softmax and SGNS skip-gram models.

	Homogeneity	Completeness	V-Measure	Adj Rand index	Adj Mutual info	Silhouette
Text vs Comb	**(+) 0.000**	(-) 0.055	**(+) 0.000**	(+) 0.342	**(+) 0.003**	**(+) 0.020**
RW vs Comb	**(+) 0.002**	**(+) 0.000**	**(+) 0.000**	**(+) 0.000**	**(+) 0.000**	(+) 0.229
NoLists vs Lists	(−) 0.342	(−) 0.970	(−) 0.418	(+) 0.659	(+) 0.358	(−) 0.362

hyperlinks structure codify complementary information for clustering purpose. However, for the Stanford website using the textual information decreases the clustering performance. The importance of combining content and hyperlink structure is confirmed by the Wilcoxon signed Rank test (see Table 3). This behaviour is quite uniform for all the evaluation measures considered.

For the last research question, results do not show a statistical contribution in the use of web lists for clustering purpose (see Table 3). This because analyzed websites are very well structured and poor of noisy links. This can be observed in Table 1, where there is not a valuable difference in terms of edges number between the real web graph and the one extracted using web lists. However, as expected the Completeness is higher for Comb-Lists, confirming that clusters have higher "precision" in the case of crawling based on web lists.

4   Conclusions and Future Works

In this paper, we have presented a new method which combines information about content, web page structure and hyperlink structure in a single vector space representation which can be used by any traditional and best-performing clustering algorithms. Experiments results show that content and hyperlink structure of web pages provide different and complementary information which can improve the efficacy of clustering algorithms. Moreover, experiments do not show statistical differences between results which use web lists and results obtained ignoring web page structure. As feature work we will run our algorithm on different domains and less structured websites in the way to observe whether web lists are really useless in the web page clustering process.

Acknowledgments. We acknowledge the support of the EU Commission through the project MAESTRA - Learning from Massive, Incompletely annotated, and Structured Data (Grant number ICT-2013-612944).

References

1. Angelova, R., Siersdorfer, S.: A neighborhood-based approach for clustering of linked document collections. In: Proceedings of CIKM 2006, pp. 778–779. ACM, New York (2006)
2. Bohunsky, P., Gatterbauer, W.: Visual structure-based web page clustering and retrieval. In: Proceedings of the 19th International Conference on World Wide Web, WWW 2010, pp. 1067–1068. ACM, New York (2010)
3. Buttler, D.: A short survey of document structure similarity algorithms. In: Proceedings of the International Conference on Internet Computing, IC 2004, Las Vegas, Nevada, USA, 21–24 June 2004, vol. 1, pp. 3–9 (2004)
4. Calado, P., Cristo, M., Moura, E., Ziviani, N., Ribeiro-Neto, B., Gonçalves, M.A.: Combining link-based and content-based methods for web document classification. In: Proceedings of the Twelfth International Conference on Information and Knowledge Management, CIKM 2003, pp. 394–401. ACM, New York (2003)
5. Campello, R.J.G.B., Moulavi, D., Sander, J.: Density-based clustering based on hierarchical density estimates. In: Pei, J., Tseng, V.S., Cao, L., Motoda, H., Xu, G. (eds.) PAKDD 2013. LNCS, vol. 7819, pp. 160–172. Springer, Heidelberg (2013). doi:10.1007/978-3-642-37456-2_14
6. Chehreghani, M.H., Abolhassani, H., Chehreghani, M.H.: Improving density-based methods for hierarchical clustering of web pages. Data Knowl. Eng. **67**(1), 30–50 (2008)
7. Crescenzi, V., Merialdo, P., Missier, P.: Clustering web pages based on their structure. Data Knowl. Eng. **54**(3), 279–299 (2005)
8. Fathi, M., Adly, N., Nagi, M.: Web documents classification using text, anchor, title and metadata information. In: Proceedings of the International Conference on Computer Science, Software Engineering, Information Technology, e-Business and Applications, pp. 1–8 (2004)
9. Firth, J.: A synopsis of linguistic theory 1930-55. In: Palmer, F.R. (ed.) Selected Papers of J.R. Firth 1952-59, pp. 168–205. Longmans, London (1968)
10. Fumarola, F., Weninger, T., Barber, R., Malerba, D., Han, J.: Hylien: a hybrid approach to general list extraction on the web. In: Proceedings of the 20th International Conference on World Wide Web, WWW 2011, Hyderabad, India, 28 March - 1 April 2011 (Companion Volume), pp. 35–36 (2011)
11. Gornerup, O., Gillblad, D., Vasiloudis, T.: Knowing an object by the company it keeps: a domain-agnostic scheme for similarity discovery. In: Proceedings of the 2015 IEEE International Conference on Data Mining (ICDM), ICDM 2015, pp. 121–130. IEEE Computer Society, Washington, DC (2015)
12. Halko, N., Martinsson, P.G., Tropp, J.A.: Finding structure with randomness: probabilistic algorithms for constructing approximate matrix decompositions. SIAM Rev. **53**(2), 217–288 (2011)
13. Haveliwala, T.H., Gionis, A., Klein, D., Indyk, P.: Evaluating strategies for similarity search on the web. In: Proceedings of WWW 2002, pp. 432–442. ACM, New York (2002)
14. Hubert, L., Arabie, P.: Comparing partitions. J. Classif. **2**(1), 193–218 (1985)
15. Lanotte, P.F., Fumarola, F., Ceci, M., Scarpino, A., Torelli, M.D., Malerba, D.: Automatic extraction of logical web lists. In: Andreasen, T., Christiansen, H., Cubero, J.-C., Raś, Z.W. (eds.) ISMIS 2014. LNCS, vol. 8502, pp. 365–374. Springer, Cham (2014). doi:10.1007/978-3-319-08326-1_37

16. Lin, C.X., Yu, Y., Han, J., Liu, B.: Hierarchical web-page clustering via in-page and cross-page link structures. In: Zaki, M.J., Yu, J.X., Ravindran, B., Pudi, V. (eds.) PAKDD 2010. LNCS (LNAI), vol. 6119, pp. 222–229. Springer, Heidelberg (2010). doi:10.1007/978-3-642-13672-6_22
17. Mikolov, T., Sutskever, I., Chen, K., Corrado, G.S., Dean, J.: Distributed representations of words and phrases and their compositionality. Adv. Neural Inf. Process. Syst. **26**, 3111–3119 (2013)
18. Perozzi, B., Al-Rfou, R., Skiena, S.: Deepwalk: online learning of social representations. In: ACM SIGKDD 2014, KDD 2014, pp. 701–710. ACM, New York (2014)
19. Qi, X., Davison, B.D.: Knowing a web page by the company it keeps. In: Proceedings of the 15th ACM International Conference on Information and Knowledge Management, CIKM 2006, pp. 228–237. ACM, New York (2006)
20. Rosenberg, A., Hirschberg, J.: V-measure: a conditional entropy-based external cluster evaluation measure. In: EMNLP-CoNLL 2007, pp. 410–420 (2007)
21. Sahlgren, M.: The distributional hypothesis. Ital. J. Linguist. **20**(1), 33–54 (2008)
22. Tang, J., Qu, M., Wang, M., Zhang, M., Yan, J., Mei, Q.: Line: large-scale information network embedding. In: Proceedings of the 24th International Conference on World Wide Web, WWW 2015, New York, NY, USA, pp. 1067–1077 (2015)
23. Turney, P.D., Pantel, P.: From frequency to meaning: vector space models of semantics. J. Artif. Int. Res. **37**(1), 141–188 (2010)
24. Weninger, T., Johnston, T.J., Han, J.: The parallel path framework for entity discovery on the web. ACM Trans. Web **7**(3), 16:1–16:29 (2013)

Concept-Enhanced Multi-view Co-clustering of Document Data

Valentina Rho$^{(\boxtimes)}$ and Ruggero G. Pensa**ID**

Department of Computer Science, University of Torino, Turin, Italy
{valentina.rho,ruggero.pensa}@unito.it

Abstract. The maturity of structured knowledge bases and semantic resources has contributed to the enhancement of document clustering algorithms, that may take advantage of conceptual representations as an alternative for classic bag-of-words models. However, operating in the semantic space is not always the best choice in those domain where the choice of terms also matters. Moreover, users are usually required to provide a valid number of clusters as input, but this parameter is often hard to guess, due to the exploratory nature of the clustering process. To address these limitations, we propose a multi-view co-clustering approach that processes simultaneously the classic document-term matrix and an enhanced document-concept representation of the same collection of documents. Our algorithm has multiple key-features: it finds an arbitrary number of clusters and provides clusters of terms and concepts as easy-to-interpret summaries. We show the effectiveness of our approach in an extensive experimental study involving several corpora with different levels of complexity.

Keywords: Co-clustering · Semantic enrichment · Multi-view clustering

1 Introduction

Clustering is a widely used tool in text document analysis. Due to its unsupervised nature, it takes part in a wide range of information retrieval applications, including summarization [20], query expansion [12] and recommendation [22], but it is also employed as a first exploratory tool in analyzing new text corpora. The principle of clustering is simple: it aims at grouping together similar documents into groups, called clusters, while keeping dissimilar documents in different clusters. The way similar documents are grouped together strongly depends on the clustering algorithm [1], but the notion of similarity itself is not straightforward. Documents can be viewed as bags of words, thus classic similarity functions (usually, the cosine similarity) can be applied on word vectors; however, they are not sufficient to capture the semantic relationship between two documents, since they do not deal with problems like synonymy (different terms with the same meaning) and polysemy (same term with multiple meanings). Moreover,

© Springer International Publishing AG 2017
M. Kryszkiewicz et al. (Eds.): ISMIS 2017, LNAI 10352, pp. 457–467, 2017.
DOI: 10.1007/978-3-319-60438-1_45

the document-term matrix (the matrix describing the frequency of terms that occur in a collection of documents) used as input for the clustering algorithm is usually very sparse and high-dimensional, leading to the well-studied problem of the curse of dimensionality, which in turn results in meaningless cluster structures. To mitigate these problems, semantic approaches can be applied to the document-terms matrix prior to clustering. For instance, Latent Semantic Analysis (LSA) [13] is a dimensionality reduction technique, based on singular-value decomposition (SVD), that provides a set of latent factors related to documents and terms, assuming that words with similar meanings occur in similar portions of text. Then clustering can be executed on a reduced document-factor matrix, rather than on the whole document-term matrix. Although these approaches provide effective solutions to some of the aforementioned problems, they suffer from some limitations weakening their exploitation in many clustering applications. First, polysemy is not handled. Second, the latent factors have no interpretable meaning in natural language, therefore they cannot be used to directly describe clustering results. Yet, cluster interpretation is fundamental in many exploratory applications. Third, the number of latent dimensions is a required parameter of the SVD algorithm performing LSA, and a wrong choice of this parameter may lead to poor clustering results.

With the evolution of structured knowledge bases (e.g., Wikipedia) and semantic resources (e.g., WordNet and BabelNet), in the last decade, new alternative approaches to the semantic enhancement of document clustering algorithms have been proposed. A first class of methods uses semantic resources to create new feature spaces [5,21]. A second group of algorithms leverages the semantic representation to reduce data dimensionality [19]. Finally, other methods define new similarity measures that take into account the semantic relations between concepts [8,9,21]. However, operate solely in the semantic space is not always the best choice for document clustering: even though the same concept can be expressed by different terms, sometimes each term is specific to a particular domain or language register. For instance, the terms *latent class analysis* and *clustering* sometimes refer to the same concept, but the former is used prevalently by statisticians, while the latter is preferred by machine learning experts. In these cases the chosen term is as important as its meaning.

To address all these limitations, we propose a multi-view clustering approach that processes simultaneously two representations of the same collection of documents: a classic document-term matrix and an enhanced document-concept representation. In our work, concepts are *abstract representations* of terms and are extracted from the document collection by means of a conceptualization approach that combines entity linking and word sense disambiguation, two natural language processing (NLP) techniques aiming at recognizing all concepts mentioned in a text. The two views are processed with a multi-view co-clustering approach that has multiple key-features: *(i)* it takes into account the peculiarity of the statistical distribution in each view, by implementing an iterative star-structure optimization approach; *(ii)* it provides an arbitrary number of clusters thanks to the adoption of an association function whose optimization does not depends on the number of clusters; *(iii)* it provides clusters of terms and concepts that can be used as easy-to-interpret summaries of the document clusters in both representation spaces. Our approach, then, also transforms texts into their direct conceptual representation, but, differently from the aforementioned methods, we embed this new representation into a 2-view setting in which both terms and concepts contribute to the clusters generation process. We show the effectiveness of our approach in an experimental study involving several corpora with different levels of complexity.

The remainder of the paper is organized as follows: we present the theoretical details of our approach in Sect. 2; in Sect. 3 we report some experimental results and discuss them; finally, we end up with some concluding remarks and ideas for future work in Sect. 4.

2 Combining Words and Concepts

We present a clustering approach that combines the expressive power of both terms and concepts to provide meaningful clusters of documents and an associated collection of clusters of features. First, we introduce a sketch of the overall clustering approach; then, we describe a possible way to extract a collection of concepts from a given text corpus. Finally, we provide more details on the multi-view co-clustering algorithm that we use to address our clustering problem.

2.1 Overall Clustering Approach

This section aims at describing the overall clustering approach, shown in Fig. 1, called *CVCC* (Concept-enhanced multi-View Co-Clustering). For a given collection of documents, represented by both a term and a conceptual view, *CVCC* partitions documents into an arbitrary number of clusters by also providing two related partitions of terms and concepts. Before entering the details of the approach, we first introduce some useful notation.

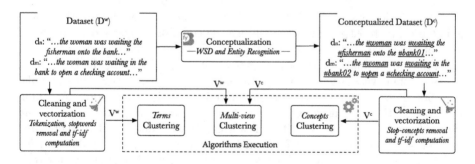

Fig. 1. A graphic overview of the overall *CVCC* clustering approach.

The input of *CVCC* is a dataset D, defined as a set of raw textual documents $\{d_1, \ldots, d_n\}$, each one represented by the sequence of words $<w_1, w_2, \ldots>$

that occur in it. The first step is to apply a *conceptualization* process on D, to obtain the *conceptualized* dataset D^c. Each document $d^c \in D^c$ is the conceptual representation of the corresponding document d in D and is defined as the sequence of concept identifiers $<c_1, c_2, \ldots>$ that occur in d. Then, we define three cleaning sets for each dataset D in order to ignore information that are considered not relevant to our purposes: $S = \{w_1^s, \ldots, w_p^s\}$ where each w_i^s is a word that is considered very common in the considered language (stopwords); $F = \{w_1^f, \ldots, w_q^f\}$, where each w_i^f is a word that occurs in more than t_f documents in D; $U = \{w_1^u, \ldots, w_r^u\}$, where each w_i^u is a word that occurs in less than t_u documents in D; t_f and t_u are two threshold values given in input to the pipeline. The last two sets are also computed on D^c, obtaining respectively F^c (too frequent concepts) and U^c (too rare concepts).

A preprocessing step applied to both D and D^c allow us to generate respectively a bag-of-words dataset V^w and a bag-of-concepts dataset V^c that will be the input of our clustering algorithm. In the former case, V^w, is represented as a $|D| \times |W|$ matrix, where W is defined as $\{w_j \mid \exists d_i \in D \wedge w_j \in d_i \wedge w_j \notin \{S \cup F \cup U\}\}$; each element v_{ij}^w of V^w is a numerical value representing the relevance of the term w_j in the document d_i. This numerical value could be computed with the well-known *tf-idf* (term frequency-inverse document frequency) function. In a similar way, we can define the preprocessed bag-of-concepts dataset V^c as a $|D| \times |C|$ matrix. In this case C is defined as $\{c_j \mid \exists d_i^c \in D^c \wedge c_j \in d_i^c \wedge w_j^c \notin S \wedge c \notin \{F^c \cup U^c\}\}$, where w_j^c is the term associated to the concept c in the corresponding document in D. Notice that the two sets are created independently. In fact, too rare (resp. too frequent) words may refer to more frequent (resp. unfrequent) concepts and vice versa, due to synonymy and polysemy. V^w and V^c are the two representations that feed the clustering algorithm used to compute the partitions on D, W and C.

2.2 Conceptualization Process

Many interpretations of what a *concept* is have been proposed during years, the most generic defining it as a high level representation of a set of items that share common characteristics. Here we embrace the commonly accepted definition of concept as an *abstract representation* of something in one's mind.

In document analysis, there are several advantages in using the conceptual representation with respect to the standard bag of words one. For example, concepts allow: to distinguish different meanings of the same word, by taking advantage of the context (polysemy, e.g. *bank* as *financial institution* or as a *land alongside water*); to aggregate different words with the same meaning (synonymy, e.g. *film* and *movie*); to identify named entities (e.g. *pink* as a color or *Pink* as the singer); to automatically consider *n*-grams instead of single terms (e.g. *United States*). In addition, another key point to consider when dealing with concepts is that, as they are *abstract*, they are language-insensitive: the same abstract concept labeled *#dog01* represents words *dog*, *cane*, *chien*, *hund* and so on, allowing us to work with multi-language text corpora.

In order to transform a generic document represented as a sequence of terms, into its conceptual representation we have to face the nontrivial issues of *entity linking* (assigning each word to the correct concept) and word-sense disambiguation (deciding which is the correct sense of each word, depending on its context). To address these issues, we make use of *Babelfy* [16], a multi-lingual semantic resource that aims at performing both entity linking and word-sense disambiguation on generic sentences. Babelfy is grounded on BabelNet, a multilingual encyclopedic resource created by the automatic integration of other well-known resources, e.g. WordNet and Wikipedia [17]. In practice, in our approach, concepts are intended as BabelNet identifiers. The conceptualization process transforms a sentence, represented as a sequence of words, into a list of concepts. We let the reader refer to [16] for more details about Babelfy.

2.3   Clustering Algorithm

We define our clustering approach as a 2-view co-clustering problem on the two matrices V^w and V^c. The goal of the 2-view co-clustering approach is to compute a set of n document clusters $X = \{x_1, \ldots, x_n\}$ on D, a set of l word clusters $Y^w = \{y_1^w, \ldots, y_l^w\}$ on W and a set of m concept clusters $Y^c = \{y_1^c, \ldots, y_m^c\}$ on C. X is such that $\bigcap_{k=1}^{n} x_k = \varnothing$ and $\bigcup_{k=1}^{n} x_k = D$. Y^w and Y^c are subject to similar constraints. Differently from most document clustering problems, n, m and l are not provided as input, i.e., our clustering approach is able to identify partitions with an arbitrary non predefined number of clusters. To achieve this goal, similarly to [11], we adopt an optimization function that is independent on the number of clusters: the Goodman and Kruskal's $\tau_{X_1|X_2}$ association measure [6]. It estimates the association between two categorical variables X_1 and X_2 by the proportional reduction of the error in predicting X_1 knowing or not the variable X_2. This measure requires that partitions X, Y^w and Y^c are defined as discrete random variables. Variable Y^w has l categories $y_1^w, \ldots, y_l^w$, corresponding to the l word clusters, with probabilities $q_1^w, \ldots, q_l^w$. Variable Y^c is defined similarly, while variable X has n categories $x_1, \ldots, x_n$ corresponding to n document clusters. However, for each view, the n categories of X have different probabilities $p_1^w, \cdots, p_n^w$, and $p_1^c, \cdots, p_n^c$. Moreover, the joint probabilities between X and Y^w (resp. Y^c) are denoted by r_{st}^w (resp r_{st}^c). All probabilities are computed directly from matrices V^w and V^c. As an example, the joint probabilities r_{st}^w between X and Y^w are computed as follows:

$$r_{st}^w = \frac{\sum_{d_i \in x_s} \sum_{w_j \in y_t^w} \overline{v}_{ij}^w}{\sum_i \sum_j \overline{v}_{ij}^w}$$

where $x_s \in X$, $y_t^w \in Y^w$, and $\overline{v}_{ij}^w$ is the value of v_{ij}^w normalized by sum of all elements in V^w.

The 2-view co-clustering problem can be defined as a multi-objective optimization problem defined over the following Goodman and Kruskal's τ coefficients, depending on which variable is considered as independent:

$$\tau_{X|Y^w,Y^c} = \frac{e_X - E[e_{X|Y^w,Y^c}]}{e_X}, \ \tau_{Y^w|X} = \frac{e_{Y^w} - E[e_{Y^w|X}]}{e_{Y^w}}, \ \tau_{Y^c|X} = \frac{e_{Y^c} - E[e_{Y^c|X}]}{e_{Y^c}} \quad (1)$$

where e_X (resp., e_{Y^w}, e_{Y^c}) is the sum of the errors over the independent variables Y^w and Y^c (resp. X). $E[e_{X|Y^w,Y^c}]$ (resp. $E[e_{Y^w|X}]$, $E[e_{Y^c|X}]$) is the expectation of the conditional error taken with respect to the distributions of Y^w and Y^c (resp. X). To optimize the objective functions we use the star-structure multi-objective optimization approach proposed in [11] which iteratively optimizes the three partitions X, Y^w and Y^c based on Goodman-Kruskal's τ measure using Eq. 1. The reader may refer to [11] for further algorithmic details.

3   Experiments

In this section we report the results of the experiments that we conducted to evaluate the performances of our document clustering approach. We first describe the datasets adopted and how we processed them. Then we introduce the algorithms involved in our comparative analysis and provide the details of the experimental protocol. Finally, we present the results and discuss them.

3.1   Datasets

The experiments are conducted on two well-known document corpora: Reuters-21578[1] and 20-Newsgroups[2]. For both datasets, categories are given that describe the content of each document. However, while 20-Newsgroups contains equally distributed disjoint categories, in Reuters-21578 corpus categories are not equally distributed and often cover very similar topics. Moreover, documents may belong to more than one category. For these reasons, we manually aggregated some of the original Reuters categories to create more homogeneous and semantically correlated groups (see Table 1 for the result of this process). Categories *earn* and *acq* are used as is. For both datasets, we prepared three reduced datasets, consisting of four categories each, as shown in Table 2. These three datasets are created to represent different complexity levels for the document clustering perspective: *level 1* (easy) datasets contain well-separated categories, *level 2* (medium) datasets contain two semantically similar categories and two different ones, *level 3* (hard) datasets are composed by two pairs of similar categories. Table 2 shows a detailed description of each considered dataset.

3.2   Experimental Settings

To evaluate *CVCC*, we compared its performances with those of three well-known algorithms: Non-negative Matrix Factorization (*NMF*), *K-Means* and *EBC*. *NMF* [3,14] is a dimensionality reduction algorithm that has been proved to be useful in different tasks, included document clustering. *K-Means* [15] is a popular clustering algorithm; in our setting we preprocess the data using Latent Semantic Analysis (LSA), in order to reduce their sparsity and improve clustering performances. The last competitor, *EBC* [18], is a very recent improvement of the

---

[1] http://www.nltk.org/book/ch02.html#reuters-corpus.
[2] http://scikit-learn.org/stable/datasets/twenty_newsgroups.html.

Table 1. Reuters-21578 aggregated categories. In **bold** the name of the resulting category, followed by the names of the Reuters categories that compose it.

Economic-indices ipi, wpi, jobs, trade, gnp, bop, cpi, income	**Money** yen, money-fx, interest, dlr	**Energy** crude, gas, fuel, propane, ship, nat-gas, naphtha, pet-chem, heat

Cereals oat, sorghum, oilseed, coconut-oil, sun-oil, rye, grain, sunseed, corn, wheat, palm-oil, barley, soybean, rice, cotton-oil, cotton, rapeseed, rape-oil, veg-oil, soy-oil

Table 2. Datasets composition and statistics, in terms of no. of documents, features and density. T and C columns refers to term and concept matrices, respectively. For included categories the number of elements of each category is reported in parentheses; pairs of semantically similar categories within each dataset are highlighted in *italic*.

Dataset		Doc.	Features		Density		Included categories
			T	C	T	C	
20-newsg.[a]	L1	2025	9523	9767	0.45%	0.44%	hardware (590), autos (594), religion (377), politics (546)
	L2	2058	10128	9478	0.37%	0.39%	*windows* (591), religion (377), autos (594), *hardware* (590)
	L3	2293	11099	10521	0.39%	0.41%	*windows* (591), crypt (595), *hardware* (590), electronics (591)
reuters-21k	L1	1495	5925	6206	0.77%	0.76%	cereals (400), energy (400), money (400), earn (400)
	L2	1501	6523	6874	0.79%	0.77%	cereals (400), energy (400), *money* (400), *acq* (400)
	L3	1555	5453	6001	0.83%	0.78%	earn (400), economic-indices (400), *acq* (400), *money* (400)

[a] 20-Newsgroups categories have been renamed for the sake of readability, as follows: comp.sys.ibm.pc.hardware as *hardware*, comp.os.ms-windows.misc as *windows*, soc.religion.christian as *religion*, rec.autos as *autos*, talk.politics.guns as *politics*, sci.crypt as *crypt*, sci.electronics as *electronics*

well-known Information-Theoretic Co-clustering algorithm [4] and it is proven to perform well with large sparse data matrices [18].

The experiments were conducted as follows. First of all, for each dataset described in Sect. 3.1 we compute matrices V^w and V^c, as shown in Sect. 2.1. We run the selected algorithms in three different configurations: *(i)* using only the **terms** matrix, in order to assess the capabilities of $CVCC$ with respect to the competitors in a standard setting; *(ii)* using only the **concepts** matrix, to evaluate the performances of all algorithms when moving from a lexical perspective to a more semantically enhanced interpretation of documents; *(iii)* using **both** terms and concepts (hereafter *both* configuration), to test $CVCC$ multi-view approach dealing with two representations of the same documents; in this last case, for single-view algorithms, we consider the hybrid matrix $[V^w, V^c]$ as the concatenation of the two original terms and concepts matrices, while for *NMF* we execute a recent co-regularized version (*CoNMF*) [7] that extends *NMF* for multi-view clustering. Additionally, when using *NMF*, we apply Non-negative Double Singular Value Decomposition (NNDSVD) [2] to preprocess sparse input matrices. Since all competitors require the number of clusters to find as input parameter, we set this value to four, that is the "correct" number of embedded clusters in our datasets; we let *CVCC* algorithm adapt this value autonomously. The number of iterations of *CVCC* has been configured, for each dataset, to $20 \times (n\_documents + n\_features)$, rounded to the nearest thousand[3].

To measure the performance of each algorithm, we adopt the Adjusted Rand Index (ARI) [10]. It measures the agreement of two different partitions of the same set, but, differently from other common statistics like Purity or Rand Index, it is not sensitive to group imbalance and allows the comparison of partitions with different number of clusters. Here, we use it to compare the cluster assignments proposed by each algorithm with the original assignment provided by the given *true* categories. As all algorithms are nondeterministic we perform 30 executions for each considered configuration and compute the ARI mean and standard deviation. All algorithms are written in Python and executed on a server with 16 3.30 GHz Xeon cores, 128 GB RAM, running Linux.

3.3   Results and Discussion

The results of the experiments are shown in Table 3 and two different aspects of our approach are highlighted. First, considering each algorithm independently, the best representation of each dataset is formatted in *italics*. Then, the best algorithm for each dataset representation is highlighted in **bold**.

As a general observation, the configurations that consider either the concepts view or the combination of terms and concepts often lead to the best results with very few exceptions, independently from the considered clustering algorithm. This result confirms that, in most contexts, the classical terms based approaches do not capture the embedded cluster structure sufficiently. Moreover, regarding the second evaluated aspect, *CVCC* almost always performs the best with 20-Newsgroups and exhibits significant differences with the other algorithms regardless of the complexity level of the dataset.

With the low complexity instance (L1) of Reuters-21578 corpus, instead, the differences among the four algorithms are less marked, with our algorithm providing always the second best results. This behavior is confirmed with L2, but in this case, the LSA-enhanced version of *K-means* performs significantly better than any other competitor. This is probably due to a minor contribution of polysemy in these two version of the dataset, which also explains the exceptional outperformances of the conceptual representation with *CVCC*. However, with the L3 instance of Reuters data, *CVCC* outperforms all other competitors by

---

[3] An iteration in *CVCC* corresponds to a single object movement [11].

Table 3. Mean and standard deviation of Adjusted Rand Index. The best ARI value for each experimental setting is highlighted in **bold**, while the best representation for each algorithm is formatted in *italic*.

Dataset	View	No. clusters	Adjusted Rand Index (ARI)			
			CVCC	EBC	LSA-KM	(Co)NMF
20ng-l1	terms	8.3 (3.01)	**0.53 (0.11)**	0.28 (0.10)	0.23 (0.03)	0.28 (0.00)
	conc	10.1 (4.21)	**0.44 (0.09)**	0.25 (0.10)	0.21 (0.02)	0.41 (0.00)
	both	5.8 (2.26)	*0.54 (0.06)*	*0.32 (0.08)*	*0.24 (0.01)*	*0.48 (0.07)*
20ng-l2	terms	10.5 (2.7)	**0.46 (0.04)**	0.21 (0.08)	0.19 (0.03)	0.23 (0.00)
	conc	9.9 (2.36)	**0.42 (0.04)**	0.17 (0.07)	0.14 (0.02)	0.31 (0.00)
	both	8.3 (1.32)	*0.47 (0.03)*	*0.25 (0.06)*	*0.20 (0.02)*	*0.36 (0.06)*
20ng-l3	terms	17.9 (7.99)	*0.30 (0.05)*	*0.23 (0.06)*	*0.21 (0.01)*	0.26 (0.00)
	conc	13 (6.01)	**0.25 (0.03)**	0.21 (0.05)	0.19 (0.01)	0.21 (0.00)
	both	9.5 (3.46)	0.28 (0.04)	*0.23 (0.07)*	*0.21 (0.01)*	***0.31 (0.03)***
reut-l1	terms	7.2 (2.41)	0.43 (0.11)	0.41 (0.09)	**0.45 (0.02)**	0.18 (0.00)
	conc	9.9 (2.43)	*0.54 (0.12)*	0.39 (0.11)	***0.55 (0.07)***	*0.45 (0.00)*
	both	7.6 (2.14)	**0.52 (0.17)**	*0.42 (0.12)*	0.45 (0.03)	0.25 (0.13)
reut-l2	terms	13.4 (1.6)	0.54 (0.06)	0.34 (0.11)	**0.57 (0.15)**	0.51 (0.00)
	conc	12.2 (1.1)	*0.64 (0.07)*	0.36 (0.10)	***0.71 (0.05)***	*0.52 (0.00)*
	both	11.9 (1.12)	0.61 (0.06)	*0.40 (0.11)*	***0.71 (0.10)***	0.49 (0.09)
reut-l3	terms	2.4 (0.95)	**0.43 (0.01)**	*0.37 (0.10)*	0.39 (0.11)	0.18 (0.00)
	conc	7.4 (2.54)	**0.48 (0.06)**	0.36 (0.09)	0.43 (0.07)	*0.21 (0.00)*
	both	3.1 (0.89)	*0.51 (0.06)*	*0.37 (0.10)*	*0.44 (0.10)*	0.20 (0.03)

far, thus confirming that in more complex scenarios our multi-view co-clustering approach shows its effectiveness compared to other approaches.

Finally, it is worth noting that, in general, the adoption of a two-view schema has two positive effects on the number of discovered clusters: not only does it better approach the correct number of categories with respect to single-view representations, but it also becomes more stable.

4   Conclusions

We presented a novel multi-view approach to semantically enhanced document co-clustering. Our algorithm can simultaneously process multiple representations of the same document. In particular, in the current setting we consider two views: document-term and document-concept. In the majority of cases, the results showed a clear advantage in using this strategy, compared to other well-known methods for document clustering. As future work, we plan to expand the conceptual representation with the inclusion of semantically related information in order to take advantage of relations between concepts. Finally, we will inspect the performances of our approach on different domains, e.g., image data or geographically annotated data, in which elements can be represented by additional views, e.g., SIFT and georeferred features.

Acknowledgments. The work is supported by Compagnia di San Paolo foundation (grant number Torino_call2014_L2_157).

References

1. Aggarwal, C.C., Zhai, C.: A survey of text clustering algorithms. In: Aggarwal, C.C., Zhai, C. (eds.) Mining Text Data, pp. 77–128. Springer, Heidelberg (2012)
2. Boutsidis, C., Gallopoulos, E.: SVD based initialization: a head start for nonnegative matrix factorization. Pattern Recogn. **41**(4), 1350–1362 (2008)
3. Cichocki, A., Phan, A.H.: Fast local algorithms for large scale nonnegative matrix and tensor factorizations. IEICE Trans. **92–A**(3), 708–721 (2009)
4. Dhillon, I.S., Mallela, S., Modha, D.S.: Information-theoretic co-clustering. In: Proceedings of ACM SIGKDD 2003, pp. 89–98. ACM (2003)
5. Gabrilovich, E., Markovitch, S.: Feature generation for text categorization using world knowledge. In: Proceedings of IJCAI 2005, pp. 1048–1053 (2005)
6. Goodman, L.A., Kruskal, W.H.: Measures of association for cross classification. J. Am. Stat. Assoc. **49**, 732–764 (1954)
7. He, X., Kan, M., Xie, P., Chen, X.: Comment-based multi-view clustering of web 2.0 items. In: Proceedings of WWW 2014, pp. 771–782 (2014)
8. Hu, J., Fang, L., Cao, Y., Zeng, H., Li, H., Yang, Q., Chen, Z.: Enhancing text clustering by leveraging wikipedia semantics. In: Proceedings of SIGIR 2008, pp. 179–186. ACM (2008)
9. Huang, A., Milne, D., Frank, E., Witten, I.H.: Clustering documents using a wikipedia-based concept representation. In: Theeramunkong, T., Kijsirikul, B., Cercone, N., Ho, T.-B. (eds.) PAKDD 2009. LNCS (LNAI), vol. 5476, pp. 628–636. Springer, Heidelberg (2009). doi:10.1007/978-3-642-01307-2_62
10. Hubert, L., Arabie, P.: Comparing partitions. J. Classif. **2**(1), 193–218 (1985)
11. Ienco, D., Robardet, C., Pensa, R.G., Meo, R.: Parameter-less co-clustering for star-structured heterogeneous data. Data Min. Knowl. Discov. **26**(2), 217–254 (2013)
12. Kalmanovich, I.G., Kurland, O.: Cluster-based query expansion. In: Proceedings of ACM SIGIR 2009, pp. 646–647. ACM (2009)
13. Landauer, T.K., Foltz, P.W., Laham, D.: An introduction to latent semantic analysis. Discourse Process. **25**(2–3), 259–284 (1998)
14. Lin, C.: Projected gradient methods for nonnegative matrix factorization. Neural Comput. **19**(10), 2756–2779 (2007)
15. Lloyd, S.P.: Least squares quantization in PCM. IEEE Trans. Inf. Theory **28**(2), 129–136 (1982)
16. Moro, A., Raganato, A., Navigli, R.: Entity linking meets word sense disambiguation: a unified approach. Trans. ACL **2**, 231–244 (2014)
17. Navigli, R., Ponzetto, S.P.: Babelnet: the automatic construction, evaluation and application of a wide-coverage multilingual semantic network. Artif. Intell. **193**, 217–250 (2012)
18. Percha, B., Altman, R.B.: Learning the structure of biomedical relationships from unstructured text. PLoS Comput. Biol. **11**(7), e1004216 (2015)
19. Recupero, D.R.: A new unsupervised method for document clustering by using wordnet lexical and conceptual relations. Inf. Retr. J. **10**(6), 563–579 (2007)
20. Shen, C., Li, T., Ding, C.H.Q.: Integrating clustering and multi-document summarization by bi-mixture probabilistic latent semantic analysis (PLSA) with sentence bases. In: Proceedings of AAAI 2011, pp. 914–920. AAAI Press (2011)

21. Wei, T., Lu, Y., Chang, H., Zhou, Q., Bao, X.: A semantic approach for text clustering using wordnet and lexical chains. Expert Syst. Appl. **42**(4), 2264–2275 (2015)
22. West, J.D., Wesley-Smith, I., Bergstrom, C.T.: A recommendation system based on hierarchical clustering of an article-level citation network. IEEE Trans. Big Data **2**(2), 113–123 (2016)

Big Data Analytics and Stream Data Mining

Scalable Framework for the Analysis of Population Structure Using the Next Generation Sequencing Data

Anastasiia Hryhorzhevska[1]([✉]), Marek Wiewiórka[1],
Michał Okoniewski[2], and Tomasz Gambin[1]

[1] Institute of Computer Science, Warsaw University of Technology,
00-665 Warsaw, Poland
anastasiia.Hryhorzhevska@gmail.com
[2] Scientific IT Services, ETH Zurich, 8092 Zurich, Switzerland

Abstract. Genomic variant data obtained from the next generation sequencing can be used to study the population structure of
the genotyped individuals. Typical approaches to ethnicity classifica-
tion/clustering consist of several time consuming pre-processing steps,
such as variant filtering, LD-pruning and dimensionality reduction of
genotype matrix. We have developed a framework using R programming language to analyze the influence of various pre-processing methods
and their parameters on the final results of the classification/clustering
algorithms. The results indicated how to fine-tune the pre-processing
steps in order to maximize the supervised and unsupervised classification performance. In addition, to enable efficient processing of large
data sets, we have developed another framework using Apache Spark.
Tests performed on 1000 Genomes data set confirmed the efficiency and
scalability of the presented approach. Finally, the dockerized version of
the implemented frameworks (freely available at: https://github.com/
ZSI-Bio/popgen) can be easily applied to any other variant data set,
including data from large scale sequencing projects or custom data sets
from clinical laboratories.

1 Introduction

Understanding of the genomic basis of diseases has become a central part of
human and molecular genetics. It motivates novel biological hypothesis, teaches
the things about epidemiology, causal risk factors and relations between different
parts of biological system and enables to develop new tools that can be used
in diagnosis and treatment. Identification of novel disease genes is a challenging
task that requires large scale case-control studies, which often involve individuals
from various human populations.

Genome-wide association studies (GWAS) and Rare variant association studies (RVAS) give the ability to exam markers across whole genomes simultaneously
and test hundreds of thousands allele variants to find an associations between

© Springer International Publishing AG 2017
M. Kryszkiewicz et al. (Eds.): ISMIS 2017, LNAI 10352, pp. 471–480, 2017.
DOI: 10.1007/978-3-319-60438-1_46

genotype in the markers and likelihood of disease. GWAS have focused on the analysis of common variants, however most of genetic heritability remained unexplained [13]. Next generation sequencing enabled to sequence whole genomes and explore the entire spectrum of allele frequencies, including rare variants. Currently, most studies focuses on variants with low minor-allele frequency ($0.5\% <$ MAF $< 5\%$) or rare variants (MAF $< 0.5\%$) [16,20].

Population structure retrieving and inferences are critical in association studies, in which population stratification (i.e., the presence of genotypic differences among different groups of individuals) can lead to inferential errors. Genotype-based clustering of individuals is an important way of summarizing the genetic similarities and differences between individuals of different ancestry. A number of methods have been proposed to deal with the problem such as principal component analysis (PCA) [22], multidimensional scaling (MDS) [18], linkage disequilibrium-based approach [15]. However, these algorithms need to be well-tuned to complete the learning process in the best possible way and considerably low computation time.

At the same time, the whole genome sequencing data generated in large-scale sequencing projects, increases the number of sequenced individuals and the feature space (the number of unique allele variants) by orders of magnitude. This sharp increase in both sample numbers and features per sample requires a massively parallel approach for data processing [25]. Traditional parallelization strategies implemented e.g. in PLINK [23] cannot scale with variable data sizes at runtime [21].

To address this issue, we developed two frameworks. The first one was developed in R programming language and was built on the top of a SNPRelate toolset [10]. It was designed for fine-tuning the quality-control (QC) and the classification model parameters. It provides also the graphical representation of the results. The second one is a distributed computing framework for scalable stratification analysis in Apache Spark [3] using ADAM [6] in combination with machine learning (ML) libraries Spark MLlib [9], H2O [8] and Apache System ML [4]. It is more effective in processing of datasets containing large number of observations (i.e. number of samples >10,000) comparing to the first framework.

2   Methods

2.1   Dataset

To test our approach we have used the 1000 Genomes Project genotype data [1] released in the Variant Call Format (VCF) [11]. The dataset contains variants obtained from sequencing of 2,504 individuals (observations) across more than 30,000,000 alleles (features). The individuals are distributed across five super populations, i.e.: African (AFR), Mixed American (AMR), East Asian (EAS), South Asian (SAS) and European (EUR) and 26 sub-populations [1].

Our aim was to test the performance of ML algorithms to reconstruct both super- and sub-populations. To do this, the classification/clustering models were built either separately, i.e. for either super or sub-population groups or hierarchically, starting from super-populations and drilling down to the sub-populations.

2.2 Frameworks for Inferring the Population Structure

Due to the limitation of the current implementation of PCA in Apache Spark, which does not allow to process matrices with more than 65,535 columns [2], we implemented two separate frameworks. The first framework implemented in R provides an efficient way for testing and fine-tuning QC, PCA and machine learning parameters on the relatively small number of observations (up to several thousands). The second framework implemented in Apache Spark, besides its limitation of the number of features deals gracefully with a large number of observations (i.e. hundred of thousands of individuals).

Since the original data are stored in VCFs, the first task was to transform the data into more efficient format. In case of the first framework, we used *gdsfmt* R package that provides an interface to CoreArray Genomic Data Structure (GDS) files designated for storing SNP genotypes. In this format each byte encodes up to four SNPs genotypes reducing the file size and access time [7]. In the second framework, we used ADAM format [6] to store and process genomic variant data in Hadoop Distributed File System (HDFS). The original 1000 Genomes data was transformed from VCF files to the respective ADAM format. Then, using ADAM Parquet file, we read the genotypes into Spark data frame and extracted the information that is required for further processing.

Next, we perform three pre-processing steps, including: (i) missing values treatment, (ii) feature selection, and (iii) dimensionality reduction. These steps are done to simplify the data analysis, reduce the noise in the data and increase the accuracy of ML algorithms, and are particularly important in case of high dimensional datasets.

Finally, we apply classification and clustering methods to the reduced dataset and present algorithms quality and efficient performance for different sets of parameters and subsets of the data. Figure 1 indicates the alternative steps in the analysis process.

2.3 Missing Values Handling

In the high dimensional genomic data it is likely that some genotypes are missing, e.g. due to low coverage of sequencing reads in selected samples. The default strategy used in the population stratification is to drop features containing missing values. However, sometimes (e.g. in case of limited genotype datasets from small gene panels) it could be desirable to test on "predictive missingness", that is, what dependency the response may have upon missing values. Therefore, in our framework we implemented both approaches. In the first one we drop all variants (features) with any missing data. In the second approach we filter out variants with the proportion of missing values exceeding the given threshold, defined as P_1.

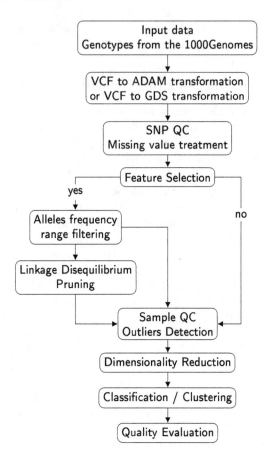

Fig. 1. The workflow for both supervised and unsupervised classification of genomes variant data

2.4   Feature Selection

Alleles Frequency Range Filtering. To reduce the number of variants in the data, we first compute the frequencies of alternate alleles across all individuals in the dataset. For further analysis we select variants with allele frequencies within a certain range, defined as P_2.

Linkage Disequilibrium Pruning. Linkage disequilibrium (LD) is a non-random association of alleles at different genomic positions (loci) [24]. LD pruning on the variants is suggested before running dimensionality reduction methods because the LD blocks can decrease the ability of the algorithms to separate populations [27]. Such blocks should be removed from the analysis.

In our framework we implemented three methods for calculation of LD values. The first algorithm calculates so called composite coefficient that is estimated from di-locus counts and sample allele frequencies [26]. The second LD calculation algorithm uses a common standardization method [17] that is a relative measure of disequilibrium compared to its maximum D' [14]. Third method involves so called R coefficient that is computed using expectation maximization algorithm under assumption of Hardy-Weinberg Equilibrium (HWE).

The LD pruning algorithm recursively removes SNPs that are greater than a LD threshold (defined as parameter P_3) within a sliding window based on the pairwise genotypic correlation.

2.5   Dimensionality Reduction

To correct the population stratification and detect true population structure we implemented PCA [19]. The PCA algorithm is carried out on a set of possibly collinear features and performs a transformation to produce a new set of uncorrelated features. Although the Spark MLlib [9] library provides methods for principal components (PCs) computation, these implementations could not handle 1000 genomes variant dataset and therefore this step was performed using parallelized algorithm implemented in SNPRelate R package.

The number of PCs that are further used for classification/clustering is selected according to the percentage of cumulative variance of the first n PCs. We tested the optimal number of PCs, defined as an input parameter P_4.

2.6   Unsupervised Learning Algorithms

To group individuals and recover the population structure three algorithms from the Spark MLlib library [9] (K-means, Bisecting K-means and Gaussian Mixture), and three implemented in R libraries (Hierarchical, K-means and Expectation-Maximization) are used. The quality of clustering are assessed using both external (Purity, Adjusted Rand Index [ARI]) and internal (Dunn Index [DI], Calinski-Harabasz Index [CH]) evaluation criteria.

2.7   Supervised Learning Algorithms

In our framework we test three algorithms for supervised classification from Spark MLlib and Spark ML, Apache System ML and R, including: Support Vector Machine (SVM), Random Forests, and Decision Trees.

To evaluate the quality of classification with respect to every class in the dataset, we build confusion matrix and compute common classification metrics such as accuracy, precision, recall, and the F1-Score (i.e. harmonic mean of precision and recall). Identification of ethnic group for an individual is a multi-class classification problem. Therefore in addition to per-class measures we average them over all the classes resulting in macro-averaged precision, recall, F1-Score. In addition, we compute the Kappa statistic, which is a measure of agreement between the predictions and the actual labels. It is interpreted as a comparison of the overall accuracy to the expected random chance accuracy.

3   Results and Discussion

To compared the quality of classifiers implemented in our frameworks, we performed all experiments on chromosome 22 from 1000 genomes dataset.

Results of Unsupervised Classification. First, we applied QC procedures to exclude problematic SNPs from the analysis. Since there was no missing values, the P_1 was set to zero in all of the experiments. Next, allele frequency filtering, LD pruning, dimensionality reduction and clustering/classification algorithms were performed. We repeated experiments for different input parameters (P_2, P_3, P_4) and investigated the clustering quality for the three methods by comparing the annotated super-population label (AMR, EUR, AFR, EAS, SAS) for each individual in the dataset to the label assigned. Table 1 presents the final selection of the fine-tuned QC parameters for each unsupervised method and the quality performance of these methods.

Table 1. Summary of quality control parameters and performance of the unsupervised classifiers

Method	MAF	LD	$nPCs$	Purity	ARI	CH	DI
Hierarchical	(0.005; 0.05)	0.2	5	97.96	95.75	1472.83	0.02
K-means	(0.005; 0.05)	0.2	4	92.21	86.89	7911.91	0.03
E-M	(0.005; 0.1)	0.2	4	96.21	91.76	3232.82	0.01

The best result was obtained using hierarchical method (95.75% of ARI). However, in case of K-means the CH index is much higher, i.e. the clusters are more compact and the distance between groups are longer than in case of other models. Visualization of the results of hierarchical algorithm for the fine-tuned parameters is presented in Fig. 2 and corresponding confusion matrix is shown in Table 2. As expected, AMR and EUR are tended to be clustered into one group, which indicates the similarities between these two populations.

Our frameworks can be applied to cluster sub-populations. The performed experiments indicate that the analysis of sub-populations is more challenging

Table 2. Confusion matrix of the hierarchical clustering, $P_2 \in (0.005; 0.05)$, $P_3 = 0.2$, $P_4 = 5$

	1	2	3	4	5
AFR	0	0	4	656	1
AMR	33	1	302	11	0
EAS	0	504	0	0	0
EUR	502	0	1	0	0
SAS	0	0	0	0	489

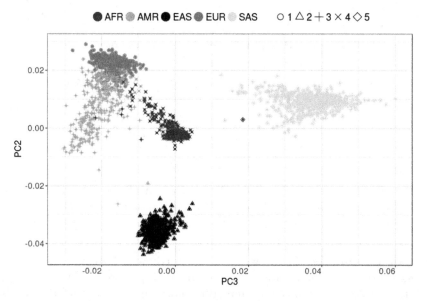

Fig. 2. Results of the hierarchical clustering. Colors indicate true populations, whereas groups discovered by clustering are marked using different shapes. Hierarchical clustering was performed on dataset pre-processed using the following thresholds: $P_2 \in (0.005; 0.05)$, $P_2 = 0.2$, $P_4 = 5$.

and requires to include more variation to obtain satisfactory results, i.e. use wider MAF interval and increased number of PCs.

Results of Supervised Classification. The process of building supervised classifiers consisted of two steps. First, we fixed classification model parameters and tuned QC parameters. Second, we estimated classifier parameters for the QC parameters obtained in the first step. Final fine-tuned QC and classification model parameters are summarized in Table 3. The results indicate that the QC parameters should to be tuned for each supervised method individually. Comparing with unsupervised learning, much larger number of PCs is required to build the best-performing model. Corresponding classification models' prediction

Table 3. Summary of quality control and classifiers parameters

Method	MAF	LD	$nPCs$	C	γ	$split$	cp	$ntree$
SVM linear	(0.005; 0.05)	0.2	60	50	-	-	-	-
SVM quadratic	(0.005; 0.05)	0.2	50	60	0.1	-	-	-
SVM radial	(0.005; 0.05)	0.2	70	20	10	-	-	-
Decision trees	(0.005; 0.05)	0.2	60	-	-	$gain$	0.01	-
Random forest	(0.01; 0.05)	0.2	60	-	-	-	-	109

Table 4. Prediction models performance of classification of super population groups

Method	Precision	Recall	F1-score	Accuracy	Kappa
SVM linear	96.43	93.63	94.75	96.25	0.95
SVM quadratic	96.72	94.40	95.38	96.61	0.96
SVM radial	98.57	98.78	98.67	98.91	0.99
Decision trees	98.14	95.53	96.60	97.70	0.97
Random forest	99.05	99.33	99.17	99.31	0.99

performances are presented in Table 4. Comparison of five classifiers revealed that SVM with radial kernel (F1 = 98.91%) and Random Forest (F 1 = 99.31%) outperformed Decision Trees and SVM with either linear or quadratic kernel.

Tests of Distributed Framework. Our distributed framework enables to process the large amount data in parallel on multiple nodes, which not only allow to overcome the problem of limited memory on a single machine but also significantly reduce the overall computation time.

The tests were done for the data on which all pre-processing steps were performed. The dataset consisted of 2,504 human individuals across 5 super-populations assayed for 5,000 SNPs. We recorded time execution of both frameworks performing PCA on the datasets generated by duplicating the original 1000 Genomes data several times as required [12]. For datasets with the number of samples greater than 20,000 non-distributed framework could not complete computation in a reasonable amount of time, and therefore for larger dataset we report the results of distributed (MLlib-based) implementation only.

Table 5. Time execution performance for distributed versus non-distributed framework on increasing datasets derived from 1000 Genomes data, using 5,000 SNPs (total number of executors is 128, executor memory is 12 GB)

Number of observations	Distributed [hh:mm:ss]	Non-distributed [hh:mm:ss]
2,504	00:13:06	00:00:12
5,008	00:13:10	00:01:12
10,016	00:13:37	00:12:18
15,024	00:13:52	00:41:06
20,048	00:14:11	01:37:06
50,080	00:15:40	>18 h
100,160	00:16:29	-
250,400	00:19:03	-
500,800	00:22:43	-

Table 5 shows that non-distributed framework deals better with a small number of observations (and thus is more suitable for fine-tuning performed on small datasets), whereas distributed implementation is much faster for a large number of observations, i.e. $N > 10,000$. In particular, Spark MLlib was able to process a dataset of 500,000 samples in less than 23 min, while SNPRelate did not complete for a dataset of 50,000 samples after 18 h of computing.

4 Conclusions

Our framework for the analysis of the population structure implemented in R programming language provides automated estimation of the optimal QC's and machine learning parameters, and ensures the high-quality performance of both clustering $(ARI = 95.75\%)$ and classification $(F1 = 98.9\%)$ algorithms for analysis of super-populations. Importantly, it allows for efficient comparison of different machine learning solutions and test for a wide range of input parameters and pre-processing strategies in order to fine-tune clustering/classification methods. Furthermore, the obtained results show that our framework gives better quality performance of clustering of super-population groups $(ARI = 95.75\%)$ than other approaches, e.g. VariantSpark $(ARI = 84\%)$ [5].

Our second framework has been developed to deal with dataset containing large number of samples (e.g. >10,000 individuals). Both dimensionality reduction and machine learning methods have been implemented using Apache Spark. The results of performed experiments on 1000 Genomes data set demonstrate that the tool can handle the population genetic analysis on the data from large sample size whole-genome sequencing cohorts and process it 100x faster than single node solution.

Acknowledgments. This work has been supported by the Polish National Science Center grants: Opus 2014/13/B/NZ2/01248 and Preludium 2014/13/N/ST6/01843.

References

1. The 1000 genomes project. http://www.internationalgenome.org/
2. Apache Spark. RowMatrix. https://github.com/apache/spark
3. Apache Spark^{TM}. http://spark.apache.org/
4. Apache SystemML - Declarative Large-Scale Machine Learning. https://systemml.apache.org/
5. BauerLab/VariantSpark. https://github.com/BauerLab/VariantSpark
6. Big Data Genomics. http://bdgenomics.org/
7. Bioconductor - gdsfmt. http://bioconductor.org/packages/gdsfmt
8. H2o.ai. http://www.h2o.ai/download/sparkling-water/
9. MLlib—Apache Spark. http://spark.apache.org/mllib/
10. SNPRelate. http://bioconductor.org/packages/SNPRelate/
11. The variant call format specification. https://github.com/samtools/hts-specs
12. Abraham, G., Inouye, M.: Fast principal component analysis of large-scale genome-wide data. PLoS ONE **9**(4), e93766 (2014)

13. Auer, P.L., Lettre, G.: Rare variant association studies: considerations, challenges and opportunities. Genome Med. **7**(1), 16 (2015)
14. Hamilton, D.C., Cole, D.E.C.: Standardizing a composite measure of linkage disequilibrium. Ann. Hum. Genet. **3**, 234–239 (2004)
15. Hinrichs, A.L., Larkin, E.K., Suarez, B.K.: Population stratification and patterns of linkage disequilibrium. Genet. Epidemiol. **33**(Suppl 1), S88–S92 (2009)
16. Lee, S., Abecasis, G., Boehnke, M., Lin, X.: Rare-variant association analysis: study designs and statistical tests. Am. J. Hum. Genet. **95**(1), 5–23 (2014)
17. Lewontin, R.C.: The interaction of selection and linkage. I. General considerations; heterotic models. Genetics **49**(1), 49–67 (1964)
18. Li, Q., Yu, K.: Improved correction for population stratification in genome-wide association studies by identifying hidden population structures. Genet. Epidemiol. **32**(3), 215–226 (2008)
19. Liu, L., Zhang, D., Liu, H., Arendt, C.: Robust methods for population stratification in genome wide association studies. BMC Bioinform. **14**, 132 (2013)
20. Manolio, T.A., Collins, F.S., Cox, N.J., Goldstein, D.B., Hindorff, L.A., Hunter, D.J., McCarthy, M.I., Ramos, E.M., Cardon, L.R., Chakravarti, A., Cho, J.H., Guttmacher, A.E., Kong, A., Kruglyak, L., Mardis, E., Rotimi, C.N., Slatkin, M., Valle, D., Whittemore, A.S., Boehnke, M., Clark, A.G., Eichler, E.E., Gibson, G., Haines, J.L., Mackay, T.F.C., McCarroll, S.A., Visscher, P.M.: Finding the missing heritability of complex diseases. Nature **461**(7265), 747–753 (2009)
21. O'Brien, A.R., Saunders, N.F.W., Guo, Y., Buske, F.A., Scott, R.J., Bauer, D.C.: VariantSpark: population scale clustering of genotype information. BMC Genom. **16**, 1052 (2015)
22. Price, A.L., Patterson, N.J., Plenge, R.M., Weinblatt, M.E., Shadick, N.A., Reich, D.: Principal components analysis corrects for stratification in genome-wide association studies. Nature Genet. **38**(8), 904–909 (2006)
23. Purcell, S., Neale, B., Todd-Brown, K., Thomas, L., Ferreira, M., Bender, D., Maller, J., Sklar, P., de Bakker, P., Daly, M., Sham, P.: PLINK: a tool set for whole-genome association and population-based linkage analyses. Am. J. Hum. Genet. **81**(3), 559–575 (2007)
24. Slatkin, M.: Linkage disequilibrium - understanding the evolutionary past and mapping the medical future. Nat. Rev. Genet. **9**(6), 477–485 (2008)
25. Stein, L.D.: The case for cloud computing in genome informatics. Genome Biol. **11**(5), 207 (2010)
26. Weir, B.S.: Genetic Data Analysis. Sinauer Associates, Inc., Sunderland (1996)
27. Zou, F., Lee, S., Knowles, M.R., Wright, F.A.: Quantification of population structure using correlated SNPs by shrinkage principal components. Hum. Hered. **70**(1), 9–22 (2010)

Modification to K-Medoids and CLARA for Effective Document Clustering

Phuong T. Nguyen[1(✉)], Kai Eckert[2], Azzurra Ragone[3], and Tommaso Di Noia[4]

[1] Duy Tan University, 182 Nguyen Van Linh, Da Nang, Vietnam
phuong.nguyen@duytan.edu.vn
[2] Stuttgart Media University, Nobelstr. 10, 70569 Stuttgart, Germany
eckert@hdm-stuttgart.de
[3] University of Milano-Bicocca, Piazza Dell Ateneo Nuovo 1, 20126 Milano, Italy
azzurra.ragone@unimib.it
[4] SisInf Lab, Polytechnic University of Bari, Via Orabona 4, 70125 Bari, Italy
tommaso.dinoia@poliba.it

Abstract. Document clustering plays an important role in several applications. K-Medoids and CLARA are among the most notable algorithms for clustering. These algorithms together with their relatives have been employed widely in clustering problems. In this paper we present a solution to improve the original K-Medoids and CLARA by making change in the way they assign objects to clusters. Experimental results on various document datasets using three distance measures have shown that the approach helps enhance the clustering outcomes substantially as demonstrated by three quality metrics, i.e. *Entropy*, *Purity* and *F-Measure*.

1 Introduction

Document analysis accounts for a crucial part in many research fields such as Data Mining and Information Retrieval. The boom of social networks in the recent years has made document analysis become even more important. Among others, document clustering algorithms have come under the spotlight. Clustering can be used to assist document browsing and produce document summary [1]. For intelligent systems and social networks, clustering is utilized in generating forecasts and recommendations [4]. The K-Means algorithm has been applied widely in document clustering due to its simplicity. However the algorithm is susceptible to noise and outliers [7,10]. K-Medoids was proposed and helps overcome the weakness of K-Means. An improved version of K-Medoids, CLARA has been derived to solve the problem of clustering big datasets. In the assignment step, both K-Medoids and CLARA attach an object to the cluster whose the medoid is closest to the object. However, we see that this may not help optimize the *compactness* of the objects within a cluster since the distance of the new object to the others is not considered. In this paper we present an approach to improve K-Medoids and CLARA. By making change in the way objects are assigned to cluster we are able to increase the overall effectiveness of the clustering. For comparison, we used some *de facto* standard document sets

M. Kryszkiewicz et al. (Eds.): ISMIS 2017, LNAI 10352, pp. 481–491, 2017.
DOI: 10.1007/978-3-319-60438-1_47

as the input data. Through the use of *Entropy*, *Purity* and *F-Measure* as quality metrics, we saw that the amendment facilitates better clustering solutions. The main contributions of this paper are: (1) Proposing a modification to enhance the effectiveness of K-Medoids and CLARA; and (2) Evaluating the performance of some clustering algorithms on standard document sets.

2   Modified K-Medoids and Modified CLARA

2.1   K-Medoids and CLARA

K-Medoids. The K-Medoids algorithm groups a set of n data objects to a pre-defined number of clusters κ. Medoids are the reference point for the assignment of objects. First, a set of initial medoids is generated randomly, then a medoid is selected as the object in the cluster that has minimum average distance to all objects in the cluster. Objects are assigned to the cluster with the closest medoid. K-Medoids is explained by means of the following greedy strategy:

- Step 1: Populate initial medoids by randomly selecting κ objects.
- Step 2: Assign each of the remaining objects to the cluster with the nearest medoid.
- Step 3: Calculate a new set of medoids. For each cluster, promote the object having the smallest average distance to the other objects in the cluster to the new medoid. If there are no changes in the set of medoids, stop the execution and return the resulting clusters in Step 2. Otherwise go back to Step 2.

CLARA. CLARA (Clustering LARge Applications) was designed to deal with large datasets [6]. It draws different samples of objects and applies K-Medoids on these samples to find the best set of medoids. The remaining objects are then assigned to the closest medoid. This helps save processing time by finding medoids from subsets of data objects. CLARA is briefly recalled as follows [6]:

Set $minD \leftarrow N$, N is a large enough number. Repeat the following steps for 5 times:

- Step 1: Choose randomly a set of $40 + 2\kappa$ objects. Apply K-Medoids on this set to find κ medoids.
- Step 2: Assign each of the remaining objects to the cluster with the nearest medoid.
- Step 3: Compute $avgD$ the average distance of the clustering solution in Step 2. If $avgD < minD$ then $minD \leftarrow avgD$ and select κ medoids in Step 1 as the current medoids. Go back to Step 1.

2.2   Modification to K-Medoids and CLARA

A clustering algorithm aims to minimize the intra-distance within a cluster and maximize the inter-distance to other clusters at the same time. With K-Medoids and CLARA, the assignment of an object to the closest medoid helps reduce the distance to the medoid. However, the average distance of a newly joining object to the other objects of the cluster might be high since the assignment does not take this into account. Objects in a cluster can be close to the medoid but they may be well apart among themselves. Given the circumstance, the assignment cannot minimize the average distance from the new object to the objects of the cluster. In this sense, we see that there is room for improvement. We propose making an amendment to the original K-Medoids (**oKM**) and the original CLARA (**oCLARA**) as follows. An object is assigned to a cluster iff the average distance from the object to all objects of the cluster is minimal. We retain the steps of **oKM** and **oCLARA** except for Step 2:

- Step 2: Assign each of the remaining objects **to the cluster with which it has the smallest average distance to all objects of the cluster.**

The proposed amendment is applied to K-Medoids and CLARA with the aim of improving their effectiveness. Two algorithms are derived, namely modified K-Medoids (**mKM**) and modified CLARA (**mCLARA**). In the succeeding sections, we are going to investigate whether the modification is beneficial by comparing the performance of **oKM**, **mKM**, **oCLARA** and **mCLARA**.

3 Document Clustering

We consider the problem of document clustering where a set of n documents needs to be grouped into different clusters. Based on the relationship among the input documents, a clustering algorithm distributes them to independent groups in a way that both the similarity among the members of a group as well as the dissimilarity among groups can be maximized.

3.1 Extraction of Document Features

To serve as the input for the clustering process, it is necessary to calculate the distance between each pair of documents. In the first place, a document needs to be represented in a mathematically computable form. We adopted the vector representation [2,5]. There, a document is modeled as a feature vector where each element corresponds to the weight of a term in the document [1,11]. If we consider a set of documents D and a set of terms $t = (t_1, t_2, .., t_r)$ then the representation of a document $d \in D$ is vector $\delta = (w_1^d, w_2^d, .., w_r^d)$ where w_l^d is the weight of term l in d and computed using the *term frequency-inverse document frequency* function with f_l^d being the frequency of t_l in d [8]:

$$w_l^d = tf \cdot idf(l, d, D) = f_l^d \cdot log \frac{n}{|\{d \in D : t_l \in d\}|} \qquad (1)$$

3.2   Distance Measures

Considering two documents d and e represented by feature vectors $\delta = \{\delta_l\} \mid_{l=1,..,r}$ and $\epsilon = \{\epsilon_l\} \mid_{l=1,..,r}$, the following metrics are utilized to calculate distance:

Cosine Similarity. A set of terms $t = (t_1, t_2, .., t_r)$ forms an r-dimension space and for each pair of two vectors δ and ϵ there is an angle between them. Intuitively, the cosine similarity metric measures the similarity as the cosine of the corresponding angle between the two vectors. And the distance between them is equivalent to the dissimilarity:

$$D_C(d, e) = 1 - SIM_C(d, e) = 1 - \frac{\sum_{l=1}^{r} \delta_l \cdot \epsilon_l}{\sqrt{\sum_{l=1}^{r}(\delta_l)^2} \cdot \sqrt{\sum_{l=1}^{r}(\epsilon_l)^2}}$$

Tanimoto Coefficient. The similarity defined by Tanimoto coefficient:

$$SIM_T(d, e) = \frac{\sum_{l=1}^{r} \delta_l \cdot \epsilon_l}{\sum_{l=1}^{r}(\delta_l)^2 + \sum_{l=1}^{r}(\epsilon_l)^2 - \sum_{l=1}^{r} \delta_l \cdot \epsilon_l}$$

And the distance between d and e is:

$$D_T(d, e) = 1 - SIM_T(d, e) \tag{2}$$

Euclidean Distance. Euclidean distance computes the geometric distance between d and e in the r-dimension space as given below [5].

$$D_E(d, e) = (\sum_{l=1}^{r} |w_l^d - w_l^e|^2)^{\frac{1}{2}} \tag{3}$$

4   Evaluation

Once the distance between every pair of documents has been identified, we are able to cluster a set of documents. In order to validate the proposed hypotheses, we compared the clustering performance of **mKM** and **mCLARA** with that of **oKM** and **oCLARA** with reference to a set of *de facto* standard document sets. We examined if the modification helps improve the effectiveness of clustering. We used Latent Dirichlet Allocation (**LDA**) as baseline for our evaluation. **LDA** deals with the modeling of topics, however it can also be utilized in clustering documents [3]. Due to space limitation, the algorithm is not recalled in this paper, interested readers are referred to the original paper for further detail [7].

Experiments on various datasets have been performed by using the distance measures in Sect. 3.2. We chose some of the datasets available at the website of Karypis Lab[1] for the experiments. There, pre-processing stages had been conducted to extract terms from documents and term frequencies were then

---

[1] http://glaros.dtc.umn.edu/gkhome/fetch/sw/cluto/datasets.tar.gz.

calculated and saved into text files [11]. Furthermore, category for each document, e.g. *Sport, Financial, Foreign*, has been identified and can be read from the provided files [12]. A summary of the datasets is given in Table 1.

Table 1. Datasets used for evaluation.

	la1	la2	re0	re1	tr31	tr41	tr45	wap
# of docs (n)	3204	3075	1504	1657	927	878	690	1560
# of terms (r)	31472	31472	2886	3758	10128	7454	8261	8460
# of classes (κ)	6	6	13	25	7	10	10	20
# of weights	484024	455383	77808	87328	248903	171509	193605	220482

4.1 Evaluation Metrics

Every document in the datasets has already been classified into a pre-defined category. In the following we call the categories in a dataset $C = (C_1, C_2, .., C_k)$ as *classes*, being C_i the set of documents whose category is i, and the resulting groups of the clustering process $\mathscr{C} = (\mathscr{C}_1, \mathscr{C}_2, .., \mathscr{C}_\kappa)$ as *clusters*. We performed *external evaluation* to measure the extent to which the produced clusters match the classes [9]. Given a clustering solution $\mathscr{C}$, the task is to compare the relatedness of the clusters in $\mathscr{C}$ to the classes in C. Three evaluation metrics *Entropy*, *Purity* and *F-Measure* were chosen to analyze the clustering solutions. The rationale for this selection is that the metrics have been widely utilized in evaluating clustering and appear to be effective [5,9,11,12]. The metrics are briefly recalled as follows.

Entropy. This metric gauges the relatedness of a cluster to the classes by measuring the presence of the classes in a cluster. If p_{ij} is the probability that a member of class i is found in cluster j then Entropy for cluster j is computed according to the probability of the existence of all classes in j:

$$E_j = -\sum_{i=1}^{\kappa} p_{ij} \cdot log(p_{ij}) = -\sum_{i=1}^{\kappa} \frac{|C_i \cap \mathscr{C}_j|}{|\mathscr{C}_j|} \cdot log(\frac{|C_i \cap \mathscr{C}_j|}{|\mathscr{C}_j|})$$

The Entropy value for a clustering solution is weighted across all clusters:

$$E = \sum_{j}^{\kappa} \frac{|\mathscr{C}_j|}{n} \cdot E_j \tag{4}$$

Purity. It is used to evaluate how well a cluster matches a single class in C:

$$P_j = \frac{1}{|\mathscr{C}_j|} \cdot \max_{i} \{|C_i \cap \mathscr{C}_j|\} \qquad \text{and} \qquad P = \sum_{j}^{\kappa} \frac{|\mathscr{C}_j|}{n} \cdot P_j \tag{5}$$

F-Measure. By this metric *Precision* and *Recall* are utilized as follows:

$$F_{ij} = \frac{2 \cdot precision_{ij} \cdot recall_{ij}}{precision_{ij} + recall_{ij}} \qquad \text{and} \qquad F = \sum_i^\kappa \frac{|C_i|}{n} \cdot \max_j \{Fij\} \quad (6)$$

Precision is the fraction of documents of class i in cluster j, whereas Recall is the fraction of documents of cluster j in class i:

$$precision_{ij} = \frac{|C_i \cap \mathscr{C}_j|}{|\mathscr{C}_j|} \qquad \text{and} \qquad recall_{ij} = \frac{|C_i \cap \mathscr{C}_j|}{|C_i|}$$

If an ideal clustering solution is found, i.e. $\mathscr{C}_i \equiv C_i \mid_{i=1,..,\kappa}$ then Entropy is equal to 0, whilst both Purity and F-Measure are 1.0. If the clusters are completely different to the classes then Purity and F-Measure are equal to 0. This implies that a good clustering solution has low Entropy but high Purity and F-Measure. It is expected that the modification helps increase Purity, F-Measure but decrease Entropy at the same time.

4.2   Results

Using the datasets described in Table 1, we performed 24 independent experiments, each corresponds to applying one distance measure to a document set. From the distance scores, **oKM**, **mKM**, **oCLARA**, and **mCLARA** were applied to produce clusters. To aim for randomness, every clustering experiment was run in 100 trials. For the experiments with Cosine Similarity, the outcomes were visualized by sketching out Entropy, Purity and F-Measure. The boxplot diagram in Fig. 1 shows the Entropy scores for the datasets using Cosine Similarity. By all datasets, **mKM** and **mCLARA** help obtain a much better Entropy than **oKM** and **oCLARA** do. Similarly, as seen in Fig. 2, the Purity values for the clustering with **mKM** and **mCLARA** are much better. **mKM** produces a clearly better Purity for **la1**. Fig. 3 shows the F-Measure scores obtained from performing the algorithms on all datasets. Equation 6 suggests that a good clustering solution has high F-Measure score. Back to the figures, we see that by all datasets, the F-Measure scores produced by **mKM** and **mCLARA** are of higher quality. These demonstrate that using Cosine Similarity on the datasets, the modified versions of the algorithms produce better results.

For other clustering experiments with Tanimoto and Euclidean, we also performed 100 trials each. Furthermore, **LDA** was used as a baseline for comparison. By **LDA**, a document can be assigned to each cluster with a specific probability and we attach the document to the cluster with the highest probability. A series of Entropy, Purity and F-Measure values have been derived from the clustering solutions. Due to space limitation the results for all trials are averaged out and the means are shown in Tables 2, 3 and 4. In Table 2, we see that the Entropy scores using Tanimoto are similar to those of Cosine Similarity. However, with Euclidean distance the results obtained from **mKM** are the best among others whereas those between **oCLARA**, **mCLARA** and **LDA** are comparable.

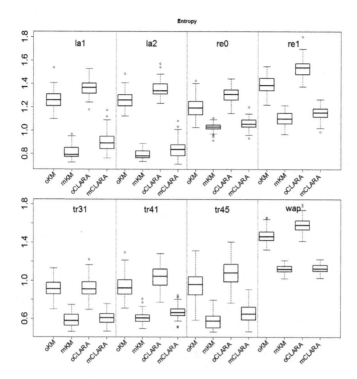

Fig. 1. Entropy of the clustering solutions using cosine similarity on the datasets

With Purity by Tanimoto, we witness the same pattern for these results as with Entropy, i.e. clustering with **mKM** and **mCLARA** yields superior Purity scores in all experiments. Again, with Euclidean distance **mKM** brings the best Purity while **oCLARA, mCLARA, LDA** possess similar F-Measure.

The F-Measure scores for experiments with Tanimoto and Euclidean are shown in Table 4, together with F-Measure produced by **LDA**. With Tanimoto Coefficient, F-Measure scores by **mKM** and **mCLARA** are always superior to those of **oKM** and **oCLARA**. With Euclidean distance, the difference among all algorithms is marginal. F-Measure of **LDA** is inferior to that of the others.

To sum up, we see that applying the proposed modification to **oKM** and **oCLARA** in clustering appears to be effective as demonstrated by all evaluation metrics. It is evident that **mKM** and **mCLARA** are suitable for use in combination with Cosine Similarity and Tanimoto. Taken all metrics into consideration, i.e. Entropy, Purity and F-Measure, we see that **LDA** does not produce good outcomes compared with other algorithms. As already suggested in [11], we are confident that **mKM** and **mCLARA** are the best algorithms in the given context. Under the circumstances, we come to the conclusion that the modification is highly beneficial to clustering on the observed datasets.

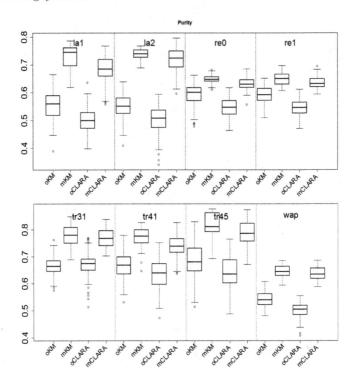

Fig. 2. Purity of the clustering solutions using cosine similarity on the datasets

5   Related Work

In [5], an evaluation of the influence of similarity measures on clustering is presented. Using various distance measures, the effectiveness of various distance measure for document clustering is compared using Purity and Entropy. [12] presents a comprehensive study of partitional and agglomerative algorithms that use different criterion functions and merging schemes. The results demonstrate that partitional algorithms always lead to better solutions than agglomerative algorithms. This is also confirmed in [7]. The authors in [11] demonstrate a comparison for two document clustering techniques, namely agglomerative hierarchical clustering and K-Means. In this work, documents are modeled as feature vectors and Entropy, F-Measure and Overall Similarity are used as the metrics for evaluation. The work in [1] provides a comprehensive survey of text clustering. There, the key methods and their advantages of the clustering problem applied to the text domain are discussed. Furthermore, the potential of clustering for social networks and Linked Open Data is also mentioned.

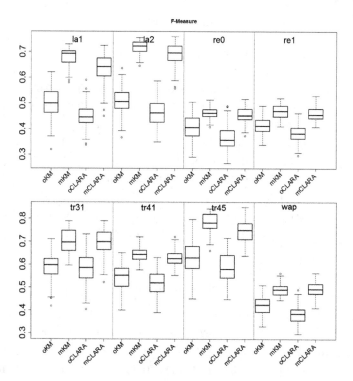

Fig. 3. F-Measure of the clustering solutions using cosine similarity on the datasets

Table 2. Entropy

		Dataset							
		la1	la2	re0	re1	tr31	tr41	tr45	wap
Tanimoto	oKM	1,33	1,33	1,26	1,43	0,94	1,02	1,05	1,46
	mKM	**1,03**	**0,95**	**1,11**	**1,14**	**0,61**	**0,69**	**0,65**	**1,10**
	CLARA	1,60	1,38	1,60	2,34	1,39	1,73	1,21	2,24
	mCLARA	1,06	1,00	1,13	1,16	0,62	0,74	0,78	1,11
Euclidean	oKM	1,26	**1,67**	1,19	1,39	0,91	0,93	1,05	1,47
	mKM	**0,81**	**1,67**	**1,02**	**1,10**	**0,58**	**0,61**	**0,63**	**1,12**
	CLARA	1,68	1,67	1,70	2,41	1,44	1,82	1,92	2,47
	mCLARA	1,68	1,67	1,77	2,42	1,46	1,85	2,00	2,40
LDA		1,68	1,67	1,74	2,39	1,51	1,85	2,01	2,40

Table 3. Purity

		Dataset							
		la1	la2	re0	re1	tr31	tr41	tr45	wap
Tanimoto	oKM	0,52	0,50	0,56	0,58	0,69	0,64	0,66	0,54
	mKM	**0,61**	**0,64**	**0,61**	**0,63**	**0,76**	**0,73**	**0,79**	**0,65**
	CLARA	0,34	0,48	0,44	0,31	0,44	0,35	0,59	0,30
	mCLARA	**0,60**	**0,63**	**0,61**	**0,63**	**0,77**	**0,71**	**0,74**	**0,64**
Euclidean	oKM	0,55	**0,29**	0,59	0,59	0,66	0,66	0,66	0,54
	mKM	**0,73**	**0,29**	**0,64**	**0,64**	**0,77**	**0,77**	**0,80**	**0,64**
	CLARA	0,30	0,30	0,43	0,29	0,43	0,32	0,29	0,25
	mCLARA	0,29	0,30	0,42	0,28	0,42	0,30	0,26	0,25
LDA		0,29	0,29	0,40	0,24	0,37	0,28	0,23	0,22

Table 4. F-measure

		Dataset							
		la1	la2	re0	re1	tr31	tr41	tr45	wap
Tanimoto	oKM	0,48	0,46	0,37	0,40	0,58	0,51	0,59	0,41
	mKM	**0,56**	**0,61**	**0,41**	**0,45**	**0,72**	**0,61**	**0,74**	**0,48**
	CLARA	0,31	0,44	0,32	0,21	0,43	0,31	0,53	0,21
	mCLARA	**0,55**	**0,60**	**0,41**	**0,45**	**0,72**	**0,60**	**0,69**	**0,49**
Euclidean	oKM	0,32	**0,33**	0,35	**0,23**	**0,40**	**0,31**	0,59	**0,19**
	mKM	**0,32**	**0,33**	**0,36**	0,22	**0,40**	**0,30**	**0,75**	**0,19**
	CLARA	**0,32**	**0,33**	**0,36**	**0,23**	0,39	**0,30**	0,26	**0,19**
	mCLARA	**0,32**	**0,33**	**0,36**	0,22	0,39	**0,30**	0,24	**0,19**
LDA		0,20	0,19	0,17	0,09	0,22	0,18	0,17	0,11

6   Conclusion

Based on the observation that **oKM** and **oCLARA** may not minimize the average distance of a newly joining object to the other objects of the cluster, we proposed modified version of the algorithms and we applied it to the use case of document clustering. A document is assigned to the cluster with which it has the smallest average distance to all existing objects. Some *de facto* standard document datasets were utilized in the evaluation. Using *Entropy, Purity* and *F-Measure* as the quality metrics, we saw that the modified versions produce better outcome compared to the original ones. For future work, we plan to examine the performance of **mKM** and **mCLARA** for a variety of documents using more distance measures. Finally, we consider using other evaluation metrics to better study the performance of the proposed algorithms.

References

1. Aggarwal, C.C., Zhai, C.: A survey of text clustering algorithms. In: Aggarwal, C.C., Zhai, C. (eds.) Mining Text Data, pp. 77–128. Springer, Heidelberg (2012). doi:10.1007/978-1-4614-3223-4_4
2. Basu, T., Murthy, C.: A similarity assessment technique for effective grouping of documents. Inf. Sci. **311**(C), 149–162 (2015)
3. Blei, D.M., Ng, A.Y., Jordan, M.I.: Latent Dirichlet allocation. J. Mach. Learn. Res. **3**, 993–1022 (2003)
4. DuBois, T., Golbeck, J., Kleint, J., Srinivasan, A.: Improving recommendation accuracy by clustering social networks with trust, New York, NY, USA (2009)
5. Huang, A.: Similarity measures for text document clustering, pp. 49–56 (2008)
6. Kaufman, L., Rousseeuw, P.J.: Finding Groups in Data: An Introduction to Cluster Analysis. Wiley, New York (1990)
7. Ng, R.T., Han, J.: Clarans: a method for clustering objects for spatial data mining. IEEE Trans. Knowl. Data Eng. **14**(5), 1003–1016 (2002)
8. Reed, J.W., Jiao, Y., Potok, T.E., Klump, B.A., Elmore, M.T., Hurson, A.R.: TF-ICF: a new term weighting scheme for clustering dynamic data streams. In Proceedings of the 5th International Conference on Machine Learning and Applications, ICMLA 2006, Washington, DC, USA, pp. 258–263. IEEE Computer Society (2006)
9. Rendón, E., Abundez, I., Arizmendi, A., Quiroz, E.M.: Internal versus external cluster validation indexes. Int. J. Comput. Commun. **5**, 27–34 (2011)
10. Rokach, L., Maimon, O.: Clustering methods. In: Maimon, O., Rokach, L. (eds.) Data Mining and Knowledge Discovery Handbook, pp. 321–352. Springer, Boston (2005)
11. Steinbach, M., Karypis, G., Kumar, V.: A comparison of document clustering techniques. In: 6th ACM SIGKDD, World Text Mining Conference (2000)
12. Zhao, Y., Karypis, G., Fayyad, U.: Hierarchical clustering algorithms for document datasets. Data Min. Knowl. Discov. **10**, 141–168 (2005)

Supporting the Page-Hinkley Test with Empirical Mode Decomposition for Change Detection

Raquel Sebastião[(✉)] and José Maria Fernandes

Institute of Electronics and Informatics Engineering of Aveiro (IEETA),
Department of Electronics, Telecommunications and Informatics (DETI),
University of Aveiro, 3810-193 Aveiro, Portugal
{raquel.sebastiao,jfernan}@ua.pt

Abstract. In the dynamic scenarios faced nowadays, when handling non stationary data streams it is of utmost importance to perform change detection tests. In this work, we propose the Intrinsic Page Hinkley Test (iPHT), which enhances the Page Hinkley Test (PHT) eliminating the user-defined parameter (the allowed magnitude of change of the data that are not considered real distribution change of the data stream) by using the second order intrinsic mode function (IMF) which is a data dependent value reflecting the intrinsic data variation. In such way, the PHT change detection method is expected to be more robust and require less tunes. Furthermore, we extend the proposed iPHT to a blockwise approach. Computing the IMF over sliding windows, which is shown to be more responsive to changes and suitable for online settings. The iPHT is evaluated using artificial and real data, outperforming the PHT.

Keywords: Data streams · Page Hinkley Test · Empirical Mode Decomposition · Personalized approach · Change detection · Sliding windows

1 Introduction

Nowadays information is gathered as a continuous flow of data streams. A data stream is a sequence of information in the form of transient data that arrives continuously (possibly at varying times) and is potentially infinite. Data streams may not be strictly static as underlying properties evolve over time presenting non-stationary distributions and demanding the detection of changes. The survey [2] discusses several detection approaches, such as, ADaptive WINdowing, Drift Detection Method, Early Drift Detection Method.

The PHT [6] is a sequential analysis technique typically used for monitoring change detection in the average of a Gaussian signal. However, it is relatively robust in the face of non normal distributions and has been recently applied for concept drift detection in data streams [3]. In this work we propose the iPHT: a change detection approach enhancing the PHT through the elimination of the

© Springer International Publishing AG 2017
M. Kryszkiewicz et al. (Eds.): ISMIS 2017, LNAI 10352, pp. 492–498, 2017.
DOI: 10.1007/978-3-319-60438-1_48

user-defined parameter (the allowed magnitude of change) by using the second order IMF.

The IMF are the result of the Empirical Mode Decomposition (EMD) of a signal, which is an adaptive and efficient decomposition method that handles data on time domain [4]. Relying on time scale characteristics of the data, the EMD can successfully handle non stationary and nonlinear data [4]. Moreover, unlikely other signal processing decomposition methods, EMD makes no assumptions on the incoming data. This means that iPHT can be applied without any prior hypothesis on the data as both PHT and EMD can be applied over an arbitrary data stream without any constraints. Furthermore, we extend the iPHT to a blockwise approach. Computing the IMF over sliding windows advances a contribution to an online strategy.

2   Proposed Strategy

The Page-Hinkley Test requires an input parameter and an input threshold, which are defined by the user according to the data properties and application purposes. By supporting the PHT with the EMD we can replace the input parameter, which controls the allowed magnitude of change, by an IMF.

2.1   Page Hinkley Test

The two-sided PHT [3,6] detects both increases and decreases in the mean of a sequence. For testing online changes it runs two tests in parallel, considering a cumulative variable defined as the accumulated difference between the observed values and their mean until the current moment.

2.2   Empirical Mode Decomposition

The EMD is being successfully used in real-time applications, such as biomedical problems [7] and image processing [5], among others. There are several free implementations of EMD available: the one used in this work was developed by Alan Tan[1].

The EMD consists of successively decomposing the original data $x(t)$ into IMF $c_i(t), i = 1, \ldots, n$ and into the monotonic residual $r(t)$. Once the first IMF is removed from the original data, the procedure is successively applied to the residual. This process will decompose the original data into the highest frequency component (c_1) to the lowest frequency component (c_n), until the residual $r(t)$ is a monotonic function from which no more IMF can be extracted: $x(t) = \sum_{i=1}^{n} c_i(t) + r(t)$, where n is the number of IMF.

---

[1] available at http://www.mathworks.com/matlabcentral/fileexchange/19681-hilbert-huang-transform (accessed in March 17th 2016).

2.3   Intrinsic Page Hinkley Test

The lower IMF orders are related with higher frequency of the data and higher IMF orders describe the data baseline wander. We are interested in using the lower IMF orders to establish the magnitude of the changes that are allowed. In this context the PHT is supported through the EMD, by replacing the δ parameter with an IMF. Hence we are loosing the user-defined parameter of the PHT by using a component that will depend on the data itself. The iPHT tests are the following:

For increase cases:

$iU_0 = 0$

$iU_T = (iU_{T-1} + x_T - \bar{x}_T - IMF_T)$

($\bar{x}_T$ is the mean until the current sample.)

$m_T = min(iU_t, t = 1 \ldots T)$

$iPH_U = iU_T - m_T$

If $iPH_U > \lambda$: change detected

For decrease cases:

$iL_0 = 0$

$iL_T = (iL_{T-1} + x_T - \bar{x}_T + IMF_T)$

$M_T = max(iL_t, t = 1 \ldots T)$

$iPH_L = M_T - iL_T$

If $iPH_L > \lambda$: change detected

The threshold λ depends on the admissible false alarm rate. Increasing λ entails fewer false alarms, but might miss or delay changes. The order of the IMF used to replace δ depends on the relation between the several IMF orders and the original data. The information shared decrease with the IMF order: the first IMF shares more information with the original data. As δ corresponds to the magnitude of the changes that are allowed, the original data and the IMF order must be positive related. On the other side, they must not share too much information. Therefore, the IMF order will be studied, in the Sect. 3, evaluating the covariance and the Spearman correlation between the data and several IMF orders.

The EMD requires the knowledge of the entire data stream to compute the IMF. To overcome this drawback, we propose a blockwise approach for the extraction of the IMF: the IMF are computed only with the data that is inside a sliding window. The main difficulty is how to select the appropriate length of the window, establishing a trade-off between good stability for the computation of the IMF and good adaptability to evolving scenarios.

3   Results and Discussion

3.1   Artificial Data Design

The artificial data sets were generated according to normal distributions, varying only the mean parameter and setting the standard deviation parameter equal to 1. Each data stream consists of 2 parts, each of with size $N = 5000$. Different changes were simulated by varying among 3 levels of magnitude and 3 rates (or speed) of change, obtaining a total of 9 types of changes. For high, medium and low magnitude levels, μ changed from 0 to 5, 3 and 2, respectively. The rates were defined assuming that the samples from the first part are from the old distribution and the $5000 - ChangeLength$ last samples are from the new distribution. The rates of change were defined as high, medium and low, for a $ChangeLength$ of $1, 0.25 * 5000$ and $0.5 * 5000$, respectively.

For each type of changes, 10 data streams were generated with different seeds. Two data sets, with similar characteristics, were generated, one to train and adjust the parameters and another to test and evaluate the proposed iPHT.

Choosing the Order of the IMF. Figure 1 shows the covariance between the original data and the correspondent five first IMF orders (average with error bars for 10 runs on data generated with different seeds). The Spearman correlation presented a similar behavior, and for both measures, the higher values are obtained for the IMF_1. When comparing the original data with the IMF_1 and with the remain IMFs, it can be observed an abrupt decay and a repeated pattern along the different changes. Considering these results on the shared information, IMF_2 was chosen to replace the user-defined parameter δ of the PHT.

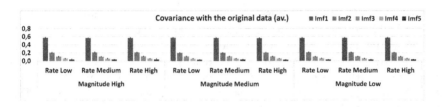

Fig. 1. Covariance between the data and the five first IMF orders

Sensibility of the λ Threshold. The λ change detection threshold was adjusted using the training set and evaluating the total number of false alarms (FA), the total number of missed detections (MD) and the detection delay time (DDT). The results are shown in Table 1. Independently of the λ, the proposed iPHT detects all the changes. This is relevant as the MD are a major concern: missing a change, besides inducing a bias on the model, would prevent the system to perform the appropriate action. The number of FA (desirable to be null) can be controlled by the change detection threshold. It can be observed the absence of FA for λ greater than 300. Considering that the DDT was quite similar for $\lambda = 400$ and for $\lambda = 500$, and following a conservative approach, we decided on establishing the $\lambda = 500$. As expected, the DDT when using $\lambda = 500$ is greater for changes with low rate and magnitude.

Comparison with the PHT. This experiment assesses the advantage of iPHT over PHT. For PHT, the λ was set to 500 and δ varied from 0.0001 to 1. Independently of the δ, both approaches detected all the changes without FA. The DDT of the iPHt was 722 ± 71 (average and standard deviation of 10 runs for the 9 types of changes). For the PHT, as the δ varied from 0.0001 to 1, the DDT varied from 723 ± 69 to 1818 ± 31 (average and standard deviation of 10 runs for the 9 types of changes). These results show that the iPHT and PHT presented similar performance, but the iPHT has the advantage of adjusting only the change detection threshold.

Table 1. DDT when varying the λ: average and standard deviation of 10 runs. In parenthesis is the number of runs, if any, where the iPHT presents FA or MD.

Magnitude	Rate	λ			
		100	300	500	700
High	Low	341 ± 174 (14;0)	628 ± 118 (0;0)	901 ± 75(0;0)	1112 ± 55 (0;0)
	Medium	225 ± 66 (18;0)	456 ± 98 (0;0)	647 ± 65 (0;0)	789 ± 53 (0;0)
	High	15 ± 4 (19;0)	45 ± 13 (1;0)	80 ± 19 (0;0)	120 ± 19 (0;0)
Medium	Low	476 ± 139 (16;0)	860 ± 154 (0;0)	1201 ± 127 (0;0)	1592 ± 113 (0;0)
	Medium	399 ± 94 (15;0)	641 ± 79 (0;0)	869 ± 54 (0;0)	1056 ± 47 (0;0)
	High	23 ± 8 (18;0)	60 ± 28 (0;0)	127 ± 30 (0;0)	196 ± 29 (0;0)
Low	Low	611 ± 176 (20;0)	1131 ± 99 (0;0)	1532 ± 92 (0;0)	1849 ± 71 (0;0)
	Medium	398 ± 145 (19;0)	735 ± 139 (0;0)	1027 ± 105 (0;0)	1263 ± 87 (0;0)
	High	36 ± 17 (15;0)	108 ± 30 (0;0)	210 ± 31 (0;0)	314 ± 30 (0;0)

Extension to Online Processing. This experiment evaluates the relation between DDT and the window length (from 100 to 1000) used to compute the IMF of the iPHT. A large window, while providing more data to compute more accurate IMF, will contribute to higher delays, since the testing for changes is performed within large time steps. A smaller window will, at a first glance, allow to detect changes earlier. However, it compromises the stability of the IMF, which may preclude the detection of changes.

The DDT is shown in Fig. 2. This experiment stands out that choosing correctly the length of the sliding window, computing the IMF in blockwise does not impair the detection delay time. Indeed, using a sliding window of length 100 the DDT is similar when computing the IMF with the entire data stream. It also can be observed that, an increase of the length of the sliding window leads to higher DDT. One result that must be underlined is that using the entire data stream or using sliding windows, the iPHT does not presented FA nor MD. It can also be observed that the advantage of using a smaller sliding window is strengthened when detecting abrupt changes, which is explained by the ability of a window with small length to better catching abrupt than smooth changes.

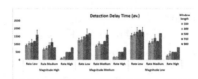

Fig. 2. DDT for different lengths of the sliding window (average and standard deviation of 10 runs of the training set)

Robustness to Detect Changes in the Presence of Noise. In this experiment noisy data was generated by adding different percentages of pink noise to the test data set. Figure 3 shows the obtained results by varying the amount of

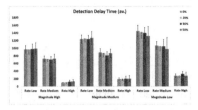

Fig. 3. DDT (average of 10 runs) of the iPHT with different amounts of noise and computed over a sliding window of length 100 (blockwise)

noise from 0% to 50%. The DDT was obtained when performing the iPHT over a sliding window of length 100. The iPHT presents a regular performance along the different amounts of noise, which are similar to the DDT in the absence of noise. This experiment sustains the argument that the iPHT, when performed in blockwise with a sliding window of length 100, is robust against noise while effectively detects changes in the data. It should be noted that, even in the presence of great amount of noise, the proposed iPHT did not miss any change. Although, with an amount of 50% of noise, it presents a FA for changes with high magnitude and low rate and with medium magnitude and rate.

3.2   Industrial Data

This industrial data set was obtained within the scope of the work presented in [1], with the objective of designing different machine learning classification methods for predicting surface roughness in high-speed machining. Data was obtained by performing tests in a Kondia HS1000 machining center equipped with a Siemens 840D open-architecture CNC. The tests were done with different cutting parameters, using sensors for registry vibration and cutting forces. For change detection purposes, the measurements of the cutting speed on X axes from 7 tests were joined sequentially in order to have only one data set with 6 changes with different magnitudes and sudden and low rates, as shown in Fig. 4. For PHT, δ was set to 1 and the iPHT used the IMF_2. The λ was set to 2000 (for both tests). Although presenting 8 FA against 7 of the PHT, the iPHT detects all the 6 changes with mean DDT of 3512 against only 5 detected changes with mean DDT of 8458 of the PHT.

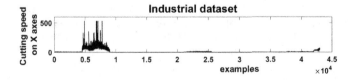

Fig. 4. The cutting speed on X axes from 7 tests sequentially joined

4 Conclusions

This paper proposes the iPHT, a test for change detection through the enhancement of the PHT with the IMF, eliminating the user-defined parameter which controls the allowed magnitude of change.

The results over the designed experiments have shown that the iPHT is a feasible approach to detect changes under different evolving scenarios, even in the presence of noise. The obtained results also sustain its application over a sliding window, which is an advantage that leads future research on this approach for online settings. While presenting similar performance to the PHT (regarding DDT, FA and MD), the iPHT has the advantage of adjusting only the change detection threshold.

With these properties, the iPHT has potential for detecting changes in several interesting scenarios, like internet of things and sensor networks context, for human and environmental monitoring. Our next step is to apply the iPHT in a physiological monitoring setting over heart rate. In such cases, a change can be seen a biomarker for emotion changes and changes in fatigue or stressful conditions. Furthermore, iPHT must be evaluated and compared to other methods in terms of computational time and memory usage.

Acknowledgments. This work was supported by the Portuguese Science Foundation (FCT) through national funds, and co-funded by the FEDER, within the PT 2020 Partnership Agreement and COMPETE2020 under projects IEETA (UID/CEC/ 00127/2013) and VR2market (funded by the CMU Portugal program, CMUP-ERI/FIA/0031/2013). Raquel Sebastião acknowledges her Post-Doc grant (BPD/ UI62/6777/2015).

References

1. Correa, M., Bielza, C., Pamies-Teixeira, J.: Comparison of bayesian networks and artificial neural networks for quality detection in a machining process. Expert Syst. Appl. **36**(3), 7270–7279 (2009). doi:10.1016/j.eswa.2008.09.024
2. Gama, J., et al.: A survey on concept drift adaptation. ACM Comput. Surv. **46**(4) (2014). doi:10.1145/2523813
3. Gama, J., Sebastião, R., Rodrigues, P.: On evaluating stream learning algorithms. Mach. Learn. **90**(3), 317–346 (2013). doi:10.1007/s10994-012-5320-9
4. Huang, N.E., et al.: The empirical mode decomposition and the hilbert spectrum for nonlinear and non-stationary time series analysis. Proc. R. Soc. Lond. A: Math. Phys. Eng. Sci. **454**(1971), 903–995 (1998). doi:10.1098/rspa.1998.0193
5. Linderhed, A.: Variable sampling of the empirical mode decomposition of two-dimentional signals. Int. J. Wavelets Multiresolut. Inf. Process. **03**(03), 435–452 (2005). doi:10.1142/S0219691305000932
6. Page, E.S.: Continuous inspection schemes. Biometrika **41**(1–2), 100–115 (1954). doi:10.1093/biomet/41.1-2.100
7. Santillan-Guzman, A., Fischer, M., Heute, U., Schmidt, G.: Real-time empirical mode decomposition for EEG signal enhancement. In: 2013 Proceedings of the 21st European Signal Processing Conference (EUSIPCO), pp. 1–5 (2013)

Co-training Semi-supervised Learning for Single-Target Regression in Data Streams Using AMRules

Ricardo Sousa[1]([⊠]) and João Gama[1,2]

[1] LIAAD/INESC TEC, Universidade do Porto, Porto, Portugal
rtsousa@inesctec.pt, jgama@fep.up.pt
[2] Faculdade de Economia, Universidade do Porto, Porto, Portugal

Abstract. In a single-target regression context, some important systems based on data streaming produce huge quantities of unlabeled data (without output value), of which label assignment may be impossible, time consuming or expensive. Semi-supervised methods, that include the co-training approach, were proposed to use the input information of the unlabeled examples in the improvement of models and predictions. In the literature, the co-training methods are essentially applied to classification and operate in batch mode.

Due to these facts, this work proposes a co-training online algorithm for single-target regression to perform model improvement with unlabeled data. This work is also the first-step for the development of online multi-target regressor that create models for multiple outputs simultaneously. The experimental framework compared the performance of this method, when it rejects unalabeled data and when it uses unlabeled data with different parametrization in the training.

The results suggest that the co-training method regressor predicts better when a portion of unlabeled examples is used. However, the prediction improvements are relatively small.

Keywords: Single-target regression · Semi-supervised learning · Co-training · Data streams

1 Introduction

The importance of prediction has increased in online data streams context [1,2]. In fact, several domains (where data is obtained through data streams) rely on the ability of making accurate predictions for decision making, planning, strategy development and reserve determination which depend on models produced by data analysis [3].

In this context, data streams produce massive quantities of data of which label assignment may be impossible, time consuming or expensive. Unlabeled data (without output values) is usually present in sensor malfunction or database failure. In addition, labels may be omitted when data is sensitive (e.g.,

© Springer International Publishing AG 2017
M. Kryszkiewicz et al. (Eds.): ISMIS 2017, LNAI 10352, pp. 499–508, 2017.
DOI: 10.1007/978-3-319-60438-1_49

privacy preservation) or may be not obtained due to labeling cost [4]. Unlabeled data usually appear in a wide range of contexts such as Engineering Systems (video object detection) [5], Physics (weather forecasting and ecological models) [6], Biology (model of cellular processes) [7] and Economy/Finance (stock price forecasting) [3]. In most of these areas, data from streams are obtained and processed in real time [4].

Semi-supervised Learning (SSL) methodology has been developed to utilize the input information for accuracy improvement of the regression model by artificial labeling [4]. These methodologies become useful when the unlabeled data is significantly more abundant than labeled data [8]. However, these methodologies may introduce errors by propagating the inaccurate artificial labels [8].

Formally, let $\mathcal{S} = \{..., (\mathbf{x}_1, y_1), (\mathbf{x}_2, y_2), ..., (\mathbf{x}_i, y_i), ...\}$ denote an unlimited stream of data examples, where $\mathbf{x}_i = [x_{i,1} \cdots x_{i,j} \cdots x_{i,M}]$ is a vector of descriptive variables and y_i is a scalar output variable (label) of the ith example (considering one example with the index of reference). The unlabeled example is represented with an empty label $y_i = \emptyset$. The aim of SSL consists of using examples $(\mathbf{x}_i, \emptyset)$ to improve the regression model $y_i \leftarrow f(\mathbf{x})$ and reduce the mean error of prediction for both labeled and unlabeled examples. Most of existent SSL methodologies are performed in batch mode with large amount of computational resources [2]. Moreover, these methodologies are often applied to classification and cannot be directly applied to regression [8].

Co-training has been showing promising results, among the SSL methods [4]. This method consists of creating two or more diversified models by using different input variables, different regressors or the same regressors with different parametrization. In the training stage, the regressors yield predictions that are processed (e.g., mean of all predictions or selection of the best prediction according to a criterion) to produce an artificial label in order to be used in the training of models. In the prediction stage, the regressors also yield predictions that are combined (e.g., weighted mean)to produce a final prediction.

The main propose of this paper is to apply and adapt the co-training method to online single-target regression context. This work also prospects the extension to online multi-target regression using the (Adaptive Model Rules)AMRules algorithm [2].

This paper is organized as follows. Section 2 presents a brief review on the principles of SSL and methods of co-training. Section 3 describes the adaptation of the co-training to online learning and regression. Section 4 explains in detail the evaluation method. Finally, the results are presented and discussed in Sect. 5 and the main conclusions are summarised in Sect. 6.

2   Related Work

In this section, the main principles of co-training and some existent co-training methods are briefly reviewed. Since no online version of co-training methods were found in the literature, the most prominent co-training batch mode methods are presented. Despite these facts, the methods are fair starting points for the development of online regression methods since they exhibit promising principles and results.

Co-training basically consists of two or more models training but with different aspects that allows to create diversity (different inputs, different regressors, different parametrization,...). The common trait is that a regressor algorithm is trained with examples (previously unlabeled) artificially labeled by other complementary regressors. These regressors are assumed to predict reliably which makes the co-training confidence driven. This method relies on several assumptions such as consensus, complementary, sufficiency, compatibility and conditional independence.

- **Consensus** assumption states that the error reduction of labeled examples prediction and the increase of the unlabeled example prediction agreement lead to more precise models [9].
- **Complementarity** assumption states that each input group contains information that the other groups do not contain. Hence, the use of multiple input groups increase the amount of information to construct more accurate models [9].
- **Sufficiency** assumption considers that each group of inputs is adequate to build a model by proper training.
- **Compatibility** assumption implies that the output predictions from the models are very similar with high probability for the simultaneous input values of the respective groups.
- **Conditional independence** assumption allows the possibility of at least one of the model to produce less errors and teach the other models the correct prediction [10]. This assumption is essential for co-training, however it is very strong. To overcome this problem, similar but less demanding assumptions were consider. **Weak dependence** assumption, where some dependence exists between inputs was proven to work [11]. **Large diversity** assumption considers that independence can be achieved by using different algorithms or the same algorithm with different parametrization [12].

The main drawbacks of co-training are related to the inaccuracy of the artificially labeled examples that convey error to the models. In addiction, the artificially labeled examples may not carry the needed information to the regressor [8]. The co-training variants may present different strategies to artificially label the unlabeled examples or may present a criterion to discard the damaging artificially labeled examples. The prediction function generally combines the predictions of the models according to a criterion to produce the final prediction [8].

In this work the Co-training regression (COREG), Co-training by Committee for Regression (CoBCReg) and Co-regularised least squares regression (coRLSR) were studied. COREG uses two k-Nearest Neighbours (kNN) regressors [4]. Initially, the labeled and unlabeled examples are separated in two sets. For each regressor, the (k-NN) is used to construct a set of labeled examples which input vector is close to the input vector of the unlabeled example by using a distance metric (user defined). Each regressor predicts a value to artificially label the example and uses it to re-train the models with all labeled examples. Mean Squared Error (MSE) variation is computed between the scenarios with and without the artificially labeled examples. If MSE is reduced, the artificially labeled example is joined to the labeled examples set. The process stops when none of unlabeled examples is interchanged between labeled and unlabeled sets. The final prediction is obtained by averaging the predictions of the two regressors. CoBCReg is based on Radial Basis Functions regressors (with a Gaussian basis function that uses the Minkowski distance) and Bagging. This algorithm implies that diversity must exist between the elements of the ensemble of regressors, which is achieved by different input subsets random initialization. In this method, each regressor selects the unlabeled examples that are more relevant for the respective model [13]. coRLSR (Co-regularised least squares regression) formulates into a regularised risk in Hilbert spaces minimisation problem [14]. It aims to find the models that minimizes the error of all models and the disagreement on unlabeled examples predictions.

3   Online Co-training Regression

This section presents the proposed co-training method by showing the main adaptations to the online and regression context. This section also presents a small description of the underlying algorithm regressor AMRules.

The proposed co-training method, at the initialization stage, divides randomly the input variables of the incoming example into two groups and produce two example types with different input variables but with equal labels (labels of the incoming example). In this step, weak dependence is assumed. The two groups may overlap some inputs randomly selected by a pre-defined overlap percentage. Posteriorly, two AMRules complementary regressors yield predictions for each examples. The initial models are obtained previously in a training stage using a small dataset. Considering an incoming unlabeled example, a score is computed in order to evaluate the benefit/confidence of the prediction to be used in the artificial labeling for models training. The score is the relative error compared to maximum absolute value of the output found in the stream. If the score is lower than a pre-defined threshold, the predictions are used to train the complementary regressor. Otherwise, the artificially labeled example is discarded. The consensus assumption is used in this step. If the example is labeled, the mean error is computed for each regressor and the example is used for both regressor training. Algorithm 1 explains the training procedure of the proposed method.

For prediction, combination of the regressors are made by using weights. These weights are inversely proportional to the error produced by labeled examples previously used in the training stage. This strategy gives more credit to the regressor that produces less errors. Algorithm 2 explains the procedure of label prediction.

The AMRules regressor was used as the underlying algorithm in the training of the models and for output prediction of the unlabeled examples. The

Algorithm 1. Co-training algorithm training

1: **Initialization:**
2:   α − *Overlap percentage* s − *Score Threshold*
3:   *Random input allocation and overlapping into the two groups using* α
4: **Input:** *Example* $(\mathbf{x}_i, y_i) \in \mathcal{S}$
5: **Output:** *Updated Models*
6: *Divide* $\mathbf{x}_i$ *into* $\mathbf{x}_i^1$ *and* $\mathbf{x}_i^2$
7: $\hat{y}_i^1 = PredictModel1(\mathbf{x}_i^1); \ \hat{y}_i^2 = PredictModel2(\mathbf{x}_i^2)$
8: **if** $(y_i = \emptyset)$ **then**
9:    **if** $(|\hat{y}_i^1 - \hat{y}_i^2|/|y_{max}| < s)$ **then**
10:       $TrainModel1((\mathbf{x}_i^1, \hat{y}_i^2)); \ TrainModel2((\mathbf{x}_i^2, \hat{y}_i^1));$
11: **else**
12:    $\bar{e}_1 = Update\ the\ mean\ error\ of\ Model1(\hat{y}_i^1, y_i)$
13:    $\bar{e}_2 = Update\ the\ mean\ error\ of\ Model2(\hat{y}_i^2, y_i)$
14:    $TrainModel1((\mathbf{x}_i^1, \hat{y}_i^2)); \ TrainModel2((\mathbf{x}_i^2, \hat{y}_i^1))$

Algorithm 2. Co-training algorithm prediction

1: **Input:** *Example* $(\mathbf{x}_i, y_i) \in \mathcal{S}$
2: **Output:** *Example prediction* $\hat{y}_i$
3: *Divide* $\mathbf{x}_i$ *into* $\mathbf{x}_i^1$ *and* $\mathbf{x}_i^2$
4: $\hat{y}_i^1 = PredictModel1(\mathbf{x}_i^1); \ \hat{y}_i^2 = PredictModel2(\mathbf{x}_i^2)$
5: $w_1 = \bar{e}_2/(\bar{e}_1 + \bar{e}_2); \ w_2 = \bar{e}_1/(\bar{e}_1 + \bar{e}_2);$
6: $\hat{y}_i = w_1 * \hat{y}_i^1 + w_2 * \hat{y}_i^2$

AMRules is a multi-target algorithm (predicts several outputs) that is based on rule learning [2]. AMRules partionates the input space and creates local models for each partition. The local models are trained using a single layer perceptron. Its main advantages are models simplicity, low computational cost and low error rates [2]. This algorithm presents convenient properties such as the modularity property that allows the construction of models for particular input variables regions (defined by the rule). It uses anomaly and change detection to increase resilience to data outliers and data changes on the stream. AMRules algorithm benefits from unlabeled examples since it prunes the input partitions.

4   The Evaluation Method

The proposed co-training algorithm was evaluated by simulating a data stream with artificial and real datasets. A percentage of 30% of each dataset first examples (30% of the first examples of the stream) were used to create an initial consistent model and the remaining examples were used in the testing stage.

In the testing stage, a binary Bernoulli random process with a probability p was applied to assign an example as labeled or unlabeled. If the example is assigned as unlabeled, the true output value is omitted from the algorithm. The p probabilities of being unlabeled were 50%, 80%, 90%, 95% and 99%. Considering the algorithm parametrization, the score threshold values were 1×10^{-4}, 5×10^{-4}

0.001, 0.005, 0.01, 0.05, 0.1, 0.5 and 1. The diversity of these values allow to observe the behaviour of the algorithm in different scales. The overlap percentage values were 0%, 10%, 30%, 50%, 70% and 90%. Prequential mode was used in evaluation. This mode first predicts the label and then train the model for both labeled and artificially labeled examples [15].

Five real world and four artificial datasets were used. The real world datasets were House8L (Housing Data Set), House16L (Housing Data Set), CASP (Physicochemical Properties of Protein Tertiary Structure Data Set), California, blogDataTrain and the artificial datasets were 2dplanes, fried, elevators and ailerons. These datasets contain a single-target regression problem and are available at UCI repository [16]. Table 1 shows the features of the real world and artificial data sets used in the method evaluation.

Table 1. Real world datasets description

Dataset	# Examples	# Inputs
House8L	22784	8
House16H	22784	16
calHousing	20640	7
CASP	45730	9
blogDataTrain	52472	281
2dplanes	40768	10
fried	40768	10
ailerons	13750	41
elevators	8752	18

As performance measures, the mean relative error (MRE) and the mean percentage of accepted unlabeled examples (MPAUE) in the training were used. Finally, the error reduction was measured by using the relative error (in percentage) between the reference scenario (no unlabeled examples used) E_0 and the case with the parametrization that lead to the lowest error E_{lowest} (includes the reference case E_0). Equation 1 defines the Error Reduction.

$$Error\ Reduction = \frac{|E_0 - E_{lowest}|}{E_0} \tag{1}$$

If the reference case yields the lowest error, then the Error Reduction is zero, which means that the algorithm is not useful for that particular scenario.

The algorithm was developed in the Massive Online Analysis (MOA) platform where the AMRules was developed [17]. Its an open source platform of Machine Learning and Data Mining algorithms applied to data streams. This platform was implemented in JAVA programming language.

5 Results

In this section, the evaluation results are presented and discussed. Some scenarios examples plots of MRE and MPAUE for overlap percentage and score threshold combination are presented. The reference curve (Ref) corresponds to the scenario where no unlabeled example is used in the training. Since the inputs are selected randomly, 10 runs and respective averaging of the MRE and MPAUE were performed in order to obtain more consistent values. This section also presents the error reduction for each dataset and stream unlabeled examples percentage simulation.

Figure 1 presents the plots of the MRE and MPAUE of a successful case for 80% of unlabeled examples stream. The curves on the left reveals that there exists some cases (combination of overlap percentage and score threshold) that lead to beneficial use of unlabeled examples. For the case of overlap of 50% and score threshold of 0.001, the use of 13.4% of the unlabeled examples in the training lead to reduction of 5,3% of the MRE in average. In general, it also observed that the overlapping decrease the MRE.

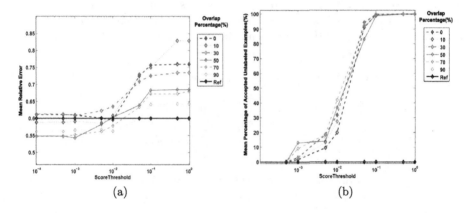

(a)   (b)

Fig. 1. Mean relative error (a) and mean percentage of accepted unlabeled examples (b) for a data stream with 80% of unlabeled examples. The examples are from the House8L dataset.

Figure 2 shows a case where the algorithm does not present any combination of overlap percentage and score thresholds that lead to model improvement. In this case, most of unlabeled examples contributed to model damage and the artificial labels conveyed significant errors (all curves are above the reference curve). This fact suggests that the dataset characteristics (e.g., inputs variables distributions) may influence the performance. The error propagation through the model lead to worst predictions in the artificial labeling. This effect leads to a cycle that reinforce the error on each unlabeled example processing. In fact, the more unlabeled examples arrive the higher is the error.

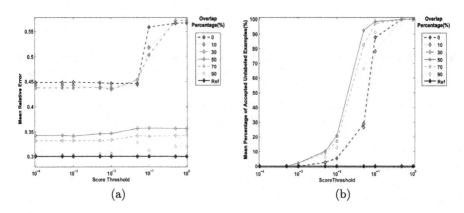

(a)   (b)

Fig. 2. Mean relative error (a) and mean percentage of accepted unlabeled examples (b) for a data stream with 80% of unlabeled examples. The examples are from the calHousing dataset.

Table 2 presents the error reduction of the experiments on real world datasets for each chosen unlabeled examples probabilities. When the value is zero, it means that there was not any combination of overlap percentage and score threshold that improved the model and the reference scenario is the best.

Table 2. Error reduction (%) for real world datasets

Datasets	Unlabeled examples probabilities				
	50%	80%	90%	95%	99%
House8L	2,33	5,31	0,26	0,00	0,00
House16H	1,01	0,01	0,01	0,90	0,00
calHousing	1,11	0,00	1,62	0,01	0,00
CASP	0,8	3,45	1,06	0,00	0,00
blogDataTrain	2,54	1,56	0,26	0,00	0,00

According to Table 2, the algorithm seems to benefit most part of the scenarios. However, the benefits are in general relatively small. As expected, the more elevated the probability of unlabeled example is, the less is the relative error reduction.

Table 3 presents the error reduction for real artificial datasets in a similar way as the real world datasets presented in Table 2. The artificial datasets also present the same trend of error reduction when the probability of unlabeled example incoming increases. The error reduction is also frequently small.

In essence, the MRE curves and the error reduction tables support the view that the algorithm leads to an error reduction (despite being small) by using labeled examples in most cases. It was observed that none of the tested scenarios

Table 3. Error reduction (%) for artificial datasets

Datasets	Unlabeled examples probabilities				
	50%	80%	90%	95%	99%
2dplanes	1,48	0,00	1,70	0,02	0,00
fried	4,64	3,21	1,04	0,70	0,00
ailerons	1,83	0,08	0,00	1,25	0,00
elevators	3,21	0,60	1,48	0,93	0,00

worked for simulated streams with 99% of unlabeled examples. In fact, this scenario is an extreme case where the model is trained essentially with artificially labeled examples and the error propagation can easily occur.

6 Conclusion

In this paper, an online semi-supervised single-target algorithm for regression based on co-training is addressed. This work prospects the development of multi-target regression algorithm that performs semi-supervised learning by co-training.

In general this co-training approach reduces the prediction error with the proper parameters calibration. The mean relative error is reduced by using a portion of unlabeled examples in most of evaluation experimental scenarios. However, the error reduction is relatively small and the parametrization depends on the dataset. It can be also conclude that, in order to obtain model improvement, only a small amount of unlabeled examples are used in the training.

As future work, this method will be extended to multi-target regression. The fact that very few unlabeled examples can lead to some improvement may suggest the study of the conditions that lead to this improvement. In order to increase the algorithm validity, the evaluation tests will be performed using a higher number of real world datasets with a significant amount of examples.

Acknowledgements. This work is financed under the project "NORTE-01-0145-FEDER-000020" funded by the North Portugal Regional Operational Programme (NORTE 2020), under the PORTUGAL 2020 Partnership Agreement, and through the European Regional Development Fund (ERDF).

References

1. Li, C., Dong, W., Liu, Q., Zhang, X.: MORES: online incremental multiple-output regression for data streams. CoRR, abs/1412.5732 (2014)
2. Duarte, J., Gama, J.: Multi-target regression from high-speed data streams with adaptive model rules. In: IEEE Conference on Data Science and Advanced Analytics (2015)

3. Ariyo, A.A., Adewumi, A.O., Ayo, C.K.: Stock price prediction using the ARIMA model. In: Proceedings of the 2014 UKSim-AMSS 16th International Conference on Computer Modelling and Simulation, UKSIM 2014, Washington, DC, USA, pp. 106–112. IEEE Computer Society (2014)
4. Zhou, Z.H., Li, M.: Semi-supervised regression with co-training style algorithms. IEEE Trans. Knowl. Data Eng. 19(11), 1479–1493 (2007)
5. Rosenberg, C., Hebert, M., Schneiderman, H.: Semi-supervised self-training of object detection models. In: Proceedings of the Seventh IEEE Workshops on Application of Computer Vision (WACV/MOTION 2005) - Volume 1 - Volume 01, WACV-MOTION 2005, Washington, DC, USA, pp. 29–36. IEEE Computer Society (2005)
6. Chalabi, Z., Mangtani, P., Hashizume, M., Imai, C., Armstrong, B.: Article: time series regression model for infectious disease and weather. Int. J. Environ. Res. 142, 319–327 (2015)
7. Uslana, V., Seker, H.: Article: quantitative prediction of peptide binding affinity by using hybrid fuzzy support vector regression. Appl. Soft Comput. 43, 210–221 (2016)
8. Kang, P., Kim, D., Cho, S.: Semi-supervised support vector regression based on self-training with label uncertainty: an application to virtual metrology in semiconductor manufacturing. Expert Syst. Appl. 51, 85–106 (2016)
9. Xu, C., Tao, D., Xu, C.: A survey on multi-view learning. CoRR, abs/1304.5634 (2013)
10. Blum, A., Mitchell, T.: Combining labeled and unlabeled data with co-training. In: Proceedings of the Eleventh Annual Conference on Computational Learning Theory, COLT 1998, pp. 92–100. ACM, New York (1998)
11. Abney, S.P.: Bootstrapping. In: Proceedings of the 40th Annual Meeting of the Association for Computational Linguistics, 6–12 July 2002, Philadelphia, PA, USA, pp. 360–367 (2002)
12. Goldman, S., Zhou, Y.: Enhancing supervised learning with unlabeled data. In: Proceedings of the 17th International Conference on Machine Learning, pp. 327–334 (2000)
13. Abdel Hady, M.F., Schwenker, F., Palm, G.: Semi-supervised learning for regression with co-training by committee. In: Alippi, C., Polycarpou, M., Panayiotou, C., Ellinas, G. (eds.) ICANN 2009. LNCS, vol. 5768, pp. 121–130. Springer, Heidelberg (2009). doi:10.1007/978-3-642-04274-4_13
14. Brefeld, U., Gärtner, T., Scheffer, T., Wrobel, S.: Efficient co-regularised least squares regression. In: Proceedings of the 23rd International Conference on Machine learning, ICML 2006, pp. 137–144. ACM, New York (2006)
15. Gama, J., Sebastião, R., Rodrigues, P.P.: On evaluating stream learning algorithms. Mach. Learn. 90(3), 317–346 (2013)
16. Bache, K., Lichman, M.: UCI machine learning repository (2013)
17. Bifet, A., Holmes, G., Kirkby, R., Pfahringer, B.: MOA: massive online analysis. J. Mach. Learn. Res. 11, 1601–1604 (2010)

Time-Series Data Analytics Using Spark and Machine Learning

Patcharee Thongtra$^{(\boxtimes)}$ and Alla Sapronova

Uni Research Computing, Uni Research, Bergen, Norway
{Patcharee.Thongtra,Alla.Sapronova}@uni.no

Abstract. This work presents a scalable architecture capable to provide real-time analysis over large-scale time-series data. Spark streaming, Spark MLlib and machine learning methods are combined to process and analyse the data streams. A high performance training model is automatically built and applied for the time-series forecasting. In order to validate the proposed architecture, authors developed a prototype system to predict the average energy consumption at real-time (estimated from 6 K Irish home- and business consumers) from 30 to 90 min ahead. The results show the best prediction was done with a convolutional neural network model, where the Mean Absolute Error and Root Mean Square Error were 7.5% and 10.5% correspondingly.

1 Introduction

A time-series data is a collection of observations of well-defined data items obtained through repeated measurements over time. One defining characteristic of time series is that ordering of data items is very important because there is dependency, and changing the order could change the meaning of the data [1]. The basic objective of time-series analysis is to determine a model that describes the pattern of the time series, and thus the model can be used to for example forecast future values of the series, explain how the past effects the future and explain how two time-series interact.

Data scientists have been able to solve these time-series analysis problems using machine learning models for a long time. The familiar and popular tools are such as R, Python and Weka Framework software [2]. A large number of useful modules or packages have been well developed and tested. However, nowadays a great amount of time-series data can be generated from a variety of application areas including finance, economics, communication, automatic control, social networks, the Internet of Things (IoT), etc. The traditional tools are not sufficient enough, as mainly because they are designed to process data on a single machine with limited resources. Fast and scalable frameworks and techniques able to cope with the large-scale datasets in a small time period (or near real time) is imperative.

Apache Spark [3] has emerged as a widely used open-source framework for general big data processing. Spark is a lightning fast, fault-tolerant cluster computing system that supports in-memory parallel operations. It possesses a rich set of APIs in Java, Scala, Python, and R. Spark streaming [4] is an extension of the core Spark API that enables scalable, high-throughput data stream processing. It can consume data stream

© Springer International Publishing AG 2017
M. Kryszkiewicz et al. (Eds.): ISMIS 2017, LNAI 10352, pp. 509–515, 2017.
DOI: 10.1007/978-3-319-60438-1_50

from various sources such as HDFS, Flume, Kafka [5], Twitter, ZeroMQ and TCP sockets. MLlib [6] is Spark's machine learning library that is designed for simplicity, scalability and easy integration with other tools, especially the other Spark libraries (Spark SQL, Streaming and GraphX). Spark MLlib provides multiple types of machine learning algorithms, including classification, regression, clustering, and collaborative filtering, as well as supporting functionality such as model evaluation and data import.

We propose an architecture that combines the big data technologies with stable, mature machine learning methods and libraries to provide real-time ingestion, processing, storage and analysis for large-scale time-series. This paper focuses on its data processing and analysis features that are realized by the integration of Spark Streaming, Spark MLlib and Weka Framework software. In our work, a set of training models are automatically built, and then the best performance model is selected and applied to get the prediction values.

The rest of this paper is organized as follows, Sect. 2 discusses related studies. In Sect. 3, the proposed architecture is described and a scenario of smart home energy consumption are under Sect. 4. Finally, conclusions and some future directions are given in Sect. 5.

2   Related Work

Batch and real-time processing of time-series have been widely investigated. Some open-source technological frameworks for data stream processing have been discussed [7, 8]. Compared to the others streaming frameworks like Storm [9] and Flink [10], Spark Streaming guarantees the exactly-once message delivery semantics; no data will be lost and no data will be processed multiple-times. Due to its micro-batch processing, Spark Streaming has higher latency. However, it can be efficiently integrated with the other Spark libraries to provide a unifying programming model.

For machine learning models for the time-series forecasting, [11] studied eight different models and compared their performance. They found that the preprocessing can have a significant impact on the performance. The large-scale data streams (pre-) processing however requires fast and scalable frameworks. [12] evaluates machine learning tools with big data in the Hadoop ecosystem including Mahout [13], MLlib and SAMOA. Similar to our work, MLlib is used for the prediction in many research works, for example in [14, 15]. But as mentioned already our architecture provides a feature to automatically select the best performance training model for each data stream, and this is not mentioned explicitly in the other works.

3   Architecture

In this paper, time-series data is focused, however the proposed architecture support general data streams. The architecture shown in Fig. 1 is composed of a streaming module, processing module, data analysis module, storage, and visualization. Kafka publish-subscribe messaging system is used in our proposed model to receive stream data from the sources. All Kafka messages are organized into topics. A topic can be

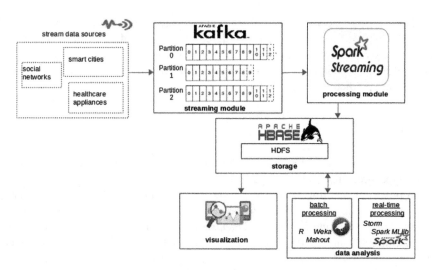

Fig. 1. The proposed architecture

divided into a number of partitions placed on multiple nodes to allow for multiple producers/consumers to submit to/read from a topic in parallel. A stream coming to Kafka is appended to a commit log and Kafka consumers can read it from this log.

For the processing part, Spark Streaming consumes the stream and perform operations on the stream. Spark Streaming provides two approaches to consume data from Kafka; Receiver-based and Receiver-less direct. By the first approach data is received from Kafka through receivers which are implemented using the Kafka high-level consumer API, and data is stored in Spark executors. In the latter there is no receiver and Spark Streaming queries Kafka periodically for the latest offsets. After Spark Streaming processes the data, it is persisted in HBase [16] tables. HBase is a NoSQL database based on Googles's BigTable that sits on top of HDFS. In HBase, data are logically organized into tables, rows and columns. Columns can have multiple versions of the same row key.

The data analysis module provides both batch and real-time processing. Batch processing is suitable for tasks that are not time-critical as it processes large volumes of static data which are already in the HDFS. These tasks are for example computing data distributions, performing data clustering, and building machine learning models. R, Weka Framework software and Mahout are softwares that are suitable to handle these. In contrast, real-time processing requires that data must be processed in a small time period. In most cases, it will take advantage of machine learning models built earlier by the batch processor. Example tasks are prediction, fraud detection and user recommendation. Storm and Spark MLlib are capable of handling these real-time machine learning tasks.

Finally the analysis results from both batch and real-time processing stored in the HDFS are summarized and visualized in a intuitive graphical way. The REST API is used to give the end users access to this graphical results.

In this work a time-series analysis is performed and a set of predictive models (with customizable parameters) are built, as described in Sect. 4. The models are built automatically every defined time interval by Weka. Then the best performance model, evaluated by prediction accuracy and processing time, will be selected and applied for the real-time prediction by Spark MLlib.

4   Use Case

A smart home scenario for energy consumption is used to evaluate our model. A prototype streaming system based on the proposed architecture has been built. This system processes real-time energy usage and makes a future prediction of the average usage up to 90 min ahead. The dataset used in this scenario is the meter data obtained from the Smart Metering Electricity Customer Behaviour Trials (CBTs) [17] that took place during 2009 and 2010 over 6 K Irish home- and business participants. The meter read data contains 4 digits of Meter ID, 5 digits of Datetime code and Electricity consumed in during an interval in kWh. The physical location of the meter can be identified by the Meter ID. The size of data from a meter at a time is 4 KB. During the system execution the data is received in a second window.

4.1   Energy Consumption Streaming System

The system is developed and run on a cluster of six nodes and 10 physical disks. Each node is equipped with Intel Xeon CPU E5-2650 v2 (32 core) 2.60 GHz, 125.99 GB RAM. These nodes runs CentOS release 6.7 and are connected to 10 Gbps Ethernet. All of them belong to the same rack. For HDFS configurations, a block size of 128 MB and $3\times$ replication are used. For the software and libraries, Hadoop version 2.7.3, Spark version 2.1.0 and Kafka version 0.10.1.0 are installed in the cluster.

A Kafka topic is created with 10 partitions. This follows Kafka documentation that recommends having about one partition per physical disk. Six spark executors are running for the processing task and 1 GB memory is assigned per executor. Spark Streaming consumes data by using the Receiver-less direct approach. It does a simple map, reduce ByKey and for each RDD operation to transform the data, group the usage into each city and save into a HBase table. In order to evaluate the efficiency of the system we also simulated data of 10 K, 50 K, 100 K and 500 K meters which each meter's usage follows a normal distribution.

Table 1. shows the average and median times measured in seconds as well as the standard deviation when the system received different numbers of datasets in a time window (1 s) and processed them. As it can be seen, except the 500 K datasets case average times are below 1 s, thus providing acceptable response time. Also, the system is response time-predictable as shown by the small- standard deviation and differentiation between the average and the mean. However for the 500 K datasets, there is a need to improve the system performance. The performance tuning can be done by increasing the number of partitions in Kafka topic or increasing the number of workers and memory for Spark streaming execution, for future work.

Table 1. Average, median times (in sec) as well as standard deviation for the streaming and processing services for different number of datasets in one sec

Number of datasets	Average	Median	Std. dev.
6 K	0.38	0.4	0.095
10 K	0.43	0.4	0.085
50 K	0.65	0.6	0.112
100 K	0.91	0.9	0.068
500 K	3.04	3	0.123

4.2   Predictive Model for Energy Consumption

In the present work a time-series prediction is made with machine learning algorithms. In particular, 30-minutes averaged energy consumption data is used as inputs to artificial neural networks (ANN) of various architecture. Time window constructing of historical data varies from 2 to 10 intervals. The prediction horizon is set up to 3 intervals (or 90 min) ahead.

Weka Framework software provides several types of ANN to chose from. The predictive modeling here is made with MLPRegressor, MultilayerPerceptron, MultilayerPerceptronCS, RBFNetwork and NeuralNetwork (convolutional) algorithms. The LinearRegression algorithm's output is used as a benchmark. In Table 2 the performance for time series prediction is summarized.

Table 2. Mean Absolute Error and Root Mean Square Error of algorithms used

Prediction error (%):	3 steps ahead		2 steps ahead		1 step ahead	
Method:	Mean Absolute Error	Root Mean Square Error	Mean Absolute Error	Root Mean Square Error	Mean Absolute Error	Root Mean Square Error
MLPRegressor	21	29.9	9.8	14	9.8	14
MultilayerPerceptron	27	38	19	29	19	29
MultilayerPerceptronCS	27	37	19	29	19	29
RBFNetwork	20	24	19	23	19	23
Convolutaional neural network	10.5	14.4	7.5	10.5	7.5	10.4
LinearRegression	48	84	17	26	17	25

As seen in Table 2, convolutional neural network algorithm shows best performance as well as most stable for various predicting horizons. The training time for this algorithm was the highest. But trained model applied by Spark MLlib performs the prediction at the same time as other models, including LinearRegression.

In Fig. 2 below the prediction for 3 steps ahead (90 min ahead) for a validation data set (that is 30% from the entire data set) is shown and the model's confidence intervals (set as 95%) are marked.

The model's prediction follows the actual (observed) data quite precise, as the actual data always stays well within the model's confidence intervals. The prediction of the energy consumption change is picked by the model well, especially the increase of the power consumption.

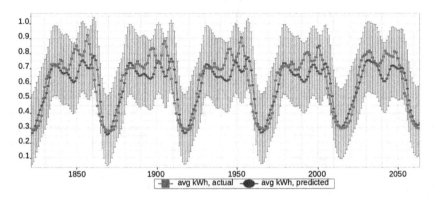

Fig. 2. The prediction for 3 steps ahead (90 min ahead)

5   Conclusions

The time-series data is ingested, processed, stored and analyzed by the proposed framework in real-time. In the use case scenario, the average energy consumption of each city in 30- to 90-minutes ahead is forecasted. Highest prediction accuracy is obtained when convolutional neural network is employed. This type of neural network also shows best generalization ability with the difference between prediction error for training and validation data sets not greater than 0.1%. While training (and re-training on the provided feedback) of ANN architecture is usually more time consuming than others architectures, the employment of Spark Streaming and Spark MLlib allows the framework to overcome this limitation. This work is going to be extended by improving the data processing performance as well as applying automatic search for the best predictive machine learning methods based on different criteria.

References

1. Applied Time Series Analysis Learning Online: https://onlinecourses.science.psu.edu/stat510/node/47. Accessed 3 Feb 2017
2. Witten, I.H., Frank, E., Hall, M.A., Pal, C.J.: Data Mining: Practical Machine Learning Tools and Techniques. Morgan Kaufmann, Burlington (2016)
3. Zaharia, M., Chowdhury, M., Franklin, M.J., Shenker, S., Stoica, I.: Spark: cluster computing with working sets. In: Proceedings of the 2nd USENIX conference on Hot topics in cloud computing (2010)

4. Zaharia, M., Das, T., Li, H., Hunter, T., Shenker, S., Stoica, I.: Discretized streams: fault-tolerant streaming computation at scale. In: Proceedings of the 24th ACM Symposium on Operating Systems Principles, pp. 423–438. ACM (2013)
5. Apache Kafka: A Distributed Streaming Platform. http://kafka.apache.org. Accessed 3rd Feb 2017
6. Meng, X., Bradley, J., Yavuz, B., Sparks, E., Venkataraman, S., Liu, D., et al.: MLlib: machine learning in apache spark. J. Mach. Learn. Res. 17(34), 1–7 (2016)
7. Namiot, D.: On big data stream processing. Int. J. Open Inf. Technol. 3(8), 48–51 (2015)
8. García, S., Ramírez-Gallego, S., Luengo, J., Benítez, J.M., Herrera, F.: Big data preprocessing: methods and prospects. Big Data Anal. 1(1), 9 (2016)
9. Apache Storm: http://storm.apache.org. Accessed 3rd Feb 2017.7
10. Apache Flink: http://flink.apache.org. Accessed 3rd Feb 2017
11. Ahmed, N.K., Atiya, A.F., Gayar, N.E., El-Shishiny, H.: An empirical comparison of machine learning models for time series forecasting. Econom. Rev. 29(5-6), 594–621 (2010)
12. Landset, S., Khoshgoftaar, T.M., Richter, A.N., Hasanin, T.: A survey of open source tools for machine learning with big data in the hadoop ecosystem. J. Big Data 2(1), 24 (2015)
13. Apache Mahout: http://mahout.apache.org. Accessed 3rd Feb 2017
14. Perez-Chacon, R., Talavera-Llames, R.L., Martinez-Alvarez, F., Troncoso, A.: Finding electric energy consumption patterns in big time series data. In: Omatu, S., et al. (eds.) Distributed Computing and Artificial Intelligence, 13th International Conference. AISC, vol. 474, pp. 231–238. Springer, Cham (2016). doi:10.1007/978-3-319-40162-1_25
15. Gachet, D., de la Luz Morales, M., de Buenaga, M., Puertas, E., Muñoz, R.: Distributed big data techniques for health sensor information processing. In: García, C., Caballero-Gil, P., Burmester, M., Quesada-Arencibia, A. (eds.) UCAmI 2016. LNCS, vol. 10069, pp. 217–227. Springer, Cham (2016). doi:10.1007/978-3-319-48746-5_22
16. George, L.: HBase: The Definitive Guide: Random Access to Your Planet-Size Data. O'Reilly Media Inc., Newton (2011)
17. Smart meter data source: http://www.ucd.ie/issda/data/commissionforenergyregulationcer/. Accessed 3rd Feb 2017

Granular and Soft Clustering for Data Science

Scalable Machine Learning with Granulated Data Summaries: A Case of Feature Selection

Agnieszka Chądzyńska-Krasowska[1]([✉]), Paweł Betliński[2], and Dominik Ślęzak[2]

[1] Polish-Japanese Academy of Information Technology,
ul. Koszykowa 86, 02-008 Warsaw, Poland
honzik@pjwstk.edu.pl
[2] Institute of Informatics, University of Warsaw,
ul. Banacha 2, 02-097 Warsaw, Poland

Abstract. We investigate how to use the histogram-based data summaries that are created and stored by one of the approximate database engines available in the market, for the purposes of redesigning and accelerating machine learning algorithms. As an example, we consider one of popular minimum redundancy maximum relevance (mRMR) feature selection methods based on mutual information. We use granulated data summaries to approximately calculate the entropy-based mutual information scores and observe the mRMR results compared to the case of working with the actual scores derived from the original data.

Keywords: Data granulation · Approximate query · Feature selection

1 Introduction

The considered approximate database engine captures information upon data ingest. This is done instantaneously in the form of single- and multi-column data summaries. The engine composes groups of data rows (so called *packrows*) incoming into a given data table and summaries are built for each group separately. The engine stores and processes only summaries without assuming any access to the original data. The SQL select statements are executed directly on summaries, i.e., each data operation (filtering, joining, grouping, etc.) *transforms* summaries of its input directly into summaries of its output [1,2].

The goal of the considered engine is to work on petabytes of summarized data. It allows users to achieve analytical insights 100–1000 times faster than traditional databases, by performing on structures that are orders of magnitude smaller than the atomic data. This is important as one can observe a significant increase in data sizes that organizations need to deal with on daily basis[1]. This proliferation of data is a result of an increase in the number of devices, services and people connected in an increasingly complex environment.

---

[1] One of the current deployments of the considered new engine assumes working with 30-day periods, wherein there are over 10 billions of new data rows coming every day and ad-hoc analytical queries are required to execute in 2 s.

© Springer International Publishing AG 2017
M. Kryszkiewicz et al. (Eds.): ISMIS 2017, LNAI 10352, pp. 519–529, 2017.
DOI: 10.1007/978-3-319-60438-1_51

In this paper, we investigate whether the considered granulated data summaries can serve as an efficient input for machine learning methods. This idea follows a popular trend of mining massive data streams based on precalculated data cluster descriptions [3]. Actually, one may say that approximate database engines are a perfect match for implementations, which construct decision models heuristically based on aggregated information derived using SQL [4]. Yet another approach is to derive required information directly from data summaries without a need of going through SQL-level interfaces. In both cases, it is expected that approximate and exact calculations of heuristic functions provide comparable basis for the decision model optimization. This is related to a general intuition that heuristic decisions can be based on approximate information.

As an example, we consider the problem of feature selection [5]. We focus on the minimum redundancy maximum relevance (mRMR) approach, where features are added to the resulting set iteratively by examining their relationships with both the dependent variable and features that were added in previous steps [6]. Relationships are modeled by means of the entropy-based mutual information measure that can be computed from the original data or, alternatively, approximated using the considered data summaries. In experiments, we analyze a data set that includes several millions of network transmissions. This is actually a fragment of the data obtained from a company that develops the tools for early detection of viruses and worms in the network. In this particular case, the goal is to identify features characterizing suspiciously big data transfers.

The paper is organized as follows. Section 2 contains basic information about the considered approximate query engine architecture. Section 3 presents quantization and rank-based algorithms that construct granulated summaries of ingested data sets. Section 4 recalls the mRMR approach to feature selection and provides foundations for approximating mutual information. Section 5 compares the results of feature selection obtained by utilizing standard and approximated mutual information computations. Section 6 concludes our studies.

2   The Engine Basics

The foundations of the approach considered in this paper relate to the earlier-developed relational database engine, now available as *Infobright DB*[2]. On data load, Infobright DB clusters the incoming rows into *packrows*, additionally decomposing each packrow onto *data packs* gathering values of particular columns [7]. The contents of data packs are described by simple summaries accessible independently from the underlying data packs. Infobright DB combines the ideas taken from modern database technologies and the theory of rough sets [8], by means of using summaries to quickly classify data packs as irrelevant, relevant or partially relevant to particular queries. All together, Infobright DB processes data according to the four following principles: storing data in data packs, creating approximate summaries for each of data packs, conducting approximate

---

[2] Formerly known as *Brighthouse* and *Infobright Community/Enterprise Edition*.

computations on summaries, and, whenever there is no other way to finish query execution, iteratively accessing the contents of some of data packs.

When compared to Infobright DB, the new approximate engine referred in this paper operates with richer summaries. For single columns, it uses enhanced histogram representations. For each original data pack, its histogram contains information about dynamically derived range-based bars and *special values* that differ from their neighboring values of the corresponding column by means of their frequencies in the corresponding packrow. The engine also stores information about the most significant *gaps*, i.e., the areas where there are no values occurring. Histogram ranges and special values are further used as the foundation of multi-dimensional summaries, which capture packrow-specific co-occurrences of values on different columns. The engine decides what to store based on the methods that rank the significance of detected co-occurrences.

Data summaries are stored in binary files accessible by the approximate query execution methods via internal interfaces. From a logical perspective, the contents of those files can be represented as a collection of tables displayed in Fig. 1. Actually, for diagnostic purposes, we implemented a kind of converter that transforms the contents of binary files into such explicit relational form [9]. One can utilize standard PostgreSQL environment to access the obtained data summary tables storing information about histogram bars' frequencies and ranges, special values, gaps, and bar/value co-occurrences, per each database, each data table, each data column, and each ingested packrow. In particular, the feature

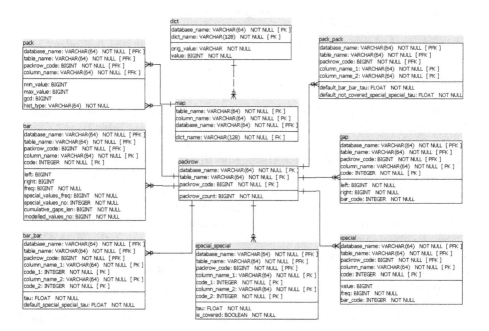

Fig. 1. Relational schema of the designed granulated data summary tables

Table 1. Short descriptions of data summary tables displayed in Fig. 1

Table name	Description of table contents
packrow	Number of rows gathered in particular packrows (2^{16} by default)
pack	Basic information about values occurring in particular data packs
bar	Information about particular bars – their supports, ranges, etc.
special	Information about special values, including bars they belong to
gap	Information about gaps – their ranges and bars they belong to
pack_pack	Compressed information about co-occurrence ratios that are not represented in tables *bar_bar* and *special_special* (see Sect. 3)
bar_bar	Information about co-occurrence ratios for the most interesting pairs of bars and compressed information about co-occurrence ratios of pairs of special values belonging to the considered pairs of bars that are not represented in table *special_special*
special_special	Co-occurrence ratios for the most interesting special value pairs
dict	Dictionary codes assigned to values of alphanumeric data columns
map	The mapping between the contents of table *pack* and table *dict*

selection approach reported in further sections has been implemented as a simple PL/pgSQL script working directly on such summary tables (Table 1).

· Once information about original data has been captured, then – for each incoming SQL select statement – the engine *transforms* data summaries iteratively to build a summary of the query result. Different transformation mechanisms are dedicated to different operations. For example, the operation of filtering requires updating frequencies of histogram bars and special values for columns relevant for further query execution stages basing on the WHERE conditions specified over other columns. This is done by adapting the methods of belief propagation [10], where – for each packrow separately – the stored most significant co-occurrences of bars and special values of different columns are used as partial information about data-driven joint probability distributions.

Once the query outcome summary is created, the engine translates it into the standard SQL select statement result format. If one interprets data ingestion as a step of information granulation [11], then the final stage of translating query result summaries into actual approximate results can be interpreted as information *degranulation*. Prior to that, information transformed through query execution stages is highly condensed, requiring only a fraction of resources of a traditional database to produce the results. This is especially interesting to compare with approximate query techniques based on data sampling [12], whereby – even though samples may contain a low percentage of original rows – the whole computational process remains at the original atomic data level.

3   Discretization and Co-Occurrence Rankings

The construction of histograms representing particular columns within particular packrows can be compared to the problem of data discretization [4]. We combined two standard discretization approaches that, up to now, yielded the best quality of approximate queries executed against granulated data sets with truly complex column domains. Namely, we applied an equal-length discretization in order to create 8 (by default) roughly equal-length buckets, and split each of them onto shorter intervals such that the amounts of rows with a given column's values belonging to particular intervals are approximately uniform, finally resulting in a total number of 64 (by default) buckets. A more detailed description of the utilized dynamic histogram generation method can be found in [9].

One may also consider another way of splitting histogram ranges, e.g., onto shorter intervals containing roughly uniform amount of distinct values of a given column occurring in a given packrow. In Sect. 4 (Table 2), these two methods of splitting will be referred to as *support* and *distinct*, respectively.

One more aspect is how to automatically choose a set of special values, which are somehow *outlying*. This topic has been already studied while attempting to introduce approximate querying into Infobright DB [7], whereby we were identifying and in some sense 'neglecting' outliers in order to produce more compact summaries. Quite oppositely, in the new engine considered in this paper, a specified number of *interesting* values per data pack (100 by default) is represented explicitly. Currently, such special values are chosen so as to minimize a variance related to summarizing other values by histogram bars [2]. Another way would be, e.g., to weight rows in order to equalize the importance of values in different equal-length intervals. In Sect. 4, these two methods of extracting special values will be referred to as *standard* and *weighted*, respectively.

The remaining stage of data ingestion is to describe co-occurrences between values of different columns. To keep low footprint of data summaries, we store co-occurrence-related information only for a limited amount (by default 128 × the number of columns) of pairs of histogram bars and special values. For packrow t and columns a and b, let us refer to a's and b's histogram bars using iterators i and j, respectively. Let us denote by $p_t(i)$, $p_t(j)$, and $p_t(i,j)$ the normalized frequencies of occurrence of a's values within its i-th bar, occurrence of b's values within its j-th bar, and joint occurrence of pairs of a's values within its i-th bar and b's values within its j-th bar, respectively. Currently, the engine uses the following function for expressing the importance of pairs of bars:

$$rank_t^{|\cdot|}(i,j) = p_t(i)p_t(j)\left|\frac{p_t(i,j)}{p_t(i)p_t(j)} - 1\right| \tag{1}$$

Formula (1) measures how much accuracy we might lose when basing on $p_t(i)$ $p_t(j)$ instead of $p_t(i,j)$. For a given packrow t, we use $rank_t^{|\cdot|}$ to jointly evaluate all pairs of bars for all pairs of columns. This way, we can devote more footprint to column pairs, which seem to be more correlated than others. After selecting a fixed number of the most important pairs of bars, for given columns a and b, we store two types of information. For the selected pairs of bars (i,j), we store ratios $\tau_t(i,j) = \frac{p_t(i,j)}{p_t(i)p_t(j)}$. For not selected pairs of bars, we store the default ratio (denoted by $default\_bar\_bar\_tau$ in Fig. 1) defined as[3]:

$$\tau_t(a,b) = \frac{1 - \sum_{i,j:\tau_t(i,j)\in\tilde{t}} p_t(i)p_t(j)\tau_t(i,j)}{1 - \sum_{i,j:\tau_t(i,j)\in\tilde{t}} p_t(i)p_t(j)} \tag{2}$$

Analogous default ratios are stored for not selected pairs of special values (fields $default\_special\_special\_tau$ and $default\_not\_covered\_special\_special\_tau$, for the cases of stored/not stored pairs of special values' parent bars).

4 mRMR Feature Selection and Mutual Information

Feature selection addresses the task of extracting subsets of columns (variables, attributes) that are sufficient for the purposes of building various kinds of decision models [5]. Feature selection algorithms rely on different types of information gain and/or preserving criteria, derived from a number of theories [8]. In this paper, we consider the widely known *minimum redundancy maximum relevance* (mRMR) method [6], presented in its simplified form as Algorithm 1.

As in our comparative experiments we are mainly interested in investigating an ordering of additions of features into the resulting feature set, Algorithm 1 does not include any stopping condition. As function $\phi : A \times A \cup \{d\} \to [0, +\infty)$ evaluating relationships between features, we employ well-known mutual information. Our goal is to study: (1) how to utilize data summaries to approximate mutual information and (2) how different outcomes may be expected when feeding Algorithm 1 with approximations instead of thorough calculations.

Let us consider columns a and b. For simplicity, let us assume that they do not have special values. (The case with special values is quite analogous.) As the mutual information approximation, we put $\tilde{I}(a,b) = \frac{1}{N} \sum_{t=1}^{N} I_t(a,b)$, where N denotes the amount of packrows and $I_t(a,b)$ is defined as follows[4]:

$$I_t(a,b) = \sum_{i,j:\tau_t(i,j)\in\tilde{t}^+} p_t(i)p_t(j)\tau_t(i,j)\log\tau_t(i,j) + \alpha_t(a,b)\log\tau_t(a,b) \tag{3}$$

where

$$\alpha_t(a,b) = 1 - \sum_{i,j:\tau_t(i,j)\in\tilde{t}} p_t(i)p_t(j)\tau_t(i,j) \tag{4}$$

In the above formulas, we keep notation introduced in Sect. 3. For a pair of i-th and j-th bars such that $\tau_t(i,j) \in \tilde{t}$, the quantity $p_t(i)p_t(j)\tau_t(i,j)\log\tau_t(i,j)$ can be rewritten as $p_t(i,j)\log\frac{p_t(i,j)}{p_t(i)p_t(j)}$, which is identical to the standard way

---

[3] $\tau_t(i,j) \in \tilde{t}$ means that information about $\tau_t(i,j)$ is stored by the engine.
[4] $\tau_t(i,j) \in \tilde{t}^+$ means that $\tau_t(i,j)$ is stored by the engine and there is $\tau_t(i,j) > 0$.

Algorithm 1. Simplified mRMR feature selection method

Input: the set of features A and the dependent variable d;
$\phi : A \times A \cup \{d\} \rightarrow [0, +\infty)$ function for measuring pairwise dependencies;
Output: the ordered set of features $A' = A$

1: **begin**
2:   $A' \leftarrow \arg\max_{a \in A} \phi(a, d); \; A \leftarrow A \backslash A';$
3:   **while** $A \neq \emptyset$ **do**
4:     $\bar{a} \leftarrow \arg\max_{a \in A} (\phi(a, d) - avg_{b \in A'} \phi(a, b));$
5:     $A' \leftarrow A' \cup \{\bar{a}\}; \; A \leftarrow A \backslash \{\bar{a}\};$
6:   **end while**
7: **end**

---

of computing mutual information. For all other pairs of a's and b's bars, the ratio under the logarithm can be approximated by $\tau_t(a, b)$. The total normalized frequency of such cases equals to $\alpha_t(a, b)$. The only remaining corner case is $\tau_t(a, b) = 0$. However, one can see that this would imply equality $\alpha_t(a, b) = 0$. In such situations, we can assume that $\alpha_t(a, b) \log \tau_t(a, b) = 0$.

The following result shows that $\tilde{I}$ has analogous properties when comparing it to standard mutual information. A simplified proof is included in order to better illustrate the meaning of the considered data summaries.

Proposition 1. *For any columns a and b of a data table T, there is $\tilde{I}(a, b) \geq 0$, where equality holds, if and only if the considered approximate query engine does not store information about any co-occurrences involving a and b.*

Proof. We will show $-I_t(a, b) \leq 0$ for every packrow t. For simplicity, let us assume that columns a and b do not have special values. (For the case with special values the proof is analogous.) We know that for $x > 0$ there is $\log(x) \leq x - 1$, where equality holds, if and only if $x = 1$. Hence, $-I_t(a, b) =$

$$= \sum_{i,j:\tau_t(i,j)\in\tilde{t}^+} p_t(i,j) \log \tfrac{1}{\tau_t(i,j)} + \alpha_t(a,b) \log \tfrac{1}{\tau_t(a,b)}$$
$$\leq \sum_{i,j:\tau_t(i,j)\in\tilde{t}^+} p_t(i,j) \left(\tfrac{1}{\tau_t(i,j)} - 1 \right) + \alpha_t(a,b) \left(\tfrac{1}{\tau_t(a,b)} - 1 \right)$$
$$= \sum_{i,j:\tau_t(i,j)\in\tilde{t}^+} p_t(i)p_t(j) - \sum_{i,j:\tau_t(i,j)\in\tilde{t}} p_t(i,j) + \tfrac{\alpha_t(a,b)}{\tau_t(a,b)} - \alpha_t(a,b)$$
$$\leq \sum_{i,j:\tau_t(i,j)\in\tilde{t}} p_t(i)p_t(j) - \sum_{i,j:\tau_t(i,j)\in\tilde{t}} p_t(i,j) + \tfrac{\alpha_t(a,b)}{\tau_t(a,b)} - \alpha_t(a,b)$$

By definition of $\alpha_t(a, b)$, we know that $\alpha_t(a, b) + \sum_{i,j:\tau_t(i,j)\in\tilde{t}} p_t(i,j) = 1$ and $\tfrac{\alpha_t(a,b)}{\tau_t(a,b)} + \sum_{i,j:\tau_t(i,j)\in\tilde{t}} p_t(i)p_t(j) = 1$. Thus, $-I_t(a, b) \leq 0$ and equality holds, if and only if: (1) $\sum_{i,j:\tau_t(i,j)\in\tilde{t}^+} p_t(i)p_t(j) = \sum_{i,j:\tau_t(i,j)\in\tilde{t}} p_t(i)p_t(j)$ and (2) for each case of $\tau_t(i,j) \in \tilde{t}^+$, there is $\tau_t(i, j) = 1$. The first condition means that there are no cases of $\tau_t(i, j) = 0$. When combining with the second condition, it means that for each $\tau_t(i, j) \in \tilde{t}$, there is $\tau_t(i, j) = 1$. Given the way of ranking pairs of histogram bars, this means that there is no need to store information about any pairs of a's and b's bars in the summary of packrow t.

5 Experimental Results

The data set used in our experiments includes 100×2^{16} rows (split onto 100 packrows) and 17 columns reflecting the network traffic. The columns represent typical information about transmissions, protocols, services, servers, as well as the source, destination and monitor characteristics. (Column abbreviations are visible in Table 3). As the dependent variable, we use column *transferred bytes*. Our choice is inspired by observation that a network monitoring tool developed by this particular data provider – one of the companies deploying the new engine considered in this paper – is generating a number of queries attempting to characterize the largest-in-size transmissions using other columns.

Table 2 reports a summary of our results. Its first column describes the way of deriving histograms and special values, as well as – analogously – discretizing the original data set as a prerequisite for standard mutual information calculations. Settings *support* versus *distinct* indicate two alternative methods of constructing histogram bars (see the second paragraph of Sect. 3). Settings *standard* versus *weighted* indicate two methods of identifying special values (see the third paragraph of Sect. 3). Clearly, these settings can be applied both to particular packrows ingested by the engine and to the original data set as a whole.

The second column in Table 2 refers to calculations over a 15% data sample. The reported quantities equal to normalized L_1 distances between vectors of columns' ordinal numbers resulting from Algorithm 1 when computing ϕ as mutual information over the whole discretized data set and over its discretized sample. One can see that discretization over a sample leads toward less similar feature selection outputs (comparing to mRMR executed on the original data set as a baseline) for the *weighted* technique of selecting special values.

Further columns in Table 2 correspond to four different ways of ranking co-occurrences of pairs of histogram bars and special values. (Settings $|\cdot|/hierarchy$ in combination with *support/standard* represent the current production version of the considered approximate query engine.) Like above, we report distances between data columns' ordinal numbers resulting from Algorithm 1 when computing ϕ as mutual information over the original data set versus computing it using formula (3). Labels $|\cdot|/hierarchy$, $|\cdot|/flat$, $(\cdot)/hierarchy$, and $(\cdot)/flat$ refer to four strategies of choosing co-occurrences that should be stored in the engine. $|\cdot|$ means applying function $rank_t^{|\cdot|}$ while $(\cdot)$ means replacing it with

$$rank_t^{(\cdot)}(i,j) = p_t(i)p_t(j)\left(\frac{p_t(i,j)}{p_t(i)p_t(j)} - 1\right) \tag{5}$$

The idea behind $rank_t^{(\cdot)}$ is to devote the multi-column footprint to *positive* co-occurrences, i.e., ratios $\tau_t(i,j)$ that are significantly greater than 1.

Settings *hierarchy* and *flat* refer to two ways of looking at pairs of special values. Let us denote by k_i and l_j special values that drop into a's i-th histogram range and b's j-th histogram range, respectively. The special-value-related ranking corresponding to the *hierarchy* setting (used currently in the engine) compares special values' ratios to the ratios of their parents[5]:

$$rank_t^{|\cdot|}(k_i, l_j) = p_t(k_i)p_t(l_j) \left| \frac{p_t(k_i, l_j)}{p_t(k_i)p_t(l_j)} - \frac{p_t(i,j)}{p_t(i)p_t(j)} \right| \tag{6}$$

As for the *flat* setting, $\frac{p_t(k_i, l_j)}{p_t(k_i)p_t(l_j)}$ is put against 1 instead of $\frac{p_t(i,j)}{p_t(i)p_t(j)}$. (In order to keep a kind of co-occurrence ranking consistency, pairs of histogram bars are then evaluated by means of their frequencies excluding special values.)

Table 2. Distances between mRMR rankings obtained using the whole data versus mRMR rankings obtained based on a data sample and mRMR rankings based on data summaries captured using various settings within the considered engine

Discretization	Sampled	$\|\cdot\|$/hierarchy	$\|\cdot\|$/flat	$(\cdot)$/hierarchy	$(\cdot)$/flat
support/standard	1.25	2.00	2.25	1.38	2.25
support/weighted	3.38	2.13	3.13	2.63	2.50
distinct/standard	0.63	2.63	1.25	1.88	1.63
distinct/weighted	4.50	3.63	4.25	3.63	3.88

Table 2 provides some insights with regard to relationships between different settings. Let us note that $(\cdot)$ seems to be a better choice than $|\cdot|$, if and only if we use it together with *hierarchy/standard* or *flat/weighted* strategies responsible for selecting (pairs of) special values. For the *standard* setting, sampled discretization seems to be more accurate than the considered granulated approximations. This requires further investigation with respect to both the quality of data summaries and a way of using them to approximate mutual information. Nevertheless, our approach has a huge advantage with regards to the speed of calculations when comparing to both standard and sampled ways of running mRMR feature selection. This is similar to our already-mentioned observations related to sampling-based approximate query solutions [12].

Table 3 illustrates an example of more detailed experimental outcomes. One can see that all versions of collecting information about co-occurrences provide us with results that are quite similar to the ordering in the first column (starting with column *p_element* as the first choice in all cases). In practice, a lot depends on the mRMR stopping condition [6]. For instance, if the algorithm is set up to select three columns, then the baseline feature subset would take a form of {*p_element, service, trans_type*}. In this particular situation, the $\tilde{I}$-driven computations would lead toward (almost) the same subsets. This is actually truly encouraging given the fact that the considered approximate engine stores quite limited information about correlations between columns.

---

[5] In (6), we combine settings *hierarchy* and $|\cdot|$. The case of $(\cdot)$ is analogous.

Table 3. mRMR orderings of columns in the network traffic data set discretized using *support/weighted* settings for bars and special values (refer to Table 2)

Standard	Sampled	| · |/hierarchy	| · |/flat	(·)/hierarchy	(·)/flat
p_element	p_element	p_element	p_element	p_element	p_element
service	trans_type	service	service	service	service
trans_type	d_class	trans_type	trans_type	s_address	trans_type
server	s_port	d_address	s_address	trans_type	s_class
d_address	service	s_vrf	s_vrf	d_port	server
protocol	s_class	s_port	d_class	s_port	d_class
s_port	d_port	s_address	s_port	d_address	s_port
monitor	server	d_port	d_port	s_vrf	d_port
d_port	s_address	protocol	protocol	protocol	s_address
s_vrf	d_address	monitor	s_class	monitor	protocol
d_class	protocol	d_class	d_address	d_class	s_vrf
s_address	d_interface	s_class	monitor	s_class	monitor
d_interface	s_vrf	d_interface	d_interface	s_interface	d_address
s_class	monitor	s_interface	s_interface	d_interface	d_interface
m_address	s_interface	m_address	m_address	m_address	s_interface
s_interface	m_address	server	server	server	m_address

6   Conclusions

We developed a novel approach to feature selection, relying on the minimum redundancy maximum relevance (mRMR) algorithm fed by mutual information score approximations. In our computations, we used granulated data summaries produced by a new approximate database engine available in the market. The outcomes – taking a form of orderings of features selected by mRMR – were compared to the analogous procedure executed in a standard way. Experimental results confirm that such *approximate* and *standard* orderings are quite similar to each other, while approximate calculations are incomparably faster. This leads to two observations: (1) It is possible to extend the functionality of the considered new engine – focused up to now on approximate execution of SQL – by some elements of scalable granular-style machine learning, and (2) It is possible to extend the current engine's testing environment by comparing the outputs of standard and granular versions of machine learning algorithms.

References

1. Glick, R.: Current Trends in Analytic Databases - Keynote Talk at FedCSIS (2015). https://fedcsis.org/2015/keynotes/rick_glick
2. Chądzyńska-Krasowska, A., Kowalski, M.: Quality of histograms as indicator of approximate query quality. In: Proceedings of FedCSIS, pp. 9–15 (2016)

3. Aggarwal, C.C. (ed.): Data Streams: Models and Algorithms. Springer, New York (2007)
4. Nguyen, H.S.: Approximate boolean reasoning: foundations and applications in data mining. In: Peters, J.F., Skowron, A. (eds.) Transactions on Rough Sets V. LNCS, vol. 4100, pp. 334–506. Springer, Heidelberg (2006). doi:10.1007/11847465_16
5. Guyon, I., Elisseeff, A.: An introduction to variable and feature selection. J. Mach. Learn. Res. **3**, 1157–1182 (2003)
6. Peng, H., Long, F., Ding, C.: Feature Selection based on mutual information criteria of max-dependency, max-relevance, and min-redundancy. IEEE Trans. Pattern Anal. Mach. Intell. **27**(8), 1226–1238 (2005)
7. Ślęzak, D., Eastwood, V.: Data warehouse technology by infobright. In: Proceedings of SIGMOD, pp. 841–846 (2009)
8. Pawlak, Z.: Some issues on rough sets. In: Peters, J.F., Skowron, A., Grzymała-Busse, J.W., Kostek, B., Świniarski, R.W., Szczuka, M.S. (eds.) Transactions on Rough Sets I. LNCS, vol. 3100, pp. 1–58. Springer, Heidelberg (2004). doi:10.1007/978-3-540-27794-1_1
9. Chądzyńska-Krasowska, A., Stawicki, S., Ślęzak, D.: A metadata diagnostic framework for a new approximate query engine working with granulated data summaries. In: Proceedings of IJCRS (2017)
10. Neapolitan, R.E.: Learning Bayesian Networks. Prentice Hall, Upper Saddle River (2003)
11. Pal, S.K., Meher, S.K., Skowron, A.: Data science, big data and granular mining. Pattern Recogn. Lett. **67**, 109–112 (2015)
12. Mozafari, B., Niu, N.: A handbook for building an approximate query engine. IEEE Data Eng. Bull. **38**(3), 3–29 (2015)

Clustering Ensemble for Prioritized Sampling Based on Average and Rough Patterns

Matt Triff[(⊠)], Ilya Pavlovski, Zhixing Liu, Lori-Anne Morgan,
and Pawan Lingras

Mathematics and Computing Science, Saint Mary's University, Halifax, Canada
{matt.triff,pawan}@cs.smu.ca

Abstract. This paper proposes a clustering ensemble for prioritized sampling to tackle a big data problem. The proposal first creates separate clustering schemes of objects using different dimensions of the dataset. These clustering schemes are then combined to create a representative sample based on all the possible combinations of profiles. The resulting clustering ensemble will help system developers to reduce the number of objects that need to be analyzed while making sure that all the profile combinations are comprehensively covered. The proposal further ranks the objects in the sample based on their ability to capture important aspects of each of the criteria. The proposed approach can be used to provide a priority based analysis/modelling over an extended period of time. The prioritized analysis/models will be available for use in a reasonably short period of time. The quality of the analysis/modelling will continuously improve as more and more objects in the sample are processed according to their rank in the sample. The proposal is applied to a large set of weather stations to create a ranked sample based on hourly and monthly variations of important weather parameters, such as temperature, solar radiation, wind speed, and humidity. The experiments also demonstrate how a combination of average and rough patterns help in creating more meaningful profiles.

Keywords: Clustering · Rough patterns · Sampling · Weather

1 Introduction

Usually, analysis and modelling of most engineering systems require weeks or months of careful work. Such a customized approach may not be cost-effective in many applications. An alternative is to maintain a knowledge base of analysis/models for all possible states of the system. For any new potential system, the knowledge base can provide a reasonably approximate analysis/model by combining analysis/models of similar systems stored in the knowledge base. In many real-world problems, it is not possible to analyze and model all the possible states of a system. A possible solution is to create a sampling process that comprehensively covers all the important aspects of a possible state. For example, if we were to install a facility that is affected by the weather at different

© Springer International Publishing AG 2017
M. Kryszkiewicz et al. (Eds.): ISMIS 2017, LNAI 10352, pp. 530–539, 2017.
DOI: 10.1007/978-3-319-60438-1_52

locations in the world, we are potentially looking at millions of locations. The problem is further compounded by the fact that there may be hundreds of different variations of the facility itself. This will lead to hundreds of millions of possible variations of a facility.

A possible solution is to use unsupervised learning to group the states of the system and create a list of representative buildings that are distributed over the entire spectrum of possibilities. However, even such a sample may be too large. In today's cloud based computing environment, it is also possible to continue the modelling process for an indefinitely long period of time. This would allow the available models can be put to use as soon as possible while the knowledge base constantly improves in performance with the new and refined models of more and more facilities in the sample, based on a prescribed priority. In order to institute such a knowledge cloud the system states in the sample will have to be prioritized. This paper proposes a clustering ensemble technique that not only creates a sample that is distributed over the entire spectrum of possibilities, but also ranks the states based on their ability to cover a broad spectrum.

The proposed approach is demonstrated for a reasonably large weather dataset, consisting of three years worth of hourly variations in temperature, solar radiation, wind speed and humidity. In order to develop the most meaningful profile descriptions, the weather stations are first represented using average hourly variations for each month of the year. However, a quick analysis of the resulting clustering suggests that the average values do not capture the extreme fluctuations in some of the more severe climates. Therefore, the representations of the weather stations is further enhanced by adding rough hourly patterns for each month of the year. A rough pattern consists of low and high values of a quantity [5]. It is especially relevant in knowledge representations where variation in the values is as important as the average values. For example, two places may have the same average temperature of say $25\,^\circ\mathrm{C}$ but one of them may have a low of 10° and high of 30° as opposed to another where the temperature varies only by $\pm 5^\circ$.

2 Review of Literature

2.1 Clustering

First, we describe the notations that will appear in this section. Let $X = \{x_1, \ldots, x_n\}$ be a finite set of objects, where the objects in X are represented by m-dimensional vectors. A classifying scheme classifies n objects into k categories $C = \{c_1, \ldots, c_k\}$. We use the term $category$ instead of class or cluster to emphasize the fact that it can be used in supervised and unsupervised learning. For a clustering scheme (CS), such as crisp clustering and rough clustering, C is the set of clusters. And each of the clusters c_i is represented by an m-dimensional vector, which is the centroid, or mean, vector for that cluster. We will use K-means clustering, which is one of the most popular statistical clustering techniques [1].

2.2   Rough Patterns

Pawlak [6] proposed the rough set theory for describing sets using lower and upper approximations. The concept gained significant momentum due to its usefulness in expert systems, rule generation, and classification [12]. While the rough concept is mostly used in the context of set theory, [9–11] have shown the flexible interpretation of the rough concept can be useful in a wider context. We will focus on the numerical rough representations in this section.

Pawlak [7] proposed the concept of rough real functions, which can be useful for rough controllers. The rough real functions use a generalized definition of approximation space. Let $P = y_1, y_2, y_3, ..., y_n$ be a sequence, where $y_i < y_{i+1}, i = 1, \cdots, n-1$. A function $y = f(x)$ may not be able to correspond to a single value of y in P. [7] proposed a rough function in such a case as follows:

$$\underline{f}(x) = \sup\{y \in P : y \leq f(x)\}, \tag{1}$$

and

$$\overline{f}(x) = \inf\{y \in P : y \geq f(x)\}. \tag{2}$$

If $\underline{f}(x) = \overline{f}(x)$, then f is exact. Otherwise, f is inexact or *rough* in x. The *error of approximation* of f in x is defined as the value $\overline{f}(x) - \underline{f}(x)$.

The notion of rough real functions was defined as an approximate value in case exact values cannot be represented by the given set of values. However, the notion can be used in a broader context. Lingras [2] used the rough values to develop supervised and unsupervised neural networks [3,8] and genetic algorithms [4]. This section describes rough values and patterns.

In some cases, a precise value of an entity may not be available. For example, one may estimate the current temperature to be between $20\,°C$ and $25\,°C$. In other cases, it may not even be possible to state a precise value. Many spatial (rainfall in Nova Scotia) or temporal (daily temperature) variables fall in this category. We cannot associate a precise value for daily temperature, only a range of values using the highest and lowest temperatures recorded on that day. We use rough or interval values to measure such quantities. For continuous variables, rough values are special cases of intervals as they focus on the end points. However, unlike intervals, rough values can also be used to represent a set of discrete values using the minimum and maximum values in the set. Let $Y = \{y_1, y_2, ..., y_n\}$ be a set of values collected for a variable such as daily temperature or stock market index. Each *rough value* y is denoted by a pair $(\underline{y}, \overline{y})$:

$$\underline{y} = \inf\{y \in Y\}, \tag{3}$$

and

$$\overline{y} = \sup\{y \in Y\}. \tag{4}$$

Here, sup is defined as the maximum value from the set, while inf corresponds to the minimum value. The definitions of inf and sup can be modified to exclude outliers. For example, one could use the bottom 5th percentile value for $\underline{y}$ and top 5th percentile value for $\overline{y}$. The above definition by Pawlak accommodates sets with continuous as well as discrete values. If the values are continuous, the set will be infinite and the resulting rough values correspond to the conventional notion of interval.

Rough patterns are sequences of rough or interval values [2]. We will look at a real-world example of a rough pattern using weather parameters in the next section.

3   Study Data and Knowledge Representation

The study data consists of weather data collected from weather stations in North America. For the purposes of this study, 1000 weather stations were randomly selected. The weather stations cover Canada and the United States of America. The weather data was recorded from each weather station at an interval of once per hour, over the course of three years. This resulted in raw data for each weather station having approximately 24 h * 365 days * 3 years = 26280 records.

The data collected from the weather stations includes parameters for temperature, solar radiation, humidity and wind speed. In order to create a more concise knowledge representation for clustering, the values were aggregated. Each weather station has four different knowledge representations. Each knowledge representation corresponds to one of the four parameters. Each month of weather data for each parameter is aggregated into average and rough hourly patterns. For example, the temperature between 00:00 and 01:00 in the month of January is aggregated as the minimum, average, and maximum temperature values that occurred throughout the entire month, during that hour. This means that each parameter measured by a single weather station has three values for every hour of every month for every year collected. A single weather station is therefore represented by 3 values * 24 h * 12 months * 3 years = 2592 values for each parameter.

Figures 1a and b show examples of two such weather locations. Both figures show lines plotted for the high, average, and lows for each hour of the day, accumulated over the same hour for the entire month. Between the two figures, we can see distinct behaviours, with the temperature being much lower in Regina. We also see that the temperature is warmer over a shorter period of time, rising and falling quickly.

4   Prioritized Clustering Ensemble

We will describe the proposed clustering ensemble based on the weather clustering problem. However, it can be generalized to any similar criterion. The steps for the proposed technique are as follows:

K1. Collect a large number of weather files from weather stations spread across the region.

K2. Extract average, min., and max. hourly patterns for all the twelve months for important weather parameters such as temperature, solar radiation, wind speed, and humidity.

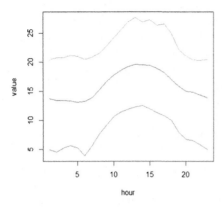 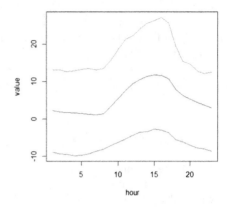

(a) Halifax, NS, Canada in September   (b) Regina, SK, Canada in October

Fig. 1. Plotted knowledge representation of the temperature parameter, for two weather stations

K3. For the dataset obtained by K2, determine the appropriate number of clusters for each of the weather parameters. Create clustering schemes such as $T = t_1, t_2, ..., t_5$ for temperature, $S = s_1, s_2, ..., s_{10}$ for solar radiation, $W = w_1, w_2, ..., w_5$ for wind speed, $H = h_1, h_2, ..., h_5$ for humidity.

K4. Find the best representative weather station for each cluster, such as t_i, s_j, w_k, h_m, where $i, j, k, m = 1, ..., 5$ or 10. Find a representative for each cluster combination such as $t_i - s_j - w_k - h_m$. The representative can be selected by any preferred criterion, such as the geographical medoid, medoid of all clustered criteria, etc.

K5. Rank the cluster combination (and their representatives from step K4) based on the size of sample they are representing.

When a new object is added to the dataset, it can now be quickly compared to the representive objects found by step K4. Based on some similarity function, the object can quickly be assigned to an approximate cluster, without reprocessing the entire dataset. To further decrease processing time, only the top objects can be used in the comparison (i.e. the top 20 representative objects from K5). Later, after some specified time or some specified number of objects have been added to the dataset, the above process can be repeated to ensure the set of representative objects remain current.

5   Results and Discussions

Once the data has been converted to our specified knowledge representation, it is ready to be clustered. The data for each parameter is first clustered separately using the K-means algorithm. In order to determine the best clustering scheme for the data, we first determine an optimal number of clusters for our dataset (Fig. 2).

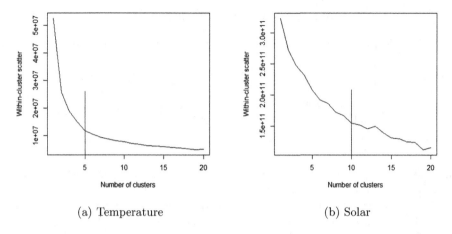

(a) Temperature   (b) Solar

Fig. 2. Sample of within-cluster scatter by number of clusters plots

We define the optimal number of clusters for a given dataset by the optimization of within cluster scatter. Within cluster scatter is the average distance
between two points within a cluster. Too few clusters and too many objects
will be grouped together, resulting in a very large within cluster scatter. Too
many clusters and the clusters become less useful as each object trends towards
receiving it's own cluster. Therefore, to determine the optimal number of clusters
we compare the results of clustering with various number of clusters. By plotting the within cluster scatter versus the number of clusters, we can determine
where the "knee" of the curve is. This "knee" defines the point where the clusters are relatively compact, but have enough members to still be valuable. The
results for each of the parameters were plotted, the figures are not shown due
to space limitations. Based on those plots, the optimal number of clusters was
found to be five for all parameters except solar, where the optimal number was
ten. This method of determining the optimal number of clusters is admittedly
slightly subjective. In general, having an additional one or two clusters would
not significantly impact the results.

Each weather station was plotted by location and labelled based on its cluster.
These figures are also not shown due to space limitations. There are some general
patterns that can be observed in the clustering of all the different parameters.
First, it is clear that the weather patterns detected by these weather stations
is largely regional. In most cases, the clusters are grouped together not only
by their similar weather parameter (solar, temperature, humidity, or wind), but
also by their physical location.

Some interesting patterns can also be observed from viewing these clusters.
For example, Western and Central Canada typically have a similar weather patterns. Coastal weather stations are the most similar, typically belonging to the
same cluster on both the East and West coasts. Solar has the most variation, as
can be inferred from the higher optimal number of clusters. The solar clusters

Table 1. Cluster sizes for each parameter

Cluster	1	2	3	4	5	6	7	8	9	10
Temperature	198	233	342	83	144					
Solar	63	44	91	136	90	69	165	107	75	160
Wind	308	251	94	45	302					
Humidity	344	156	65	138	297					

are therefore also smaller, with more regional variation in the United States, particularly in the Midwestern and Southern United States.

Table 1 shows the size of the clusters for each of the four parameters. In general, the parameters with five clusters each have two to three large clusters of approximately 250–350 objects. These clusters typically correspond to the United States Mid-West and East coast, where there is a higher density of weather stations. Each parameter also has a comparatively small cluster, with around 45–80 objects. These clusters correspond with a number of different areas, such as Alaska, the coasts, and the American Sun Belt. Solar, with a larger number of overall clusters, has the smallest range of all clusters, from 44 to 165 objects. This again shows the smaller regional variations that can be observed in solar patterns.

Monthly average patterns of various clusters can be observed in Fig. 3a. Although the temperature varies throughout the day, and by the month, the patterns of the fluctuation can define our clusters. In Fig. 3a, we see that Cluster 4 has relatively constant temperature throughout the entire day. For Cluster 2, we see that the temperature falls significantly overnight. When comparing these results to the map of weather stations labelled by temperature cluster it was observed that weather stations in Cluster 4 are typically in coastal regions, whereas the weather stations in Cluster 2 are typically inland, and thus we would expect to see more "continental" weather patterns.

From Fig. 3b, we can see the yearly temperature patterns for each of the five clusters. In this case, we can see that clusters 1–3 and 5 have similar weather patterns throughout the year, cold during the winter months of the Northern hemisphere, and steadily rising through Spring, peaking in Summer, and falling in the Fall. The difference between the weather stations in each cluster is where the range of temperatures lie. Additionally, we can see the range is greatest in Cluster 5, where the curve is steepest. Finally, Cluster 4 is the outlier, in that the temperature throughout the entire year remains relatively constant, with only a (relatively) slight increase over the Summer months.

The average solar patterns from the ten different clusters are shown in Fig. 4. Here the patterns are more difficult to discern between clusters. In general, all clusters follow the same pattern of having the most sunlight in the Summer months, and the least in Winter months. However, some weather stations have identifiable peaks and valleys from month to month, perhaps most notably in Cluster 10. Similarly to the temperature clusters, the clusters vary in the range of

Table 2. Cluster sizes for the top 10 cluster combinations for the weather patterns

Weather cluster combination	Temperature cluster	Solar cluster	Wind cluster	Humidity cluster	Size
o_{w1}	t_3	s_4	w_2	h_5	32
o_{w2}	t_2	s_{10}	w_5	h_1	30
o_{w3}	t_1	s_1	w_5	h_1	29
o_{w4}	t_3	s_4	w_2	h_1	28
o_{w5}	t_4	s_5	w_4	h_4	27
o_{w6}	t_3	s_9	w_1	h_5	23
o_{w7}	t_3	s_7	w_5	h_1	22
o_{w8}	t_2	s_7	w_5	h_1	21
o_{w9}	t_3	s_8	w_1	h_5	21
o_{w10}	t_1	s_{10}	w_5	h_1	20

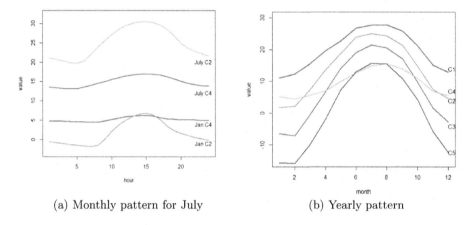

(a) Monthly pattern for July   (b) Yearly pattern

Fig. 3. Sample of average temperature patterns, by cluster

the amount of light they receive. Weather stations in Cluster 5 will almost always receive the least light. Likewise, weather stations in Cluster 3 often receive the most light. Based on the plot of weather stations by solar clusters, we observed that weather stations belonging to Cluster 3 appear in California and the American Sun Belt. Weather stations in Cluster 5 are mainly located in Alaska and the Canadian Arctic, where daylight hours are more limited throughout the year due to some winter days of complete darkness, and typically foggy or cloudy summers.

Once the clusters for the individual parameters have been determined, meta-clusters can be created. Meta-clusters are combinations of the parameters clusters. Simply, these cluster combinations are the set of weather stations that belong to a given set of parameter clusters (t_i, s_j, w_k, h_m). For all possible

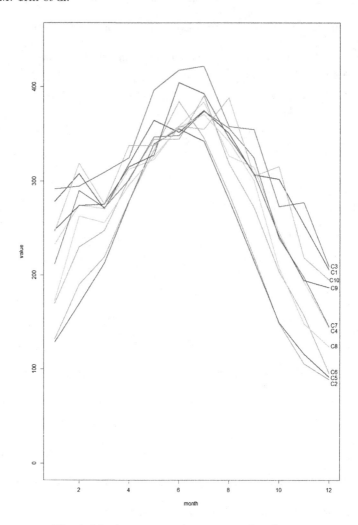

Fig. 4. Yearly average solar patterns, by cluster

cluster combinations, we rank them by the number of weather stations that belong to the cluster combination i.e. the size. From our data, we show the top 10 cluster combinations in Table 2. Out of all 1250 possible cluster combinations, only 201 had at least one weather station. Of the 201 cluster combinations, only 62 had at least five weather stations. All of the valid cluster combinations can serve as our prototypical weather profiles. Based on these clusters, future weather stations can be compared to their centroids to be quickly assigned.

Analyzing the top 10 cluster combinations, we can see that their sizes are reasonably distributed. We do not see a significant drop in size within the top 10. In total, the top 10 clusters encompass 253 of the 1000 weather stations analyzed. The largest cluster combination consists primarily of a set of weather stations within the Great Lakes region. The second largest cluster combination consists primarily of a set of weather stations within the American South. In analyzing these cluster combinations, we see that the location of their composite weather stations are in physically similar locations. This similarity can occur in proximity or topography.

6   Summary and Conclusions

Objects in a dataset can be combined with a number of different criteria. Usually, such a profiling begins with determining an appropriate knowledge representation to capture the essence of the criteria. For example, in this paper 1000 randomly selected weather stations from North America are profiled based on temperature, humidity, solar radiation and wind. For each of these criteria, we use annual patterns of low and high values. Such patterns are termed rough annual patterns. Each criterion results in a different grouping of weather stations. We want to use these profiles based on a different set of criteria to select a sample that covers all possible types of weather patterns. The paper demonstrates how one can assign priority to different weather stations based on their ability to cover as much of the most common weather variations possible. The weather stations with higher priorities can be subject to early engineering analysis and modelling.

References

1. Hartigan, J.A., Wong, M.A.: Algorithm AS 136: a k-means clustering algorithm. J. R. Stat. Soc. Ser. C (Appl. Stat.) **28**(1), 100–108 (1979). http://www.jstor.org/stable/2346830
2. Lingras, P.: Applications of rough patterns. In: Polkowski, L., Skowron, A. (eds.) Rough Sets in Data Mining and Knowledge Discovery 2, pp. 369–384. Springer, Heidelberg (1998)
3. Lingras, P.: Fuzzy-rough and rough-fuzzy serial combinations in neurocomputing. Neurocomput. J. **36**, 29–44 (2001)
4. Lingras, P., Davies, C.: Applications of rough genetic algorithms. Comput. Intell.: Int. J. **17**(3), 435–445 (2001)
5. Lingras, P., Butz, C.J.: Rough support vector regression. Eur. J. Oper. Res. **206**(2), 445–455 (2010)
6. Pawlak, Z.: Rough Sets: Theoretical Aspects of Reasoning about Data. Kluwer Academic Publishers, Berlin (1992)
7. Pawlak, Z.: Rough real functions (1994). http://citeseer.ist.psu.edu/105864.html
8. Peters, J.F., Han, L., Ramanna, S.: Rough neural computing in signal analysis. Comput. Intell. **17**(3), 493–513 (2001)
9. Polkowski, L., Skowron, A.: Rough mereology: a new paradigm for approximate reasoning. J. Approx. Reason. **15**(4), 333–365 (1996)
10. Skowron, A., Polkowski, L.: Rough mereological foundations for design, analysis, synthesis and control in distributed systems. Inf. Sci. **104**(1–2), 129–156 (1998)
11. Yao, Y.Y.: Constructive and algebraic methods of the theory of rough sets. Inf. Sci. **109**, 21–47 (1998)
12. Ziarko, W.: Variable precision rough set model. J. Comput. Syst. Sci. **46**(1), 39–59 (1993)

C&E Re-clustering: Reconstruction of Clustering Results by Three-Way Strategy

Pingxin Wang[1,2,3($\boxtimes$)], Xibei Yang[2,4], and Yiyu Yao[2]

[1] School of Science, Jiangsu University of Science and Technology,
Zhenjiang 212003, China
pingxin_wang@hotmail.com
[2] Department of Computer Science, University of Regina,
Regina, SK S4S 0A2, Canada
[3] College of Mathematics and Information Science,
Hebei Normal University, Shijiazhuang 050024, China
[4] School of Computer Science, Jiangsu University of Science and Technology,
Zhenjiang 212003, China

Abstract. Many existing approaches to clustering are based on a two-way strategy that does not adequately show the fact that a cluster may not have a well-defined boundary. In this paper, we propose a Contraction and Expansion Re-clustering (C&E Re-clustering for short) model based on three-way strategy. The model utilizes the ideas of erosion and dilation in mathematical morphology. Contraction is used to shrink clusters while expansion is used to stretch clusters obtained by using an existing clustering algorithm. The difference between the results of contraction and expansion is regarded as the fringe region of the specific cluster. Therefore, a three-way explanation of the cluster is naturally formed. A C&E NJW re-clustering algorithm is proposed and the results on synthetic data set show that such a strategy is effective in improving the structure of clustering results.

Keywords: Three-way clustering · Contraction · Expansion

1 Introduction

Clustering aims to find k groups based on a measure of similarity such that similarities between items in the same group are high while similarities between items in different groups are low. A common assumption underlying two-way clustering methods is that a cluster can be represented by a set with a crisp boundary. That is, a data point is either in or not in a specific cluster. The requirement of a sharp boundary leads to easy analytical results, but may not be good enough for characterizing the ambiguity that can be seen everywhere in real-world applications. For example, in Fig. 1, there are two highly concentrated areas and three discrete points x_1, x_2 and x_3. If a two-way clustering technique is used, x_1 needs to be assigned to either the right or the left clusters. It seems unreasonable to add x_1 into either cluster because the distances between x_1

© Springer International Publishing AG 2017
M. Kryszkiewicz et al. (Eds.): ISMIS 2017, LNAI 10352, pp. 540–549, 2017.
DOI: 10.1007/978-3-319-60438-1_53

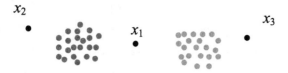

Fig. 1. A schematic diagram of data set

and two clusters are almost the same. On the other hand, although there is no difficulties in choosing clusters for x_2 and x_3, respectively, their inclusion in a cluster would weaken the structure of the cluster.

The concept of three-way decisions was first proposed by Yao [13–15] to interpret rough set three regions. The positive, negative and boundary regions are viewed, respectively, as the regions of acceptance, rejection, and noncommitment in a ternary classification. The positive and negative regions can be used to induce rules of acceptance and rejection. Whenever it is impossible to make an acceptance or a rejection decision, the third noncommitment decision is made. Motivated by the three-way strategy, Yu et al. [16,17] proposed a three-way clustering framework that uses two regions to represent a cluster, i.e., core region (Co) and fringe region (Fr) rather than one set. The core region is an area where the elements are highly concentrated of a cluster and fringe region is an area where the elements are loosely concentrated. There may be common elements in the fringe region among different clusters. This paper aims at presenting an three-way clustering method by re-constructing the results of two-way clustering. In the proposed method, each cluster is represented by a core region and a fringe region, revealing a better cluster structure.

Mathematical morphology was introduced by Matheron [7] to study porous media using set theory. Erosion and dilation are two primary morphological operations used to shrink and stretch an image according to the structuring element. Similar to the idea of erosion and dilation operations for image processing, we can shrink and stretch each cluster obtained from a two-way clustering algorithm based on a pair of morphological operations. We use contraction to indicate the deletion of objects while expansion to indicate the adding of more objects with respect to a specific cluster. Through operations of contraction and expansion, cluster may shrink to be a core region and stretch to be an expanded region. Immediately, a fringe region of the specific cluster is derived by the difference between the result of contraction and expansion operations. This reconstruction of clusters is referred to as C&E Re-clustering.

In Fig. 2, C_1 and C_2 are two clusters obtained by a two-way clustering technique on data given in Fig. 1. Each object is either in C_1 or not in C_1; The same is true about C_2. If we apply the proposed contraction and expansion method on C_1 and C_2, we obtain the results in Fig. 3. We can see that core region is more compact and the structure is better than clusters in Fig. 2.

The rest of the paper is organized as follows. Section 2 briefly introduces the background knowledge. Section 3 proposes a C&E strategy that aims to

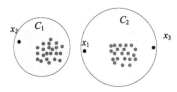
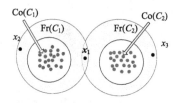

Fig. 2. Clustering results by two-way clustering

Fig. 3. Clustering results by C&E clustering

transform two-way clusters into three-way based clusters. A C&E NJW re-clustering algorithm is designed in Sect. 4. Experiment results are reported in Sect. 5.

2   Three-Way Clustering and Mathematical Morphology

Traditional two-way crisp boundary clustering is too restrictive in many areas such as network structure analysis, wireless sensor networks and biological processing. In order to overcome this problems, many authors have proposed methods to improve the traditional two-way strategy. Hoppner et al. [3] studied fuzzy clustering that represents a cluster by using a fuzzy set with a gradually changing boundary. Lingras et al. [4,5] presented and systematically investigated rough clustering and interval set clustering. Takaki et al. [12] produced a method of overlapping clustering for network structure analysis. Aydin et al. [1] introduced an overlapping clusters algorithm and applied it to mobile Ad hoc networks. Three-way clustering proposed by Yu et al. [16,17] is a most recent proposal for modeling clusters with an unsharp boundary. Mathematical morphology may provide a sound basis for such three-way approaches to clustering [11].

2.1   Three-Way Clustering

The concept of three-way decisions plays an important role in many real world decision-making problems. One usually makes a decision based on available information and evidence. When the evidence is insufficient or too weak, it might be impossible to make either a positive or a negative decision. One chooses an alternative decision that is neither yes nor no. By combining three-way decisions [13–15] with cluster analysis, Yu et al. [16,17] presented a framework of three-way clustering which represents a cluster by a pair of sets called the lower and upper bounds.

We summarize the basic concepts of three-way clustering [16]. Assume $C = \{C_1, \cdots, C_k\}$ is a family clusters of universe $V = \{v_1, \cdots, v_n\}$. A two-way clustering requires that C satisfies the following conditions:

(i) $C_i \neq \phi, (i = 1, \cdots, k)$,   (ii) $\bigcup_{i=1}^{k} C_i = V$,   (ii) $C_i \bigcap C_j = \phi \ (i \neq j)$.

Property (i) states that each cluster cannot be empty. Properties (ii) and (iii) state that every $v \in V$ belongs to only one cluster. In this case, C is a partition of the universe. Three-way clustering represents a cluster as an interval set. That is, C_i is represented by an interval set $C_i = [C_i^l, C_i^u]$, where C_i^l and C_i^u are the lower bounds and upper bounds of the cluster C_i, respectively. There are different requirements on C_i^l and C_i^u according to different applications. In this paper, we adopt the following properties:

$$\text{(I)} \ \ C_i^l \neq \phi, (i = 1, \cdots, k), \quad \text{(II)} \ \ \bigcup_{i=1}^{k} C_i^u = V, \quad \text{(III)} \ \ C_i^l \bigcap C_j^l = \phi \ (i \neq j).$$

Property (I) demands that each cluster cannot be empty. Property (II) states that it is possible that an element $v \in V$ belongs to more than one cluster. Property (III) requires that the lower bounds of clusters are pairwise disjoint. The elements in C_i^l can be interpreted as typical elements of the cluster C_i, which we called the core region $(\text{Co}(C_i))$ of a cluster. The elements of core region cannot, at the same time, be a typical elements of another cluster. The elements in $C_i^u - C_i^l$ are the fringe elements and the region is called fringe region $(\text{Fr}(C_i))$ of the cluster.

Base on the above discussion, we can also use $(\text{Co}(C_i), \text{Fr}(C_i))$ to represent the i-th cluster. That is, we use the following family of clusters to represent a family of three-way clusters:

$$\mathbb{C} = \{(\text{Co}(C_1), \text{Fr}(C_1)), (\text{Co}(C_2), \text{Fr}(C_2)), \cdots, (\text{Co}(C_k), \text{Fr}(C_k))\}.$$

Note that the properties of $\mathbb{C}$ can be obtained from properties (I)–(III).

2.2   Erosion and Dilation in Mathematical Morphology

In mathematical morphology, an image is viewed as a subset of the Euclidean space $\mathbb{R}^d$ or the integer grid $\mathbb{Z}^d$. Its basic purpose is to transform an image by using a simple, pre-defined shape, according to how this shape fits or misses the shapes in the image. A pre-defined shape is called a structuring element, and is itself a binary image (i.e., a subset of the space). There are two morphological operations called erosion and dilation. Erosion is to remove some pixels from the object's boundary while dilation is to add some pixels to the object's boundary. The erosion and dilation of a binary image A in integer grid $\mathbb{Z}^2$ by the structuring element B is illustrated in Fig. 4.

Definition 1. *Let E be Euclidean space $\mathbb{R}^d$ or the integer grid $\mathbb{Z}^d$, A is a binary image in E and B is a structuring element. The erosion of the binary image A by the structuring element B is defined by [10]:*

$$A \ominus B = \{x \in E \mid B_x \subseteq A\},$$

The dilation of A by the structuring element B is defined by:

$$A \oplus B = \{x \in E \mid \check{B}_x \cap A \neq \phi\},$$

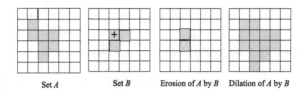

Fig. 4. Erosion and dilation of set A by the structuring element B

where B_x denotes the translation of the structuring element at a point x and $\check{B}$ denotes the symmetrical transformation of B with respect to the origin.

Erosion has the effect of shrinking the image and the dilation operation has the effect of enlarging the image if the origin, denoted by $+$ in Fig. 4, is in the structuring element B. The results of erosion and dilation depend on the choice of structuring element. Shape and size are two main characteristics that are directly related to structuring elements. Different structuring elements will produce different results of erosion and dilation.

3   A C&E Re-clustering Method

3.1   Framework of C&E Re-clustering Method

Suppose that a set $V = \{v_1, \cdots, v_n\}$ has been divided into k clusters $C = \{C_1, \cdots, C_k\}$ by a two-way clustering method. We present a strategy to transform two-way cluster C_i into $(\mathrm{Co}(C_i), \mathrm{Fr}(C_i))$. The idea origins from the dilation and erosion operations in mathematical morphology. As we have pointed out in Sect. 2, erosion operation has the effect of shrinking the image and the dilation operation has the effect of enlarging the image if the origin is in the structuring element. Motivated by the erosion and dilation of an image, we can shrink or stretch a cluster. In order to extend erosion and dilation operation to a general set, we introduce the concept of structure-generating function.

Definition 2. *Suppose V is a nonempty set. A mapping $S: V \to 2^V$ is called a structure-generating function and $S(x)$ is called a structuring element of $x \in V$. A structure-generating function S is reflexive if S satisfies the condition that $x \in S(x)$ for any $x \in U$.*

Similar to the erosion and dilation of an image, we can shrink or stretch a set by structure-generating function. We use contraction to indicate the deletion of objects while expansion to indicate the adding of more objects in a set.

Definition 3. *Suppose V is a nonempty set equipped with a pair of reflexive structure-generating functions (S_1, S_2) and A is subset of V. The contraction of A by structure-generating function S_1 is defined by:*

$$A \ominus S_1 = \{x \mid S_1(x) \subseteq A\}.$$

The expansion of A by structure-generating function S_2 is defined by:

$$A \oplus S_2 = \{x \mid S_2(x) \cap A \neq \phi\}.$$

From Definition 3, we can see that contraction deletes the elements of A that do not satisfy the condition $S_1(x) \subseteq A$, and expansion adds those elements outside of A that satisfy the condition $S_2(x) \cap A \neq \phi$. Through contraction and expansion, a set may shrink to a small one and stretch to an expanded one. We therefore can transform a two-way cluster into a three-way cluster through contraction and expansion operations by choosing a pair of appropriate (S_1, S_2).

Definition 4 *C&E reclustering*. *Suppose* $C = \{C_1, \cdots, C_k\}$ *is a two-way clustering result of a universal set* $V = \{v_1, \cdots, v_n\}$ *and* (S_1, S_2) *is a pair of reflexive structure-generating functions on* V. *We define the core region and fringe region* C_i *as,*

$$Co(C_i) = C_i \ominus S_1 = \{v \mid S_1(v) \subseteq C_i\},$$
$$Fr(C_i) = C_i \oplus S_2 - Co(C_i),$$

where $C_i \oplus S_2 = \{v \mid S_2(v) \cap C_i \neq \phi\}$.

There are many structure-generating functions for contraction and expansion operations and different structure-generating functions will produce different results of contraction and expansion. All of these methods by contraction and expansion operations on the results of two-way clusters are called C&E re-clustering method. An essential issue of C&E re-clustering method is to choose appropriate reflexive structure-generating functions (S_1, S_2).

3.2 Neighbor-Based C&E Re-clustering Method

In this subsection, we present a neighbor-based C&E re-clustering method based on the framework of C&E re-clustering. We use $Neig_q(v)$ to represent the q-nearest points of v, where q is a given parameter. Before introducing the reflexive structure-generating functions for contraction and expansion operations, we investigate the characteristics of points in V. Let us take Fig. 1 as an example. We can easily divide these objects of Fig. 1 into two clusters except the three discrete points x_1, x_2 and x_3. As for x_1, it is difficult to decide which cluster it belongs to because the point is between the two clusters. As for x_2, although there is no difficulties in choosing cluster, the distance between x_2 and the center of the cluster is significantly greater than other point-to-center distance. The same case is true for x_3. Therefore, the $Co(C_1)$ should be $C_1 - \{x_2\}$ and x_1 should be belong to $Fr(C_1)$. This procedure can be explained by using the contraction and expansion operations for set C_1.

Extending the above discussion to set V, we can classify all objects of V into three types based on the results of a two-way clustering:

Type I $= \{v \mid \exists i \neq j, Neig_q(v) \cap C_i \neq \phi \wedge Neig_q(v) \cap C_j \neq \phi\}$,

Type II $= \{v \mid \exists i, Neig_q(v) \subset C_i \wedge d > \rho d_i\}$,

Type III $= \{v \mid \exists i, Neig_q(v) \subset C_i \wedge d < \rho d_i\}$, where $\rho > 1$ is a parameter, d is the mean distance between v and $Neig_q(v)$ and d_i is the mean value of d for all $v \in C_i$. We use the following strategy to reconstruct the clusters: the points in Type I and Type II should be assigned to the fringe region and the points in Type III should be assigned to the core region of the corresponding cluster. Based on the above strategy, we can define the following structure-generating functions S_1 for contraction and S_2 for expansion on cluster C_i. If V is divided into k clusters $C = \{C_1, \cdots, C_k\}$ by a two-way clustering method, we have:

$$S_1(v) = \begin{cases} V, & d > \rho d_i \\ Neig_q(v), & d \le \rho d_i \end{cases}, \quad S_2(v) = Neig_q(v),$$

where v is any element of V, $\rho > 1$ is a given parameter, d is the mean distance between v and $Neig_q(v)$ and d_i is the mean value of d for all $v \in C_i$. It can be easily seen that S_1 and S_2 are reflexive. Using the definitions of contraction and expansion, we express the core region and the fringe region of cluster C_i as,

$$\text{Co}(C_i) = C_i \ominus S_1 = \{v | S_1(v) \subseteq C_i\},$$
$$\text{Fr}(C_i) = C_i \oplus S_2 - \text{Co}(C_i),$$

where $C_i \oplus S_2 = \{v | S_2(v) \cap C_i \ne \phi\}$.

The procedure of the reconstruction of clusters is referred as neighbor-based C&E re-clustering. An algorithm of neighbor-based C&E re-clustering of two-way clustering can be designed as given in Algorithm 1.

Algorithm 1. Algorithm of neighbor-based C&E re-clustering

Input: A set of points $V = \{v_1, \cdots, v_n\}$, the number of clusters k and parameter q and ρ.

Output: $\mathbb{C} = \{(\text{Co}(C_1), \text{Fr}(C_1)), (\text{Co}(C_2), \text{Fr}(C_2)), \cdots, (\text{Co}(C_k), \text{Fr}(C_k))\}$.

1. Divide the data set $V = \{v_1, \cdots, v_n\}$ into k clusters $C = \{C_1, \cdots, C_k\}$ by two-way clustering method.
2. Given a cluster C_i, for a point $v \notin C_i$, $Neig_q(v)$ is the q-nearest points of v. If $Neig_q(v) \bigcap C_i \ne \phi$, then $v \in \text{Fr}(C_i)$.
3. Given a cluster C_i, for every point $v_j \in C_i$, compute the average distance d_j between v_j and its q-nearest neighborhood $Neig_q(v)$ and the mean d_i of all d_j for $v_j \in C_i$. If $d_j < \rho d_i$, then $v \in \text{Co}(C_i)$, If $d_j > \rho d_i$, then $v_j \in \text{Fr}(C_i)$.
4. Return $\{(\text{Co}(C_1), \text{Fr}(C_1)), (\text{Co}(C_2), \text{Fr}(C_2)), \cdots, (\text{Co}(C_k), \text{Fr}(C_k))\}$.

---

4   Neighbor-Based C&E Spectral Re-clustering

Spectral clustering [2,6,9] origins from the theory of graph partitioning. It divides the vertices of a graph into disjoint clusters to minimize the total cost of an edge cut induced by clusters. Spectral clustering does not make any assumption on the form of the data clusters, is very simple to implement, and can be solved efficiently by standard linear algebra methods. A review of spectral clustering can be found in [6].

Most of the spectral clustering algorithms can be divided to three steps. The first step is to construct the graph and the similarity matrix representing the data sets, the second step is to compute eigenvalues and eigenvectors of the Laplacian matrix and map each point to a lower-dimensional representation based on one or more eigenvectors, and the last step is to assign points to two or more classes, based on the new representation. Taking NJW algorithm [8] for example, we give the procedure of spectral clustering in Algorithm 2.

Algorithm 2. NJW algorithm

Input: $V = \{v_1, \cdots, v_n\} \in R^l$, the number of clusters k and parameter σ.
Output: Clusters $C = \{C_1, C_2, \cdots, C_k\}$.
1. Form the similarity matrix W defined by $w_{ij} = \exp(-\frac{d^2(v_i, v_j)}{2\sigma^2})$, $i \neq j$, and $w_{ii} = 0$.
2. Define D to be the diagonal matrix whose (i, i)-element is the sum of $A's$ i-th row, and construct the matrix $L = D^{-\frac{1}{2}} W D^{\frac{1}{2}}$.
3. Find the first k eigenvectors of L (chosen to be orthogonal to each other in the case of repeated eigenvalues), and form the matrix U by stacking the eigenvectors in columns $X = (x_1, x_2, \cdots, x_k) \in R^{n \times k}$.
4. Form the matrix Y from X by normalizing each of $X's$ rows to have unit length, $Y_{ij} = \frac{X_{ij}}{\sum_j X_{ij}^2}$.
5. Treat each row of Y as a point in R^k and classify them into k classes via k-means algorithm.
6. Assign the original points v_i to cluster j if and only if row i of the matrix Y was assigned to cluster j.

We introduce a neighbor-based C&E NJW re-clustering algorithm by combining Algorithm 1 and Algorithm 2. The algorithm has two main steps. The first one is to spectral clustering with the data and the second steps is to compute the core region and fringe region of each cluster. Finally, a significance neighbor-based C&E NJW re-clustering can be designed as Algorithm 3.

In Algorithm 3, Step 1 produces an initial clustering results from spectral clustering. Step 2 and Step 3 find the core region and the fringe region of each cluster.

5 Experimental Illustration

To illustrate the effectiveness of the neighbor-based C&E NJW re-clustering algorithm, one synthetic two dimensions data set with 1590 points is employed. All codes are run in Matlab R2010b on a personal computer and the parameters

---

Algorithm 3. Neighbor-based C&E NJW re-clustering algorithm

Input: $V = \{v_1, \cdots, v_n\} \in R^l$, the number of clusters k and parameters σ, q, ρ.
Output: $\mathbb{C} = \{(\mathrm{Co}(C_1), \mathrm{Fr}(C_1)), (\mathrm{Co}(C_2), \mathrm{Fr}(C_2)), \cdots, (\mathrm{Co}(C_k), \mathrm{Fr}(C_k))\}$.

1. Execute steps 1–7 of Algorithm 2 and obtained k clusters $C = \{C_1, \cdots, C_k\}$ by spectral clustering method.
2. Given a cluster C_i, for a point $v \notin C_i$, $Neig_q(v)$ is the q-nearest points of v. If $Neig_q(v) \bigcap C_i \neq \phi$, then $v \in \mathrm{Fr}(C_i)$.
3. Given a cluster C_i, for every point $v_j in C_i$, compute the average distance d_j between v_j and its q-nearest neighborhood $Neig_q(v)$ and the mean d_i of all d_j for $v_j \in C_i$. If $d_j > \rho d_i$, then $v_j \in \mathrm{Fr}(C_i)$. If $d_j < \rho d_i$, then $v \in \mathrm{Co}(C_i)$.
4. Return $\mathbb{C} = \{(\mathrm{Co}(C_1), \mathrm{Fr}(C_1)), (\mathrm{Co}(C_2), \mathrm{Fr}(C_2)), \cdots, (\mathrm{Co}(C_k), \mathrm{Fr}(C_k))\}$.

---

are $q = 15$ and $\rho = 1.5$. The results of spectral clustering and neighbor-based C&E NJW re-clustering are depicted in Figs. 5 and 6, respectively.

From Fig. 5, we can see that these points are clustered into two clusters and each point is belong to only one cluster by spectral clustering. Figure 6 shows that neighbor-based C&E NJW re-clustering finds the core region and the fringe region of every cluster and the overlapping parts of two classes. There are 80 points in the fringe region of clusters C_1 and 90 points in the fringe region of clusters C_2, where 17 points in overlap of two clusters.

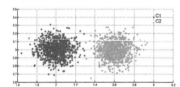

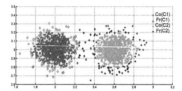

Fig. 5. Spectral clustering result

Fig. 6. C&E spectral re-clustering result

6   Conclusion

In this paper, we developed a general framework of C&E re-clustering method by employing the ideas of corrosion and expansion in mathematical morphology. Overlapping points and discrete points will be assigned to the fringe region of one or more clusters according to their q-nearest neighborhoods. A neighbor-based C&E NJW re-clustering algorithm is provided and experimental results demonstrate that the new algorithm can significantly improve the structure of classification, compared with the traditional spectral clustering algorithm. The present study is the first step for the research of three-way clustering. How to determine the parameter and the number of clusters is an interesting topic to be addressed for further research.

Acknowledgements. This work was supported in part by National Natural Science Foundation of China (Nos. 61503160 and 61572242), Natural Science Foundation of the Jiangsu Higher Education Institutions of China (No. 15KJB110004), and a Discovery Grant from NSERC, Canada.

References

1. Aydin, N., Naït-Abdesselam, F., Pryyma, V., Turgut, D.: Overlapping clusters algorithm in ad hoc networks. In: 2010 IEEE Global Telecommunications Conference (2010)
2. Donath, W., Hoffman, A.J.: Lower bounds for the partitioning of graphs. IBM J. Res. Dev. **17**, 420–425 (1973)
3. Hoppner, F., Klawonn, F., Kruse, R., Runkler, T.: Fuzzy Cluster Analysis: Methods for Classification, Data Analysis and Image Recognition. Wiley, Chichester (1999)
4. Lingras, P., Hogo, M., Snorek, M., West, C.: Temporal analysis of clusters of supermarket customers: conventional versus interval set approach. Inf. Sci. **172**, 215–240 (2005)
5. Lingras, P., West, C.: interval set clustering of web users with rough k-means. J. Intell. Inf. Syst. **23**, 5–16 (2004)
6. Luxburg, U.: A tutorial on spectral clustering. Stat. Comput. **17**, 395–416 (2007)
7. Matheron, G.: Random Sets and Integral Geometry. Wiley, New York (1975)
8. Ng, A., Jordan, M., Weiss, Y.: On spectral clustering: analysis and an algorithm. In: Dietterich, T., Becker, S., Ghahramani, Z. (eds.) Advances in Neural Information Processing Systems 14, pp. 849–856. MIT Press, Cambridge (2002)
9. Shi, J., Malik, J.: Normalized cuts and image segmentation. IEEE Trans. Pattern Anal. Mach. Intell. **22**, 888–905 (2000)
10. Serra, J.: Image Analysis and Mathematical Morphology. Academic Press, London (1982)
11. Stell, J.G.: Relations in mathematical morphology with applications to graphs and rough sets. In: Winter, S., Duckham, M., Kulik, L., Kuipers, B. (eds.) COSIT 2007. LNCS, vol. 4736, pp. 438–454. Springer, Heidelberg (2007). doi:10.1007/978-3-540-74788-8_27
12. Takaki, M., Tamura, K., Mori, Y.: A extraction method of overlapping cluster based on network structure analysis. In: IEEE/WIC/ACM International Conferences on Web Intelligence and Intelligent Agent Technology, pp. 212–217 (2007)
13. Yao, Y.Y.: Three-way decisions with probabilistic rough sets. Inf. Sci. **180**, 341–353 (2010)
14. Yao, Y.: An outline of a theory of three-way decisions. In: Yao, J.T., Yang, Y., Słowiński, R., Greco, S., Li, H., Mitra, S., Polkowski, L. (eds.) RSCTC 2012. LNCS (LNAI), vol. 7413, pp. 1–17. Springer, Heidelberg (2012). doi:10.1007/978-3-642-32115-3_1
15. Yao, Y.Y.: Three-way decisions and cognitive computing. Cogn. Comput. **8**, 543–554 (2016)
16. Yu, H., Zhang, C., Wang, G.: A tree-based incremental overlapping clustering method using the three-way decision theory. Knowl.-Based Syst. **91**, 189–203 (2016)
17. Yu, H., Jiao, P., Yao, Y.Y., Wang, G.: Detecting and refining overlapping regions in complex networks with three-way decisions. Inf. Sci. **373**, 21–41 (2016)

Multi-criteria Based Three-Way Classifications with Game-Theoretic Rough Sets

Yan Zhang$^{(\boxtimes)}$ and JingTao Yao

Department of Computer Science, University of Regina, Regina, SK S4S 0A2, Canada
{zhang83y,jtyao}@cs.uregina.ca

Abstract. Three-way classifications divide the universe of objects into three regions based on a given concept. Rough sets and its extensions provide effective ways to construct three-way classifications. When multiple criteria are involved to determine three-way classifications, the problem of determining three-way regions can be formulated as a typical multi-criteria decision making (MCDM) problem. In this paper, we use game-theoretic rough set model (GTRS) to solve and address the multi-criteria based three-way classifications constructed in the context of rough sets. GTRS implement competitive games amongst multiple criteria in order to obtain a compromise between criteria by finding an equilibrium of the games. Applying GTRS in MCDM consists of three stages, namely, competitive game formulation, repetition learning process, and decision making based on equilibria. The advantage of applying GTRS is twofold. GTRS do not require the predefined weights for criteria or compound decision objectives. GTRS are inherently suitable for a competitive environment in which the involved criteria maximize their own benefits and the payoff of each criterion is influenced by other's strategies.

Keywords: Game-theoretic rough sets · Multi-criteria decision making · Three-way classifications · Competitive games

1 Introduction

Three-way classifications are constructed based on the notions of acceptance, rejection and non-commitment if the classifications are used for decision making [14,15]. Given U as a finite nonempty set of objects and C as a undefinable target concept, the aim of three-way classifications is to partition U based on C into three disjoint regions [14]. The three-way classifications can be formulated from different models or theories, such as rough sets and its extensions, fuzzy sets, shadow sets, interval sets [15]. We focus on the three-way classifications in the context of rough sets in this research. The determination of rough sets based three-way classification regions is a critical research question. In order to solve this question, we need some criteria or measures, such as accuracy, confidence, generality, coverage, cost, and uncertainty, to evaluate the three-way classifications [2,20]. These criteria evaluate three-way classifications from different views.

© Springer International Publishing AG 2017
M. Kryszkiewicz et al. (Eds.): ISMIS 2017, LNAI 10352, pp. 550–559, 2017.
DOI: 10.1007/978-3-319-60438-1_54

When multiple criteria are involved to determine three-way classifications, especially, the involved criteria are conflicting to each other, we have to find an approach to balance them or to reach a compromise among them. Determining three-way classifications based on multiple criteria can be formulated as a typical multiple criteria decision making (MCDM) problem. MCDM deals with complicated decision making problems which involve multiple conflicting criteria [19]. It aims to obtain a suitable solution among various options based on the performance of evaluation on multiple criteria, and the selected solution can represent a tradeoff among the involved criteria. Many concepts, techniques and approaches have been proposed to resolve MCDM problems, including: analytic hierarchy process [9], dominance-based rough set approach [4], weighted sum model [11], and others [10]. Typically, an MCDM problem contains four basic elements: a set of criteria, a set of alternatives or actions, the outcomes of involved criteria, and a preference structure of criteria [18]. The set of alternatives represents different decision options. The set of criteria is used to evaluate alternatives. The outcomes of involved criteria mean the evaluations for the set of alternatives based on multiple criteria. The preference structure guides in selecting an appropriate alternative [18].

Game theory is a mathematical tool to study the conflict and cooperation among decision makers [6]. MCDM problems can be formulated as competitive games with multiple players and strategies. Using game theory to solve MCDM problems has been attracting much attention [3,7]. Game-theoretic rough set model (GTRS) is an advancement in determining suitable partition of three-way regions by formulating competitive or coordinative games between multiple measures [12,13]. The essential idea of GTRS is to implement games to obtain suitable three-way classifications in the rough set context when multiple measures are involved to evaluate three regions [1]. GTRS combine game theory and rough sets to solve MCDM problems in rough sets and three-way classifications. In this problem solving process, the selection of game players, the configuration of strategies and the definition of payoff functions are investigated in detail [5,13]. The aim of GTRS is to improve the rough sets based decision making by finding a compromise among the involved measures.

In this paper, we apply GTRS to determine three-way classifications with the presence of multiple criteria. This process contains three stages, i.e., competitive game formulation, repetition learning, and decision making based on equilibrium. Compared with the conventional MCDM methods, the advantage of applying GTRS in multi-criteria based three-way classifications is twofold. On the one hand, GTRS do not require the predefined weights for criteria or compound decision objectives as the conventional MCDM methods. On the other hand, GTRS are inherently suitable for a competitive environment where involved criteria maximize their benefits and the payoff of each criterion is influenced by other's strategies, while the conventional MCDM methods ignore that the behaviors of other decision makers may effect their payoffs.

2 Background Knowledge

In this section, we briefly introduce the background concepts about three-way classifications in the context of rough sets, and the criteria of evaluating three-way classifications.

2.1 Rough Sets Based Three-Way Classifications

The three-way classification theory was outlined by Yao in [14,15]. It classifies the universe of objects U into three disjoint regions based on a undefinable target concept C. We can construct three-way classifications with rough sets and its extensions. Suppose the universe of objects U is a finite nonempty set. Let $E \subseteq U \times U$ be an equivalence relation on U, where E is reflexive, symmetric, and transitive [8]. For an element $x \in U$, the equivalence class containing x is given by $[x] = \{y \in U | xEy\}$. The family of all equivalence classes defines a partition of the universe and is denoted by $U/E = \{[x] | x \in U\}$, that is the intersection of any two elements is an empty set and the union of all elements are the universe U [8]. For an undefinable target concept $C \subseteq U$, probabilistic rough sets utilize conditional probability $Pr(C|[x])$ as evaluation function and thresholds (α, β) to define three-way regions of C, i.e., positive, negative, and boundary regions of the concept C [16]:

$$POS_{(\alpha,\beta)}(C) = \bigcup\{[x] \mid [x] \in U/E, Pr(C|[x]) \geq \alpha\},$$
$$NEG_{(\alpha,\beta)}(C) = \bigcup\{[x] \mid [x] \in U/E, Pr(C|[x]) \leq \beta\},$$
$$BND_{(\alpha,\beta)}(C) = \bigcup\{[x] \mid [x] \in U/E, \beta < Pr(C|[x]) < \alpha\}. \tag{1}$$

These three-way regions are pair-wise disjoint and their union is the universe of objects U according to Eq. (1). They form a tripartition of the universe U. The family of the three regions constitute a three-way classification model:

$$\pi_{(\alpha,\beta)}(C) = \{POS_{(\alpha,\beta)}(C), BND_{(\alpha,\beta)}(C), NEG_{(\alpha,\beta)}(C)\} \tag{2}$$

When we use this three-way classification to classify objects, acceptance and rejection decision rules can be induced from positive and negative regions, respectively.

2.2 Evaluating Three-Way Classifications

Given a target concept C and a pair of probabilistic thresholds (α, β), we can obtain a three-way classification according to Eq. (1). Many criteria or measures have been proposed to evaluate three-way classifications [17,20]. The accuracy and coverage are two most commonly used measures to evaluate the performance of three-way classifications [21].

The criterion accuracy intends to capture the degree of classification correctness of three-way classifications. It is calculated as the ratio of the number of correctly classified objects by a three-way classification model and the number of objects that can be classified by this model. The range of an accuracy value is between 0 and 1. The accuracy of a three-way classification $\pi_{(\alpha,\beta)}(C)$ is defined as:

$$Acc(\pi_{(\alpha,\beta)}(C)) = \frac{|C \cap POS_{(\alpha,\beta)}(C)| + |C^c \cap NEG_{(\alpha,\beta)}(C)|}{|POS_{(\alpha,\beta)}(C)| + |NEG_{(\alpha,\beta)}(C)|}. \tag{3}$$

The criterion coverage intends to express the applicability of three-way classifications. It is the ratio of the number of objects that can be classified by a three-way classification to the number of objects in the universe U. It expresses the proportions of objects that can be classified by a three-way classification model. The coverage of a three-way classification $\pi_{(\alpha,\beta)}(C)$ is defined as:

$$Cov(\pi_{(\alpha,\beta)}(C)) = \frac{|POS_{(\alpha,\beta)}(C)| + |NEG_{(\alpha,\beta)}(C)|}{|U|}. \tag{4}$$

We would like to obtain a high level of classification accuracy and a high level of classification coverage for three-way classifications. A high level of accuracy means we can make more accurate classification decisions based on this three-way classification model. A high level of coverage means we can classify more objects by using this three-way classification model. In general, a more accurate three-way classification model tends to be weaker in applicability or coverage. Similarly, three-way classifications with a high coverage level may not be very accurate. It may not be wise to consider only one criterion and ignore the others in order to obtain suitable three-way classifications. Please note that three-way classifications can be constructed from many different methods or theories, and it is independent on rough sets and its extensions. The measures discussed here can be used to evaluate three-way classifications formulated by other theories.

3   Applying GTRS in Multi-criteria Based Three-Way Classifications

Determining three-way classifications based on multiple criteria can be formulated as a multi-criteria decision making problem. In order to make the problem easy to understand and solve, assuming that two criteria c_1 and c_2 are considered to determine the partition of three regions in a three-way classification. When more than two criteria are involved, the problem solving process is similar. The four elements of this MCDM problem are:

- The set of criteria contains two criteria c_1 and c_2;
- The set of alternatives contains all possible probabilistic threshold pairs $(\alpha_1, \beta_1), (\alpha_2, \beta_2), ..., (\alpha_k, \beta_k)$, where $0 \le \beta_i \le 0.5 \le \alpha_i \le 1$ and $1 \le i \le k$;
- The outcomes of criteria with different alternatives represent the values of two criteria c_1 and c_2 under all possible probabilistic threshold pairs;
- The preference of decision makers is to maximize their own criteria values.

These MCDM elements can match the elements of a game defined in GTRS. Applying GTRS in multi-criteria based three-way classifications includes three stages, namely competitive game formulation, repetition learning process, and decision making based on equilibrium, respectively. Each stage contains different tasks and some of tasks are covered in two stages, as shown in Fig. 1.

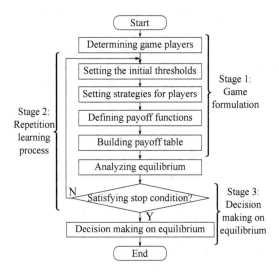

Fig. 1. The three stages of applying GTRS

3.1 Competitive Game Formulation

In game-theoretic rough sets, there are three elements when formulating a game G, that is, game players set O, strategy sets performed by the players S, and the payoff set of players u, and $G = \{O, S, u\}$ [5]. In the game formulation stage, the three elements of a game are specified. In fact, the elements of a game correspond to the basic elements of MCDM problems.

Game Players. The set of game players contains two criteria c_1 and c_2, that is $O = \{c_1, c_2\}$. The game players represent the criteria that are used to evaluate three-way classifications. Each player evaluates three-way classifications from its own perception, and they have opposite interests when evaluating a three-way classification.

Initial Thresholds. The initial thresholds (α, β) are defined and they can be any thresholds values that satisfy the constraint $0 \leq \beta \leq 0.5 \leq \alpha \leq 1$. For example, we can set initial thresholds as $(\alpha, \beta) = (0.5, 0)$. Two players start from the initial thresholds and perform their strategies to change the initial thresholds. The setting of initial thresholds can directly influence the strategies of players. For example, when initial thresholds are set as $(\alpha, \beta) = (0.5, 0)$, which means α and β get the minimum values in their own limits, the strategies can only be set as increasing α and β. Similarly, if initial thresholds are set as $(\alpha, \beta) = (1, 0.5)$, which means α and β get the maximum values in their own limits, the strategies can only be set as decreasing α and β.

Strategies. The set of strategies or actions S contains two sets of strategies performed by the game players, i.e., $S = \{S_1, S_2\}$, where $S_1 = \{s_1, s_2, ..., s_{k1}\}$ is a set of possible strategies or actions for player c_1, and $S_2 = \{t_1, t_2, ..., t_{k2}\}$ is a set of possible strategies or actions for player c_2. All these strategies are the changes of probabilistic thresholds α and β. Two players may have different strategies. For example, the player c_1's strategies can be the increase of α, $S_1 = \{\alpha$ increases 0.05, α increases 0.1, α increases $0.15\}$. The player c_2's strategies can be the increase of β, $S_2 = \{\beta$ increases 0.05, β increases 0.1, β increases $0.15\}$.

A strategy profile $p = \{p_1, p_2\}$ is a particular play of a game, in which player c_1 performs the strategy or action p_1 and player c_2 performs the strategy or action p_2, here $p_1 \in S_1$ and $p_2 \in S_2$.

Payoff Functions. The set of payoff functions results from players performing strategies $u = \{u_1, u_2\}$. The payoff of player c_1 under the strategy profile $p = \{p_1, p_2\}$ is denoted as $u_1(p) = u_1(p_1, p_2)$. The payoff of player c_2 under the strategy profile $p = \{p_1, p_2\}$ is denoted as $u_2(p) = u_2(p_1, p_2)$. The payoff functions of two players are defined by the criteria they are representing. The payoff of each player depends on the strategies or actions performed by both game players. The strategy performed by one game player can influence the payoff of the other player. For example, if the initial thresholds are $(\alpha, \beta) = (0.5, 0)$, the player c_1 performs the strategy s_1, i.e., α increases 0.05 and the player c_2 performs the strategy t_3, i.e., β increases 0.15. The probabilistic thresholds under the strategy profile $p = \{s_1, t_3\}$ is $(\alpha, \beta) = (0.55, 0.15)$. The payoffs of players are the values of the criteria they represent when $(\alpha, \beta) = (0.55, 0.15)$.

Payoff Tables. For a two-player game, we can build a payoff table to represent the game. The Table 1 shows a payoff table. Each cell of the payoff table corresponds to a strategy profile and contains a pair of payoff values based on that strategy profile.

Table 1. A payoff table of a two-player game

		c_2		
		t_1	t_2	......
c_1	s_1	$\langle u_1(s_1, t_1), u_2(s_1, t_1) \rangle$	$\langle u_1(s_1, t_2), u_2(s_1, t_2) \rangle$	......
	s_2	$\langle u_1(s_2, t_1), u_2(s_2, t_1) \rangle$	$\langle u_1(s_2, t_2), u_2(s_2, t_2) \rangle$	......
	s_3	$\langle u_1(s_3, t_1), u_2(s_3, t_1) \rangle$	$\langle u_1(s_3, t_2), u_2(s_3, t_2) \rangle$	......
	......	......	......	......

3.2   Repetition Learning Mechanism

The second stage formulates competitive games repeatedly with different initial thresholds and strategies to approach a balanced solution. In other words, if the values of thresholds obtained in the current game are not good enough to apply in decision making, we can repeat the game with the new thresholds. This stage first analyzes pure strategy equilibrium of the game and then check if the stop condition is satisfied.

Pure Strategy Equilibrium. The game solution of pure strategy Nash equilibrium is typically used to determine possible game outcomes in GTRS. In a two-player game, the strategy profile (s_i, t_j) is a pure strategy Nash equilibrium, if for players c_1 and c_2, s_i and t_j are the best responses to each other. This is expressed as [6],

$$\forall s_i' \in S_1, \quad u_1(s_i, t_j) \geqslant u_1(s_i', t_j), \quad \text{where } s_i \in S_1 \text{ and } s_i' \neq s_i,$$
$$\forall t_j' \in S_2, \quad u_2(s_i, t_j) \geqslant u_2(s_i, t_j'), \quad \text{where } t_j \in S_2 \text{ and } t_j' \neq t_j. \quad (5)$$

Equation (5) may be interpreted as a strategy profile such that no player would like to change his strategy or they would loss benefit if deriving from this strategy profile, provided this player has the knowledge of other player's strategies.

After analyzing the equilibrium, we check if the stop condition is satisfied. If the stop condition is not satisfied, new thresholds values will be set as initial thresholds and another game will be formulated. For example, the initial thresholds are (α, β), equilibrium analysis shows that the result thresholds are (α', β') and the stop condition is not satified. In the subsequent iteration of the game, the initial thresholds will be set as (α', β').

3.3   Making Decision Based on Equilibrium

The last stage contains setting the stop conditions and making the final decision based on the equilibrium.

Stop Conditions. We need to stop the iterations at the proper time in order to obtain the balanced thresholds. This requires proper stop conditions be defined. There are many possible stop conditions, for example, thresholds (α, β) violate the constraint $0 \leq \beta \leq 0.5 \leq \alpha \leq 1$, the payoffs of players are beyond some specific values, a subsequent iteration does not improve previous configurations, or the gain of one player's payoff is less than the loss of the other player's payoff in the current game.

Decision Making. The final three-way classification can be defined by the initial thresholds used in the last game.

4   Illustrative Example

In this section, we present an example to demonstrate that a balanced three-way classification can be obtained by applying GTRS when two criteria are involved to evaluate this three-way classification. In the experiment, we use a random generator to generate the probabilistic information of experimental data. The probability $Pr(X_i)$ is a random number between 0.001 and 0.1, and the sum of all $Pr(X_i)$ is 1. The condition probability $Pr(C|X_i)$ is a random number between 0 and 1. Table 2 summarizes probabilistic data about a concept C. There are 20 equivalence classes denoted by $X_i (i = 1, 2, ..., 20)$, which are listed in a decreasing order of the conditional probabilities $Pr(C|X_i)$ for convenient computations.

Table 2. Summary of the experimental data

	X_1	X_2	X_3	X_4	X_5	X_6	X_7	X_8	X_9	X_{10}
$Pr(X_i)$	0.083	0.077	0.07	0.066	0.06	0.051	0.04	0.025	0.021	0.011
$Pr(C|X_i)$	1	0.98	0.93	0.89	0.81	0.77	0.71	0.66	0.64	0.58

	X_{11}	X_{12}	X_{13}	X_{14}	X_{15}	X_{16}	X_{17}	X_{18}	X_{19}	X_{20}
$Pr(X_i)$	0.013	0.019	0.028	0.041	0.049	0.055	0.061	0.069	0.074	0.087
$Pr(C|X_i)$	0.51	0.49	0.44	0.39	0.31	0.26	0.21	0.11	0.08	0

Two criteria accuracy and coverage are used to evaluate three-way classifications defined by different probabilistic thresholds. We can set these two criteria accuracy and coverage as game players, i.e., $O = \{acc, cov\}$. The strategy set is $S = \{S_{acc}, S_{cov}\}$. The possible strategies of two players are the changes of thresholds. The initial thresholds are $(\alpha, \beta) = (1, 0.5)$. The player acc tries to decrease β, so its strategy set is $S_{acc} = \{\beta$ doesn't change, β decreases 0.05, β decreases 0.1$\}$. Under these strategies, the corresponding β values are 0.5, 0.45, and 0.4, respectively. The player cov tries to decrease α, so its strategy set is $S_{cov} = \{\alpha$ doesn't change, α decreases 0.05, α decreases 0.1$\}$. The corresponding α values are 1, 0.95, and 0.9, respectively. The payoff functions of two players are defined by the criteria they are representing, i.e., the Eqs. (3) and (4).

Table 3. The payoff table

		cov		
		$\alpha = 1$	$\alpha = 0.95$	$\alpha = 0.9$
acc	$\beta = 0.5$	<0.8349, 0.569>	<0.8523, 0.643>	<0.8599, 0.713>
	$\beta = 0.45$	<0.8462, 0.547>	<0.8627, 0.624>	<0.8695, 0.694>
	$\beta = 0.4$	<0.8617, 0.519>	<0.8769, 0.596>	**<0.8825,0.666>**

The payoff table is shown in Table 3. The strategy profile (β decreases 0.1, α decreases 0.1) is the equilibrium. We set the stop condition as the gain of one player's payoff is less than the loss of the other player's payoff in the current game. When the thresholds change from $(1, 0.5)$ to $(0.9, 0.4)$, the accuracy increases from 0.8349 to 0.8825 and the coverage increases from 0.566 to 0.666. We repeat the game by setting $(\alpha, \beta) = (0.9, 0.4)$ as initial thresholds. In the second iteration of the game, the initial thresholds is $(\alpha, \beta) = (0.9, 0.4)$, two players' strategy sets are $S_{acc} = \{\beta$ doesn't change $, \beta$ decreases $0.05, \beta$ decreases $0.1\}$ and $S_{cov} = \{\alpha$ doesn't change $, \alpha$ decreases $0.05, \alpha$ decreases $0.1\}$, respectively. The competition will be repeated three times. The result is shown in Table 4. In the third iteration, we can see that the gain of the payoff values of player acc is 0.0021 which is less than the loss of payoff values of player cov 0.025. The repetition of game is stopped and the final result is the initial thresholds of the third competitive game $(\alpha, \beta) = (0.8, 0.3)$.

Table 4. The repetition of games

	Initial (α, β)	Result (α, β)	Payoffs	
1	(1, 0.5)	(0.9, 0.4)	<0.8825, 0.666>	+0.1476
2	(0.9, 0.4)	(0.8, 0.3)	<0.9064, 0.702>	+0.0559
3	(0.8, 0.3)	(0.7, 0.2)	<0.9085, 0.677>	−0.0229

5   Conclusion

We use game-theoretic rough sets to determine multi-criteria based three-way classifications. When multiple criteria are involved to determine three-way classifications, we formulate this problem as a multi-criteria decision making problem. Within the GTRS framework, multiple criteria that are used to evaluate three-way classifications are set as game players. The competitive games are formulated to solve the conflict among criteria. The strategies of both players are the changes of thresholds. Both players can gradually approach the balanced probabilistic thresholds by repeatedly modifying the initial thresholds and finding the pure strategy equilibrium of repeated games. GTRS provide a feasible and effective method for multi-criteria based three-way classifications in the context of rough sets. GTRS accommodate and meet the rough sets related special requirements when formulating games between criteria. GTRS do not rely on any predefined knowledge about criteria or compound objective functions. Moreover, GTRS are more suitable to solve a competitive situation where involved criteria have opposite interest and the payoff of each criterion is influenced by other criteria's strategies.

Acknowledgements. This work is partially supported by a Discovery Grant from NSERC Canada, the University of Regina Verna Martin Memorial Scholarship, and a Mitacs Accelerate program.

References

1. Azam, N., Zhang, Y., Yao, J.T.: Evaluation functions and decision conditions of three-way decisions with game-theoretic rough sets. Eur. J. Oper. Res. **261**(2), 704–714 (2017)
2. Deng, X.F., Yao, Y.Y.: A multifaceted analysis of probabilistic three-way decisions. Fundam. Inf. **132**(3), 291–313 (2014)
3. Deng, X.Y., Zheng, X., Su, X.Y., Chan, F.T., Hu, Y., Sadiq, R., Deng, Y.: An evidential game theory framework in multi-criteria decision making process. Appl. Math. Comput. **244**, 783–793 (2014)
4. Greco, S., Inuiguchi, M., Slowiński, R.: Dominance-based rough set approach. Multi-Objective Program. Goal Program.: Theory Appl. **21**, 129 (2013)
5. Herbert, J.P., Yao, J.T.: Game-theoretic rough sets. Fundam. Inf. **108**(3–4), 267–286 (2011)
6. Leyton-Brown, K., Shoham, Y.: Essentials of Game Theory: A Concise Multidisciplinary Introduction. Morgan & Claypool Publishers, Williston (2008)
7. Madani, K., Lund, J.R.: A monte-carlo game theoretic approach for multi-criteria decision making under uncertainty. Adv. Water Resour. **34**(5), 607–616 (2011)
8. Pawlak, Z.: Rough Sets: Theoretical Aspects of Reasoning About Data. Kluwer Academatic Publishers, Boston (1991)
9. Saaty, T.L., Peniwati, K.: Group Decision Making: Drawing Out and Reconciling Differences. RWS Publications, Pittsburgh (2013)
10. Stewart, T.J.: A critical survey on the status of multiple criteria decision making theory and practice. Omega **20**(5–6), 569–586 (1992)
11. Triantaphyllou, E.: Multi-Criteria Decision Making Methods: A Comparative Study. Kluwer Academic Publishers, Boston (2000)
12. Yao, J.T., Herbert, J.P.: A game-theoretic perspective on rough set analysis. J. Chongqing Univ. Posts Telecommun. **20**(3), 291–298 (2008)
13. Yao, J.T., Azam, N.: Web-based medical decision support systems for three-way medical decision making with game-theoretic rough sets. IEEE Trans. Fuzzy Syst. **23**(1), 3–15 (2015)
14. Yao, Y.: An outline of a theory of three-way decisions. In: Yao, J.T., Yang, Y., Słowiński, R., Greco, S., Li, H., Mitra, S., Polkowski, L. (eds.) RSCTC 2012. LNCS, vol. 7413, pp. 1–17. Springer, Heidelberg (2012). doi:10.1007/978-3-642-32115-3_1
15. Yao, Y.: Rough sets and three-way decisions. In: Ciucci, D., Wang, G., Mitra, S., Wu, W.-Z. (eds.) RSKT 2015. LNCS, vol. 9436, pp. 62–73. Springer, Cham (2015). doi:10.1007/978-3-319-25754-9_6
16. Yao, Y.Y.: Probabilistic rough set approximations. Int. J. Approx. Reason. **49**(2), 255–271 (2008)
17. Yao, Y.Y.: The superiority of three-way decisions in probabilistic rough set models. Inf. Sci. **181**(6), 1080–1096 (2011)
18. Yu, P.L.: Multiple-criteria Decision Making: Concepts, Techniques, and Extensions, vol. 30. Springer Science & Business Media, Philadelphia (2013)
19. Zeleny, M., Cochrane, J.L.: Multiple Criteria Decision Making. University of South Carolina Press, Columbia (1973)
20. Zhang, Y., Yao, J.T.: Rule measures tradeoff using game-theoretic rough sets. In: Zanzotto, F.M., Tsumoto, S., Taatgen, N., Yao, Y. (eds.) BI 2012. LNCS, vol. 7670, pp. 348–359. Springer, Heidelberg (2012). doi:10.1007/978-3-642-35139-6_33
21. Zhang, Y., Yao, J.T.: Gini objective functions for three-way classifications. Int. J. Approx. Reason. **81**, 103–114 (2017)

Theoretical Aspects of Formal Concept Analysis

A Formal Context for Acyclic Join Dependencies

Jaume Baixeries[(✉)]

Departament de Ciències de la Computació,
Universitat Politècnica de Catalunya, 08024 Barcelona, Catalonia, Spain
jbaixer@cs.upc.edu

Abstract. Acyclic Join Dependencies (AJD) play a crucial role in database design and normalization. In this paper, we use Formal Concept Analysis (FCA) to characterize a set of AJDs that hold in a given dataset. This present work simplifies and generalizes the characterization of Multivalued Dependencies with FCA.

1 Introduction and Motivation

In database theory, a **dependency** expresses a relationship between sets of attributes in a dataset. There are numerous types of dependencies: conditional dependencies, sequential dependencies, order dependencies, to name just a few of them (see [13] for a more detailed survey). Dependencies may have different semantics, and may express relationships of different nature: equality, similarity, order, distance, etc.

However, not all of those dependencies have been equally popular. The most common dependencies in the relational database model are **functional dependencies** (FD's), which have been widely studied in the field of database theory [16]. The reason of this success may be twofold: on the one hand, their semantics is very simple and intuitive, on the other hand, they have been proven to be very versatile, since they can be used for database design, database validation, and, also, data cleaning [9]. They play a key role to explain the normalization of a database scheme in the relational database model.

Another type of dependencies that have been relevant in database theory are **multivalued dependencies** (MVDs) [16]. These dependencies are a generalization of functional dependencies, and their semantics is capable of expressing how a table can be split into two different tables such that their join is exactly the original table. This procedure is of key importance for database normalization and design.

Acyclic Join Dependencies (AJDs) [17] are a generalization of multivalued dependencies. AJDs are of critical importance in the decomposition method [10], which is a method for designing database schemes. An AJD specifies a lossless decomposition of a dataset, which is the decomposition into different (smaller) datasets such that their composition restores the original dataset. The interest of this decomposition is that it allows the dataset to be in the so-called 4th-normal form (4NF), which prevents redundancy and update errors. We discuss AJDs in more detail in Sect. 2.1.

M. Kryszkiewicz et al. (Eds.): ISMIS 2017, LNAI 10352, pp. 563–572, 2017.
DOI: 10.1007/978-3-319-60438-1_55

Formal Concept Analysis (FCA) is a simple and elegant lattice-oriented mathematical framework that is strongly connected to lattice theory. The uses of FCA are manyfold, as, for instance, knowledge discovery and machine learning [14], among many others. We discuss FCA in more detail in Sect. 2.2.

Formal Concept Analysis has been widely used to characterize and compute different types of dependencies. We present the most relevant work on this subject in Sect. 2.3.

In this paper, we deal with the characterization of a set of AJDs that hold in a given dataset, using the formalism of FCA. The goals of this paper are twofold: on the one hand, extend previous work that dealt with the characterization of functional and multivalued dependencies. On the other hand, propose FCA as a tool to compute sets of AJDs that hold in a dataset, compute the closure of a set of AJDs and explore the possibility of computing minimal bases for this kind of dependencies.

This paper starts with the Notation section, where we explain the basics of AJDs, FCA and also, previous work linking FCA and data dependencies. In the Results section, we present a new formal context for AJDs. We also present an example in a separate section to illustrate the results. Finally, we present the conclusions and future work.

2   Notation

The primary objects with which we deal in this paper are a set of **attributes** and a **dataset**. A dataset T is a set of tuples: $T = \{t_1, \ldots, t_N\}$ (we use indistinctively dataset and set of tuples as equivalent terms) and a set of attributes $\mathcal{U} = \{a, b, \ldots\}$ (commonly known as column names). Each tuple has a value associated to each attribute. We use non capital letters for single elements of the set of attributes, starting with $a, b, c, \ldots$, and capital letters $X, Y, Z, \ldots$ for subsets of $\mathcal{U}$. We drop the union operator and use juxtaposition to indicate set union. For instance, instead of $X \cup Y$ we write XY. If the context allows, we drop the set notation, and write abc instead of $\{a, b, c\}$.

id	a	b	c	d
t_1	1	1	1	1
t_2	1	2	1	1
t_3	1	1	2	2
t_4	1	2	2	2

Fig. 1. Example dataset.

For instance, Fig. 1 is an example of a dataset $T = \{t_1, t_2, t_3, t_4\}$, with its set of attributes $\mathcal{U} = \{a, b, c, d\}$. We use the notation $t(X)$ to indicate the **restriction** of a tuple t to the set of attributes X. In this same example, $t_2(\langle b, c \rangle) = \langle 2, 1 \rangle$. It is necessary to note that when the values of the tuple are given, some order must be implicit, because a value that appears in a tuple is always related to an attribute. We also use juxtaposition for the composition of tuples. For instance, $t(\langle a,b \rangle)t(\langle c,d \rangle)$ is the tuple $t(\langle a,b,c,d \rangle)$. We also have that $t(\langle a,b \rangle)t(\langle c,d \rangle)$ is equivalent to $t(\langle c,d \rangle)t(\langle a,b \rangle)$, assuming that we have a total order on the attributes set. We use the notation $\varPi_X(S)$, where $X \subseteq \mathcal{U}$ and $S \subseteq T$ as the set $\varPi_X(S) = \{\, t(X) \mid t \in S \,\}$, this is, the set of restrictions of all tuples in S to the set of attributes X. In this same example, $\varPi_{\langle a,d \rangle}(T) = \{\, \langle 1,1 \rangle, \langle 1,2 \rangle \,\}$.

Given a set S, we define also the set of its splits and pairs:

Definition 1. *The set* Split(S) *is the set of partitions of S of size 2. For instance, if $S = \{\, a,b,c,d \,\}$, then,* Split($\mathcal{U}$) $= \{\, [a \mid bcd], [b \mid acd], [c \mid abd], [d \mid abc], [ab \mid bc], [ac \mid bd], [ad \mid cd] \,\}$.

Definition 2. *The set* Pair(S) *is the set of pairs of elements of S, modulo reflexivity and commutativity. For instance, if $S = \{\, a,b,c,d \,\}$, then,* Pair(S) $= \{\, (a,b), (a,c), (a,d), (b,c), (b,d), (c,d) \,\}$.

Finally, we define the **join** operator $\bowtie$ on the datasets R and S as follows: $R \bowtie S = \{\, r \cup s \mid r \in R \,\wedge\, s \in S \,\wedge\, r(X) = s(X) \,\}$, where X is the set of attributes that are common to both R and S. If they have no common attributes, then, this operation becomes a cartesian product.

2.1 Acyclic Join Dependencies

An acyclic join dependency is a special case of a join dependency. We provide first the definition of a join dependency, and then, we describe the restriction that applies to acyclic join dependencies.

Definition 3. *Let T be a set of tuples and let $\mathcal{U}$ be its attribute set. A **join dependency** $R = [R_1, \ldots, R_N]$ is a set of sets of attributes such that:*

1. $R_i \subseteq \mathcal{U}, \forall i : 1 \le i \le N$.
2. $\mathcal{U} = \bigcup\limits_{1 \le i \le N} R_i$, this is, all attributes are present in R.

A join dependency R holds in T if and only if:

$$T = \varPi_{R_1}(T) \bowtie \ldots \bowtie \varPi_{R_N}(T)$$

The intuition behind a join dependency is that the set of tuples T can be decomposed into different smaller (with less attributes and, maybe, tuples) sets of tuples, such that their composition according to R is lossless, this is, no information is lost [7]. **Acyclic join dependencies** are join dependencies that hold in a set of tuples according to the condition in Definition 3. However, they have some syntactical restrictions that make them more tractable than join dependencies, both in terms of axiomatization and computational complexity [12]. A deep discussion on the differences between both join and acyclic join dependencies is far beyond the scope of this paper, and we will provide only the basic definition and properties of AJDs that are relevant to this paper.

The notion of a join dependency is closely related to that of a **hypergraph** [8]. Let S be a set of vertices, a hypergraph extends the notion of a graph in the sense that an edge in a hypergraph is not limited to two vertices, as in a graph, but to any number of vertices. In a join dependency $R = [R_1, \ldots, R_N]$, the set of vertices would be $\mathcal{U}$, and R would be the set of edges. Also in hypergraphs, there is the notion of **acyclicity**, but this notion is not as intuitive as in the case of graphs. In fact, there are different definitions of acyclicity, some more restrictive than others (in [6] some are discussed). In this case, we use the definition of acyclicity that appears in [6]. But, again, this definition can be enunciated in as many as 12 different equivalent ways, but we just use one of them which is necessary to understand the results in this paper. With this definition, we proceed to define acyclic join dependencies.

Definition 4 [6]. *Let $R = [R_1, \ldots, R_N]$ be a join dependency. A **join tree** JT for R is a tree such that the set of nodes is the same as R and:*

1. *Each edge (R_i, R_j) is labeled with $R_i \cap R_j$.*
2. *For any pair $R_i, R_j \in R$, where $i \neq j$, we take the only path $P = \{R_i, R_{i+1} \ldots R_j\}$ between R_i and R_j. For all edges (R_{i+k}, R_{i+k+1}) we have that: $R_i \cap R_j \subseteq R_{i+k} \cap R_{i+k+1}$).*

*A join dependency R is an **acyclic join dependency** (AJD) if and only if it has a join tree. If an acyclic join dependency has only two components, then, it is called a **multivalued dependency**.*

Example 1. For instance, if we have that $\mathcal{U} = \{a, b, c, d, e, f, g\}$ and the AJD $R = [abc, ace, abfg, abd]$, a join tree for R is in Fig. 2. We can see, for instance, that in the path from node ace to node abd, the intersection $ace \cap abd = a$ appears in all the edges of the path.

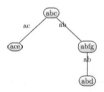

Fig. 2. Join tree for R.

Therefore, an acyclic join dependency is a join dependency that has a join tree or, equivalently, is an acyclic hypergraph. For a given AJD, there can be more than one possible join trees, since, for instance, the choice of a root is completely arbitrary (although switching the root is not the only possible way to have different join trees).

One important and useful property of AJDs is that a single AJD is equivalent to a set of multivalued dependencies. We remind the reader that a MVD is a special case of AJD with cardinality 2. Definition 5 is a technicality that helps to understand how to compute the set of MVDs that are equivalent to an AJD in Proposition 2.

Definition 5. *Let* $JT = \langle R, E \rangle$ *be the join tree of an AJD R. We define the function* Removal *that returns a partition of the set of attributes* $\mathcal{U}$ *into three classes according to an edge of JT.*
Let $C_1 = \langle V_1, E_1 \rangle, C_2 = \langle V_2, E_2 \rangle$ *be the two connected components that appear after the removal of an edge* $(R_i, R_j) \in E$ *in JT. Let* $C_i = \langle V_i, E_i \rangle$, *then* attrib$(C_i)$ *returns the set of attributes that appear in that connected component, this is:* attrib$(C_i) = \bigcup_{X \in V_i} X$. *The function* Removal$(\langle R, E \rangle, (R_i, R_j))$ *returns the triple:*

$$\text{Removal}(\langle R, E \rangle, (R_i, R_j)) := \langle R_i \cap R_j, \quad \text{attrib}(C_1) \setminus (R_i \cap R_j), \quad \text{attrib}(C_2) \setminus (R_i \cap R_j) \rangle$$

Proposition 1. *Let R be an AJD and let* $JT = (\mathcal{U}, R)$ *be its join tree. The function* Removal *returns a partition of the set* $\mathcal{U}$.

Proof. Since R is an AJD, then, R contains all the attributes in $\mathcal{U}$. Lets assume that we are removing the edge (R_i, R_j) After splitting JT we have two connected components: $C_1 = \langle V_1, E_1 \rangle, C_2 = \langle V_2, E_2 \rangle$. Therefore, all the attributes are either in V_1 or in V_2. This implies that all attributes will appear in one of the classes returned by Removal, since either this attribute will be in V_1, or in V_2 or in $R_i \cap R_j$. Now we need to prove that an attribute will not appear in more than one class. Because $R_i \cap R_j$ is substracted in V_1 and V_2, if there is a repeated attribute, it will only be in V_1 and V_2. Let us suppose, by way of contradiction, that there is an attribute that is in attrib$(C_1) \setminus (R_i \cap R_j)$ and attrib$(C_2) \setminus (R_i \cap R_j)$, this is, in $R_k \in E_1$ and $R_j \in E_2$. Since JT is a join tree, it means that this attribute will appear in all the edges in the path from R_k to R_j, also in $R_i \cap R_j$ because the edge (R_i, R_j) is necessarily in that path, which yields a contradiction.

Example 2. We return to Example 1. We have the AJD $R = [ace, abc, abfg, abd]$ and the corresponding join tree $JT = \langle AJD, (ace, abc), (abc, abfg), (abfg, abd) \rangle$. We have that:

$$\text{Removal}(JT, (abc, abfg)) = \langle ab, ce, dfg \rangle$$

$$\text{Removal}(JT, (ace, abc)) = \langle ac, e, bdfg \rangle$$

We now define the set of MVDs that are equivalent to a single AJD.

Proposition 2. *Let R be an AJD, and let* $JT = \langle R, E \rangle$ *be its join tree. R holds in a table T if and only if* $\forall (R_i, R_j) \in E : [XY, XZ]$ *holds in T, where* $\langle X, Y, Z \rangle := \text{Removal}(JT, (R_i, R_j))$. *The proof is in Theorem 7.1 in [6].*

2.2 Formal Concept Analysis

In this brief account of Formal Concept Analysis (FCA), we use standard definitions from [11]. Let G and M be arbitrary sets and $I \subseteq G \times M$ be a binary relation between G and M. The triple (G, M, I) is called a formal context. Each $g \in G$ is interpreted as an object, each $m \in M$ is interpreted as an attribute. The statement $(g, m) \in I$ is interpreted as "g has attribute m". The two following derivation operators $(\cdot)'$:

$$A' = \{m \in M \mid \forall g \in A : gIm\} \qquad for\ A \subseteq G,$$
$$B' = \{g \in G \mid \forall m \in B : gIm\} \qquad for\ B \subseteq M$$

define a Galois connection between the powersets of G and M. The derivation operators $\{(\cdot)', (\cdot)'\}$ put in relation elements of the lattices $(\wp(G), \subseteq)$ of objects and $(\wp(M), \subseteq)$ of attributes and reciprocally.

2.3 Previous Work

The characterization of dependencies (in a generic sense) and FCA can be roughly divided into two different parts: a syntactical characterization and a semantical characterization. In a **syntactical** characterization, a formal context characterizes the closure of a set of dependencies. This is, given a set of dependencies, what is the maximal set of dependencies that can be derived according to the axioms for these dependencies?

The **semantical** characterization of a set of dependencies takes into account the definition of these dependencies with respect to a dataset. In this line of work, a formal context is constructed, such that the set of objects is related to a dataset, and the set of attributes is related to the set of attributes of the dataset, which is a parameter of the formal context. In this paper, we continue this latter like of work.

The definition of a formal context to characterize the functional dependencies that hold in a dataset can be found in [1,15,18]. These dependencies are also characterized with pattern structures (a generalization of FCA) in [5]. Degenerate multivalued dependencies are characterized in [2], multivalued dependencies in [3], and similarity dependencies are characterized in terms of pattern structures in [4].

In all previous cases, the characterization in terms of FCA (or pattern structures) allows the formal context to answer the question: does a specific dependency hold in that dataset? It can also compute the whole set of dependencies that hold in that dataset. As we have mentioned, previously, the defined context takes into account the set of tuples and the set of attributes of a given dataset. As an example, we describe the formal context that was defined in [3] to characterize the set of multivalued dependencies that hold in a dataset T. This formal context was defined as $\mathbb{K}_T = (\text{Part}(T), \text{Part}(\mathcal{U}), I)$ where $\text{Part}(T)$ is the set of all partitions that can be formed with the set of tuples T, and $\text{Part}(\mathcal{U})$ is the set of all partitions of the set of attributes $\mathcal{U}$. We mention this example because multivalued dependencies are a special case of acyclic join dependencies, which are treated in this paper. In the next section, we define a new formal context for acyclic join dependencies that will be simpler (in size) than this previous context, and that will characterize acyclic join dependencies, as well as multivalued dependencies, as a special case.

3   Results

In this section we present the main result in this paper. Given a dataset, we define a formal context that can be used to check if an AJD holds in that dataset. Eventually, this formal context could also compute the set of all AJDs that hold in that dataset, although this is not discussed in this paper.

Before defining the formal context, we need to define the binary relation it will contain.

Definition 6. *Let T be a set of tuples, and $\mathcal{U}$ its set of attributes. Let $S = [X \mid Y] \in \mathrm{Split}(\mathcal{U})$ and let $(t_i, t_j) \in \mathrm{Pair}(T)$. We define the relation $I \subseteq \mathrm{Split}(\mathcal{U}) \times \mathrm{Pair}(T)$ as follows: $[X \mid Y]$ is related to the pair of tuples (t_i, t_j) if and only if the tuples $t_i(X)t_j(Y)$ and $t_j(X)t_i(Y)$ are in T. More formally:*

$$[X \mid Y] \ \ I \ \ (t_i, t_J) \Leftrightarrow t_i(X)t_j(Y), t_j(X)t_i(Y) \in T$$

We are now ready to define the following formal context:

Definition 7. *Let T be a set of tuples, and $\mathcal{U}$ its set of attributes. The AJD-formal context for the set of tuples T is:*

$$\mathbb{K}_T = (\mathrm{Split}(\mathcal{U}), \mathrm{Pair}(T), I)$$

We now use this formal context in order to check if an AJD holds in T. We have seen that, according to Proposition 2, in order to check if an AJD holds in a dataset, we need to check if the equivalent set of MVDs hold as well. This is the base for our next theorem.

Theorem 1. *Let T be a dataset, and let $R = \{R_1, \ldots, R_N\}$ be an acyclic join dependency, and let $JT = \langle R, E \rangle$ the corresponding join tree of R. R holds in T if and only if:*

$$\bigwedge_{e \in E} \{ [X \mid YZ] \}' = \{ [Y \mid XZ], [Z \mid XY] \}'$$

holds in the formal context $\mathbb{K}_T = (\mathrm{Split}(\mathcal{U}), \mathrm{Pair}(T), I)$, where $\langle X, Y, Z \rangle = \mathrm{Removal}(JT, e)$.

Proof. The base of this proof is to check if this condition holds for all the MVDs that are equivalent to an AJD according to Proposition 2.

($\Rightarrow$) We assume that R holds in T, we need to prove that

$$\bigwedge_{e \in E} \{ [X \mid YZ] \}' = \{ [Y \mid XZ], [Z \mid XY] \}'$$

where $\langle X, Y, Z \rangle = \text{Removal}(JT, e)$. We take an arbitrary $e \in E$ and a pair of tuples $t_i, t_j \in T$. We need to prove now two things:

1. If $(t_i, t_j) \in \{ [X \mid YZ] \}'$, then, we have that $(t_i, t_j) \in \{ [Y \mid XZ], [Z \mid XY] \}'$. Since $(t_i, t_j) \in \{ [X \mid YZ] \}'$ tuples $t_i(X)t_j(YZ)$ and $t_j(X)t_i(YZ)$ are in T. Therefore, we have that T contains, at least:

$$
\begin{array}{ll}
t_i(X)t_i(Y)t_i(Z) & t_j(X)t_j(Y)t_j(Z) \\
t_i(X)t_j(Y)t_j(Z) & t_j(X)t_i(Y)t_i(Z)
\end{array}
$$

Now, if $(t_i, t_j) \in \{ [Y \mid XZ] \}'$, we need the following tuples to be in T: $t_i(X)t_j(Y)t_i(Z)$ and $t_j(X)t_i(Y)t_j(Z)$ and for $(t_i, t_j) \in \{ [Z \mid XY] \}'$, we need the following tuples: $t_i(X)t_i(Y)t_j(Z)$ and $t_j(X)t_j(Y)t_i(Z)$. By Proposition 2, we have that $[XY, XZ]$ holds in T. This implies that the following tuples are also in T:

$$
\begin{array}{ll}
t_i(X)t_i(Y)t_j(Z) & t_i(X)t_j(Y)t_i(Z) \\
t_j(X)t_j(Y)t_i(Z) & t_j(X)t_i(Y)t_j(Z)
\end{array}
$$

2. We need to prove the inverse condition: if $(t_i, t_j) \in \{ [Y \mid XZ], [Z \mid XY] \}'$, then, we have that $(t_i, t_j) \in \{ [X \mid YZ] \}'$. Since $(t_i, t_j) \in \{ [Y \mid XZ], [Z \mid XY] \}'$, we have, at least, the following tuples in T:

$$
\begin{array}{ll}
t_i(X)t_i(Y)t_i(Z) & t_j(X)t_j(Y)t_j(Z) \\
t_i(X)t_j(Y)t_i(Z) & t_j(X)t_i(Y)t_j(Z) \\
t_i(X)t_i(Y)t_j(Z) & t_j(X)t_j(Y)t_i(Z)
\end{array}
$$

If we want $(t_i, t_j) \in \{ [X \mid YZ] \}'$ we need the following tuples to be in T as well: $t_i(X)t_j(Y)t_j(Z)$ and $t_j(X)t_i(Y)t_i(Z)$. Since $[XY, XZ]$ holds in T, we also have the following tuples in T: $t_i(X)t_j(Y)t_j(Z)$ and $t_j(X)t_i(Y)t_i(Z)$.

($\Leftarrow$) We now assume that $\bigwedge_{e \in E} \{ [X \mid YZ] \}' = \{ [Y \mid XZ], [Z \mid XY] \}'$ where $\langle X, Y, Z \rangle = \text{Removal}(JT, e)$ holds, we need to prove that R holds in T. For that, we use Proposition 2, and, therefore, we prove that all the MVDs that are equivalent to R hold in T.

We prove that an arbitrary MVD $[XY, XZ]$, where $\langle X, Y, Z \rangle = \text{Removal}$ (JT, e), holds if $\{ [X \mid YZ] \}' = \{ [Y \mid XZ], [Z \mid XY] \}'$. We take a pair of tuples (t_i, t_j) such that $t_i(X) = t_j(X)$. In order to prove that $[XY, XZ]$ holds, we need to prove that the following tuples are in T as well: $t_i(X)t_i(Y)t_j(Z)$ and $t_i(X)t_j(Y)t_i(Z)$. Clearly, $(t_i, t_j) \in \{ [X \mid YZ] \}'$, and by the hypothesis, we have that $(t_i, t_j) \in \{ [Y \mid XZ], [Z \mid XY] \}'$. This means that the following tuples are in T: $t_i(X)t_i(Y)t_j(Z)$ and $t_i(X)t_j(Y)t_i(Z)$.

□

4   Example

We provide a running example in order to illustrate and clarify the results that are contained in the previous section. From the dataset in Example 1, we define the formal context $\mathbb{K}_T = (\mathrm{Split}(\mathcal{U}), \mathrm{Pair}(T), I)$ in Table 1.

Table 1. Formal context $\mathbb{K}_T = (\mathrm{Split}(\mathcal{U}), \mathrm{Pair}(T), I)$

$\mathbb{K}$	(t_1,t_2)	(t_1,t_3)	(t_1,t_4)	(t_2,t_3)	(t_2,t_4)	(t_3,t_4)
$[a \mid bcd]$	$\times$	$\times$	$\times$	$\times$	$\times$	$\times$
$[b \mid acd]$	$\times$	$\times$	$\times$	$\times$	$\times$	$\times$
$[c \mid abd]$	$\times$					$\times$
$[d \mid abc]$	$\times$					$\times$
$[ab \mid cd]$	$\times$	$\times$	$\times$	$\times$	$\times$	$\times$
$[ac \mid bd]$	$\times$					$\times$
$[ad \mid bc]$	$\times$					$\times$

The AJD $R = [\,ab, bc, cd\,]$ holds in T because $T = \Pi_{ab}(T) \bowtie \Pi_{bc}(T) \bowtie \Pi_{cd}(T)$. We check that in our context we also have this result. Let $E = \{\,(ab, bc), (bc, cd)\,\}$ and $JT = \langle R, E \rangle$ be the join tree of R. We verify that $\bigwedge_{e \in E} \{\,[X \mid YZ]\,\}' = \{\,[Y \mid XZ], [Z \mid XY]\,\}'$, where $\langle X, Y, Z \rangle = \mathrm{Removal}(JT, e)$. This is:

$$\{\,[b \mid acd]\,\}' = \{\,[a \mid bcd], [ab \mid cd]\,\}' \wedge \{\,[c \mid abd]\,\}' = \{\,[ab \mid cd], [abc \mid d]\,\}' \Leftrightarrow$$
$$\{\,(t_1,t_2), (t_1,t_3), (t_1,t_4), (t_2,t_3), (t_2,t_4), (t_3,t_4)\,\} =$$
$$\{\,(t_1,t_2), (t_1,t_3), (t_1,t_4), (t_2,t_3), (t_2,t_4), (t_3,t_4)\,\} \cap \{\,(t_1,t_2), (t_1,t_3), (t_1,t_4), (t_2,t_3), (t_2,t_4), (t_3,t_4)\,\}$$
$$\wedge\{\,(t_1,t_2), (t_3,t_4)\,\} = \{\,(t_1,t_2), (t_1,t_3), (t_1,t_4), (t_2,t_3), (t_2,t_4), (t_3,t_4)\,\} \cap \{\,(t_1,t_2), (t_3,t_4)\,\}$$

which is true.

5   Conclusions and Future Work

We have presented a new formal context for acyclic join dependencies. This context generalizes a previous approach for multivalued dependencies simply because these dependencies are a special case, and it simplifies it because the formal context has a smaller size. Acyclic join dependencies are of capital importance in database design and validation, among many others.

This result is just a first step towards (1) a more complete characterization of acyclic join dependencies or join dependencies within the formal concept analysis framework, (2) the application of FCA algorithms to compute the set of AJDs that hold in a dataset, (3) the computation of minimal bases for AJDs, and (4) using other FCA-related formalisms for the same purpose, as, for instance, pattern structures.

Acknowledgments. This research work has been supported by the SGR2014-890 (MACDA) project of the Generalitat de Catalunya, and MINECO project APCOM (TIN2014-57226-P).

References

1. Baixeries, J.: A formal concept analysis framework to model functional dependencies. In: Mathematical Methods for Learning (2004)
2. Baixeries, J., Balcázar, J.L.: Characterization and armstrong relations for degenerate multivalued dependencies using formal concept analysis. In: Ganter, B., Godin, R. (eds.) ICFCA 2005. LNCS (LNAI), vol. 3403, pp. 162–175. Springer, Heidelberg (2005). doi:10.1007/978-3-540-32262-7_11
3. Baixeries, J., Balcázar, J.L.: A lattice representation of relations, multivalued dependencies and armstrong relations. In: ICCS, pp. 13–26 (2005)
4. Baixeries,J., Kaytoue, M., Napoli, A.: Computing similarity dependencies with pattern structures. In: Ojeda-Aciego, M., Outrata, J. (eds.) CLA. CEUR Workshop Proceedings, vol. 1062, pp. 33–44 (2013). http://ceur-ws.org/
5. Baixeries, J., Kaytoue, M., Napoli, A.: Characterizing functional dependencies in formal concept analysis with pattern structures. Ann. Math. Artif. Intell. **72**(1–2), 129–149 (2014)
6. Beeri, C., Fagin, R., Maier, D., Yannakakis, M.: On the desirability of acyclic database schemes. J. ACM **30**(3), 479–513 (1983)
7. Beeri, C., Vardi, M.Y.: Formal systems for join dependencies. Theoret. Comput. Sci. **38**, 99–116 (1985)
8. Berge, C.: Hypergraphs. North-Holland Mathematical Library, vol. 45. North-Holland, Amsterdam (1989)
9. Bohannon, P., Fan, W., Geerts, F., Jia, X., Kementsietsidis, A.: Conditional functional dependencies for data cleaning. In: ICDE, pp. 746–755 (2007)
10. Codd, E.F.: Further normalization of the data base relational model. IBM Research Report, San Jose, California, RJ909 (1971)
11. Ganter, B., Wille, R.: Formal Concept Analysis. Springer, Berlin (1999)
12. Gyssens, M.: On the complexity of join dependencies. ACM Trans. Database Syst. **11**(1), 81–108 (1986)
13. Kanellakis, P.C.: Elements of relational database theory. In: van Leeuwen, J. (ed.) Handbook of Theoretical Computer Science (Vol. B), pp. 1073–1156. MIT Press, Cambridge (1990)
14. Kuznetsov, S.O.: Machine learning on the basis of formal concept analysis. Autom. Remote Control **62**(10), 1543–1564 (2001)
15. Lopes, S., Petit, J.-M., Lakhal, L.: Functional and approximate dependency mining: database and fca points of view. J. Exp. Theor. Artif. Intell. **14**(2–3), 93–114 (2002)
16. Maier, D.: The Theory of Relational Databases. Computer Science Press, Rockville (1983)
17. Malvestuto, F.: A complete axiomatization of full acyclic join dependencies. Inf. Process. Lett. **68**(3), 133–139 (1998)
18. Medina, R., Nourine, L.: Conditional functional dependencies: an FCA point of view. In: Kwuida, L., Sertkaya, B. (eds.) ICFCA 2010. LNCS (LNAI), vol. 5986, pp. 161–176. Springer, Heidelberg (2010). doi:10.1007/978-3-642-11928-6_12

On Containment of Triclusters Collections Generated by Quantified Box Operators

Dmitrii Egurnov[1]([⊠]), Dmitry I. Ignatov[1], and Engelbert Mephu Nguifo[2]

[1] National Research University Higher School of Economics, Moscow, Russia
egurnovd@yandex.ru, dignatov@hse.ru
[2] Université Clermont Auvergne, CNRS, LIMOS, Clermont-Ferrand, France
engelbert.mephu_nguifo@uca.fr,mephu@isima.fr

Abstract. Analysis of polyadic data (for example n-ary relations) becomes a popular task nowadays. While several data mining techniques exist for dyadic contexts, their extensions to triadic case are not obvious. In this work, we study development of ideas of Formal Concept Analysis for processing three-dimensional data, namely OAC-triclustering (from Object, Attribute, Condition). We consider several similar methods, study relations between their outputs and organize them in an ordered structure.

Keywords: Three-way data mining · Triadic Formal Concept Analysis · Triclustering · OAC-triclustering

1 Introduction

Triadic Formal Concept Analysis (3-FCA) was introduced by Lehman and Wille [1] and is aimed at analysis of object-attribute-condition relational data. However, in some cases its strict requirements may be relaxed, that is we could search for less dense structures called triclusters instead of triconcepts [2,3]. Several authors proposed n-ary extensions of FCA for pattern mining with closed n-sets including exact [4,5] and approximate ones [6,7].

The goal of this work is to prove important properties of the previously introduced OAC-triclustering [8].

The paper shortly recalls basic notions of Triadic Formal Concept Analysis in Sect. 2. Then general approach to triclustering is discussed in Sect. 3. Section 4 is dedicated to OAC-triclustering and its methods. Finally, in Sect. 5 we investigate relations between OAC-triclustering operators.

We consider formulas, definitions and statements only for object dimension due to the space limits. Missing formulations can be developed in a similar way due to intrinsic symmetry between the dimensions.

2 Triadic Formal Concept Analysis

In 3-FCA we deal with triadic formal contexts. They are very similar to traditional formal contexts, but include three dimensions, or modalities: objects, attributes and conditions [1].

© Springer International Publishing AG 2017
M. Kryszkiewicz et al. (Eds.): ISMIS 2017, LNAI 10352, pp. 573–579, 2017.
DOI: 10.1007/978-3-319-60438-1_56

Definition 1. *Let G, M and B be arbitrary sets. Subset of their Cartesian product defines a triadic relation $I \subseteq G \times M \times B$. The quadruple $\mathbb{K} = (G, M, B, I)$ is called a triadic formal context, or tricontext. The sets G, M and B are called set of objects, set of attributes, and set of conditions, respectively.*

For each triple $(g, m, b) \subseteq I$, where $g \in G$, $m \in M$, and $b \in B$, it is said that "object g has attribute m under condition b".

Like in Formal Concept Analysis we use concept-forming (prime or derivation) operators, but because of extra modalities there exists six variations of those operators:

$$\theta_G : 2^G \to 2^M \times 2^B, \qquad \theta_{G,M} : 2^G \times 2^M \to 2^B,$$
$$\theta_M : 2^M \to 2^G \times 2^B, \qquad \theta_{G,B} : 2^G \times 2^B \to 2^M,$$
$$\theta_B : 2^B \to 2^G \times 2^M, \qquad \theta_{M,B} : 2^M \times 2^B \to 2^G.$$

These variations are called triadic concept-forming (or prime) operators. They are assorted by the number of sets in their input into two groups: 1-set (triadic) prime operators and 2-set (triadic) prime operators.

For example, if $X \subseteq G$, $Y \subseteq M$ for a given tricontext $\mathbb{K} = (G, M, B, I)$, then $\theta_G(X) = (X)' = \{(m, b) \in M \times B \mid \forall g \in X : (g, m, b) \in I\}$ and $\theta_{G,M}(X, Y) = (X, Y)' = \{b \in B \mid \forall (g, m) \in X \times Y : (g, m, b) \in I\}$. Remaining operators are defined similarly. Further, we use the same prime-based notation for all these operators: $(\cdot)'$.

Therefore, triadic formal concept is defined in the following way:

Definition 2. *A triple of sets (X, Y, Z), where $X \subseteq G$, $Y \subseteq M$, $Z \subseteq B$, is called a triadic formal concept (or a triconcept) iff three conditions hold: $(X, Y)' = Z$, $(X, Z)' = Y$ and $(Y, Z)' = X$.*

The first and the second elements of the triple inherit their names from dyadic case, which are respectively extent and intent, while the third component is called modus (modi in plural). As well as in dyadic case, triadic formal concept can be interpreted as a maximal cuboid of positive values (or crosses) in Boolean matrix representation of the formal context, possibly under suitable permutations of elements of dimensions. In set notation, the statement is equivalent to maximality of $X \times Y \times Z \subseteq I$ w.r.t. set inclusion order over G, M, and B, respectively.

The set of all triadic formal concepts of a triadic formal context $\mathbb{K}$ can be organized into a structure called concept trilattice $\mathfrak{T}(\mathbb{K})$. However unlike in dyadic case, extents (intents and modi, respectively) do not form a closure system since two different triconcepts may have the same extent, but their intent and modus components may be incomparable in terms of set inclusion.

3   Triclustering

Methods of 3-FCA have found numerous successful applications in different fields, but sometimes they may be too strict to operate on real-world data.

Big datasets are prone to contain missing values, errors and noise, which may lead to losing some part of relevant information in the output. The clustering approach studied in this section is supposed to be more flexible and applicable to corrupted or sparse data.

A formal definition of tricluster is usually adjusted for each specific problem as it depends on the problem statement, data types, the chosen triclustering technique, and the desired output. In this section we give the most general definition.

Definition 3. *Let $A_{n \times k \times l}$ be a three-dimensional binary matrix. Let sets $G = \{g_1, g_2, \ldots, g_n\}$, $M = \{m_1, m_2, \ldots, m_k\}$, and $B = \{b_1, b_2, \ldots, b_l\}$ be index sets of A. Then for some arbitrary sets $X \subseteq G$, $Y \subseteq M$ and $Z \subseteq B$ submatrix $A_{XYZ} = \{a_{xyz} \mid x \in X, y \in Y, z \in Z\}$ is called a tricluster. Sets X, Y and Z are called respectively extent, intent and modus of the tricluster.*

Triadic formal concepts may be considered as a special case of triclusters, because they form a three-dimensional structure without zeros ("holes") in triadic formal context. Therefore, 3-FCA methods are part of triclustering methods. However, most datasets have little distortions, that may affect output of the algorithms, so it is considered useful to relax absolute density condition, allowing some number of empty cells in clusters.

Various constraints may be applied to triclusters. Usually they feature structure requirements, cardinality restrictions on extent, intent and modus, and limitations of other parameters. For example, the most common conditions, which eliminate small and meaningless structures from the output, are minimal support condition ($|X| \geq s_X, |Y| \geq s_Y, |Z| \geq s_Z$) and minimum density threshold:

$$\rho(A_{XYZ}) = \frac{\sum_{i=1}^{|X|} \sum_{j=1}^{|Y|} \sum_{k=1}^{|Z|} a_{x_i y_j z_k}}{|X||Y||Z|} \geq \rho_{min}.$$

4   OAC-triclustering

The main triclustering method studied within FCA framework previously is OAC-triclustering [6]. It is the result of development of the OA-biclustering algorithm [9] to triadic case. Two existing variations of this method are very similar, but one relies on box operators, while another uses prime operators [6]. Let us shortly describe them.

4.1   Box Operator Based OAC-triclustering

The box operator based variant of OAC-triclustering was chronologically the first. It was introduced in [3]. The method utilizes the following idea:

Let $\mathbb{K} = (G, M, B, I)$ be a triadic context. For triples $g \in G$, $m \in M$, $b \in B$ of the context, where $(g, m, b) \in I$, we generate triclusters by applying so-called box operators. Here we rely on prime operators, defined in Sect. 2. Further, we simplify set notation for both prime and box operators and write $a^{\square}$ and a'

instead of $\{a\}^{\square}$ and $\{a\}'$, but readers should keep in mind, that the operators can also take sets as input.

For a fixed triple $(\tilde{g}, \tilde{m}, \tilde{b}) \in I$ the box operators are defined in the following way:

$$\tilde{g}^{\square} := \left\{ g \mid \exists m : (g, m) \in \tilde{b}' \vee \exists b : (g, b) \in \tilde{m}' \right\}.$$

The output sets of box operators are called box sets.

Definition 4. *For a triple* $(\tilde{g}, \tilde{m}, \tilde{b}) \in I$ *a triple of sets* $T = (\tilde{g}^{\square}, \tilde{m}^{\square}, \tilde{b}^{\square})$ *is called a box operator based OAC-tricluster. The components of the triple are called extent, intent, and modus, respectively. The triple* $(\tilde{g}, \tilde{m}, \tilde{b})$ *is called the generating triple of the tricluster* T.

Figure 1 illustrates the addition condition for an element $g \in G$ to be included into $\tilde{g}^{\square}$. The element is added to a box set, if the gray zone contains at least one cross.

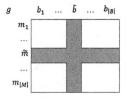

Fig. 1. Addition of g to $\tilde{g}^{\square}$

The method consists of iterating over all of the triples in the relation I of a triadic context $\mathbb{K}$, generation of box operator based triclusters for each of them, and storing the triclusters in a set $\mathcal{T}$. The tricluster T should be added to the set $\mathcal{T}$ only if it has not been discovered earlier, as some generators may result in the same triclusters.

4.2 Alternative Box Operator Definition

An alternative definition for box operators utilizes possible change of logical operators for obtaining different sets and, therefore, different triclusters. The general definition of box operators has the following form:

$$\tilde{g}^{\square} := \left\{ g \mid \exists\, m : (g, m) \in \tilde{b}' \,\lozenge\, \exists\, b : (g, b) \in \tilde{m}' \right\}.$$

Here the symbol $\lozenge$ signifies variability between logical operators $\wedge$ and $\vee$.

This yields two variations of box operators. They are distinguished by characters, which substitute optionality symbol, and are called respectively $\vee$-box and $\wedge$-box operators. Let us investigate their structure.

1. V-box variation is the simplest one, as it is identical to the ordinary box operators. An element is added to the box set, if the slice corresponding to the element contains at least one cross in the gray area, as shown in Fig. 2a. The aforementioned general definition of the box operators contains 1-set prime operators, which makes it difficult to use in theoretical work. For this reason, we introduce a detailed definition for V-box operators:

$$\tilde{g}_{\lor}^{\square} := \left\{ g \mid \exists\, m : (g, m, \tilde{b}) \in I \;\lor\; \exists\, b : (g, \tilde{m}, b) \in I \right\}.$$

2. ∧-box operators impose stricter conditions for an element to be included in a box set. The corresponding slice needs to have at least one cross in both gray row and gray column (see Fig. 2b). This way of forming triclusters is supposed to give higher density. The detailed definition is as follows:

$$\tilde{g}_{\land}^{\square} := \left\{ g \mid \exists\, m : (g, m, \tilde{b}) \in I \;\land\; \exists\, b : (g, \tilde{m}, b) \in I \right\}.$$

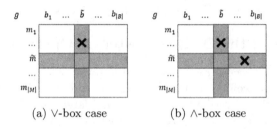

(a) V-box case   (b) ∧-box case

Fig. 2. ◇-box addition conditions

4.3   Prime Operator Based OAC-triclustering

The prime-based algorithm [8] uses a simpler way of generating triclusters. It corresponds to dyadic method of mining OA-biclusters described in [9].

Like in the previous method, triclusters are generated by applying 2-set prime operators, as defined in Sect. 2, to pairwise combinations of components of each triple in context. The motivation for development of this method was to improve quality of resulting triclusters in terms of density and simplicity of their structure. Outputs of these operators are called prime sets.

Definition 5. *For a triple* $(\tilde{g}, \tilde{m}, \tilde{b}) \in I$ *a triple of sets* $T = ((\tilde{m}, \tilde{b})', (\tilde{g}, \tilde{b})', (\tilde{g}, \tilde{m})')$ *is called a prime operator based OAC-tricluster. The components of the triple are called extent, intent, and modus, respectively. The triple* $(\tilde{g}, \tilde{m}, \tilde{b})$ *is called the generating triple of tricluster* T.

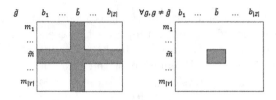

Fig. 3. Prime operator based tricluster structure

Again, these triclusters have a star-like structure in the 3-dimensional table representation of the formal context (under appropriate permutation of rows, columns and layers).

Figure 3 shows the structure of a prime operator based tricluster $T = (X, Y, Z)$, generated from a triple $(\tilde{g}, \tilde{m}, \tilde{b}) \in I$. The gray zone contains crosses. The content of the white zone is unknown and define density (quality) of the tricluster.

The algorithm is identical to the box operator based one: it iterates over all the triples of the context $\mathbb{K}$, generates triclusters T-s and adds them to the resulting set $\mathcal{T}$ if they were not found on the previous steps.

5 Relations of the OAC-triclustering Operators

In this section, we discuss relations between variations of box operators and prime operators, as well as triclusters that they produce. We propose several ordering lemmas that describe the relations, and later are used to organize the operators into a linear order with respect to set inclusion.

Lemma 1. *A box set generated by the* ∨*-box operator contains as a subset the box set generated by corresponding* ∧*-box operator from the same generating triple* $(\tilde{g}, \tilde{m}, \tilde{b}) \in I$, *but the contrary is not always true.*

Lemma 2. *A box set generated by* ∧*-box operator contains as a subset the prime set generated by corresponding prime operator from the same generating triple* $(\tilde{g}, \tilde{m}, \tilde{b}) \in I$, *but the contrary is not always true.*

Theorem 1 (Nesting Order of tricluster components generated from the same generating triple). *The box and prime sets generated by prime operators and variations of box operators from the same generating triple* $(\tilde{g}, \tilde{m}, \tilde{b}) \in I$ *are ordered in the following way with respect to set inclusion:*

$$(\tilde{m}, \tilde{b})' \subseteq \tilde{g}_\wedge^\square \subseteq \tilde{g}_\vee^\square.$$

Corollary 1. *The triclusters built with box and prime sets generated from the same generating triple* $(\tilde{g}, \tilde{m}, \tilde{b}) \in I$ *inherit the same nesting order with respect to component-wise inclusion:*

$$T' \sqsubseteq T_\wedge^\square \sqsubseteq T_\vee^\square.$$

These findings may help to analyze and explain the changes in number and quality of triclusters found by different methods in the same triadic context.

6   Conclusion

This work is dedicated to methods of OAC-triclustering. We described two kinds of tricluster generating operators. One of them has an alternative definition that yields two different variations of the operator. We investigated the relations between these operators and proposed a set of ordering lemmas that establish a nesting order of triclusters generated by the operators. This finding theoretically proves advantages of one operators over the others.

Acknowledgments. We would like to thank our colleagues, B. Ganter, S. Kuznetsov, B. Mirkin, R. Missaoui, L. Cerf, J.-F. Boulicaut, A. Napoli, M. Kaytoue and S. Ben Yahia for their piece of advice and useful prior communication. The paper was prepared within the framework of the Basic Research Program at HSE and supported within the framework of a subsidy by the Russian Academic Excellence Project "5–100". The second co-author was partially supported by Russian Foundation for Basic Research. This work was also partially supported by the French LabEx project IMobS3.

References

1. Lehmann, F., Wille, R.: A triadic approach to formal concept analysis. In: Proceedings of the Third International Conference on Conceptual Structures: Applications, Implementation and Theory, ICCS 1995, Santa Cruz, California, USA, 14–18 August 1995, pp. 32–43 (1995)
2. Jäschke, R., Hotho, A., Schmitz, C., Ganter, B., Stumme, G.: TRIAS - an algorithm for mining iceberg tri-lattices. In: Proceedings of the 6th IEEE International Conference on Data Mining (ICDM 2006), Hong Kong, China, pp. 907–911 (2006)
3. Ignatov, D.I., Kuznetsov, S.O., Magizov, R.A., Zhukov, L.E.: From triconcepts to triclusters. In: Kuznetsov, S.O., Ślęzak, D., Hepting, D.H., Mirkin, B.G. (eds.) RSFDGrC 2011. LNCS (LNAI), vol. 6743, pp. 257–264. Springer, Heidelberg (2011). doi:10.1007/978-3-642-21881-1_41
4. Cerf, L., Besson, J., Robardet, C., Boulicaut, J.: Closed patterns meet n-ary relations. TKDD **3**(1), 3:1–3:36 (2009)
5. Jelassi, M.N., Yahia, S.B., Nguifo, E.M.: Towards more targeted recommendations in folksonomies. Soc. Netw. Anal. Min. **5**(1), 68:1–68:18 (2015)
6. Ignatov, D.I., Gnatyshak, D.V., Kuznetsov, S.O., Mirkin, B.G.: Triadic formal concept analysis and triclustering: searching for optimal patterns. Mach. Learn. **101**(1–3), 271–302 (2015)
7. Cerf, L., Besson, J., Nguyen, K., Boulicaut, J.: Closed and noise-tolerant patterns in n-ary relations. Data Min. Knowl. Discov. **26**(3), 574–619 (2013)
8. Gnatyshak, D., Ignatov, D.I., Kuznetsov, S.O.: From triadic FCA to triclustering: experimental comparison of some triclustering algorithms. In: Proceedings of the Tenth International Conference on Concept Lattices and Their Applications, La Rochelle, France, pp. 249–260 (2013)
9. Ignatov, D.I., Kuznetsov, S.O., Poelmans, J.: Concept-based biclustering for internet advertisement. In: 12th IEEE International Conference on Data Mining Workshops, ICDM Workshops, Brussels, Belgium, 10 December 2012, pp. 123–130 (2012)

The Inescapable Relativity of Explicitly Represented Knowledge: An FCA Perspective

David Flater$^{(\boxtimes)}$

National Institute of Standards and Technology, Gaithersburg, USA
david.flater@nist.gov

Abstract. Knowledge models are supposed to capture knowledge of lasting value in a reusable form. However, reuse of these models is hampered by arbitrary and application-specific constraints; any constraints that conflict with a new application must be altered or removed before the models can be reused. This article explores seven facets of conceptual relativity that would impact the use of Formal Concept Analysis formalisms to represent knowledge, demonstrating that the capture of application-specific constraints is inextricable from the modelling process.

1 Introduction

Knowledge models are intended to capture knowledge of lasting value in a reusable form. They can model the concepts relevant to a software project, a domain of discourse, or any world view of whatever scope. Their range of representations includes specialized ontology languages such as Web Ontology Language (OWL) [9], general-purpose modelling languages such as Unified Modeling Language (UML) [13], and the formalisms of Formal Concept Analysis (FCA) [5].

Reuse of knowledge models is hampered by arbitrary and application-specific constraints. Any constraints that conflict with a new application must be altered or removed before the models can be reused. For this reason, the modeller seeks to avoid mingling arbitrary and application-specific constraints with those believed to be universal. In a typical application of FCA, the algorithmically identified formal concepts would promulgate application-specific constraints with no warning to the practitioner. However, the problem actually is endemic to the process, and no language or technique can avoid it entirely. It is therefore important to be aware of the dangers.

This article explores seven facets of relativity in knowledge models, demonstrating that the capture of application-specific constraints is inextricable from the modelling process. The facets of conceptual relativity to be discussed are shown in Fig. 1.

The term *intent* is already well-established in the knowledge modelling domain. The terms *essence*, *identity*, and *unity* were previously introduced to

© Springer International Publishing AG 2017
M. Kryszkiewicz et al. (Eds.): ISMIS 2017, LNAI 10352, pp. 580–586, 2017.
DOI: 10.1007/978-3-319-60438-1_57

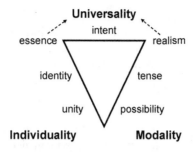

Fig. 1. Facets of conceptual relativity

the knowledge modelling community by Guarino and Welty [6,7]. The remaining terms have been introduced in parallel fashion to complete the framework needed to discuss conceptual relativity. The dilemmas confounding knowledge modellers today were first identified and explored by philosophers, so the corresponding terms from philosophical references have been applied.

As the diagram suggests, the facets are not independent; neither do they fall neatly into categories. However, generally speaking, essence, identity, and unity have to do with individuality—the factoring of the domain of discourse into separate things. Possibility, tense, and realism have to do with modality—the factoring of the domain of discourse into different ways of existing. Essence, realism, and intent have to do with universality—the determination of what is held constant. Essence is an individual perspective on universality (what universality means for individuals); realism is a modal perspective on universality (what universality means for existence).

The following sections examine the seven facets of conceptual relativity as they would impact the use of FCA formalisms to represent knowledge. The formal contexts and concept lattices of examples that have been redacted due to the page limit can be found in an unabridged technical report [3].

2   Essence

A property of an entity is essential to that entity if it must hold for it. This is a stronger notion than one of permanence, that is, a property of an entity is not essential if it just happens to be true of it, accidentally [6].

In the context of this discussion, to refer to properties as being essential to an *entity* means, more accurately, that those properties are deemed logically necessary in some intensional classification or identification of that entity. Essence identifies conditions that are necessary; it does not address sufficiency.

The appropriateness of any given classification is relative to the application. For example, consider the classification of chemical elements. A research institute would need to identify different isotopes of helium in nuclear physics experiments, but in many industrial applications, it goes without saying that the helium cylinder contains mostly ^{4}He. Consequently, FCA of data from these different applications would lead to different sets of attributes for a concept called 'helium.' Manual reconciliation would conclude that a neutron count of 2 is essential to the industrial concept of helium, even if it was never stated explicitly in the data, while it is non-essential to the nuclear physics concept.

3   Identity

When something undergoes a change, whether or not it is considered the *same* thing afterwards depends on how that thing is identified.

Using nominal scales to transform many-valued attributes, the formal context shown in Table 1 models the decay of a ^{5}He atom at time t_1 to ^{4}He at time t_2. As yet, no commitment has been made regarding whether the ^{4}He is the same atom as the ^{5}He. This formal context merely models observations of phenomena (1 and 2).

The decision to identify the two observations with the same atom (one that mutated) can be modelled using the extension of FCA called Temporal Concept Analysis [14], which adds a time relation between the two observations (called "actual objects"). If one chooses instead to view the ^{4}He as a different atom (the product of the previous atom's decay), the time relation is deleted, but the formal contexts are *structurally identical*. This suggests, accurately, that the decision to identify the two observations with the same atom or with different atoms is somewhat arbitrary. It is a subjective interpretation of objective phenomena. In contrast, some modelling environments treat identity as a transcendental, non-qualitative property ('haecceity') and cannot accurately represent the relationships among alternate identities for the same object.

That the appropriateness of any given selection of identity criteria is relative to the application is easily shown using the classic 'ship of Theseus' example. A ship that has been repaired and restored is the same ship as far as navigation is concerned, but it is not the same from the forensic perspective, e.g., in an investigation of whether the ship was in compliance with regulations when an incident occurred. In the latter application, the ships "before the repair" and "after the repair" are treated as different objects.

Table 1. Formal context for ^{5}He decay

	t_1	t_2	2 protons	2 neutrons	3 neutrons
1	×		×		×
2		×	×	×	

4   Unity

Unity relates to the philosophical notion of boundaries: one can define a thing by selecting spatial and temporal boundaries, deciding what is part of it and what is not. When formalized, the process is analogous to the example in the previous section: one subjectively picks objects out of a soup of observations.

It is often if not always the case that the spatial and/or temporal boundaries of a thing as people conceive of it are *vague*. Any precise model of such boundaries, including one using FCA, necessarily adds arbitrary and application-specific constraints. However, one can use conceptual scaling to minimize the impact. For example, again considering the process of radioactive decay, it is unlikely that the precise instant at which an atom decayed (or the instants at which the process of decay began and ended) would be known. However, it would be known with some certainty that it had not yet decayed at time t_1, and that it had already decayed at time t_2. With appropriate scaling, one need only consider those time granules for which precise knowledge is available.

5   Possibility

Possible things is a way of saying "things that might actually exist, but that we do not *know* exist;" alternately, "things that could potentially exist someday, but that do not exist now."

The uncertainty about possible things is epistemic in nature. Any hypothesized class that does not intend a logical contradiction could possibly have instances. This uncertainty is not directly modelled by FCA extensions that apply possibility theory and deal with missing/unknown values (e.g., Ref. [1]) because the objects themselves are hypothetical and thus entirely missing from the data.

One way to model possible things in FCA is to represent existential assertions *about* hypothetical objects as objects, and then use attribute scaling to capture all of the modality of the truth values of those statements. Figure 2 shows the concept lattice for an epistemic modal logic that enables an accurate description of the state of knowledge or belief about an assertion. The names of the attributes have been prefixed by "know" to emphasize that a statement can be true without one knowing that it is true, but not vice-versa.

A context that does not support *unknown* clearly leaves the modeller with little choice other than to make invalid substitutions. But too many *unknowns* results in a vacuous model—all things are possible; nothing can ever be ruled out. The treatment of possible things thus becomes a compromise between the desire for a generally valid model and the desire for a model that is constrained enough to enable the application for which it was built.

6   Tense

Since it is the goal of modellers to capture knowledge of lasting value, the question of how to model the past and future as distinct from the present is often ignored. The resulting model is timeless in the sense of having no concept of time whatsoever. Things simply are as they are, unchanging; or, if things do change, the result is a *different* model. There is no formal connection between

584   D. Flater

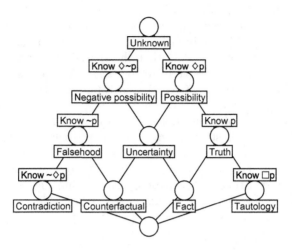

Fig. 2. Concept lattice for epistemic modal logic

the old and the new. This suffices if, in the application of interest, no contradictions result from instantiating all things past, present, and future as if they were contemporaries. However, it does not suffice for any application that needs to deal with change.

Those modellers who do model time choose to structure it in different ways. Tense logic structures time in terms of past, present, and future (a.k.a. the A-series or the tenser approach) [4]. UML sequence diagrams structure time in terms of earlier and later (a.k.a. the B-series or the detenser approach). Process Specification Language (PSL) [12] structures time in terms of reified time points (a.k.a. the four-dimensional approach). In FCA, these different structures of time can be integrated through appropriate scaling of attributes that indicate the times at which an object existed.

7   Realism

Classifying things is a process of abstraction, but some reject the claim that classes are abstractions of a higher level than the things classified.

Some modelling architectures segregate levels of abstraction (*fixed architecture*); others do not (*flat architecture*). The desirability of strict separation is a topic of debate. There are intuitively attractive notions that cannot be rendered faithfully using a fixed architecture. For example, consider the class that is called *class*. Intuitively, *class* is an instance of itself. But proponents of fixed architecture argue that it is confused thinking to identify any instance of a class with the class itself, and that doing so produces a model that has no sensible interpretation by man or machine [8,10], or at best an unconventional interpretation that does not integrate readily with conventional logic [11]. In any event, it certainly invalidates the set-theoretic interpretation of classification.

The decision to use a fixed architecture or not is a technical one, influenced by the relative expedience of expressing the concepts needed to serve particular applications. However, the barriers to translation between fixed and flat architectures are significant. In FCA, the concept whose intent consists of all attributes of an object g and whose extent consists of all objects having those attributes is called the *object concept* of the object g and is denoted by γg [5, Definition 22]. Formally, g and γg are not even comparable; one is an object, the other is a concept. FCA formalisms thus are best suited for a fixed architecture representation of knowledge, but this introduces constraints that some see as arbitrary.

8   Intent

In a static universe, extensional definitions are sufficient. If two concepts have the same extent, then they are interchangeable within the scope of the static universe and there is no value in distinguishing them. The value of intent is in making statements regarding possible and future individuals. Whether or not it is necessary to make statements about possible and future individuals or to distinguish concepts having the same extents is clearly application-specific, as is the selection of essential properties for intensional definitions.

FCA reduces the intensional/extensional dichotomy to a mathematical extreme, defining formal concepts in such a way that intensional definitions (in terms of necessary and sufficient attributes) and extensional definitions (in terms of objects) entail one other via a formal mapping. The intent that is derived from available data is forced to change when the data do. In knowledge representation generally, intensional definitions are invalidated when some possible individual that breaks the assumptions of the modeller becomes actual or becomes known. This problem is not intractable; it can be avoided by sacrificing the ability to map between intent and extent [2].

9   Conclusion

This article explored seven facets of relativity that would impact the use of FCA formalisms to represent knowledge, demonstrating that the capture of application-specific constraints is inextricable from the modelling process.

The semantic differences between models built for different applications can also be modelled and analyzed using FCA. By analyzing those differences, one can formally determine whether the applications are sufficiently compatible at the conceptual level to enable integration. That next step is explored in the unabridged technical report [3].

Acknowledgments. Thanks to all whose reviews and suggestions have improved this article, including Edward Barkmeyer, Peter Becker, Joachim Hereth Correia, Steven Fenves, Michael Grüninger, and the reviewers for the special session on knowledge discovery with FCA and related formalisms (FCA4KD++).

References

1. Ait-Yakoub, Z., Djouadi, Y., Dubois, D., Prade, H.: From a possibility theory view of Formal Concept Analysis to the possibilistic handling of incomplete and uncertain contexts. In: 5th International Workshop "What can FCA do for Artificial Intelligence?" August 2016. http://oatao.univ-toulouse.fr/17213/
2. Flater, D.: A logical model of conceptual integrity in data integration. J. Res. Natl. Inst. Stand. Technol. **108**(5), 395–402 (2003). doi:10.6028/jres.108.034
3. Flater, D.: Relativity of explicit conceptual models. NIST IR 7148, National Institute of Standards and Technology, 100 Bureau Drive, Gaithersburg, MD 20899, July 2004. doi:10.6028/NIST.IR.7148
4. Galton, A.: Temporal logic. In: Zalta, E.N. (ed.) Stanford Encyclopedia of Philosophy. Metaphysics Research Lab, Stanford University, winter 2003 edn., December 2003. https://plato.stanford.edu/archives/win2003/entries/logic-temporal/
5. Ganter, B., Wille, R.: Formal Concept Analysis: Mathematical Foundations. Springer, Heidelberg (1999)
6. Guarino, N., Welty, C.: Evaluating ontological decisions with OntoClean. Commun. ACM **45**(2), 61–65 (2002)
7. Guarino, N., Welty, C.: Identity and subsumption. In: Green, R., Bean, C.A., Myaeng, S.H. (eds.) The Semantics of Relationships: An Interdisciplinary Perspective. Information Science and Knowledge Management, pp. 111–126. Springer, Dordrecht (2002). doi:10.1007/978-94-017-0073-3_7
8. Korzybski, A.: Science and Sanity: An Introduction to Non-Aristotelian Systems and General Semantics, 5th edn. Institute of General Semantics, Englewood (1994)
9. OWL 2 Web Ontology Language document overview, 2nd edn. W3C recommendation, December 2012. https://www.w3.org/TR/owl2-overview/
10. Pan, J.Z., Horrocks, I.: Metamodeling architecture of web ontology languages. In: Proceedings of the Semantic Web Working Symposium, pp. 131–149, July 2001
11. Pan, J.Z., Horrocks, I.: RDFS(FA) and RDF MT: two semantics for RDFS. In: Fensel, D., Sycara, K., Mylopoulos, J. (eds.) ISWC 2003. LNCS, vol. 2870, pp. 30–46. Springer, Heidelberg (2003). doi:10.1007/978-3-540-39718-2_3
12. PSL Core Ontology (2004). http://www.mel.nist.gov/psl/psl-ontology/pslcore_page.html
13. Unified Modeling Language specification, version 2.5. OMG document formal/2015-03-01, Object Management Group, June 2015. http://www.omg.org/spec/UML/2.5/
14. Wolff, K.E., Yameogo, W.: Time dimension, objects, and life tracks. A conceptual analysis. In: Ganter, B., de Moor, A., Lex, W. (eds.) ICCS-ConceptStruct 2003. LNCS, vol. 2746, pp. 188–200. Springer, Heidelberg (2003). doi:10.1007/978-3-540-45091-7_13

Blocks of the Direct Product of Tolerance Relations

Christian Jäkel$^{(\boxtimes)}$ and Stefan E. Schmidt

Technische Universität Dresden, 01062 Dresden, Germany
christian.jaekel@tu-dresden.de, midt1@msn.com

Abstract. The blocks of a tolerance relation generalize equivalence classes of equivalence relations and are in one to one correspondence with certain set coverings. We will, within the framework of formal concept analysis, analyse the blocks of the direct product of two tolerance relations. The question is how blocks of the direct product are related to the structure of the factors. It turns out that directly induced and non-induced blocks exist.

For tolerance relations, the problem of detecting the blocks of the direct product can be seen as a special instance of the task to determine the blocks of the union of two tolerance relations. In general, the blocks of the union of two tolerance relations are not directly derived from blocks of the unions components. Furthermore, we will apply our results to factor analysis and discuss open problems.

Keywords: Tolerance relation · Block · Formal concept analysis · Direct product · Tensor product · Factor analysis

1 Introduction

A *tolerance relation* or simply a *tolerance* is a reflexive and symmetric binary relation τ on a non-empty finite set U. The pair (U, τ) is called a *tolerance space*. An introduction to tolerance spaces, together with applications in image processing can be found in [10]. In [1], tolerances are applied to formal concept analysis, in [8] to rough sets and in [11] to linguistics.

For a tolerance τ on U, a non-empty subset $T \subseteq U$ is called a τ-*preblock* if $T \times T$ is contained in τ. A maximal τ-preblock with respect to set inclusion is called a τ-*block*. In other words, this means that a τ-block $T \subseteq U$ defines a non-enlargeable *square* $T \times T \subseteq \tau$. In the sequel, we will only say *(pre-)block* when there is no risk of ambiguity.

The set of all blocks of τ is denoted by $\mathrm{BL}(\tau)$ and determines the tolerance τ, that is $(u_1, u_2) \in \tau$ if and only if there is a block T with $u_1, u_2 \in T$. This can be written as $\tau = \bigcup \{T \times T \mid T \in \mathrm{BL}(\tau)\}$. Furthermore, for every $u \in U$, let $\mathrm{BL}(u) := \{T \in \mathrm{BL}(\tau) \mid u \in T\}$ be the set of all blocks containing u.

For two tolerances τ_1 and τ_2 on U, let $\tau = \tau_1 \cap \tau_2$ be their intersection. In [3], it is shown that every block of τ is an intersection of one block from τ_1 and on from τ_2:

© Springer International Publishing AG 2017
M. Kryszkiewicz et al. (Eds.): ISMIS 2017, LNAI 10352, pp. 587–596, 2017.
DOI: 10.1007/978-3-319-60438-1_58

$$\mathrm{BL}(\tau_1 \cap \tau_2) = \{T \subseteq U \mid \exists T_1 \in \mathrm{BL}(\tau_1), T_2 \in \mathrm{BL}(\tau_2) : T = T_1 \cap T_2\}.$$

On the other hand (see [3]), a block of the union $\tau_1 \cup \tau_2$ is generally not a union of one block from τ_1 and one from τ_2.

In this paper, we will treat the union of two tolerance relations by means of formal concept analysis. Therefore, Sect. 2 provides a short introduction to formal concept analysis. Especially, we want to emphasize that every formal context $\mathbb{K}$ can be transformed into a tolerance space, the blocks of which are in one to one correspondence with the concepts of $\mathbb{K}$. Section 3 shows how formal concept analysis applies to tolerance spaces. The fourth section treats the direct product, Sect. 5 discusses open problems regarding factor analysis and the last section gives a conclusion.

2   Basics of Formal Concept Analysis

We assume the reader to be familiar with the basics of formal concept analysis (see [6]). Still, we will provide the definitions and facts that will be used in the sequel. A *formal context* is a triple $\mathbb{K} = (G, M, I)$, where $I \subseteq G \times M$ is a binary relation. For $A \subseteq G$ and $B \subseteq M$, we define two derivation operators:

$$A^I := \{m \in M \mid \forall a \in A : (a, m) \in I\} = \bigcap_{a \in A} \{a\}^I,$$

$$B_I := \{g \in G \mid \forall b \in B : (g, b) \in I\} = \bigcap_{b \in B} \{b\}_I.$$

If $A^I = B$ and $B_I = A$, the pair (A, B) is called a *formal concept* with *extent* A and *intent* B. The set of all formal concepts of $\mathbb{K}$ is denoted by $\mathfrak{B}(\mathbb{K})$ and defines the concept lattice $\underline{\mathfrak{B}}(\mathbb{K})$, via the order $(A_1, B_1) \leq (A_2, B_2) :\Longleftrightarrow A_1 \subseteq A_2$.

Using notation from [4], for two contexts $\mathbb{K}_1 = (G_1, M_1, I_1)$ and $\mathbb{K}_2 = (G_2, M_2, I_2)$, we define the *direct product* $\mathbb{K}_1 \stackrel{\vee}{\times} \mathbb{K}_2 := (G_1 \times G_2, M_1 \times M_2, I_1 \stackrel{\vee}{\times} I_2)$ with

$$((g, h), (m, n)) \in I_1 \stackrel{\vee}{\times} I_2) :\Longleftrightarrow (g, m) \in I_1 \text{ or } (h, n) \in I_2,$$

and the *cardinal product* $\mathbb{K}_1 \stackrel{\wedge}{\times} \mathbb{K}_2 := (G_1 \times G_2, M_1 \times M_2, I_1 \stackrel{\wedge}{\times} I_2)$ with

$$((g, h), (m, n)) \in I_1 \stackrel{\wedge}{\times} I_2) :\Longleftrightarrow (g, m) \in I_1 \text{ and } (h, n) \in I_2.$$

The product relations $I_1 \stackrel{\vee}{\times} I_2$ and $I_1 \stackrel{\wedge}{\times} I_2$ can be expressed as:

$$I_1 \stackrel{\vee}{\times} I_2 = (G_1 \times M_1) \stackrel{\wedge}{\times} I_2 \cup I_1 \stackrel{\wedge}{\times} (G_2 \times M_2), \tag{1}$$

$$I_1 \stackrel{\wedge}{\times} I_2 = (G_1 \times M_1) \stackrel{\wedge}{\times} I_2 \cap I_1 \stackrel{\wedge}{\times} (G_2 \times M_2). \tag{2}$$

If $A \subseteq G_1$, $B \subseteq G_2$ and $M_1^{I_1} = M_2^{I_2} = \emptyset$, the derivation of $A \times B$ (see [13]) with respect to $I_1 \stackrel{\vee}{\times} I_2$ is:

$$(A \times B)^{I_1 \stackrel{\vee}{\times} I_2} = A^{I_1} \times M_2 \cup M_1 \times B^{I_2}. \tag{3}$$

In the sequel, we will use the fact that the derivation of a union is the intersection of each components derivation.

$$(A \cup B)^I = A^I \cap B^I. \tag{4}$$

For two complete lattices L_1 and L_2 the *tensor product* $L_1 \otimes L_2$ is the concept lattice $\underline{\mathfrak{B}}(L_1 \overset{\vee}{\times} L_2)$, where L_1 and L_2 are regarded as formal contexts with respect to their order relations. It is shown in [6] that the concept lattice of the direct product is isomorphic to the tensor product of the factors concept lattices:

$$\underline{\mathfrak{B}}(\mathbb{K}_1 \overset{\vee}{\times} \mathbb{K}_2) \cong \underline{\mathfrak{B}}(\mathbb{K}_1) \otimes \underline{\mathfrak{B}}(\mathbb{K}_2). \tag{5}$$

The next aspect of formal concept analysis that we want to consider is *factor analysis* (see [2]). For this purpose, let $\mathbb{K} = (G, M, I)$ with $|G| = m$ and $|M| = n$, and a subset of formal concepts $\mathcal{F} = \{(A_1, B_1), ..., (A_l, B_l)\} \subseteq \mathfrak{B}(\mathbb{K})$ be given. The set $\mathcal{F}$ is called a *factorization* if $\bigcup_{k=1}^{l} A_k \times B_k = I$. Let $A_{\mathcal{F}}$ and $B_{\mathcal{F}}$ denote Boolean $m \times l$ and $l \times n$ matrices, defined by $(A_{\mathcal{F}})_{ik} := \delta_{A_k}(i)$ and $(B_{\mathcal{F}})_{kj} := \delta_{B_k}(j)$, where δ_X denotes the characteristic function of X. It follows that I, considered as a Boolean matrix, is equal to the Boolean matrix product $A_{\mathcal{F}} \circ B_{\mathcal{F}}$. Such a *decomposition* of I in terms of formal concepts is optimal in the sense of the following theorem.

Theorem 1 ([2]). *Let $\mathbb{K} = (G, M, I)$ be as above. Let $I = A \circ B$ for $m \times l$ and $l \times n$ Boolean matrices A and B. Then there exists a subset $\mathcal{F} \subseteq \mathfrak{B}(\mathbb{K})$ with $|\mathcal{F}| \leq l$ and $I = A_{\mathcal{F}} \circ B_{\mathcal{F}}$.*

Lastly, we show how to interpret a formal context as a tolerance space. Therefore, let $\mathbb{K} = (G, M, I)$ be a formal context with $G \cap M = \emptyset$. We define the *dual context* $\mathbb{K}^d := (M, G, I^{-1})$, where $I^{-1} := \{(m, g) \in M \times G | (g, m) \in I\}$. The *symmetrization* of $\mathbb{K}$, as a special case of the direct sum (see [7]), is defined as $\mathbb{K}^s = (G^s, M^s, I^s) := (G \cup M, G \cup M, I \cup I^{-1} \cup (G \times G) \cup (M \times M))$ (Fig. 1).

I^s	G	M
G	$G \times G$	I
M	I^{-1}	$M \times M$

Fig. 1. Relation I^s of the symmetrization $\mathbb{K}^s$.

The blocks of I^s correspond to the formal concepts of $\mathbb{K}$.

Theorem 2. *It holds that* $\mathrm{BL}(I^s) \cong \mathfrak{B}(\mathbb{K})$ *as sets.*

Proof. Obviously, the map $\mathfrak{B}(\mathbb{K}) \to \mathrm{BL}(I^s)$, $(A, B) \mapsto A \cup B$ has the map $\mathrm{BL}(I^s) \to \mathfrak{B}(\mathbb{K})$, $T \mapsto (T \cap G, T \cap M)$ as inverse.

3   Tolerance Relations as Formal Contexts

In the sequel, a tolerance space (U, τ) will be interpreted as a formal context (U, U, τ). Since τ is symmetric, there is only one derivation operator $(-)^\tau : 2^U \to 2^U$. The corresponding concept lattice is denoted by $\mathfrak{B}(U, \tau)$. It is according to [6] a polarity lattice (that is a complete lattice with an involutory antiautomorphism) and the polarity p maps a concept (A, B) to the concept (B, A).

From the definition of blocks, it follows that a block is a fixed point of the derivation operator:

$$T \in \mathrm{BL}(\tau) \iff T^\tau = T.$$

Hence, each block T corresponds to a *square concept* (T, T) and this square is a fixed point of the polarity map.

Theorem 3. *Let (U, τ) be a tolerance space. For every $u \in U$, the attribute concept $(u^\tau, u^{\tau\tau})$ equals $(\bigcap \mathrm{BL}(u), \bigcup \mathrm{BL}(u))$. Consequently, the object concept $(u^{\tau\tau}, u^\tau) = (\bigcup \mathrm{BL}(u), \bigcap \mathrm{BL}(u))$.*

Proof. Since $(u, \tilde{u}) \in \tau$ if and only if there exists $T \in \mathrm{BL}(\tau)$ with $u, \tilde{u} \in T$, it follows that $u^\tau = \bigcap \mathrm{BL}(u)$. Since it always holds that

$$\bigcup \mathrm{BL}(u) \subseteq (\bigcap \mathrm{BL}(u))^\tau = (\bigcup \mathrm{BL}(u))^{\tau\tau},$$

it suffices to show that $\bigcup \mathrm{BL}(u) \supseteq (\bigcap \mathrm{BL}(u))^\tau$.

Suppose for a contradiction that $\tilde{u} \in (\bigcap \mathrm{BL}(u))^\tau \setminus \bigcup \mathrm{BL}(u)$ exists. To shorten the notation, set $\mathfrak{S} := \bigcap \mathrm{BL}(u)$. Since τ is a tolerance and $\{\tilde{u}\} \times \mathfrak{S} \subseteq \tau^1$, we have:

$$(\{\tilde{u}\} \cup \mathfrak{S}) \times (\{\tilde{u}\} \cup \mathfrak{S}) \subseteq \{(\tilde{u}, \tilde{u})\} \cup \{\tilde{u}\} \times \mathfrak{S} \cup \mathfrak{S} \times \{\tilde{u}\} \cup \mathfrak{S} \times \mathfrak{S} \subseteq \tau.$$

Consequently, $\{\tilde{u}\} \cup \mathfrak{S}$ is a preblock and there exists a block $T \in \mathrm{BL}(\tau)$ with $\{\tilde{u}\} \cup \mathfrak{S} \subseteq T$. As $T \in \mathrm{BL}(u)$, we get the contradiction $\tilde{u} \in T \subseteq \bigcup \mathrm{BL}(u)$.

Corollary 1. *The square concepts of a tolerance space (U, τ) generate the concept lattice $\mathfrak{B}(U, \tau)$.*

Next, we will provide two examples which will be used later on.

Example 1. *We consider the tolerance space $(\{v_1, v_2\}, \tau)$, together with the respective concept lattice and blocks $T_1 = \{v_1\}$ and $T_2 = \{v_2\}$ (Fig. 2).*

Example 2. *We consider the tolerance space $(\{u_1, u_2, u_3, u_4\}, \sigma)$. The blocks of σ are $\{S_1, S_2, S_3\}$ with $S_1 = \{u_1, u_2\}$, $S_2 = \{u_2, u_3\}$ and $S_3 = \{u_4\}$ (Fig. 3).*

Lastly, regarding factor analysis of tolerances, we recall a result from [12].

Theorem 4. *Let (U, τ) be a tolerance space and let $\mathcal{T} \subseteq \mathfrak{B}(U, \tau)$ be the set of all square concepts. It holds that $\tau = (A_{\mathcal{T}})^t \circ A_{\mathcal{T}}$, where $(-)^t$ denotes the transpose.*

---

[1] Generally: $X \subseteq Y^I \Leftrightarrow Y \subseteq X^I \Leftrightarrow X \times Y \subseteq I$.

τ	v_1	v_2
v_1	1	0
v_2	0	1

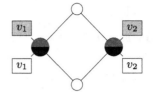

Fig. 2. The concept lattice of τ.

σ	u_1	u_2	u_3	u_4
u_1	1	1	0	0
u_2	1	1	1	0
u_3	0	1	1	0
u_4	0	0	0	1

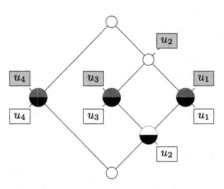

Fig. 3. The concept lattice of σ.

4 The Cardinal- and Direct Product of Tolerance Spaces

Let (U, σ) and (V, τ) be tolerance spaces. We assume that the sets of blocks $\mathrm{BL}(\sigma)$ and $\mathrm{BL}(\tau)$ are known. Furthermore, since the deletion of full rows and full columns does not change the structure of the concept lattice, both tolerances are assumed to have no full rows and no full columns, that is $U^\sigma = \emptyset$ and $V^\tau = \emptyset$.

Definition 1. *A block $X \subseteq U \times V$ of the cardinal or direct product of (U, σ) and (V, τ) is called* induced block *if $X = S \times V$ for some $S \in \mathrm{BL}(\sigma)$ or there exists $T \in \mathrm{BL}(\tau)$ with $X = U \times T$. If this is not the case, then X is a* non-induced block. *Furthermore, an induced block gives rise to an* induced formal square *and a non-induced block to a* non-induced formal square.

From Identity 2, it follows that the cardinal product of (U, σ) and (V, τ) is constructed by first lifting (U, σ) and (V, τ) to $U \times V$ and then intersecting these liftings. From [3], we conclude:

Theorem 5. *Let (U, σ) and (V, τ) be tolerance spaces. Every block of their cardinal product is the intersection of an induced block from σ and an induced block from τ.*

However, the blocks of the direct product are not always a union of induced blocks from each factor, even though Identity 1 implies that the direct product is constructed by first lifting (U, σ) and (V, τ) to $U \times V$ and then unite these liftings.

But every block in a factor induces a block in the direct product $(U \times V, \sigma \stackrel{\vee}{\times} \tau)$.

Proposition 1. *Let (U, σ) and (V, τ) be tolerance spaces. If $S \in \mathrm{BL}(\sigma)$, then $S \times V \in \mathrm{BL}(\sigma \stackrel{\vee}{\times} \tau)$ and if $T \in \mathrm{BL}(\tau)$, then $U \times T \in \mathrm{BL}(\sigma \stackrel{\vee}{\times} \tau)$.*

Proof. Let $S \in \mathrm{BL}(\sigma)$. We use Identity 3 to calculate:

$$(S \times V)^{\sigma \stackrel{\vee}{\times} \tau} = S^{\sigma} \times V \cup U \times V^{\tau} = S \times V \cup U \times \emptyset = S \times V.$$

The corresponding calculation for $U \times T$ is analogues. Hence, $S \times V$ and $U \times T$ are fixed points of $(-)^{\sigma \stackrel{\vee}{\times} \tau}$ and therefore blocks of $\sigma \stackrel{\vee}{\times} \tau$.

If the blocks of σ and τ do not overlap, which means that σ and τ are equivalence relations, then each block of their direct product is induced by a factor.

Theorem 6. *Let (U, σ) and (V, τ) be tolerance spaces with equivalence relations σ and τ. Every block of their direct product $(U \times V, \sigma \stackrel{\vee}{\times} \tau)$ is an induced block.*

Proof. As clarification (that is, the deletion of duplicate rows and columns) of (U, σ) and (V, τ) does not change the structure of their concept lattices, we assume them to be clarified. Consequently, σ and τ only have singleton blocks (Fig. 4).

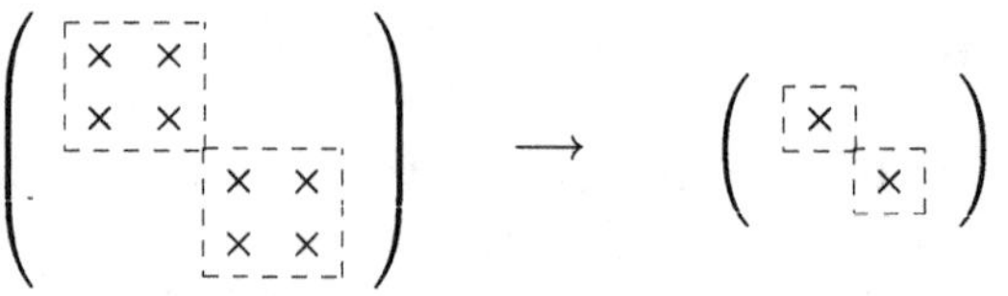

Fig. 4. Simplification of an equivalence relation.

Let $X \in \mathrm{BL}(\sigma \stackrel{\vee}{\times} \tau)$ be a non-induced block, *i.e.*, there is no $u \in U$ with $X = \{u\} \times V$ and no $v \in V$ with $X = U \times \{v\}$. Hence, there exist distinct $u_1, u_2 \in U$ and $v_1, v_2 \in V$, such that $(u_1, v_1), (u_2, v_2) \in X$. However, this would imply that $((u_1, v_1), (u_2, v_2)) \in \sigma \stackrel{\vee}{\times} \tau$ holds (see Sect. 1), but neither is $(u_1, u_2) \in \sigma$ nor $(v_1, v_2) \in \tau$.

Next, we give an example for the occurrence of non-induced blocks.

Example 3. *This example shows the direct product $\sigma \stackrel{\vee}{\times} \tau$ of the tolerance spaces from Examples 1 and 2. Instead of (u, v), we write uv. The non-induced blocks are $X_1 = \{u_1v_1, u_2v_1, u_2v_2, u_3v_1\}$ and $X_2 = \{u_1v_2, u_2v_1, u_2v_2, u_3v_2\}$, and the induced ones are $X_3 = S_1 \times V$, $X_4 = S_2 \times V$, $X_5 = S_3 \times V$, $X_6 = U \times T_1$ and $X_7 = U \times T_2$. For $i = 1, .., 7$, each block $X_i \subseteq U \times V$ is visualized as an adjacency matrix of a binary relation. Furthermore, note that each row of X_i is an extent from a concept of $\underline{\mathfrak{B}}(U, \sigma)$ and every column one from $\underline{\mathfrak{B}}(V, \tau)$ (see "dual bonds" in [6]) (Fig. 5).*

$\sigma \check{\times} \tau$	u_1v_1	u_2v_1	u_3v_1	u_4v_1	u_1v_2	u_2v_2	u_3v_2	u_4v_2
u_1v_1	1	1	1	1	1	1	0	0
u_2v_1	1	1	1	1	1	1	1	0
u_3v_1	1	1	1	1	0	1	1	0
u_4v_1	1	1	1	1	0	0	0	1
u_1v_2	1	1	0	0	1	1	1	1
u_2v_2	1	1	1	0	1	1	1	1
u_3v_2	0	1	1	0	1	1	1	1
u_4v_2	0	0	0	1	1	1	1	1

X_1	v_1	v_2
u_1	1	0
u_2	1	1
u_3	1	0
u_4	0	0

X_2	v_1	v_2
u_1	0	1
u_2	1	1
u_3	0	1
u_4	0	0

X_3	v_1	v_2
u_1	1	1
u_2	1	1
u_3	0	0
u_4	0	0

X_4	v_1	v_2
u_1	0	0
u_2	1	1
u_3	1	1
u_4	0	0

X_5	v_1	v_2
u_1	0	0
u_2	0	0
u_3	0	0
u_4	1	1

X_6	v_1	v_2
u_1	1	0
u_2	1	0
u_3	1	0
u_4	1	0

X_7	v_1	v_2
u_1	0	1
u_2	0	1
u_3	0	1
u_4	0	1

Fig. 5. Induced and non-induced blocks of $\sigma \check{\times} \tau$.

The next theorem states three closed formulas for non-induced blocks of the direct product.

Theorem 7. *Let (U, σ) and (V, τ) be tolerance spaces and $(U \times V, \sigma \check{\times} \tau)$ their direct product.*

1. *For all formal concepts $(A, B) \in \mathfrak{B}(U, \sigma)$, blocks $S \in \mathrm{BL}(\sigma)$ with $A \subseteq S \subseteq B$ and for all $(C, D) \in \mathfrak{B}(V, \tau)$ with $C \subseteq D$, it holds that*

$$(A \times V) \cup (S \times D) \cup (B \times C) \in \mathrm{BL}(\sigma \check{\times} \tau).$$

2. *For all blocks $S_1, S_2 \in \mathrm{BL}(\sigma)$ and for all formal concepts $(C, D) \in \mathfrak{B}(V, \tau)$ with $C \cap D = \emptyset$, it holds that*

$$((S_1 \cap S_2) \times V) \cup (S_1 \times D) \cup (S_2 \times C) \in \mathrm{BL}(\sigma \check{\times} \tau).$$

3. *We consider formal concepts $(A_1, B_1), (A_2, B_2) \in \mathfrak{B}(U, \sigma)$ and a block $S \in \mathrm{BL}(\sigma)$, and $(C_1, D_1), (C_2, D_2) \in \mathfrak{B}(V, \tau)$ and a block $T \in \mathrm{BL}(\tau)$. It holds that $A_1 \subseteq S \subseteq B_1$, $A_2 \subseteq S \subseteq B_2$, $A_1 \cap A_2 = \emptyset$ and $B_1 \cap B_2 = S$. Similarly, it holds that $C_1 \subseteq T \subseteq D_1$, $C_2 \subseteq T \subseteq D_2$, $C_1 \cap C_2 = \emptyset$ and $D_1 \cap D_2 = T$. Then,*

$$(A_1 \times D_1) \cup (A_2 \times D_2) \cup (B_2 \times C_1) \cup (B_1 \times C_2) \cup (S \times T) \in \mathrm{BL}(\sigma \check{\times} \tau).$$

In statements 1 and 2, the role of (U, σ) and (V, τ) can be interchanged. This yields two more similar formulas for non-induced blocks.

Proof. We use again the fact that a fixed point of the derivation operator is a block and show that the sets given in 1 and 2 are fixed points of $(-)^{\sigma \check{\times} \tau}$. Statement 3 can be shown in a similar way.

For the following calculations, we will repeatedly use 3 and 4.

$$(A \times V \cup S \times D \cup B \times C)^{\sigma \check{\times} \tau}$$
$$= (A^{\sigma} \times V \cup U \times V^{\tau}) \cap (S^{\sigma} \times V \cup U \times D^{\tau}) \cap (B^{\sigma} \times V \cup U \times C^{\tau})$$
$$= B \times V \cap (S \times V \cup U \times C) \cap (A \times V \cup U \times D)$$
$$= ((B \cap S) \times V \cup (B \cap U) \times (V \cap C)) \cap (A \times V \cup U \times D)$$
$$= (S \times V \cup B \times C) \cap (A \times V \cup U \times D)$$
$$= A \times V \cup S \times D \cup A \times C \cup B \times C$$
$$= A \times V \cup S \times D \cup B \times C.$$

$$((S_1 \cap S_2) \times V \cup S_1 \times D \cup S_2 \times C)^{\sigma \check{\times} \tau}$$
$$= ((S_1 \cup S_2)^{\sigma\sigma} \times V) \cap (S_1 \times V \cup U \times C) \cap (S_2 \times V \cup U \times D)$$
$$= (S_1 \times V \cup (S_1 \cup S_2)^{\sigma\sigma} \times C) \cap (S_2 \times V \cup U \times D)$$
$$= (S_1 \cap S_2) \times V \cup S_1 \times D \cup S_2 \times C \cup ((S_1 \cup S_2)^{\sigma\sigma} \times (C \cap D)$$
$$= (S_1 \cap S_2) \times V \cup S_1 \times D \cup S_2 \times C.$$

Now, we are able to describe the non-induced blocks from Example 3.

Example 4. *By putting* $(A, B) = (\{u_2\}, \{u_1, u_2, u_3\})$, $S = \{u_1, u_2\}$ *and* $(C, D) = (\{v_1\}, \{v_1\})$ *or* $(C, D) = (\{v_2\}, \{v_2\})$, *the non-induced blocks* X_1, X_2 *from Example 3 can be expressed by part 1 of Theorem 7. This expression can be further simplified.*

$$X_1 = \{u_2\} \times V \cup \{u_1, u_2\} \times \{v_1\} \cup \{u_1, u_2, u_3\} \times \{v_1\}$$
$$= (S_1 \cap S_2) \times T_2 \cup (S_1 \cup S_2) \times T_1.$$
$$X_2 = \{u_2\} \times V \cup \{u_1, u_2\} \times \{v_2\} \cup \{u_1, u_2, u_3\} \times \{v_2\}$$
$$= (S_1 \cap S_2) \times T_1 \cup (S_1 \cup S_2) \times T_2.$$

5   Open Problems with Regard to Factor Analysis

In this section, we will state a conjecture about minimal factorizations of direct products of tolerance spaces and their size. First, the previous section suggests the formulation of the following open problem.

Problem 1. Given two tolerance spaces (U, σ) and (V, τ). Can we fully determine the blocks of their direct product $(U \times V, \sigma \check{\times} \tau)$?

With growing size of U and V, it is unlikely possible to find general formulas in terms of Theorem 7. However, one can hope for generalized construction principles or informations about the nature of non-induced blocks.

Due to Identity 5, the theory of lattice tensor products could be used to achieve this. Several characterizations of extents of tensor products are given in [5,9,13]. This might be an indication how extents which belong to a square concept can be characterized.

Regarding factor analysis, Theorem 4 showed that the set of all formal squares is always a factorization. But, as Example 3 suggests, not all formal squares are necessary. This raises the question of whether factorizations made of squares are always more efficient than those made of formal concepts, in the sense that they have less factors.

Problem 2. Let (U, σ) be a tolerance space and $\mathcal{T} \subseteq \mathfrak{B}(U, \tau)$ a minimal factorization with formal squares. Let $\mathcal{F} \subseteq \mathfrak{B}(U, \tau)$ be an arbitrary minimal factorization. Does it hold that $|\mathcal{T}| \leq |\mathcal{F}|$?

In other words, is a minimal covering of τ with maximal squares more "economical" than a minimal covering with maximal rectangles? Intuitively, we think this is true, but lack a proof of this statement. Eventually, the fact that $\mathfrak{B}(U, \tau)$ is a polarity lattice can be utilized, since it implies that to every minimal factorization $\mathcal{F}$, there exists a *dual* minimal factorization $p[\mathcal{F}]$. As formal squares are fixed points of p, it follows that $\mathcal{T} = p[\mathcal{T}]$, for every minimal factorization with formal squares $\mathcal{T}$.

Finally, it is natural to ask about the factorizations of the direct product of two tolerance spaces. Obviously, the set of all induced maximal squares covers $\sigma \check{\times} \tau$, but minimal covers of σ and τ are sufficient. We think that minimal factorizations with induced formal squares are already optimal.

Conjecture 1. *For tolerance spaces (U, σ) and (V, τ), let $\mathcal{S}$ and $\mathcal{T}$ be minimal factorizations with formal squares. For every minimal factorization $\mathcal{F}$ of the direct product $(U \times V, \sigma \check{\times} \tau)$ it holds that $|\mathcal{F}| = |\mathcal{S}| + |\mathcal{T}|$.*

In order for this conjecture to be true, Problem 2 must have a positive answer. Informations about the structure of non-induced blocks (Problem 1) are necessary too, so that it can be shown that non-induced formal squares do not provide a covering of $\sigma \check{\times} \tau$ with less elements than a covering with induced formal squares only. In case of two equivalence relations, Problem 2 has a positive answer and due to Theorem 6, a positive answer to Problem 2 in general would imply Conjecture 1.

6 Conclusion

In our current work we investigate certain building blocks of the direct product of tolerance relations and focus on the construction of blocks of the direct product when the blocks of each factor are known. Our main tool is formal concept analysis by considering the fact that blocks can be characterized as fixed points of the derivation operator. In Sect. 4, a description for some blocks of the direct product of two tolerances in terms of the blocks of each factor can be found. Further research should aim for a general description or construction principle for the blocks of the direct product.

Since, unlike equivalence classes of equivalence relations, the blocks of tolerances may overlap, the question of minimality (*i.e.*, minimal sets of squares that suffice to cover and thereby factorize the tolerance relation) arose. In Sect. 5, we discuss the connection between the structure of blocks of the direct product and minimal coverings by formal squares. This discourse leads to a conjecture on the size of factorizations of direct products.

Acknowledgments. Finally, we want to express our thanks to the anonymous referees for their valuable suggestions to improve our paper.

References

1. Assaghir, Z., Kaytoue, M., Kuznetsov, S.O., Napoli, A.: Embedding tolerance relations in formal concept analysis: an application in information fusion. In: Proceedings of the 19th International Conference on Information and Knowledge Management (2010)
2. Belohlavek, R., Vychodil, V.: Discovery of optimal factors in binary data via a novel method of matrix decomposition. JCSS **76**, 3–20 (2010)
3. Chajda, I., Niederle, J., Zelinka, B.: On existence conditions for compatible tolerances. Czechoslov. Math. J. **26**(101) (1976)
4. Deiters, K., Erné, M.: Sums, products and negations of contexts and complete lattices. Algebra Univers. **60**, 469–496 (2009)
5. Erné, M.: Tensor products of contexts and complete lattices. Algebra Univers. **31**, 36–65 (1994)
6. Ganter, B., Wille, R.: Formal Concept Analysis: Mathematical Foundations, 284 p. Springer, New York (1997). doi:10.1007/978-3-642-59830-2
7. Garanina, N.O., Grebeneva, J.V., Shilov, N.V.: Towards description logic on concept lattices. In: Proceedings of the Tenth International Conference CLA, pp. 287–292 (2013)
8. Järvinen, J., Radeleczki, S.: Rough sets determined by tolerances. Int. J. Approx. Reason. **55**(6), 1419–1438 (2014)
9. Krötzsch, M., Malik, G.: The tensor product as a lattice of regular galois connections. In: Missaoui, R., Schmidt, J. (eds.) ICFCA 2006. LNCS, vol. 3874, pp. 89–104. Springer, Heidelberg (2006). doi:10.1007/11671404_6
10. Peters, J.F., Wasilewski, P.: Tolerance spaces: origins, theoretical aspects and applications. Inf. Sci. **195**, 211–225 (2012)
11. Pogonowski, J.: Tolerance spaces with applications to linguistics. University Press, Institute of Linguistics, Adam Mickiewicz University, 103 p. (1981)
12. Schmidt, G.: Relational Mathematics. Cambridge University Press, Cambridge (2011)
13. Wille, R.: Tensorial decomposition of concept lattices. Order **2**, 81–95 (1985)

Viewing Morphisms Between Pattern Structures via Their Concept Lattices and via Their Representations

Lars Lumpe[(✉)] and Stefan E. Schmidt

Technische Universität Dresden, 01062 Dresden, Germany
larslumpe@gmail.com, midt1@msn.com

Abstract. In continuation of [13], we investigate pattern structures and their morphisms aiming to provide a theoretical background for complexity reduction. Our results follow a top-down strategy starting with a general setup of poset adjunctions; then we specify the situation for pattern structures and their representations. In particular, we discuss the situation where morphisms between pattern structures induce morphisms between their representations.

Morphisms between adjunctions turn out to be of crucial interest for a better understanding of morphisms between concept lattices of pattern structures.

1 Introduction

In [4], pattern structures have been introduced within the scope of formal concept analysis. They have turned out to be a powerful tool for analysing various real-world applications (cf. [4,7–10]). On the theoretical side, however, there has been only recently progress in a better understanding of the role of pattern morphisms. Our approach is a continuation of [13] and discusses the theoretical framework for pattern structures and their morphisms with respect to their concept lattices and their representations. In particular, we investigate the possible complexity reductions beyond projections and o-projections as studied in [2,11].

As a novel idea we present the theoretical conditions of complexity reduction for adjunctions and, subsequently, for pattern morphisms and their representations. In particular, our Theorem 7 clarifies the theoretical background of Theorem 2 in [4] and its adjustments (e.g. see [7] Theorem 3).

A major motivation for our work on pattern structures and their morphisms lies in the discovery that results on pattern structures as gained in [2,4,7] are largely based on hands-on definitions and constructions (e.g. projections, o-projections, representation contexts). In contrast to this, the investigation of pattern morphisms, closely connected with adjunctions, provides a theoretical framework for new insights and in addition proving old results on pattern structures.

For a better understanding, our results are guided by diagrams. This is a formal similarity to [3], however in a diffferent context.

© Springer International Publishing AG 2017
M. Kryszkiewicz et al. (Eds.): ISMIS 2017, LNAI 10352, pp. 597–608, 2017.
DOI: 10.1007/978-3-319-60438-1_59

2   Preliminaries

The fundamental order theoretic concepts of our paper are nicely presented in the book on *Residuation Theory* by Blyth and Janowitz (cf. [1]). Also see [5,6].

Definition 1 (Adjunction). Let $\mathbb{P} = (P, \leqslant)$ and $\mathbb{L} = (L, \leqslant)$ be posets; furthermore let $f : P \to L$ and $g : L \to P$ be maps.

(1) The pair (f, g) is an **adjunction** w.r.t. $(\mathbb{P}, \mathbb{L})$ if $fx \leqslant y$ is equivalent to $x \leqslant gy$ for all $x \in P$ and $y \in L$. In this case, we will refer to $(\mathbb{P}, \mathbb{L}, f, g)$ as a **poset adjunction**.

(2) f is **residuated** from $\mathbb{P}$ to $\mathbb{L}$ if the preimage of a principal ideal in $\mathbb{L}$ under f is always a principal ideal in $\mathbb{P}$, that is, for every $y \in L$ there exists $x \in P$ s.t.
$$f^{-1}\{t \in L \mid t \leqslant y\} = \{s \in P \mid s \leqslant x\}.$$

(3) g is **residual** from $\mathbb{L}$ to $\mathbb{P}$ if the preimage of a principal filter in $\mathbb{P}$ under g is always a principal filter in $\mathbb{L}$, that is, for every $x \in P$ there exists $y \in L$ s.t.
$$g^{-1}\{s \in P \mid x \leqslant s\} = \{t \in L \mid y \leqslant t\}.$$

(4) The dual of $\mathbb{L}$ is given by $\mathbb{L}^{\mathrm{op}} = (L, \geqslant)$ *with* $\geqslant := \{(x, t) \in L \times L \mid t \leqslant x\}$. The pair (f, g) is a **Galois connection** w.r.t. $(\mathbb{P}, \mathbb{L})$ if (f, g) is an **adjunction** w.r.t. $(\mathbb{P}, \mathbb{L}^{\mathrm{op}})$.

The following well-known facts are straightforward (cf. [1]).

Proposition 1. *Let* $\mathbb{P} = (P, \leqslant)$ *and* $\mathbb{L} = (L, \leqslant)$ *be posets.*

(1) A map $f : P \to L$ is residuated from $\mathbb{P}$ to $\mathbb{L}$ iff there exists a map $g : L \to P$ s.t. (f, g) is an adjunction w.r.t. $(\mathbb{P}, \mathbb{L})$.

(2) A map $g : L \to P$ is residual from $\mathbb{L}$ to $\mathbb{P}$ iff there exists a map $f : P \to L$ s.t. (f, g) is an adjunction w.r.t. $(\mathbb{P}, \mathbb{L})$.

(3) If (f, g) and (h, k) are adjunctions w.r.t. $(\mathbb{P}, \mathbb{L})$ with $f = h$ or $g = k$ then $f = h$ and $g = k$.

*(4) If f is a residuated map from $\mathbb{P}$ to $\mathbb{L}$, then there exists a unique residual map f^+ from $\mathbb{L}$ to $\mathbb{P}$ s.t. (f, f^+) is an adjunction w.r.t. $(\mathbb{P}, \mathbb{L})$. In this case, f^+ is called the **residual map** of f.*

*(5) If g is a residual map from $\mathbb{L}$ to $\mathbb{P}$, then there exists a unique residuated map g^- from $\mathbb{P}$ to $\mathbb{L}$ s.t. (g^-, g) is an adjunction w.r.t. $(\mathbb{P}, \mathbb{L})$. In this case, g^- is called the **residuated map** of g.*

(6) A residuated map f from $\mathbb{P}$ to $\mathbb{L}$ is surjective iff $f \circ f^+ = id_L$ iff f^+ is injective.

(7) A residuated map f from $\mathbb{P}$ to $\mathbb{L}$ is injective iff $f^+ \circ f = id_P$ iff f^+ is surjective.

Definition 2. Let $\mathcal{P} := (\mathbb{P}, \mathbb{S}, \sigma, \sigma^+)$ and $\mathcal{Q} := (\mathbb{Q}, \mathbb{T}, \tau, \tau^+)$ be poset adjunctions. Then a pair (α, β) forms a morphism from $\mathcal{P}$ to $\mathcal{Q}$ if $(\mathbb{P}, \mathbb{Q}, \alpha, \alpha^+)$ and $(\mathbb{S}, \mathbb{T}, \beta, \beta^+)$ are poset adjunctions satisfying

$$\tau \circ \alpha = \beta \circ \sigma$$

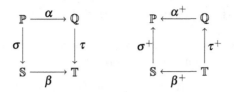

Fig. 1. Illustration of commutative diagrams

Remark: This implies $\alpha^+ \circ \tau^+ = \sigma^+ \circ \beta^+$, that is, the diagrams in Fig. 1 are commutative:

Definition 3 (Concept Poset). *For a poset adjunction $\mathcal{P} = (\mathbb{P}, \mathbb{S}, \sigma, \sigma^+)$ let*

$$B\mathcal{P} := \{(p, s) \in P \times S \mid \sigma p = s \wedge \sigma^+ s = p\}$$

denote the set of **(formal) concepts** in $\mathcal{P}$. Then the **concept poset** of $\mathcal{P}$ is given by $\mathbb{B}\mathcal{P} := (\mathbb{P} \times \mathbb{S}) \mid B\mathcal{P}$, that is, $(p_0, s_0) \leqslant (p_1, s_1)$ holds iff $p_0 \leqslant p_1$ (iff $s_0 \leqslant s_1$), for all $(p_0, s_0), (p_1, s_1) \in B\mathcal{P}$. If (p, s) is a formal concept in $\mathcal{P}$ then p is referred to as **extent** in $\mathcal{P}$ and s as **intent** in $\mathcal{P}$.

From [12] we point out Theorem 1, which provides a fundamental construction:

Theorem 1. *Let (α, β) be a morphism from a poset adjunction $\mathcal{P} = (\mathbb{P}, \mathbb{S}, \sigma, \sigma^+)$ to a poset adjunction $\mathcal{Q} = (\mathbb{Q}, \mathbb{T}, \tau, \tau^+)$. Then $(\mathbb{B}\mathcal{P}, \mathbb{B}\mathcal{Q}, \Phi_{(\alpha,\beta)}, \Phi^+_{(\alpha,\beta)})$ is a poset adjunction for*

$$\Phi_{(\alpha,\beta)} : B\mathcal{P} \to B\mathcal{Q}, (p, s) \mapsto (\tau^+ \beta s, \beta s)$$

and

$$\Phi^+_{(\alpha,\beta)} : B\mathcal{Q} \to B\mathcal{P}, (q, t) \mapsto (\alpha^+ q, \sigma \alpha^+ q).$$

*We call $(\mathbb{B}\mathcal{P}, \mathbb{B}\mathcal{Q}, \Phi_{(\alpha,\beta)}, \Phi^+_{(\alpha,\beta)})$ the **concept poset adjunction** induced by (α, β). Figure 2 shows an illustration of a concept poset adjunction.*

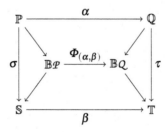

Fig. 2. Diagram of a concept poset adjunction

3   Morphisms Between Pattern Structures

From [12,13] we quote the following fundamental definitions of pattern structures, including their morphisms and representations.

Definition 4. A triple $\mathcal{G} = (G, \mathbb{D}, \delta)$ is a **pattern setup** if G is a set, $\mathbb{D} = (D, \sqsubseteq)$ is a poset, and $\delta : G \to D$ is a map. In case every subset of $\delta G := \{\delta g \mid g \in G\}$ has an infimum in $\mathbb{D}$, we will refer to $\mathcal{G}$ as **pattern structure**. Then the set

$$\mathbb{C}_{\mathcal{G}} := \{\inf_{\mathbb{D}} \delta X \mid X \subseteq G\}$$

forms a closure system in $\mathbb{D}$ and furthermore $\mathbb{C}_{\mathcal{G}} := \mathbb{D}|\mathbb{C}_{\mathcal{G}}$ forms a complete lattice. If $\mathcal{G} = (G, \mathbb{D}, \delta)$ and $\mathcal{H} = (H, \mathbb{E}, \varepsilon)$ each is a pattern setup, then a pair (f, φ) forms a **pattern morphism** from $\mathcal{G}$ to $\mathcal{H}$ if $f : G \to H$ is a map and φ is a residual map from $\mathbb{D}$ to $\mathbb{E}$ satisfying $\varphi \circ \delta = \varepsilon \circ f$, that is, the diagram in Fig. 3 is commutative.

Fig. 3. Commutative diagram of a pattern morphism

In the sequel we show how our previous considerations apply to pattern structures, therefore we need to clarify how to compose morphisms.

Definition 5. Let (α, β) be a morphism from a poset adjunction $\mathcal{P} = (\mathbb{P}, \mathbb{S}, \sigma, \sigma^+)$ to a poset adjunction $\mathcal{Q} = (\mathbb{Q}, \mathbb{T}, \tau, \tau^+)$ and (φ, ψ) be a morphism from $\mathcal{Q}$ to a poset adjunction $\mathcal{Q}' = (\mathbb{Q}', \mathbb{T}', \tau', \tau'^+)$ then the **concatenation** of (α, β) with (φ, ψ) is given by $(\varphi, \psi) \circ (\alpha, \beta) := (\varphi \circ \alpha, \psi \circ \beta)$ (Fig. 4).

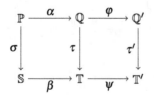

Fig. 4. Illustration of a concatenation

Remark: The concatenation of morphisms between poset adjunctions is again a morphism between poset adjunctions.

Definition 6. If $\mathcal{P}$, $\mathcal{Q}$, $\mathcal{P}'$, $\mathcal{Q}'$ are poset adjunctions and $\mathcal{P} \xrightarrow{a} \mathcal{Q}$, $\mathcal{P}' \xrightarrow{a'} \mathcal{Q}'$, $\mathcal{P} \xrightarrow{b} \mathcal{P}'$ and $\mathcal{Q}' \xrightarrow{c} \mathcal{Q}'$ are morphisms such that $c \circ a = a' \circ b$ holds, we will say that (a, c, b, a') is a **commutative square** of morphisms between poset adjunctions (Fig. 5).

Fig. 5. Commutative square of morphisms between poset adjunctions

To properly understand our Theorems 5 and 6 we have to recap the constructions of Theorems 3 and 4 in [13].

Construction 1. Let (f, φ) be a pattern morphism from a pattern structure $\mathcal{G} = (G, \mathbb{D}, \delta)$ to a pattern structure $\mathcal{H} = (H, \mathbb{E}, \varepsilon)$. To apply Theorem 1, we provide the following construction: f gives rise to an adjunction (α, α^+) between the power set lattices $2^G := (2^G, \subseteq)$ and $2^H := (2^H, \subseteq)$ via

$$\alpha : 2^G \to 2^H, X \mapsto fX \text{ and } \alpha^+ : 2^H \to 2^G, Y \mapsto \{x \in G | fx \in Y\}.$$

Further let φ^- denote the residuated map of φ w.r.t. $(\mathbb{E}, \mathbb{D})$, that is, $(\mathbb{E}, \mathbb{D}, \varphi^-, \varphi)$ is a poset adjunction. Then, obviously, $(\mathbb{D}^{op}, \mathbb{E}^{op}, \varphi, \varphi^-)$ is a poset adjunction too.

For pattern structures the following operators are essential:

$$\diamond : 2^G \to D, X \mapsto \inf_{\mathbb{D}} \delta X; \quad {}^{\bullet} : D \to 2^G, d \mapsto \{g \in G \mid d \sqsubseteq \delta g\} \text{ and}$$
$$\square : 2^H \to E, Z \mapsto \inf_{\mathbb{E}} \varepsilon Z; \quad {}^{\bullet} : E \to 2^H, e \mapsto \{h \in H \mid e \sqsubseteq \varepsilon h\}$$

It now follows that (α, φ) forms a morphism from the poset adjunction

$$\mathcal{P} = (2^G, \mathbb{D}^{op}, \diamond, {}^{\bullet}) \text{ to the poset adjunction } \mathcal{Q} = (2^H, \mathbb{E}^{op}, \square, {}^{\bullet}).$$

For the following we recall that the concept lattice of $\mathcal{G}$ is given by $\mathbb{B}\mathcal{G} := \mathbb{B}\mathcal{P}$ and the concept lattice of $\mathcal{H}$ is $\mathbb{B}\mathcal{H} := \mathbb{B}\mathcal{Q}$.
Then Theorem 1 yields that the quadruple $(\mathbb{B}\mathcal{G}, \mathbb{B}\mathcal{H}, \Phi_{(f,\varphi)}, \Phi^+_{(f,\varphi)})$ is an adjunction for (Fig. 6)

$$\Phi_{(f,\varphi)} : \mathbb{B}\mathcal{G} \to \mathbb{B}\mathcal{H}, (X, d) \mapsto ((\varphi d)^{\bullet}, \varphi d) \text{ and}$$
$$\Phi^+_{(f,\varphi)} : \mathbb{B}\mathcal{H} \to \mathbb{B}\mathcal{G}, (Z, e) \mapsto (f^{-1}Z, (f^{-1}Z)^{\diamond}).$$

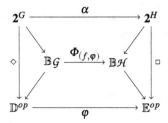

Fig. 6. Diagram associated with a pattern morphism

Remark: To apply Theorem 1 to Theorem 2 we point out that $\Phi_{(f,\varphi)} = \Phi_{(\alpha,\varphi)}$ holds.

Theorem 2. *Let $\mathcal{G} = (G, \mathbb{D}, \delta)$ and $\mathcal{H} = (H, \mathbb{E}, \epsilon)$ be pattern structure. And let $\mathcal{G}^{\bullet} = (G, \mathbb{C}_G, \delta^{\bullet})$ be the pattern structure induced by $\mathcal{G}$ via $\delta^{\bullet} : G \to C_G$, $g \mapsto \delta g$. This implies that $\mathbb{B}\mathcal{G}^{\bullet} = \mathbb{B}\mathcal{G}$. Further let (f, φ) be a pattern morphism from $\mathcal{G}^{\bullet}$ to $\mathcal{H}$. Then with the notation introduced in the previous theorem, the map $\Phi_{(f,\varphi)}$ from $\mathbb{B}\mathcal{G}$ to $\mathbb{B}\mathcal{H}$ is residuated. If f is surjective then so is $\Phi_{(f,\varphi)}$, if φ is injective then so is $\Phi_{(f,\varphi)}$. If f is surjective and φ is injective then $\Phi_{(f,\varphi)}$ is an isomorphism from $\mathbb{B}\mathcal{G}$ to $\mathbb{B}\mathcal{H}$.*

In the following theorems we investigate the theoretical background of a reduction of complexity for pattern structures and, even more, of pattern morphisms. Our results follow a top-down strategy starting with a general setup of poset adjunctions (see Theorem 3); then we specify the situation for pattern structures (see Theorem 4) and their representations. In particular, we discuss the situation how morphisms between pattern structures induce morphisms between their representations (see Theorem 5). Our Theorem 6 clarifies the theoretical background of Theorem 2 in [4] and its adjustments (e.g. see [7] Theorem 3).

Theorem 3. *A commutative square of morphisms between poset adjunctions induces a commutative square of residuated maps between their concept posets. More explicitly, let $\mathcal{P} = (\mathbb{P}, \mathbb{S}, \sigma, \sigma^+)$, $\mathcal{Q} = (\mathbb{Q}, \mathbb{T}, \tau, \tau^+)$, $\mathcal{P}' = (\mathbb{P}', \mathbb{S}', \sigma', \sigma'^+)$ and $\mathcal{Q}' = (\mathbb{Q}', \mathbb{T}', \tau', \tau'^+)$ poset adjunctions and let Fig. 7 be a commutative square of morphisms between poset adjunctions, that is, the six sides of Fig. 8 consist of commutative squares of residuated maps. Then the diagram in Fig. 9 of induced residuated maps between the concept posets of the involved poset adjunctions (as constructed in Theorem 1) is commutative. In particular the residuated maps shown in Fig. 9 form a commutative square between concept posets.*

Proof. Our claim is $\Phi_{(\varphi 2, \varphi 4)} \circ \Phi_{(\alpha, \beta)} = \Phi_{(\alpha', \beta')} \circ \Phi_{(\varphi 1, \varphi 3)}$. For the proof of this we refer to Fig. 10. So let $(p, s) \in \mathbb{B}\mathcal{P}$. Then we have

$$\Phi_{(\varphi 2, \varphi 4)}(\Phi_{(\alpha, \beta)}(p, s)) = \Phi_{(\varphi 2, \varphi 4)}(\tau^+ \beta s, \beta s) = (\tau'^+ \varphi 4 \beta s, \varphi 4 \beta s) \text{ and}$$
$$\Phi_{(\alpha', \beta')}(\Phi_{(\varphi 1, \varphi 3)}(p, s)) = \Phi_{(\alpha', \beta')}(\sigma^+ \varphi 3 s, \varphi 3 s) = (\tau'^+ \beta' \varphi 3 s, \beta' \varphi 3 s).$$

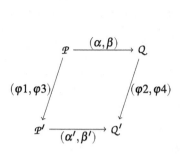

Fig. 7. Commutative square of morphisms between poset adjunctions

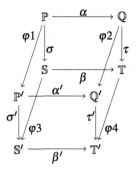

Fig. 8. Cube of commutative squares

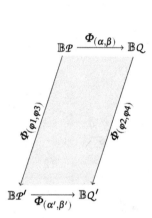

Fig. 9. Commutative square

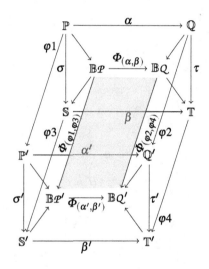

Fig. 10. Induced residuated maps

Since $\varphi4 \circ \beta = \beta' \circ \varphi3$ we conclude

$$(\Phi_{(\varphi2,\varphi4)} \circ \Phi_{(\alpha,\beta)})(p, s) = (\tau'^{+}\varphi4\beta s, \varphi4\beta s)$$
$$= (\tau'^{+}\beta'\varphi3 s, \beta'\varphi3 s)$$
$$= (\Phi_{(\alpha',\beta')} \circ \Phi_{(\varphi1,\varphi3)})(p, s)$$

$\square$

Next we show the remarkable impact of the previous theorem on pattern morphisms.

Theorem 4. *A commutative square of pattern morphisms between pattern structures induces a commutative square of residual maps between their concept lattices.*

More explicitly, let $G = (G, \mathbb{D}, \delta)$, $\mathcal{H} = (H, \mathbb{E}, \varepsilon)$, $G' = (G', \mathbb{D}', \delta')$ and $\mathcal{H}' = (H', \mathbb{E}', \varepsilon')$ be pattern structures and let a commutative square of pattern morphisms be given as in Fig. 11. This means that the six sides of the cube in Fig. 12 are commutative squares. Then it follows that Fig. 13 forms a commutative square of residuated maps between concept lattices of pattern structures.

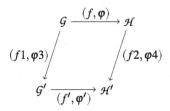

Fig. 11. Commutative square of pattern morphisms

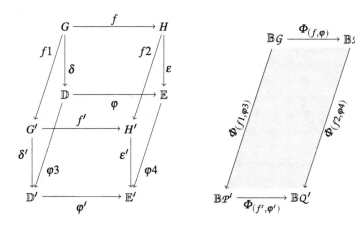

Fig. 12. Cube of commutative squares

Fig. 13. Commutative square of resid-uated maps

Proof. The six sides of the cube in Fig. 14 are commutative squares of residuated maps. Now we apply Theorem 3 and receive

$$\Phi_{(\varphi 2, \varphi 4)} \circ \Phi_{(\alpha, \varphi)} = \Phi_{(\alpha', \varphi')} \circ \Phi_{(\varphi 1, \varphi 3)}, \text{ that is,}$$
$$\Phi_{(f 2, \varphi 4)} \circ \Phi_{(f, \varphi)} = \Phi_{(f', \varphi')} \circ \Phi_{(f 1, \varphi 3)}$$

□

Definition 7. Let $G = (G, \mathbb{D}, \delta)$ be a pattern structure and M be a subset of D. Then the *M-representation* of G is defined as the pattern structure $\mathcal{P}(G, M) := (G, 2^M, \underline{\delta})$ with $\underline{\delta} : G \rightarrow 2^M, g \mapsto_M^\downarrow \delta g := \{m \in M | m \sqsubseteq \delta g\}$, that is, $\underline{\delta}g$ is the set of all subpatterns of $g \in G$ which are contained in M.

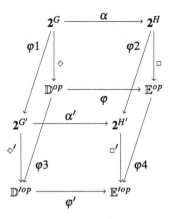

Fig. 14. Cube of commutative squares

Remark:(1) It is worth mentioning that for every formal context its concept lattice coincides with the concept lattice of its associated pattern structure. More precisely, for a formal context $\mathbb{K} = (G, M, I)$ its **associated pattern structure** is given by $\mathcal{P}\mathbb{K} := (G, 2^M, \delta : G \to 2^M, g \mapsto \{m \in M \mid gIm\})$. Then obviously, the concept lattices of $\mathbb{K}$ and $\mathcal{P}\mathbb{K}$ are equal.

(2) For a pattern structure $\mathcal{G} = (G, \mathbb{D}, \delta)$ and a subset M of D the **representation context** of $\mathcal{G}$ over M (as defined in [13]) is given by $\mathbb{K}(\mathcal{G}, M) := (G, M, I)$ with $I := \{(g, m) \in G \times M \mid m \sqsubseteq \delta g\}$. Then the concept lattices of $\mathcal{P}(\mathcal{G}, M)$ and $\mathbb{K}(\mathcal{G}, M)$ coincide and will be denoted by $\mathbb{B}(\mathcal{G}, M)$.

We want to point out that representation contexts as defined in [4,13] are in this paper discussed within the framework of M-representations of pattern structures. The next result gives new insight how pattern morphisms, which allow a representation, lead to a commutative square of residuated maps between the associated concept lattices.

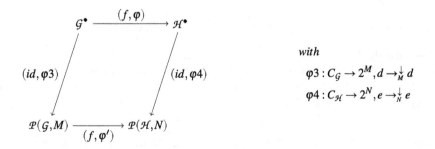

Fig. 15. Commutative square of pattern morphisms

Theorem 5. *Let* $\mathcal{G} = (G, \mathbb{D}, \delta)$ *and* $\mathcal{H} = (H, \mathbb{E}, \varepsilon)$ *be pattern structures, furthermore let* $M \subseteq D$ *and* $N \subseteq E$. *Also let Fig. 15 be a commutative square of pattern morphisms.*

Then Fig. 16 is a commutative square of residuated maps between concept lattices of pattern structures.

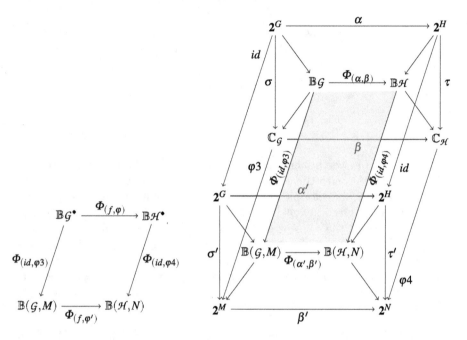

Fig. 16. Commutative square between concept lattices

Fig. 17. Induced residuated maps

Proof. The proof follows from Theorem 4 by considering the commutative diagram in Fig. 17.   □

Our next result is a representation theorem, which clarifies Theorem 2 in [4].

Theorem 6. *Let* $\mathcal{G} = (G, \mathbb{D}, \delta)$ *and* $\mathcal{H} = (H, \mathbb{D}, \varepsilon)$ *be pattern structures, furthermore let* M, N *be sets with* $N \subseteq M \subseteq D$. *Also let Fig. 18 be a commutative square of pattern morphisms. Then Fig. 19 is a commutative square of residuated maps between concept lattices of pattern structures.*

Proof. The proof follows from Theorem 4 by considering the diagram in Fig. 20.
□

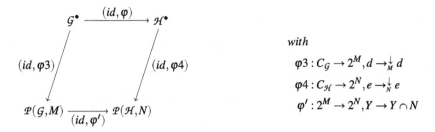

Fig. 18. Commutative square of pattern morphisms

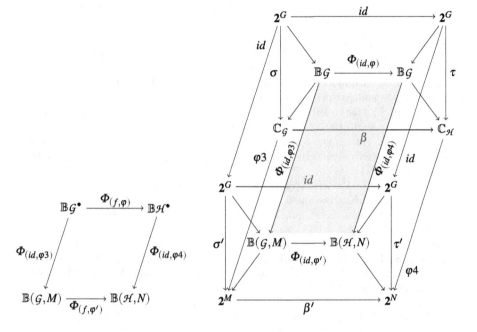

Fig. 19. Commutative square between concept lattices

Fig. 20. Induced residuated maps

4   Conclusion

We introduced a novel idea for complexity reduction for adjunctions which refers to a generalized concept of "projection" applied not only to objects but also to arrows. Also, we provide a new framework of how the corrections (like Theorem 3 in [7]) of Theorem 2 in [4] can be better understood via morphisms and their representations. Beyond this, we show that poset adjunctions and their morphisms provide an elegant theoretical tool for discussing possibilities of complexity reduction of pattern structures and their morphisms. As a major insight, morphisms between adjunctions are apparently highly relevant for the analysis of morphisms between concept lattices of pattern structures.

References

1. Blyth, T.S., Janowitz, M.F.: Residuation Theory, pp. 1–382. Pergamon Press, Oxford (1972)
2. Buzmakov, A., Kuznetsov, S.O., Napoli, A.: Revisiting pattern structure projections. In: Baixeries, J., Sacarea, C., Ojeda-Aciego, M. (eds.) ICFCA 2015. LNCS, vol. 9113, pp. 200–215. Springer, Cham (2015). doi:10.1007/978-3-319-19545-2_13
3. Dubois, D., Prade, H., Rico, A.: The cube of opposition: a structure underlying many knowledge representation formalisms. In: IJCAI 2015, pp. 2933–2939 (2015)
4. Ganter, B., Kuznetsov, S.O.: Pattern structures and their projections. In: Delugach, H.S., Stumme, G. (eds.) ICCS-ConceptStruct 2001. LNCS, vol. 2120, pp. 129–142. Springer, Heidelberg (2001). doi:10.1007/3-540-44583-8_10
5. Erné, M., Koslowski, J., Melton, A., Strecker, G.E.: A primer on Galois connections. Ann. N. Y. Acad. Sci. **704**(1), 103–125 (1993)
6. Erné, M.: Adjunctions and Galois connections. In: Galois Connections and Applications, pp. 1–138 (2004)
7. Kaiser, T.B., Schmidt, S.E.: Some remarks on the relation between annotated ordered sets and pattern structures. In: Kuznetsov, S.O., Mandal, D.P., Kundu, M.K., Pal, S.K. (eds.) PReMI 2011. LNCS, vol. 6744, pp. 43–48. Springer, Heidelberg (2011). doi:10.1007/978-3-642-21786-9_9
8. Kaytoue, M., Kuznetsov, S.O., Napoli, A., Duplessis, S.: Mining gene expression data with pattern structures in formal concept analysis. Inf. Sci. (Elsevier) **181**, 1989–2001 (2011)
9. Kuznetsov, S.O.: Pattern structures for analyzing complex data. In: Sakai, H., Chakraborty, M.K., Hassanien, A.E., Ślęzak, D., Zhu, W. (eds.) RSFDGrC 2009. LNCS (LNAI), vol. 5908, pp. 33–44. Springer, Heidelberg (2009). doi:10.1007/978-3-642-10646-0_4
10. Kuznetsov, S.O.: Scalable knowledge discovery in complex data with pattern structures. In: Maji, P., Ghosh, A., Murty, M.N., Ghosh, K., Pal, S.K. (eds.) PReMI 2013. LNCS, vol. 8251, pp. 30–39. Springer, Heidelberg (2013). doi:10.1007/978-3-642-45062-4_3
11. Lumpe, L., Schmidt, S.E.: A note on pattern structures and their projections. In: Baixeries, J., Sacarea, C., Ojeda-Aciego, M. (eds.) ICFCA 2015. LNCS, vol. 9113, pp. 145–150. Springer, Cham (2015). doi:10.1007/978-3-319-19545-2_9
12. Lumpe, L., Schmidt, S.E.: Pattern structures and their morphisms. In: CLA 2015, pp. 171–179 (2015)
13. Lumpe, L., Schmidt, S.E.: Morphisms between pattern structures and their impact on concept lattices. In: FCA4AI, pp. 25–33 (2016)

Formal Concept Analysis for Knowledge Discovery

On-Demand Generation of AOC-Posets: Reducing the Complexity of Conceptual Navigation

Alexandre Bazin[1], Jessie Carbonnel[2]([⊠]), and Giacomo Kahn[1]

[1] LIMOS and Université Clermont Auvergne, Clermont-Ferrand, France
contact@alexandrebazin.com, giacomo.kahn@isima.fr
[2] LIRMM, CNRS and Université de Montpellier, Montpellier, France
jessie.carbonnel@lirmm.fr

Abstract. Exploratory search allows to progressively discover a dataspace by browsing through a structured collection of documents. Concept lattices are graph structures which support exploratory search by conceptual navigation, i.e., navigating from concept to concept by selecting and deselecting descriptors. These methods are known to be limited by the size of concept lattices which can be too large to be efficiently computed or too complex to be browsed intelligibly. In this paper, we address the problem of providing techniques that reduce the complexity of FCA-based exploratory search. We show the suitability of AOC-posets, a condensed alternative structure to achieve conceptual navigation. Also, we outline algorithms to enable an on-demand generation of AOC-posets. The necessity to devise more flexible methods to perform product selection in software product line engineering is what motivates our work.

Keywords: Formal concept analysis · AOC-poset · Concept navigation · Software product line engineering · Product selection

1 Introduction

Exploratory search is an information retrieval strategy that aims at guiding the user into a space of existing documents to help him select the one that best suits his needs. This process is particularly adapted to situations where a user is unfamiliar with the dataspace, or when the data is too large to be known entirely. Lattice structures were among the first structures used to support information retrieval processes [9], and their usage was later generalised to Formal Concept Analysis (FCA) theory [8]. The concept lattice offers a convenient structure to do exploratory search, where navigating from concept to concept by selecting or deselecting attributes emulates iterative modifications of the document descriptor selection, and thus of the current research state. Exploratory search by conceptual navigation has been used in several applications, for instance querying web documents [4] or browsing a collection of images [6]. However, FCA-based

© Springer International Publishing AG 2017
M. Kryszkiewicz et al. (Eds.): ISMIS 2017, LNAI 10352, pp. 611–621, 2017.
DOI: 10.1007/978-3-319-60438-1_60

exploratory search raises some problems, mainly because of the size (in terms of number of concepts) of lattices, which are well known to grow exponentially with the size of the input data. Computing the whole concept lattice can take time and it needs adapted algorithms to be efficiently used in applications. Moreover, a user can rapidly get disoriented while navigating in such a large and convoluted structure. Therefore, several ways to overcome these limitations have been studied in the literature [7,11,13].

In this paper, we propose a new and more scalable approach to perform exploratory search by conceptual navigation, that relies on local generation of AOC-posets, a partial sub-order of concept lattices. Unlike concept lattices, which depict all possible queries a user can formulate, this alternative conceptual structure represents and structures the minimal set of queries that are necessary to perform conceptual navigation, and therefore permit navigation through a less complex structure. Also, to avoid generating the whole AOC-poset, we only generate the current concept and its neighbourhood, represented by its direct sub-concepts and super-concepts. In fact, even though an AOC-poset is smaller than its associated concept lattice, it still can be advantageous to only generate the parts of the structure we are interested in, especially in large datasets. We outline algorithms to identify neighbour concepts in AOC-posets, i.e., determining upper and lower covers of a given concept in an AOC-poset. An application of exploratory search in the field of software product line engineering, for an activity called product selection, is what motivates our work.

The remainder of this paper is organised as follows. In Sect. 2, we present our motivations in the domain of software product line engineering. In Sect. 3, we study AOC-posets to perform exploratory search. We then propose algorithms to compute upper and lower covers of a concept in AOC-posets in Sect. 4, and we test our approach on existing datasets. Related work is discussed in Sect. 5, and Sect. 6 concludes and presents some future work.

2   Motivation

Software product line engineering (SPLE) [14] is a development paradigm that aims to efficiently create and manage a collection of related software systems. SPLE is based on factorisation and exploitation of a common set of artifacts, organised around a generic architecture from which several software variants can be derived. A central point of SPLE is the modelisation of the common parts and the variants contained in the related software systems, called the *variability* of the software product line (SPL). This variability is represented by variability models, which are the traditional starting points to perform information retrieval operations on SPLs, including product selection, an important task that consists in guiding the user into selecting the functionalities he wants in the final derived software system. The most prevalent approach to model variability relies on *feature models* (FMs) [12], a family of visual languages that describe a set of features (i.e., main characteristics) and dependencies between them. Figure 1 depicts an FM representing an SPL about cell phones. A combination of features

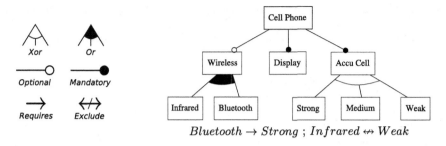

Fig. 1. Excerpt of a feature model representing an SPL about cell phones

respecting all the constraints expressed in the FM is called a *valid configuration*, and corresponds to a derivable software system.

Current approaches for product selection rely on the feature model hierarchy to automatically deploy configurators; however, these methods are too stiff considering that it does not allow the user to change his final configuration without having to start again the product selection, or to see which other configurations are similar to his. We propose to apply exploratory search in the context of product selection to complement these methods and offer a more flexible selection. In fact, conceptual navigation allows a user to start from an existing or partial configuration, explore similar ones, and be informed on how he can select or deselect features to obtain these configurations. It is noteworthy that the number of valid configurations depicted by an FM grows exponentially with its number of features. To be able to conceive applications using conceptual navigation in this context, reducing the complexity of the underlying conceptual structure along with its generation time is crucial.

3   AOC-Poset: A Condensed Structure for Conceptual Navigation

Formal Concept Analysis (FCA) [8] is a mathematical framework that structures a set of objects described by attributes depending on the attributes they share. As input, FCA takes a *formal context* $K = (\mathcal{O}, \mathcal{A}, \mathcal{R})$, where $\mathcal{O}$ is the set of objects, $\mathcal{A}$ the set of attributes and $\mathcal{R} \subseteq \mathcal{O} \times \mathcal{A}$ a binary relation. A pair (a, o) from $\mathcal{R}$ states that "the object o possesses the attribute a". Table 1 presents the formal context representing the 7 valid configurations of the FM of Fig. 1.

The application of FCA permits to extract from a context K a finite set of *formal concepts* through the use of two *derivation operators* $(\cdot)'$; $(\cdot)' : 2^{\mathcal{O}} \mapsto 2^{\mathcal{A}}$, and $(\cdot)' : 2^{\mathcal{A}} \mapsto 2^{\mathcal{O}}$. Thus, $O' = \{a \in \mathcal{A} \mid \forall o \in O, (o, a) \in \mathcal{R}\}$ and $A' = \{o \in \mathcal{O} \mid \forall a \in A, (o, a) \in \mathcal{R}\}$. A formal concept C is a pair (E, I) with $E \subseteq \mathcal{O}$ and $I \subseteq \mathcal{A}$, representing a maximal set of objects that share a maximal set of common attributes. $E = I'$ is the concept's *extent* (denoted $Ext(C)$), and $I = E'$ is the concept's *intent* (denoted $Int(C)$). The set of all concepts extracted from K together with the extent set-inclusion order forms a lattice structure called

Table 1. Formal context depicting the 7 configurations of the SPL about cell phones

	Cell phone	Wireless	Infrared	Bluetooth	Display	Accu cell	Strong	Medium	Weak
c_1	x				x	x	x		
c_2	x				x	x		x	
c_3	x				x	x			x
c_4	x	x	x		x	x	x		
c_5	x	x	x		x	x		x	
c_6	x	x		x	x	x	x		
c_7	x	x	x	x	x	x	x		

a *concept lattice*. Figure 2 (left) presents the concept lattice associated with the formal context of Table 1. We simplify the representation of intents and extents in the lattice by displaying each attribute (resp. object) only once in the structure, in the lowest (resp. the greatest) concept having this attribute (resp. object). We say that these concepts *introduce* an element. The attributes of a concept are inherited from top to bottom, and the objects from bottom to top.

A concept introducing at least an attribute is called an *attribute-concept* (AC), and a concept introducing at least an object is called an *object-concept* (OC). A concept can introduce both an attribute and an object (*attribute-object-concept* (AOC)), or it can introduce neither of them (*plain-concept*). Plain-concepts appear in the lattice as concepts with empty extents and intents. In some types of applications, it is not necessary to take these concepts into account. For instance, this is the case when the lattice is only used as a support to organise objects and their attributes (therefore represented by their introducer concepts), and not to highlight maximal groups of elements. In these particular cases, one can benefit from only generating the sub-order restricted to the *introducer* concepts instead of the whole concept lattice. This smaller structure (in terms of number of concepts) is called an *Attribute-Object-Concept partially ordered set* (AOC-poset) [10]. Figure 2 (right) presents the AOC-poset associated with the context from Table 1: it corresponds to the partial order of concepts from Fig. 2 (left), minus Concept_0 and Concept_7. While a concept lattice can have up to $2^{min(|\mathcal{A}|,|\mathcal{O}|)}$ concepts, the associated AOC-poset cannot exceed $|\mathcal{O}| + |\mathcal{A}|$ concepts.

AOC-posets can be used as a smaller alternative to concept lattices to (1) structure a collection of objects depending on the attributes they share and (2) navigate through this collection by selecting and deselecting attributes. But, if concept lattices represent all possible queries a user can formulate, AOC-posets restrict this set to the minimal queries required to perform conceptual navigation. As we have seen before, neighbour concepts in a concept lattice represent minimal possible modifications a user can make to the current query and therefore offer a dataspace in which one can navigate in minimal steps. This means that concept lattices allow to select and deselect non-cooccurrent attributes one by one. AOC-posets do not preserve the minimal step query refinement/enlargement

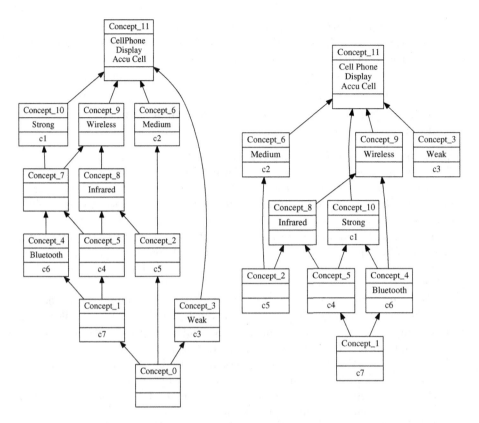

Fig. 2. Concept lattice (left) and AOC-poset (right) associated with the formal context of Table 1

property, but factorise the possible query modification steps to keep the most prevalent ones. For instance, in the concept lattice of Fig. 2 (left), if a user has selected Concept_4 as the current concept, he can choose to deselect attribute *Bluetooth* and thus move to Concept_7. From this concept, he can now choose to deselect either *Strong* or *Wireless* and move respectively to Concept_9 or Concept_10. In AOC-posets, because plain-concepts, playing the role of "transition steps", are not present, selection/deselection choices to move from concept to concept are condensed. This time, in Fig. 2 (right), if a user want to enlarge its query from Concept_4, he can either deselect both *Bluetooth* and *Strong* in one step to move to Concept_9, or deselect both *Bluetooth* and *Wireless* to reach Concept_10.

4   Partial Generation of AOC-Poset

On-demand, or local, generation consists in generating only the part of the structure we are interested in, and has already been applied to concept lattices, with algorithms such as nextClosure [8]. To our knowledge, several algorithms exist to build AOC-posets: Ares, Ceres, Pluton and Hermes [3]. However, none of them perform on-demand generation of AOC-posets. In what follows, we outline algorithms to retrieve the neighbourhood of a given concept in an AOC-poset.

4.1   Computing Upper and Lower Covers of a Concept in the AOC-Poset

Exploration can start from the top concept (i.e., the most general query), in the case where the user wants to make a software configuration from scratch. But, it is possible that the user already has partial knowledge of the configuration he wants, and it is then necessary to be able to start from any concept in the AOC-poset. As the concept corresponding to the (potentially partial) configuration that the user has in mind does not necessarily introduce an object or an attribute, we suppose the input of the exploration is a formal concept, plain or not. The problem is thus to compute the upper and lower covers of a given concept C_i in the AOC-poset.

Let us start with computing the upper covers. We are looking for the smallest ACs or OCs greater than the input. We start out by computing the smallest ACs greater than C_i. They can be obtained by computing the concepts $(\{a\}', \{a\}'')$ for each attribute $a \in Int(C_i)$. We remark that $(\{a_1\}', \{a_1\}'') \geq (\{a_2\}', \{a_2\}'')$ if and only if $a_1 \in \{a_2\}''$. As such, the smallest ACs are the ones that are computed from attributes that do not appear in the closures of other attributes. Once we have the smallest ACs, we want to compute the smallest OCs that are between them and C_i. This means that we are looking for concepts $(\{o\}'', \{o\}')$ such that o is in the extent of one of the ACs we have and $\{o\}' \subset Int(C_i)$. We remark once again that $(\{o_1\}', \{o_1\}'') \geq (\{o_2\}', \{o_2\}'')$ implies $o_2 \in \{o_1\}''$ and that the closures of some objects give us information on OCs that can't be minimal.

Algorithm 1 computes the upper neighbours of the input concept in the AOC-poset. The first loop computes the closure of single attributes. Each closure allows us to remove attributes that correspond to non-minimal ACs. The resulting set R contains the intents of the ACs that are both super-concepts of C_i and minimal for this property. The second loop constructs the set O of objects that are in the extent of an element of R but not in the extent of C_i. The third loop removes the objects of O that cannot possibly be introduced by a superset of C_i and, finally, the fourth loop removes the objects of O that produce non-minimal OCs. The ACs that are no longer minimal are also removed. Therefore, considering the initial configuration, the OCs introduce the most similar and more generalised configurations, and the ACs show the factorised possible attribute de-selections the user can make.

Computing the lower covers is done using the same algorithm, exchanging the roles of attributes and objects. This time, OCs present the most similar and more specialised configurations, and the ACs the possible condensed attribute selections.

Algorithm 1. UPPER COVERS

Input: A concept C_i
Output: The upper covers of C_i in the AOC-poset

1   $A \leftarrow Int(C_i)$
2   **foreach** $a \in A$ **do**
3 | $Y \leftarrow \{a\}''$
4 | $A \leftarrow A \setminus \{Y \setminus \{a\}\}$
5   $R \leftarrow \{\{a\}''|a \in A\}$
6   $O \leftarrow \emptyset$
7   **forall** $S \in R$ **do**
8 | $X \leftarrow S'$
9 | $O \leftarrow O \cup (X \setminus Ext(C_i))$
10 **forall** $o \in O$ **do**
11 | **if** $o' \not\subseteq Int(C_i)$ **then**
12 | | $O \leftarrow O \setminus \{o\}$
13 **forall** $o \in O$ **do**
14 | $T = \{S|(S \in R) \wedge (o \in S')\}$
15 | $R \leftarrow R \setminus T$
16 | $Y \leftarrow \{o\}''$
17 | **if** $\exists p \in O$ *such that* $p \in Y$ **then**
18 | | $O \leftarrow O \setminus \{o\}$
19 $R \leftarrow \{(\{o\}'', \{o\}') \mid o \in O\}$
20 **return** R

4.2 Implementation

We have implemented our algorithms, and we tested them on SPL datasets extracted from the SPLOT repository[1]. SPLOT (for *Software Product Line Online Tools*) is an academic website providing a repository of feature models along with a set of tools to create, edit and perform automated analysis on them. We have selected 13 representative feature models which describe SPLs as e-shops, cell phones or video games, from small sizes (13 configurations) to larger ones (4774 configurations). To test our method on data extracted from feature models, we first create a formal context $configurations \times features$ for each one of them. Then, from a context, our implementation permits to find a concept corresponding to a subset of features and compute its conceptual neighbourhood in AOC-posets. In this experiment, we assume that a user will not exceed 50 navigation steps, as he wants to be familiarised to the similar valid configurations around his initial selection of features. To measure the gain of our method in terms of number of computed concepts, we compare for each context the number of computed concepts for 50 navigation steps (i.e., a concept and its neighbourhood) to the total number of concepts in the associated AOC-poset and concept lattice. The results are presented in Fig. 3. The concept lattice curve permits to

---

[1] http://www.splot-research.org/.

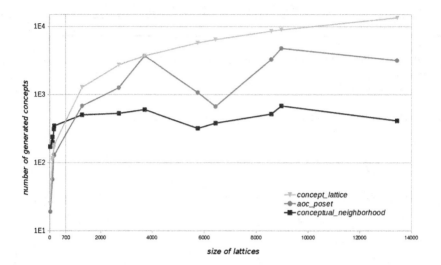

Fig. 3. Number of generated concepts for AOC-posets and conceptual neighbourhood for 50 navigation steps, depending on the size of their associated concept lattices (logarithmic scale)

indicate an "upper-bound" to visualise more easily the gain of AOC-posets and local-generation of AOC-posets comparing to concept lattices.

Figure 3 shows that both aspects of our method are useful for conceptual structures with a size around 700+ concepts. AOC-posets are smaller than concept lattices, but the gap between the two conceptual structures can grow exponentially with their sizes. In fact, the difference when the structures are small is not very important (e.g., 19 concepts against 25, 131 against 166), but AOC-posets can become very interesting with larger structures (e.g., 1074 concepts against 5761, 669 against 6430). Performing several navigation steps in a small structure makes re-computation of same concepts more likely. In these cases, it is preferable to compute the whole AOC-poset from the beginning. Our experiment shows that this is the case when the initial conceptual structure possesses less than around 700 concepts, and with 50 navigation steps.

5   Related Work

Several methods have been proposed through the literature to reduce the complexity of conceptual navigation. In [9], the authors choose not to show the whole concept lattice to the user, but only a part of it, restricted to a focus concept and its neighbourhood. This navigation approach, which we study and apply in this paper, can be found in several works [1,5,6]. In [13], the authors propose two methods to extract trees from concept lattices and use them as less complex structures for browsing and visualisation. The difference between the two methods lies in the way the "best" parent for each concept in the tree is assigned: the first one is based on the selection of one parent per layer, and the second one on conceptual indexes. They then simplify again the final structure by applying two reduction methods based on fault-tolerance and clustering on the extracted trees. In [2], the authors propose a tool to build and visualise formal concept trees. Carpineto and Romano [4] allowed the user to bound the information space by dynamically applying constraints during the search to prune the concept lattice. Bounding allows to reduce the explorable dataspace and help the user focus on the parts he is interested in. Following the same idea, iceberg concept lattices [15] are pruned structures that only show the top-part of concept lattices which can be used to perform conceptual navigation. By comparison, we use a partial sub-order of concept lattices. In [5], the authors present `SearchSleuth`, a tool for local analysis of web queries based on FCA, that derives a concept and its neighbourhood from a query. Because the domain cannot be computed entirely, it generates a new formal context at each navigation step: for each user query, it retrieves the list of results, extracts the relevant terms, and builds a context from these terms and their associated documents. The navigation is managed through an interface which suggests terms, making implicit the underlying graph structure and its complexity. Alam et al. [1] present a tool, `LatViz`, that provides several operations to reduce the information space. One of them facilitates the visualisation and the navigation. The authors propose to display the concept lattice *level-wise*: selecting a concept at a level n displays all its sub-concept at level $n - 1$. Another functionality allows to prune the concept lattice by restricting navigation in sub-concepts and/or super-concepts of concepts selected by the user, in the same way as in [4]. Also, the tool permits to compute AOC-posets to support conceptual navigation: the authors describe AOC-posets as the "core" of their corresponding concept lattices. However, they compute the whole structure using the Hermes algorithm and do not propose an on-demand generation. Greene and Fischer [11] discuss refinement and enlargement (broadening) approaches which are not restricted to neighbour concepts, and therefore allow navigation by non-minimal steps to ease exploratory search in large information spaces.

6   Conclusion and Future Work

In this paper, we address the problem of providing scalable and praticable techniques to perform conceptual navigation with formal concept analysis. Product selection, a software product line engineering task that can benefit from exploratory search by conceptual navigation to complement current methods which lack of flexibility is the motivation of our work. We used AOC-posets, the concept lattice sub-hierarchy restricted to introducer concepts, as a smaller, condensed alternative to concept lattices that preserve objects and attributes taxonomy. We show that AOC-posets depict the minimal set of queries necessary to browse the dataspace, as they "factorise" the possible attribute selection and deselection steps. To avoid generating the whole sub-hierarchy, we outline algorithms to enable on-demand generation of AOC-posets by computing the neighbourhood of any concept in the AOC-poset. We implemented these algorithms to test our approach on a dozen of SPLs extracted from the SPLOT repository. These experiments reveal that our method provides a gain in terms of number of generated concepts when used instead of concept lattices, when the concept lattice has a size larger than about 700 concepts to perform 50 navigation steps.

In the future, we are considering further experiments on different datasets to provide a more complete evaluation of the gain of the proposed approach. From a more theoretical point of view, we plan to extend exploratory search by conceptual navigation to relational data using Relational Concept Analysis.

References

1. Alam, M., Le, T.N.N., Napoli, A.: LatViz: a new practical tool for performing interactive exploration over concept lattices. In: 13th International Conference on Concept Lattices and Their Applications (CLA), pp. 9–20 (2016)
2. Andrews, S., Hirsch, L.: A tool for creating and visualising formal concept trees. In: 5th Conceptual Structures Tools and Interoperability Workshop (CSTIW 2016) Held at the 22nd International Conference on Conceptual Structures (ICCS 2016), pp. 1–9 (2016)
3. Berry, A., Huchard, M., McConnell, R.M., Sigayret, A., Spinrad, J.P.: Efficiently computing a linear extension of the sub-hierarchy of a concept lattice. In: Ganter, B., Godin, R. (eds.) ICFCA 2005. LNCS (LNAI), vol. 3403, pp. 208–222. Springer, Heidelberg (2005). doi:10.1007/978-3-540-32262-7_14
4. Carpineto, C., Romano, G.: Exploiting the potential of concept lattices for information retrieval with CREDO. J. Univers. Comput. Sci. **10**(8), 985–1013 (2004)
5. Ducrou, J., Eklund, P.W.: SearchSleuth: the conceptual neighbourhood of an web query. In: 5th International Conference on Concept Lattices and Their Applications (CLA) (2007)
6. Ducrou, J., Vormbrock, B., Eklund, P.: FCA-based browsing and searching of a collection of images. In: Schärfe, H., Hitzler, P., Øhrstrøm, P. (eds.) ICCS-ConceptStruct 2006. LNCS, vol. 4068, pp. 203–214. Springer, Heidelberg (2006). doi:10.1007/11787181_15
7. Ferré, S.: Efficient browsing and update of complex data based on the decomposition of contexts. In: Rudolph, S., Dau, F., Kuznetsov, S.O. (eds.) ICCS-ConceptStruct 2009. LNCS, vol. 5662, pp. 159–172. Springer, Heidelberg (2009). doi:10.1007/978-3-642-03079-6_13
8. Ganter, B., Wille, R.: Formal Concept Analysis - Mathematical Foundations. Springer, Heidelberg (1999). doi:10.1007/978-3-642-59830-2
9. Godin, R., Gecsei, J., Pichet, C.: Design of a browsing interface for information retrieval. In: 12th International Conference on Research and Development in Information Retrieval (SIGIR), pp. 32–39 (1989)
10. Godin, R., Mili, H.: Building and maintaining analysis-level class hierarchies using Galois lattices. In: 8th Conference on Object-Oriented Programming Systems, Languages, and Applications (OOPSLA), pp. 394–410 (1993)
11. Greene, G.J., Fischer, B.: Single-focus broadening navigation in concept lattices. In: 3rd Workshop on Concept Discovery in Unstructured Data, Co-located with the 13th International Conference on Concept Lattices and Their Applications (CLA), pp. 32–43 (2016)
12. Kang, K.C., Cohen, S.G., Hess, J.A., Novak, W.E., Peterson, A.S.: Feature-oriented domain analysis (FODA) feasibility study. Citeseer (1990)

13. Melo, C.A., Grand, B.L., Aufaure, M.: Browsing large concept lattices through tree extraction and reduction methods. Int. J. Intell. Inf. Technol. (IJIIT) **9**(4), 16–34 (2013)
14. Pohl, K., Böckle, G., van der Linden, F.: Software Product Line Engineering - Foundations, Principles, and Techniques. Springer, Heidelberg (2005). doi:10.1007/3-540-28901-1
15. Stumme, G., Taouil, R., Bastide, Y., Pasquier, N., Lakhal, L.: Computing iceberg concept lattices with Titanic. Data Knowl. Eng. (DKE) **42**(2), 189–222 (2002)

From Meaningful Orderings in the Web of Data to Multi-level Pattern Structures

Quentin Brabant[(⊠)], Miguel Couceiro, Amedeo Napoli, and Justine Reynaud

LORIA (CNRS - Inria Nancy Grand Est - Université de Lorraine), BP 239,
54506 Vandoeuvre-les-Nancy, France
{quentin.brabant,miguel.couceiro,amedeo.napoli,justine.reynaud}@loria.fr

Abstract. We define a pattern structure whose objects are elements of a supporting ontology. In this framework, descriptions constitute trees, made of triples subject-predicate-object, and for which we provide a meaningful similarity operator. The specificity of the descriptions depends on a hyperparameter corresponding to their depth. This formalism is compatible with ontologies formulated in the language of RDF and RDFS and aims to set up a framework based on pattern structures for knowledge discovery in the web of data.

Keywords: Formal concept analysis · Pattern structure · Ontology · Knowledge discovery

1 Introduction

The recent development of the web of data made available an increasing amount of ontological knowledge in various fields of interest. This knowledge is mainly expressed as set out in the specifications of the Resource Description Framework (RDF)[1] and RDF Schema (RDFS)[2].

This preliminary study aims to provide a knowledge discovery framework that produces an ordered structure summarizing and describing objects from a given ontology. The definition of ontology that we use can be seen as an abstraction of the data model defined by RDF and RDFS standards. In this paper, by an ontology we mean a structure that combines two types of knowledge about a set of objects. The first type is *subsumptional knowledge*: it corresponds to the ordering of objects in terms of their specificity, which is given by a subsumption relation (also sometimes referred to as *IS A* relation). The second type is *predicate knowledge*, thought of as a set of relations that occur between objects. The latter kind of knowledge can be naturally represented by an oriented labelled multigraph where nodes are objects and where the labels of arcs indicate the kind of relation that occurs between the origin and the target objects of the arc.

Our approach is rooted in Formal Concept Analysis (FCA) [7] and pattern structures [6], which are two frameworks for knowledge discovery that allow the

---

[1] https://www.w3.org/RDF/.

[2] https://www.w3.org/TR/rdf-schema/.

© Springer International Publishing AG 2017
M. Kryszkiewicz et al. (Eds.): ISMIS 2017, LNAI 10352, pp. 622–631, 2017.
DOI: 10.1007/978-3-319-60438-1_61

comparison and the classification of objects with respect to their descriptions. In this article we propose a framework to integrate knowledge lying in ontologies into the knowledge discovery process. More precisely, we define a pattern structure in which objects are described according to the information available in the ontology. This work follows recent propositions made in [1,10], with the goal of extracting knowledge from the ontology with as much accuracy as possible.

In Sect. 2 we survey basic background on order theory and introduce the basic structures that constitute ontology in our framework. In Sect. 3 we recall basic notions and terminology of FCA and pattern structures. In Sect. 4 we define a pattern structure in which knowledge from the ontology is used to build the descriptions of objects. The potential of this framework in knowledge discovery is then briefly discussed in Sect. 5.

2 Preliminaries

2.1 Subsumption Relations and Posets

In this paper we will make use of several order relations on differents sets, that we think of as subsumption relations. In other words, they order objects from the most specific to the most general. With no danger of ambiguity, they all will be denoted by the same symbol $\sqsubseteq$. For a set X endowed with $\sqsubseteq$ and $a, b \in X$, the fact that a is subsumed by b will be denoted by $a \sqsubseteq b$, with the meaning that a is more specific than b or, equivalently, that b is more general than a.

The subsumption relations considered hereinafter have the property that any two elements from a given set have a unique *least common subsumer*, in other words, they constitute semilattice orders. This enables us to define a *similarity* (or *meet*) operator $\sqcap$ by, for $a, b \in X$, $a \sqcap b$ is the least common subsumer of a and b. Note that

$$a \sqsubseteq b \quad \Leftrightarrow \quad a \sqcap b = b. \tag{1}$$

Note that since we only consider finite sets, X necessarily has a most general element. As for subsumption relations, we will denote all similarity operators by the same symbol $\sqcap$ when there is no danger of ambiguity.

Also, for $A \subseteq X$, we will denote by $mins(A)$ the subset of minimal values of A, i.e.,

$$mins(A) = \{c \in A \mid \nexists c' \in A : c' \sqsubseteq c \text{ and } c \neq c'\}.$$

In this paper, we will use the term *partially ordered set* (or simply *poset*) for any set X of elements ordered by a subsumption relation $\sqsubseteq$.

2.2 Ontological Knowledge

In this section we aim to define an ontology that combines information of two kinds, namely *subsumptional knowledge* and *predicate knowledge* as explained below.

Throughout this paper we will consider two disjoint sets, a set V of *objects* and a set P of *predicates*. We will also consider a subsumption order on each of these sets, such that $(P, \sqsubseteq)$ and $(V, \sqsubseteq)$ are posets with most general elements denoted by $\top_P$ and $\top_V$, respectively.

Now let $\mathcal{G} = (V, E)$ be an oriented multigraph with labelled arcs, where V is a set of vertices and $E \subseteq V \times P \times V$ is a set of triples such that $(s, p, o) \in E$ if s is connected to o by an arc labelled by p. Note that an element $s \in V$ can be connected to $o \in V$ by several arcs with different labels, however there can be at most one arc with a label p connecting s to o.

In our framework, $(P, \sqsubseteq)$ and $(V, \sqsubseteq)$ provide the *subsumptional knowledge* of the ontology while the graph $\mathcal{G}$ provides its *predicate knowledge*.

Example 1. The following example illustrates the kind of knowledge that can be captured by such structures. Abbreviated namings of objects and predicates are given between parentheses. Let

$$V = \{\text{beef}, \text{dessert}, \text{dish}, \text{egg}, \text{ice-cream(i-c.)}, \text{onion tortilla(o. tortilla)},$$
$$\text{menu1}(\text{m}_1), \text{menu2}(\text{m}_2), \text{onion}, \text{potato}, \text{steak and chips(s\&c)},$$
$$\text{strawberry i-c.}, \text{tortilla}, \text{vanilla i-c.}, \top_V\},$$
$$P = \{\text{has component(hcomp)}, \text{has ingredient(hing)}, \top_P\}.$$

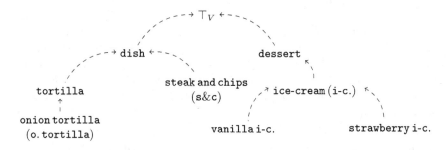

Fig. 1. The poset of objects given by $(V, \sqsubseteq)$. Dashed arrows represent subsumption relations from their origin to their target. All elements of V that do not appear in the figure are subsumed by $\top_V$. This poset indicates for example that $\text{tortilla} \sqsubseteq \text{dish}$, and that $\text{tortilla} \sqcap \text{s\&c} = \text{dish}$.

Figures 1, 2 and 3 represent respectively the posets of objects, the poset of predicates and the graph $\mathcal{G}$ describing relations between objects through predicates.

The pair of subsumptional and predicate knowledge is commonly used to describe objects within the same field of interest (e.g., class diagrams in object oriented programming). Ontologies using RDF and RDFS formats also describe these types of knowledge, but with a clear separation made between *classes* and

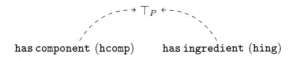

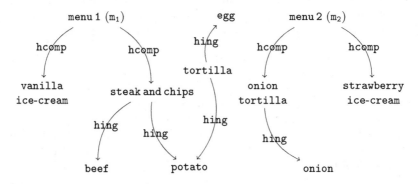

Fig. 2. The poset of predicates given by $(P, \sqsubseteq)$. Dashed arrows represent a subsumption relation between their target and origin.

Fig. 3. The relation graph $\mathcal{G}$. Vertices are elements of V, while arcs are labelled with predicates of P. Arcs provide information about the relations between vertices. This graph represents, for instance, the facts that tortilla is made of eggs and potatoes, and that menu 1 contains vanilla ice-cream. Isolated vertices (i.e. objects that share no arcs with any other object) are not represented in the graph.

instances. Indeed, RDF and RDFS specifications constitute a framework for describing *resources*, which are divided into three categories: classes, instances and *properties*. Instances can be related to each other via properties, while classes and properties are organized into so-called *hierarchies*. Hierarchies, as well as relation between instances, are expressed through a set of triples of the form resource-property-resource. Class and properties hierarchies are built using the properties `rdfs:subClassOf` and `rdfs:subProperty`, respectively. Instances are affected to classes using the `rdf:type` property.

A natural correspondence between RDFS and our formalism is then the following. The graph $\mathcal{G}$ represents the set of RDF triples (except those triples describing hierarchical relations), the set V represents the set of all classes and instances in the ontology, and the subsumption relation on V summarizes the information given by `rdfs:subClassOf` and `rdf:type` properties. Moreover, $(P, \sqsubseteq)$ corresponds to the RDFS property hierarchy that is expressed through `rdfs:subPropertyOf`.

Remark 1. RDFS hierarchies are not necessarily semilattices, since two elements do not necessarily have a unique least common subsumer. However, it is always possible to embed a poset into a lattice of subsets (see, e.g., [3,4]).

In the following section we briefly recall the basics of FCA and pattern structures, which can be used to explore ontological knowledge.

3   FCA and Pattern Structures

Formal Concept Analysis (FCA) is a mathematical framework used in classification and knowledge discovery. The basic framework considers a set of objects, a set of attributes, and an incidence relation between the two. More precisely a *formal context* is a triple (G, M, I) where G is the set of objects, M is the set of attributes and $I \subseteq G \times M$ is an incidence relation between G and M. Here $(g, m) \in I$ is interpreted as the object g has the attribute m. In this setting two derivation operators $.' : 2^G \to 2^M$ and $.' : 2^M \to 2^G$ are usually defined by

$$A' = \{m \in M \mid \forall g \in A : (g, m) \in I\},$$
$$B' = \{g \in G \mid \forall m \in B : (g, m) \in I\},$$

for every $A \subseteq G$ and $B \subseteq M$. A *formal concept* is then a pair (A, B) such that $A = B'$ and $B = A'$. In other words, A is the set of all objects that possess all attributes in B and, dually, B is set of all attributes common to all objects in A. The concept lattice associated to the formal context is then the set of all formal concepts ordered by

$$(A_1, B_1) \leq (A_2, B_2) \quad \text{if and only if} \quad A_1 \subseteq A_2 \quad \text{or, equivalently,} \quad B_2 \subseteq B_1.$$

FCA is fully detailed in [7].

FCA only deals with binary attribute values, i.e. objects are described by binary $|M|$-sequences where each $0/1$ component expresses the presence/absence of the corresponding attribute in the object. In the pattern structure framework, these descriptions become richer, the so-called *patterns*, and they may take values other than $0/1$.

In this framework, the set of patterns is usually denoted by D and endowed with a similarity operator $\sqcap$ that defines a semilattice $\underline{D} = (D, \sqcap)$. Note that with this similarity operator can define a subsumption relation $\sqsubseteq$ given by (1). In this sense we say that a description d is more *general* than a description d' if $d' \sqsubseteq d$. Descriptions of objects $g \in G$ are given by a mapping $\delta : G \to D$. The corresponding pattern structure is then $(G, \underline{D}, \delta)$.

As in FCA, two derivation operators are considered, namely, $.^\diamond : 2^G \to D$ and $.^\diamond : D \to 2^G$, and defined by

$$A^\diamond = \bigsqcap_{g \in A} \delta(g) \quad \text{and} \quad d^\diamond = \{g \in G \mid d \sqsubseteq \delta(g)\},$$

for every $A \subseteq G$ and every $d \in D$. A *pattern concept* is then a pair (A, d) such that $A = d^\diamond$ and $d = A^\diamond$. The set A is the set of objects whose descriptions are more general than d, and d is the similarity of descriptions of objects in A.

Note that a pattern structure can be translated into the FCA formalism [6]. A *representation context* of a pattern structure $(G, \underline{D}, \delta)$ is a formal context (G, M, I) where the set of attributes M is a subset of the set of patterns D, and the incidence relation I is defined by: $(g, m) \in I$ if $m \sqsubseteq \delta(g)$. A representation context can be seen as a formal context where patterns are translated in terms of binary attributes values.

4   A Pattern Structure for Objects in an Ontology

In this section we discuss the construction of a pattern structure whose objects are the vertices of $\mathscr{G}$, that is, where $G \subseteq V$ (see Sect. 2).

4.1   Simple Descriptions

The description of an object $g \in G$ is an element of the set of patterns that represents the characteristics of g. To give a meaningful description to g, we make use of the two types of knowledge present in the ontology, described in Sect. 2, namely, subsumptional knowledge which is given by the position of g in the poset, and the predicate knowledge which is provided by the multigraph $\mathscr{G}$.

Example 2. Consider the two structures given in Example 1. From the poset of objects we know that "onion tortilla" is a tortilla, a dish, and a thing ($\top_V$). From $\mathscr{G}$ we know that an onion tortilla is made of onion and, since it is a kind of tortilla, of potatoes and eggs. We know that "steak and chips" is a dish and a thing, and that it is made of beef and potatoes. We can extract common characteristics about "onion tortilla" and "steak and chips": both are dishes made of potatoes.

It should be noticed in this example that the predicate knowledge about a subsumer of g also applies to g. We can represent the predicate knowledge about a g by the set

$$\{(p, o) \mid \exists s : (s, p, o) \in E \text{ and } g \sqsubseteq s\}.$$

Each couple in this set corresponds to a knowledge unit about g, thought of as one of its characteristics. Some of these characteristics can be more specific than others. For instance, if we have "menu 1 is composed of strawberry ice-cream", then we also have that "menu 1 is composed of ice-cream", while the converse is not true. Thus we say that (hcomp, strawberry i-c.) is a more specific characteristic than (hcomp,i-c.). This intuition is formalized by the subsumption relation on $P \times V$ given by

$$(p_1, o_1) \sqsubseteq (p_2, o_2) \text{ if } p_1 \sqsubseteq p_2 \text{ and } o_1 \sqsubseteq o_2, \tag{2}$$

for $p_1, p_2 \in P$ and $o_1, o_2 \in V$. With this definition, $(p_1, o_1) \sqsubseteq (p_2, o_2)$ expresses the fact that (p_1, o_1) gives a more specific information than (p_2, o_2). Moreover, for $A \subseteq P \times V$, $mins(A)$ is the smallest set (w.r.t the subsumption relation on $P \times V$) that describes the same characteristics as A.

We are now able to provide a description function for elements of G, based on both subsumptional and predicate knowledge. The description function $\delta^1 :$ $G \to D^{(1)}$, where $D^{(1)} = V \times 2^{P \times V}$, is defined by

$$\delta^1(g) = \ <g, mins(\{(p, o) \mid \exists s : (s, p, o) \in E \text{ and } g \sqsubseteq s\})>,$$

where $mins$ is defined in terms of the order $\sqsubseteq$ over $P \times V$ given by (2). Note that $D^{(1)}$ is now used as the set of patterns. The similarity between two descriptions $d_1 = \,<s_1, e_1>$ and $d_2 = \,<s_2, e_2>$ belonging to $D^{(1)}$ is given by

$$d_1 \sqcap d_2 = \,<s_1 \sqcap s_2, mins(\{(p_1 \sqcap p_2, o_1 \sqcap o_2) \mid (p_1, o_1) \in e_1 \text{ and } (p_2, o_2) \in e_2\})>.$$

The description $d_1 \sqcap d_2$ represents the characteristics that are common to both d_1 and d_2. We can see that the first component of the similarity is given by $s_1 \sqcap s_2$, namely the least common subsumer of s_1 and s_2 in $(V, \sqsubseteq)$. For the second component, we have all possible overlappings between couples from e_1 and e_2.

This similarity operator endows $D^{(1)}$ with a semilattice structure whose most general element is $<\top_V, \{\}>$. Therefore $\mathcal{P}^1 = (G, (D^{(1)}, \sqcap), \delta^1)$ is a pattern structure.

Example 3. To illustrate, let us compute the descriptions of menu 1 (m_1) and menu 2 (m_2).

$$\delta^1(m_1) = \,<m_1, \{(\texttt{hcomp}, \texttt{vanilla i-c.}), (\texttt{hcomp}, \texttt{s\&c})\}>,$$
$$\delta^1(m_2) = \,<m_2, \{(\texttt{hcomp}, \texttt{strawberry i-c.}), (\texttt{hcomp}, \texttt{salad})\}>,$$

The similarity between both menus is

$$\delta^1(m_1) \sqcap \delta^1(m_2) = \,<\top_V, \{(\texttt{hcomp}, \texttt{i-c.}), (\texttt{hcomp}, \texttt{dish})\}>.$$

From this similarity we obtain that menu 1 and menu 2 are two things ($\top_V$) that are composed of an ice-cream and a dish.

It could be argued that the pattern structure described in this subsection is not fully satisfying since the description function δ^1 may fail to capture some characteristics that appear "deeper" in the graph $\mathcal{G}$. This problem is already illustrated in Example 3, where we see that the fact that both menus contain a dish made of potatoes is not taken into account. This is due to the fact that the description of each menu expresses "composed of a dish that is steak and chips" or "composed of a dish that is onion tortilla", but not "composed of a dish that has potato as ingredient". To arrive at such level of specificity, in the next section we will examine deeper connections in the graph $\mathcal{G}$ through multi-level descriptions.

4.2 Multi-level Descriptions

The principle of multi-level descriptions is to nest the descriptions of neighbors of g into the description of g. For instance, the multi-level description of a menu should contain the descriptions of the dishes that compose it. We define k-*level description functions* $\delta^k : G \to D^{(k)}$ where $D^{(k)} = V \times 2^{P \times D^{(k-1)}}$ and $k > 1$ is the maximal number of nestings that are allowed in the description, by:

$$\delta^k(g) = \,<g, mins(\{(p, \delta^{k-1}(o)) \mid \exists s : (s, p, o) \in E \text{ and } g \sqsubseteq s\})>.$$

Note that $\delta^{k-1}(o)$ is the description of level $(k-1)$ of the neighbors of g in $\mathcal{G}$. The similarity between two descriptions $d_1 = <s_1, e_1>$ and $d_2 = <s_2, e_2>$ is given by

$$d_1 \sqcap d_2 = <s_1 \sqcap s_2, mins(\{(p_1 \sqcap p_2, d_1' \sqcap d_2') \mid (p_1, d_2') \in e_1 \text{ and } (p_2, d_2') \in e_2\})>.$$

Again, this similarity operator endows $D^{(k)}$ with a semilattice structure whose most general element is $<\top_V, \{\}>$. Therefore $\mathcal{P}^k = (G, (D^{(k)}, \sqcap), \delta^k)$ is a pattern structure.

Example 4. Continuing from Example 3, the 2-level descriptions of menu 1 (m_1) and menu 2 (m_2) are

$$\delta^2(m_1) = <m_1, \{(\texttt{hcomp}, <\texttt{vanilla i-c.}, \{\}>),$$
$$(\texttt{hcomp}, <\texttt{s\&c}, \{(\texttt{hing, potato}), (\texttt{hing, beef})\}>)\}>,$$
$$\delta^2(m_2) = <m_1, \{(\texttt{hcomp}, <\texttt{strawberry i-c.}, \{\}>),$$
$$(\texttt{hcomp}, <\texttt{o.tortilla}, \{(\texttt{hing, potato}), (\texttt{hing, egg}), (\texttt{hing, onion})\}>)\}>,$$

and the 2-level similarity between these descriptions is

$$\delta^2(m_1) \sqcap \delta^2(m_2) = <\top_V, \{(\texttt{hcomp, i-c.}), (\texttt{hcomp}, <\texttt{dish}, \{(\texttt{hing, potato})\}>)\}>.$$

The similarity between the 2-level descriptions of m_1 and m_2 expresses the fact that both are things composed of ice-cream and of a dish in which potato is an ingredient.

Interestingly, due to the recursive nature of k-level descriptions, they can be represented as trees of depth at most k. The tree corresponding to a description $d = <s, e> \in D^{(k)}$ can be drawn through the following steps:

1. The root of the tree has value s.
2. For all $(p, d') \in e$, create a branch labelled by p, leading to a child that is the subtree given by d'.

Example 5. The trees corresponding to $\delta^2(m_1)$, $\delta^2(m_2)$ and the similarity between these descriptions are given in Fig. 4.

This pattern structure $\mathcal{P}^k$ is based on a similarity operation that the extraction of common characteristics of descriptions in a meaningful manner. Moreover the parameter k can be chosen by the user to control the deepness of descriptions. As we have seen, deeper descriptions allow a more complete detection of common characteristics.

Even though the number of levels considered can be as high as desired, the size of descriptions could grow rapidly with the number of levels. This could constitute a drawback as such pattern structures could deemed to be unusable in practice. However, it is reasonable to assume that most relevant information about an object can be found in its vicinity, that is for small values of k.

630   Q. Brabant et al.

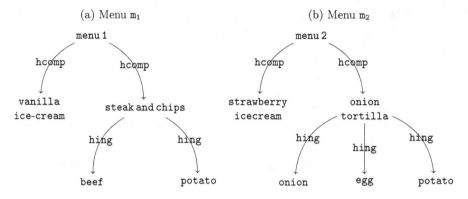

(a) Menu m₁   (b) Menu m₂

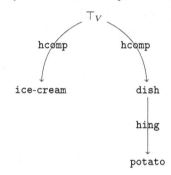

(c) Similarity between the descriptions of m₁ and m₂.

Fig. 4. Trees representing the 2-level descriptions of m₁ (a) and m₂ (b), and the similarity of their descriptions (c).

5   Discussion and Perspectives

In this paper we proposed a method for building a pattern structure that exploits ontological knowledge in object descriptions, and that opens new perspectives in the classification of complex and structured data. Indeed, since pattern structures identify common characteristics of subsets of objects (through the similarity operation), they can be used to learn intentional definitions of target classes (see [2]). Since we are mainly interested in the classification of instances, the relational graph is only used as a source of information from which we build descriptions of the instances, that are represented as trees. This makes our approach different from those of [5,11], that are based on FCA and pattern structures, but where the comparisons are made between graphs.

Also, in [8,9] the authors propose a setting for ordinal classification, based on the theory of rough sets, and that makes use of ontological knowledge. This approach considers a generality/specificity relation between objects for aiding the preference prediction. However, this approach has two main limitations: it implicitly forces a correspondence between the preference and the generality relations, and it does not take into account all kinds of predicates available in the ontology. Our framework brings an alternative to this rough set approach that can even be used in situations where objects are relational structures such as trees, ordered structures or even graphs. For instance, when dealing with menus (seen as trees), the classification task could be that of classifying menus into preference classes, or with respect to suitable diets. Future work will also focus on application and empirical aspects, in order to evaluate the efficiency of this formalism in knowledge discovery and classification tasks.

References

1. Alam, M., Napoli, A.: Interactive exploration over RDF data using formal concept analysis. In: Proceedings of IEEE International Conference on Data Science and Advanced Analytics, Paris, France, August 2015
2. Belohlavek, R., De Baets, B., Outrata, J., Vychodil, V.: Inducing decision trees via concept lattices 1. Int. J. Gen. Syst. **38**(4), 455–467 (2009)
3. Caspard, N., Leclerc, B., Monjardet, B.: Finite Ordered Sets: Concepts, Results and Uses, no. 144. Cambridge University Press, Cambridge (2012)
4. Davey, B.A., Priestley, H.A.: Introduction to Lattices and Order. Cambridge University Press, Cambridge (2002)
5. Ganter, B., Grigoriev, P.A., Kuznetsov, S.O., Samokhin, M.V.: Concept-based data mining with scaled labeled graphs. In: Wolff, K.E., Pfeiffer, H.D., Delugach, H.S. (eds.) ICCS-ConceptStruct 2004. LNCS, vol. 3127, pp. 94–108. Springer, Heidelberg (2004). doi:10.1007/978-3-540-27769-9_6
6. Ganter, B., Kuznetsov, S.O.: Pattern structures and their projections. In: Delugach, H.S., Stumme, G. (eds.) ICCS-ConceptStruct 2001. LNCS, vol. 2120, pp. 129–142. Springer, Heidelberg (2001). doi:10.1007/3-540-44583-8_10
7. Ganter, B., Wille, R.: Formal Concept Analysis: Mathematical Foundations, 1st edn. Springer, Heidelberg (1997). doi:10.1007/978-3-642-59830-2
8. Pancerz, K.: Decision rules in simple decision systems over ontological graphs. In: Burduk, R., Jackowski, K., Kurzynski, M., Wozniak, M., Zolnierek, A. (eds.) Proceedings of the 8th International Conference on Computer Recognition Systems CORES 2013. Advances in Intelligent Systems and Computing, vol. 226, pp. 111–120. Springer, Heidelberg (2013). doi:10.1007/978-3-319-00969-8_11
9. Pancerz, K., Lewicki, A., Tadeusiewicz, R.: Ant-based extraction of rules in simple decision systems over ontological graphs. Int. J. Appl. Math. Comput. Sci. **25**(2), 377–387 (2015)
10. Reynaud, J., Toussaint, Y., Napoli, A.: Contribution to the classification of web of data based on formal concept analysis. In: What Can FCA Do for Artificial Intelligence (FCA4AI) (ECAI 2016), La Haye, Netherlands, August 2016
11. Soldano, H.: Extensional confluences and local closure operators. In: Baixeries, J., Sacarea, C., Ojeda-Aciego, M. (eds.) ICFCA 2015. LNCS, vol. 9113, pp. 128–144. Springer, Cham (2015). doi:10.1007/978-3-319-19545-2_8

On Locality Sensitive Hashing for Sampling Extent Generators

Victor Codocedo$^{(\boxtimes)}$ and My Thao Tang

Inria Chile, Santiago, Chile
victor.codocedo@inria.cl, mythao.tang@inria.cl

Abstract. In this article we introduce a method for *sampling* formal concepts using locality sensitive hashing (LSH). LSH is a technique used for finding approximate nearest neighbours given a set of hashing functions. Through our approach, we are able to predict the probability of an extent in the concept lattice given set of objects and their similarity index, a generalization of the Jaccard similarity between sets. Our approach allows defining a lattice-based amplification construction to design arbitrarily discriminative sampling settings.

1 Introduction

Formal Concept Analysis (FCA) [5] practitioners are often faced with computational complexity problems when dealing with large datasets. The most efficient algorithms for mining formal concepts (Fast CbO, AddIntent, InClose, etc. [8]) are all limited by the fact that enumerating all formal concepts on a given formal context is an NP-hard problem. For this reason, a rich branch of the FCA domain is dedicated on finding better strategies to work with the same algorithms. While mining only frequent patterns is the most traditional one, other strategies for restricting the search space such as scoring, projecting and constraining formal concepts (k-patterns, stability, o-patterns, etc.) [2,6,7] can be found in the literature. The contribution of this article is the proposition of a novel strategy to sample extent generators (object sets with a non-empty set of common attributes) based on a probabilistic approach, namely locality sensitive hashing (LSH) [9].

2 Theoretical Background

2.1 Formal Concept Analysis and Pattern Structures

For the sake of brevity, we do not provide a description in depth of the FCA framework. Nevertheless, we introduce some notations taken from [5] that will be necessary to understand our formalizations. We denote a formal context as $\mathcal{K} = (G, M, I)$ where G is the set of objects, M is the set of attributes and $I \subseteq G \times M$ is an incidence relation (an example is provided in Table 1). The derivation operators for an object set $A \subseteq G$ and an attribute set $B \subseteq M$ in $\mathcal{K}$ are A^I and

© Springer International Publishing AG 2017
M. Kryszkiewicz et al. (Eds.): ISMIS 2017, LNAI 10352, pp. 632–641, 2017.
DOI: 10.1007/978-3-319-60438-1_62

Table 1. Formal context and example hashing functions f and π

	m_1	m_2	m_3	m_4	m_5	m_6	m_7	m_8	m_9	m_{10}	m_{11}	m_{12}	$f(g^I)$	$\pi(g^I)$
g_1	×	×	×		×				×				5	1
g_2	×			×	×	×	×		×				6	1
g_3		×	×	×	×			×					5	2
g_4	×		×		×			×					4	1
g_5			×			×	×						3	4
g_6								×			×		2	9
g_7									×	×			2	10
g_8									×	×	×		3	10
g_9											×	×	2	11

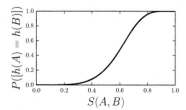

Fig. 1. Amplified probability for a minhash family $r = 5, b = 8$

B^I, respectively. A formal concept is a pair (A, B) where $A = B^I$ and $B = A^I$. An object set A_1 is said to be a generator of A iff $A_1^{II} = A$. The object concept of object $g \in G$ is defined as $\gamma(g) = ((g^I)^I, g^I)$. A projection $\psi : \wp(M) \to \wp(M)$ is a monotone, idempotent and contractive mapping defined over the complete lattice of attributes $(\wp(M), \cap)$ [4].

2.2   Locality Sensitive Hashing

LSH is a probabilistic approach to the problem of k-nearest neighbours (KNN), relaxed in a problem known as rNN. While in KNN we are interested in finding the *k closest nearest neighbours* to any given point in a metric space, in rNN we are interested in finding points in the space that are at distance at most r from any given point. In LSH this is achieved by means of a *family of locality sensitive hashing functions* specially designed for grouping points which are near to each other w.r.t. a distance metric. In the following, we provide a brief introduction to LSH for sets using the notation of FCA.

Definition 1 (Jaccard Similarity & Distance). *The Jaccard similarity S between any two sets A and B (s.t. $A \cup B \neq \emptyset$) is defined as:*

$$S(A, B) = \frac{|A \cap B|}{|A \cup B|}$$

Analogously we can define the *Jaccard distance J* as the complement of the Jaccard similarity, i.e. $J(A, B) = 1 - S(A, B)$. In this work we will use both notions when convenient. Let us consider a space of sets D. It has been showed that the Jaccard distance J is a metric in D, as it verifies the conditions of non-negativity, identity of indiscernibles, symmetry and triangle inequality. Thus, D and J define a metric space in which any two sets $A, B \in D$ can be regarded as points in the space. Obviously, we are interested in the case $D = \wp(M)$, where M is a set of attributes in a formal context. The Jaccard similarity can be applied to two object representations in the formal context $S(g_i^I, g_j^I)$ where $g_i, g_j \in G$. For example, consider the formal context in Table 1, we have: $S(g_7^I, g_8^I) = 2/3$ and $S(g_7^I, g_9^I) = 1/3$

Definition 2 (Hashing). *A hash function* $h : D \rightarrow B$ *is a surjection* $(|B| < |D|)$ *that assigns a bucket* $b \in B$ *to every point* $A \in D$.

The role of a hashing function is to group points into buckets given some arbitrary condition usually encoded in the function itself. For example, we can consider a hashing function $f(B) = |B|$ that applied to object representations $\mathbf{g}^{\mathrm{I}}$ in Table 1 yield the values in the first grey column of the same table. Notice that in this case $|B| = 12$ (12 different possible cardinalities) whereas $|\wp(M)| = 2^9 = 512$. The hashing function f groups together sets of the same size. Similarly, we can define a hashing function $\pi(B) = x; \forall \mathbf{m}_x, \mathbf{m}_y \in B, x \leq y$ assigning the lowest element index in B (in this case $|B| = 12$ as well). Values for this hash are also provided in the second grey column of Table 1.

The usefulness of hashing functions becomes evident when considering the problem of grouping together sets of the same size. A naive strategy would imply comparing all pairs of sets to decide whether they have the same cardinality. This requires $|G|^2$ comparisons (recall that in the case of the formal context, we are interested in comparing object representations). Instead, by using hashing function f defined above we can solve the problem with $|G|$ evaluations of f.

Similarly, let us consider finding groups of *similar* object representations in G w.r.t. the Jaccard similarity. In fact, this is a problem of object clustering for which we could consider the technique of *hierarchical clustering*. In order to find similar groups we need to compare at least the distances between each possible pair of object representations in G (the complexity of hierarchical clustering is at least $|G|^2$). Instead, using a hash-based solution, we would like to define a hashing that, given two similar sets, it would assign them to the same bucket (*hashing collision*). This is the main idea behind locality sensitive hashing.

Definition 3 (LSH family). *Let* (D, d) *be a metric space where* d *is a distance in the space* D. *Let* d_1 *and* d_2 *be two arbitrary distance values s.t.* $d_1 \ll d_2$, *and* $p_1, p_2 \in [0, 1]$ *be two arbitrary probability values s.t.* $p_1 \gg p_2$. *A family of hashing functions* H *is called a* (d_1, d_2, p_1, p_2)-*sensitive LSH family if for a function* $h \in H$ *and a pair of points* $A, B \in D$ *we have:*

$$d(A, B) \leq d_1 \implies P([h(A) = h(B)]) \geq p_1 \qquad (1)$$
$$d(A, B) \geq d_2 \implies P([h(A) = h(B)]) \leq p_2 \qquad (2)$$

Intuitively, we can understand distance values d_1 and d_2 as the *fuzzy* concepts *near* and *far*, respectively. Similarly, probabilities p_1 and p_2 can be understood as *high* and *low*. Thus, we can read the (1) as indicating that given two points which are *near* in the space, the probability of hashing collision is *high*. Similarly, (2) indicates that given two points which are *far* in the space, the probability of hashing collision is *low*. Different LSH families have been proposed for different metric spaces. In this work we will only focus in the metric space of sets.

Consider hashing function π in Table 1. Disregarding the particular order assigned to attributes in Table 1 (their indices), we can calculate the probability of hashing collision ($P([\pi(\mathbf{g}_i^{\mathrm{I}}) = \pi(\mathbf{g}_j^{\mathrm{I}})])$) as follows. If $\pi(\mathbf{g}_i^{\mathrm{I}}) = \pi(\mathbf{g}_j^{\mathrm{I}})$, then there exists an element $\mathbf{m}_x$ in $\mathbf{g}_i^{\mathrm{I}} \cap \mathbf{g}_j^{\mathrm{I}}$ s.t. $\pi(\mathbf{g}_i^{\mathrm{I}}) = \pi(\mathbf{g}_j^{\mathrm{I}}) = x$. In the case that

$\pi(\mathbf{g}_i^{\mathrm{I}}) \neq \pi(\mathbf{g}_j^{\mathrm{I}})$, then either $\pi(\mathbf{g}_i^{\mathrm{I}}) < \pi(\mathbf{g}_j^{\mathrm{I}})$ s.t. $\mathbf{m}_x \in \mathbf{g}_i^{\mathrm{I}}, \mathbf{m}_x \notin \mathbf{g}_j^{\mathrm{I}}$ or $\pi(\mathbf{g}_i^{\mathrm{I}}) > \pi(\mathbf{g}_j^{\mathrm{I}})$ s.t. $\mathbf{m}_x \in \mathbf{g}_j^{\mathrm{I}}, \mathbf{m}_x \notin \mathbf{g}_i^{\mathrm{I}}$. Thus, the probability can be derived as:

$$P([\pi(\mathbf{g}_i^{\mathrm{I}}) = \pi(\mathbf{g}_j^{\mathrm{I}})]) = \frac{|\mathbf{g}_i^{\mathrm{I}} \cap \mathbf{g}_j^{\mathrm{I}}|}{|\mathbf{g}_i^{\mathrm{I}} \cap \mathbf{g}_j^{\mathrm{I}}| + |\mathbf{g}_j^{\mathrm{I}} \backslash \mathbf{g}_i^{\mathrm{I}}| + |\mathbf{g}_i^{\mathrm{I}} \backslash \mathbf{g}_j^{\mathrm{I}}|} = \frac{|\mathbf{g}_i^{\mathrm{I}} \cap \mathbf{g}_j^{\mathrm{I}}|}{|\mathbf{g}_i^{\mathrm{I}} \cup \mathbf{g}_j^{\mathrm{I}}|} = S(\mathbf{g}_i^{\mathrm{I}}, \mathbf{g}_j^{\mathrm{I}})$$

Actually, $P([\pi(\mathbf{g}_i^{\mathrm{I}}) = \pi(\mathbf{g}_j^{\mathrm{I}})])$ is the Jaccard *similarity* between $\mathbf{g}_i^{\mathrm{I}}$ and $\mathbf{g}_j^{\mathrm{I}}$. It is worth noticing that this probability does not depend on one particular order among the attribute indices and thus, we can create new hashing functions using random permutations of the original order. Function π is called the *minhash* function [1] and has been extensively used for document indexing. A *minhash* function defines a $(d_1, d_2, (1-d_1), (1-d_2))$-LSH family for two arbitrary distances $0 \le d_1 < d_2 \le 1$. For example, for $d_1 = 0.2$ and $d_2 = 0.7$, the $(0.2, 0.7, 0.8, 0.3)$-LSH family indicates that two points with a Jaccard distance of at most 0.2 have at least probability 0.8 of hashing collision. Inversely, points at distance at least 0.7 have at most probability 0.3 of being hashed into the same bucket.

LSH families allow for probability amplification using more than one hashing function. In general terms, consider hashing functions h_1 and h_2 and two points $A, B \in D$. If $d(A, B) \le d_1$ then the probability of both hashing functions *simultaneously* assigning the same bucket to A and B is p_1^2. Doing the same for r hashing functions, we can build a (d_1, d_2, p_1^r, p_2^r)-LSH family. This is known as an *AND* construction (as it requires r simultaneous collisions) and it allows getting p_2 (*low* probability of *minhash* collisions for dissimilar objects) very close to 0. Analogously, we can build an *OR* construction with b different hashing functions (requiring at least 1 collision over b hashings) yielding a $(d_1, d_2, 1 - (1 - p_1)^b, 1 - (1 - p_2)^b)$-LSH family which takes p_1 (*high* probability for *minhash* collisions of similar objects) closer to 1. Finally, we can combine both constructions into an *AND-OR* amplification yielding a $(d_1, d_2, 1 - (1 - p_1^r)^b, 1 - (1 - p_2^r)^b)$-LSH family.

A *minhash* family of functions can be *amplified* using *AND* and *OR* constructions in order to create a $(d_1, d_2, 1 - (1 - (s_1)^r)^b, 1 - (1 - (s_2)^r)^b)$-LSH, where $s_1 = 1 - d_1$ and $s_2 = 1 - d_2$ are Jaccard similarities. Considering $r = 5$ and $b = 8$ for the same values of d_1 and d_2 we have a $(0.2, 0.7, 0.95, 0.01)$-LSH family. Figure 1 shows probabilities of hashing collision w.r.t. the Jaccard similarity with $r = 5$ and $b = 8$. It can be observed the characteristic S *shape* of a logistic regresion.

3 LSH and FCA

In this section we introduce our approach for using LSH to sample extent generators. In a nutshell, given a formal context $\mathcal{K} = (\mathbf{G}, \mathbf{M}, \mathbf{I})$, we derive a new formal context $\mathcal{K}_\Pi = (\mathbf{G}, \mathbf{T}_\Pi, \mathbf{I}_\Pi)$ where the probability of $(\mathbf{A}^{\mathrm{I}_\Pi \mathrm{I}_\Pi}, \mathbf{A}^{\mathrm{I}_\Pi})$ is known for a given set of objects $\mathbf{A} \in \mathbf{G}$, and it is a function of the *similarity index* of $\mathbf{A}$ (we define the *similarity index* as a generalization of the Jaccard Similarity over the set of object representations taken from $\mathbf{A}$). We begin by calculating the probability of the *join* of two object concepts in Sect. 3.1 and generalize this notion to extent generators in Sect. 3.2.

636   V. Codocedo and M.T. Tang

3.1 Hashed Formal Concepts

Let us consider a context $\mathcal{K} = (G, M, I)$ and a *minhash* function $\pi_a \colon G \to T_{\pi_a}$ for a given permutation of the attributes in M. We define a new formal context associated to π_a as $\mathcal{K}_\pi = (G, T_{\pi_a}, I_{\pi_a})$ such that $T_{\pi_a} = \{(a, \pi_a(g^I)) \mid \forall g \in G\}$ is a set of attributes, and $(g, (a, \pi_a(g^I))) \in I_\pi$. Recall that $\pi_a(g^I) = x$; $\forall m_x, m_y \in g^I, x < y$ (e.g. $\pi_a(g_1^I) = 1$ in formal context of Table 1 when the permutation is the same order as the one given in M). Trivially, the set of formal concepts of $\mathcal{K}_{\pi_a}$ is an anti-chain with $|T_{\pi_a}|$ elements $(A, (a, x))$ s.t. $\forall g \in A$, $\pi_a(g^I) = x$ (since π_a generates an equivalence relation in G). Using the results from Sect. 2.2, we can calculate the probability that for any two objects, $g_i, g_j \in G$, the *join* of their object concepts is not the *suprema* of the concept lattice, i.e. $\gamma(g_i) \vee \gamma(g_j) = (\{g_i, g_j\}^{I_\pi I_\pi}, g_i^{I_\pi} \cap g_j^{I_\pi}) \neq \top$ where $\top = (G, \emptyset)$, or $P[\pi(g_i^I) = \pi(g_j^I)] = S(g_i^I, g_j^I)$. We will refer to this as the *"the probability of $\gamma(g_i)$ and $\gamma(g_j)$"*. It is worth noticing that $g_i^{I_\pi} \cap g_j^{I_\pi} \neq \emptyset \implies g_i^I \cap g_j^I \neq \emptyset$, meaning that $(\{g_i, g_j\}^{II}, g_i^I \cap g_j^I)$ *is necessarily a formal concept in $\mathcal{K}$.*

For example, given minhash π in Table 1 we can derive the following formal concepts: $(\{g_1, g_2, g_4\}, \{(\_, 1)\})$, $(\{g_3\}, \{(\_, 2)\})$, $(\{g_5\}, \{(\_, 4)\})$, $(\{g_6\}, \{(\_, 9)\})$, $(\{g_7, g_8\}, \{(\_, 10)\})$ and $(\{g_9\}, \{(\_, 11)\})$.

In order to extend this result, let us consider now a family of n minhash functions $\Pi = \{\pi_1, \pi_2, \ldots, \pi_n\}$ (for different random attribute permutations). A new formal context can be built for Π by the apposition [5] of the formal contexts associated to each $\pi \in \Pi$. This is given by:

$$\mathcal{K}_\Pi = \mathcal{K}_{\pi_1} \mid \mathcal{K}_{\pi_2} \mid \ldots \mid \mathcal{K}_{\pi_n} = (G, T_\Pi, I_\Pi) = (G, \bigcup_{\pi \in \Pi} T_\pi, \bigcup_{\pi \in \Pi} I_\pi)$$

In $\mathcal{K}_\Pi$, we have that $\gamma(g_i) \vee \gamma(g_j) \neq \top \iff \exists \pi \in \Pi, \pi(g_i^I) = \pi(g_j^I)$. This has an associated probability of $P[\exists \pi \in \Pi, \pi(g_i^I) = \pi(g_j^I)] = 1 - (1 - S(g_i^I, g_j^I))^n$, where $n = |\Pi|$. Notice that this corresponds to an *OR* construction for an amplified LSH family. Figure 2 depicts the probability of $\gamma(g_i) \vee \gamma(g_j) \neq \top$ given the Jaccard similarity of two objects when $n = 1$ (one minhash function, diagonal) and when $n = 10$ (ten minhash functions, convex curve). We can observe that when $n = 10$, we have a more tolerant framework where the *join* of object concepts of less similar objects occurs with higher probability, e.g. with $n = 1$ the probability of the *join* $\gamma(g_i) \vee \gamma(g_j)$ where $S(g_i^I, g_j^I) = 0.6$ is 0.6, while with $n = 10$, the probability is close to 1.

While Fig. 2 shows a more tolerant probability distribution, it is also less discriminative as the probability of the *join* of object concepts with low similarity is also increased. Indeed, we would prefer to have a probability distribution more similar to that of Fig. 1 than that of Fig. 2.

Proposition 1. *Let $\mathcal{K}_\Pi = (G, T_\Pi, I_\Pi)$ be a formal context created from n minhash functions. The function $\psi_k : \wp(T_\Pi) \to \wp(T_\Pi)$ defined as:*

$$\psi_k(B) = \begin{cases} \emptyset & |B| < k \\ B & |B| \geq k \end{cases}$$

with $k \in [1, n]$ is a projection in $\mathcal{K}_\Pi$.

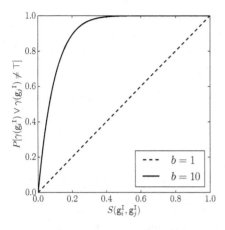
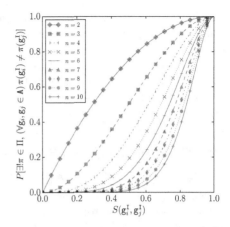

Fig. 2. Probability of the *join* of object concepts w.r.t. their Jaccard similarity using 1 (diagonal) and 10 (convex curve) minhash functions

Fig. 3. Lattice-based amplification of the probability of a join when $k = (n-1)$ for $n \in [2, 10]$

The proof of Proposition 1 is straightforward since ψ_k is clearly monotone ($\mathsf{B}_1 \subseteq \mathsf{B}_2 \implies \psi_k(\mathsf{B}_1) \subseteq \psi_k(\mathsf{B}_2)$), idempotent ($\psi_k(\psi_k(\mathsf{B})) = \psi_k(\mathsf{B})$) and contractive ($\psi_k(\mathsf{B}) \subseteq \mathsf{B}$) for any $k \in [1, n]$. Intuitively, ψ_k *removes* the *join* of object concepts that have less than k minhash collisions (out of the n possible collisions). For example, let us consider a $k = 2$. For any two objects $\mathsf{g}_i, \mathsf{g}_j \in \mathsf{G}$ we have that $\psi_k(\mathsf{g}_i^{I_\Pi} \cap \mathsf{g}_j^{I_\Pi}) \neq \emptyset$ iff there exist *at least* two minhashes $\pi_1, \pi_2 \in \Pi$ such that $\pi_1(\mathsf{g}_i^I) = \pi_1(\mathsf{g}_j^I)$ *and* $\pi_2(\mathsf{g}_i^I) = \pi_2(\mathsf{g}_j^I)$.

Parameter k in projection ψ_k has a similar role to parameter r in an *AND* amplification construction for an LSH family (Sect. 2.2), this is, it restricts the *join* of object concepts of dissimilar objects. Nevertheless, there is a very important difference. In an actual *AND-OR* construction, the r minhash collisions that should occur *simultaneously* are split into b groups (or bands) that can occur *independently*. Instead, in $\mathcal{K}_\Pi$, the k minhash collisions that should occur *simultaneously* may be *any* of the n minhash functions available. This changes the probability function given that groups of k (or more) minhash collisions are not independent events. Consequently, it becomes hard to derive a formula for the probability given a combinatorial explosion in the number of variables required for even a few minhash functions. Nevertheless, the probability function remains being a polynomial with degree n. Actually, it is possible to derive a simple function for the probability of the *join* of two objects when $k = (n - 1)$ as $n \cdot S(\mathsf{g}_i^I, \mathsf{g}_j^I)^{n-1} - (n - 1) \cdot S(\mathsf{g}_i^I, \mathsf{g}_j^I)^n$.

Furthemore, it is possible to show that this function is monotonically increasing for $S(\mathsf{g}_i^I, \mathsf{g}_j^I) \in [0, 1]$ and for $n \in \mathbb{N}^+$. For the sake of brevity we do not provide the development of this probability. Our main concern is to show that it is also a function of the Jaccard similarity between objects and the parameter $k \in [1, n]$ chosen for the projection. We call this construction a *lattice-based amplification* of the LSH family. Figure 3 depicts the probability distribution when $k = (n-1)$

for $n \in [2, 10]$. Lattice-based amplification provides a steeper probability function than an *AND-OR* construction with less minhash functions. For example, consider the *AND-OR* construction depicted in Fig. 1 with a less restrictive function, i.e. less similar pairs of objects are more likely to be hashed together. It takes $b \times r = 45$ minhash functions to build compared to the quite more restrictive function of a lattice-based construction with $k = n - 1$ requiring just 10 minhash functions.

3.2   The Probability of a Set of Objects

So far we have only considered the probability that firstly, the *join* of the object concepts $\gamma(g_1)$ and $\gamma(g_2)$ exists and secondly, the probability that $\gamma(g_1) \vee \gamma(g_2)$ is not projected to the top concept. We want to extend both of these notions to the *join* of any set of object concepts. Luckily, this extension is straightforward. Let us consider the probability that a minhash function π collides for all the objects in a set $A \subseteq G$. A similar analysis to that of Sect. 2.2 yields that:

$$P[\forall g_i, g_j \in A, \pi(g_i^I) = \pi(g_j^I)] = \frac{|\bigcap_{g \in A} g^I|}{|\bigcup_{g \in A} g^I|} = S(A)$$

We will overload S to also mean that $S(A)$ is *the similarity index* of the set of objects A. Knowing this allows deriving the probability of $(A^{I_\pi I_\pi}, A^{I_\pi})$ existing in different settings using the same rationale as before. In $\mathcal{K}_\pi$, the probability of $A^{I_\pi I_\pi}$ existing (given A) is $P[A^{I_\pi I_\pi}|A] = S(A)$. For a set Π of n minhash functions, the probability of $A^{I_\Pi I_\Pi}$ in $\mathcal{K}_\Pi$ (given A) is $P[A^{I_\Pi I_\Pi}|A] = (1 - (1 - S(A))^n)$. Finally, $P[\psi_k(A^{I_\Pi}) \neq \emptyset]$ when $k = n - 1$ is $n \cdot S(A)^{(n-1)} - (n-1) \cdot S(A)^n$. It is important to notice that the first two probabilities are conditional. This is due to the fact that $P[A^{I_\Pi I_\Pi}]$ depends on the conditional probability of each of the generators of $A^{I_\Pi I_\Pi}$ which are unknown previous to the calculation of the formal concepts in $\mathcal{K}_\Pi$.

Final Remarks. Let us consider the function $\omega(g^I) = \langle (i, \pi_i(g^I)) \rangle_{\pi_i \in \Pi}$ where Π is a set of r minhash functions. We will consider $\omega(g^I)$ to be a categorical atomic value, i.e. it is not a set of values but a *signature*. Clearly, the probability of signature collision for two different objects, i.e. $P[\rho(g_i^I) = \rho(g_j^I)] = S(g_i^I, g_j^I)^r$, corresponds to that of an *AND* amplification construction for r minhash functions. The inclusion of this kind of amplification in our framework is straightforward. We just need to define the formal context $\mathcal{K}_\omega = (G, T_\omega, I_\omega)$ for a given conjunction of r random minhashes and the formal context $\mathcal{K}_\Omega = (G, T_\Omega, \psi_k \circ I_\Omega)$ built by apposition of a set of formal contexts derived for n conjunctions of r random minhashes. $P[\psi_k(A^{I_\Omega}) \neq \emptyset]$ is the same as before, however it has a degree of $r \cdot n$ making the curve more steep and consequently, more selective.

4   Understanding the Model - Experiments and Discussion

To illustrate our previous findings, we provide a small experimental evaluation of the probabilities calculated for the formal context in Table 1. The experiment consists in sampling formal concepts by generating 10000 lattices. For each formal concept $(\mathtt{A}, \mathtt{B})$ we present their extents, associated frequencies and similarity indices using 4 minhash functions $(n = 4)$ and a projection created for $k = 3$ and $k = 2$ in Tables 2a and b, respectively. Table 2c presents the results for $k = 2$ using an AND amplification construction with $r = 2$. The polynomial for the probability $P[\psi_k(\mathtt{A}^{I_n}) \neq \emptyset]$ in each setting is shown at the bottom of each table along with a small illustration of the probability distribution function (x-axis corresponds to $S(\mathtt{A})$ and y-axis, to $P[\psi_k(\mathtt{A}^{I_n}) \neq \emptyset]$). Tables are ordered by the frequency of appearance of each extent.

It is worth noticing that the order of entries in Tables 2a, b and c is not strictly increasing w.r.t. the similarity index, while the probability function *is monotonically increasing* w.r.t. it. This is not an error. We recall that the function actually represents *the probability of a set of objects appearing within the projected lattice*, or what is the same, the probability of an extent in the projected lattice containing the set of objects. In the following, we discuss this and other characteristics of our approach.

Let us focus in the case $n = 4, k = 3$ since it corresponds to a $k = n - 1$ situation. We can calculate $P[\psi_k(\{\mathtt{g_7}, \mathtt{g_8}, \mathtt{g_9}\}^{I_n}) \neq \emptyset]$ using $4 \cdot S(\mathtt{A})^{(3)} - 3 \cdot S(\mathtt{A})^4$ with $S(\mathtt{A}) = 0.33$ as indicated by Table 2a. This yields 0.108 which is near the 0.111 frequency reported for this extent. Notice that the probability of $\{\mathtt{g_7}, \mathtt{g_8}, \mathtt{g_9}\}^{I_n I_n}$ being something else than $\{\mathtt{g_7}, \mathtt{g_8}, \mathtt{g_9}\}$ is zero (keeping the fact that $\psi_k(\{\mathtt{g_7}, \mathtt{g_8}, \mathtt{g_9}\}^{I_n}) \neq \emptyset$). In contrast, $P[\psi_k(\{\mathtt{g_7}, \mathtt{g_8}\}^{I_n}) \neq \emptyset]$ yields 0.599 which is far from the 0.530 reported in Table 2a. This is due to the fact that in some occasions $\{\mathtt{g_7}, \mathtt{g_8}\}^{I_n I_n} \neq \{\mathtt{g_7}, \mathtt{g_8}\}$. For example in Table 2a we have that $\{\mathtt{g_7}, \mathtt{g_8}\}^{I_n I_n}$ may be $\{\mathtt{g_7}, \mathtt{g_8}, \mathtt{g_9}\}$ or $\{\mathtt{g_1}, \mathtt{g_7}, \mathtt{g_8}\}$. Clearly, the sum associated to all these sets is not 0.599 either. This is explained because, as we have previously indicated, these events are not independent and may occur simultaneously in any possible arrangement. What $P[\psi_k(\{\mathtt{g_7}, \mathtt{g_8}\}^{I_n}) \neq \emptyset]$ indicates is that *there is an extent containing $\mathtt{g_7}, \mathtt{g_8}$ in the lattice* with a probability of 0.599. In the experiment, this actually occurs with a frequency of 0.5958 as predicted (not reported in the Table).

Another interesting observation from the results is that $\{\mathtt{g_1}, \mathtt{g_2}\}$ and $\{\mathtt{g_2}, \mathtt{g_4}\}$ occur less frequently than $\{\mathtt{g_1}, \mathtt{g_2}, \mathtt{g_4}\}$ when the similarity index of the latter (0.20) is lower than the indices of the first two extents (0.22 and 0.25, respectively). Indeed, this is a very good example of how minhash functions group objects together in the concept lattice. Firstly, notice that in $\mathcal{K}$ in Table 1, both $\{\mathtt{g_1}, \mathtt{g_2}\}$ and $\{\mathtt{g_2}, \mathtt{g_4}\}$ are generators of $\{\mathtt{g_1}, \mathtt{g_2}, \mathtt{g_4}\}$, i.e. $\{\mathtt{g_1}, \mathtt{g_2}\}^{II} = \{\mathtt{g_2}, \mathtt{g_4}\}^{II} = \{\mathtt{g_1}, \mathtt{g_2}, \mathtt{g_4}\}$. Consequently, a minhash function π that collides for $\mathtt{g_1}$ and $\mathtt{g_2}$ is likely to collide for $\mathtt{g_1}$ and $\mathtt{g_4}$ (consequently for $\mathtt{g_2}$ and $\mathtt{g_4}$). If this is not the case, then $\pi(\mathtt{g_4^I}) = x$ and $\mathtt{m_x} \in \mathtt{g_4^I}$ and $\mathtt{m_x} \notin \mathtt{g_1^I} \cup \mathtt{g_2^I}$. Since the difference between $\mathtt{g_4^I}$ and $\mathtt{g_1^I} \cup \mathtt{g_2^I}$ is only one element and their intersection contains two elements, the probability that π groups g_1 and g_2 without g_4 is lower than the probability that it groups them all together. This probability gets amplified by the arrange of n minhash functions, thus maintaining the relation and generating the results we observe in Table 2a. If we consider $\{g_2, g_4\}$, we can observe that g_1^I has two elements of difference with $g_2^I \cup g_4^I$ (maintaining an intersection of two elements). Consequently, the frequencies of $\{g_2, g_4\}$ and $\{g_1, g_2, g_4\}$ in Table 2a are comparable.

Table 2b has a similar arrangement of extents to that of Table 2a. Since $k = 2$ yields a less restrictive projection than $k = 3$, the frequency of occurrence of each extent is higher. Differently, in Table 2c frequencies are higher for extents with high Jaccard indices (the first two) and lower for extents with low Jaccard indices. This construction yields a more *selective* sampling of extents, i.e. those with higher Jaccard indices are more likely to occur, while those with low Jaccard indices are less likely to occur. Nevertheless, this construction needs twice ($r = 2$) as many minhash functions to represent objects.

Table 2. (a) $n = 4, k = 3$. (b) $n = 4, k = 2$. (c) $n = 4, k = 2, r = 2$

Toy Example - Sampled formal concepts from Table 1 in 10000 concept lattices. Figures at the bottom of each Table depict the probability function given by the polynomial

(a)

$A^{I_\Pi I_\Pi}$	Frequency	$S(A)$
g_2, g_3, g_4	0.003	0.11
g_1, g_2, g_3, g_4	0.004	0.10
g_2, g_3, g_5	0.004	0.11
g_1, g_7	0.006	0.17
g_3, g_5	0.007	0.14
g_1, g_2	0.010	0.22
g_1, g_7, g_8	0.010	0.14
g_2, g_6	0.011	0.14
g_2, g_4	0.023	0.25
g_2, g_3	0.025	0.22
g_1, g_2, g_4	0.026	0.20
g_6, g_8, g_9	0.050	0.25
g_6, g_9	0.067	0.33
g_1, g_3, g_4	0.070	0.29
g_7, g_8, g_9	0.111	0.33
g_1, g_3	0.148	0.43
g_1, g_4	0.250	0.50
g_3, g_4	0.265	0.50
g_2, g_5	0.310	0.50
g_8, g_9	0.497	0.67
g_7, g_8	0.530	0.67

$P = 4 \cdot S(A)^3 - 3 \cdot S(A)^4$

(b)

$A^{I_\Pi I_\Pi}$	Frequency	$S(A)$
g_2, g_3, g_4	0.014	0.11
g_1, g_7	0.039	0.17
g_1, g_2	0.041	0.22
g_3, g_5	0.042	0.14
g_1, g_2, g_3, g_4	0.054	0.10
g_2, g_3, g_5	0.062	0.11
g_2, g_4	0.091	0.25
g_1, g_7, g_8	0.102	0.14
g_2, g_6	0.104	0.14
g_2, g_3	0.121	0.22
g_1, g_2, g_4	0.142	0.20
g_6, g_9	0.197	0.33
g_6, g_8, g_9	0.263	0.25
g_1, g_3, g_4	0.290	0.29
g_1, g_3	0.355	0.43
g_7, g_8, g_9	0.403	0.33
g_1, g_4	0.449	0.50
g_3, g_4	0.512	0.50
g_2, g_5	0.668	0.50
g_8, g_9	0.676	0.67
g_7, g_8	0.741	0.67

$P = 6S(A)^2 - 8S(A)^3 + 3S(A)^4$

(c)

$A^{I_\Omega I_\Omega}$	Frequency	$S(A)$
g_2, g_3, g_4	0.000	0.11
g_2, g_3, g_5	0.001	0.11
g_1, g_2, g_3, g_4	0.001	0.10
g_1, g_7, g_8	0.002	0.14
g_3, g_5	0.002	0.14
g_1, g_7	0.002	0.17
g_2, g_6	0.003	0.14
g_1, g_2	0.005	0.22
g_1, g_2, g_4	0.008	0.20
g_2, g_4	0.011	0.25
g_2, g_3	0.012	0.22
g_6, g_8, g_9	0.021	0.25
g_1, g_3, g_4	0.035	0.29
g_6, g_9	0.044	0.33
g_7, g_8, g_9	0.062	0.33
g_1, g_3	0.126	0.43
g_1, g_4	0.227	0.50
g_3, g_4	0.240	0.50
g_2, g_5	0.264	0.50
g_8, g_9	0.566	0.67
g_7, g_8	0.573	0.67

$P = 6S(A)^4 - 8S(A)^6 + 3S(A)^8$

5   Conclusions and Perspectives

In this article we have introduced a framework to sample extent generators from a given formal context using a family of locality sensitive hashing functions, namely *minhash* functions. We have shown that through our framework we can predict the probability of occurrence of a set of objects within the extent of one formal concept in the concept lattice. Our approach introduces three parameters to tune this probability function, namely the number of minhashes to apply to object representations n, the number of minhash functions to group in a conjunctive signature r, and the minimal number of minhash collisions to build a projection over the concept lattice k. We have named this approach a *lattice-based amplification* for an LSH family.

We consider this approach to be the groundwork of a larger framework to actually sample patterns from a formal context or a pattern structure. For example, we can consider an approach based on heterogeneous pattern structures [3] where objects are dually represented in the space of minhash functions and the original space of attributes. Heterogeneous concepts would contain both, the patterns sought and a known probability of occurrence. In the same context, we consider the extension of our approach to other families of object representations. For example when considering *interval pattern structures*, sampling concepts is an important part of the mining process given the overwhelming number of interval pattern concepts generated for even small datasets of vectors. In this new setting, we can consider the corresponding LSH family to estimate the Euclidean distance of points in the space, plugging in the associated hashing collision probabilities to our model accordingly.

References

1. Broder, A.Z.: On the resemblance and containment of documents. In: Proceedings, Compression and Complexity of Sequences (1997)
2. Buzmakov, A., Kuznetsov, S.O., Napoli, A.: Revisiting pattern structure projections. In: Baixeries, J., Sacarea, C., Ojeda-Aciego, M. (eds.) ICFCA 2015. LNCS, vol. 9113, pp. 200–215. Springer, Cham (2015). doi:10.1007/978-3-319-19545-2_13
3. Codocedo, V., Napoli, A.: A proposition for combining pattern structures and relational concept analysis. In: Glodeanu, C.V., Kaytoue, M., Sacarea, C. (eds.) ICFCA 2014. LNCS, vol. 8478, pp. 96–111. Springer, Cham (2014). doi:10.1007/978-3-319-07248-7_8
4. Ganter, B., Kuznetsov, S.O.: Pattern structures and their projections. In: Delugach, H.S., Stumme, G. (eds.) ICCS-ConceptStruct 2001. LNCS, vol. 2120, pp. 129–142. Springer, Heidelberg (2001). doi:10.1007/3-540-44583-8_10
5. Ganter, B., Wille, R.: Formal Concept Analysis: Mathematical Foundations. Springer, Heidelberg (1999)
6. Han, J., Wang, J., Lu, Y., Tzvetkov, P.: Mining top-k frequent closed patterns without minimum support. In: IEEE International Conference on Data Mining (2002)
7. Kuznetsov, S.O.: On stability of a formal concept. Ann. Math. Artif. Intell. **49**(1–4), 101–115 (2007)
8. Kuznetsov, S.O., Obiedkov, S.A.: Comparing performance of algorithms for generating concept lattices. J. Exp. Theor. Artif. Intell. **14**, 189–216 (2002)
9. Rajaraman, A., Ullman, J.D.: Mining of Massive Datasets. Cambridge University Press, New York (2011)

An Application of AOC-Posets: Indexing Large Corpuses for Text Generation Under Constraints

Alain Gutierrez, Michel Chein, Marianne Huchard$^{(\boxtimes)}$, and Pierre Pompidor

LIRMM, CNRS and Montpellier University, Montpellier, France
{Alain.Gutierrez,michel.chein,marianne.huchard,pierre.pompidor}@lirmm.fr
http://www.lirmm.fr

Abstract. In this paper, we describe the different ingredients of the COGITEXT tool which can be used for building, editing, and using large corpuses for text generation under constraints *à la* ALAMO. In COGI-TEXT, AOC-posets are used as indexes that give information about the shape of the corpuses and that help to efficiently find terms for the text creation process. We give some figures about their size and the needed time for computing them and for making a specific text creation.

Keywords: Formal concept analysis · AOC-poset · Text generation with constraints · ALAMO

1 Introduction

OuLiPo [9] is a literary approach founded in 1960 by Raymond Queneau and François Le Lionnais that aims to create literary text with constraints in writing. In 1981, members of OuLiPo created ALAMO [1], which is, as indicated by its name, a *Workshop (Atelier in french) of Literature Assisted by Mathematics and Computers (Ordinateurs in french)*. Several tools were designed to assist this approach. In this paper, we introduce COGITEXT, which can be considered as the continuation of LAPAL (the last tool for automatic literary text creation developed within the framework of ALAMO). COGITEXT contains tools for building, editing, or using large corpuses. For instance, the examples given in the paper are using corpuses built from DELA [5] for substantives (nouns) and adjectives, and from Morphalou [7] for verbs. Besides classical attributes (e.g. gender) associated to each corpus item, phonetics has been obtained with an original software and can be used for computing metric properties (e.g. syllable number) or consonance properties (e.g. rhyme). For dealing with the large size of these corpuses, an original indexing method based on AOC-poset has been built. Another specificity is the use of a knowledge representation system enabling different ways to describe constraints: graphical interface, Datalog rules or Beanshell scripts. COGITEXT = *CoGui + Text*, where CoGui is a visual tool for building knowledge bases [4].

© Springer International Publishing AG 2017
M. Kryszkiewicz et al. (Eds.): ISMIS 2017, LNAI 10352, pp. 642–652, 2017.
DOI: 10.1007/978-3-319-60438-1_63

Section 2 illustrates and outlines the approach, by showing an example of writing under constraints. Section 3 defines the corpuses and the phonetics. Production schemes are presented in Sect. 4. The use of AOC-posets for efficient indexing is detailed in Sect. 5. Section 6 describes the implementation and gives figures on the computation time of the AOC-posets and of the text production. We conclude in Sect. 7 with a few prospects of this work.

2   Motivation and Outline of the Approach

In this section, we illustrate the purpose of CogiText with a simple example. Let us assume that an author would like to produce a parody of the Jean de La Fontaine[1] fable "le corbeau et le renard" ("the fox and the crow"). The title and the first two lines of the original text used to exemplify the approach are shown in the upper part of Table 1. The design of this parody here is based on the definition of a *production scheme* including a *production template* and a *constraint set*. We describe them here in a textual form, but a user interface is provided to assist the writer (see Sect. 6).

Table 1. Constraints inspired by the Jean de La Fontaine fable "Le corbeau et le renard". rhyme3 (resp. rhyme2) stands for rhymes with 3 (resp. 2) phonemes.

First lines of the original text (French)	Translation (English)
Le corbeau et le renard.	The crow and the fox.
Maître Corbeau,	Mister Crow,
sur un arbre perché,	perched on a tree,
Tenait en son bec un fromage.	was holding in his beak a cheese.
Production template (french)	**Production template (transposed)**
Le $\{X_1.txt\}$ et le $\{X_2.txt\}$.	The $\{X_1.txt\}$ and the $\{X_2.txt\}$.
Maître $\{X_1.txt\}$,	Mister $\{X_1.txt\}$,
sur un $\{X_3.txt\}$ $\{X_4.txt\}$	$\{X_4.txt\}$ on a $\{X_3.txt\}$,
Tenait en son $\{X_5.txt\}$ un $\{X_6.txt\}$	was holding in its $\{X_5.txt\}$ a $\{X_6.txt\}$.
Constraint set (french)	**Constraint set (transposed)**
X_1=element(corpusNoun);	X_1=element(corpusNoun);
X_1.rhyme3="Rbo";	X_1.rhyme2="oW";
X_1.nbsyl=2;	X_1.nbsyl=1;
X_1.gender="_masculine";	X_1.gender="_neuter";
X_1.number="_singular";	X_1.number="_singular";
X_2=element(corpusNoun);	X_2=element(corpusNoun);
X_2.rhyme3="naR";	X_2.rhyme1="oX";
X_2.nbsyl=3;	X_2.nbsyl=1;
X_3.gender=X_4.gender;	X_3.gender=X_4.gender;
X_3.number=X_4.number; ...	X_3.number=X_4.number; ...

The central part of Table 1 shows a production template for the parody. In the production template of the example, several words have been replaced by

---

[1] Jean de la Fontaine is a famous French fabulist of the 17th century.

Table 2. Text productions for "Le corbeau et le renard". The french text is automatically produced by CogiText. The english text is manually composed using the rhyming dictionary Rhymer [10] for making the technique understandable by english-speaking readers.

Production (french)	Production (transposed)
Le barbot et le fouinard.	The woe and the vox.
Maitre barbot,	Mister woe,
sur un marbre torché,	lurched on a knee,
Tenait en son bec un dommage	Holds in his creek a sneeze.

expressions referring to variables: here the text of the variable will be used to replace the initial word. Constraints are properties that the variables appearing in the production template should satisfy. The bottom part of Table 1 shows a few constraints for "the fox and the crow" parody (see Appendix 2 for a more complete production scheme[2]). From the *production scheme*, and corpuses described in Sect. 3, CogiText builds *text productions* as the one shown in Table 2 (left-hand-side). In productions, the expressions on variables are replaced by values. Here, the values are the texts of randomly chosen corpus elements that satisfy the constraints. For example, expression $\{X_1.txt\}$ is replaced by "barbot" which is a term found in the DELA lexicon [5], satisfying the constraints: noun, rhyme in "Rbo", 2 syllables, singular, masculine.

The approach is outlined in Fig. 1: (1) the writer chooses corpuses among CogiText corpuses or builds his own corpuses (each term being equipped with

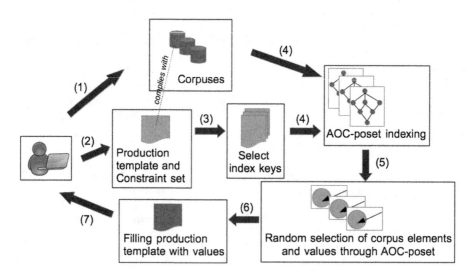

Fig. 1. Outline of the approach

---

[2] Appendix 1 and Appendix 2 are available at: http://www.lirmm.fr/~huchard/Documents/Papiers/appendix12.pdf.

attributes, key/value pairs, used in production schemes); (2) the writer types the production template and the constraints, that have to comply with the selected corpuses; (3) the system extracts the relevant key/value pairs on corpus elements (like gender="_masculine", or rhyme3="Rbo") from the production scheme; (4) these key/value pairs are used to build indexes (with an AOC-poset structure) on corpuses; (5) corpus elements and values are randomly selected through the AOC-poset; (6) the production is built by filling the expressions with the chosen element values; (7) the production is returned to the user.

3   Corpuses and Phonetic Information

The approach requires significant linguistic resources to serve its purpose: corpuses that include a large set of terms with attributes, especially phonetics in our example since it concerns poetry.

CogiText *Corpuses.* CogiText is designed to work with any text corpus or lexicon, provided it is equipped with the structure that follows (examples of this paragraph are translated in English). A *corpus* is a set of elements. A *corpus element* is a set of (key, value) pairs. For example, an element can be elem1= (txt, "home"), (rhyme2, "oM"), (gender, "_neutral"), (nbsyl, 3), (syn, ["residence", "house"]). A *key* is simply a string, as txt (for "text"), rhyme2, gender, nbsyl (for "number of syllables"), syn (for "synonyms"). For example, for elem1, "home" is the value of key txt, 3 is the value of nbsyl. The possible types of *values* are primitive types (string, integer, float, boolean) or arrays of these primitive types. Strings may include several words, lines, spaces or punctuation marks. Thus, if in our example the CogiText corpuses are lexicons, in other applications they can be corpuses in the usual linguistic meaning. A *corpus schema* is a structure for a corpus. It is composed of a set of (key, type) pairs. For the previous example, the corpus schema contains (txt, string), (rhyme2, string), (gender, string), (nbsyl, integer), (syn, array of string). A corpus element e complies with a corpus schema S if every (key, value) pair (k, v) is such that: if k appears in S, then it appears in at most one pair of e, and v is a value of its associated type in S (a corpus element does not necessarily contain all the keys of the schema). Several schemas can be associated with a given corpus. A *corpus mapping* associates a computed value with a corpus element. The mappings that return the values associated with a key are predefined. For example, e.txt is the mapping which associates to an element e the value associated with its key txt, e.g. elem1.txt="home". The null value is returned when the corpus element does not own this key. Other mappings can be defined by the user, in a dedicated language. For example, one can develop a mapping cutReturn6 that returns, for elements whose text has at least 6 characters, the string obtained by splitting the text into two parts and reversing these parts; e.g. if $e.txt =$ "*congratulate*" e.cutReturn6 = "tulatecongra".

At this point CogiText contains three corpuses built from the french lexicons DELA (for nouns and adjectives) and Morphalou (for verbs). They are equipped with corpus schemas. DELA contains 102 073 lemmas, giving 683 824 inflected forms (including plural forms for example), and Morphalou introduces 8790 verb lemmas. An example of a DELA entry for a lemma is [précepteur, 1.N36(Hum)], where "précepteur" is the canonical form, "1" is the lexical level, N36 is a morphological code which allows to calculate the different inflected forms, and "Hum" indicates that it applies to a human. An entry for an inflected form is: [préceptrice,précepteur. N36(Hum):fs], where "préceptrice" is the inflected form, "précepteur" is the canonical form, "fs" is the gender and the number (feminine, singular). Such information will appear in two corpus elements: p1 = (txt, "précepteur"), (gender, "_masculine"), (nbsyl, 3), ... and p2 = (txt, "préceptrice"), (gender, "_feminine"), (nbsyl, 3), ... The same principle applies to other categories. Details about the CogiText corpuses built from DELA and Morphalou are shown in Appendix 1.

Phonetic and Syllable Information. Phonetic syllabification of french words plays an important role in applications dealing with literary texts. The resources are composed of 641 handmade phonetic rules which come from the lexicon Descartes analysis [8] and a handmade lexicon of 1399 lemmas which have exceptional phonetics due to their exogenous origin. The tool has a few limitations, including: recognizing between the different forms of weak/mute "e" in french language is difficult; some ambiguous cases that would require a syntactic analysis are not considered; the pronounced liaison between the words is hard to know; and of course, the prosody is absent. Nevertheless 98.5% of the words are correctly pronounced. We illustrate our method on the verbal form "accéléraient" (as in "they accelerated" in English). The analysis is achieved in four steps:

– match with the longest phonetic suffix: ent| (matching with a verbal form) => . ("e" is mute in "ent", thus this is not considered as a phoneme)
– match with the longest phonetic prefix: |acc => ak-s+
– find intercalated phonemes: é => e, l => l+, é => e, r => r+, ai => E
– produce the final result: ak-s+|e|l+|e|r+|E|. => ak se le rE.

4   Production Scheme

A *production scheme* is composed of a *production template* and *constraints*.

A simple production template is a sequence of strings and expressions of the form {corpus_variable.corpus_mapping}. A *corpus variable* is an identifier which represents an unknown corpus element. Each corpus variable refers to a specific corpus, for example a variable may refer to the common noun corpus while another refers to the verb corpus. For example, production pattern "Le $\{X_1.\texttt{txt}\}$ et le $\{X_2.\texttt{txt}\}$." is a sequence composed of string "Le", expression "$\{X_1.\texttt{txt}\}$", string "et le", expression $\{X_2.\texttt{txt}\}$, and string ".". The fact that variable X_1 refers to the common noun corpus is the first constraint as noted in the constraint part of Table 1.

A constraint is a property that variables appearing in a production template have to satisfy. They can be applied to a single variable as $\{X_1.\texttt{rhyme3="Rbo"}\}$ or they can apply to several variables as $\{X_1.\texttt{nbsyl} = X_2.\texttt{nbsyl}\}$. A constraint that applies to a single variable is a unary constraint. A unary constraint defines a part of the corpus, for example $\{X_1.\texttt{nbsyl=2}\}$ amounts to considering only the common nouns that have two syllables, acting as a filter and limiting the number of the corpus elements that have to be considered. The binary constraints may involve different mappings, such as $\{X_1.\texttt{nbsyl} = X_2.\texttt{nbvowels}\}$ and variables may refer to different corpuses: e.g. X_1 may refer to a common noun and X_2 may refer to a verb. We call *simple* constraints the constraints involving a relation $=$, $\neq$, $<$, or $>$ between two unary mappings. *Complex* constraints can be defined (by programming) as $\{X_1.\texttt{nbsyl} + X_2.\texttt{nbsyl} + X_3.\texttt{nbsyl} = 12\}$. The language is rather rich, enabling the use of standard functions that manipulate strings, as well as any user-defined function. The expression enclosed in braces $\{\}$ can be: an expression that can be evaluated as a string, like $\{\texttt{"hello"}\}$, $\{\texttt{""+}X_1.\texttt{nbsyl+"syllabes"}\}$, $\{\texttt{1+8}\}$, $\{\texttt{9}\}$, or $\{\texttt{"9"}\}$; or the body of a function that returns a string, like $\{\texttt{if("\_masculine".equals(X1.gender)) return "Le"; else return "La";}\}$.

These templates and constraints are written using CoGui [4], which is a knowledge representation language. CoGui provides a graphical language for building conceptual graph knowledge bases [3] and allows us to define the constraints as predicates similar to Datalog rules.

5   Efficient Indexing and Text Production with AOC-Posets

A crucial step for efficiency of CogiText is a rapid access to the corpuses to find relevant corpus elements to be assigned to the corpus variables. This is achieved in three main steps: (1) An offline building of AOC-posets associated with the involved corpuses. These AOC-posets will be used as an index on the corpuses; (2) A computation of the needed key-value pairs for corpus variables from the constraints; (3) An assignment of corpus elements to variables using the AOC-posets of the variable corpus.

Offline Building of AOC-Posets. For each corpus, a general index is built, which takes the form of an AOC-poset associated with a formal context $K = (O, A, R)$, where formal objects O are the corpus elements, formal attributes A are all the possible key-value pairs according to the corpus schema, and $R \subseteq O \times A$ associates a corpus element to a key-value pair it owns. The concepts built on top of K are pairs of sets (E, I) such that $E \subseteq O$ and $I \subseteq A$. E is a maximal set of formal objects (extent) associated with the maximal set I of formal attributes (intent) they share [6]. They are organized by inclusion of their extent in the concept lattice, giving a specialisation order $\leq$. $C_1 \leq C_2$ if and only if $E_1 \subseteq E_2$ and $I_2 \subseteq I_1$. C_1 is a subconcept of C_2 and C_2 is a superconcept of C_1. A concept C introduces a formal attribute a (resp. a formal object o)

if C is the highest (resp. lowest) concept in $\leq$ with a in its intent (resp. o in its extent). The AOC-poset is the suborder of the concept lattice restricted to the concepts that introduce at least one formal attribute, or one formal object. Specialized algorithms for building AOC-posets are presented in [2]. Compared to the concept lattice, whose concept number can reach $2^{min(|O|,|A|)}$, the AOC-poset concept number is limited to $|O| + |A|$.

Table 3. Partial formal context for the corpus built upon DELA nouns

Offset (text) × key-value	gender _masculine	number _singular	nbsyl 1	nbsyl 2	nbsyl 3	rhyme3 naR	rhyme3 Rbo	rhyme3 RbR	rhyme3 maZ	rhyme3 bEk	...
164555 (renard)	x	x		x		x					...
348 (fouinard)	x	x		x		x					...
110976 (corbeau)	x	x		x			x				...
345724 (barbot)	x	x		x			x				...
734657 (turbo)	x	x		x			x				...
12456 (arbre)	x	x		x				x			...
78347 (marbre)	x	x		x				x			...
1110723 (bec)	x	x	x							x	...
34677 (fromage)	x	x			x				x		...
125044 (dommage)	x	x			x				x		...
.....											...

Table 3 shows a part of the formal context for the corpus built upon the DELA nouns. The shown part of the table focuses on some corpus elements and key-value pairs useful to illustrate our approach. Figure 2 (left-hand side) shows a table (restricted to the key-value pairs used in "Le corbeau et le renard" example: e.g. `gender="_masculine"`, `number="_singular"`, `nbsyl=1`), which allows a rapid access to the AOC-poset. Figure 2 (central part) shows the AOC-poset associated with the shown part of the formal context of Table 3. For the whole corpus, the AOC-poset is of course larger and has a different shape. Offsets are pointers towards the data files that enable to efficiently access from a concept extent to corpus data files (links from center to right-hand side of Fig. 2).

Computation of Key-Value Pairs for Corpus Variables. The second step computes the key-value pairs for the corpus variables of the production scheme. The equalities are used to group equal expressions. For example if the constraints are X_1.`gender`=X_2.`gender` and X_1.`gender="_masculine"`, then X_1.`gender`, X_2.`gender` and `"_masculine"` are grouped. If the group contains a single value, this value is assigned to the expressions. If the group contains several different values, the group is inconsistent and no solution can be found. This may happen if, for example, to the previous constraints we add: X_2.`gender="_feminine"`.

A group may only contain expressions (and no fixed value). For example, a hypothetical group may only contain X_3.`nbsyl` and X_4.`nbvowels`, due to a given constraint X_3.`nbsyl` = X_4.`nbvowels` and the fact that no other constraint gives a value neither to X_3, nor to X_4. In this case, a value has to be randomly chosen. Each expression of the group has a set of possible values in the associated variable corpus, for example X_3 could come from the DELA noun corpus, and

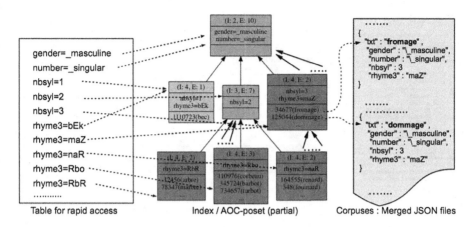

Table for rapid access   Index / AOC-poset (partial)   Corpuses : Merged JSON files

Fig. 2. Partial AOC-poset for partial formal context of Table 3. "I" stands for intent and is followed by the intent size (number of formal attributes); "E" stands for extent and is followed by the extent size (number of formal objects).

X_3.nbsyl can take values in $\{1, 2, ...14\}$, while X_4 could come from the DELA adjective corpus and X_4.nbvowels can take values in $\{1, 2, ...6\}$. The intersection of the value sets is computed. For the example, this intersection is $\{1, 2, ...14\} \cap \{1, 2, ...6\} = \{1, 2, ...6\}$. The AOC-posets associated with the corpuses allow to count how many corpus elements own each value of the intersection. A value is randomly chosen with a weighted sampling based on the number of relevant corpus element tuples. E.g. for computing this number in the previous example, to each value x of $\{1, 2, ...6\}$, we associate the number of (noun, adjective) pairs such that the noun has x syllables and the adjective has x vowels.

Assignment of Corpus Elements to Variables. After the previous step, all expressions relative to a variable have a value. These key-value pairs determine one or more concepts (the highest concepts containing all these key-value pairs) which exist if corpus elements exist with these values. For each variable, a corpus element is randomly selected in the union of the extents of the concepts that own all the key-value pairs associated with this variable. In the simplified example of "Le corbeau et le renard" parody, for each set of constraints on a variable, a single concept (introducing the initial word) has the whole set of key-value pairs of the variable, but this is a specific case. A selection in this case is simple, e.g. the noun corpus element with text "barbot" can be randomly chosen in the extent of the concept introducing "corbeau" and assigned to X_1 of Table 1.

6   Implementation

Implementation Framework. CogiText enhances the Cogui environment and provides a graphical interface for easy typing of production templates and constraints (see Fig. 3 which shows a graphical window for constraint editing).

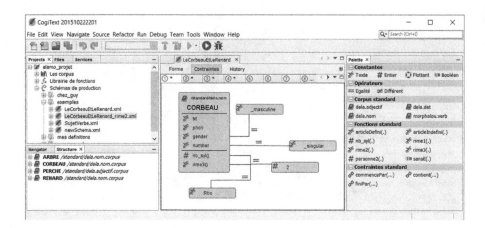

Fig. 3. Graphical description of the variable X_1 (called CORBEAU in the interface) constraints.

The two corpuses DELA and Morphalou are equipped with corpus schemas as explained in Sect. 3 and are encoded in JSON (JavaScript Object Notation) which is a lightweight data-interchange format, readable by humans and easy to automatically parse. The corpuses are stored in concatenated JSON files to ensure an efficient access to corpus elements via integers that serve as pointers.

Size and Computation Time. Table 4 shows the size and the computation time for AOC-posets associated with the "Le corbeau et le renard" parody and the DELA noun corpus. The AOC-posets are built without filtering, or with a filtering which consists in keeping the key-value pairs useful for answering to the query and the corpus elements which have at least one of these key-value pairs.

Several algorithms have been applied, using two different Java implementations for each of them: one using the Java `BitSet` data structure and one using the Java `HashSet` data structure. For these data and the `BitSet` implementation, CERES is the most efficient, e.g. running within 1 s for the filtered data (3-phonemes) and within 4 min for the whole data (3-phonemes). For the `HashSet` implementation, CERES remains competitive, but in the filtered cases, PLUTON is the best.

The AOC-posets are built offline. For our example, AOC-posets are built for the DELA noun and adjective corpuses.

Then the time t for a text production is:

- For a 3-phonemes search (#possibilities: X1:7, X2:27, X3:2, X4:3, X5:1, X6:27)
  - with the filtered data: $t = 787$ ms, including 3 ms needed to traverse the AOC-posets (for getting the whole concept extents in which a corpus element is randomly chosen).
  - with the non filtered data: $t = 1571$ ms, including a 37 ms traversal.
- For a 2-phonemes search (#possibilities: X1:37, X2:494, X3:16, X4:27, X5:13 X6:682)
  - with the filtered data: $t = 761$ ms, including a 12 ms traversal.
  - with the non filtered data: $t = 692$ ms, including again a 12 ms traversal.

Table 4. Size and computation time of AOC-posets for DELA nouns

key-value pairs for **rhyme3+nbsyl+gender+number with** filtering

#elements	#key-value pairs	density	building matrix ex. time	#concepts
137276	10	0.17	50s	56
Time	Ceres (ms)	Pluton (ms)	Hermes (ms)	Ares (ms)
BitSet	1229	2057	3124	26445
HashSet	1327	425	85887	36186

key-value pairs for **rhyme2+nbsyl+gender+number with** filtering

#elements	#key-value pairs	density	building matrix ex. time	#concepts
137413	10	0.17	50s	85
Time	Ceres (ms)	Pluton (ms)	Hermes (ms)	Ares (ms)
BitSet	1315	2427	3478	31350
HashSet	1329	388	78500	31671

key-value pairs for **rhyme3+nbsyl+gender+number without** filtering

#elements	#key-value pairs	density	building matrix ex. time	#concepts
160268	4800	8.32E-4	50s	33669
Time	Ceres (ms)	Pluton (ms)	Hermes (ms)	Ares (ms)
BitSet	216152	1884040	1422808	4018082
HashSet	138069	400936	580275	3635452

key-value pairs for **rhyme2+nbsyl+gender+number without** filtering

#elements	#key-value pairs	density	building matrix ex. time	#concepts
160268	627	6.32E-3	50s	7999
Time	Ceres (ms)	Pluton (ms)	Hermes (ms)	Ares (ms)
BitSet	7839	145148	179065	455852
HashSet	3229	22700	122855	729366

7   Conclusion

We presented an approach that assists the generation of literary texts with the support of corpuses equipped with corpus schemas, production schemes (composed of production patterns and constraints) and AOC-posets that provide information on the corpus structure (e.g. for choosing values for non-valued variable keys) and allow an efficient access for filling the corpus variables and the production patterns. The results show the benefits of the approach in a realistic case. As a future work, we would like to consider more complex constraints, e.g. inequalities or differences between variable key-values and also constraints expressed as predicates satisfying a set of DATALOG+ rules. We also would like to investigate a process systematically including an on-the-fly generation of AOC-posets specialized for a specific production scheme.

Acknowledgments. The authors warmly thank Guy Chaty who introduced them in the ALAMO world.

References

1. ALAMO: Workshop (Atelier in French) of Literature Assisted by Mathematics and Computers (Ordinateurs in French) (1981). http://www.alamo.free.fr/. Accessed 01 Jan 2017
2. Berry, A., Gutierrez, A., Huchard, M., Napoli, A., Sigayret, A.: Hermes: a simple and efficient algorithm for building the AOC-poset of a binary relation. Ann. Math. Artif. Intell. **72**(1–2), 45–71 (2014)
3. Chein, M., Mugnier, M.L.: Graph-Based Knowledge Representation: Computational Foundations of Conceptual Graphs, 1st edn. Springer Publishing Company (2008, incorporated)
4. CoGui: Visual tool for building conceptual graph knowledge bases (2008). http://www.lirmm.fr/cogui/. Accessed 01 Jan 2017
5. DELA: (dictionnaires/lexicons). LADL (Laboratoire d'Automatique Documentaire et Linguistique)- now in Institut Gaspard Monge (IGM). http://infolingu.univ-mlv.fr/. Accessed 01 Jan 2017
6. Ganter, B., Wille, R.: Formal Concept Analysis - Mathematical Foundations. Springer, Heidelberg (1999)
7. Morphalou: (lexique/lexicon). laboratoire ATILF (Nancy Université - CNRS). http://www.cnrtl.fr/lexiques/morphalou/LMF-Morphalou.php. Accessed 01 Jan 2017
8. New, B., Pallier, C., Brysbaert, M., Ferrand, L.: Lexique 2: a new French lexical database. Behav. Res. Methods Instrum. Comput. **36**(3), 516–524 (2004)
9. OuLiPo: Ouvroir de Littérature Potentielle ("workshop of potential literature") (1961). http://www.oulipo.net/. Accessed 01 Jan 2017
10. Rhymer: Rhyming Dictionary, WriteExpress. http://www.rhymer.com/RhymingDictionary/. Accessed 01 Jan 2017

On Neural Network Architecture Based on Concept Lattices

Sergei O. Kuznetsov, Nurtas Makhazhanov$^{(\boxtimes)}$, and Maxim Ushakov

Faculty of Computer Science, Department of Data Analysis and Artificial
Intelligence, National Research University Higher School of Economics,
Kochnovskiy pr. 3, Moscow, Russia
{skuznetsov,nmahaghanov}@hse.ru, mnushakov_1@edu.hse.ru
https://www.hse.ru/

Abstract. Selecting an appropriate network architecture is a crucial
problem when looking for a solution based on a neural network. If the
number of neurons in network is too high, then it is likely to overfit.
Neural networks also suffer from poor interpretability of learning results.
In this paper an approach to building neural networks based on concept
lattices and on lattices coming from monotone Galois connections is proposed in attempt to overcome the mentioned difficulties.

Keywords: Neural network architecture · Formal concept analysis ·
Optimal NN architecture · Lattice-based NN

1 Introduction

Neural Networks (NN) is one of the most popular approaches to Machine Learning [15]. Fitting of NN architecture to a dataset under study is a standard
procedure that major researchers apply to obtain the best performance results.
Matching appropriate architecture is important, because if we take too large network for training, the possibility of getting over-trained model increases. Because
of too much redundant connections, NN may learn to select required outcome for
each example in the training dataset. So, if the trained network is applied on an
object with unobserved set of attributes, it is unable to make a correct classification of it. An over-trained network loses generalizing ability, however if one takes
a too small network, one misses the ability to detect possible non-linearities in
data. Hence, selecting an appropriate topology and a size of neural network is
crucial for its performance. Another problem of NNs is a "black box"-problem.
Even if the system with NN shows good classification results, one cannot give
an intuitively clear explanation of this.

The first attempts to relate FCA and Neural Networks were done in [13–
15]. In [2] authors apply FCA for interpretation of neural codes. In this article
we propose an approach to generating neural network architecture based on the
covering relation (graph of the diagram) of a lattice coming from antitone Galois

© Springer International Publishing AG 2017
M. Kryszkiewicz et al. (Eds.): ISMIS 2017, LNAI 10352, pp. 653–663, 2017.
DOI: 10.1007/978-3-319-60438-1_64

connections (concept lattice) [6] or monotone Galois connections [1]. The motivation for that is two-fold: First, vertices of such neural networks are related to sets of similar objects with similarity given by their common attributes, so easily interpretable. The edges between vertices are also easily interpretable in terms of concept generality (bottom-up) or conditional probability (top-bottom). Second, many well-known indices of concept quality can be used to select "most interesting" concepts [10], thus reducing the size of the resulting network. In this paper we study two different types of lattices underlying network architecture: standard concept lattices [6] and lattices based on monotone Galois connections [1]. The sets of attributes closed wrt. the latter have "disjunctive meaning" in contrast to "conjunctive meaning" of standard FCA intents (sets of attributes closed wrt. antitone Galois connection). This disjunctive understanding of a set of attributes might fit better the principle of threshold function underlying the standard model of a neuron.

2   Formal Concept Analysis and Hypotheses

First of all, let us recall the basic definitions of Formal Concept Analysis [6]. We consider a set G of objects, a set M of attributes and a binary relation $I \subseteq G \times M$ such that $(g, m) \in I$ iff object g has the attribute m. Such a triple $K = (G, M, I)$ is called a *formal context*. Using the *derivation operators*, defined for $A \subseteq G$, $B \subseteq M$ by

$$A' = \{m \in M \mid gIm \, for \, all \, g \in A\},$$
$$B' = \{g \in G \mid gIm \, for \, all \, m \in B\},$$

we can define a *formal concept* of the context K as a pair (A, B) such that $A \in G$, $B \in M$, $A' = B$, $B' = A$. A is called the *extent* B is called the *intent* of the concept (A, B). These concepts, ordered by

$$(A_1, B_1) \geq (A_2, B_2) \iff A_1 \supseteq A_2$$

form a complete lattice, called *the concept lattice of $K = (G, M, I)$*.

Next, we recall concept-based hypotheses [4,5,7–9], which originate from JSM-hypotheses [3]. A *target attribute* $\omega \notin M$ partitions the set G of all objects into $c + 1$ subsets, where c is the number of values of the target attribute. The set G_i of those objects that are known to belong to ω_i class. The set $G_\tau \subseteq G$ consists of undetermined examples, i.e., of those objects, for which it is unknown what class they belong to.

Further on we consider the case of binary target and two respective classes, which we call positive and negative. By G_+ we denote the set of objects that are known to have the property of ω and G_-, the set of objects for which it is known that they do *not* have the target attribute ω, and G_τ is the set of objects for which it is not known if they have or do not have target attribute ω. Then, $K_+ = (G_+, M, I_+)$, $K_- = (G_-, M, I_-)$, and $K_\tau = (G_\tau, M, I_\tau)$ are positive, negative, and undefined contexts, respectively. Consider an example in Table 1.

Table 1. Example of formal context. Here we use the following abbreviations: "w" for white, "y" for yellow, "g" for green, "b" for blue, "s" for smooth, "r" for round, "f" for firm, and we use "$\bar{m}$" to denote negation of a binary attribute m.

Context	N	G/M	w	y	g	b	r	$\bar{r}$	f	$\bar{f}$	s	$\bar{s}$	Target
K_+	1	Apple		×			×		×	×			×
	2	Grapefruit		×			×		×		×		×
	3	Kiwi			×		×		×		×		×
	4	Plum				×	×		×	×			×
K_-	5	Toy cube		×			×	×		×			
	6	Egg	×				×	×		×			
	7	Tennis ball	×				×		×		×		
K_τ	8	Mango		×			×		×	×			?

3 Neural Network Based on Concept Lattice

In this section we introduce an algorithm for the construction of neural network architecture from the diagram of formal concept lattice. The first step is generating the set of concepts. For the generation of the covering relation of the concept lattice we apply *Add Extent* algorithm, a dual version of *Add Intent* from [12].

The first k upper layers of the lattice can be used as a neural network, where concepts stay for neurons and links between them stay for network connections. By adding extra output layer for classes we obtain an architecture of the neural network. The last hidden layer (with most specific concepts) is connected with the output layer, so that the activations of this layer determine the class of the undefined object. We call the upper part of the lattice diagram with intents of size $\leq k$ the diagram of level k. To attain a less number of classification errors by the network, one can try to select the "best concepts," retain the corresponding vertices in the network and discard the remaining ones. There are many ways to define what are "best concepts". Here we consider two measures, assuming that higher values of them correspond to "better concepts."

– F-value:
$$F = 2 \cdot \frac{Precision \cdot Recall}{Precision + Recall}$$

– Score accounted on Precision and Recall:
$$Score(h_i) = \alpha Precision(h_i) + (1 - \alpha)Recall(h_i),$$

where $\alpha \in [0, 1]$. We will consider different values of α.
– purity of formal concept, which is the maximal part of objects from one class in the concept extent;

To form the set of best concepts H, we start from the empty set, and then iteratively add a concept with the highest score (one of those given above) calculated on examples uncovered by H. This procedure stops when the set H covers all examples from the training set.

In [11,16] authors proposed their approaches for applying Concept Lattices to constructing NN's architecture. Here, we will describe another possible implementation of building interpretable NN using FCA. Previously, we build a diagram of level k, and after that we discard all formal concepts from the last hidden layer that are not considered to be good classifiers. All the neurons above that are not connected with the remaining concepts in the last hidden layer are also removed from the network. We connect the remaining part of the diagram with the input and output layers, obtaining the final architecture of the neural network.

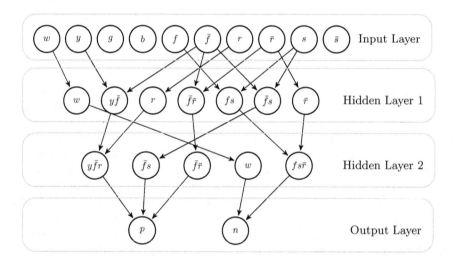

Fig. 1. Architecture of feedforward neural network based on a concept lattice.

Assume that we have a context $K = (G, M, I)$ and diagram D of its concept lattice. Then the network structure can be build as follows:

- The input layer Inp consists of neurons which represent attributes $m \in M$ of the context K.
- Hidden layers Hid_i consist of neurons which represent formal concepts of the context K. The connections between neurons in the hidden layers are the same as in the diagram D (so, two neurons are connected if the corresponding formal concepts are neighbours in the diagram of formal concept lattice). The lower (most specific) concepts are connected to neurons staying for classes. In Fig. 1 you can see the architecture of NN that has been constructed by the method described above.

- The output layer *Out* consists of neurons representing classes. It is connected to the last hidden layer representing the most specific concepts.

In Fig. 1 you can see that in the output layer every neuron is related to each class. There are neurons related to formal concepts in the last hidden layer. We can interpret each weight on the link connecting a concept from the last hidden layer with the output layer as importance of this concept. Here, we want to note that one has to see the proportion of such weights in each particular output neuron. So, these weights do not allow us to compare the concepts related to different classes.

4 Neural Networks Based on Monotone Galois Connections

In this section we describe an approach to constructing neural networks from lattices arising from monotone Galois connections between powersets of objects and attributes. The motivation for this approach comes from the basic properties of standard formal concepts and related closed sets of attributes (intents).

This properties may result in problems when we take intents for nodes of a neural network and covering relation of the concept lattice for connections (arcs) of the neural network. For example, consider a neural network with node C, which have two parents, node A and node B, connected with corresponding weights w_{AC} and w_{BC}. Assume that for some object g neuron A is activated, but neuron B is not. According to neural network model, neuron C also will be activated: $w_{AC} \cdot 1 + w_{BC} \cdot 0 = w_{AC}$ (here we use linear activation function). However, if the given neurons are formal concepts with intents $\hat{A}$, $\hat{B}$ and $\hat{C}$, respectively, then object g does not have some attributes from $\hat{B}$ (as B is not activated), and, as a result, object g does not have some attributes from $\hat{C}$. The latter means that object g does not belong to the extent of concept C, but neuron C is activated for g. This difference may lead to a problem when neurons, which are not supposed to be activated, have significant weights in the constructed neural network.

In this section we will use monotone Galois connection [1,17,18] for building network architecture, which will help us to deal with the problem of 'conjunctivity' of formal concepts, i.e., the property of the concept intent that one has to have all attributes when having the intent.

Consider a formal context (G, M, I), then monotone Galois connections are defined as

$$A' = \{b \mid \nexists a \in G \setminus A \quad such \quad that \quad aIb\},$$
$$B' = \{a \mid \exists b \in B \quad such \quad that \quad aIb\},$$

where $A \subseteq G$ is a set of objects and $B \subseteq M$ is a set of attributes. Further we will call a pair (A, B), where $A' = B$, $B' = A$, a *disjunctive formal concept*.

As in the case of standard formal concepts, disjunctive formal concepts form a lattice with operation $\cup$, defined as $(A_1, B_1) \cup (A_2, B_2) = (A_1 \cup A_2, (B_1 \cup B_2)'')$.

So, we can use the diagram of the concept lattice for building neural network (in the same way as we did it in the previous section).

To show advantages of this approach, consider the previous example, but with disjunctive formal concepts instead of the standard ones. Consider that neuron A is activated, but neuron B is not. So, object g has some attributes from $\hat{A}$, and does not have any attribute from $\hat{B}$. As $\hat{A} \subset \hat{C}$, then object g has some attributes from $\hat{C}$ too. Thus, object g belongs to the extent of $\hat{C}$, so neuron C is activated (we do not have contradictions between disjunctive formal concepts and neural network model).

In order to compute the set of disjunctive formal concepts in a context $K = (G, M, I)$, we need to perform three steps:

1. Compute the complement of the initial relation (replacing all zeros by ones and vice versa);
2. Run *Add Extent* algorithm (a dual version of *Add Intent* from [12]) to compute formal concepts and the covering relation on them;
3. Replace all extents A by $G \setminus A$.

Besides determining the network structure, to train the network one needs to set initial weights on neuron connections. Here we consider several ways of initializing weights.

First, consider the following assignment of weights:

1. $w_1((A_1, B_1), (A_2, B_2)) = \frac{|A_1|}{|A_2|}$ The weights of this kind just give the confidence of the association rule $B_1 \to B_2$.
   Second, one can use similar weights, but with the zero mean:
2. $w_2((A_1, B_1), (A_2, B_2)) = \frac{|A_1|}{|A_2|} - 0.5$.
   The idea of the following kind of weights is quite simple: the more general the concept, the less important a connection from any concept to it:
3. $w_3((A_1, B_1), (A_2, B_2)) = \frac{1}{|A_2|}$.

Here $w((A_1, B_1), (A_2, B_2))$ is the weight of the connection from concept (A_1, B_1) to (A_2, B_2). For edges connecting the last hidden layer and the output layer we initialize weights $w((A, B), i)$ as follows:

$$w((A, B), i) = \frac{|\{a : a \in A, l(a) = i\}|}{|A|},$$

where i is a class, and $l(\cdot)$ is a function that takes any object g to its class $l(g)$.

5   Experiments

We have performed experiments with the following six datasets from the open source UCI Machine Learning Repository (http://archive.ics.uci.edu/ml/index. html) (Table 2):

1. Breast Cancer

2. Credit Card Default
3. Heart Disease
4. Mammographic Mass Data
5. Seismic Bumps.

Table 2. Basic characteristics of the datasets

Dataset	Train sample	Test sample	Number of variables	Number of classes in target variable
Breast cancer	512	57	30	2
CreditCard default	27000	3000	23	2
Heart disease	273	31	13	5
Mammographic mass data	865	96	5	2
Seismic bumps	2326	258	18	2

5.1 Experiments with Different ML Methods

First, we consider the performance of Neural Networks without any prior information about dataset. For all of the five datasets above we have learnt neural networks with different architectures. We apply Adam stochastic optimization strategy because of small number of parameters required for tuning.

As you can see in Fig. 2, neural networks are very unstable wrt. size. For each dataset, it is required to select its own best performing architecture. Even small modifications in the network structure can dramatically affect the results.

Now, let us compare the performance of various ML methods and FCA-based NN algorithms. We constructed a network with one hidden layer and random initial weights. As you can see in Table 3, on Breast Cancer and on Mammographic Mass datasets we achieve same results or that comparable with other algorithms. On Credit Card Default and Seismic Bumps datasets the network based on antitone Galois connections perform better than other ML methods. On the other hand, FCA-based NN models shows worse performance on Heart Disease dataset. At the same time, you can see that the performance of fully-connected NN on this sample is higher than ours. If we consider FCA-based NN as a simple neural network without redundant connections, then we can suppose that it can achieve performance comparable with fully-connected neural networks. The reason why we have obtained worse results may reside in poor selection of the "best concepts". Nevertheless, the main goal of this work was to construct neural network architecture, which can give interpretable results. In future work we would like to attain performance of lattice-based neural networks close to results obtained with simple feed-forward neural networks.

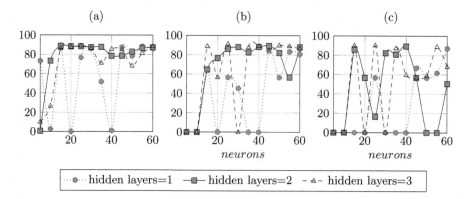

Fig. 2. Performance of NNs with different architectures on breast cancer data. On x-scale number of neurons in each hidden layer, on y-scale F-values. Each NN were learned applying Adam stochastic optimization strategies with initial learning rates equal: (a) 0.01, (b) 0.001, (c) 0.0001.

Table 3. Performance of machine learning methods.

Method	F-value				
	Breast cancer	CreditCard default	Heart disease	Mammogr mass	Seismic bumps
Nearest neighbour	89.0%	27.6%	7.8%	77.2%	5.5%
Decision tree	92.4%	40.5%	34.7%	76.7%	18.5%
Random forest	91.9%	41.9%	28.2%	80.0%	13.2%
Neural network (for the best architecture)[a]	92.9%	37.1%	48.5%	81.7%	12.7%
FCA with AGC based NN	91.8%	50.1%	37.3%	83.4%	23.2%
FCA with MGC based NN	91.8%	34.6%	39.2%	80.5%	10.2%

[a]For Breast Cancer and CC Default datasets: 2 layers, 25 neurons; for Heart Disease: 3 layers, 10 neurons; for Mammographic Mass data: 1 layer, 40 neurons; for Seismic Bumps: 2 layers, 10 neurons

5.2 Comparing Different Methods of Pretraining Neural Network

In Sect. 4 we have proposed three methods of pretraining initial weights of the model based on the properties of disjunctive formal concepts. We have compared their efficiency on Car Evaluation dataset for different activation functions. To this end, we have constructed the network architecture from the 4-level diagram of disjunctive formal concepts, by initializing weights according to the formulas above and training the model. The table below shows the results of the experiments:

As you can see, the first method of initializing weights for disjunctive concepts $w_1((A_1, B_1), (A_2, B_2))$ shows the worst results, almost like coming from random choice. The reason can be in high values of preinitialized weights, which result in high values of activation function in class nodes, so it is difficult for neural network to fit the data (Table 4).

This model is significantly less accurate than the previous one. The reason of such difference may reside in the monotone nature of disjunctive concepts (the size of extents increases with the size of intent). The top level of the lattice gives very general concepts with big extents and intents, which are not so good for classifying objects.

Table 4. Performance of disjunctive formal concepts with various initial weights and activation functions

Activation function	Predefined weights		
	w_1	w_2	w_3
Sigmoid	32.7%	43.1%	41%
Rectify	33.8%	45.1%	41.9%
Softmax	29.2%	30.9%	31.2%

6   Conclusion

In this paper we have proposed an approach for constructing neural networks based on lattices coming from antitone Galois connections (standard concept lattices) and monotone Galois connections.

Neural networks that are based on concept lattices are very sparse compared to standard fully-connected networks. All neurons in the last hidden layer are related to concepts coming from the dataset. Another advantage of neural networks based on concept lattices is their interpretability, which is very significant in domains like medical decision making and credit scoring. One can both predict the probability of default of applicants, but also implement specific rules and then weight them according to their importance for predicting target variable. Thus, NNs based on concept lattices can be implemented in domains where it is important to explain why objects are assigned to particular classes.

We have presented some results of experiments with different heuristics and parameters of the model. We have calculated performances of simple neural networks with different number of hidden layers and neurons. Also, we have evaluated performance of other learning algorithms in order to compare them with neural networks based on concept lattices for different datasets. We can conclude that on some datasets we have achieved results comparable with those obtained by other learning approaches. Our further research will be on the study of methods for selecting best concepts for better network performance.

Acknowledgments. The paper was prepared within the framework of the Basic Research Program at the National Research University Higher School of Economics (HSE) and supported within the framework of a subsidy by the Russian Academic Excellence Project 5-100.

References

1. Düntsch, I., Gediga, G.: Approximation operators in qualitative data analysis. In: Swart, H., Orłowska, E., Schmidt, G., Roubens, M. (eds.) Theory and Applications of Relational Structures as Knowledge Instruments. LNCS, vol. 2929, pp. 214–230. Springer, Heidelberg (2003). doi:10.1007/978-3-540-24615-2_10
2. Endres, D., Foldiak, P.: Interpreting the neural code with formal concept analysis. In: Koller, D., Schuurmans, D., Bengio, Y., Bottou, L. (eds.) Advances in Neural Information Processing Systems 21, pp. 425–432. MIT Press, Cambridge (2009)
3. Finn, V.K.: Plausible reasoning in systems of JSM type. Itogi Nauki i Tekhniki, Seriya Informatika, Moscow (1991, in Russian)
4. Ganter, B., Kuznetsov, S.O.: Hypotheses and version spaces. In: Ganter, B., De Moor, A., Lex, W. (eds.) ICCS-ConceptStruct 2003. LNCS, vol. 2746, pp. 83–95. Springer, Heidelberg (2003). doi:10.1007/978-3-540-45091-7_6
5. Ganter, B., Kuznetsov, S.O.: Formalizing hypotheses with concepts. In: Ganter, B., Mineau, G.W. (eds.) ICCS-ConceptStruct 2000. LNCS, vol. 1867, pp. 342–356. Springer, Heidelberg (2000). doi:10.1007/10722280_24
6. Ganter, B., Wille, R.: Formal Concept Analysis: Mathematical Foundations. Springer, Heidelberg (1999)
7. Kuznetsov, S.O.: Mathematical aspects of concept analysis. J. Math. Sci. **80**(2), 1654–1698 (1996)
8. Kuznetsov, S.O.: Machine learning and formal concept analysis. In: Eklund, P. (ed.) ICFCA 2004. LNCS (LNAI), vol. 2961, pp. 287–312. Springer, Heidelberg (2004). doi:10.1007/978-3-540-24651-0_25
9. Kuznetsov, S.O.: Fitting pattern structures to knowledge discovery in big data. In: Cellier, P., Distel, F., Ganter, B. (eds.) ICFCA 2013. LNCS (LNAI), vol. 7880, pp. 254–266. Springer, Heidelberg (2013). doi:10.1007/978-3-642-38317-5_17
10. Kuznetsov, S.O., Makhalova, T.P.: On interestingness measures of formal concepts. Inf. Sci. (2017) (accepted for publication)
11. Nguifo, E.M., Tsopze, N., Tindo, G.: M-CLANN: multiclass concept lattice-based artificial neural network. In: Franco, L., Elizondo, D.A., Jerez, J.M. (eds.) Constructive Neural Networks. Studies in Computational Intelligence, vol. 258, pp. 103–121. Springer, Heidelberg (2009)
12. Merwe, D., Obiedkov, S., Kourie, D.: AddIntent: a new incremental algorithm for constructing concept lattices. In: Eklund, P. (ed.) ICFCA 2004. LNCS (LNAI), vol. 2961, pp. 372–385. Springer, Heidelberg (2004). doi:10.1007/978-3-540-24651-0_31
13. Norris, E.M.: Maximal rectangular relations. In: Karpiński, M. (ed.) FCT 1977. LNCS, vol. 56, pp. 476–481. Springer, Heidelberg (1977). doi:10.1007/3-540-08442-8_118

14. Rudolph, S.: Using FCA for encoding closure operators into neural networks. In: Priss, U., Polovina, S., Hill, R. (eds.) ICCS-ConceptStruct 2007. LNCS, vol. 4604, pp. 321–332. Springer, Heidelberg (2007). doi:10.1007/978-3-540-73681-3_24
15. Shavlik, W.J., Towell, G.G.: KBANN: knowledge based artificial neural networks. Artif. Intell. **70**, 119–165 (1994)
16. Tsopze N., Nguifo, E.M., Tindo G., CLANN: concept-lattices-based artificial neural networks. In: Proceedings of 5th International Conference on Convcept Lattices and Applications (CLA 2007), pp. 157–168, Montpellier, France, 24–26 October 2007
17. Vimieiro, R., Moscato, P.: Disclosed: an efficient depth-first, top-down algorithm for mining disjunctive closed itemsets in high-dimensional data. Inf. Sci. **280**, 171–187 (2014)
18. Zhao, L., Zaki, M.J., Ramakrishnan, N.: BLOSOM: a framework for mining arbitrary Boolean expressions. In: KDD 2006, Philadelphia USA (2006)

Query-Based Versus Tree-Based Classification: Application to Banking Data

Alexey Masyutin[(✉)] and Yury Kashnitsky

National Research University Higher School of Economics, Moscow, Russia
alexey.masyutin@gmail.com, ykashnitsky@hse.ru

Abstract. The cornerstone of retail banking risk management is the estimation of the expected losses when granting a loan to the borrower. The key driver for loss estimation is probability of default (PD) of the borrower. Assessing PD lies in the area of classification problem. In this paper we apply FCA query-based classification techniques to Kaggle open credit scoring data. We argue that query based classification allows one to achieve higher classification accuracy as compared to applying classical banking models and still to retain interpretability of model results, whereas black-box methods grant better accuracy but diminish interpretability.

Keywords: PD · Classification · Kaggle · FCA · Credit scoring

1 Introduction

From the 1960s, banks have started to adopt statistical scoring systems that were trained on datasets of applicants, consisting of their socio-demographic and loan specific features [3]. The aim of those systems was to support decision making whether to grant a loan for an applicant or not. As far as mathematical models are concerned, they were typically logistic regressions run on selected sets of attributes. The target variable was defined as a binary logical value, one if default occurs, zero otherwise. Typical scorecard is built in several steps. The first step is so-called WOE-transformation [1], which transforms all numerical and categorical variables into discrete numerical variables. For continuous variables the procedure, in effect, breaks the initial variable into several ranges, for categorical ones - the procedure regroups the initial categories. The second step is single factor analysis, when significant attributes are selected. The commonly used feature selection method is then based on either information value, or Gini coefficient calculation [1]. With the most predictive factors included into the model, they are further checked for pairwise correlations and multicollinearity. Features with high observed correlation are excluded. As soon as single-factor analysis is over, logistic regression is run taking the selected transformed attributes as input. The product of beta-coefficient and WOE value of the particular category produces the score for that particular category. The sum of variable scores produces the final score for the loan application. Finally, the cutoff score is selected based on the revenue and loss in the historical portfolio. When the scorecard is launched

© Springer International Publishing AG 2017
M. Kryszkiewicz et al. (Eds.): ISMIS 2017, LNAI 10352, pp. 664–673, 2017.
DOI: 10.1007/978-3-319-60438-1_65

into work, the loan application immediately receives its score which is compared to the cutoff point. In case the score is lower than cutoff value, the application is rejected, otherwise it is approved. It has to be mentioned that despite its simple mathematical approach scorecards were incredibly attractive for lending institutions for several reasons. First of all, new loan application received score for each of its attributes, which provided clarity: in case of rejection the reason, why the final score was lower than cutoff, can be retrieved. The discriminative power of the models, however, is still at the moderate level. The Gini coefficient for the application scorecards varies from 45% to 55%, and for the behavioral scorecards the range is from 60% to 70% [4]. Apparently, a considerable amount of research was done in the field of alternative machine learning techniques seeking the goal to improve the results of the wide-spread scorecards [5–7].

The methods of PD estimation can either produce so-called "black box" models with limited interpretability of model result, or, on the contrary, provide interpretable results and clear model structure. The key feature of risk management practice is that, regardless of the model accuracy, it must not be a black box. That is why methods such as neural networks and SVM classifiers did not earn much trust within banking community [14].

On the contrary, alternative methods such as associative rules and decision trees provide the user with easily interpretable rules which can be applied to the loan application. FCA-based algorithms also belong to the second group since they use concepts in order to classify objects. The intent of the concept can be interpreted as a set of rules that is supported by the extent of the concept. However, for non-binary context the computation of the concepts and their relations can be very time-consuming. In case of credit scoring we deal with numerical context, as soon as categorical variables can be transformed into set of dummy variables. Lazy classification [13] seems to be appropriate to use in this case since it provides the decision maker with the set of rules for the loan application and can be easily parallelized.

In this paper, we test query-based classification framework on Kaggle open data contest.[1] The contest was held in 2011 and provided credit scoring data to test different classification algorithms. We compare results of query-based classification with classical methods adopted in banks and black-box methods. We argue that query-based classification allows one to achieve higher accuracy than classical methods and still to retain interpretability of model results, whereas black-box methods grant better accuracy but diminish interpretability.

2   Main Definitions

First, we recall some standard definitions related to Formal Concept Analysis, see e.g. [8].

Let G be a set (of objects), let $(D, \sqcap)$ be a meet-semi-lattice (of all possible object descriptions) and let δ: $G \rightarrow D$ be a mapping. Then $(G, \underline{D}, \delta)$, where

---

[1] https://www.kaggle.com/c/GiveMeSomeCredit.

$\underline{D} = (D, \sqcap)$, is called a *pattern structure* [9], provided that the set $\delta(G) := \{\delta(g) | g \in G\}$ generates a complete subsemilattice $(D_\delta, \sqcap)$ of $(D, \sqcap)$, i.e., every subset X of $\delta(G)$ has an infimum $\sqcap X$ in $(D, \sqcap)$. Elements of D are called *patterns* and are naturally ordered by *subsumption* relation $\sqsubseteq$: given $c, d \in D$ one has $c \sqsubseteq d \leftrightarrow c \sqcap d = c$. Operation $\sqcap$ is also called a *similarity operation*. A pattern structure $(G, \underline{D}, \delta)$ gives rise to the following *derivation operators* $(\cdot)^\diamond$:

$$A^\diamond = \prod_{g \in A} \delta(g) \qquad \text{for } A \in G,$$

$$d^\diamond = \{g \in G \mid d \sqsubseteq \delta(g)\} \qquad \text{for } d \in (D, \sqcap).$$

These operators form a Galois connection between the powerset of G and $(D, \sqcap)$. The pairs (A, d) satisfying $A \subseteq G$, $d \in D$, $A^\diamond = d$, and $A = d^\diamond$ are called *pattern concepts* of $(G, \underline{D}, \delta)$, with *pattern extent* A and *pattern intent* d. Operator $(\cdot)^{\diamond\diamond}$ is an algebraical closure operator on patterns, since it is idempotent, extensive, and monotone [8].

The concept-based learning model for standard object-attribute representation (i.e., formal contexts) is naturally extended to pattern structures. Suppose we have a set of positive examples G_+ and a set of negative examples G_- w.r.t. a target attribute, $G_+ \cap G_- = \emptyset$, objects from $G_\tau = G \setminus (G_+ \cup G_-)$ are called undetermined examples. A pattern $c \in D$ is an α - weak positive premise (classifier) iff:

$$\frac{|c^\diamond \cap G_-|}{|G_-|} \leq \alpha \text{ and } \exists A \subseteq G_+ : c \sqsubseteq A^\diamond$$

A pattern $h \in D$ is an α - weak positive premise iff:

$$\frac{|h^\diamond \cap G_-|}{|G_-|} \leq \alpha \text{ and } \exists A \subseteq G_+ : h = A^\diamond$$

In case of credit scoring we work with pattern structures on intervals as soon as a typical object-attribute data table is not binary, but has many-valued attributes. Instead of binarizing (scaling) data, one can directly work with many-valued attributes by applying interval pattern structure [16]. For two intervals $[a_1, b_1]$ and $[a_2, b_2]$, with $a_1, b_1, a_2, b_2 \in \mathbb{R}$ the *meet operation* is defined as [12]:
$[a_1, b_1] \sqcap [a_2, b_2] = [min(a_1, a_2), max(b_1, b_2)]$.

The original setting for lazy classification with pattern structures can be found in [10, 11].

3   Loan Default Prediction in Banking: Scorecards

The event of default in retail banking is defined as more than 90 days of delinquency within the first 12 months after the loan origination. Defaults are divided into fraudulent cases and ordinary defaults. The default is told to be a fraudulent case when delinquency starts at one of the three first months. It means that when submitting a credit application, the borrower did not even intend to pay back. Otherwise, the default is ordinary when the delinquency starts after the first three months on book. That is why scorecards are usually divided into fraud and application scorecards. In fact the only difference is the target variable definition, while the sets of predictors and the data mining techniques remain the same. The default cases are said to be "bad", and the non-default cases are said to be "good". Banks and credit organizations have been traditionally using scorecards to predict whether a loan applicant is going to be bad or good.

Mathematical architecture of scorecards is based on a logistic regression, which takes the transformed variables as an input. The transformation of the initial variables is known as WOE-transformation [1]. It is wide-spread in credit scoring to apply such a transformation to the input variables as soon as it accounts for non-linear dependencies and provides certain robustness coping with potential outliers. The aim of the transformation is to divide each variable into no more than k categories. At step 0, all the continuous variables are binned into 20 quantiles, the nominal and ordinal variables are either left untouched or are one-hot encoded. Now, when all the variables are categorized, the odds ratio is computed for each category.

$$odds_{ij} = \frac{\%goods_{ij}}{\%bads_{ij}}$$

Then for each predictor variable X_i ($i = 1...n$) non-significant categories are merged. Significance is measured by standard chi-square test for differences in odds with p-value threshold up to 10%. So, for each feature the following steps are done:

1. If X_i has 1 category only, stop and set the adjusted p-value to be 1.
2. If X_i has k categories, go to step 7.
3. Else, find the allowable pair of categories of (an allowable pair of categories for ordinal predictor is two adjacent categories, and for nominal predictor is any two categories) that is least significantly different (i.e. most similar) in terms of odds. The most similar pair is the pair whose test statistic gives the largest p-value with respect to the dependent variable Y.
4. For the pair having the largest p-value, check if its p-value is larger than a user-specified alpha-level merge. If it does, this pair is merged into a single compound category. Then a new set of categories of is formed. If it does not, then if the number of categories is less or equal to user-specified minimum segment size, go to step 6, else merge two categories with highest p-value.
5. Go to step 2.
6. (Optional) Any category having too few observations (as compared with a user-specified minimum segment size) is merged with the most similar other category as measured by the largest of the p-values.
7. The adjusted p-value is computed for the merged categories by applying Bonferroni adjustments [2]. Having accomplished the merging steps, we acquire categorized variables instead of the continuous ones.

When each variable X_i $(i = 1...n)$ is binned into a certain number of categories (k_i), one is able to calculate the odds for each category j $(j = 1...k_i)$, the weight of evidence for each category.

$$WOE_{ij} = ln(odds_{ij})$$

The role of the WOE-transformation is that, instead of initial variables, logistic regression receives WOE features as input. So, each input variable is a discrete transformed variable, which takes values of WOE. When estimating the logistic regression, the usual maximum likelihood is applied.

4   Query-Based Classification Algorithm

Query-based classification is in effect an approach proposed in [15] with certain voting scheme applied to predict the test object class (positive or negative). The idea behind the algorithm is to check whether it is positive or negative context that test object is more similar to. The similarity is defined as a total support of α - weak positive (negative) premises that contain the description of test object. The algorithm uses three parameters: subsample size, number of iterations and alpha-threshold. The first parameter is expressed as percentage of the observations in the context. At each step the subsample is extracted and the descriptions of the objects in subsample are intersected with the description of test object. As subsample size grows, the resulting intersection $\delta(g_1) \sqcap ... \sqcap \delta(g_k) \sqcap \delta(g)$ becomes more generic and it is more frequently falsified by the objects from the opposite context. We randomly take the chosen number of objects from positive (negative) context as candidates for intersection with the test object. The number of times (i.e. number of iterations) we randomly extract a subsample from the context is the second parameter of the algorithm, which is also tuned through grid search. Intuition says, the higher the value of the parameter the more premises should be mined from the data. However, the obvious penalty for increasing the value of this parameter is time required for computing intersections. As we mentioned, the greater the subsample size, the more it is likely that $(\delta(g_1) \sqcap ... \sqcap \delta(g_k) \sqcap \delta(g))^\diamond$ contains the object of the opposite class. In order to control this issue, we add third parameter which is alpha-threshold. If the percentage of objects from the positive (negative) context that falsify the premise $\delta(g_1) \sqcap ... \sqcap \delta(g_k) \sqcap \delta(g)$ is greater than alpha-threshold of this context than the premise will be considered as falsified, otherwise the premise will be α-weak and, thereafter, used in classification of the test object. These steps are performed for each test object for positive and negative contexts seprately, producing a set of positive and negative α-weak premises. The final output for the test object we used was a difference between the total number of objects from positive context supporting the set of positive premises and the total number of objects from negative context supporting the set of negative premises.

Table 1. Kaggle data description

Variable name	Description	Type
SeriousDlqin2yrs	Person experienced 90 days past due delinquency or worse	Y/N
RevolvingUtilization OfUnsecuredLines	Total balance on credit cards and personal lines of credit except real estate and no installment debt like car loans divided by the sum of credit limits	Percentage
Age	Age of borrower in years	Integer
NumberOfTime30-59DaysPastDueNotWorse	Number of times borrower has been 30–59 days past due but no worse in the last 2 years	Integer
DebtRatio	Monthly debt payments, alimony, living costs divided by monthly gross income	Percentage
MonthlyIncome	Monthly income	Real
NumberOfOpenCredit LinesAndLoans	Number of Open loans (installment like car loan or mortgage) and Lines of credit (e.g. credit cards)	Integer
NumberOfTimes90Days Late	Number of times borrower has been 90 days or more past due	Integer
NumberRealEstateLoans OrLines	Number of mortgage and real estate loans including home equity lines of credit	Integer
NumberOfTime60-89DaysPastDueNotWorse	Number of times borrower has been 60–89 days past due but no worse in the last 2 years	Integer
NumberOfDependents	Number of dependents in family excluding themselves (spouse, children etc.)	Integer

5 Data and Experiments

We decided to retrieve open dataset devoted to the credit scoring. We considered the "Give Me Some Credit" contest held in 2012[2]. The data has a binary target variable (class label) whether the borrower defaulted or not. However, it is not specified whether the default event was ordinary or fraudulent. We develop a scorecard and examine its accuracy via out-of-sample validation with provided target variable. The validation process requires calculation of performance metrics (ROC AUC and Gini coefficient) of the model based on the data sample that was retrieved from the same distribution but was not used to develop the model itself. This approach allows the user to check for accuracy and stability of the model. In order to train the models we extracted 1000 good loans and 1000 bad loans. The size of the validation set was 300 observations. All these

[2] https://www.kaggle.com/c/GiveMeSomeCredit.

observations were randomly extracted from the contest dataset. Our aim was to compare classical scorecard versus black-box models such as boosting versus query-based classification approach based on interval patterns. We implemented the query-based classification algorithm using R, which is a flexible tool for statistical analysis. The R language is becoming more recognizable in the banking sphere as well. The features for loan default prediction are presented in Table 1:

First, we concluded that the variable distributions might be not very appropriate for applying trees-like transformations. The values of features are evenly distributed across wide ranges both for good and bad loans, therefore applying cutpoint does not perform well to distinguish among loan applicants. Examples of such distributions are presented below (Fig. 1):

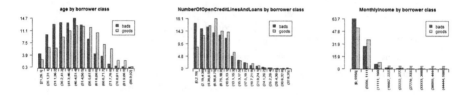

Fig. 1. Age distribution by goods and bads (left), number of open credit lines and loans by goods and bads (middle), and monthly applicant income by goods and bads (right)

In order to build scorecard we applied WOE-transformation to the variables (using rpart and smbinning packages in R) on training sample. The WOE-transformation was controlled for maximum number of observations in the final nodes of one-factor trees in order to escape overfitting at the starting point. Therefore, variables were binned into two to four categories. The examples of variable binning are provided in Fig. 2.

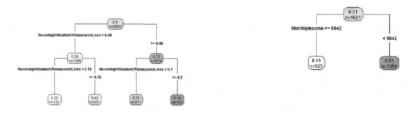

Fig. 2. One-Factor Trees for WOE-transformation of Revolving Utilization of Unsecured Lines (left) and Monthly Income (right)

As soon as we have transformed the factors, the individual Gini coefficients were calculated to assess the predictive power of the coefficients. We excluded variables that have shown dramatic drop in Gini on validation sample. The rest were fed to logistic regression and the final model included the features presented in Table 2.

Table 2. Logistic regression output

Feature	Estimate	Std. Error	t-stat	P-value
(Intercept)	−0.56881	0.05002	−11.371	<2e−16***
trscr_RevolvingUtilizationOfUnsecuredLines	0.73361	0.04317	16.992	<2e−16***
trscr_age	0.39750	0.08257	4.814	1.59e−06***
trscr_NumberOfTime3059DaysPastDueNotWorse	0.55770	0.05593	9.971	<2e−16***
trscr_NumberOfTime6089DaysPastDueNotWorse	0.44882	0.06373	7.043	2.58e−12***

After training the scorecard we applied query-based classification to the validation set. The algorithm QBCA (for "Query-Based Classification Algorithm") defines number of iterations, *alpha*-level and *subsample-size* parameters upon algorithm tuning. Finally, the algorithms were compared on a validation set by plotting ROC curves and calculating Gini coefficients achieved.

Table 3. Experimental results: cross-validation and validation Gini coefficients for 3 models. "Scorecard" stands for logistic regression with WOE-transformed features, and "QBCA" designates the query-based classification algorithm

Metric\algo	Scorecard	QBCA	Xgboost
Valid. Gini	0.5806	0.6624	0.708

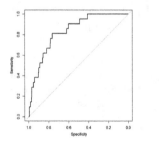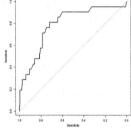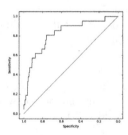

Fig. 3. ROC curves for QBCA (left), Scorecard (middle) and Xgboost (right)

Finally, we applied the Xgboost[3] gradient boosting algorithm to the same data to estimate the classification accuracy achievable with the "black-box" model. The parameters were tuned via 5-fold stratified cross-validation. The results (cross-validation and validation Gini) for 3 tested algorithms are given in Table 3. The ROC curves for validation set are presented in Fig. 3. As we can see, Xgboost performs best in terms of Gini. However, its results are not

---

[3] https://github.com/dmlc/xgboost.

interpretable, and the best explanation for classification that we one can extract from the trained Xgboost model is the estimated feature importance, based on the number of times splits in trees were done with each feature.

On the contrary, it is interesting to realize that certain patterns can be extracted from the QBCA model. We can observe rules such as if a loan applicant's age is greater than 50 and there was no delinquency in the past and the overall revolving utilization of unsecured lines was less than 11%, then the probability of default is almost 4 times lower than average. On the other side applicants younger than 30 and having revolving utilization of unsecured lines greater than 72% will default 1.5 times more frequent than on average. This is where we enjoy the advantage of interval pattern structures: they represent the rules that can be easily interpreted, and at the same time they make prediction for each new object in validation dataset individually, which allows to improve classification accuracy over the default scorecard model.

6   Conclusion

We considered three approaches to modeling probability of default in the problem of credit scoring. All approaches were tested on the random sample from Kaggle dataset. The first was testing classical methods of scorecard, which is easily interpretable but provides limited predictive accuracy. The second, was query-based classification algorithm on interval pattern structures, which provides higher predictive performance, and still keeps the interpretability clear. The third, was a black-box algorithm represented by Xgboost, which showed best predictive ability but nevertheless did not allow one to extract interesting client insights from the data. Therefore, we argue that FCA based classification algorithms can compete with ordinary statistical instruments adopted in banks and still provide the sets of rules which were relevant for particular loan applicant.

Acknowledgments. The paper was prepared within the framework of the Basic Research Program at the National Research University Higher School of Economics (HSE) and supported within the framework of a subsidy by the Russian Academic Excellence Project '5-100'.

References

1. Bigss, D., Ville, B., Suen, E.: A method of choosing multiway partitions for classification and decision trees. J. Appl. Stat. **18**(1), 49–62 (1991)
2. Bonferroni, C.E.: Teoria statistica delle classi e calcolo delle probabilitá. In: Pubblicazioni del R Istituto Superiore di Scienze Economiche e Commerciali di Firenze, vol. 8, pp. 3–62 (1936)
3. Naeem, S.: Credit Risk Scorecards: Developing and Implementing Intelligent Credit Scoring. SAS Publishing, Cary (2005)
4. Baesens, B., Gestel, T.V., Viaene, S., Stepanova, M., Suykens, J.: Benchmarking state-of-the-art classification algorithms for credit scoring. J. Oper. Res. Soc. **54**(6), 627–635 (2003)

5. Yu, L., Wang, S., Lai, K.K.: An intelligent agent-based fuzzy group decision making model for financial multicriteria decision support: the case of credit scoring. Eur. J. Oper. Res. **195**, 942–959 (2009)
6. Gestel, T.V., Baesens, B., Suykens, J.A., Van den Poel, D., Baestaens, D.E., Willekens, B.: Bayesian kernel based classification for financial distress detection. Eur. J. Oper. Res. **172**, 979–1003 (2006)
7. Kumar, P.R., Ravi, V.: Bankruptcy prediction in banks and firms via statistical and intelligent techniques - a review. Eur. J. Oper. Res. **180**(1), 1–28 (2007)
8. Ganter, B., Wille, R.: Formal Concept Analysis: Mathematical Foundations. Springer-Verlag New York Inc., New York (1997)
9. Ganter, B., Kuznetsov, S.O.: Pattern structures and their projections. In: Delugach, H.S., Stumme, G. (eds.) ICCS-ConceptStruct 2001. LNCS, vol. 2120, pp. 129–142. Springer, Heidelberg (2001). doi:10.1007/3-540-44583-8_10
10. Kuznetsov, S.O.: Scalable knowledge discovery in complex data with pattern structures. In: Maji, P., Ghosh, A., Murty, M.N., Ghosh, K., Pal, S.K. (eds.) PReMI 2013. LNCS, vol. 8251, pp. 30–39. Springer, Heidelberg (2013). doi:10.1007/978-3-642-45062-4_3
11. Kuznetsov, S.O.: Fitting pattern structures to knowledge discovery in big data. In: Cellier, P., Distel, F., Ganter, B. (eds.) ICFCA 2013. LNCS (LNAI), vol. 7880, pp. 254–266. Springer, Heidelberg (2013). doi:10.1007/978-3-642-38317-5_17
12. Kaytoue, M., Duplessis, S., Kuznetsov, S.O, Napoli, A.: Mining gene expression data with pattern structures in formal concept analysis. Inf. Sci. (2011). Special Issue: Lattices
13. Aha, D.W. (ed.): Lazy Learning. Kluwer Academic Publishers, Berlin (1997)
14. Li, X., Zhong, Y.: An overview of personal credit scoring: techniques and future work. Int. J. Intell. Sci. **2**(4A), 182–189 (2012)
15. Masyutin, A., Kashnitsky, Y., Kuznetsov, S.O.: Lazy classification with interval pattern structures: application to credit scoring. In: Kuznetsov, S.O., Napoli, A., Rudolph, S. (eds.) Proceedings of the International Workshop "What can FCA do for Artificial Intelligence?", FCA4AI at IJCAI 2015, pp. 43–54. Buenos Aires, Argentina (2015)
16. Kaytoue, M., Kuznetsov, S.O., Napoli, A.: Revisiting numerical pattern mining with formal concept analysis. In: IJCAI 2011, pp. 1342–1347 (2011)

Using Formal Concept Analysis for Checking the Structure of an Ontology in LOD: The Example of DBpedia

Pierre Monnin[1]([⊠]), Mario Lezoche[2], Amedeo Napoli[1], and Adrien Coulet[1]

[1] LORIA (CNRS, Inria NGE, Université de Lorraine), Vandœuvre-lès-Nancy, France
`pierre.monnin@loria.fr`
[2] CRAN (CNRS, Université de Lorraine), Vandœuvre-lès-Nancy, France

Abstract. Linked Open Data (LOD) constitute a large and growing collection of inter-domain data sets. LOD are represented as RDF graphs that allow interlinking with ontologies, facilitating data integration, knowledge engineering and in a certain sense knowledge discovery. However, ontologies associated with LOD are of different quality and not necessarily adapted to all data sets under study. In this paper, we propose an original approach, based on Formal Concept Analysis (FCA), which builds an optimal lattice-based structure for classifying RDF resources w.r.t. their predicates. We introduce the notion of lattice annotation, which enables comparing our classification with an ontology schema, to confirm subsumption axioms or suggest new ones. We conducted experiments on the DBpedia data set and its domain ontologies, DBpedia Ontology and YAGO. Results show that our approach is well-founded and illustrates the ability of FCA to guide a possible structuring of LOD.

Keywords: Linked Open Data · Formal Concept Analysis · Classification · Ontology engineering

1 Introduction

Linked Open Data (LOD) are resulting from a community effort for building a web of data, where data resources may be freely accessed and interpreted by human or software agents for various problem-solving activities. LOD are composed of a large and growing collection of data sets represented within Semantic Web standards that include the use of RDF (Resource Description Framework) and URIs (Uniform Resource Identifiers). In LOD, resources are identified by a URI and can represent entities of the real world (*e.g.*, persons, organizations, places). RDF statements link resources to other resources or to literals (*e.g.* strings, numbers) using properties, also called predicates. Predicates can be used to link resources from the same data set, from different data sets, or from a data set and an ontology [3]. Here, an ontology is a formal representation of a particular domain consisting of classes and relationships between classes.

© Springer International Publishing AG 2017
M. Kryszkiewicz et al. (Eds.): ISMIS 2017, LNAI 10352, pp. 674–683, 2017.
DOI: 10.1007/978-3-319-60438-1_66

Indeed, resources of a data set may be typed by ontology classes, resulting in class instantiation.

In this paper, we consider DBpedia, an RDF data set of the LOD, built from Wikipedia [12]. Wikipedia consists of a very large number of articles, mostly composed of plain text, but that may also include structured data as ones in *infoboxes*. For example, in the Wikipedia article about the French philosopher Voltaire[1], the infobox includes his birth date, his nationality and his occupation. The DBpedia project aims at extracting this structured content and making it available respecting Semantic Web standards. In DBpedia, each Wikipedia article is represented as a *DBpedia page*. Each page can instantiate multiple classes from several ontologies of the DBpedia data set.

Because of their rapid growth and popularity, LOD data sets and their associated ontologies are of various quality and completeness [16]. Consequently, several research papers aim at correcting and completing LOD data sets and ontologies [14]. Some of the proposed approaches extract association rules that express relations between variables, *e.g.*, implications (see [13] for more details on association rule mining with FCA). For example, d'Amato *et al.* mine both data and an associated ontology to learn assertional axioms (*e.g.*, class instantiations) using association rule mining [5]. AMIE and AMIE+ use Inductive Logic Programming to mine association rules over large knowledge bases composing the LOD [6,7]. Alternatively, Alam *et al.* use FCA to mine DBpedia and obtain implications, subsequently transformed into definitions of ontology classes [1].

In this paper, we compare the structure of an ontology with a hierarchical structure built from RDF resources and the predicates they are subject of. Thanks to this comparison, we are able to check if the structure of the ontology is in agreement with regularities in RDF data. For this purpose, we use Formal Concept Analysis (FCA), which is a mathematical framework well adapted to data analysis and knowledge engineering purposes [2,4]. First, we propose applying standard FCA to classify RDF resources within a concept lattice w.r.t. their predicates. Second, we propose to extend FCA with the definition of the *annotation* of a concept, which associates concepts of the lattice with classes of an ontology. Finally, because they have similar properties to intents, annotations yield rules that we use for checking axioms in ontologies. In the following, we assume that the reader is familiar with the basics of FCA that can be found in [9].

The paper is organized as follows. Section 2 introduces LOD, ontologies and DBpedia. Section 3 presents the method we propose for checking the structure of an ontology. Section 4 details an experimentation with DBpedia data. Finally, we conclude in Sects. 5 and 6 with a discussion on how FCA could contribute to the structuring of LOD.

---

[1] Wikipedia article about Voltaire: https://en.wikipedia.org/wiki/Voltaire.

2   Preliminaries

2.1   Linked Open Data (LOD) and Ontologies

Semantic web standards facilitate the publication of LOD online and their connection to existing data sets [3]. Indeed, LOD are represented in the form of graphs encoded using RDF. Atomic elements of an RDF graph are triples denoted by:

$$\langle subject, predicate, object \rangle \in (U \cup B) \times (U \cup B) \times (U \cup B \cup L)$$

where U is the set of URIs, L is the set of literals and B represent blank nodes. We note that B is absent in DBpedia.

Ontologies consist of classes and relationships between these classes [10]. In this paper, we are interested in the subsumption relation. It is a transitive relation denoted by $\sqsubseteq$, where $c \sqsubseteq d$ means that c is a subclass of d. LOD resources can be linked to several ontologies, *i.e.*, a subject can instantiate one or several classes from one or several ontologies. The semantics of the subsumption relation as well as the semantics of the instantiation of a class by a resource depend on the representation formalism of the ontology. Indeed, taxonomies expressed with SKOS (Simple Knowledge Organization System) use the `skos:broader` predicate to express subsumption relations and `dcterms:subject` to express instantiations. Alternatively, ontologies using RDFS (RDF Schema) express subsumption relations using the `rdfs:subClassOf` predicate and instantiations using the `rdf:type` predicate. For example, the DBpedia page `Voltaire` is typed with the class `Writer` from the DBpedia Ontology with the following triple:

$\langle$http://dbpedia.org/page/Voltaire, `rdf:type`,
http://dbpedia.org/ontology/Writer$\rangle$

In this paper, we consider that we have at our disposal an abstract ontology $\mathcal{O}$ using `abstract:subClassOf` to express subsumption relations and `abstract:type` to express instantiations. We denote $\mathcal{C}_{\mathcal{O}}$ the set of classes of $\mathcal{O}$. We define the *type* of a resource e as the set of classes of $\mathcal{O}$ that e instantiates. Formally,

$$\text{type}(e) = \{c \in \mathcal{C}_{\mathcal{O}} \mid \langle e, \texttt{abstract:type}, c \rangle\}$$

Accordingly to the definition of the subsumption, an instance of a class c is also an instance of all the classes subsuming c in $\mathcal{C}_{\mathcal{O}}$. Thus, we define the *extended type* of e as the whole set of superclasses of the classes of type(e). Formally,

$$\text{extdtype}(e) = \text{type}(e) \cup \{d \in \mathcal{C}_{\mathcal{O}} \mid \exists c \in \text{type}(e), \ c \sqsubseteq d\}$$

2.2   DBpedia

Our study focuses on DBpedia as it is a large LOD data set and as it is associated with ontologies. Particularly, we are interested in DBpedia pages and their classifications w.r.t. available ontologies in DBpedia.

Each DBpedia page is a RDF resource representing an article from Wikipedia. DBpedia pages instantiate classes from multiple ontologies encoded with various languages such as SKOS or RDFS. We consider, in our experiments, the DBpedia Ontology [12] and YAGO [15], which are both encoded in RDFS. The DBpedia Ontology has been manually created based on the most common structured contents available in Wikipedia. YAGO facts have been automatically extracted from Wikipedia and WordNet and their accuracy has been manually evaluated.

3   Bridging a Concept Lattice and an Ontology

Although our experiments were performed on DBpedia, we explain our approach considering the example of an abstract RDF data set. It is composed of the RDF triples from Table 1 containing 4 different resources that are subjects of various triples. We also consider an abstract ontology $\mathcal{O}$ (Fig. 2) composed of *(i)* classes that are instantiated by the subjects of the triples and *(ii)* subsumption relationships between these classes.

3.1   Construction of the Concept Lattice

We build a formal context (G, M, I), where formal objects G are subjects of triples in our data set and attributes M are predicates of the same triples. The incidence relation I indicates that there exists at least one triple where a subject and a predicate appear simultaneously. Here, only two elements of the triples are considered: the subject and the predicate, while the object is not taken into account. Then, we apply standard FCA on this context to build a concept lattice. For example, Table 1 shows RDF triples and the associated formal context. The resulting formal lattice is displayed in Fig. 1.

Table 1. Example of RDF triples and the associated formal context. Triples are represented using the Turtle syntax.

e_1	rdf:type	k_1, k_2 .
e_1	$pred_1$	o_1 .
e_1	$pred_2$	o_2 .
e_2	rdf:type	k_1, k_2, k_4, k_5 .
e_2	$pred_1$	o_3 .
e_2	$pred_2$	o_4 .
e_2	$pred_3$	o_5 .
e_3	rdf:type	k_1, k_2 .
e_3	$pred_1$	o_6 .
e_4	rdf:type	k_1, k_2, k_5 .
e_4	$pred_2$	o_7 .
e_4	$pred_3$	o_8 .

	rdf:type	$pred_1$	$pred_2$	$pred_3$
e_1	×	×	×	
e_2	×	×	×	×
e_3	×	×		
e_4	×		×	×

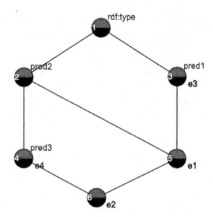

Fig. 1. Line diagram representing the concept lattice built from the formal context in Table 1. The lattice is displayed using the reduced notation. Following this notation, formal objects, depicted in black, are associated with one concept and, implicitly, with all its superconcepts. Attributes, in grey, are associated with one concept and, implicitly, with all its subconcepts. Formal concepts are arbitrarily numbered from 1 to 6.

3.2   Annotation of Formal Concepts

In the lattice in Fig. 1, each extent contains subjects and each intent contains predicates. To compare the lattice with the ontology $\mathcal{O}$, we need to "link" formal concepts with ontology classes. To do so, given $A \subseteq G$ and $K \subseteq \mathcal{C}_\mathcal{O}$, we define two new dual derivation operators denoted by $(\cdot)^\diamond : 2^G \to 2^{\mathcal{C}_\mathcal{O}}$ and $(\cdot)^\diamond : 2^{\mathcal{C}_\mathcal{O}} \to 2^G$ such as:

$$A^\diamond = \bigcap_{e \in A} \text{extdtype}(e) \quad \text{and} \quad K^\diamond = \{e \in G \mid K \subseteq \text{extdtype}(e)\}$$

$A^\diamond$ contains the ontology classes shared by the extended types of all the subjects in A. $K^\diamond$ represents the set of subjects having all the ontology classes of K in their extended type. We prove in Appendix that $(\cdot)^{\diamond\diamond} : 2^{\mathcal{C}_\mathcal{O}} \to 2^{\mathcal{C}_\mathcal{O}}$ and $(\cdot)^{\diamond\diamond} : 2^G \to 2^G$ are *closure operators*.

Given a formal concept (A, B), we call $A^\diamond$ the *annotation* of the concept and we define the triple $(A, B, A^\diamond)$ as an *annotated concept*. For example, consider the concept 4 ($\{e_2, e_4\}, \{\text{rdf:type}, \text{pred}_2, \text{pred}_3\}$) in Fig. 1. Based on the ontology in Fig. 2, we have $\{e_2, e_4\}^\diamond = \{k_1, k_2, k_3, k_5\}$. So, the annotated concept is ($\{e_2, e_4\}, \{\text{rdf:type}, \text{pred}_2, \text{pred}_3\}, \{k_1, k_2, k_3, k_5\}$). Accordingly, we define the *annotated lattice* as the lattice where each concept is replaced by its corresponding annotated concept.

Given two concepts such as $(A_1, B_1) \leqslant (A_2, B_2)$. As $A_1 \subseteq A_2$, we have $A_2^\diamond \subseteq A_1^\diamond$ (see Eq. (1) in Appendix). Therefore, we also extend the reduced notation to represent an annotated lattice: extents and intents are depicted as usual and annotations are depicted showing only the new classes that are not already in the annotations of the superconcepts. For example, on the annotated lattice in Fig. 3, the annotation of the concept 4 shows only k_5 because k_1, k_2 and k_3 are already in the annotation of its superconcepts.

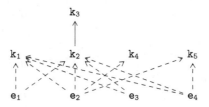

Fig. 2. Example of ontology classes being instantiated by subjects. Dotted arrows represent instantiations whereas solid arrows represent subsumption relations. For example, e_1 instantiates k_1 and k_2, and k_2 is subsumed by k_3.

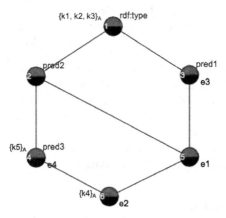

Fig. 3. Line diagram representing the annotated lattice based on the concept lattice in Fig. 1 and the ontology in Fig. 2. It is displayed using our extension of the reduced notation. Formal objects are depicted in black, attributes in grey and annotations are depicted by $\{\cdot\}_A$.

3.3 Comparing Rules Extracted in the Lattice and Axioms in the Ontology

In the annotated lattice in Fig. 3, extents contain subjects, intents contain predicates and annotations contain classes of $\mathcal{O}$. To check the structure of the ontology, in other words, to check the subsumption axioms existing between classes of $\mathcal{O}$, we have to extract rules from the lattice that can be considered similar to subsumption axioms. To do so, we consider each pair composed of an annotated concept $(A, B, A^\diamond)$ and one of its covering concept $(E, F, E^\diamond)$, *i.e.* one of its direct superconcepts, such as the pair composed of concepts 6 and 4 in Fig. 3.

Similarly to Fig. 3, we denote $A_A^\diamond = \{x_1, x_2, \ldots, x_p\}$ and $E_A^\diamond = \{y_1, y_2, \ldots, y_q\}$ the reduced notation of the annotations $A^\diamond$ and $E^\diamond$.

Then, we consider each pair of ontology classes (x_i, y_j) from $A_A^\diamond \times E_A^\diamond$. Considering concepts 6 and 4, the only pair that can be considered is (k_4, k_5). Because $A \subseteq E$ and $E^\diamond \subseteq A^\diamond$, we can say that all subjects typed by x_i are also typed by y_j. Moreover, y_j can type subjects that are not typed by x_i. Hence, x_i can be viewed as "more specialized" than y_j. This can be interpreted as follows: the annotated lattice "suggests" x_i as a subclass of y_j. Because we use the cover relation, we consider this suggestion as a direct subclass suggestion. Here, k_4 is suggested as a direct subclass of k_5.

This rule extracted from the lattice is then compared with the subsumption axioms defined in $\mathcal{O}$:

- If x_i is already explicitly declared as a subclass of y_j, we call this rule a *confirmed axiom*.
- If x_i is an indirect subclass of y_j, *i.e.* x_i is not explicitly stated as a subclass of y_j but it can be inferred from $\mathcal{O}$ thanks to the transitivity of the subsumption relation, we call the rule an *inferable axiom*.
- Otherwise, the rule does not exist in the ontology and we consider it as a potential *new axiom*, proposed to improve the ontology.

In our example, according to the ontology in Fig. 2, there is no link existing between k_4 and k_5. Thus, $k_4 \sqsubseteq k_5$ is a new axiom proposal.

If we consider the pair composed of concepts 6 and 5 in Fig. 3, we notice that the reduced notation of the annotation of concept 5 is empty. Thus, no rule can be extracted from this pair of concepts. Similarly, no rule involving k_1, k_2 and k_3 can be extracted with the current approach because the reduced notation of the annotations of concepts 2 and 3 are empty.

4   Experimentation

We experimented this approach, classifying 904 pages selected from the DBpedia 2014 data set that describe persons who died between 01/01/2000 and 07/01/2000 (included). The resulting lattice consisted of 15,234 concepts. Axioms from the DBpedia Ontology (11) and YAGO (199) were retrieved among the rules and, therefore, were considered as confirmed axioms. For example, the DBpedia Ontology class Boxer was found as a subclass of Athlete. Inferable axioms (2,250) as well as new axioms (1,372) were only found for YAGO. This may be due to the small size of the DBpedia Ontology which only contains 683 classes. Therefore, pages have more similar extended types than with a more fine-grained ontology such as YAGO. Consequently, annotations of formal concepts are also similar, leading to empty reduced notation of annotations from which no rule is extracted (see the example with concepts 5 and 6 in Fig. 3). For example, the lattice we obtained from our subset of DBpedia pages suggested to add BostonUniversityAlumni as a subclass of Scholar110557854. Nevertheless, this subsumption could be inferred from the ontology. Another example was the suggestion of `FilipinoChildActors` as a subclass of `FilipinoActors`. It was neither explicitly stated in the ontology nor inferable and seemed to make sense in a general case.

5   Discussion

Regarding the results of the experiments, when classifying RDF subjects w.r.t. their related predicates, we are able to suggest direct subclass relations between ontology classes typing these subjects. Some of the suggestions already exist as direct subclass relations in the considered ontology. Thus, predicates allow to classify subjects in a lattice whose structure is similar to some parts of the existing ontology. Consequently, it seems that some predicates applied to subjects are specific of the classes typing these subjects. Therefore, predicates could be considered as "indicators" of the classification of subjects by ontology classes. Some of the suggested axioms indirectly exist in the ontology, meaning that the structure of the lattice is not as fine-grained as the ontology. This may be due to the fact that predicates applied to subjects are not the only indicators of their classification by classes of an ontology. In the annotated lattice in Fig. 3, we also notice that the reduced notation of the annotation is empty for concepts 2, 3 and 5 because these concepts are not closed w.r.t. the $(\cdot)^{\diamond\diamond}$ operator. Therefore, we do not extract rules from them, in addition the classes k_1, k_2 and k_3 never appear in rules extracted from the lattice.

As future work, we will consider additional features in classification to provide more granularity. For example, the formal context could be built using predicates, $(predicate, \text{extdtype}(object))$ and $(predicate, object)$ pairs as attributes. Pattern Structures [8] could be used on extended types as there is a partial order on classes of an ontology. Because $(\cdot)^{\diamond\diamond}$ is a closure operator, we could also use the triadic approach to FCA [11] to model the three sets that describe each of our concepts (extents, intents and annotations). Another perspective of the current work is to reduce the lattice by keeping only the closed concepts relatively to the $(\cdot)^{\diamond\diamond}$ operator. This will increase the number of suggestions. For example, in Fig. 3, removing concepts 2, 3 and 5 leads to suggest k_5 and k_4 as subclasses of k_1, k_2 and k_3. Finally, in the case of subjects typed by two (or more) ontologies, two annotations can be computed for each concept of the lattice, providing potential relationships between independent ontologies. In this case, the lattice can be seen as a pivot structure for ontology matching experiments.

6   Conclusion

In this paper, we apply FCA to classify RDF subjects w.r.t. their related predicates in a concept lattice. A method is proposed to annotate this lattice with ontology classes used to type these subjects. Based on the structure of the resulting lattice, subsumption axioms between the ontology classes are suggested and then compared with existing axioms in the ontology. Running this method on a subset of pages from DBpedia and considering two associated reference ontologies (DBpedia Ontology and YAGO), we are able to suggest axioms that already exist. This means that the structure of the annotated lattice is similar to some parts of the existing ontologies. Consequently, predicates applied to pages can be considered as indicators of classes instantiated by these pages. New axioms are also proposed. One of the next challenges resides in performing a qualitative study of these new axioms.

Appendix: Proving that $(\cdot)^{\diamond\diamond}$ is a Closure Operator

An operator on a partial ordered set is a closure operator if it is monotonous, extensive and idempotent. Firstly, let us prove that:

$$X_1 \subseteq X_2 \Rightarrow X_2^{\diamond} \subseteq X_1^{\diamond} \tag{1}$$

To do so, let's consider $E_1 \subseteq G$ and $E_2 \subseteq G$ such as $E_1 \subseteq E_2$ and, dually, $K_1 \subseteq \mathcal{C}_{\mathcal{O}}$ and $K_2 \subseteq \mathcal{C}_{\mathcal{O}}$ such as $K_1 \subseteq K_2$. We have:

$$E_2^{\diamond} = \bigcap_{e \in E_2} \text{extdtype}(e) = (\cap_{e \in E_1} \text{extdtype}(e)) \bigcap (\cap_{e \in E_2 \setminus E_1} \text{extdtype}(e))$$

$$= E_1^{\diamond} \bigcap (\cap_{e \in E_2 \setminus E_1} \text{extdtype}(e)) \subseteq E_1^{\diamond}$$

$$K_2^{\diamond} = \{e \in G \mid K_2 \subseteq \text{extdtype}(e)\}$$

$$= \{e \in G \mid K_1 \subseteq \text{extdtype}(e) \wedge K_2 \setminus K_1 \subseteq \text{extdtype}(e)\}$$

$$= \{e \in G \mid K_1 \subseteq \text{extdtype}(e)\} \bigcap \{e \in G \mid K_2 \setminus K_1 \subseteq \text{extdtype}(e)\}$$

$$= K_1^{\diamond} \bigcap \{e \in G \mid K_2 \setminus K_1 \subseteq \text{extdtype}(e)\} \subseteq K_1^{\diamond}$$

Monotonicity. Considering $X_1 \subseteq X_2$. Because of (1), we have $X_2^{\diamond} \subseteq X_1^{\diamond}$ and then $X_1^{\diamond\diamond} \subseteq X_2^{\diamond\diamond}$.

Extensivity. Considering $E \subseteq G$, we have $E^{\diamond} = \bigcap_{e \in E} \text{extdtype}(e)$. So, $E^{\diamond\diamond} = \{f \in G \mid \bigcap_{e \in E} \text{extdtype}(e) \subseteq \text{extdtype}(f)\}$. Therefore, $E \subseteq E^{\diamond\diamond}$. Dually, considering $K \subseteq \mathcal{C}_{\mathcal{O}}$, we have $K^{\diamond} = \{e \in G \mid K \subseteq \text{extdtype}(e)\}$. So, $K^{\diamond\diamond} = \bigcap_{f \in \{e \in G \mid K \subseteq \text{extdtype}(e)\}} \text{extdtype}(f)$. Therefore , $K \subseteq K^{\diamond\diamond}$.

Idempotence. Because of the extensivity, we know that $X \subseteq X^{\diamond\diamond}$. Therefore, because of (1), $X^{\diamond\diamond\diamond} \subseteq X^{\diamond}$ and $X^{\diamond\diamond} \subseteq X^{\diamond\diamond\diamond\diamond}$. Because of the extensivity, $X^{\diamond} \subseteq X^{\diamond\diamond\diamond}$. So because of (1), $X^{\diamond\diamond\diamond\diamond} \subseteq X^{\diamond\diamond}$. Consequently, $X^{\diamond\diamond} = X^{\diamond\diamond\diamond\diamond}$.

References

1. Alam, M., Buzmakov, A., Codocedo, V., Napoli, A.: Mining definitions from RDF annotations using formal concept analysis. In: Proceedings of IJCAI 2015, pp. 823–829 (2015)
2. Bendaoud, R., Napoli, A., Toussaint, Y.: Formal concept analysis: a unified framework for building and refining ontologies. In: Gangemi, A., Euzenat, J. (eds.) EKAW 2008. LNCS (LNAI), vol. 5268, pp. 156–171. Springer, Heidelberg (2008). doi:10.1007/978-3-540-87696-0_16
3. Bizer, C., Heath, T., Berners-Lee, T.: Linked data - the story so far. Int. J. Semant. Web Inf. Syst. **5**(3), 1–22 (2009)
4. Cimiano, P., Hotho, A., Staab, S.: Learning concept hierarchies from text corpora using formal concept analysis. J. Artif. Intell. Res. (JAIR) **24**, 305–339 (2005)
5. d'Amato, C., Staab, S., Tettamanzi, A.G., Minh, T.D., Gandon, F.: Ontology enrichment by discovering multi-relational association rules from ontological knowledge bases. In: Proceedings of ACM SAC 2016, pp. 333–338 (2016)
6. Galárraga, L., Teflioudi, C., Hose, K., Suchanek, F.M.: Fast rule mining in ontological knowledge bases with AMIE+. VLDB J. **24**(6), 707–730 (2015)
7. Galárraga, L.A., Teflioudi, C., Hose, K., Suchanek, F.: AMIE: association rule mining under incomplete evidence in ontological knowledge bases. In: WWW 2013, pp. 413–422 (2013)
8. Ganter, B., Kuznetsov, S.O.: Pattern structures and their projections. In: Delugach, H.S., Stumme, G. (eds.) ICCS-ConceptStruct 2001. LNCS, vol. 2120, pp. 129–142. Springer, Heidelberg (2001). doi:10.1007/3-540-44583-8_10
9. Ganter, B., Wille, R.: Formal Concept Analysis: Mathematical Foundations. Springer, Berlin (1999)
10. Gruber, T.R., et al.: A translation approach to portable ontology specifications. Knowl. Acquis. **5**(2), 199–220 (1993)
11. Lehmann, F., Wille, R.: A triadic approach to formal concept analysis. In: Ellis, G., Levinson, R., Rich, W., Sowa, J.F. (eds.) ICCS-ConceptStruct 1995. LNCS, vol. 954, pp. 32–43. Springer, Heidelberg (1995). doi:10.1007/3-540-60161-9_27
12. Lehmann, J., et al.: DBpedia - a large-scale, multilingual knowledge base extracted from Wikipedia. Semant. Web **6**(2), 167–195 (2015)
13. Pasquier, N., Bastide, Y., Taouil, R., Lakhal, L.: Efficient mining of association rules using closed itemset lattices. Inf. Syst. **24**(1), 25–46 (1999)
14. Paulheim, H.: Automatic knowledge graph refinement: a survey of approaches and evaluation methods. Semant. Web J. (to appear)
15. Suchanek, F.M., Kasneci, G., Weikum, G.: YAGO: a core of semantic knowledge unifying WordNet and Wikipedia. In: Proceedings of WWW 2007, pp. 697–706. ACM (2007)
16. Zaveri, A., Rula, A., Maurino, A., Pietrobon, R., Lehmann, J., Auer, S.: Quality assessment for linked data: a survey. Semant. Web **7**(1), 63–93 (2016)

A Proposal for Classifying the Content of the Web of Data Based on FCA and Pattern Structures

Justine Reynaud[1]([✉]), Mehwish Alam[2], Yannick Toussaint[1], and Amedeo Napoli[1]

[1] LORIA (CNRS – Inria Nancy-Grand Est – Université de Lorraine), 239, 54506 Vandoeuvre les Nancy, France
justine.reynaud@loria.fr
[2] Labortoire d'Informatique Paris Nord, Université Paris 13, Paris, France

Abstract. This paper focuses on a framework based on Formal Concept Analysis and the Pattern Structures for classifying sets of RDF triples. Firstly, this paper proposes a method to construct a pattern structure for the classification of RDF triples w.r.t. domain knowledge. More precisely, the poset of classes representing subjects and objects and the poset of predicates in RDF triples are taken into account. A similarity measure is also proposed based on these posets. Then, the paper discusses experimental details using a subset of DBpedia. It shows how the resulting pattern concept lattice is built and how it can be interpreted for discovering significant knowledge units from the obtained classes of RDF triples.

1 Introduction

The Web of Data (WOD) has become a very huge space of experimentation especially regarding knowledge discovery and knowledge engineering due to its rich and diverse nature. WOD is a database as it includes different kinds of data e.g. documents, images, videos etc. It can also be considered as a knowledge base because a major part of it relies on the Linked Data (LD) cloud. LD are further based on RDF triples of the form <subject, predicate, object> where each element in the triple denotes a resource (accessible through a URI). Moreover, the elements in a triple can be organized within partial orderings using the predefined vocabularies such as RDF Schema (RDFS), i.e. a subclass relation (`rdfs:subClassOf`) and a subproperty relation (`rdfs:subPropertyOf`, where a predicate in an RDF triple is also called a property). We rely on this double vision of WOD, as a database and as a knowledge base, for proposing a way of classifying the content of WOD thanks to Formal Concept Analysis (FCA) and its extension called as Pattern Structures. As a database, WOD can be navigated, searched, queried through SPARQL queries, and mined. As a knowledge base, WOD provides domain knowledge that can be used for guiding information retrieval, knowledge discovery and knowledge engineering. Regarding these

© Springer International Publishing AG 2017
M. Kryszkiewicz et al. (Eds.): ISMIS 2017, LNAI 10352, pp. 684–694, 2017.
DOI: 10.1007/978-3-319-60438-1_67

tasks, questions are arising, e.g. "how to organize set of RDF triples such as triples returned as answers to a SPARQL query", "how to carry on a knowledge discovery process on WOD as a database and as a knowledge base at the same time". The first question has already been investigated by some authors of the present paper (see [3]) but improvements are still needed. The second question remains a challenge since knowledge discovery does not just amount to query processing, but can take advantage of partial orderings and of knowledge repositories lying in WOD (i.e. ontologies). Databases already define a certain schema but it is usually not as elaborate as a knowledge base, mainly due to the fact that a knowledge base shapes the human perception of the world in the form that a machine can understand thanks to an expressive knowledge representation language (e.g. OWL). Moreover, a knowledge repository is based on specific resources, e.g. ontologies, and can be seen as a set of facts and partial orderings (posets) organizing concepts and properties. Then, the posets supporting knowledge repository are of first importance for knowledge discovery within WOD.

Accordingly, we present in the following a knowledge discovery process based on Formal Concept Analysis and Pattern Structures that is applied to sets of RDF triples, taking into account the context, i.e. the knowledge resources related to the components of the RDF triples. Then, one main objective is to propose an operational mining process working on RDF triples w.r.t. domain knowledge. We extend preceding approaches by defining an order product able to organize pairs of properties and objects in the triples w.r.t. related posets of properties and objects. FCA and Pattern Structures are good candidates for mining the web of data and output concept lattices that can be explored, including concepts, implications and association rules. A concept lattice resulting from the discovery process can be considered as a new knowledge repository providing a well-founded organization to the original set of triples. Finally, the concept lattice offers a point of view on data from which an analyst can discover useful and significant knowledge units otherwise hidden in the data.

The paper is organized as follows. Section 2 motivates the proposed approach and presents WOD and Pattern Structures. Section 3 presents the existing work and the extension proposed in the current study. Section 4 discusses the experimental setup and finally Sect. 5 concludes the paper.

2   Preliminaries

2.1   The Web of Data

The Content of the Web of Data. The amount of data in the WOD has increased drastically over the past 10 years. Many important on-line resources are now represented as a part of Linked Data such as DBpedia, which represents Wikipedia Infoboxes in the form of RDF. All these data sources are represented in the form node-arc-labeled graph where each resource is connected to another through internal links and each data set is connected to another resource through external links forming Linked Open Data (LOD) cloud.

More formally, WOD can be seen as an oriented multigraph[1] $G = (V, E)$, where nodes correspond to resources and edges correspond to labeled links between those resources. Each resource can be represented as a URI, Blank Node or a literal. A literal represents a value (string, integer, date, ...) whereas a blank node designates an unidentified resource[2]. As a graph, WOD can also be considered as a set of triples (s, p, o), where s and o denote vertices, and p denotes an edge between them. Multiple RDF triples connect together to form a graph.

RDF and RDFS. Resource Description Framework[3] (RDF) allows the user to represent facts as statements, where each statement is a triple. This set of facts corresponds to an ABox in description logics. For example, (ÉVARISTE_GALOIS, hasDeathPlace, PARIS) is an RDF triple which expresses a relation hasDeath- Place between the resources ÉVARISTE_GALOIS and PARIS, meaning that Galois died in Paris. RDF also proposes special predicates such as rdf:type, which links an instance to its class, e.g. (ÉVARISTE_GALOIS, rdf:type, Mathematician).

RDF Schema[4] (RDFS) is the language including constructions for ordering RDF triples into a structure that corresponds to a TBox in description logics. The relation C_1 rdfs:subClassOf C_2 corresponds to the subsumption relation in description logics. A class C_1 is said to be *more specific than* a class C_2, declared as C_1 rdfs:subClassOf C_2, if the interpretation of C_1, i.e. the set of instances of C_1, is included in the interpretation of C_2. Similarly, the relation p_1 rdfs:subPropertyOf p_2 means that if there is a relation p_1 between x and y, then there is a relation p_2 between x and y. Both relations rdfs:subClassOf and rdfs:subPropertyOf are transitive and have a logical semantics that can be operationalized as a set of inference rules [1].

The Posets of Classes and Predicates. The relations rdfs:subClassOf and rdfs:subPropertyOf are particularly interesting in the current scenario. The rdfs:subClassOf relation defines a partial order over classes, whereas the rdfs:subPropertyOf relation defines a partial order over properties.

Viewing WOD as a graph $G = (V, E)$, these two relations define two partial orders, the first over the set of vertices $(V, \leqslant_V)$ and the second over the set of edges $(E, \leqslant_E)$. In the following, we assume that both posets $(V, \leqslant_V)$ and $(E, \leqslant_E)$ are trees, an multiple inheritance will not be discussed here. We work with DBpedia, which does not make use of multiple inheritance, hence this approach is still relevant.

---

[1] Having more than one edges between two nodes.
[2] In this work we do not consider blank nodes.
[3] https://www.w3.org/TR/2014/REC-rdf11-mt-20140225/.
[4] https://www.w3.org/TR/2014/REC-rdf-schema-20140225/.

Querying Web of Data. SPARQL[5] is the standard language for querying WOD. The queries can be constructed with the help of graph patterns represented as a set of triples, formally termed as Basic Graph Patterns (BGP). The answer of such a query is the set of all subgraphs matching the BGP. Then, the variables are replaced by the resources of the graph. An example is presented in Fig. 1.

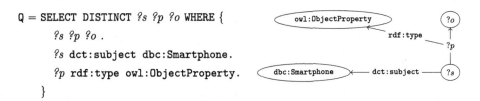

```
Q = SELECT DISTINCT ?s ?p ?o WHERE {
        ?s ?p ?o .
        ?s dct:subject dbc:Smartphone.
        ?p rdf:type owl:ObjectProperty.
    }
```

Fig. 1. Query for extracting the data and the associated basic graph pattern. Every triple extracted is connected to some subject *?s* which is an element (dc:subject) of the category which deals with smartphones (dbc:Smartphone). The prefix dbc is for all the DBpedia categories, whereas the prefix dc represents Dublin Core, a generic vocabulary for describing resources.

2.2 Pattern Structures

Pattern structures (PS) [8] are a generalization of Formal Concept Analysis[6] (FCA) [9] for dealing with complex data. While FCA is based on a binary relation between objects (G) and attributes (M), PS consider that objects in G have a description. Descriptions are partially ordered in a meet-semilattice, thanks to a subsumption relation $\sqsubseteq$ which is associated to a similarity relation denoted as $\sqcap$. More precisely, if c and d are two descriptions, then $c \sqcap d = c \Leftrightarrow c \sqsubseteq d$. Formally, a pattern structure is defined as follows:

Definition 1 (Pattern structure). *Let G be a set of objects, $(D, \sqcap)$ a semi-lattice of descriptions and $\delta : G \to D$ a mapping associating a description to an object. Then $(G, (D, \sqcap), \delta)$ is called a pattern structure. The Galois connections are the following:*

$$A^{\square} = \bigsqcap_{g \in A} \delta(g) \qquad\qquad for\ A \subseteq G$$

$$d^{\square} = \{g \in G \mid d \sqsubseteq \delta(g)\} \qquad\qquad for\ d \in D$$

As in FCA, the composition of these mappings are closure operators: given a set of objects $A \subseteq G$ we have that $A \subseteq A^{\square\square}$ and A is closed when $A = A^{\square\square}$ (the same for $d \subseteq (D, \sqcap)$, $d \subseteq d^{\square\square}$ and d is closed when $d = d^{\square\square}$).

---

[5] https://www.w3.org/TR/rdf-sparql-query/.
[6] We assume that the reader is familiar with the basics of FCA thus we directly detail the basics of pattern structures.

A pattern concept (A, d) verifies that $A^\square = d$ and $d^\square = A$ where A and d are closed. Given a set of objects $A \in G, (A^{\square\square}, A^\square)$ is a pattern concept. Similarly, if $d \in (D, \sqcap)$ is a description, $(d^\square, d^{\square\square})$ is a pattern concept. A partial order on pattern concepts is defined in a way similar to FCA: $(A_1, d_1) \leqslant (A_2, d_2) \Leftrightarrow A_1 \subseteq A_2 \Leftrightarrow d_2 \sqsubseteq d_1$. This partial order gives rise to a pattern concept lattice.

Example 1. Given the objects and their descriptions in Fig. 2, we have $\delta(g_2) = d_4$ and $\delta(g_3) = d_6$. We have $\delta(g_2) \sqcap \delta(g_3) = d_1$. Thus, $(\{g_2, g_3, g_4\}, d_1)$ is a pattern concept.

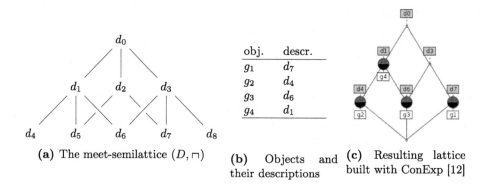

obj.	descr.
g_1	d_7
g_2	d_4
g_3	d_6
g_4	d_1

(a) The meet-semilattice $(D, \sqcap)$ **(b)** Objects and their descriptions   **(c)** Resulting lattice built with ConExp [12]

Fig. 2. Example of formal context and the resulting lattice for pattern structures.

3   Building a Pattern Structure for RDF Data

3.1   Preliminaries

FCA and patterns structures have already been used for classifying RDF data using graph structure [7,10,11] and using RDF triples [2,3]. In [2], the authors aim to provide a navigation space over RDF resources. The extent of a concept is a set of resources, and the intent is a set of pairs (predicates, objects). The similarity between two descriptions is computed pairwise. The relation rdfs:subClassOf is taken into account as domain knowledge.

The work in [2] is the starting point of the present work. We present hereafter an example to give the intuition on how RDF triples are taken into account and how we generalize the work in [2].

Example 2. The first part of this example gives an intuition of the pattern structure used in [2]. Given the example Fig. 3, we have:

$$\delta(Paris) = \{\underbrace{(cityOf, \{Europe\})}_{P1}, \underbrace{(capitalOf, \{France\})}_{P2}\}$$

$$\delta(Nancy) = \{\underbrace{(hasLocation, \{Europe\})}_{N1}, \underbrace{(cityOf, \{France\})}_{N2}\}$$

$$\delta(Paris) \sqcap \delta(Nancy) = \{P1 \sqcap N1, P1 \sqcap N2, P2 \sqcap N1, P2 \sqcap N2\}$$

According to [2], the similarity is the following:

$$\delta(Paris) \sqcap \delta(Nancy) = \{(cityOf, \{Place\})\} \qquad \text{from } P1 \sqcap N2$$

The comparison between two pairs (predicate, object) is possible only if the predicates are the same. In the following, we extend this pattern structure to take into account the `rdfs:subPropertyOf` relation, leading to this similarity:

$$\delta(Paris) \sqcap \delta(Nancy) = \{(hasLocation, \{Europe\}), (cityOf, \{France\})\}$$
$$\text{from } P1 \sqcap N1 \text{ and } P2 \sqcap N2, \text{ which are the most specific}$$

This leads to a more accurate similarity between the two descriptions. In the next section, we show how to take into account both `rdfs:subClassOf` and `rdfs:subPropertyOf`.

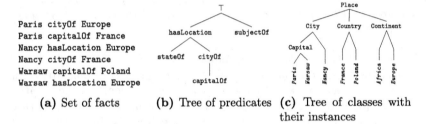

(a) Set of facts   **(b)** Tree of predicates   **(c)** Tree of classes with their instances

Fig. 3. Toy knowledge base. Subfigure (a) illustrates a set of facts. Subfigure (b) illustrates a poset of properties w.r.t. `rdfs:subPropertyOf` relation. Subfigure (c) shows a poset of classes with their instances.

3.2 A Pattern Structure for RDF Triples

In this section, we present a pattern structure to classify RDF triples, considering the posets of classes and of predicates as domain knowledge. The data set $\mathcal{B}$ is extracted from DBpedia with a SPARQL query Q: all the triples satisfying the constraints expressed in the query Q are kept.

$$\mathcal{B} = \{(s, p, o) \mid Q \vDash (s, p, o)\}$$

In order to avoid confusion between the objects in FCA and the objects in RDF, objects in FCA are called entities. Then, the set G of entities corresponds to the set of subjects in the RDF triples:

$$G = \{s \mid (s, p, o) \in \mathcal{B}\}$$

For descriptions, we have a set M of pairs (p, o) corresponding to the pairs in data set $\mathcal{B}$.

$$M = \{(p, o) \mid (s, p, o) \in \mathcal{B}\}$$

This set is structured w.r.t two partial orders, contrasting with [2] where only one order is considered. Indeed, the order on predicates and the order on classes are taken into account. The resulting poset $(V \times E, \leqslant_\pi)$ is the Cartesian product of the posets $(V, \leqslant_V)$ and $(E, \leqslant_E)$:

$$(p_i, o_i) \leqslant_\pi (p_j, o_j) \Leftrightarrow p_i \leqslant_E p_j \text{ and } o_i \leqslant_V o_j$$

We define the extended set M^* as the set of all pairs (p, o) which are in M together with all pairs (p_i, o_j) such as $(p, o) \leqslant_\pi (p_i, o_j)$:

$$M^* = M \cup \bigcup_{(p,o) \in M} \{(p_i, o_j) \mid (p, o) \leqslant_\pi (p_i, o_j)\}$$

The set M^* plays the same role as the extended set of attributes introduced in [5,6] including all attributes and their subsumers.

Example 3. Considering Fig. 3, if $(capitalOf, Country)$ is contained in M, then $(capitalOf, Place)$, $(hasLocation, Country)$ and $(hasLocation, Place)$ are included in M^*.

The descriptions of entities are mappings from G to M^*, such that if a pair (p, o) is in the description of a subject s, then (s, p, o) belongs to the data set $\mathcal{B}$. From this set, we keep only the most specific elements, i.e. if $\delta(s) = \{(p, C_1), (p, C_0)\}$ and C_1 rdfs:subClassOf C_0, then (p, C_0) follows from (p, C_1) and $\delta(s) = \{(p, C_1)\}$. Thus, the description of a subject is the antichain of the minimal pairs in its description:

$$\delta(s) = \min\{(p, o) \mid (s, p, o) \in \mathcal{B}\}$$

where min selects the pairs which are minimal w.r.t the order defined on pairs (p, o). The intuition is the following. A description in M^* is a filter, i.e. a pair (p,o) and all subsumers of (p, o) in M^*. The filter then can be "represented" by its minimal elements. The order on descriptions is written as $\delta(s_1) \sqsubseteq \delta(s_2)$ and is interpreted as "$\delta(s_1)$ is more specific than $\delta(s_2)$":

$$\delta(s_1) \sqsubseteq \delta(s_2) \Leftrightarrow \forall (p_1, o_1) \in \delta(s_1), \exists (p_2, o_2) \in \delta(s_2) \text{ s.t. } (p_1, o_1) \leqslant_\pi (p_2, o_2)$$

Since a description such as $\delta(s_1)$ or $\delta(s_2)$ is an antichain of (p, o) pairs, when a pair $(p_1, o_1) \in \delta(s_1)$ is lower than a pair $(p_2, o_2) \in \delta(s_2)$, there does not exist any pair $(p, o) \in \delta(s_2)$ which is lower than (p_1, o_1). We can now define the similarity operator as:

$$\delta(s_1) \sqcap \delta(s_2) = \min_{\substack{(p_{1_i}, o_{1_j}) \in \delta(s_1) \\ (p_{2_i}, o_{2_j}) \in \delta(s_2)}} \{(lcs_E(p_{i_1}, p_{i_2}), lcs_V(o_{j_1}, o_{j_2}))\}$$

where lcs is the *least common subsumer* of two elements in the tree of classes or of properties.

Definition 2 (Least common subsumer). *Given a tree* $(H, \leqslant)$, *the least common subsumer of two nodes* x *and* y *of that tree is the node* z *s.t.* $x \leqslant z, y \leqslant z, \not\exists z_1 \leqslant z$ *s.t.* $x \leqslant z_1$ *and* $y \leqslant z_1$.

The pair $(lcs_E(p_{i_1}, p_{i_2}), lcs_V(o_{j_1}, o_{j_2}))$ belongs to M^* since lcs_E relies on the `rdfs:subPropertyOf` relation and lcs_V relies on the `rdfs:subClassOf` relation. Finally, we have that:

Proposition 1. $\delta(s_1) \sqsubseteq \delta(s_2) \Leftrightarrow \delta(s_1) \sqcap \delta(s_2) = \delta(s_2)$.

This equation ensures that the resulting construction has all the good properties of a lattice, which is mandatory in the knowledge discovery process. It is reversed from the usual equation – $\delta(s_1) \sqsubseteq \delta(s_2) \Leftrightarrow \delta(s_1) \sqcap \delta(s_2) = delta(s_1)$, since the two connections are order-isomorphisms.

4 Experiments

This section illustrates our approach with the help of an experiment. Pattern concept lattice was built on a set of RDF triples extracted from DBpedia. The main points discussed are the construction of the data set and the construction of the pattern concept lattice.

Table 1. Statistics on DBpedia (April, 2016) and the smartphones corpus (January, 2017). The number of predicates and classes correspond to the number of nodes in each of the domain knowledge trees.

	Triples	Entities	Predicates	Classes	Concepts
DBpedia	9.5 billion	5.2 million	1103	754	–
Smartphones	566	3423	25	17	775
Toy ex.	5	54	4	13	14

Building the Data Set. DBpedia contains more than 9 billion triples. In the current work we focus on extracting domain specific subset of RDF triples about smartphones. The data set was extracted using the query given in Fig. 1, i.e. triples (s, p, o) such that s represents smartphones and p is an `objectProperty`, i.e. a property whose range is a resource (and not a literal). The extracted data set contains 3423 triples and is detailed Table 1. In order to present the resulting pattern concept lattice, we use a toy example with only 5 entities, described Table 1. The resulting pattern concept lattice is presented in Fig. 4. The experiment has also been run on the full corpus of Smartphones.

Resulting Lattice. Using the previous toy example, the pattern structure built a pattern concept lattice with 14 formal concepts. The extent of a formal concept is a set of instances occurring in a subject position of a triple. The intent of a pattern concept is a set of couples (predicate, object). From this resulting lattice, we can make several observations. First, object concepts were located at a very low level in the lattice, just above the bottom concept. Theoretically, object concepts could end up higher in the lattice, depending of course on the data set, but it is a rare occurrence. This means that the lattice cannot be used to rank answers to a SPARQL query as suggested in [5], as there is no way to decide if one instance fits the query better than another. Instead, the lattice provides context to the answers, highlighting similarities and differences between the entities.

Second, descriptions of instances through triples vary a lot from one to another and does not follow a regular schema. Thus, the *Blackberry* is located to the side of the lattice, sharing very few similarities with other smartphones. It is described as being a *multitouch screen* phone but not a *touch screen* phone while the *IPhone* is both a *multitouch* and a *touch screen* phone. This is probably due to some missing information in the data set. Moreover, some pairs (predicate,object) should be assigned to more instances. For example, (type,Merchandise), should be shared by all instances.

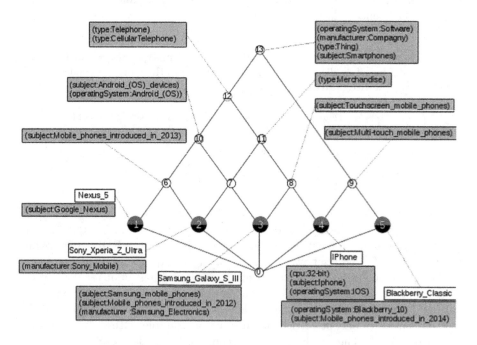

Fig. 4. Lattice built from the toy example.

We noticed that, the date of introduction is encoded in a string in such a way that we cannot formally reason about dates. However, following the informal meaning of the strings, there is a disjunction between phones introduced in 2013 and those introduced in 2014.

Another interesting observation is that, the operating systems leads to a partition of the instances and all the concepts subsumed by (operatingSystem,x) form disjoint sub-lattices with different x. This is more obvious when looking at the lattice built with the complete data set as there is currently no phone with more than one operating system.

Finally, pattern structures, like FCA, define implication rules. Thus, we found that phones introduced in 2013 are all under Android system in our toy example. We may also learn some equivalence inside a description. For example, (subject:Android_(OS)_devices) ≡ (operatingSystem:Android_(OS)). As a matter of fact, the two properties correspond to two formulations of the same property, one coming from a Wikipedia encoding, the other one from DBpedia.

From these observations, we can conclude that the lattice is of great help to add context to the data extracted from WOD using some external knowledge, giving a synthetic and structured view of the data extracted by a SPARQL query. The approach highlights some descriptions that play a major role in structuring the lattice and, conversely, highlights descriptions that are meaningless, or those that are not associated with instances where they should be. To avoid the above problems that weaken the interpretation of the lattice, two questions arise: *how can we improve data set collection from WOD to identify non-relevant or noisy properties?* and *how can we identify within the lattice, missing associations between properties and instances and then improve the data set?* One possible answer is to rely on association rules with high confidence for finding possible missing definitions. This works has been discussed in [4] and should be extended.

5   Conclusion and Future Work

In this work, we defined a pattern structure in the continuity of [2,3]. This approach is relevant for WOD, especially in the case of DBpedia, since entities (i.e. subjects of the triples) correspond to Wikipedia pages. We showed that pattern structures are relevant for taking into account domain knowledge, even with more than one partial order. Finally, we presented the resulting pattern concept lattice and discussed the observed results. An interesting extension to our work would be to consider the triples with literals having numeric values (for ages and dates). Ongoing work is using association rule mining for generating pseudo definitions using the formalism of description logics.

Acknowledgments. This work has been conducted with the support of "Région Lorraine" and "Délégation Générale de l'Armement".

694 J. Reynaud et al.

References

1. Abiteboul, S., Manolescu, I., Rigaux, P., Rousset, M.C., Senellart, P.: Web Data Management. Cambridge University Press, Cambridge (2011)
2. Alam, M., Buzmakov, A., Napoli, A., Sailanbayev, A.: Revisiting pattern structures for structured attribute sets. In: CLA Proceedings, pp. 241–252 (2015)
3. Alam, M., Napoli, A.: Interactive exploration over RDF data using formal concept analysis. In: DSAA Proceedings (2015)
4. Alam, M., Buzmakov, A., Codocedo, V., Napoli, A.: Mining definitions from RDF annotations using formal concept analysis. In: IJCAI Proceedings, pp. 823–829 (2015)
5. Carpineto, C., Romano, G.: A lattice conceptual clustering system and its application to browsing retrieval. Mach. Learn. **24**(2), 95–122 (1996)
6. Carpineto, C., Romano, G.: Concept Data Analysis: Theory and Applications. Wiley, Chichester (2004)
7. Ferré, S.: A proposal for extending formal concept analysis to knowledge graphs. In: Baixeries, J., Sacarea, C., Ojeda-Aciego, M. (eds.) ICFCA 2015. LNCS (LNAI), vol. 9113, pp. 271–286. Springer, Cham (2015). doi:10.1007/978-3-319-19545-2_17
8. Ganter, B., Kuznetsov, S.O.: Pattern structures and their projections. In: Delugach, H.S., Stumme, G. (eds.) ICCS-ConceptStruct 2001. LNCS (LNAI), vol. 2120, pp. 129–142. Springer, Heidelberg (2001). doi:10.1007/3-540-44583-8_10
9. Ganter, B., Wille, R.: Formal Concept Analysis - Mathematical Foundations. Springer, Heidelberg (1999)
10. Kötters, J.: Concept lattices of RDF graphs. In: Proceedings of FCA&A@ICFCA, pp. 81–91 (2015)
11. Kuznetsov, S.O.: Computing graph-based lattices from smallest projections. In: ICCS Proceedings, pp. 35–47 (2007)
12. Yevtushenko, S.A.: System of data analysis "concept explorer". In: Proceedings of 7th national Conference on Artificial Intelligence KII, pp. 127–134 (2000)

ISMIS 2017 Data Mining Competition on Trading Based on Recommendations

ISMIS 2017 Data Mining Competition: Trading Based on Recommendations

Mathurin Aché[1], Andrzej Janusz[2], Kamil Żbikowski[3,4], Dominik Ślęzak[2(✉)],
Marzena Kryszkiewicz[3], Henryk Rybinski[3], and Piotr Gawrysiak[3,4]

[1] Montrouge, France
mathurin.ache@outlook.fr
[2] Institute of Informatics, University of Warsaw,
ul. Banacha 2, 02-097 Warsaw, Poland
{janusza,slezak}@mimuw.edu.pl
[3] Institute of Computer Science, Warsaw University of Technology,
ul. Nowowiejska 15/19, 00-665 Warsaw, Poland
{kamil.zbikowski,mkr,hrb,p.gawrysiak}@ii.pw.edu.pl
[4] mBank S.A., ul. Senatorska 18, 00-950 Warsaw, Poland

Abstract. We describe ISMIS 2017 Data Mining Competition – "Trading Based on Recommendations" – which was held between November 22, 2016 and January 22, 2017 at the platform Knowledge Pit. We explain its scope and summarize its results. We also discuss the solution which achieved the best result among all participating teams.

Keywords: Data mining competitions · Stock market trading · Expert-based recommendations

1 Introduction

It is not an easy task to predict movements of financial markets. A lot of effort has been committed to development of methods that would persistently provide profits for investors. These methods can be divided into several categories. One of approaches is to focus on information that can be extracted from financial instruments quotations. Another approach considers micro and macro economic factors. An example of the former is a group of methods called 'technical analysis' that applies various transformations of underlying time series in order to extract patterns that are the source of various (often contradictory) investors interpretations. An example of methods belonging to the latter category can be 'fundamental analysis' of companies quoted on a stock market.

The aforementioned methods struggle to make accurate predictions of future outcomes based on the currently available data. Thus, there is a huge potential to apply machine learning algorithms to construct efficient prediction models in this area. In the literature, there are various examples of combining financial

Partially supported by mBank and Tipranks.

M. Kryszkiewicz et al. (Eds.): ISMIS 2017, LNAI 10352, pp. 697–707, 2017.
DOI: 10.1007/978-3-319-60438-1_68

market prediction methods with machine learning techniques. One can find, e.g., applications of neural networks [1], support vector machines [2] or random forests [3], in modeling and validating stock trading strategies.

In this paper, we present the results of an open online data mining competition wherein the main goal was to determine – for a preselected number of stocks and analysts – whether recommendations from an analyst do have a sufficient predictive power to make '*Buy*'/'*Hold*'/'*Sell*' decisions. The motivation for organizing this kind of competition has come from the observation of phenomena called 'mutual fund herding' that is basically a situation in which decision makers of mutual funds blindly follow analyst recommendations [4]. Another observation of a similar market inefficiency can be found in [5] where, over the period from 1986 to 1996, a portfolio of the stocks with the most favorable consensus analyst recommendations was constructed. It provided an annual abnormal gross return of 4.13%. Other examples can be found in [6].

The paper has several aims. First, in Sect. 2, we go into the details of the competition task and summarize the obtained outcomes at a general level. Second, in Sect. 3, we describe the winning solution[1] and discuss how it refers to our observations originating from competitions held in the past. Third, we analyze the competition results from the perspective of their practical applicability, as well as the richness/completeness of the underlying data.

2   ISMIS 2017 Data Mining Competition

The considered competition was held between November 22, 2016 and January 22, 2017 at the platform Knowledge Pit [7]. All data sets used in the competition are available on-line to facilitate future research on this topic (https://knowledgepit.fedcsis.org/contest/view.php?id=119). In this section, we describe the scope of the competition and the characteristics of the used data sets. We also summarize the main competition results.

2.1   Task Description

The objective of the competition was to verify whether expert recommendations can be used as a reliable basis for making informed decisions regarding investments in a stock market. The task for participants was to devise an algorithm that is able to accurately predict a class of a return rate from an investment in a stock over the next three months, using only tips given by analysts. All the recommendations included into the competition data were provided by Tipranks (https://www.tipranks.com/) – a hub for major analysts, hedge fund managers, bloggers and financial reporters who bring the most accurate and accountable financial advice to the general public.

---

[1] The first author of this paper is the competition winner. The remaining authors are the competition organizing team members including representatives of sponsors and task/data providers, competition platform creators and conference organizers.

The data sets in the competition were provided in a tabular format. The training data set contained 12,234 records that corresponded to recommendations for stock symbols at different points in time. We refer to these time points as *decision dates*. Each data record was composed of three columns. The first one gave an identifier of a stock symbol (i.e. *Symbol_ID*, the actual symbol names were hidden from participants). The second column stored an ordered list of recommendations issued by experts for a given stock during two months before the decision date. The third column gave information related to the actual stock price computed within the period of three months after the decision date, mapped into three possible values depending on what decision would turn out to be the best one: '*Buy*', '*Hold*', '*Sell*'.

In each record, the list of recommendations consisted of one or more tips from financial experts. Any single tip was expressed using four values. The first value was an identifier of an expert. The second value gave the expert recommendation. The third value expressed expert expectations regarding the future return rate. It has to be stressed that information regarding the expected return rates was sometimes inconsistent and generally less reliable than the prediction of the rating, due to different interpretations of stock quotes by experts (e.g. not considering splits and/or dividends). Moreover, some experts did not share their expectations about the returns. Such situations were denoted by *NA* values in the data. The fourth value in each tip denoted a time distance to the decision date (in days), e.g. value 5 meant that the recommendation was published five days before the decision date. The list of tips in each record was sorted by the above-mentioned time distances, thus it could be regarded as a time series. A sample of the training data is shown in Fig. 1.

SymbolID	Recommendations	Decision
S94311056	{3084,Buy,0.4603721,0}{3864,Buy,0.1864133,6}	Sell
S175871	{307,Buy,0.1847301,0}{307,Buy,0.1643529,40}	Hold
S1913464	{3888,Buy,0.1612903,0}	Sell
S3682384	{1702,Buy,0.1366866,0}	Buy
S94311056	{3110,Buy,NA,0}{3117,Buy,0.2684415,8}{4527,B	Buy
S4258978	{3490,Sell,NA,0}	Buy
S3515147	{3250,Hold,NA,0}{746,Hold,-0.07084785,0}	Sell
S3682384	{1702,Hold,0.06903425,0}	Hold
S48162117	{1346,Hold,0.1217359,0}{3519,Hold,-0.01690238	Buy

Fig. 1. A sample of data rows extracted from the original training data set for ISMIS 2017 data mining competition

In order to additionally enrich the competition data, there was provided a table that grouped experts by companies for which they work. In total, the data consisted of recommendations from 2,832 experts who were employed in 228 different financial institutions.

Table 1. The final and preliminary results of the top-ranked teams

Rank	Team name	Preliminary	Final score
1	Mathurin	0.42017	0.43751
2	Bongod	0.42824	0.43746
3	Michalm	0.42818	0.43252
...	...	...	...
	Baseline	0.42656	0.43958

The testing data consisted of 7,555 records. It had a similar format as the training data, however, excluding the third column. The task for participants was to predict the labels for the test cases. It is important to note that the training and testing data sets corresponded to different time periods and the records within both sets were given in a random order.

The submitted solutions were first evaluated on-line and the preliminary results were published on the competition public leaderboard. These scores were computed on a subset of the testing set consisting of randomly selected 1000 records, fixed for all participants. The final evaluation was performed after completion of the competition using the remaining part of the testing data. These results were also published on-line (Table 1). The whole process was conducted in accordance to typical rules of Knowledge Pit which are quite comparable to the case of other data mining competitions related to financial domain (see e.g. https://www.kaggle.com/c/informs2010#description or https://dms.sztaki.hu/ecml-pkkd-2016/#/app/home).

2.2 Summary of Results

ISMIS 2017 Data Mining Competition ended on January 22, 2017. In total, it attracted 159 participants out of whom 73 were active and submitted at least one solution to the public leaderboard. We received 2,570 valid solutions. Moreover, 25 teams decided to send us descriptions of their approaches.

The final evaluation showed that none of the teams was able to beat the baseline score corresponding to a maximum of scores associated with '*always buy*' and '*always sell*' strategies (i.e. assigning all test cases with decisions '*Buy*' and '*Sell*', respectively). This might be the main cause of a lower than expected number of submitted competition reports. It confirms the difficulty of the considered competition and can be seen as a strong argument against usefulness of the recommendation data in predicting revenue from investments in stocks. Further discussion on this issue can be found in Sect. 4.

3   Description of the Winning Submitted Solution

In this section we describe a solution submitted by the team *mathurin*, represented by the first author of this paper – the competition winner. Hereafter, we

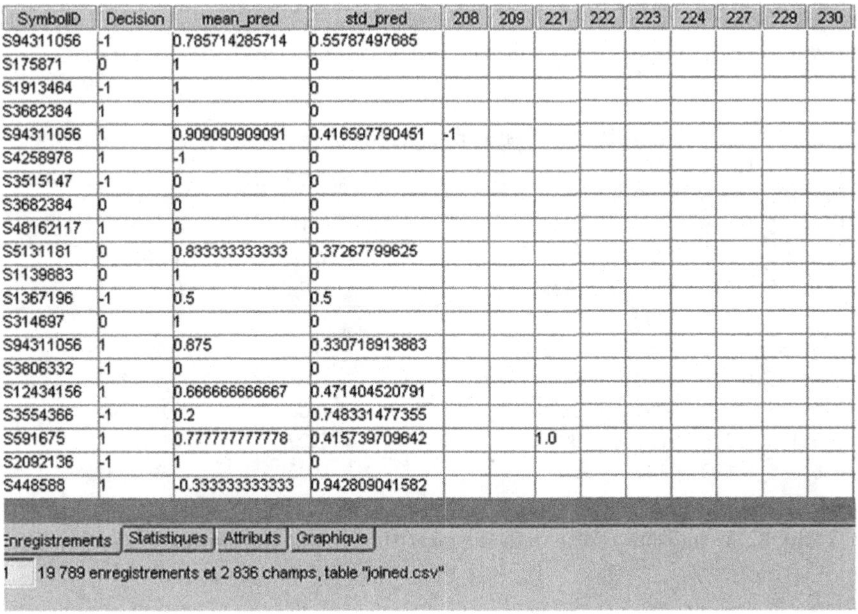

SymbolID	Decision	mean_pred	std_pred	208	209	221	222	223	224	227	229	230
S94311056	-1	0.785714285714	0.55787497685									
S175871	0	1	0									
S1913464	-1	1	0									
S3682384	1	1	0									
S94311056	1	0.909090909091	0.416597790451	-1								
S4258978	1	-1	0									
S3515147	-1	0	0									
S3682384	0	0	0									
S48162117	1	0	0									
S5131181	0	0.833333333333	0.37267799625									
S1139883	0	1	0									
S1367196	-1	0.5	0.5									
S314697	0	1	0									
S94311056	1	0.875	0.330718913883									
S3806332	-1	0	0									
S12434156	1	0.666666666667	0.471404520791									
S3554366	-1	0.2	0.748331477355									
S591675	1	0.777777777778	0.415739709642			1.0						
S2092136	-1	1	0									
S448588	1	-0.333333333333	0.942809041582									

Enregistrements | Statistiques | Attributs | Graphique

1   19 789 enregistrements et 2 836 champs, table "joined.csv"

Fig. 2. A snapshot of the data set after the first transformation phase

will name this solution as the *Mathurin method*. Apart from showing the layout of the final classifier used in the competition, we explain a rationale behind the key decisions taken during the modeling phase. In particular, we compare some aspects of the Mathurin method with our experiences from the previous data mining competitions organized at Knowledge Pit.

3.1   Data Preparation and Feature Engineering

In the Mathurin method, at the first data modeling phase, the available data set was transformed into a more convenient format. First, each recommendation was assigned to a new column, named in accordance to the expert identifier. For *time_distance* and *return_rate* variables, the average and standard deviation were computed. In the end, a newly constructed data table contained roughly 19 thousand rows and nearly 3 thousand columns corresponding to recommendations from 2834 experts (Fig. 2). All recommendations and decision classes were coded as integers, i.e. 'Buy' $= 1$, '$Hold$' $= 0$, '$Sell$' $= -1$.

In the second phase, each combination of *Symbol_ID*; *Decision_date*; *Expert_ID*; *Recommendation*; *return_rate*; *time_distance*; *Company_ID* values was transformed into a new row, following the multi-relational approach [8]. As a result, 19 thousand rows of the data (the training and testing sets) were transformed into approximately 110 thousand rows (Fig. 3). After this transformation, a custom software module was applied that allows to create 'on the fly' indicators, computed over several different dimensions of the data. These aggregations were performed by a *Symbol_ID* + *Decision_date* pair or by *Symbol_ID* to compute values corresponding to each row in the training and testing sets.

SymbolID 1	ExpertID	Recommandation	return_rate	time_distance	CompanyID
510151185_10376	277	Buy	0,33	52,00	26
510151185_10376	1700	Buy	0,21	51,00	231
510151185_10376	4157	Hold	0,10	37,00	167
510151185_10376	1700	Buy	0,16	31,00	231
510151185_10376	769	Hold	0,23	30,00	83
510151185_10376	1700	Buy	0,14	26,00	231
510151185_10376	4251	Hold	0,12	22,00	95
510151185_10376	1700	Buy	0,05	22,00	231
510151185_10376	1061	Sell	-0,08	9,00	134
510151185_10376	1700	Buy	0,22	0,00	231
510151185_10533	1115	Buy	0,05	43,00	265

Fig. 3. A snapshot of the data set after the second transformation phase

In the consecutive phase of the Mathurin method, manual and automatic feature engineering operations were conducted. First, summaries of the recommendations were created with respect to *Symbol_ID* and then *Symbol_ID* + *Decision_date*. This was done to create a sum of recommendations '*Buy*', '*Hold*' and '*Sell*', as well as a new attribute called *Top* whose value expressed the most common recommendation. This common-sense method proved to be the most discriminating information during the modeling. The *Top* attribute indicates the majority trend in expectations of the experts. There is a strong correlation between this attribute and the decision values in the training data (the value of *Top* indicates the correct decision in nearly 50% of cases).

Attributes that summarize a general attitude of experts toward the stocks express the main underlying trend of the market. For this reason, additional features were constructed, such as *Overall Appraisal Performance by Expert_ID* and by *Company_ID*. They were used to identify the recommendations that have the most predictive value.

The average accuracy of expert recommendations is nearly 40%. All experts/companies substantially exceeding this average are the ones to follow. Conversely, all experts/companies who under-perform with regard to this average value should be avoided. For example, the expert associated to *Expert_ID* = '1700', who generates the highest number of recommendations, has a success rate of 40% which is an average rating among all experts in the data, i.e. he/she is not better than a randomly selected expert. On the other hand, the expert with *Expert_ID* = '3084' has a success rate of 51% which makes him/her someone to follow. At the opposite end, the expert with *Expert_ID* = '1712' is below average (31%), so his/her recommendations should be avoided (Fig. 4).

If we check performance of recommendations aggregated by *Company_ID*, we can see that the company that corresponds to *Company_ID* = '284' is 8% above the average in recommendation accuracy, while e.g. company with *Company_ID* = '8134' performs 10% below the average.

ExpertID	% of success	Count ↓
1 700	0,40	5 679
3 084	0,51	724
1 712	0,31	627
2 208	0,38	591
1 702	0,46	572
8 593	0,38	572
1 709	0,41	540
1 156	0,29	529
4 365	0,35	518
603	0,47	501
860	0,44	496
3 813	0,29	450

CompanyID	% of success	Count ↓
231	0,39	7 747
163	0,41	2 472
134	0,38	2 284
26	0,40	2 225
95	0,40	2 208
70	0,39	1 857
167	0,41	1 802
302	0,40	1 899
254	0,43	1 627

Fig. 4. Counts and success rates of individual experts (on the left) and companies (on the right) in the training data set

In addition to manually generated features, an automatic method was used to generate 50 thousand features by aggregating the vertically extended data on different *Symbol_IDs* and *Decision_dates* (which define the aggregation level). Additionally, numerical attributes were discretized and some values of categorical data columns were grouped together in order to define new dimensions for analysis, e.g., *time_distance* = 0; *return_rate* > 5. The aggregations were corresponding to basic statistics derived from the grouped data, such as mean absolute deviation (MAD), average, standard deviation, count, distinct count, max, median, min, mode, percentile, range, skewness, kurtosis and sum.

It can be clearly seen that using the above method it is possible to create any desired number of additional features [9]. It is important to note, however, that the training and testing data sets in the competition correspond to different time periods and the records in both sets are given in a random order. Due to this fact, it was not possible to construct shift variables (lags) which could have taken into account the previous recommendations to refine the performance of the model. This seems to be a promising direction for the future study.

3.2   Concept Drift Analysis

Preliminary observation of the performance of the models developed within the Mathurin method revealed significant differences between the training and testing data sets. A similar phenomenon has been observed in the case of other competitions organized at Knowledge Pit (see e.g. [10]). During the AAIA 2015 Data Mining Competition, Boullé highlighted a distribution problem [11], often referred to as a concept drift [12]. Thus, this aspect is also worth examining with respect to ISMIS 2017 Competition.

Let us stop for a while the analysis of the winning submitted solution and take a closer look at the training and testing data sets. Following Boullé's approach developed for the purposes of previous competitions, let us merge the training and testing ISMIS 2017 Competition data sets and define an additional boolean attribute equal to 0 for the cases from the training data and 1 otherwise. We used cross-validation to check how accurately it is possible to predict whether a case belongs to the testing data. It turned out that for this task a model learnt using the XGBoost algorithm [13] had an accuracy greater than 0.82, which suggests a strong distribution shift between the training and testing sets. The two plots in Fig. 5 show a comparison between the predictive power of individual features in predicting true decision labels and their ability to discern between the training and testing data. Nearly all individual features can identify objects from the testing set (the plot on the left) while it is much harder to distinguish between any two random subsets of the training data (the plot on the right).

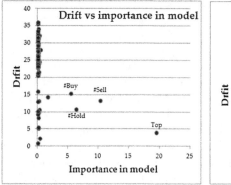 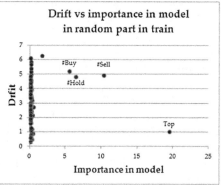

Fig. 5. Concept drift between the training and testing data (on the left) and between two randomly selected parts of the training data (on the right)

The strong concept drift confirms that the temporal shift between the training and testing sets has a significant impact on model's performance. It suggests that since financial markets are volatile [14], a context of a recommendation is needed to assess its value. On the other hand, the distribution of the recommendations did not change significantly between the training and testing sets, as if the experts did not see the change of a trend. One might even speculate that some of the experts could artificially maintain their recommendation in order to influence the market for their own profit.

3.3   Model Construction and Validation

Let us now go back to the Mathurin method where the aforementioned XGBoost algorithm was used as well. XGBoost is a scalable decision tree gradient boosting method that is known for its efficiency and robustness. Since the decision attribute in the considered competition data had three different values, two different quality metrics were tested, namely a multi-class log-loss and a weighted accuracy (ACC). 5-fold stratified cross-validation was used in a combination with the grid search to determine the best settings of parameters.

Since XGBoost struggled to converge into optimal parameter settings using ACC, after initial tests only the log-loss metric was used. For every test case, the model returned a probability distribution over the three decision values. At first, the recommendation with a maximal probability was chosen for each test case. However, the obtained preliminary result (ACC $\approx$ 0.40) was not satisfactory and significantly below the estimation of performance computed using 5-fold cross-validation on the training data (ACC $\approx$ 0.56).

For that reason and due to the significant concept drift, another solution was submitted. This time, however, the decision with the lowest probability was selected for each of the test examples. It turned out that the preliminary score of such a counter-intuitive solution was significantly higher ($ACC \approx 0.42$) and its final result was the highest among all participants of the competition. It confirms the hypothesis that the available data set is insufficient for learning a predictive model that could be used to reliably achieve better performance than the overall trend of the market. It also shows that making financial decisions based solely on recommendations from experts is burdened with a considerable risk.

4   Conclusions and Plans for the Future Work

The aim of this study was to determine whether financial analysts can provide meaningful recommendations, i.e. their predictions can constitute a solid basis for making informed decisions regarding investments in a stock market. The data mining competition that we organized to augment our research provided interesting insights toward answering this question.

During the competition, there were submitted 2570 correctly formatted solutions. The winning one did not manage to achieve results better than the baseline that was constructed as the maximum from 'always buy' and 'always sell' strategies over the testing period. This fact does not mean that the best performing models are worthless. Beating the market is a very difficult task. Models which perform slightly worse but are stable and still are able to generate profit (i.e. their expected return rate is greater than zero) have an unquestionable practical value. It is worth noticing that all of the top-ranked teams achieved significantly better results[2] than random predictions or the predictions made using some simple voting methods (e.g. the majority voting between experts). The unexpected outcome of the method that won the competition may actually be the effect of an unexpected trend reversion during the testing period. As a result, stocks that were overpriced during the training period due to the effect of positive recommendations have lost much of their value.

---

[2] The significance of differences was measured using the t-test, based on scores obtained by 1000 random classification vectors of the testing data.

One of the promising directions for the future research on this topic could be a deeper investigation of dependencies between different experts. For instance, feature clustering methods [15] can be applied to detect hidden relations between analysts. Moreover, some recently developed heuristics for finding so-called decision bireducts can be used to concurrently detect influential or reliable experts and parts of the data where the predictions are considerably more confident [16]. Finally, the impact of less restrictive cost functions on the learning performance of prediction algorithms should be investigated. It could provide valuable insight about an ability to construct more cautious classification models that prefer answering '*Hold*' when they are not confident about their predictions.

References

1. Chavarnakul, T., Enke, D.: Intelligent technical analysis based equivolume charting for stock trading using neural networks. Expert Syst. Appl. **34**(2), 1004–1017 (2008)
2. Żbikowski, K.: Using volume weighted support vector machines with walk forward testing and feature selection for the purpose of creating stock trading strategy. Expert Syst. Appl. **42**(4), 1797–1805 (2015)
3. Ładyżyński, P., Żbikowski, K., Grzegorzewski, P.: Stock trading with random forests, trend detection tests and force index volume indicators. In: Rutkowski, L., Korytkowski, M., Scherer, R., Tadeusiewicz, R., Zadeh, L.A., Zurada, J.M. (eds.) ICAISC 2013. LNCS (LNAI), vol. 7895, pp. 441–452. Springer, Heidelberg (2013). doi:10.1007/978-3-642-38610-7_41
4. Brown, N.C., Wei, K.D., Wermers, R.: Analyst recommendations, mutual fund herding and orverreaction in stock prices. Manage. Sci. **60**(1), 1–20 (2014)
5. Barber, B., Lehavy, R., McNichols, M., Trueman, B.: Can investors profit from the prophets? Security analyst recommendations and stock returns. J. Financ. **56**(2), 531–563 (2001)
6. Kim, S.T., Lin, J.C., Slovin, M.B.: Market structure, informed trading, and analysts' recommendations. J. Financ. Quant. Anal. **32**(4), 507–524 (1997)
7. Janusz, A., Ślęzak, D., Stawicki, S., Rosiak, M.: Knowledge pit - a data challenge platform. In: Proceedings of CS&P 2015, pp. 191–195 (2015)
8. Padhy, N., Panigrahi, R.: Multi relational data mining approaches: a data mining technique. CoRR abs/1211.3871 (2012)
9. Wróblewski, J.: Analyzing relational databases using rough set based methods. In: Proceedings of IPMU 2000, vol. 1, pp. 256–262 (2000)
10. Janusz, A., Ślęzak, D., Sikora, M., Wróbel, Ł.: Predicting dangerous seismic events: AAIA'16 data mining challenge. In: Proceedings of FedCSIS 2016, pp. 205–211 (2016)
11. Boullé, M.: Tagging fireworkers activities from body sensors under distribution drift. In: Proceedings of FedCSIS 2015, pp. 389–396 (2015)
12. Brzeziński, D., Stefanowski, J.: Combining block-based and online methods in learning ensembles from concept drifting data streams. Inf. Sci. **265**, 50–67 (2014)
13. Chen, T., Guestrin, C.: XGBoost: a scalable tree boosting system. In: Proceedings of KDD 2016, pp. 785–794 (2016)
14. Jeffers, E., Moyé, D.: Dow Jones, CAC 40, SBF 120: Comment Expliquer Que le CAC 40 Est le Plus Volatil? Revue d'Économie Financiere **74**(1), 203–218 (2004)

15. Janusz, A., Ślęzak, D.: Rough set methods for attribute clustering and selection. Appl. Artif. Intell. **28**(3), 220–242 (2014)
16. Stawicki, S., Ślęzak, D., Janusz, A., Widz, S.: Decision bireducts and decision reducts - a comparison. Int. J. Approx. Reason. **84**, 75–109 (2017)

Predicting Stock Trends Based on Expert Recommendations Using GRU/LSTM Neural Networks

Przemyslaw Buczkowski[1,2](✉)

[1] National Information Processing Institute, Warsaw, Poland
pbuczkowski@opi.org.pl
[2] Warsaw University of Technology, Warsaw, Poland

Abstract. Predicting the future value of the stock is very difficult task, mostly because of a number of variables that need to be taken into account. This paper tackles problem of stock market predicting feasibility, especially when predictions are based only on a subset of available information, namely: financial experts' recommendations. Analysis was based on data and results from ISMIS 2017 Data Mining Competition. An original method was proposed and evaluated. Participants managed to perform substantially better than random guessing, but no participant outperformed baseline solution.

Keywords: Stock exchange · Sequence modeling · Time series prediction · Artificial neural networks · Recurrent neural networks

1 Introduction

Prediction of stock market trends is a very difficult task. There is no substantial evidence that it is possible in long-term approach at all. The reason of that difficulty is the stock price is function of large number of variables. Financial results of the company may depend on the historical performance of that company, development direction, board changes, global economic situation, political situation and countless other factors. Those relations are not trivial to measure which leads to question about feasibility of the task. There are two opposite standpoints: "Efficient Markets Hypothesis" (EMH) and "random walk hypothesis" (RWH). Efficient Markets Hypothesis proposed by Professor Eugene Fama posits that all available information at any time is already reflected in prices. That works both ways. Any price change observed is caused by a release of new information [1]. Random walk proposed by Burton Malkiel, on the other hand, posits that changes of the prices have random nature and are not strictly correlated with information. Burton Malkiel suggests that paying financial services for prediction hurts return value.

ISMIS 2017 Data Mining Competition (www.knowledgepit.fedcsis.org) is data mining challenge organized as the special event of 23rd International Symposium on Methodologies and Intelligent Systems and sponsored by mBank

© Springer International Publishing AG 2017
M. Kryszkiewicz et al. (Eds.): ISMIS 2017, LNAI 10352, pp. 708–717, 2017.
DOI: 10.1007/978-3-319-60438-1_69

(www.mbank.pl) and TipRanks (www.tipranks.com). It is closely related to stock market prediction problem but it is constrained by rather specific subset of available information. In contrast to the most popular approach of exploiting historical price time series, ISMIS 2017 Data Mining Competition focuses on predicting class ("Buy", "Hold", "Sell") from expert recommendations. These recommendations are provided by TipRanks company which aggregates recommendations and ranks analysts. More specific information about dataset is provided in section Data. The score in this challenge is calculated according to (1).

$$score(X) = \frac{\sum_{i=1..3}(X_{i,i} \cdot C_{i,i})}{\sum_{i=1..3}\sum_{j=1..3}(X_{i,j} \cdot C_{i,j})} \tag{1}$$

X is 3×3 confusion matrix and C is 3×3 cost matrix used for scaling different types of errors and is provided by mBank employees by applying their expert knowledge of the problem. Weights $C_{i,j}$ are presented in Table 1.

Table 1. Costs $C_{i,j}$ for calculating score

Preds	Truth		
	Buy	Hold	Sell
Buy	8	4	8
Hold	2	1	2
Sell	8	4	8

There are traditionally three main groups of stock market prediction methods: fundamental analysis, technical analysis, and technological methods. Fundamental analysis uses expert knowledge about the financial situation of a company, sector of the company and global economics. In the process, the condition of the company is evaluated and prediction is made based on the difference between this evaluation ("true" value of the company) and the current price. During the process, many numeric indicators are calculated and used to support evaluations. The second group is called technical analysis. In this approach, only historical prices are analysed. According to the EMH, every available information is already reflected in current or historical price so there is no need to analyze any external sources e.g. financial reports. This group, therefore boils down to time series classification or time series prediction tasks. Both of described method groups contains method applied by a human expert rather than a computer program, although methods from both groups can be automated to some extent. Those automated equivalents fit in the third group called technological methods.

This paper describes solution created for ISMIS 2017 Data Mining Competition. A novel neural network approach was proposed in order to make use of examples from testing dataset in an unsupervised fashion.

2   Data

Challenge data consists of two separate datasets: a training set with (target classes included) with 12234 examples and test set (target classes not included) with 7555 examples. Records are anonymized so connecting stock identifiers and expert identifiers to real entities is not feasible in any simple way. A single record consists of a stock identifier, a list of expert recommendations of varying length and optional target label (class). Every recommendation is a tuple of four values: expert identifier, class predicted by expert, expert's expectation about return value and a number of days between recommendation and decision. Class predicted by an expert as well as target label is one of: "Buy", "Hold", "Sell". Third value, the return prediction, is real value giving quantitative information about a change in price. Unfortunately, challenge's website states that this value may be calculated and interpreted differently among experts. Last value is a natural number. Less value of this variable should imply higher level of certainty. A number of expert recommendation per record varies from 1 to over 100. In additional file information about affiliation of experts is provided. It is a one-to-many relation with "many" expert belonging to single company (Fig. 1).

S47757139;{4035,Hold,0.0940919,0}{2137,Buy,NA,44};Hold

Fig. 1. Random data record

The training set is unbalanced with respect to target label. There are almost two times more "Buy" labels than "Sell". This ratio for test set is unknown for now. This imbalance is addressed by batch balancing (see Method section). More concerning fact about this data is totally different distribution of label between real labels and expert's predictions. On the training set, experts are far more biased toward buying. Those differences are presented in Table 2.

Table 2. Number of label occurrences as targets and recommendations in datasets

	Training set		Test set
	Recommendation	Target	Recommendation
Buy	34306 (62%)	5722 (47%)	34166 (62%)
Hold	18388 (33%)	3403 (28%)	18751 (34%)
Sell	2564 (5%)	3109 (25%)	2287 (4%)

It looks like experts think they are experiencing stronger bull market than they really are. Distribution of expert's recommendation over test set can be calculated and is included in Table 2 as well. Differences in those (expert's prediction) distributions are surprisingly small. It would be possible if market trends didn't change for the whole period. It is said, that the periods of data acquisition for training and test set are disjoint but adjacent. But there is a reason why one can assume that bull market is no longer there during test dataset acquisition. Three benchmark solutions consisting of single label repeated 7555 times were sent for evaluation using challenge platform. Best score (0.4395) was achieved for solution containing only "Sell". It is clear indicator of change in market situation toward a bear trend. Unfortunately, experts have not notice that and proceed to favor buying over selling. This solution (let us denote is Baseline Solution (BS)) is probably the exact solution used as baseline by administrators as it yields identical pre-evaluation and final score.

To further investigate quality of data, samples from training set have been manually examined. It was found that most of the records contain contradictory data. Most common example is having some expert's predictions mismatch target label. More problematic are records without any correct recommendation. It renders impossible to predict correct labels for such records (Fig. 2).

```
S2201432;{754,Buy,NA,0};Sell
S362146;{1218,Hold,-0.0331087,0};Sell
S7291152;{4640,Buy,NA,0}{3339,Buy,0.2429892,25};Sell
S3386173;{4285,Sell,NA,0};Buy
```

Fig. 2. Records containing contradictory data

Noisy nature of data is real problem here. To evaluate quality of recommendations, confusion matrix for experts' predictions has been calculated on training set. Using this confusion matrix, plain accuracy $ACC = 0.4010$ as well as score (1) $SCORE = 0.5139$ were calculated. This accuracy should be considered poor for a general 3-way classification problem (Table 3).

Table 3. Confusion matrix of experts' predictions calculated on training set

Preds	Truth		
	Buy	Hold	Sell
Buy	16233	8511	1261
Hold	9399	5267	644
Sell	8674	4610	659

Having identifier of expert for every recommendation it is possible to calculate accuracy for every single expert. If some of them are significantly more accurate than others, a method exploiting this information might yield better result than average expert.

At this point, it is difficult to judge if task of proposing method more accurate than Baseline Solution is feasible. For many cases, it is perfectly achievable to construct strong classifier with many weak classifiers (experts in this case). In fact, this is what boosting does. The difference is that experts cannot be trained during the process. It is more about proper aggregating of their prediction.

3   Method

3.1   Data Preprocessing

It seems to be reasonable idea to provide model with precalculated information about accuracy of experts recommendations if such information is available. Using training set one can calculate accuracy for every expert in this set. This is done by grouping recommendation by expert identifier and calculating accuracy independently for every group. This is not the only grouping which can make sense. It is reasonable to presume that some expert can predict some stocks value better than other stocks. Such expert may simply have better knowledge of some economic sector. It leads to idea of grouping recommendations by expert identifier and stock identifier. Another possible grouping is performed by stock identifier alone. This describes how accurately price of that item can be predicted by any expert. Some stocks may be more unpredictable than others and this might be helpful hint for network to classify more conservatively, e.g. predict "Hold" with greater probability. The last grouping uses company identifier. The assumption was made, that "good" company employs "good" experts. Therefore one can assume that information about quality of company may be estimating (perhaps lacking) information about quality of expert. Let us denote set of all experts by E, set of all stocks by S and set of companies by C. From now on, accuracy of all recommendations of stock $s \in S$, made by expert $e \in E$, employed in company $c \in C$ will be denoted by A_{esc}. Note that in this case index c is redundant because single expert works in only one company. Finally, let us denote accuracy of recommendations across different values of any index by putting $*$ in place of single value of such index. For example A_{*sc} should be understood as accuracy of all recommendations of company c concerning stock "s". Using training set and cross-validation process those accuracies were picked as good features: A_{e**}, A_{es*}, A_{*s*} and A_{**c}. These accuracies are not used directly because some of the values calculated this way are calculated on too few examples. For example, if some expert e made only one recommendation, A_{e**} is calculated based on this single guess and therefore should not be trusted. In such case, average expert accuracy is used instead. Accuracy in every grouping is substituted by appropriate mean if there is less than ten examples in group. Second and last modification of those artificially calculated features is subtracting (appropriate) group mean accuracy so all features are concentrated around 0. Let us denote such modified accuracies with prim character A'_{xyz}. Time parameter of recommendation is divided by 60 to fit $[0, 1]$ interval for almost every

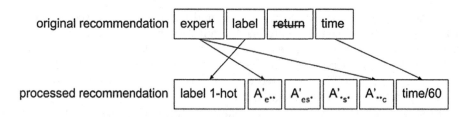

Fig. 3. Single recommendation preprocessing

recommendation. Prediction label is replaced by its one-hot encoding. Raw identifiers of stock, expert, and company are omitted as they are difficult for network to understand directly in form of integers (Fig. 3).

3.2   Model

The model proposed in this paper is quite complicated as it consists of three separate neural networks: two recurrent networks (RNN#1 and RNN#2) whose concatenated outputs is fed to the final fully connected feed-forward network. All of the networks are trained separately but FC#1 depends on results of its predecessors (Fig. 4).

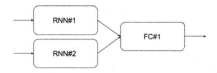

Fig. 4. Overview of model architecture

Recurrent neural networks are special kind of neural networks which maintain its state across different time steps. In this context they are not stateless and such state is stored as vector of numbers. From perspective of single (spatial) layer this can be viewed as having more input values and therefore more learnable parameters: beside current input such layer accepts its own output from previous time step as input. A recurrent neural network can be "unrolled" to become deep feed-forward networks which can be trained with variant of backpropagation [2] called backpropagation through time.

Such depth would be painful due to known problems with learning very deep networks such as vanishing gradient [3] or exploding gradient. The solution is to share weights among different temporal copies of network. Even if the gradient vanishes in distant layers weights still can be updated as gradient from closer layers will stay different from zero. However vanishing gradient is not the problem anymore preserving state vector for long dependencies is. Repeatable update of state vector is eventually killing stored information. The solution is to conditionally update state vector. Updates and readings of state vector can be driven through gates conditionally like in digital memory. But when to open or close such gates? Let us delegate this burden to trainable parameters to decide. An example of such network is Long Short-Term Memory (LSTM) [4]. There are many variations of LSTM e.g. Gated Recurrent Units (GRU) [5] which simplifies forget and update gates into single gates or others [6–8].

First network (RNN#1) is encoder-decoder LSTM network which tries to find constant length vector representation of recommendation series from both training and test set. This setup can be considered sequence-enabled equivalent of autoencoder network. Autoencoder networks are networks which try to learn compact representation of data in unsupervised fashion. It is achieved by using hourglass-like architecture of (in most cases) two layers of neurons. During the training, input of the first layer is trained to pop out from output of the second layer. To do this, the first layer must learn such transformation that will preserve essential information needed to reconstruct input while reducing dimensionality. This small intermediate representation is called embedding and is believed to contain important features from classification perspective [9].

It is possible to construct the equivalent of autoencoder for sequential data which will provide constant length embeddings summarising sequences. This approach is used for a while [5]. The easiest way to achieve this is to use one recurrent neural called encoder (which is equivalent of first layer in traditional autoencoder) to produce embedding and another RNN fed with such embedding (which is equivalent of second layer in traditional autoencoder) which reconstruct the initial sequence from embedding (Fig. 5).

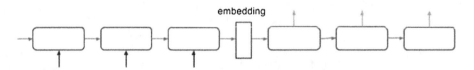

Fig. 5. A more precise depiction of RNN#1 component. The encoder outputs a vector, from which the decoder tries to reconstruct input sequence.

By using encoder-decoder setup one can utilize unlabeled recommendation sequences from a test set as well as labeled from training set. Encoder part is meant to learn good features from both sets which should, in theory, decrease overfitting measured on a test set. This network is trained with backpropagation and RMSprop update rule [11]. After training of RNN#1, decoder part is no longer needed so it is discarded. From now on, e.i. in inference mode, RNN#1 is returning only compact representation.

Second network (RNN#2) is simple GRU network trained for performing sequence classification. It is also fed with preprocessed recommendation series but it has softmax layer on its end to perform efficient classification. Softmax layer's outputs are normalized so they sum up to 1, therefore they can be interpreted as probabilities of classes. Cross-entropy cost function was used as most common in such scenario (Fig. 6).

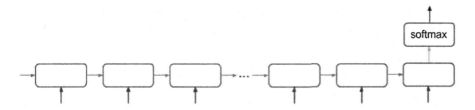

Fig. 6. A more precise depiction of RNN#2 component. Output of RNN is connected to the softmax layer.

When both RNN#1 and RNN#2 have finished learning their parameters (weights) become frozen. For training last part of the network, FC#1, record is fed to RNN#1 and RNN#2. RNN#2 produces a vector of three probabilities, one per every label. RNN#1 produces feature vector of arbitrary length, potentially useful for FC#1. Concatenating both of these vectors generate new vector which is fed to FC#1. FC#1 is simple MLP [10] with a single layer of hidden neurons followed by softmax layer of 3 neurons for final classification learned with cross-entropy cost function (Fig. 7).

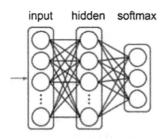

Fig. 7. A more precise depiction of FC#1 component. The last layer is the softmax layer

Due to an imbalance of target labels in training set some balancing method needs to be introduced in order to reduce prior probability bias (as distributions of target labels in training and test sets are different). Training with mini batches lets one prepare batches with equal class distribution. This method was used in this solution. In training of RNN#1 batch consisting of union of two batches of equal size was used: balanced batch from the training set and random batch from the test set.

To benchmark this model before final evaluation, the training set was randomly split into two sets $t1$ (90% examples) and $t2$ (10% examples). While training model on $t1$ only, evaluation on both $t1$ and $t2$ was periodically performed. The model proved to have enough capacity to learn (to overfit) data from $t1$ but failed to generalize well on $t2$. A problem of overfitting was addressed with attempt of dropout [12,13] usage. Dropout is a trick reducing overfitting by omitting random weights with some fixed probability what results in effect similar to averaging of many models. Experiments with different number of layers, number of neurons per layer and different keep probability were conducted but yield no significant improvement. Actual dropping of classification error on Fig. 8 is subtle and takes place in first 100 epochs which is barely visible on the left hand side of figure.

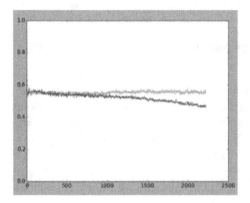

Fig. 8. Classification error during training. Error calculated on $t1$ is plotted in red, error on $t2$ – in green (Color figure online).

4   Results

Described model achieved 5th place in ISMIS 2017 Data Mining Competition with score 0.4271. A total number of 74 teams participated in this challenge. What is worth noticing, no solution managed to perform better than Baseline Solution (BS) which was constructed by predicting "Sold" label for every record.

5   Discussion

As was stated in the introduction section, predicting stock trends proved to be difficult task. The most meaningful fact is that no participant managed to outperform single-label baseline solution using recommendations from experts. There is very little information about the situation on the market during challenge data acquisition. A potential explanation of this situation is that training set was collected during a bull market and test set during a bear market. This could explain inability of models to generalize. However this story does not hold, because models fail to generalize on cross-validation as well. Considering poor performance of experts calculated on training set, one can blame little to no correlation between experts' predictions and target labels (Table 4).

Results of this challenge support claim of Burton Malkiel that paying for recommendations might actually hurt net portfolio return. However, it does not proof that predicting stock market is not a feasible task to do. Perhaps better expert selection or gathering more information can result in beneficial investing strategy.

Table 4. Final results of the competition

No	Name	Score
-	Baseline Solution (BS)	0.4395
1	mathurin	0.4375
2	bongod	0.4374
3	michalm	0.4325
4	pwbluehorizon1	0.4298
5	**boocheck**	**0.4271**
6	a.ruta	0.4242
7	mkoz	0.4234
8	amy	0.4172
9	grzegorzkozlowski	0.4164
10	vinh	0.4159

References

1. Fama, E.: Efficient capital markets: a review of theory and empirical work. J. Finan. **25**(2), 383–417 (1970)
2. LeCun, Y., Bottou, L., Orr, G.B., Müller, K.-R.: Efficient BackProp. In: Orr, G.B., Müller, K.-R. (eds.) Neural Networks: Tricks of the Trade. LNCS, vol. 1524, pp. 9–50. Springer, Heidelberg (1998). doi:10.1007/3-540-49430-8_2
3. Hochreiter, S.: The vanishing gradient problem during learning recurrent neural nets and problem solutions. Int. J. Uncertainty Fuzziness Knowl. Based Syst. **06**, 107 (1998)
4. Hochreiter, S., Schmidhuber, J.: Long short-term memory. Neural Comput. **9**, 1735–1780 (1997)
5. Cho, H.: Learning phrase representations using RNN encoder-decoder for statistical machine translation (2014)
6. Yao, K., Cohn, T.: Depth-Gated Recurrent Neural Networks (2015)
7. Greff, K., Srivastava, R.: LSTM: a search space odyssey (2015)
8. Jozefowicz, R., Zaremba, W.: An empirical exploration of recurrent network architectures (2015)
9. Geoffrey, H.: Where do features come from? Cogn. Sci. **38**, 1078–1101 (2014)
10. Rosenblatt, F.: Principles of Neurodynamics: Perceptrons and the Theory of Brain Mechanisms. Spartan Books, Washington DC (1961)
11. Tieleman, T., Hinton, G.: Lecture 6.5 - RMSProp, COURSERA: neural networks for machine learning. Technical report (2012)
12. Hinton, G., Srivastava, N., Krizhevsky, A.: Improving neural networks by preventing co-adaptation of feature detectors (2012)
13. Srivastava, N.: Improving neural networks with dropout (2013)

Using Recommendations for Trade Returns Prediction with Machine Learning

Ling Cen[1]([✉]), Dymitr Ruta[1], and Andrzej Ruta[2]

[1] EBTIC, Khalifa University, Abu Dhabi, UAE
{cen.ling,dymitr.ruta}@kustar.ac.ae
[2] ING Bank Slaski, Katowice, Poland
andrzej.ruta@ingbank.pl

Abstract. Automatically predicting stock market behavior using machine learning and/or data mining technologies is quite a challenging and complex task due to its dynamic nature and intrinsic volatility across global financial markets. Forecasting stock behavior solely based on historical prices may not perform well due to continuous, dynamic and in general unpredictable influence of various factors, e.g. economic status, political stability, voiced leaders' opinions, emergency events, etc., which are often not reflected in historic data. It is, therefore, useful to look at other data sources for predicting direction of market movement. The objective of ISMIS 2017 Data Mining Competition was to verify whether experts' recommendations can be used as a reliable basis for making informed decisions regarding investments in stock markets. The task was to predict a class of a return from an investment in different assets over the next three months, using only opinions given by financial experts. To address it, the trading prediction is formulated as a 3-class classification problem solved within supervised machine learning domain. Specifically, a hybrid classification system has been developed by combining traditional probabilistic Bayesian learning and Extreme Learning Machine (ELM) based on Feed-forward Neural Networks (NN). Assuming feature space narrowed down to just the latest experts recommendations probabilistic and ELM classifiers are trained and their outputs fed to train another baseline ELM classifier. The outputs from baseline classifiers are combined by voting at the decision level to generate final decision class. The presented hybrid model achieved the prediction score of 0.4172 yielding 8th place out of 159 teams competing in the ISMIS' 2017 competition.

Keywords: Trading prediction · Maximum a posteriori estimation · Hybrid classification · Decision level fusion · Extreme Learning Machines (ELM)

1 Introduction

Stock markets are generally non-stationary and chaotic, which are influenced by many direct or indirect variables and uncertainties, such as historical prices, economic conditions, investors' sentiments, political events, etc. Quick changes in

© Springer International Publishing AG 2017
M. Kryszkiewicz et al. (Eds.): ISMIS 2017, LNAI 10352, pp. 718–727, 2017.
DOI: 10.1007/978-3-319-60438-1_70

stock markets lead to dynamic, non-parametric, chaotic and noisy nature of stock time series resembling random processes with short term fluctuations in price and volume [1]. In order to achieve optimal capital gain and minimize investment risk, one need to be aware of stock future price movements and accordingly, identify trade opportunity. With rapid development of intelligent/smart computing, banks, trading houses, investment firms have relied heavily on algorithmic trading utilizing computer programs and mathematical models to develop and execute trading strategies. They generate model triggered automated orders that are instantly executed at financial trading platforms often at very high trading frequency (HFT) and require continuous and high bandwidth feed of data and information to trade in real-time. Apart from exploiting profit opportunities, algorithmic trading makes markets more liquid and trading more systematic mostly thanks to removing the impact of human emotions on trading activities [2]. The key question in algorithmic trading is how to accurately predict future market movements based on historic and present stock data (e.g. prices, volume, etc.) and other relevant financial information (e.g. company annual reports, expert recommendation, commodity prices, etc.).

Machine learning and data mining have gone through a rapid development in the fields of computer science, data analysis, software engineering, and artificial intelligence and have found numerous applications for predictive analytics in various areas. Discovering the hidden patterns of market behavior and automatically predicting the change of stock prices with a combination of data mining and machine learning have recently attracted growing attention, despite many challenges in handling highly dynamic nature and intrinsic volatility in price and volume flows across global markets [1,3–13]. Most of the methods reported in the literature forecast stock market behavior based on the knowledge inferred from historic data. In this work, we intend to verify whether experts' recommendations can be used as a reliable source to devise profitable trading strategies on various stock markets in the context of ISMIS 2017 Data Mining Competition. The task in the competition is to predict the class of a investment return in an asset over the next 3 months using recommendations provided by multiple financial experts. The task is formulated as a 3-class classification problem with features extracted based on expert's recommendation. A hybrid classification approach has been developed utilizing Maximum a posteriori (MAP) class estimation and Extreme Learning Machine (ELM). The presented model achieved the prediction accuracy score of 0.4172 yielding the 8th place out of 159 teams competing in the ISMIS 2017 competition. The relatively low accuracy, although still higher than chance, could be explained by the highly unstable nature of market sentiments and generally poor individual expert performances that limit the utility of their recommendations for a successful trading strategy. However, as we have demonstrated, recommendations provide a new external and beneficial financial information other than historic price/volume data that can improve the performance of stock price prediction.

The remainder of the paper is organized as follows. A literature survey on stock price prediction is given in Sect. 2. Following that, ISMIS' 2017 Data

Mining Competition is introduced, specifically describing the objective of the competition, data format, and performance evaluation function. In Sect. 4, the prediction model is presented, where the feature representation and the hybrid learning method are proposed. The experimental results obtained through model evaluation on KnowledgePit online platform within ISMIS' 2017 Data Mining Competition are summarized in Sect. 5, followed with concluding remarks provided in Sect. 6.

2   Related Work

In the literature, stock price prediction has been formulated as either classification or regression problems, and solved by using supervised learning algorithms, e.g. support vector machines (SVM), neural network (NN), linear regression, etc., based on stock historical data. In [3], classification and regression models were both created to predict the daily return of a stock from a set of direct features such like open, close, high, low, volume, etc., and indirect features corresponding to some external economic indices that could influence the return of a stock (e.g. commodity prices such as gold, crude oil, nature gas, corn or cotton). In the classification model, the trend of a stock in a specific day was predicted with logistic regression, while in the regression model the exact return of a stock was forecast by linear regression. Compared to the SVM models, linear models performed better in both classification and regression frameworks and logistic regression achieved highest success rate of 55.65% and 2000% cumulative return over 14 years. An automatic stock trading process was proposed in [4], which involved a hierarchy of a feature selecting method based on multiple indicators (e.g. Sharpe ratio, Treynor ratio, etc.) by using backward search, four types of machine-learning algorithms (i.e. locally weighted linear regression (LWLR), logistic regression, $l-1$ regularized v-support vector machine (v-SVM), and multiple additive regression trees (MART)), as well as an on-line learning mechanism. The trading was automatically executed based on the prediction of tomorrow's trend, i.e. buying if the stock price would increase tomorrow while selling if the price would decrease. Quite high average prediction accuracy up to 92.36% was achieved as reported in [4]. In [5], stock price was modeled on the daily and longer term basis, so that the prices on the next-day or the next n-days were predicted, respectively. Different types of learning algorithms such like logistic regression, quadratic discriminant analysis, and SVM were utilized in price modeling on historic data ranging from Sept. 2008 to Aug. 2013. The highest accuracy achieved in the next-day prediction model was 58.2%, while the accuracy can be as high as 79.3% in the SVM based long-term model with a time window of 44. This was further improved to 96.92% with a time window of 88 in [1] that utilized random forest classier with features extracted from multiple technical indicators (e.g. Relative Strength Index (RSI), stochastic oscillator, etc.). Neural network (NN) based approaches have also been applied in stock predictive analytics, e.g. predicting stock trend (i.e. rising or falling) using neural network ensemble [6], classifying stocks into groups of buying, holding, and selling by using one-vs-all and one-vs-one NN, etc. [7]. Recently, deep learning (DL) has achieved tremendous successes in diverse applications, e.g. visual object recognition, image processing, speech recognition, hand-writing recognition, natural language processing, information retrieval, etc. [16]. In [8], an autoencoder composed of stacked restricted Boltzmann machines (RBM) was utilized to extract features from the history of individual stock prices, which successfully enhanced the momentum trading strategy [9] without extensive hand-engineering of input features and delivered an annualized return of 45.93% over the period of 1990–2009 versus 10.53% for basic momentum. In [10], a high-frequency strategy was developed based on deep neural networks (DNN) that were trained for prediction of the next one-minute average price based on current time (hour and minute), and n-lagged one-minute pseudo-returns, price standard deviations and trend indicators, in which a stock would be bought or sold when its next predicted average price is higher/lower than its last closing price. The model achieved an accuracy of 66% with the AAPL (stock for Apple Inc.) tick-by-tick transactions during September to November of 2008.

Although these models have demonstrated more or less success in prediction of stock price/trend, they were built based solely on historic data, which contradicts a basic rule in finance known as the Efficient Market Hypothesis [11]. The hypothesis implies that if one was to gain an advantage by analyzing historical stock data, then the entire market will become aware of this advantage and as a result adjust the price of the stock to eliminate this opportunity in the future, aspects of which are still subject of debates and disputes [1]. Indeed, market value is affected by many uncertainties and factors such as economic conditions, investors' sentiments and perceptions, political events, etc. Quick changes in stock markets lead to dynamic, non-parametric, chaotic and noisy nature of stock's price series that resemble random process with short term fluctuations [1]. Modeling market behavior only on historical data may miss important factors with significant impact on market change that remain outside of market scope, i.e. exogenous to the market prices. Indeed, a very little work in the literature has attempted to deal with stock prediction problems based on information other than historical price related data in algorithmic trading. In [12], discrete stock price was predicted using a synthesis of linguistic, financial and statistical techniques to create an Arizona Financial Text System (AZFinText). Specifically, the system predicted stock price based on the presence of key terms in financial news article as well as the stock prices within short 20+ min time window after releasing, and achieved 2% higher return than the best performing quantitative funds monitoring the same securities.

Banks, trading houses, investment experts use deep human intuition and complex knowledge acquired from years' of experience to make daily trading recommendations for individual and business customers. The disturbing difficulty in adopting these recommendations for average non-professional investors is their normally poor and variable reliability, often additionally contaminated by inconsistent conflicting recommendations provided by multiple experts. In [13], the expert investors with similar investment preferences based on their publicly available portfolio were firstly matched to non-professional investors by taking advantage of social network analysis, and appropriate managed portfolio was then recommended to them according to the combined recommendations from matched financial experts. Although this work provides a good way for non-professional investors to identify appropriate experts to follow, the reliability of recommendations is not investigated. In this work, we intend to explore the reliability of expert recommendations and their feasibility for prediction of stock market change in the context of ISMIS Data Mining Competition 2017, which to our best knowledge, has not been studied in the literature.

3   ISMIS 2017 Data Mining Competition

The objective of the ISMIS' 2017 competition [14] is to verify whether experts' recommendations can be used as a reliable basis for making informed decisions regarding investments in a stock market. The trading recommendation problem has been defined as predicting best trading decisions from the set of classes: {sell,hold,buy} corresponding to the considerable negative, near zero or considerable positive return, respectively, observed for different financial assets in the subsequent 3 months, based on multiple trading recommendations made by many different financial experts in a period of up to 2 months prior to the trading decision date.

The contestants were provided with the labeled training set of 12,234 examples corresponding to expert recommendation for 489 assets at different points in time. Each example is composed of an asset ID, an ordered list of recommendations issued by experts for the asset during 2 months before the decision date, and the true return class of the stock computed over the period of 3 months after the decision date that may take one of the 3 values, i.e. "sell", "hold", and "buy" corresponding to considerably positive, close to zero, and considerably negative returns, respectively. The expert recommendations for every asset were structured as a table including expert id, number of days prior to trading decision that the recommendation was made, expected return, and suggested trading action from the same set as the target classes of {sold, hold, buy}. An unlabeled testing data set consisting of 7555 examples was been provided as well, which has a similar format as the training set. All recommendations provided were from 2832 experts who were employed in 228 different financial institutions.

To factor in uneven impact of trading decisions on corresponding return, the performance function of the trading classification system in response to a vector of features X has been defined by the cost-weighted accuracy metric, given as

$$ACC(X) = \frac{\sum_{i=1}^{3}(C_{i,i}W_{i,i})}{\sum_{i=1}^{3}\sum_{j=1}^{3}(C_{i,j}W_{i,j})} \tag{1}$$

where $C_{i,j}$ denotes the confusion matrix entry for i^{th} true and j^{th} predicted class and $W_{i,j}$ is a corresponding weight from the cost matrix W, shown in Table 1.

Table 1. Cost matrix of the weighted accuracy ACC in Eq. (1)

		Predicted	
	Sell	Hold	Buy
Sell	8	1	8
Hold	4	1	4
Buy	8	1	8

4   Prediction Model

Trading prediction has been formulated as a classification problem, in which the features are extracted from expert recommendations and the trading decisions are from the set of classes: {sell,hold,buy}. A hybrid scheme is developed, in which we combine a probabilistic Bayesian learning model and Extreme Learning Machine (ELM) in the decision-level fusion. The details are elaborated below.

4.1   Feature Representation

Based on the data provided by the competition, the features are extracted from experts' recommendations. As introduced in Sect. 3, the recommendation data contain expected returns in the subsequent 3 months from the decision date and the suggested trading actions. Considering that the information regarding the expected return rates may be inconsistent due to different interpretations of stock quotes by experts, e.g. whether or not considering splits or dividends [14], the features used in our model are constructed based solely on the suggested trading actions taking values $\{1, 2, 3\}$ corresponding to {sell,hold,buy}, respectively. If an expert gave more than one recommendations to an asset at different time points, only the latest recommendation considered as the most reliable is used.

The features are represented in the form of a matrix. Given N assets and M unique experts, the feature matrix has a dimension of $N \times M$, in which each row gives suggested trading actions for one asset and each column lists all recommendations made by one expert. Although there are in total 2832 experts, i.e. $M = 2832$, only a quite small fraction, up to 47 of 2832, gave recommendations to one asset. The value of the feature is set to be 0 if the corresponding recommendation is missing. The feature matrix obtained in this way, is extremely sparse with around 0.13% and 0.19% of non-zero values for the training and testing data sets, respectively.

4.2   Learning Algorithm

To solve the classification problem with a quite sparse feature matrix, a hybrid classification scheme that combines the advantages of a probabilistic learning model and Extreme Learning Machine (ELM) in decision level has been developed. The base classifiers and the fusion scheme are elaborated below.

It has been shown that the feature dimension is quite high, say 2832, and the matrix is extremely sparse. The sparsity is caused due to the fact that only a quite small fraction of experts published their recommendation to each of assets. Feature selection that chooses a fixed subset of features cannot make sense to all examples since their recommendations are from different experts. Features with a quite large size, whose major components are meaningless (unrecommended), however, may lead to poor performance in model learning when commonly used classification algorithms, e.g. decision trees, neural network, etc., are used. To address this, a probabilistic learning method based on Maximum a posteriori (MAP) estimation is developed with assumption that the recommendations of experts were made independently. The advantages of the probabilistic model are that it considers the reliability of each unique expert and avoids the use of meaningless feature components without the need of feature selection process.

Let the true return class be c_r and the trading decision suggested by the eth expert be $c_p^{(e)}$, where $c_r, c_p^{(e)} \in \{$sold,hold,buy$\}$. The conditional probability of the true return class being c_r given recommendations $c_p^{(e)}$ can be expressed as:

$$p(c_r|c_p^{(e)}) = \frac{p(c_p^{(e)}|c_r) \times p(c_r)}{p(c_p^{(e)})}, e \in E,$$ (2)

where E denotes the set of indices of the experts and $p(c_r)$ represents the prior probability. The value of $p(c_r|c_p^{(e)})$ can be estimated from the training data set for all combination of $c_r, c_p^{(e)} \in \{$sold,hold,buy$\}$, based on which a look-up table with a dimension of 3×3 can be created for each unique expert.

The conditional probability of c_r given a list of suggested trading actions of $c_p^{(e)}$ for all $e \in E$ can be expressed as:

$$p(c_r|c_p^{(1)}c_p^{(2)}...c_p^{(M)}) = \frac{p(c_p^{(1)}c_p^{(2)}...c_p^{(M)}|c_r)p(c_r)}{p(c_p^{(1)}c_p^{(2)}...c_p^{(M)})}$$ (3)

where $M = 2832$ is the total number of experts. Assuming that all recommendations are made independently, we have

$$p(c_p^{(1)}c_p^{(2)}...c_p^{(M)}) = p(c_p^{(1)})p(c_p^{(2)})...p(c_p^{(M)}).$$ (4)

By substituting (4) into (3), the conditional probability can be rewritten as

$$p(c_r|c_p^{(1)}c_p^{(2)}...c_p^{(M)})$$
$$= \frac{p(c_p^{(1)}|c_r)p(c_p^{(2)}|c_r)...p(c_p^{(M)}|c_r)p(c_r)}{p(c_p^{(1)})p(c_p^{(2)})...p(c_p^{(M)})}$$
$$= \frac{\prod_{e=1}^{M} p(c_p^{(e)}, c_r)}{\prod_{e=1}^{M} p(c_p^{(e)})} \times \frac{1}{p^{M-1}(c_r)}$$ (5)
$$= \prod_{e=1}^{M} p(c_r|c_p^{(e)}) \times \frac{1}{p^{M-1}(c_r)}.$$

Considering that the feature matrix is quite sparse, it may not cover all possible combinations of c_r and c_p. If a combination is not found in the training data set, it does not necessarily indicate that it is impossible to happen. In this case, the conditional probability given in (2) is set to be 1, instead of being 0, which is aimed to ignore its impact in the estimation of probability in (5).

Maximum a Posteriori (MAP) estimation principle is used to identify the best return class that yields the highest posterior class probability given the evidence in a form of the suggested trading actions and prior assumptions $p(c_r)$:

$$\hat{cr} = \arg\max_{c_r} p(c_r | c_p^{(1)} c_p^{(2)} ... c_p^{(M)}). \tag{6}$$

To boost the performance of the probabilistic learning model, it is combined at the decision level with the ELM based feed-forward neural network that is diverse to the probabilistic approach with low correlation among the outputs from the base classifiers [17]. Traditional feed-forward neural networks use slow gradient-based learning algorithms to train networks and tune the parameters iteratively, which makes their learning speeds rather slow. To overcome these drawbacks, Huang et al. proposed the ELM based feedforward neural networks, which randomly choose hidden nodes and analytically determine the output weights of the networks [15]. Compared to other computational intelligence methods such as the conventional back-propagation (BP) algorithm and SVMs, the ELM has much faster learning speeds, ease of implementation, the least human intervention, and high generalization performance. It has been reported by Huang et al. from the experimental results that the ELM can produce better generalization performance and can learn thousands of times faster than traditional learning algorithms for feed-forward neural networks [15].

Each base classifier is trained individually on the training data set, which gives one prediction label representing the forecast return class to each example. The scores of the 3 classes from both base classifiers are then used as input features feeding to the ELM again to train a mapping model that is used as one base classifier as well. The results from the 3 base classifiers are then combined together to identify the final class according to majority voting principle, i.e. the best trading class is the one with most votes.

5   Competition Results

The facility to score derived model solutions on the chunk of the testing set was provided via web-based KnowledgePit platform. Although the submissions with predicted trading labels had to be made for the whole testing set, the feedback in a form of the performance score was received based on only 10% randomly chosen testing examples, identities of which were hidden from the competitors.

The final solution that we submitted to the competition received the ACC of 0.461 based on 10% of testing examples scoring the 8th place in the preliminary stage, and that was decreased to 0.417 when evaluated on the whole set, which in the final stage also yielded the 8th place out of 159 international teams participated in ISMIS' 2017 Data Mining Competition.

Figure 1 depicts evolution of the validation and testing performance of submitted solutions, in which the green line illustrates the preliminary scores achieved based on 10% of the testing set and the red line shows the final scores received based on the whole set. It can be seen from the figure that the highest ACC score evaluated on the whole set highlighted in the Fig. 1 is 0.44 that is slightly higher than 0.4375 scoring the first place in the competition. However, this solution was not chosen as the final submission that was selected based on the score given in the preliminary stage.

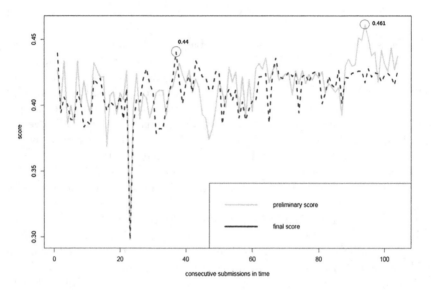

Fig. 1. Evolution of the validation and testing performance of submitted solutions

6   Conclusions

In this work, we intended to verify whether experts' recommendations can be used as a reliable basis for making informed decisions regarding investments in financial markets. To this end, prediction of the return class from stock investment over the next three months, using only tips provided by financial expert, is addressed in the context of ISMIS Data Mining Competition 2017. A hybrid classification scheme is developed by combining maximum a posteriori (MAP) estimation based probabilistic learning model and Extreme Learning Machine (ELM) in decision fusion level. The prediction accuracy measured on external testing set achieved the score of 0.4172 yielding the 8th place out of 159 international teams' competing in ISMIS Data Mining Competition 2017.

References

1. Khaidem, L., Saha, S., Dey, S.R.: Predicting the direction of stock market prices using random forest. Appl. Math. Finan. 1–20 (2016). CoRR abs/1605.00003

2. Algorithmic   Trading.   Investopedia.   www.investopedia.com/terms/a/ algorithmictrading.asp
3. Li, H., Yang, Z.J., Li, T.L.: Algorithmic Trading Strategy Based On Massive Data Mining. Stanford University, Stanford (2014)
4. Shao, C.X., Zheng, Z.M.: Algorithmic trading using machine learning techniques: final report (2013)
5. Dai, Y., Zhang, Y.: Machine Learning in Stock Price Trend Forecasting. Stanford University, Stanford (2013)
6. Giacomel, F., Galante, R., Pareira. A.: An algorithmic trading agent based on a neural network ensemble: a case of study in North American and Brazilian stock markets. In: IEEE/WIC/ACM International Conference on Web Intelligence and Intelligent Agent Technology (2015)
7. Boonpeng, S., Jeatrakul, P.: Decision support system for investing in stock market by using OAA-neural network. In: 8th International Conference on Advanced Computational Intelligence Chiang Mai, February 2016
8. Takeuchi, L., Lee, Y.A.: Applying Deep Learning to Enhance Momentum Trading Strategies in Stocks, December 2013
9. Jegadeesh, N., Titman, S.: Returns to buying winners and selling losers: implications for stock market eciency. J. Finan. **48**, 65–91 (1993)
10. Arévalo, A., Niño, J., Hernández, G., Sandoval, J.: High-frequency trading strategy based on deep neural networks. In: Huang, D.-S., Han, K., Hussain, A. (eds.) ICIC 2016. LNCS, vol. 9773, pp. 424–436. Springer, Cham (2016). doi:10.1007/978-3-319-42297-8_40
11. Malkiel, B.G., Fama, E.F.: Efficient capital markets: a review of theory and empirical work. J. Finan. **25**(2), 383–417 (1970)
12. Schumakera, R.P., Hsinchun, C.: A quantitative stock prediction system based on financial news. Inf. Process. Manag. (ELSEVIER) **45**(5), 571–583 (2009)
13. Koochakzadeh, N., Kianmehr, K., Sarraf, A., Alhajj, R.: Stock market investment advice: a social network approach. In: IEEE/ACM International Conference on Advances in Social Networks Analysis and Mining (ASONAM), pp. 71–78 (2012)
14. ISMIS Data Mining Competition: Trading based on Recommendations. https://knowledgepit.fedcsis.org/mod/page/view.php?id=1012
15. Huang, G.B., Zhu, Q.Y., Siew, C.K.: Extreme learning machine: theory and applications. Neurocomputing **70**, 489–501 (2006)
16. Deng, L.: A tutorial survey of architectures, algorithms, and applications for deep learning. APSIPA Trans. Sig. Inf. Process. **3**, 1–29 (2014)
17. Goebel, K.F., Yan, W.: Choosing classifiers for decision fusion. In: Svensson, P., Schubert, J. (eds.) Proceedings of the International Conference on Information Fusion, vol. I, pp. 563–568 (2004)

Heterogeneous Ensemble of Specialised Models - A Case Study in Stock Market Recommendations

Michał Kozielski[1]([⊠]), Katarzyna Dusza[2], Józef Flakus[2], Krzysztof Kozłowski[2], Sebastian Musiał[2], and Bartłomiej Szwej[2]

[1] Institute of Electronics, Silesian University of Technology,
Akademicka 16, Gliwice, Poland
`michal.kozielski@polsl.pl`
[2] Institute of Informatics, Silesian University of Technology,
Akademicka 16, Gliwice, Poland
`http://adaa.polsl.pl`

Abstract. The paper presents an approach to the task of a data mining competition "Trading Based on Recommendations". The approach is based on a heterogeneous ensemble of classification models, where each model was created independently by different participant of the competition. Each base-model is presented in the paper and a concept, and results of the overall solution are discussed.

Keywords: Ensemble learning · Heterogeneous ensemble · Classification

1 Introduction

Ensemble methods have become a very popular approach to data classification and regression [3,7,10] as they are able to reduce the expected error of a method. Taking an error of a model into consideration it is known that a single model has associated a certain bias with. Additionally, the learning process is never performed on a whole possible population of examples, which results in additional error called the variance of the learning method for a given problem. Therefore, the total expected error of a method is the sum of these two components: bias and variance [6].

In order to reduce the expected error of a method the ensemble methods are applied. From the perspective of the ensemble type there are homogeneous

B. Szwej—The authors would like to thank all the participants developing the models included in the presented ensemble: Karolina Adamczyk, Szymon Bartnik, Michał Bychawski, Piotr Dankowski, Tomasz Dabek, Piotr Drewniak, Michał Dziwoki, Łukasz Gawin, Krzysztof Hanzel, Karol Herok, Karol Kalaga, Mateusz Kaleta, Grzegorz Kozłowski, Robert Krupa, Mateusz Łysień, Mateusz Małota, Wojciech Niemkowski, Krzysztof Paszek, Stefania Perlak, Dawid Poloczek, Marek Pownug, Tomasz Rzepka, Artur Siedlecki, Krzysztof Śniegoń, Katarzyna Toporek.

© Springer International Publishing AG 2017
M. Kryszkiewicz et al. (Eds.): ISMIS 2017, LNAI 10352, pp. 728–734, 2017.
DOI: 10.1007/978-3-319-60438-1_71

and heterogeneous approaches. The first ones generate an ensemble of models on the basis of one algorithm. Boosting and bagging methods ([1,4] respectively) are of this type. The latter ones utilize an ensemble consisting different models generated by different algorithms. The stacking approach [11] is an example of this type. The application of ensemble as a classification task was verified for many cases [2,8,9], including stock market analysis.

The goal of this paper is to present our solution submitted to the data mining competition "Trading Based on Recommendations"[1]. The objective of the competition was to verify whether experts recommendations can be used as a reliable basis for making decisions regarding investments in a stock market.

The presented solution is an ensemble consisting of six models created independently by six teams enrolled to the competition. The teams were formed by 30 students taking part in a project within the course Social Networks and Data Mining at the Institute of Informatics, Silesian University of Technology. The goal of each team was to develop possibly the best model, as the ranking based on the initial evaluation had an impact on the course final grade. Each team created several approaches implementing different concepts of data transformation and modelling. The methods chosen to create the ensemble were the best ones according to the preliminary evaluation performed on the competition platform. Therefore, it is assumed that the ensemble consists of the specialised models that were tuned to the given task.

The presented solution is an example of the approach, where data science is performed by a group of students having limited experience in the domain and creating an ensemble of specialised models. This approach can be compared to the single models developed by experienced professionals who participated in the competition.

The structure of the paper is as follows. Section 2 presents the competition objectives and a data set available for the participants. The base-models created by the students and the created ensemble are presented in Sect. 3. Results of the analysis are presented in Sect. 4, whereas final conclusions are presented in Sect. 5.

2   Stock Market Recommendations

The data mining competition "Trading Based on Recommendations" was organised using the KnowledgePit platform [5] and sponsored by mBank S.A. and Tipranks. The task of the competition was "to devise an algorithm that would most accurately predict the class of return from an investment in a stock over the next quarter, basing on historical recommendations related to a particular stock."

The data set available to the competition participants was divided into a training and test data. Training data contained 12 234 records. Each record contained three columns named: SymbolID, Recommendations, Decision. The

---

[1] https://knowledgepit.fedcsis.org/contest/view.php?id=119.

first column is an identifier of a stock symbol. The second column is an ordered list of recommendations. Each recommendation contains a unique identifier of an expert, an expert decision (*Buy*, *Hold* or *Sell*), return rate and time distance to the decision date. The third column gives information about the true return class of the stock and it takes one of three values: *Buy*, *Hold*, *Sell*. Additionally, there was a table available that bound the experts (2 832 experts) to the companies they worked for (228 companies).

The evaluation of the participants' solutions was performed on a test set containing 7 555 examples. Before the final deadline it was possible to perform a preliminary evaluation on 1 000 test examples only. The complete evaluation was performed after the final deadline for the solutions, which were described in a report.

An initial analysis of the training data set showed that there are significant differences between the expected decisions that should have been taken and the recommendations provided by the experts. A comparison of a number of the *Buy*, *Hold* and *Sell* decisions provided as a ground truth and as the expert recommendations is presented in Table 1. This observation was utilised in further analysis.

Table 1. Cardinality of decisions provided as a ground truth and as the expert recommendations

	Buy	Hold	Sell
True decision	26 005	15 310	13 943
Expert decision	34 306	18 388	2 564

3   Ensemble Models

The created ensemble consisted of the six approaches presented in the consecutive paragraphs. Each approach was developed independently, therefore, each description consists of the presentation of the applied data transformation, the utilised model and the environment where the analysis was performed.

Model 1. Within this approach the data were transformed to the form where each example represented a decision of one expert (the information about the experts' companies was not taken under consideration). A derived variable representing correctness was added to the training data. Correctness was calculated as $c = \frac{1}{w}$, where $w \in \{1, 3, 5\}$ was a value depending on a strength of mistake made by an expert. Defining an order of the possible decisions as a sequence <Sell, Hold, Buy>, we can measure the weight of an error between the predicted decision and the correct one. Therefore, the weight of the error is set to 5 if the distance between the decisions in the sequence is 2 (e.g. Sell and Buy).

If the distance between the predicted and correct decisions is 1 (e.g. Sell and Hold) the weight of the error is set to 3; otherwise (the prediction was correct), the value of weight was set to 1.

The classification of each test data example was performed by means of two steps. At first, for each expert recommending a decision within this data example the predicted expert correctness was calculated. Then, a final decision was calculated by an ensemble of experts delivering weighted decisions, where the weights corresponded to the expert precision.

The model was developed in Python.

Model 2. Within this approach an efficiency of each expert was derived on the basis of training data set. The efficiency was calculated as a per cent of correct predictions out of all the predictions of the given expert. In this way the reliable experts, who had got the efficiency exceeding 70%, were identified.

The model that was created as an ensemble depending on a condition verifying the presence of reliable expert in a given data example. If there were the reliable experts on a list of experts contained in a data example, then the final decision was calculated as voting exclusively of these experts. Otherwise voting of all the experts was performed.

The solution was developed in a Java programming language.

Model 3. Within this approach the training data were aggregated. Therefore, real valued attributes (e.g., number of days to decision, return rate) were averaged and a resultant expert decision was calculated by majority voting. If a number of votes for different decisions was equal, then the following priority was applied: Sell, Hold, Buy. The resulting data set was characterised by the imbalanced class size. Therefore, oversampling of the minority classes (Sell and Hold) was applied, which enabled the size of all three classes to be balanced. The training data set created in this way was applied to create a model by means of a Naive Bayes approach.

The solution was developed within a RapidMiner environment.

Model 4. Within this approach the training data set was initially transformed to the form where each example represented a decision of one expert. Each example of the resulting data set was enriched with the information about the company where a given expert was working. Analysis of the created data set and preliminary evaluation of the developed solutions led to submission of a model, where the final decision was calculated as a majority voting performed on the expert recommendations contained in a given example.

The model was implemented in Python.

Model 5. Within this approach the initial data set was transformed to the form where each example represented a decision of one expert. Next, the median value of a decision date for each stock was calculated and a group of decisions referenced to this date was identified. Finally, the decisions were encoded by integer values: Buy = 2, Hold =´1, Sell = 0 and the final decision was calculated as an average of the expert decisions. The choice of the average operation is a result of the intuition that if there are opposite recommendations (Sell and Buy) provided, then the safest solution is to choose Hold option.

The solution was implemented in a RapidMiner environment.

Model 6. Within this approach information on how many days are left to decision is utilised again. It enables us to calculate statistics for each expert. The derived statistics showed the expert efficiency in a given day, where the efficiency was calculated as a per cent of correct predictions out of all the predictions in this day. Therefore, the initial data set was transformed to the form, where each example represented a decision of one expert. Next, this data was enriched by a derived attribute containing the expert efficiency in a given day. Finally, the numeric values were normalised by subtracting an average value for a given attribute and dividing a result by standard deviation for this attribute. Such training data were used to create a model based on eXtreme Gradient Boosting that predicted an expert efficiency.

The final decision for a new example was reached by majority voting of the expert decisions. If there was equal number of votes for different decisions the sum of predicted expert efficiency values was calculated for each decision class and the class label associated to a greater expert efficiency was chosen.

The solution was implemented in an R environment.

Ensemble. The six solutions presented above created a heterogeneous ensemble. It was decided not to prune the ensemble. It was also decided not to weight the solutions on the basis of the preliminary evaluation results. However, the analysis of the training data set showed the imbalanced class distribution for the expert decisions (see Table 1). Therefore, the expert recommendations contained within the training data strongly underrepresent *Sell* decision and overrepresent *Buy* decision. Thus, it was decided to weight each decision of a base-classifier depending on the value it takes. The final outcome of the ensemble for a data example is the class where the highest sum of weights (for the 6 base-classifiers) was calculated.

4   Experimental Results

Each of the models creating the ensemble was separately evaluated due to participation in the competition. Unfortunately, the students did not decide to deliver to the competition system the reports describing their approaches. Therefore, only the results of preliminary evaluation are available and they are presented in Table 2.

Results of the ensemble consisting of the described base-models are presented in Table 3. Initially, no weights were calculated neither at a model nor at a decision level. This approach resulted in a quality that was worse then the average

Table 2. Quality of the base solutions creating the presented ensemble – preliminary evaluation

Model ID	1	2	3	4	5	6
Preliminary evaluation	0.43	0.4094	0.4359	0.416	0.4224	0.4095

quality of the base-models. Next, decision weights were applied due to imbalanced class representation in recommendations (see Sects. 2 and 3). The values of the weights were adjusted experimentally. They were incremented by 0.5 starting from 1. Finally, the following values were set:

- each decision *Buy* was given a weight value 1,
- each decision *Hold* was given a weight value 1.5,
- each decision *Sell* was given a weight value 6.

Application of the weights improved the results and therefore, this approach was submitted to final evaluation.

Table 3. Quality of the presented ensemble - with and without weights

Model	Preliminary evaluation	Final evaluation
Ensemble without weights	0.4124	–
Ensemble with weights	0.4517	0.4234

5 Conclusions

The heterogeneous ensemble of the specialised models was presented in this paper as a solution taking part in the data mining competition "Trading Based on Recommendations". The models were meant to be specialised because they were developed by students as separate submissions to the competition. The models were prepared by different teams and were chosen from a set of solutions developed by each team as the best possible approach according to preliminary evaluation.

Weighting of each model decision seemed to be a key solution implemented in the final ensemble. The weights were introduced on the basis of initial data analysis showing significant differences between the expected decision values and the recommendations provided by the experts. Introduction of the weighting scheme improved the preliminary evaluation results significantly.

Another set of conclusions can be drawn considering involvement of the students into the competition. It is worth highlighting that each team consisting of 5 students developed about 3 solutions on average. Some of these solutions were fairly sophisticated and included clustering, learning of expert efficiency and weighted voting of expert decisions. Other solutions were trivial, including random generation of results, what was a sign of desperation resulting from poor performance of other methods. It was interesting to see the students involved into this kind of project. Additionally, it is interesting that an ensemble of student approaches was ranked on the 7th place in the competition. However, a conclusion that such ensemble of solutions, that were developed by students obliged to participation in the contest, can be competitive seems to be unjustified. Especially, looking at the absolute quality values of the developed solution, it is clear that the analysed problem is not trivial and the final results are far from being valuable.

Acknowledgements. The work was carried out within the statutory research project of the Institute of Electronics, Silesian University of Technology: BK_220/RAu-3/2016 (02/030/BK_16/0017).

References

1. Breiman, L.: Bagging predictors. Mach. Learn. **24**(2), 123–140 (1996). http://dx.doi.org/10.1007/BF00058655
2. Chang, P.-C., Liu, C.-H., Fan, C.-Y., Lin, J.-L., Lai, C.-M.: An ensemble of neural networks for stock trading decision making. In: Huang, D.-S., Jo, K.-H., Lee, H.-H., Kang, H.-J., Bevilacqua, V. (eds.) ICIC 2009. LNCS, vol. 5755, pp. 1–10. Springer, Heidelberg (2009). doi:10.1007/978-3-642-04020-7_1
3. Dietterich, T.G.: Ensemble methods in machine learning. In: Kittler, J., Roli, F. (eds.) MCS 2000. LNCS, vol. 1857, pp. 1–15. Springer, Heidelberg (2000). doi:10.1007/3-540-45014-9_1
4. Freund, Y., Schapire, R.E.: A desicion-theoretic generalization of on-line learning and an application to boosting. In: Vitányi, P. (ed.) EuroCOLT 1995. LNCS, vol. 904, pp. 23–37. Springer, Heidelberg (1995). doi:10.1007/3-540-59119-2_166
5. Janusz, A., Ślezak, D., Stawicki, S., Rosiak, M.: Knowledge pit-a data challenge platform. In: CS&P, pp. 191–195 (2015)
6. Kong, E.B., Dietterich, T.G.: Error-correcting output coding corrects bias and variance. In: Proceedings of the Twelfth International Conference on Machine Learning (1995)
7. Mendes-Moreira, J., Soares, C., Jorge, A.M., Sousa, J.F.D.: Ensemble approaches for regression: a survey. ACM Comput. Surv. **45**(1), 10:1–10:40. http://doi.acm.org/10.1145/2379776.2379786
8. Opitz, D., Maclin, R.: Popular ensemble methods: an empirical study. J. Artif. Intell. Res. **11**, 169–198 (1999)
9. Seker, S.E., Mert, C., Al-Naami, K., Ayan, U., Ozalp, N.: Ensemble classification over stock market time series and economy news. In: 2013 IEEE International Conference on Intelligence and Security Informatics, pp. 272–273, June 2013
10. Witten, I.H., Frank, E., Hall, M.A.: Data Mining: Practical Machine Learning Tools and Techniques. Morgan Kaufmann, Burlington (2011)
11. Wolpert, D.H.: Stacked generalization. Neural Netw. **5**(2), 241–259 (1992). http://www.sciencedirect.com/science/article/pii/S0893608005800231

Algorithmic Daily Trading Based on Experts' Recommendations

Andrzej Ruta[1]($\boxtimes$), Dymitr Ruta[2], and Ling Cen[2]

[1] ING Bank Slaski, Katowice, Poland
andrzej.ruta@ingbank.pl
[2] Emirates ICT Innovation Center, EBTIC, Khalifa University, Abu Dhabi, UAE
{dymitr.ruta,cen.ling}@kustar.ac.ae

Abstract. Trading financial products evolved from manual transactions, carried out on investors' behalf by well informed market experts to automated software machines trading with millisecond latencies on continuous data feeds at computerised market exchanges. While high-frequency trading is dominated by the algorithmic robots, mid-frequency spectrum, around daily trading, seems left open for deep human intuition and complex knowledge acquired for years to make optimal trading decisions. Banks, brokerage houses and independent experts use these insights to make daily trading recommendations for individual and business customers. How good and reliable are they? This work explores the value of such expert recommendations for algorithmic trading utilising various state of the art machine learning models in the context of ISMIS 2017 Data Mining Competition. We point at highly unstable nature of market sentiments and generally poor individual expert performances that limit the utility of their recommendations for successful trading. However, upon a thorough investigation of different competitive classification models applied to sparse features derived from experts' recommendations, we identified several successful trading strategies that showed top performance in ISMIS 2017 Competition and retrospectively analysed how to prevent such models from over-fitting.

Keywords: Algorithmic trading · Feature selection · Classification · Gradient boosting decision trees · Sparse features · K-nn

1 Introduction

Algorithmic trading utilizes computer programs and mathematical models to create and determine trading strategies and make automatic transactions in financial markets for optimal returns, which has become a modern approach to replace human manual trading. Nowadays, banks, trading houses and investment firms rely heavily on algorithmic trading in the stock markets, especially for high-frequency trading (HFT) that requires processing of large amounts of information to make instantaneous investment decisions. Algorithmic trading allows consistent and systematic execution of designed trading strategies that are free from (typically damaging) impact of human emotions. It also makes markets more liquid and efficient [1].

© Springer International Publishing AG 2017
M. Kryszkiewicz et al. (Eds.): ISMIS 2017, LNAI 10352, pp. 735–744, 2017.
DOI: 10.1007/978-3-319-60438-1_72

1.1   Related Work

The key question in algorithmic trading is how to define a set of rules or build mathematical models based on historical stock data, e.g. prices, volume, order book, as well as other available information, such as companies' annual reports, expert recommendations or commodity prices, to accurately predict market behaviour and correctly identify trading opportunities. Simple trade criteria, as an example, can be defined based on 5-day and 20-day moving averages as follows: buying 100 shares when the 5-day moving average of a stock price goes above its 20-day moving average and selling half of shares when the price's 5-day moving average goes below the 20-day moving average. Based on such rules a machine can be coded to monitor stock prices and corresponding indicators, and automatically place buy/sell orders triggered when the defined conditions are met. However, stock markets are non-stationary and chaotic and are influenced by many direct or indirect variables and uncertainties beyond trader's control or knowledge. Simple rules, as the ones above, typically do not suffice to account for all impact factors and fail to simultaneously achieve high returns at low risk of financial loss. Machine learning (ML) and data mining (DM) have undergone a rapid development in the recent years and have found numerous applications in predictive analytics across different disciplines and industries. In algorithmic trading, ML/DM can help to discover hidden patterns of market behaviour from related financial data in order to decode their complex impact on market movements or trends at different time horizons.

In [2], five types of stock analysis: typical price (TP), Bollinger bands, relative strength index (RSI), and moving average (MA), were combined together to predict the trend of closing price in the following day. With help of data mining techniques their model achieved an average accuracy of well over 50%. Application of supervised learning to determine future trends in stock market has been a subject of intense research, bulk of which focused on using well-established models, such as support vector machine (SVM), neural network (NN) or linear regression, to learn markets behaviour based on their own historical signals. In [3] classification and regression models were both designed to predict the daily stock returns from direct price/volume signals: open, close, high, low, volume, and indirect features corresponding to external economic indicators. Simple Logistic Regression (LR) predicting daily trend direction was reported to outperform SVM, yielding over 2000% cumulative return over 14 years.

An automated stock trading system was proposed in [4]. It considered a hierarchy of features and methods selected based on multiple risk-adjusted investment performance indicators. It also used backward search and four ML algorithms (linear/logistic regression, $l-1$ regularized v-SVM, and multiple additive regression tree (MART)), as was capable of online learning. The system traded automatically based on the following day's trend prediction and reported high average accuracy in excess of 90% [4].

In [5] stock price was modelled on daily or longer intervals to predict future single- or multi-day averages respectively. LR, quadratic discriminant analysis (QDA) and SVM models were tested on the historical price data over the period between 2008 and 2013. LR reported the top performance in next-day price prediction: 58.2%, further improved to 79.3% in the SVM-based long-term model with a 44-day window. In [6] a random forest classier was built on features extracted from technical analysis indicators, such as RSI or stochastic oscillator, and achieved 96.92% accuracy in 88-day window price prediction.

Neural network (NN) based approaches have also been applied to stock price trend prediction [7], classification into groups of buying, holding, and selling [8], and other related tasks. Recently, deep learning (DL) has achieved tremendous success in diverse applications, e.g. visual object recognition, speech recognition or information retrieval [9]. In [10], an auto-encoder composed of stacked restricted Boltzmann machines (RBM) was utilized to extract features from the history of individual stock prices, which successfully enhanced the momentum trading strategy [11] and delivered an annualized return of 45.93% over the period of 1990–2009 versus 10.53% for basic momentum. In [12] a high-frequency strategy was developed based on deep neural networks (DNN) that were trained for prediction of the next-minute average price based on the most recent and n-lagged one-minute pseudo-returns, price standard deviations and trend indicators. It achieved 66% accuracy on Apple Inc. stocks' tick-by-tick transactions over the period of Sep–Nov 2008.

Although these models have been successful in predicting stock price trends, they were built based solely on historical data, which contradicts a basic rule in finance known as the *Efficient Market Hypothesis* [13]. It implies that if one was to gain an advantage through historical stock data analysis, then the entire market would immediately become aware of this advantage causing correction of the stock price [6]. This dynamic and reactive nature of international financial markets combined with their high sensitivity to all kinds of micro- and macro-economic events in business, financial and geopolitical spheres, make them appear chaotic, very noisy and allegedly truly unpredictable, especially in short time horizons [6]. Very little research reported in the public domain literature has been devoted to algorithmic trading based on information other than historical price related data. In [14], discrete stock price was predicted using a synthesis of linguistic, financial and statistical techniques to create an Arizona Financial Text System (AZFinText). Specifically, the system combined stock prices and the presence of key terms in the related financial news articles within the window of up to 20 min after the release yielding 2% higher return than the best-performing quantitative funds monitoring the same securities.

Banks, trading houses and investment experts use deep human intuition and complex knowledge acquired over years of practice to make daily trading recommendations for individual and business customers. The main concern for non-professional investors to follow these recommendations is their non-guaranteed reliability, especially when inconsistent recommendations are given by various experts. In [15] the expert investors with similar investment preferences based on their publicly available portfolios were first matched to non-professional investors by taking advantage of social network analysis. Then, appropriately managed portfolios were recommended to them according to their assigned financial experts. Although the authors proposed an interesting way for non-professional investors to identify appropriate experts to follow, recommendations reliability was not investigated. In this work, in the context of ISMIS 2017 Data Mining Competition, we intend to explore the feasibility and value of expert recommendations in stock trend prediction for algorithmic daily trading utilizing various machine learning models, which has been seldom studied in the literature.

1.2   ISMIS 2017 Competition Problem Formulation

The trading recommendation problem has been defined as predicting best trading decisions from the set of classes: {sell,hold,buy}, corresponding to the considerable negative, near zero or considerable positive return observed for different financial assets in the subsequent 3 months, based on multiple trading recommendations made by many different experts in a period of up to 60 days prior to the trading decision. The expert recommendations for every asset were structured as a table including expert id, number of days prior to trading decision that the recommendation is made, expected return, and suggested trading action from the same set as the target classes: {sell,hold,buy}.

To factor in uneven impact of trading decisions on corresponding return, the performance function of the trading classification system in response to a vector of features X has been defined by the following cost-weighted accuracy metric:

$$ACC(X) = \frac{\sum_{i=1}^{3}(C_{i,i}W_{i,i})}{\sum_{i=1}^{3}\sum_{j=1}^{3}(C_{i,j}W_{i,j})} \tag{1}$$

where $C_{i,j}$ denotes the confusion matrix entry for i^{th} true and j^{th} predicted class and $W_{i,j}$ is a corresponding weight from the following cost matrix W:

Table 1. Cost matrix of the weighted accuracy ACC (1)

	Predicted Sell	Predicted Hold	Predicted Buy
Actual Sell	8	1	8
Actual Hold	4	1	4
Actual Buy	8	1	8

The contestants were provided with the labelled training set of 12234 examples as well as the facility to score their predictions on the chunk of the testing set (7555 examples) via web-based KnowledgePit platform. Although the submissions with predicted trading labels had to be made for the whole testing set, the feedback in a form of the ACC score was received based on only 10% randomly chosen testing examples identities of which were hidden from the competitors.

1.3   Market Expectation

The competition setup allowed to extract general market expectations or sentiments in the target period, in which the testing set was prepared, in terms of
the prior trading class expectations. Namely, given the performance cost matrix
W shown in Table 1 and the feedback from uniform prediction submissions: all
sell, all hold, all buy, correspondingly: ACC_S, ACC_H and ACC_B, the prior class
probabilities can be extracted using the following formula:

$$p(H) = ACC_H \qquad p(B) = \frac{ACC_H ACC_B}{2(1 - ACC_B - ACC_S)} \qquad (2)$$

$$p(S) = \frac{ACC_H ACC_S}{2(1 - ACC_B - ACC_S)}$$

derived from simple solutions of 3 unknowns with 3 equations determined by
elementwise multiplications of the confusion and cost matrices as defined in
(1). Compared to the training period for which sentiments among sell, hold
and buy were distributed as: 0.25, 0.28 and 0.47, respectively, in the testing
period extracted via (2) it changed significantly to 0.36, 0.30 and 0.34, i.e. from
strong buy to strong sell. As we show in the subsequent analysis, access to
this information significantly contributed to the model over-fitting, since the
contestants attempted to exploit this information to boost their preliminary
leaderboard position. It is also worth noting that in real trading scenario the
testing set market expectations constitute the information from the future and
will not be available to help fine-tune or calibrate the trading model.

2   Trading Classification Models

2.1   Representation and Extraction of Recommendation Features

There were a number of different choices on how to represent the recommendation features, which part of the recommendation to include in feature definitions
and how many features to choose for the model. We have adopted a simple,
sparse feature representation associating each feature with an individual expert
e. Numerous experiments were done with the aim to determine the best function
mapping the original observation x associated with a given stock to the feature
value $f_e(x)$. Possible choices involved the most recent return class suggested by
the expert, the last percentage return expected by them, and the average return
with multiple variants of temporal weighting and missing return imputation
schemes. Finally we have decided to consider two best feature families $f_e(x)$: the
most recent return class and the time weighted return from each expert:

$$f_e(x) = c(argmin(t_i)) \qquad f_e(x) = \frac{\sum_{i=1}^{k} r_i e^{-\lambda t_i}}{\sum_{i=1}^{k} e^{-\lambda t_i}} \qquad (3)$$

where r_i denotes percentage return of expert e at time point distant by t_i from
decision date and $\lambda = 0.05$. The resulting temporal weighting scheme emphasises

returns expected more recently and follows our intuition that human experts tend to discount correctly rather near- than distant-future market changes. The feature matrix $X^{[12234 \times 2832]}$ obtained this way for the training set was extremely sparse with less than 0.1% of non-zero values.

Interestingly, no other feature generation scheme appeared contributive to the task of return class prediction. We trialled a new feature space in which for every data record the transformed vector contained values of hand-crafted functions, among others: number/percentage of recommendations with specific return class label, minimum/maximum/average expected return, average lead time of the recommendation or percent of missing return expectations. The total number of data dimensions generated this way exceeded 40. In this new feature space we observed low generalization power of several state-of-the-art classifiers outperforming the best dummy (all-SELL) solution by no more than 2%.

The above disappointing results made us focus more on the recommenders rather than recommendations. Specifically, we ranked the experts according to the global accuracy of their recommendations over the training set and then retained for every data record only the {expert id, recommendation, expected return, days to decision date} entries assigned to the best-performing experts. With this modification no visible improvement was observed either.

The company membership of experts was attempted to be used for clustering and for generation of company-wide features, but also did not result in any improvement of performance compared to the sparse matrix representation.

2.2   Explored Model Design Choices

Trading recommendation was presented as a 3-class classification problem, however it is arguable whether the *hold* class should be considered a genuine independent class or it is just a state in between *sell* and *buy* classes. The argument in support of the latter could be that the *hold* occurring during transition from *sell* to *buy* (when price rises) is clearly different than the same *hold* happening during transition from *buy* to *sell* (when price drops). Given this ambiguity of the *hold* class, as well as the aforementioned discrepancy of target variable distribution between the training and the test set, we have tried various modeling approaches and obtained rather diverse results without a clear winner in terms of the consistency of the design. However, we will exploit the diversity of model solutions to refine the final prediction.

2.3   Top Baseline Classifiers and Their Fine-Tuning

We have explored a number of standard classification models with thorough parametric fine-tuning. Specifically the following models have been considered:

- *Naive Bayes* with multivariate multinomial distribution and enforced uniform prior class probabilities that achieved the performance score of 0.45,
- *Sum of votes* over selected subset of binarised expert recommendations that achieved the performance of above 0.45,

- *k-nearest neighbours* with 9 neighbours, standard Euclidean distance, uniform class priors and *hold*-class penalizing cost matrix that scored 0.47,
- *Support Vector Machine* (SVM) with linear kernel and test class priors that achieved the score of almost 0.45,
- *Boosted classifier ensemble* with decision trees acting as week learners that jointly with class distribution rebalancing reached accuracy of over 0.47 and as much as 0.49 after further "sum-of-scores" combination of three ensembles.

Below we present our modelling strategy that was ultimately adopted. We considered the most beneficial to focus on correct *sell* and *buy* classes predictions, even at the high cost of making mistakes. In principle we attempted to obtain as many correct predictions of *sell* and *buy* based on pure recommendation evidence as possible and then use the diversity of our model versions and the extracted market expectation to refine some of the extreme class instances towards *hold*.

To get a "raw" model we chose two different baseline classifiers, $k-$nearest neighbours (k-NN) and a boosted decision tree ensemble. The former appeals for its conceptual simplicity, lack of training, and the ability to generate arbitrarily complex decision boundaries. The latter is natural for robust selection of predictors out of an overcomplete feature space. Boosted tree ensemble has an additional property of providing a measure of relative importance which reflects how frequently each variable is chosen in individual splits and how much choosing it contributes to reduction of the prediction error. Initially multi-class AdaBoost [16] algorithm was used for the purpose of building the model and selecting the best predictors in parallel. Further it was replaced by Gradient Boosted Decision Trees (GBDT) [17] which offered slightly better accuracy.

While k-NN does not require explanation, we further provide some details on the boosted ensemble configuration. In GBDT weak learners, which are decision trees themselves, are combined into a strong classifier in such a way that in each subsequent round of training the new weak learner is fit to the error of the previously obtained classifier. In the provided implementation the error is expressed in terms of deviance for classification with probabilistic outputs, as in logistic regression. In addition, aiming at minimisation of model variance, the trees were randomized, i.e. we set the fraction of samples to be used for training of individual base learners to less than 1.0 and reduced the number of features to consider when looking for the best split at each level of each individual tree to $\log_2 M$, where M is the number of all features.

3 Preliminary Evaluation and Post-processing

Further refinement of the return class prediction model was based on the exploitation of market expectations extracted as shown in (2). The top baseline classifier outputs have been taken as a starting point for the refinement process and were subjected to several layers of label replacement aimed at reconstruction of the retrieved market expectations. The output class replacement followed a simple logic of identifying subsets of outputs where a pair of classes were significantly over/under represented compared to the market expectations and switching them accordingly to bring the outputs distribution closer to the reconstructed figures: $p(S) = 0.36$, $p(H) = 0.30$ and $p(B) = 0.34$ (2).

The following set of post-processing output replacement methods have been applied based on the feedback during the preliminary stage of the competition:

- **Correction based on output labels imbalance.** For various characteristics of model's output scores, as well as auxiliary expert-specific properties, such as average recommendations age or recommendations inconsistency over time, histograms of output labels were inspected and reshaped in the regions where the actual labels distribution deviated the most significantly from the expected test label proportions.
- **Correction based on candidate model outputs agreement.** The classification output correction was carried out along various levels of agreement among the candidate models' outputs. For every output class separately, the class agreement was measured as a percentage of candidate models that agreed with the final model output. Using this corrective strategy for different ranges of the agreement levels the final output class was switched from the most over-represented to the most under-represented.
- **Correction based on disagreement with baseline classifiers.** The soft outputs from auxiliary baseline classifiers were compared against the current final model outputs. In the case of significant disagreement between the two, the final predictions in the most extreme subset of up to 100 examples were replaced with the most under-represented class.
- **Correction based on submitted model versions disagreement.** The outputs from all previously submitted model versions were taken as inputs to derive a models disagreement measure. Given the input example $\mathbf{x}_i$ and its corresponding m competing outputs or votes $\mathbf{y}_i = y_{ij}$, $j = 1, .., m$ from the ensemble of m predictors, each taking values from a set of $c = 3$ classes: $\{sell, hold, buy\}$, their disagreement d was defined as the normalised set cardinality of the least popular class outputs:

$$d(\mathbf{y}_i) = \frac{c}{m} min^c_{k=1}|\{y_{ij} : y_{ij} = k\}| \qquad (4)$$

taking the values between 0 (at least one class has no votes) and 1 (votes equally distributed among classes). Outputs for cases with the extreme disagreement were replaced with the "safe" *hold* class, while subsets with continuous disagreement ranges detecting high classes imbalance compared to the expectations were re-labelled towards the most under-represented class.

The two ML models based on k-NN and GBDT and the presented post-processing scored top 1st and 2nd place during the preliminary evaluation stage.

4   Final Evaluation

Final evaluation provided surprising yet not entirely unexpected results that completely changed the leaderboard ranking. The top performer from the validation stage (second author's model) scored only $ACC = 0.387$ on the full test set yielding the 24th place while the 2nd contestant in the preliminary stage (first author's model) scored $ACC = 0.424$ (6th place). The top performer received the score of $ACC = 0.437$. At the same time organizers informed that the intermediate solution of the 2nd-ranked contestant from the preliminary stage achieved the highest test score out of all contestants. Unfortunately, that solution was not picked as final since it had not received the best score on the validation set.

Figures 1(a) and (b) show the evolution of the preliminary scores on the validation set and their corresponding final test scores that were hidden throughout the competition. Clearly the market-expectation-guided post-processing led to massive over-fitting and spoiled the well performing early model setups that could have won the contest otherwise. It can be recognised particularly by the validation and test scores evolution divergence that in both cases started to play a role prior to any relabelling attempted. 10% validation sample available to contestants at the model development stage appeared to be insufficiently representative and thus misleading.

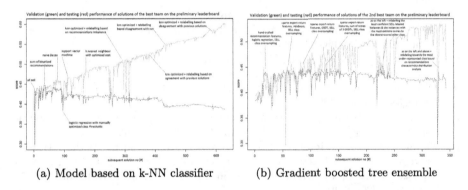

(a) Model based on k-NN classifier   (b) Gradient boosted tree ensemble

Fig. 1. Evolution of the validation and test performance of submitted solutions from the 2 top contestants in the preliminary competition leaderboard

5   Conclusions

The competition uncovered very weak predictability of the future market direction based on experts' recommendations learnt from the past observations. Isolated expert-specific features capturing the most recent experts' opinions proved to be the best at discriminating between three major asset return classes. A number of baseline classifiers, led by gradient boosted decision trees and a simple voting, resulted in the top overall accuracy in detecting optimal trading actions reported during the ISMIS 2017 Data Mining competition. The subsequent classification outputs post-processing, guided by the market feedback (unavailable in reality), turned out detrimental to the model and led to massive overfitting.

References

1. Algorithmic Trading. www.investopedia.com/terms/a/algorithmictrading.asp
2. Kannan, K.S., Sekar, P.S., Sathik, M.M., Arumugam, P.: Financial stock market forecast using data mining techniques. In: International MultiConference of Engineers and Computer Scientists, vol. 1 (2010)
3. Li, H., Yang, Z.J., Li, T.L.: Algorithmic Trading Strategy Based on Massive Data Mining. Stanford University, Stanford (2014)
4. Shao, C.X., Zheng, Z.M.: Algorithmic trading using machine learning techniques: final report (2013)
5. Dai, Y., Zhang, Y.: Machine Learning in Stock Price Trend Forecasting. Stanford University, Stanford (2013)
6. Khaidem, L., Saha, S., Dey, S.R.: Predicting the direction of stock market prices using random forest. Appl. Math. Finan. (2016)
7. Giacomel, F., Galante, R., Pareira, A.: An algorithmic trading agent based on a neural network ensemble: a case of study in North American and Brazilian stock markets. In: International Conference on Web Intelligence and Intelligent Agent Technology (2015)
8. Boonpeng, S., Jeatrakul, P.: Decision support system for investing in stock market by using OAA-neural network. In: 8th International Conference on Advanced Computational Intelligence (2016)
9. Deng, L.: Three classes of deep learning architectures and their applications: a tutorial survey. APSIPA Trans. Sig. Inf. Process. (2012)
10. Takeuchi, L., Lee, Y.A.: Applying Deep Learning to Enhance Momentum Trading Strategies in Stocks (2013)
11. Jegadeesh, N., Titman, S.: Returns to buying winners and selling losers: implications for stock market efficiency. J. Finan. **48**(1), 65–91 (1993)
12. Arévalo, A., Niño, J., Hernández, G., Sandoval, J.: High-frequency trading strategy based on deep neural networks. In: Huang, D.-S., Han, K., Hussain, A. (eds.) ICIC 2016. LNCS, vol. 9773, pp. 424–436. Springer, Cham (2016). doi:10.1007/978-3-319-42297-8_40
13. Malkiel, B.G., Fama, E.F.: Efficient capital markets: a review of theory and empirical work. J. Finan. **25**(2), 383–417 (1970)
14. Schumakera, R.P., Hsinchun, C.: A quantitative stock prediction system based on financial news. Inf. Process. Manag. **45**(5), 571–583 (2009)
15. Koochakzadeh, N., Kianmehr, K., Sarraf, A., Alhajj, R.: Stock market investment advice: a social network approach. In: International Conference on Advances in Social Networks Analysis and Mining, pp. 71–78 (2012)
16. Zhu, J., Zou, H., Rosset, S., Hastie, T.: Multi-class adaboost. Stat. Interface **2**, 349–360 (2009)
17. Friedman, J.: Greedy function approximation: a gradient boosting machine. Ann. Stat. **29**(5), 1189–1232 (2001)
18. Jegadeesh, N., Kim, J., Krische, S.D., Lee, C.M.C.: Analyzing the analysts: when do recommendations add value? J. Finan. **59**(3), 1083–1124 (2004)

Author Index

Aché, Mathurin 697
Akbas, Esra 417
Akgul, Yusuf Sinan 129
Al Bkhetan, Ziad 19
Alam, Mehwish 684
Aleyxendri, Andrea 240
Almardini, Mamoun 29
Ammar, Asma 77
Andreasen, Troels 424
Andruszkiewicz, Paweł 146
Angelastro, Sergio 37, 368
Aryal, Amar Mani 396

Baixeries, Jaume 563
Bazin, Alexandre 611
Beksa, Katarzyna 215
Bembenik, Robert 347
Ben Sassi, Imen 157
Ben Yahia, Sadok 157
Betliński, Paweł 519
Białek, Łukasz 229
Błaszczyński, Jerzy 271
Bocheński, Tomasz 146
Brabant, Quentin 622
Brzezinski, Dariusz 105
Buczkowski, Przemyslaw 708
Bujnowski, Paweł 215
Bulskov, Henrik 424

Carbonell, Jaime G. 292
Carbonnel, Jessie 611
Cardiel, Oscar 240
Ceci, Michelangelo 446
Cen, Ling 718, 735
Chądzyńska-Krasowska, Agnieszka 519
Chein, Michel 642
Cheng, Chi-Cheng 168
Chesani, Federico 3
Clark, Patrick G. 282
Codocedo, Victor 632
Couceiro, Miguel 622
Coulet, Adrien 674
Czyżewski, Andrzej 47

da Penha Natal, Igor 358
Dayioglugil, Ali Batuhan 129
de Avellar Campos Cordeiro, Rogerio 358
Di Noia, Tommaso 481
Dunin-Kęplicz, Barbara 229
Dusza, Katarzyna 728

Eckert, Kai 481
Egurnov, Dmitrii 573
Eick, Christoph F. 396
Elouedi, Zied 77
Esposito, Floriana 37, 195, 368

Ferilli, Anna Maria 37
Ferilli, Stefano 37, 195, 368
Fernandes, José Maria 492
Ferreira, Carlos Abreu 57
Flakus, Józef 728
Flater, David 580
Fumarola, Fabio 446
Fushimi, Takayasu 87

Gama, João 57, 499
Gambin, Tomasz 471
Gao, Cheng 282
Garcia, Ana Cristina Bicharra 358
Garrido, Angel Luis 240
Gawrysiak, Piotr 697
Grekow, Jacek 175
Grzymala-Busse, Jerzy W. 282
Gutierrez, Alain 642

Hartkopp, Oliver 205
Hofmockel, Julia 378
Hryhorzhevska, Anastasiia 471
Huchard, Marianne 642

Ignatov, Dmitry I. 573
Ikeda, Tetsuo 87

Jäkel, Christian 587
Jansen, Peter J. 292
Janusz, Andrzej 697

Jensen, Per Anker   424
Jeon, Heesik   215

Kahn, Giacomo   611
Kashnitsky, Yury   664
Kazama, Kazuhiro   87
KhudaBukhsh, Ashiqur R.   292
Kimura, Masahiro   116
Kłopotek, Mieczysław A.   97
Koržinek, Danijel   404
Kostek, Bożena   47
Kowalczyk, Wojtek   323
Kozielski, Michał   728
Kozłowski, Krzysztof   728
Kozlowski, Marek   435
Krachunov, Milko   251
Kryszkiewicz, Marzena   697
Kubera, Elżbieta   404
Kuranc, Andrzej   404
Kurowski, Adam   47
Kursa, Miron Bartosz   302
Kuznetsov, Sergei O.   653

Lango, Mateusz   312
Lanotte, Pasqua Fabiana   446
Latkowski, Tomasz   215
Lech, Michał   47
Lezoche, Mario   674
Ligęza, Antoni   261
Lingras, Pawan   530
Liu, Zhixing   530
Lu, Tsan-Chu   168
Lumpe, Lars   597

Makhazhanov, Nurtas   653
Malerba, Donato   446
Marasek, Krzysztof   136
Marhula, Joanna   215
Masyutin, Alexey   664
Mattfeld, Dirk C.   205
Mello, Paola   3
Mellouli, Sehl   157
Mephu Nguifo, Engelbert   573
Milani, Alfredo   185
Monnin, Pierre   674
Montali, Marco   3
Morgan, Lori-Anne   530
Morzy, Mikolaj   105
Morzy, Tadeusz   105

Motoda, Hiroshi   116
Mulawka, Jan   64
Musiał, Sebastian   728

Napierala, Krystyna   312
Napoli, Amedeo   622, 674, 684
Nguyen, Phuong T.   481
Nilsson, Jørgen Fischer   424
Nisheva, Maria   251
Niyogi, Rajdeep   185
Nogueira, Ana Rita   57

Odya, Piotr   47
Ohara, Kouzou   116
Okoniewski, Michał   471

Pavlovski, Ilya   530
Pazienza, Andrea   195
Pensa, Ruggero G.   457
Petrov, Peter   251
Piernik, Maciej   105
Pijnenburg, Mark   323
Plewczynski, Dariusz   19
Pompidor, Pierre   642
Protaziuk, Grzegorz   347

Quilez, Ruben   240

Ragone, Azzurra   481
Raś, Zbigniew W.   29
Redavid, Domenico   37, 368
Reynaud, Justine   622, 684
Rho, Valentina   457
Richter, Felix   205, 378
Rokita, Przemysław   136, 146
Ruta, Andrzej   718, 735
Ruta, Dymitr   718, 735
Rybinski, Henryk   435, 697

Saito, Kazumi   87, 116, 385
Sapronova, Alla   509
Sax, Eric   378
Schmidt, Stefan E.   587, 597
Sebastião, Raquel   492
Sikorski, Szymon   215
Ślęzak, Dominik   519, 697
Słowik, Tomasz   404
Sousa, Ricardo   499
Spaleniak, Paweł   47

Stańczyk, Urszula 333
Stefanowski, Jerzy 271, 312
Stokowiec, Wojciech 136
Szałas, Andrzej 229
Szczuko, Piotr 47
Szurmak, Przemyslaw 64
Szwaj, Jacek 347
Szwej, Bartłomiej 728

Tang, My Thao 632
Thongtra, Patcharee 509
Toussaint, Yannick 684
Triff, Matt 530
Trzciński, Tomasz 136, 146

Ushakov, Maxim 653

Vassilev, Dimitar 251

Wang, Pingxin 540
Wang, Sujing 396
Wieczorkowska, Alicja 404
Wiewiórka, Marek 471
Wołk, Krzysztof 136

Yamagishi, Yuki 385
Yang, Xibei 540
Yao, JingTao 550
Yao, Yiyu 540

Żbikowski, Kamil 697
Zembrzuski, Maciej 215
Zhang, Yan 550
Zhang, Yongli 396

Printed in the United States
By Bookmasters